Kiwi

# SIGNAL PROCESSING OF POWER QUALITY DISTURBANCES

**Books in the IEEE Press Series on Power Engineering**

*Rating of Electric Power Cables in Unfavorable Thermal Environments*
George J. Anders

*Power System Protection*
P. M. Anderson

*Understanding Power Quality Problems: Voltage Sags and Interruptions*
Math H. J. Bollen

*Electric Power Applications of Fuzzy Systems*
Edited by M. E. El-Hawary

*Principles of Electrica Machines with Power Electronic Applications*, Second Edition
M. E. El-Hawary

*Pulse Width Modulation for Power Converters: Principles and Practice*
D. Grahame Holmes and Thomas Lipo

*Analysis of Electric Machinery and Drive Systems*, Second Edition
Paul C. Krause, Oleg Wasynczuk, and Scott D. Sudhoff

*Risk Assessment for Power Systems: Models, Methods, and Applications*
Wenyuan Li

*Optimization Principles: Practical Applications to the Operations and Markets of the Electric Power Industry*
Narayan S. Rau

*Electric Economics: Regulation and Deregulation*
Geoffrey Rothwell and Tomas Gomez

*Electric Power Systems: Analysis and Control*
Fabio Saccomanno

*Electrical Insulation for Rotating Machines: Design, Evaluation, Aging, Testing and Repair*
Greg Stone, Edward A. Boulter, Ian Culbert, and Hussein Dhirani

# SIGNAL PROCESSING OF POWER QUALITY DISTURBANCES

**MATH H. J. BOLLEN**

**IRENE YU-HUA GU**

IEEE PRESS SERIES ON POWER ENGINEERING

**IEEE PRESS SERIES ON POWER ENGINEERING**
**MOHAMED E. EL-HAWARY, *SERIES EDITOR***

**IEEE**
**IEEE PRESS**

**WILEY-INTERSCIENCE**

**A JOHN WILEY & SONS, INC., PUBLICATION**

Published by John Wiley & Sons, Inc.
Published simultaneously in Canada.



***Library of Congress Cataloging-in-Publication Data is available.***

ISBN-13 978-0-471-73168-9
ISBN-10 0-471-73168-4

FSC www.fsc.org MIX Paper from responsible sources FSC® C013604

To my father and in memory of my mother (from Irene)

# CONTENTS

# PREFACE

This book originated from a few occasional discussions several years ago between the authors on finding specific signal-processing tools for analyzing voltage disturbances. These simple discussions have led to a number of joined publications, several Masters of Science projects, three Ph.D. projects, and eventually this book. Looking back at this process it seems obvious to us that much can be gained by combining the knowledge in power system and signal processing and bridging the gaps between these two areas.

This book covers two research areas: signal processing and power quality. The intended readers also include two classes: students and researchers with a power engineering background who wish to use signal-processing techniques for power system applications and students and researchers with a signal-processing background who wish to extend their research applications to power system disturbance analysis and diagnostics. This book may also serve as a general reference book for those who work in industry and are engaged in power quality monitoring and innovations. Especially, the more practical chapters (2, 5, 6, and 10) may appeal to many who are currently working in the power quality field.

The first draft of this book originated in 2001 with the current structure taking shape during the summer of 2002. Since then it took another three years for the book to reach the state in which you find it now. The outside world did not stand still during these years and many new things happened in power quality, both in research and in the development of standards. Consequently, we were several times forced to rewrite parts and to add new material. We still feel that the book can be much more enriched but decided to leave it in its current form, considering among others the already large number of pages. We hope that the readers will pick up a few open subjects from the book and continue the work. The conclusion sections in this book contain some suggestions on the remaining issues that need to be resolved in the authors' view.

Finally, we will be very happy to receive feedback from the readers on the contents of this book. Our emails are m.bollen@ieee.org and i.gu@ieee.org. If you find any mistake or unclarity or have any suggestion, please let us know. We cannot guarantee to answer everybody but you can be assured that your message will be read and it will mean a lot to us.

MATH H. J. BOLLEN
IRENE Y. H. GU

*Ludvika, Sweden*
*Gothenburg, Sweden*
*May 2006*

# ACKNOWLEDGMENTS

The authors would like to thank those who have contributed to the knowledge for the writing of this book. A main word of thanks goes to colleagues and students at Eindhoven University of Technology, Eindhoven, The Netherlands; University of Manchester Institute of Science and Technology (UMIST, currently part of University of Manchester), Manchester, United Kingdom; the University of Birmingham, Birmingham, United Kingdom; Chalmers University of Technology (Gothenburg, Sweden); STRI AB (Ludvika, Sweden); and Luleå University of Technology (Skellefteå, Sweden). Especially we would like to thank Emmanouil Styvaktakis for his contributions to bridging the gap between our research areas.

The availability of data from real power system measurements has been an important condition for allowing us to write this book. Measurement data and other power system data and information were collected through the years. Even though not all of them were used for the material presented in this book, they all contributed to our further understanding of power quality monitoring and disturbance data analysis. Therefore we would like to thank all those that have contributed their measurement data through the years (in alphabetical order): Peter Axelberg (Unipower); Geert Borloo (Elia); Larry Conrad (Cinergy); Magnus Ericsson (Trinergi); Alistair Ferguson (Scottish Power); Zhengti Gu (Shanghai, China); Per Halvarsson (Trinergi and Dranetz BMI); Mats Häger (STRI); Daniel Karlsson (Sydkraft, currently at Gothia Power); Johan Lundquist (Chalmers, currently at Sycon); Mark McGranaghan (Electrotek, currently at EPRI Solutions); Larry Morgan (Duke Power); Robert Olofsson (Göteborg Energi, currently at Metrum, Sweden); Giovanna Postiglione (University of Naples, currently at FIAT Engineering); Christian Roxenius (Göteborg Energi); Dan Sabin (Electrotek); Ambra Sannino (Chalmers, currently at ABB); Helge Seljeseth (Sintef Energy Research);

Torbjörn Thiringer (Chalmers); Erik Thunberg (Svenska Kraftnät); and Mats Wahlberg (Skellefteå Kraft).

The interesting discussions in a number of working groups and international cooperation projects also contributed to the material presented in this book. The authors especially acknowledge the contribution from fellow members in IEEE task force P1564 and CIGRE working group C4.07 (originally 36.07).

Many thanks are due to the anonymous reviewers of this book for their valuable suggestions. Thanks are also due to Peter Willett (University of Connecticut, United States) for his encouragement and very useful suggestions, and to Mats Viberg (Chalmers, Sweden) for support. A special thanks also goes to Lars Moser (KvaLita, Sweden) for invaluable encouragement and support.

A final thanks goes to Marilyn Catis and Anthony Vengraitis at IEEE Press for encouraging us to start writing this book and for the help through the whole writing process.

M. H. J. B
I. Y-H. G

# CHAPTER 1

# INTRODUCTION

This chapter introduces the subjects that will be discussed in more detail in the remainder of this book: power quality events and variations, signal processing of power quality measurements, and electromagnetic compatibility (EMC) standards. This chapter also provides a guide for reading the remaining chapters.

## 1.1 MODERN VIEW OF POWER SYSTEMS

The overall structure of the electric power system as treated in most textbooks on power systems is as shown in Figure 1.1: The electric power is generated in large power stations at a relatively small number of locations. This power is then transmitted and distributed to the end users, typically simply referred to as "loads." Examples of books explicitly presenting this model are [193, 211, 322].

In all industrialized countries this remains the actual structure of the power system. A countrywide or even continentwide transmission system connects the large generator stations. The transmission system allows the sharing of the resources from the various generator stations over large areas. The transmission system not only has been an important contributing factor to the high reliability of the power supply but also has led to the low price of electricity in industrialized countries and enabled the deregulation of the market in electrical energy.

Distribution networks transport the electrical energy from the transmission substations to the various loads. Distribution networks are typically operated radially and power transport is from the transmission substation to the end users. This

*Signal Processing of Power Quality Disturbances.* By Math H. J. Bollen and Irene Yu-Hua Gu

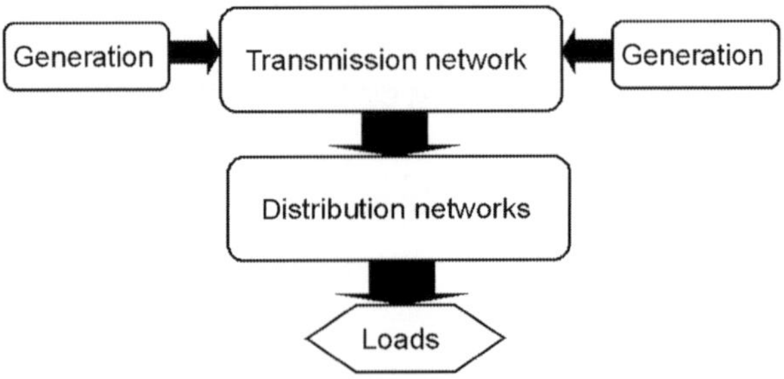


**Figure 1.1** Classical structure of power system.

allows for easy methods of protection and operation. The disadvantage is that each component failure will lead to an interruption for some end users.

There are no absolute criteria to distinguish between distribution and transmission networks. Some countries use the term *subtransmission networks* or an equivalent term to refer to the networks around big cities that have a transmission system structure (heavily meshed) but with a power transport more or less in one direction. Discussion of this terminology is however far outside the scope of this book.

Due to several developments during the last several years, the model in Figure 1.1 no longer fully holds. Even though technically the changes are not yet very big, a new way of thinking has emerged which requires a new way of looking at the power system:

- The deregulation of the electricity industry means that the electric power system can no longer be treated as one entity. Generation is in most countries completely deregulated or intended to be deregulated. Also transmission and distribution are often split into separate companies. Each company is economically independent, even where it is electrically an integral part of a much larger system.
- The need for environmentally friendly energy has led to the introduction of smaller generator units. This so-called embedded generation or distributed generation is often connected no longer to the transmission system but to the distribution system. Also economic driving forces, especially with combined heat and power, may result in the building of smaller generation units.
- Higher demands on reliability and quality mean that the network operator has to listen much closer to the demands of individual customers.

A more modern way of looking at the power system resulting from these developments is shown in Figure 1.2. The electric power network no longer transports energy from generators to end users but instead enables the exchange of energy between customers. Note that these customers are the customers of the network (company), not only the end users of the electricity.

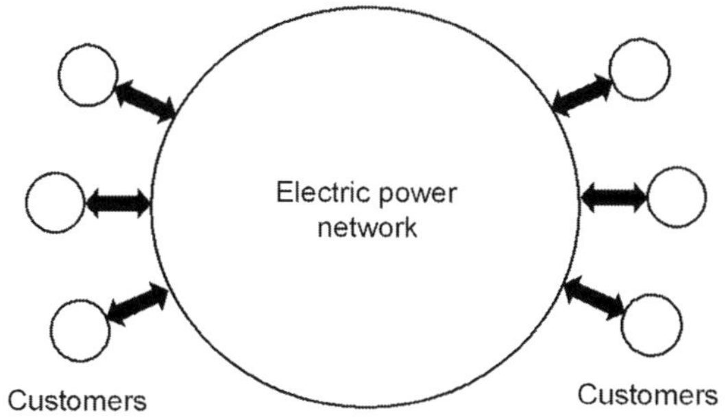


**Figure 1.2** Modern view of power system.

The actual structure of the power system is still very much as in Figure 1.1, but many recent developments require thinking in the structure of Figure 1.2. The power network in Figure 1.2 could be a transmission network, a distribution network, an industrial network, or any other network owned by a single company. For a transmission network, the customers are, for example, generator stations, distribution networks, large industrial customers (who would be generating or consuming electricity at different times, based on, e.g., the price of electricity at that moment), and other transmission networks. For a distribution network, the customers are currently mainly end users that only consume electricity, but also the transmission network and smaller generator stations are customers. Note that all customers are equal, even though some may be producing energy while others are consuming it. The aim of the network is only to transport the electrical energy, or in economic terms, to enable transactions between customers. An example of a transmission and a distribution network with their customers is shown in Figure 1.3.

The technical aim of the electric power networks in Figures 1.2 and 1.3 becomes one of allowing the transport of electrical energy between the different customers,

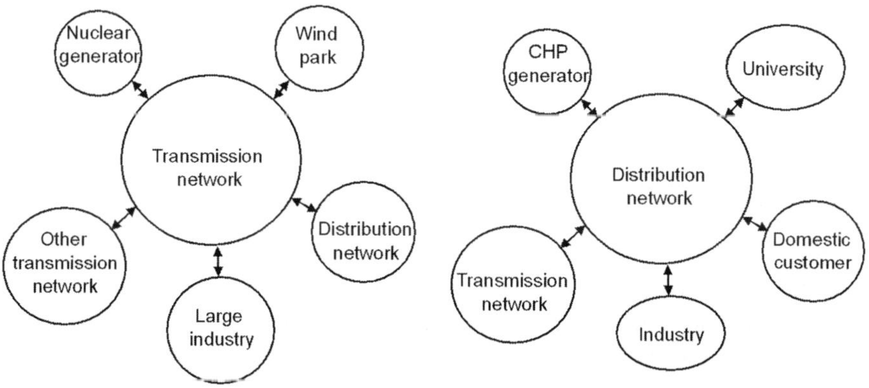


**Figure 1.3** Customers of a transmission network (left) and a distribution network (right).

guaranteeing an acceptable voltage, and allowing the currents taken by the customers. As we will see in Section 1.2.2 power quality concerns the interaction between the network and its customers. This interaction takes place through voltages and currents. The various power quality disturbances, such as harmonic distortion, of course also appear at any other location in the power system. But disturbances only become an issue at the interface between a network and its customers or at the equipment terminals.

The model in Figure 1.2 should also be used when considering the integration of renewable or other environmentally friendly sources of energy into the power system. The power system is no longer the boundary condition that limits, for example, the amount of wind power that can be produced at a certain location. Instead the network's task becomes to enable the transport of the amount of wind power that is produced and to provide a voltage such that the wind park can operate properly. It will be clear to the reader that the final solution will be found in cooperation between the customer (the owner of the wind park) and the network operator considering various technical and economic constraints.

Concerning the electricity market, the model in Figure 1.2 is the obvious one: The customers (generators and consumers) trade electricity via the power network. The term *power pool* explains rather well how electricity traders look at the power network. The network places constraints on the free market. A much discussed one is the limited ability of the network to transport energy, for example, between the different European countries. Note that under this model lack of generation capacity is not a network problem but a deficiency of the market.

## 1.2 POWER QUALITY

### 1.2.1 Interest in Power Quality

The enormous increase in the amount of activity in the power quality area can be observed immediately from Figure 1.4. This figure gives the number of papers in the INSPEC database [174] that use the term *power quality* in the title, the abstract, or the list of keywords. Especially since 1995 interest in power quality appears to have increased enormously. This means not that there were no papers on power quality issues before 1990 but that since then the term power quality has become used much more often.

There are different reasons for this enormous increase in the interest in power quality. The main reasons are as follows:

- Equipment has become less tolerant of voltage quality disturbances, production processes have become less tolerant of incorrect operation of equipment, and companies have become less tolerant of production stoppages. Note that in many discussions only the first problem is mentioned, whereas the latter two may be at least equally important. All this leads to much higher costs than before being associated with even a very short duration disturbance. The

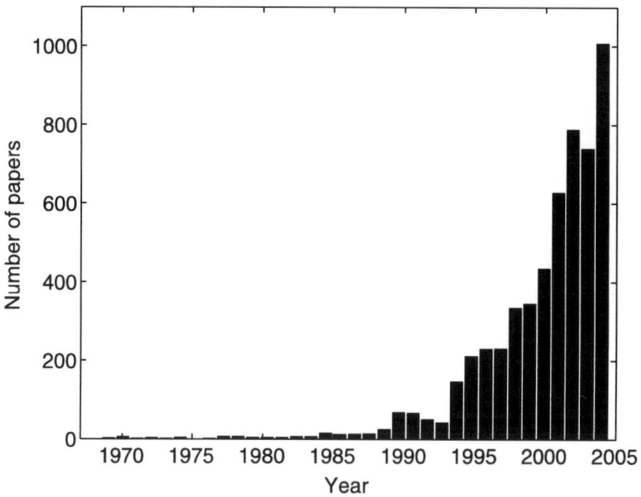


**Figure 1.4** Use of term power quality, 1968–2004.

main perpetrators are (long and short) interruptions and voltage dips, with the emphasis in discussions and in the literature being on voltage dips and short interruptions. High-frequency transients do occasionally receive attention as causes of equipment malfunction but are generally not well exposed in the literature.

- Equipment produces more current disturbances than it used to do. Both low- and high-power equipment is more and more powered by simple power electronic converters which produce a broad spectrum of distortion. There are indications that the harmonic distortion in the power system is rising, but no conclusive results are available due to the lack of large-scale surveys.
- The deregulation (liberalization, privatization) of the electricity industry has led to an increased need for quality indicators. Customers are demanding, and getting, more information on the voltage quality they can expect. Some issues of the interaction between deregulation and power quality are discussed in [9, 25].
- Embedded generation and renewable sources of energy create new power quality problems, such as voltage variations, flicker, and waveform distortion [325]. Most interfaces with renewable sources of energy are sensitive to voltage disturbances, especially voltage dips. However, such interfaces may be used to mitigate some of the existing power quality disturbances [204]. The relation between power quality and embedded generation is discussed among others in [178, Chapter 5; 99, Chapter 9; 118, Chapter 11]. An important upcoming issue is the immunity of embedded generation and large wind parks to voltage dips and other wide-scale disturbances. The resulting loss of generation as a result of a fault in the transmission system becomes a system security (stability) issue with high penetration of embedded generation.

- Also energy-efficient equipment is an important source of power quality disturbances. Adjustable-speed drives and energy-saving lamps are both important sources of waveform distortion and are also sensitive to certain types of power quality disturbances. When these power quality problems become a barrier for the large-scale introduction of environmentally friendly sources and end-user equipment, power quality becomes an environmental issue with much wider consequences than the currently merely economic issues.

### 1.2.2 Definition of Power Quality

Various sources give different and sometimes conflicting definitions of power quality. The Institute of Electrical and Electronics Engineers (IEEE) dictionary [159, page 807] states that "power quality is the concept of powering and grounding sensitive equipment in a matter that is suitable to the operation of that equipment." One could, for example, infer from this definition that harmonic current distortion is only a power quality issue if it affects sensitive equipment. Another limitation of this definition is that the concept cannot be applied anywhere else than toward equipment performance.

The International Electrotechnical Commission (IEC) definition of power quality, as in IEC 61000-4-30 [158, page 15], is as follows: "Characteristics of the electricity at a given point on an electrical system, evaluated against a set of reference technical parameters." This definition of power quality is related not to the performance of equipment but to the possibility of measuring and quantifying the performance of the power system.

The definition used in this book is the same as in [33]: Power quality is the combination of voltage quality and current quality. *Voltage quality* is concerned with deviations of the actual voltage from the ideal voltage. *Current quality* is the equivalent definition for the current. A discussion on what is ideal voltage could take many pages, a similar discussion on the current even more. A simple and straightforward solution is to define the ideal voltage as a sinusoidal voltage waveform with constant amplitude and constant frequency, where both amplitude and frequency are equal to their nominal value. The ideal current is also of constant amplitude and frequency, but additionally the current frequency and phase are the same as the frequency and phase of the voltage. Any deviation of voltage or current from the ideal is a *power quality disturbance*. A disturbance can be a voltage disturbance or a current disturbance, but it is often not possible to distinguish between the two. Any change in current gives a change in voltage and the other way around. Where we use a distinction between voltage and current disturbances, we use the cause as a criterion to distinguish between them: Voltage disturbances originate in the power network and potentially affect the customers, whereas current disturbances originate with a customer and potentially affect the network. Again this classification is due to fail: Starting a large induction motor leads to an overcurrent. Seen from the network this is clearly a current disturbance. However, the resulting voltage dip is a voltage disturbance for a neighboring customer. For the network operator this is a current disturbance, whereas it is a voltage disturbance for the neighboring customer. The fact that one underlying event (the motor start in this case) leads to different disturbances for different customers or at different locations is very common for power quality issues. This still often leads to confusing discussions and confirms the need for a new view of power systems, as mentioned in Section 1.1.

This difficulty of distinguishing between voltage and current disturbances is one of the reasons the term power quality is generally used. The term voltage quality is reserved for cases where only the voltage at a certain location is considered. The term current quality is sometimes used to describe the performance of power-electronic converters connected to the power network.

Our definition of power quality includes more disturbances than those that are normally considered part of power quality: for example, frequency variations and non-unity power factor. The technical aspects of power quality and power quality disturbances are not new at all. From the earliest days of electricity supply, power system design involved maintaining the voltage at the load terminals and ensuring the resulting load currents would not endanger the operation of the system. The main difference with modern-day power quality issues is that customers, network operators, and equipment all have changed. The basic engineering issues remain the same, but the tools have changed enormously. Power-electronic-based (low-power and high-power) equipment is behind many of the timely power quality problems. Power-electronic-based equipment is also promoted as an important mitigation tool for various power quality problems. The introduction of cheap and fast computers enables the automatic measurement and processing of large amounts of measurement data, thus enabling an accurate quantification of the power quality. Those same computers are also an essential part in power-electronic-based mitigation equipment and in many devices sensitive to power quality disturbances.

A large number of alternative definitions of power quality are in use. Some of these are worth mentioning either because they express the opinion of an influential organization or because they present an interesting angle.

Our definition considers every disturbance as a power quality issue. A commonly used alternative is to distinguish between *continuity* (or *reliability*) and *quality*. Continuity includes interruptions; quality covers all other disturbances. Short interruptions are sometimes seen as part of continuity, sometimes as part of quality. Following this line of reasoning, one may even consider voltage dips as a reliability issue, which it is from a customer viewpoint. It is interesting to note that several important early papers on voltage dips were sponsored by the reliability subcommittee of the IEEE Industrial Applications Society [e.g., 30, 75, 73].

The Council of European Energy Regulators [77, page 3] uses the term *quality of service in electricity supply* which considers three dimensions:

- *Commercial quality* concerns the relationship between the network company and the customer.
- *Continuity of supply* concerns long and short interruptions.

- *Voltage quality* is defined through enumeration. It includes the following disturbances: "frequency, voltage magnitude and its variation, voltage dips, temporary and transient overvoltages, and harmonic distortion."

It is interesting that "current quality" is nowhere explicitly mentioned. Obviously current quality is implicitly considered where it affects the voltage quality. The point of view here is again that adverse current quality is only a concern where it affects the voltage quality.

A report by the Union of the Electricity Industry (Eurelectric) [226, page 2] states that the two primary components of supply quality are as follows:

- *Continuity*: freedom from interruptions.
- *Voltage quality*: the degree to which the voltage is maintained at all times within a specific range.

Voltage quality, according to [226], has to do with "several mostly short-term and/or frequency related ways in which the supply voltage can vary in such a way as to constitute a particular obstacle to the proper functioning of some utilization equipment." The concept of voltage quality, according to this definition, is especially related to the operation of end-use equipment. Disturbances that do not affect equipment would not be part of voltage quality. Since at the measurement stage it is often not possible to know if a disturbances will affect equipment, such a definition is not practical.

Another interesting distinction is between *system quality* and *service quality*. System quality addresses the performance of a whole system, for example, the average number of short-circuit faults per kilometer of circuit. This is not a value which directly affects the customer, but as faults lead to dips and interruptions, it can certainly be considered as a quality indicator. A regulator could decide to limit the average number of faults per kilometer per circuit as a way of reducing the dip frequency. Service quality addresses the voltage quality for one individual customer or for a group of customers. In this case the number of dips per year would be a service quality indicator. Like any definition in the power quality area, here there are also uncertainties. The average number of dips per year for all customers connected to the network could be seen as a service quality indicator even though it does not refer to any specific customer. The 95% value of the number of dips per year, on the other hand, could be referred to as a system quality indicator. We will come back to this distinction when discussing site indices and system indices in Chapters 5 and 10.

Reference [260] refers in this context to *aggregate system service quality* and *individual customer service quality*. Reference [77] refers to the quality-of-supply and the quality-of-system approach of regulation. Under the *quality-of-supply approach* the quality would be guaranteed for every individual customer, whereas under the *quality-of-system approach* only the performance of the whole system would be guaranteed. An example of the quality-of-supply approach is to pay compensation to customers when they experience an interruption longer than a predefined duration (24 h is a commonly used value). Under the quality-of-system approach a network operator would, for example, have to reduce the use-of-system charges for all customers when more than 5% of customers experience an interruption longer than this predefined duration.

A term that is very much related to power quality is the term *electromagnetic compatibility* as used within IEC standards. According to IEC 61000-1-1 [148], "Electromagnetic compatibility is the ability of an equipment or system to function satisfactorily in its electromagnetic environment without introducing intolerable electromagnetic disturbances to anything in that environment." The first part of the definition, "ability ··· to function ··· in its ··· environment" fits well with the aforementioned definition of voltage quality. The second part of the definition, "introducing ··· disturbances ···" is rather similar to our term current quality. The IEC has published a whole series of standards and technical reports on EMC, most of which are part of the IEC 61000 series. Most international standards on power quality are part of this series. The most important ones are listed in Appendix A. Some aspects of EMC that are important for power quality are discussed in Section 1.4.

Within the IEC standards on EMC, a distinction is made between an (electromagnetic) disturbance and (electromagnetic) interference: "A disturbance is a phenomenon which may degrade the performance of a device, equipment or system, or adversely affect living or inert matter" [148]. In power quality terms, any deviation from the ideal voltage or current can be labeled as a disturbance. *Interference* is much stricter defined: It is the actual *degradation of a device, equipment, or system caused by an electromagnetic disturbance* [148]. The term *power quality problem* could be used as a synonym. In this book we will mainly discuss (power quality) disturbances as the term *interference* can only be used with reference to a specific piece of equipment.

### 1.2.3 Events and Variations

An important division of power quality disturbances is between variations and events. Variations are steady-state or quasi-steady-state disturbances that require (or allow) continuous measurements. Events are sudden disturbances with a beginning and an ending. Such a distinction is made in almost all publications on power quality, but the terminology differs. With reference to more classical power engineering, the measurement of variations is similar to metering of the energy consumption (i.e., continuous), whereas the measurement of events is similar to the functioning of a protection relay (i.e., triggered).

A typical example of a power quality variation is the variation of the power system frequency. Its nominal value is 50 Hz but the actual value always differs from this by up to about 1 Hz in a normal system. At any moment in time the frequency can be measured and a value will be obtained. For example, one may decide to measure the power system frequency once a second from the number of voltage zero crossings of the voltage waveform. In this way the average frequency is obtained every second. After one week this measurement will have resulted in $7 \times 24 \times 60 \times 60 = 604{,}800$ frequency values. These values can next be used to obtain information on the probability distribution, like average, standard deviation, and 99% interval (the range not exceeded by 99% of the values).

The issues to be discussed when measuring power quality variations include

- extracting the characteristics, in this case the frequency, from the sampled voltage or current waveform;
- statistics to quantify the performance of the supply at one location; and
- statistics to quantify the performance of a whole system.

These issues will be discussed in detail in the forthcoming chapters. The origin of some power quality variations is discussed in Chapter 2. Signal-processing methods for extracting characteristics from measured voltage and current waveforms are discussed in Chapters 3 and 4. Statistical methods for further processing the characteristics obtained are discussed in Chapter 5.

A typical example of a power quality event is an interruption. During an interruption the voltage at the customer interface or at the measurement location is zero. To measure an interruption, one has to wait until an interruption occurs. This is done automatically in most power quality monitors by comparing the measured voltage magnitude with a threshold. When the measured voltage magnitude is less than the threshold for longer than a certain time, the monitor has detected the start of an interruption. The end of the interruption is detected when the voltage magnitude rises above a threshold again. The duration of the interruption is obtained as the time difference between the beginning and the end of the event. This description for a rather simple event already shows the complexity in the measurement of events:

- A method has to be defined to obtain the *voltage magnitude* from the sampled waveform.
- Threshold levels have to be set for the beginning threshold and for the ending threshold. These two thresholds could be the same or different. Also a value has to be chosen for the minimum duration of an interruption.
- Characteristics have to be defined for the event, in this case the duration of the interruption.

After a sufficiently long monitoring time at a sufficiently large number of locations, it is again possible to obtain statistics. But these statistics are of a completely different nature than for power quality variations. Instead of a distribution over time, a distribution of the duration of the interruption is obtained. One may be interested in the number of interruptions lasting longer than 1 min or longer than 3 h. The average duration of an interruption no longer has any direct meaning, however. It will depend on the minimum duration of an interruption to be recorded. If only interruptions longer than 1 min are recorded, the average duration may be 25 min. If, however, all interruptions longer than 1 s are recorded, a large number of very short duration events may show up, leading to an average duration of only 20 s. The choice of the thresholds will also affect the average values with some events.

The origins of some power quality events are discussed in Chapter 6, methods for detecting events in Chapter 7, characterization of events in Chapter 8, event classification in Chapter 9, and the presentation of event statistics in Chapter 10.

The distinction between variations and events is not always easy to make. If we, for instance, consider changes in the voltage magnitude as a power quality disturbance, one may consider a voltage dip as an extreme case of a voltage magnitude variation. A unique way of defining events is by the triggering that is required to start their recording. Variations do not need triggering, events do. The difference between a voltage dip and a voltage (magnitude) variation is in the triggering. A voltage dip has a specific starting and ending instant, albeit not always uniquely defined. Both voltage dips and voltage variations use the root-mean-square (rms) voltage as their basic measurement quantity. However, for the further processing of voltage variations all values are important, whereas for the further processing of voltage dips only the rms values below a certain threshold are considered.

### 1.2.4 Power Quality Monitoring

From a pure measurement viewpoint there is no difference between power quality measurements and the measurement of voltages and currents, for example for protection or control purposes. In fact, many signal-processing tools discussed in this book have a wider application that just power quality. The difference is in the further processing and application of the measured signals. The results of power quality monitoring are not used for any automatic intervention in the system. Exceptions are the measurements as part of power quality mitigation equipment, but such equipment is more appropriately classified as protection or control equipment.

Power quality measurements are performed for a number of reasons:

- Finding the cause of equipment malfunction and other power quality problems. Finding the cause of a power quality problem is in many cases the first step in solving and mitigating the problem. The term *power quality troubleshooting* is often used for this. With these kind of measurements it is important to extract as much information as possible from the recorded voltage and current waveforms. With most existing equipment the power quality engineer directly interprets the recorded waveform or some simple characteristics such as the rms voltage versus time or the spectrum of the voltage or current. In most cases hand-held or movable equipment is used and the measurements are performed during a relatively short period. This has been the main application of power quality measurements for a long time.
- Permanent and semipermanent monitoring to get statistical information on the performance of the supply or of the equipment. An increasing number of network companies are installing permanent monitors to be able to provide information to their customers on the performance of their system. In some

cases, a national regulator demands this kind of information as well. The latter is becoming common practice for long interruptions but only very slowly taking off for other power quality disturbances.

- Permanent and semipermanent monitoring can also be used to monitor the network instead of only the voltage and current quality at the interface with the customer. A number of network companies have used voltage-dip recordings and statistics to assess the performance of the distribution system protection. Long voltage dips are often due to overly slow settings of protection relays. The resetting of protection relays resulted in a significant reduction of the number of long voltage dips and thus in an improvement of the voltage quality. But also system events that do not lead directly to problems with customer equipment provide information on the performance of the network. Examples are prestrike and restrike transients with capacitor switching and self-clearing faults in high-impedance grounded systems. Taking the right action may prevent future dips and even interruptions. The term *power quality predictive maintenance* is used in this context in [217]. Permanent power quality monitors can play an important role in reliability-centered maintenance (RCM).
- Another important application of permanent power quality monitoring is that troubleshooting no longer requires additional measurements. The moment a problem is reported, past data can be used to find the cause. When a sufficiently long data period is available, it is even possible to compare the effectiveness of different mitigation methods.
- The results of wide-scale monitoring campaigns, such as the distribution power quality (DPQ) survey in the United States, can be used to define the electromagnetic environment to which end-user equipment is subjected.
- The data obtained from permanent monitors can be used to analyze the system events that led to an interruption or blackout. Even though transmission operators have installed disturbance recorders for this purpose, power quality monitors may give important additional information. This holds to an even higher degree for public and industrial distribution systems. Knowledge about the chain of events that led to an interruption or blackout is important for preventing future events. An analysis of power quality recordings during the August 2003 blackout in the United States and Canada was published within a few days [132].

A general scheme for carrying out power quality measurements is shown in Figure 1.5. Part of the measurements take place in dedicated devices, often referred to as power quality monitors, part take place in devices that have other functions as well. The postprocessing of the data often takes place on computers far away from the monitors. The actual measurement takes place in a measurement device, which often includes the standard instrument transformers. The whole chain from the analog voltages and currents in the power system to the statistical indices resulting from the postprocessing is referred to as *power quality monitoring*.

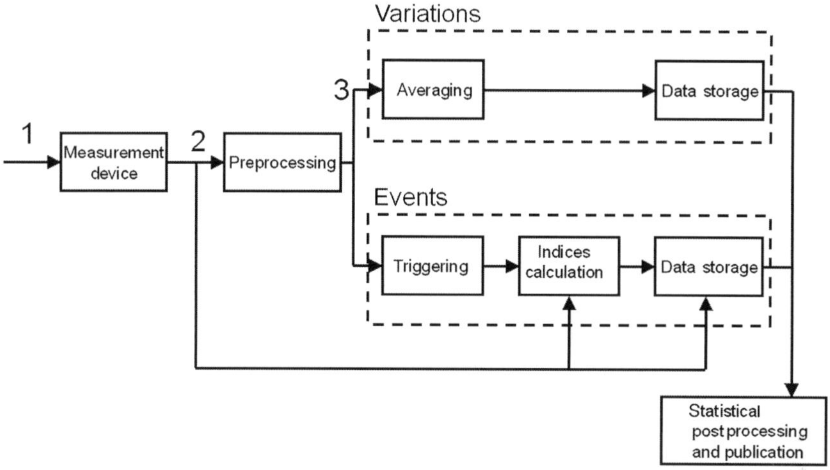


**Figure 1.5** General scheme of power quality measurements: (1) voltage or current in system; (2) sampled and digitized voltage or current; (3) quantity for further processing.

The first step in power quality monitoring is the transformation from analog voltages and currents in the power system to sampled digital values that can be processed automatically. The measurement device block in Figure 1.5 includes

- instrument transformers,
- analog anti-aliasing filters,
- sampling and digitizing, and
- digital anti-aliasing and down sampling.

Anti-aliasing is needed to prevent frequency components above the Nyquist frequency (half the sampling frequency) from showing up at low-frequency components. This is a standard part of any digital measurement device. The use of special instrument transformers is a typical power system issue. Voltages and currents in the power system are in many cases far too high to be measured directly. Therefore they are transformed down to a value that can be handled, traditionally 110 V and 1 or 5 A. These so-called instrument transformers are designed and calibrated for 50 or 60 Hz. At this frequency they have a small error. However, some of the power quality disturbances of interest require the measurement of significantly higher frequencies. For those frequencies the accuracy of the instrument transformers can no longer be taken for granted. This is especially important when measuring harmonics and transients. For some measurements special equipment such as resistive voltage dividers and Rogowski coils is being used.

In this book we will assume that the sampled and digitized voltage or current waveforms (referred to as *waveform data*) are available for processing. From the waveform data a number of characteristics are calculated for further processing. The example mentioned a number of times before is the rms voltage. The voltage waveform cannot be directly used to detect events: It would lead to the detection of 100 voltage dips per second. It would also not be very suitable to describe variations in the magnitude of the voltage. For the detection of voltage dips, the one-cycle rms voltage shall be compared with a threshold every half cycle, according to IEC 61000-4-30 [158]. Once an event is detected, its indices are calculated and stored. Some monitors not only store calculated event data but also part of the complete voltage and/or current waveform data. These data can later be used for diagnostics, for calculating additional indices, or for educational purposes. In our research groups we learned a lot about power quality and about power systems in general from the study of waveforms obtained by power quality monitors.

Note that we will refer to the whole chain, including the instrument transformers and the postprocessing outside the actual monitors, as power quality monitoring. This book will not go into further detail on the transformation from voltages and currents in the system to digital waveform data. The main theme of this book is the further processing of these digital waveform data.

The further processing of the data is completely different for variations and events. For power quality variations the first step is again the calculation of appropriate characteristics. This may be the rms voltage, the frequency, or the spectrum. Typically average values over a certain interval are used, for example, the rms voltage obtained over a 10-cycle window. The standard document IEC 61000-4-30 prescribes the following intervals: 10 or 12 cycles, 150 or 180 cycles, 10 min, and 2 h. Some monitors use different window lengths. Some monitors also give maximum and minimum values obtained during each interval. Some monitors do not take the average of the characteristic over the whole interval but a sample of the characteristic at regular intervals, for example, the spectrum obtained from one cycle of the waveform once every 5 min. Further postprocessing consists of the calculation of representative statistical values (e.g., the average or the 95 percentile) over longer periods (e.g., one week) and over all monitor locations. The resulting values are referred to as site indices and system indices, respectively.

The processing of power quality events is different from the processing of power quality variations. In fact, the difference between events and variations is in the method of processing, not necessarily in the physical phenomenon. Considering again the rms voltage, the events considered are short and long interruptions, voltage dips and swells, and (long-duration) overvoltages and undervoltages. The standard first step in their processing is the calculation of the rms voltage, typically over a one-cycle window. But contrary to power quality variations, the resulting value is normally not stored or used. Only when the calculated rms voltage exceeds a certain threshold for a certain duration does further processing start. Some typical theshold and duration values are given in Figure 1.6. These events are referred to as *voltage magnitude events* in [33] and as *rms variations* by some authors. We will refrain from using the latter term because of the potential confusion with our term, (*power quality*) *variations*.

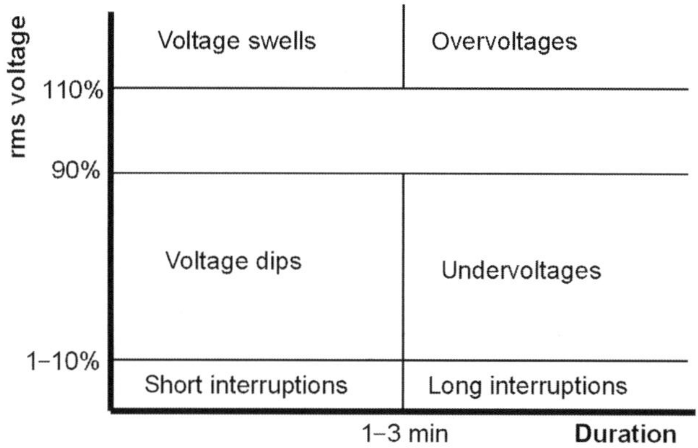


**Figure 1.6** Examples of threshold values triggering further processing of events based on rms voltage.

The vertical axis of Figure 1.6 gives the threshold value as a percentage of a reference voltage. Typically the nominal voltage is used as a reference, but sometimes the average voltage over a shorter or longer period before the event is used as a reference. The horizontal axis gives the time during which the rms voltage should exceed the threshold before further processing of the event starts. Further processing of a voltage-dip event is triggered whenever the rms voltage drops below the voltage-dip threshold (typically 90%), whereas further processing of a long interruption is triggered when the rms voltage drops below the interruption threshold (typically 1 or 10%) for longer than 1 to 3 min. Different values are used for the border between dips and interruptions and for the border between short and long interruptions. A further discussion on triggering of power quality events can be found in Chapter 7.

The triggering levels in Figure 1.6 are often referred to as "definitions" for these events. This is, for example, the case in IEEE standard 1159[165]. The authors are of the opinion that this is strictly speaking not correct. The thresholds are aimed at deciding which voltage-dip events require further processing (e.g., to be included in voltage-dip statistics). Any temporary reduction in rms voltage, no matter how small, is a voltage dip, even if there is no reason to record the event.

The further processing of a power quality event consists of the calculation of various indices. The so-called single-event indices (also known as *single-event characteristics*) typically include a duration and some kind of magnitude. The actual processing differs for different types of events and may include use of the sampled waveform data. Statistical processing of power quality events consists of the calculation of site indices (typically number of events per year) and system events (typically number of events per site per year). The calculation of single-event indices will be discussed in further detail in Chapter 8, the calculation of site and system indices in Chapter 10.

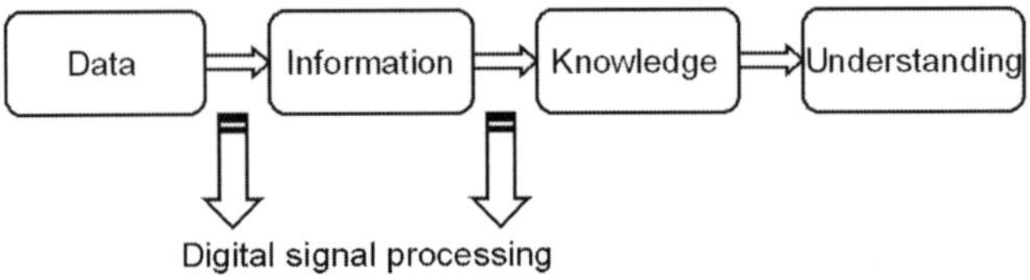


**Figure 1.7** Role of signal processing in extraction of information from power quality data.

## 1.3 SIGNAL PROCESSING AND POWER QUALITY

Digital signal processing, or signal processing in short, concerns the extraction of features and information from measured digital signals. As a research area signal processing covers any type of signal, including electrocardiogram (ECG) and electroencephalogram (EEG) signals, infrared pictures taken from fields suspected of containing land mines, radio waves from distant galaxies, speech signals transmitted over telephone lines, and remote-sensing data. A wide variety of signal-processing methods have been developed through the years both from the theoretical point of view and from the application point of view for a wide range of signals. In this book, we will study the application of some of these methods on voltage and current waveforms. The processing of power quality monitoring data can be described by the block diagram in Figure 1.7. Data are available in the form of sampled voltage and/or current waveforms. From these waveforms, information is extracted, for example, the retained voltage and duration of a voltage dip (see Section 8.6). Signal-processing tools play an essential role in this step. To extract knowledge from the information (e.g., the type and location of the fault that caused the voltage dip), both signal-processing tools and power system knowledge are needed. Having enough knowledge will in the end lead to understanding, for example, that dips occur more during the summer because of lightning storms and to potential mitigation methods. See also the discussion on the reasons for power quality monitoring at the start of Section 1.2.4.

We will not discuss the details of the difference between data, information, knowledge, and understanding. In fact, there are no clear boundaries between them. Having lots of data and the ability to assess them may give the impression of understanding. What is important here is that each step in Figure 1.7 is a kind of refinement of the previous step. One may say that signal processing extracts and enhances the information that is hidden or not directly perceivable.

### 1.3.1 Monitoring Process

The main emphasis in the signal-processing parts of this book will be on the analysis and extraction of relevant features from sampled waveforms. The process of power quality monitoring involves a number of steps that require signal processing:

- Characterizing a variation is done by defining certain features. The choice of features is often very much related to the essence of the variation: What is actually varying? For example, voltage variations concern variations in the magnitude of the voltage waveform. But such a definition is not sufficient to quantify the severity of the voltage variation through measurements. There are a number of features that can be used to quantify this magnitude: the absolute value of the complex voltage, the rms voltage, and the peak voltage. Other choices have to be made as well: the sampling frequency, the length of the window over which the characteristic is extracted, the repetition frequency of the measurement, and the way of processing a series of values. The choice of measurement methods may impose influence on the resulting value, which in turn may be decisive for compliance with a standard.

An excellent example of the use of signal-processing tools to extract and analyze features is the flickermeter standard for characterization of voltage fluctuations. See Section 2.4 for a detailed discussion on the flickermeter algorithm.

- Distinguishing between a variation and an event, a triggering mechanism is needed. The most commonly used method compares a sliding-window rms value with a threshold value. This again requires the definition of a number of values, such as the size of the window, the overlap between successive windows, and the choice of the threshold. However, other triggering methods may be more appropriate. A discussion on triggering methods can be found in Chapter 7.
- Characterizing each event through a number of parameters once an event is detected (or captured). This again involves the extraction of one or more features. For voltage dips the event characterization is very much related to the characterization of voltage variations. There is a reasonable amount of agreement in the power quality field on how this should be done. But for voltage and current transients no standardized method is currently available. We will discuss the characterization of voltage dips and transients in more detail in Sections 8.6 and 8.10, respectively.
- Classifying each event according to its underlying causes from the extracted features. This can often be considered as the final aim of the analysis. One of the essential issues is to choose between the categories of classification methods, for example, linear or nonlinear classifier, depending on the signal characteristics. Next, in each category, a number of possible candidates can be selected. For example, a Newman–Pearson method is selected if one needs to maximize the classification rate while the false-alarm rate of classification should be below a certain threshold. Or, one may choose a support vector machine where the learning complexity of the machine is a practical issue of concern and the performance of the classifier to the testing data needs to be guaranteed.

Although the fundamental signal-processing techniques used in practical power quality monitoring have been the discrete Fourier transform (DFT) and the rms, many more have been proposed in the literature. This book will discuss recent developments in more detail.

### 1.3.2 Decomposition

Analyzing the sampled voltage or current waveforms offers quantitative descriptions of power quality, for example, the dominant harmonic components and their associated magnitudes, the points where disturbances start and end, and the block of data where different system faults led to the disturbances. Many signal-processing methods can be applied for such purposes. As we will describe later, a signal-processing method could be very good for one application but not very suitable for another application. For example, the wavelet transform may be very attractive for finding the transitions while it could be unattractive to harmonic analysis [129]. For each application, a set of methods can be chosen as candidates, and each may offer different performance and complexity; again it is a matter of trade-off. Below we try to roughly summarize the types of signal-processing methods that may be attractive to power quality analysis. It should be mentioned that the list of methods below is far from complete. They can be roughly categorized into two classes: transform or subband filter-based methods and model-based methods.

1. *Data Decomposition Based on Transforms or Subband Filters* These methods decompose the measurement into components. Depending on the stationarity of the measurement data (or data blocks), one may choose frequency- (or scale-) domain analysis or time–frequency- (or time–scale-) domain analysis.
   - *Frequency-Domain Analysis* If the measurement data (or block of the data) are stationary, frequency-domain decomposition of the data is often desirable. A standard and commonly preferred method is the DFT or its fast algorithm, the fast Fourier transform (FFT).

     Wavelet transform is another transform closely related to frequency-domain analysis. Wavelet transform decomposes data to scale-domain components where scales are related to frequencies in logarithmic scales.
   - *Time–Frequency- (or Time–Scale-) Domain Analysis* If the measurement data are nonstationary, it is desirable that they are decomposed into time-dependent frequency components. To obtain the time–frequency representation of data, a commonly used method is the short-time Fourier transform (STFT) or a set of sliding-window FFTs.

     The STFT can be explained equivalently by a set of bandpass filters with an equal bandwidth. The bandwidth is determined by the selected window and the size of the window.

     Another way to implement time–frequency representation of data is to use time–scale analysis by discrete wavelet filters. This is mostly done by successively applying wavelet transforms to decompose the low-pass-filtered data (or the original data) into low-pass and high-pass bands. This is equivalently described by a set of bandpass filters with octave bandwidth. The advantages are the possibility to trade off between time resolution and frequency resolution given a fixed joint time–frequency resolution value constrained under the uncertainty principle.

2. *Data Analysis Using Model-Based Methods* Another important set of signal-processing methods for power system data analysis are the model-based methods. Depending on the prior knowledge of systems, one may assume that the data sequences are generated from certain models, for example, sinusoidal models, autoregressive models, or state-space models. One of the advantages of model-based methods is that if the model is correctly chosen, it can achieve a high-frequency resolution as compared with filter-bank and transform-based methods. Conversely, if an incorrect model is applied, the performance is rather poor.
   - *Sinusoidal Models* The signal is modeled as the sum of a finite number of frequency components in white noise. The number of components is decided beforehand and the frequencies and (complex) magnitudes are estimated by fitting the measured waveform to the model. Three estimation methods—multiple signal classification (MUSIC), estimation of signal parameters via rotational invariance techniques (ESPRIT), and Kalman filters—are discussed in detail in Chapters 3 and 4.
   - *Other Stochastic Models* We limit ourselves to the autoregressive (AR), autoregressive moving-average (ARMA), and state-space models. In the models, the signal is modeled as the response of a linear time-invariant system with white noise as the input, where the system is modeled by a finite number of poles or poles and zeros. The AR and ARMA models are both discussed in detail in Chapters 3 and 4. For state-space modeling essential issues include predefining state variables and formulating a set of state and observation equations. Although Kalman filters are employed for estimating harmonics in the examples, their potential applications for power system disturbance analysis are much broader depending on how the state space is defined.

### 1.3.3 Stationary and Nonstationary Signals

As far as the signals are concerned, we can roughly classify them into two cases: stationary and nonstationary signals. The signal-processing methods introduced in Chapter 3 are for stationary signals. However, strictly stationary signals do not exist in real-life power systems: Both small and large statistical changes occur in the signal parameters. The presence of small and relatively slow statistical changes is addressed through so-called block-based methods. The signal is assumed stationary over a short duration of time (or window), a so-called block of data; the signal features (or characteristics or attributes) are estimated over this window. Next the window is shifted in time and the calculations are repeated for a new block of data. The resulting estimated features become a function of time depending on the location of the window. Apart from these block-based signal-processing methods, Kalman filters offer non-block-based processing which can be directly applied to nonstationary signal processing. The different aspects of block-based (batch processing) and non-block-based (iterative processing) signal-processing methods are discussed in detail in Chapter 4. A more sophisticated segmentation method which automatically uses an adaptive block size is described in Chapter 7.

### 1.3.4 Machine Learning and Automatic Classification

A logical next step after quantifying and characterizing the data is to classify, diagnose, and mitigate the disturbances. Appropriate tools to achieve this include machine learning and automatic classification and diagnostics. Classification methods use features (or attributes or characteristics) of data as the input and the designated class label of the data as the output. A classification process usually consists of steps such as feature extraction and optimization, classifier design that finds a mapping function between the feature space and decision space, supervised or nonsupervised machine learning, validation, and testing. In Chapter 9 we will describe some frequently used linear and nonlinear classification methods where static features are considered. Further, in Chapter 9 a simple rule-based expert system for data classification will be described through a number of examples. A number of statistical-based classification methods such as Bayesian learning, Newman–Pearson methods, support-vector machines, and artificial neural networks together with practical application examples will also be described in Chapter 9.

## 1.4 ELECTROMAGNETIC COMPATIBILITY STANDARDS

### 1.4.1 Basic Principles

All communication between electrical devices is in the form of electromagnetic waves ruled by Maxwell's equations. This holds for intentional as well as unintentional communication. Electromagnetic waves are responsible for the power supply to the equipment (note that for the power supply in most cases Kirchhoff's equations are used as a low-frequency approximation of Maxwell's equations) and for the exchange of information between equipment. They are also responsible for all kinds of disturbances that may endanger the correct operation of the equipment. These so-called electromagnetic disturbances may reach the equipment trough metallic wires (so-called conducted disturbances) or in the form of radiation (radiated disturbances). Note that there is no difference between intentional information exchange and a disturbance from an electromagnetic viewpoint.

The general approach is to achieve *electromagnetic compatibility* between equipment. Electromagnetic compatibility is defined as "the ability of a device, equipment or system to function satisfactorily in its electromagnetic compatibility without introducing intolerable electromagnetic disturbances to anything in that environment" [148]. The IEC has published several standard documents (mainly in the 61000 series) which define various aspects of EMC. These include a number of standards on power quality issues like harmonics and voltage dips. An overview of all relevant EMC standards is given in Appendix A.

The principle of the EMC standards can best be explained by considering two devices: one which produces an electromagnetic disturbance and another that may be adversely affected by this disturbance. In EMC terms, one device (the "emitter") emits an electromagnetic disturbance; the other (the "susceptor") is susceptible to this disturbance. Within the EMC standards there is a clear distinction in meaning between (electromagnetic) "disturbance" and (electromagnetic) "interference." An electromagnetic disturbance is any unwanted signal that may lead to a degradation of the performance of a device. This degradation is referred to as electromagnetic interference. Thus the disturbance is the cause, the interference the effect.

The most obvious approach would be to test the compatibility between these two devices. If the one would adversely affect the other, there is an EMC problem, and at least one of the two needs to be improved. However, this would require testing of each possible combination of two devices, and if a combination would fail the test, it would remain unclear which device would require improvement.

To provide a framework for testing and improving equipment, the concept of *compatibility level* is introduced. The compatibility level for an electromagnetic disturbance is a reference value used to compare equipment emission and immunity. From the compatibility level, an emission limit and an immunity limit are defined. The immunity limit is higher than or equal to the compatibility level. The emission limit, on the other hand, is lower than or equal to the compatibility level (see Fig. 1.8). Immunity limit, compatibility level, and emission limit are

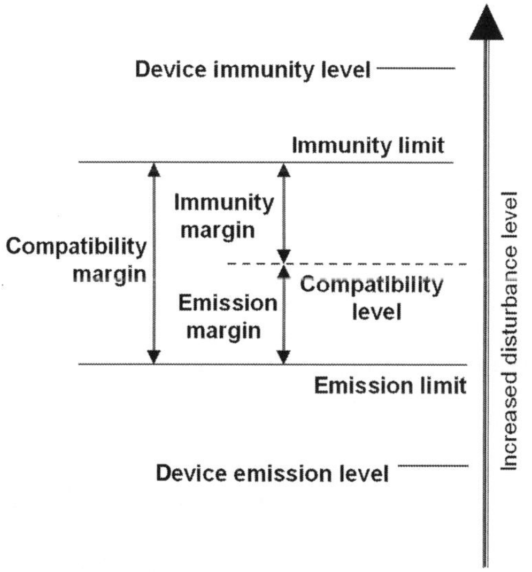


**Figure 1.8** Various levels, limits, and margins used in EMC standards.

defined in IEC standards. The ratio between the immunity limit and the compatibility level is called the *immunity margin*; the ratio between the compatibility level and the emission level is referred to as the *emission margin*. The value of these margins is not important in itself, as the compatibility level is just a predefined level used to fix emission and immunity limits. Of more importance for achieving EMC is the *compatibility margin*: the ratio between the immunity limit and the emission limit. Note that the compatibility margin is equal to the product of the emission margin and the immunity margin. The larger the compatibility margin, the smaller the risk that a disturbance from an emitter will lead to interference with a susceptor.

The testing of equipment involves comparing the amount of electromagnetic disturbance produced by the equipment (the *emission level*) with the emission limit and the maximum amount of disturbance for which the equipment functions normally (the *immunity level*) with the immunity limit. To pass the test, the emission level should be less than the emission limit and the immunity level should be higher than the immunity limit. The result of this is obvious: When two devices both pass the test, they will not adversely affect each other.

In the next section we will discuss how to choose the ratio between compatibility level and maximum emission level. This, however, does not say where to start when deciding on emission and immunity levels.

In case both emission and immunity levels can be freely chosen, the compatibility level can be freely chosen. A too high compatibility level would lead to high costs for equipment immunity; a too low level would lead to high costs for limiting equipment emission. The compatibility level should be chosen such that the sum of both costs (for all devices, thus the total costs to society) is minimal. With conducted disturbances that can be attributed to equipment such a trade-off is in principle possible. However, in practice the existing disturbance levels are often used as a basis for determining the compatibility level.

For some disturbances the emission level cannot be affected. One may think of (variations in) Earths magnetic field or cosmic radiation as an example in which it is impossible to affect the source of the disturbance. But also disturbances due to events in the power system (faults, lightning strokes, switching actions) are treated like this in the EMC standards even though it is possible to affect the source of the disturbance. This way of treating the power system as something that cannot be affected is again related to the fact that EMC standards apply to equipment only. There is no technical argument for this. An often-used argument is that voltage dips cannot be prevented because lightning strokes (leading to faults, leading to dips) are part of nature. (The very inappropriate term "acts of God" is sometimes used.) Even though it is not possible to prevent lightning, it is technically very well possible to limit the number of faults due to lightning strokes to overhead lines. Shielding wires, higher insulation levels, and underground cables are possible options. The prohibiting costs associated with some of these improvements would be a more valid argument.

Finally there are disturbances for which it is not possible (or not practical) to affect the immunity of the equipment. With voltage fluctuation the "equipment" is the combination of our eye with our brain. The susceptibility of the eye–brain combination to light intensity fluctuations is not possible to affect. Therefore the compatibility level is determined by the immunity limit.

### 1.4.2 Stochastic Approach

From the previous section the reader could get the impression that a very small margin between immunity and emission limits would be sufficient to guarantee that no device is affected adversely by another device. This would be the case if the tests could reproduce all possible operating conditions for all possible devices. But that would require an almost infinite number of tests to be applied on all devices. In practice a limited number of tests under strictly defined laboratory conditions are applied to a small number of devices (the so-called type testing). If these devices pass the test, all other devices of the same type are supposed to have passed the test and will get the appropriate certification.

A number of phenomena contribute to the uncertainty in immunity and emission limits[148], all in some way related to the fact that it is not possible to test each device for each possible situation:

- *Relevance of Test Methods* Each device to be tested is subjected to a number of well-defined tests in a laboratory environment. The actual operating conditions of the device are likely to deviate from the laboratory environment. This could lead to a device emitting more disturbance than under test conditions or to a susceptor being more susceptible than under test conditions. For example, the harmonic current emission of a device is tested for a voltage waveform of a given magnitude with small distortion and for a well-defined source impedance. Any difference in voltage magnitude, voltage distortion, and/or source impedance will affect the current distortion (i.e., the emission). The emission or immunity of a device may also be affected by the outside temperature, by its loading, or by the state of the device in its duty cycle.

  The tests have to be done under well-defined conditions to guarantee that they are reproducible. A test performed by one testing laboratory should give the same results as the same test performed by another laboratory. The result of this is however that the variations in ambient conditions and state of the device cause a spread in immunity and emission levels when the device is used in the real world.
- *Normal Spread in Component Characteristics* As said before, only a small number of devices are subjected to the tests. The characteristics of different devices will show slightly different characteristics, even if they are from the same type. An example is the second-harmonic current taken by a power-electronic converter. Second-harmonic current is due to the difference in component parameters between the diodes or transistors within one pair. This level will likely vary a lot between different devices.
- *Superposition Effects and Multidimensional Characteristics* During the tests the device is subjected to one disturbance at a time. When a device is subjected

to different disturbances at the same time, its susceptibility may be significantly different than for the individual disturbances. A classical example is the combination of voltage dips and harmonic distortion. Most voltage distortion in low-voltage networks is such that the peak value of the voltage shows a reduction without affecting the rms voltage; the voltage waveform becomes more of a trapezoidal wave instead of a sine wave. This is due to single-phase rectifiers all together drawing current during the maximum voltage. The effect of this reduction in crest factor is that the voltage at the direct current (dc) bus inside the single-phase rectifiers gets smaller; the dc bus voltage is more or less proportional to the crest factor. This reduction in dc bus voltage makes the rectifiers more susceptible to voltage dips.

Another example concerns the susceptibility of rotating machines for voltage unbalance. The machine can tolerate a higher unbalance for a nondistorted voltage waveform than for a distorted waveform. In the same way, its susceptibility against harmonic distortion is affected by the voltage unbalance.

- *Multiple-Emitting Sources* The problem of multiple-emitting sources was mentioned already before as it is one of the contributing factors to the uncertainty in emission level. The discussion before on the choice of emission and immunity limits was based on the assumption of one emitter and one susceptor. In reality there are in most cases a number of emitters and a number of susceptors. The presence of multiple susceptors does not affect the earlier reasoning very much. The susceptibility of a device to an electromagnetic disturbance is not affected by the susceptibility of a neighboring device. The neighboring device may affect the disturbance level and in that way the operation of the device. But the susceptibility in itself is not affected nor is the susceptibility of the neighboring device of any influence. The effect of neighboring devices on the disturbance is included in the statistical uncertainty in emission and immunity levels.

  The presence of multiple emitters will of course also not affect the susceptibility of a device. But it will have such a large effect on the electromagnetic environment that it cannot be treated as just another statistical variation. The concept for setting the immunity limit, as shown in Figure 1.8, can still be applied when the total emission (the resulting disturbance level) is used instead of the emission level. A method similar to the hosting-capacity approach (Section 1.7.1) may be used when the number of emitters is not known.

In the IEC EMC standards and in most publications on EMC it is stated that emission and immunity limits should be chosen such that the probability of electromagnetic interference is sufficiently small. This assumes of course that the distributions are known, which is generally not the case. The various contributions given here are such that it is very hard to bring then into a probability distribution. The result is that a large amount of engineering judgment is needed when deciding about emission and immunity limits. With almost any of the phenomena adjustments to the limits and/or to the tests have been shown to be needed. The EMC standards are not as static as one would expect. With new types of equipment, new types of emission and new types of susceptibility will be introduced. This means that EMC will remain an interesting area for many years to come.

### 1.4.3 Events and Variations

A distinction was made before between *events* and *variations*. This distinction also appears in the EMC standards, or more precisely in the lack of standard documents on power quality events. For example, the European voltage characteristics documents EN 50160 [106] gives useful information for variations but nothing for events [33].

As we saw before, the EMC standards are developed around the concept of compatibility level. For variations, which are measured continuously, a probability distribution can be obtained for each location. The 95% values for time and location can be used to obtain the compatibility level. It will be rather difficult to perform measurements for each location, but even an estimated probability distribution function will do the job. For events the situation becomes completely different, as it is no longer possible to obtain a value that is not exceeded 95% of the time. This requires a reevaluation of the concept of compatibility level. One can no longer define the requirement (emission limits and immunity limits) by a probability that interference will occur. Instead the setting of limits should be ruled by the number of times per year that interference will occur. This part of the EMC standards is not very well defined yet. We will come back to the statistical processing of events in Chapter 10.

A place where the distinction between variations and events becomes very clear is with voltage magnitude (rms voltage). Small deviations from the nominal voltage are called *voltage variations* or *voltage fluctuations*; large deviations are called *voltage dips*, *overvoltages*, or *interruptions*. Without distinguishing between variations and events, the somewhat strange situation would arise that the small deviations are regulated and the large ones are not.

### 1.4.4 Three Phases

Most power systems consist of three phases. But neither the EMC standards nor the power quality literature make much mention of this. There are a number of reasons for the lack of attention for three-phase phenomena:

- In normal operation the system voltages are almost balanced. For a balanced system a single-phase approach (more exactly the positive sequence) is sufficient. Most variations concern normal operation so that this approach is also generally deemed sufficient here.
- The EMC standards apply to devices, most of which are single phase.
- Three-phase models increase the complexity of the approach, in some cases significantly. This makes it harder to get a standard document accepted.

The only phenomenon that is treated in a three-phase sense is *unbalance*, where the negative-sequence voltage is divided by the positive-sequence voltage to quantify the unbalance. For most other disturbances the individual phase voltages are treated independently, after which the worst phase is taken to characterize the disturbance level.

However, most disturbances are not balanced as they are deviations from the ideal (constant and balanced) situation. The fact that *three-phase unbalance* is addressed in the EMC standards further emphasizes that even for variations the voltages are not fully balanced. During events (e.g., voltage dips) the deviation from the balanced case can be very large. During a phase-to-phase fault the magnitudes of positive-sequence and negative-sequence voltages become equal at the fault location. For fundamental frequency disturbances such as voltage fluctuations and voltage dips, a symmetrical-component approach seems the most appropriate. Also for harmonic distortion symmetrical-component methods have been proposed. For transients a different approach may have to be developed. We will come back to this discussion at various places in the forthcoming chapters.

## 1.5 OVERVIEW OF POWER QUALITY STANDARDS

The main set of international standards on power quality is found in the IEC documents on EMC. The IEC EMC standards consist of six parts, each of which consists of one or more sections:

- *Part 1: General* This part contains for the time being only one section in which the basic definitions are introduced and explained.
- *Part 2: Environment* This part contains a number of sections in which the various disturbance levels are quantified. It also contains a description of the environment, classification of the environment, and methods for quantifying the environment.
- *Part 3: Limits* This is the basis of the EMC standards where the various emission and immunity limits for equipment are given. Standards IEC 61000-3-2 and IEC 61000-3-4 give emission limits for harmonic currents; IEC 61000-3-3 and IEC 61000-3-5 give emission limits for voltage fluctuations.
- *Part 4: Testing and Measurement Techniques* Definition of emission and immunity limits is not enough for a standard. The standard must also define standard ways of measuring the emission and of testing the immunity of equipment. This is taken care of in part 4 of the EMC standards.
- *Part 5: Installation and Mitigation Guidelines* This part gives background information on how to prevent electromagnetic interference at the design and installation stage.
- *Part 6: Generic Standards* Emission and immunity are defined for many types of equipment in specific product standards. For those devices that are not covered by any of the product standards, the generic standards apply.

The most noticeable non-IEC power quality standard is the voltage characteristics document EN 50160 published by Cenelec. This document will be discussed in Section 5.6.3. Several countries have written their own power quality documents, especially on harmonic distortion. The IEEE has published a significant number of standard documents on power quality, with the harmonics standard IEEE 519 (see Section 5.6.6) probably being the one most used outside of the United States. A more recent document that has become a de facto global standard is IEEE 1366 defining distribution reliability indices (see Section 10.1.2). Other IEEE power quality standard documents worth mentioning are IEEE 1346 (compatibility between the supply and sensitive equipment, see Section 10.2.4), IEEE 1100 (power and grounding of sensitive equipment), IEEE 1159 (monitoring electric power quality), and IEEE 1250 (service to sensitive equipment). The relevant IEC and IEEE standard documents on power quality are listed in Appendixes A and B, respectively. Also the as-yet not-officially-published work within task forces 1159 (power quality monitoring) and 1564 (voltage-sag indices) is already widely referenced.

## 1.6 COMPATIBILITY BETWEEN EQUIPMENT AND SUPPLY

The interest in power quality started from incompatibility issues between equipment and power supply. Therefore it is appropriate to spend a few lines on this issue, even though this is not the main subject of this book. The distinction between voltage and current quality originates from these compatibility issues. We will mainly discuss voltage quality here but will briefly address current quality later.

Voltage quality, from a compatibility viewpoint, concerns the performance of equipment during normal and abnormal operation of the system. What matters to the equipment are only the voltages. In Section 1.2.3 we introduced the distinction between variations and events. This distinction is also important for the compatibility between equipment and supply. Events will further be divided into "normal events" and "abnormal events." In the forthcoming paragraphs guidelines are given for the compatibility between equipment and supply. It is thereby very important to realize that ensuring compatibility is a joined responsibility of the network and the customer. This responsibility sharing plays an important part in the discussion below.

### 1.6.1 Normal Operation

The voltage as experienced by equipment during normal operation corresponds to what we refer to here as *voltage variations*. Voltage variations will lead to performance deterioration and/or accelerated aging of the equipment. We distinguish between three levels of voltage variations:

- voltage variations that have no noticeable impact on equipment,
- voltage variations that have a noticeable but acceptable impact on equipment, and
- voltage variations that have an unacceptable impact on equipment, which includes malfunction and damage of equipment.

The design of the system and the design of the equipment should be coordinated in such a way that the third level is never reached during normal operation and the time spent at the second level is limited. In practice this means that the design of equipment should be coordinated with the existing level of voltage variations. A good indication of the existing level of voltage variations can in part be obtained from such documents as EN 50160 and IEC 61000-2-2 (see Sections 5.6.3 and 5.6.4, respectively). The alternative, to perform local measurements, is not always feasible. Any further discussion on the appropriateness of these documents is beyond the scope of this chapter.

The responsibility of the network operator is to ensure that the voltage quality does not deteriorate beyond a mutually agreed-upon level. Again the limits as given in EN 50160 are an appropriate choice for this. Several countries have regulations in place or are developing such regulations, in many cases based on the EN 50160 limits.

### 1.6.2 Normal Events

For design purposes it is useful to divide power quality events into *normal events* and *abnormal events*. Normal events are switching events that are part of the normal operation of the system. Examples are tap changing, capacitor switching, and transformer energizing as well as load switching. If the resulting voltage events are too severe, this will lead to equipment damage or malfunction. If there are too many events, this will cause unacceptable aging of equipment. The same approach may be used as for normal operation: Equipment design should be coordinated with the existing voltage quality. There are, however, two important differences. The first difference is in the type of limits. For events limits are in the form of a maximum severity for individual events and in a maximum number of events. (In Chapter 10 we will refer to these two types of event limits as single-event indices and single-site indices, respectively.) The second difference with normal operation is that there is no document that gives the existing levels. Some normal events are discussed in EN 50160, but the indicated levels are too broad to be of use for equipment design. What is needed is a document describing the existing and acceptable voltage quality with relation to switching actions in the system (i.e., normal events).

Fortunately, normal events rarely lead to problems with equipment. The main recent exception are capacitor-energizing transients. These have caused erroneous trips for many adjustable-speed drives. The problem is solved by a combination of system improvement (synchronized switching) and improved immunity of equipment. In terms of responsibility sharing, the network operator should keep the severity and frequency of normal events below mutually agreed-upon limits; the customer should ensure that the equipment can cope with normal events within those limits.

### 1.6.3 Abnormal Events

Abnormal events are faults and other failures in the system. These are events that are not part of normal operation and in most cases are also unwanted by the network operator. Voltage dips and interruptions are examples of voltage disturbances due to abnormal events in the system. It is not possible to limit the severity of abnormal events and thus also not possible to ensure that equipment can tolerate any abnormal event. A different design approach is needed here.

When the performance of the supply is known, an economic optimization can be made to determine an appropriate immunity of the equipment. A higher immunity requirement leads to increased equipment costs but reduced costs associated with downtime of the production. The total costs can be minimized by choosing the appropriate immunity level. This approach is behind the *voltage-sag coordination chart* as defined in IEEE 1346 [74]. Such an optimization is, however, only possible when detailed information is available on system performance and is therefore difficult to apply for domestic and commercial customers.

An alternative approach is to define a *minimum equipment immunity*. The requirements placed by normal operation and normal events already place a lower limit on the equipment immunity. This lower limit is extended to include also *common abnormal events*. An example of such a curve is the Information Technology Industry Council (ITIC) curve for voltage dips and swells. Although the origins of these curves are different, they may all be used as a minimum immunity curve. (Note that the term *voltage tolerance curve* is more commonly used than *immunity curve*.) The practical use of such a curve only makes sense when the number of events exceeding the curve are limited. This is where the responsibility of network operators comes in. The responsibility sharing for abnormal events such as voltage dips is as follows: The network operator should ensure a limited number of events exceeding a predefined severity; the customer should ensure that all equipment will operate as intended for events not exceeding this predefined severity. In the current situation a large compatibility gap is present between immunity requirements placed on equipment and regulatory requirements placed on the network operator. This compatibility gap is shown in Figure 1.9. The immunity curve shown is according to the class 3 criteria in Edition 2 of IEC 61000-4-11. Regulatory requirements are available in some countries for long interruptions, typically starting at durations between 1 and 5 min (3 min duration and 10% residual voltage have been used for the figure). The range in dips between the two curves is regulated on neither the equipment side nor the network side.

A more desirable situation is shown in Figure 1.10: A mutually agreed-upon curve defines both the minimum equipment immunity and the range of events that occur only infrequently. A regulatory framework may be needed to ensure the latter. As part of the regulation, the frequency of occurrence of events below the curve should be known. This allows an economic optimization to be performed by those customers that require a higher reliability than the standard one.

The existing compatibility gap is even larger than would follow from Figure 1.9. The IEC immunity standard refers to equipment performance, not to the performance of a production process. If a piece of equipment safely shuts down during an event, this may classify as compliance with the standard (this is the case for the drive standard IEC 61800-3). The process immunity curve is thus located toward the left of the equipment immunity curve.

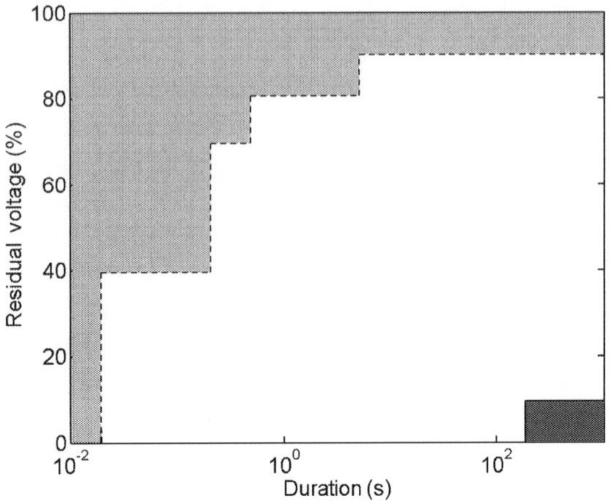


**Figure 1.9** Compatibility gap with IEC 61000-4-11 class 3 criteria toward left and area to which regulation most commonly applies to right.

Power quality also has a current quality side, which requires design rules in the same way as voltage quality. There are two reasons for limiting the severity and frequency of current disturbances. Current disturbances should not lead to damage, malfunction, or accelerated aging of equipment in the power system. The design rules should be the same as for normal operation and normal events as discussed before. The only difference is that the network operator is now on the receiving end of the disturbance. The second reason for limiting current disturbances is that they cause voltage disturbances, which are in turn limited. The limits placed by

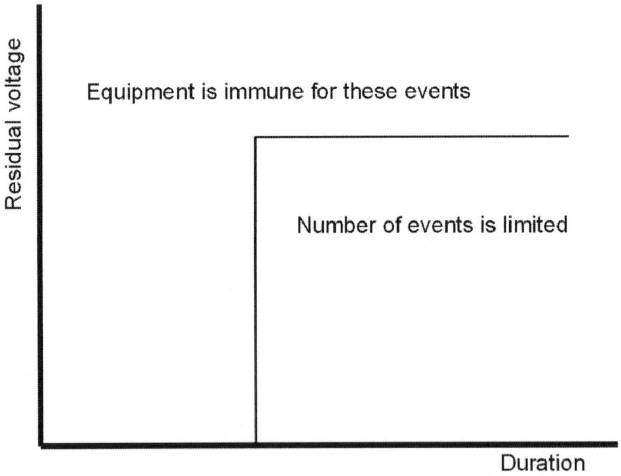


**Figure 1.10** Responsibility sharing between equipment and network operator: Compatibility gap has disappeared.

the network operator on the current quality for customers and equipment should correspond with the responsibility of the network operator to limit voltage disturbances.

## 1.7 DISTRIBUTED GENERATION

In Section 1.2.2 power quality was defined as the electrical interaction between the electricity grid and its customers. These costumers may be consumers or generators of electrical energy. The interaction was further divided into *voltage quality* and *current quality*, referring to the way in which the network impacts the customer and the way in which the customer impacts the network, respectively. When considering systems with large amounts of distributed generation, power quality becomes an important issue that requires a closer look. Three different power quality aspects are considered in [40]:

- Distributed generation is affected by the voltage quality in the same way as all other equipment is affected. The same design rules hold as in Section 1.6. An important difference between distributed generation and most industrial installations is that the erroneous tripping of the generator may pose a safety risk: The energy flow is interrupted, potentially leading to overspeed of the machine and large overvoltages with electronic equipment. This should be taken into consideration when setting immunity requirements for the installations.
- Distributed generation affects the current quality and through the grid also the voltage quality as experienced by other customers. The special character of distributed generation and its possible wide-scale penetration require a detailed assessment of this aspect. We will discuss this in detail below.
- A third and more indirect aspect of the relation between distributed generation and power quality is that the tripping of a generator may have adverse consequences on the system. Especially when large numbers of generators trip simultaneously, this can have an adverse impact on the reliability and security of the system.

### 1.7.1 Impact of Distributed Generation on Current and Voltage Quality

The impact of distributed generation on power quality depends to a large extent on the criteria that are considered in the design of the unit. When the design is optimized for selling electricity only, massive deployment of distributed generation will probably adversely impact quality, reliability, and security. Several types of interfaces are however capable of improving the power quality. In a deregulated system this will require economic incentives, for example, in the form of a well-functioning ancillary services market.

To quantify the impact of increasing penetration of distributed generation on the power system, the hosting capacity approach is proposed in [40] and [272]. The basis of this approach is a clear understanding of the technical requirements that the customer places on the system (i.e., quality and reliability) and the requirements that the system operator may place on individual customers to guarantee a reliable and secure operation of the system. The hosting capacity is the maximum amount of distributed generation for which the power system operates satisfactorily. It is determined by comparing a performance index with its limit. The performance index is calculated as a function of the penetration level. The hosting capacity is the penetration level for which the performance index becomes less than the limit. A hypothetical example is shown in Figure 1.11.

The calculation of the hosting capacity should be repeated for each different phenomenon in power system operation and design: The hosting capacity for voltage variations is different from the hosting capacity for frequency variations. Even for one phenomenon the hosting capacity is not a fixed value: It will depend on many system parameters, such as the structure of the network, the type of distributed generation (e.g., with or without storage; voltage/power control capability), the kind of load, and even climate parameters (e.g., in case of wind or solar power). For studying the impact of distributed generation on power quality phenomena the indices introduced in Chapters 5 and 10 should be used. Note that the "ideal" value of many of those indices is zero, so that the hosting capacity is reached when the index value exceeds the limit.

By using the hosting capacity approach, the issue of power quality and distributed generation has now been reduced to a discussion of the acceptable performance of a power system. This may not always be an easy discussion, but it is at least one that leads to quantifiable results. Obviously what is acceptable to one customer may not be acceptable to another customer and here some decisions may have to be made.

Figure 1.12 gives an example of how to implement this method for the overvoltages due to the injection of active power by distributed generation units. This

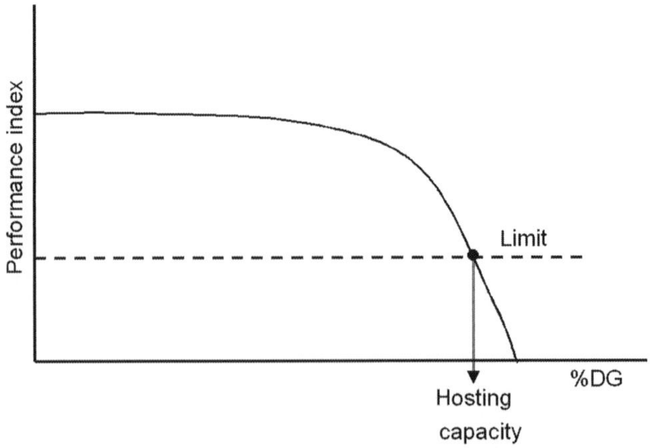


**Figure 1.11** Definition of hosting capacity approach for distributed generation.

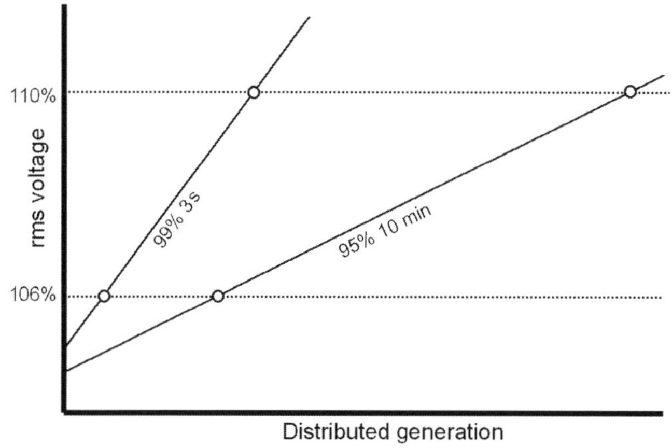


**Figure 1.12** Example of hosting capacity approach as applied to voltage variations; two different limits and two different indices result in four different values for hosting capacity.

is a standard example in many studies. In the figure two different indices are used, both based on the rms voltage. One index uses the 95 percentile of the 10-min rms values, whereas the other one uses the 99 percentile of the 3-s rms values. The figure also shows two different limits: 106 and 110% of the nominal voltage. The choice of two limits and two indices results in four values for the hosting capacity. The hosting capacity depends strongly on the choice of index and the choice of limit. The amount of distributed generation that can be accepted by the system depends on the performance requirements placed on the system. Note that this is exactly the same discussion as the coordination between current quality and voltage quality in Section 1.6. By quantifying the responsibility of the network operator for voltage quality, the hosting capacity for distributed generation is also determined.

Distributed generation may have an adverse influence on several power quality variations. The most-discussed issue is the impact on voltage variations. The injection of active power may lead to overvoltages in the distribution system (see Section 2.2.3.3). Also, increased levels of harmonics and flicker are mentioned as potential adverse impacts of distributed generation. But distributed generation can also be used to mitigate power quality variations. This especially holds for power-electronic interfaces that can be used to compensate voltage variations, flicker, unbalance, and low-frequency harmonics. The use of power-electronic interfaces will however lead to high-frequency harmonics being injected into the system. These could pose a new power quality problem in the future.

### 1.7.2 Tripping of Generator Units

As we mentioned at the start of this section, there is a third power quality aspect related to distributed generation. With large penetration of distributed generation, their tripping is an issue not only for the generator owner but also for the system operator and other customers. The tripping of one individual unit should not be a concern to the system, but the simultaneous tripping of a large number of units is a serious concern. Seen from the network this is a sudden large increase in load. Simultaneous tripping occurs due to system events that exceed the immunity of the generator units. As discussed in the previous section, we assume that distributed generator units will not trip for normal events such as transformer or capacitor energizing. Their behavior for abnormal events such as faults (voltage dips) and loss of a large generator unit (frequency swings) is at first a matter of economic optimization of the unit.

A schematic diagram linking a fault at the transmission level with a large-scale blackout is shown in Figure 1.13. The occurrence of a fault will lead to a voltage dip at the terminals of distributed generation units. When the dip exceeds the immunity level of the units, they will disconnect, leading to a weakening of the system. For a fault at the transmission system, clearing the fault will also lead to a weakening of the system. The safety concerns and the loss of revenue are a matter for the unit operator; they will be taken care of in a local economic optimization. Of importance for the system is that the loss of generation potentially leads to instability and component overload. The voltage drop resulting from the increased system loading may lead to further tripping of distributed generation units.

The most threatening event for the security of a system with a large penetration of distributed generation is the sudden loss of a large (conventional) generation unit. This will lead to a frequency swing in the whole system. The problem is more severe in Scandinavia and in the United Kingdom than in continental Europe because of the size of the interconnected systems. The problem is most severe on small islands. Even with low penetration it is recommended that all generation units remain online with the tripping of a large power station, as such events may happen several times a week. Larger frequency swings occur only once every few

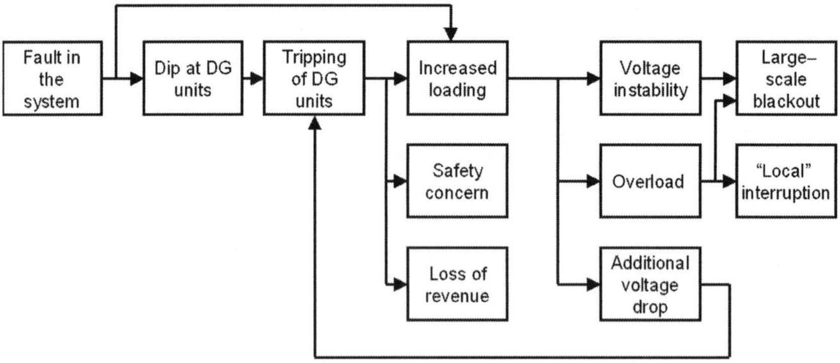


**Figure 1.13** Potential consequences of fault or loss of generation in system with large penetration of distributed generation.

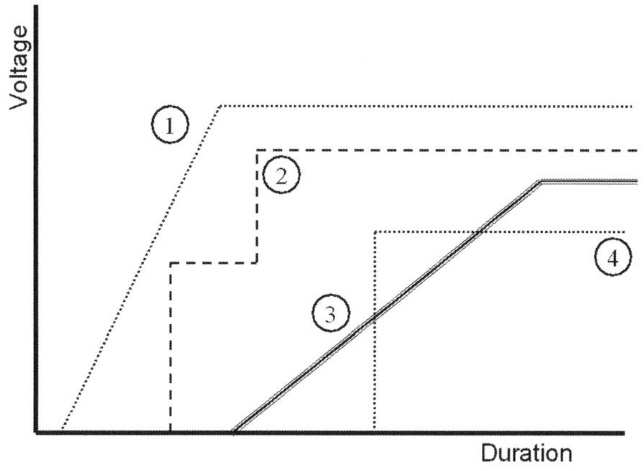


**Figure 1.14** Voltage tolerance curves for distributed generation.

years, and there is no economical need for the generator operator to be immune against such a disturbance. However, with a large penetration of distributed generation tripping due to severe events will increase the risk of a blackout.

Several network operators place protection requirements on distributed generation for the maximum tripping time with a given undervoltage. Preventing islanding and the correct operation of the short-circuit protection are typically behind such requirements. Network operators may also prescribe immunity requirements based on system security and reliability concerns. The different types of voltage tolerance curves are plotted together in Figure 1.14:

1. A (future) immunity requirement set by the transmission system operator or by an independent system operator to guarantee that the generators remain connected to the system during a fault in the (transmission) system. This curve is a minimum requirement: The unit is not allowed to be disconnected for any disturbance above or toward the left of this curve.
2. The actual immunity of the generator determined by the setting of the protection. The operator of the unit is free to set this curve within the limitations posed by curves 1, 3, and 4.
3. The limits set by the physical properties of the generator components: thermal, dielectrical, mechanical, and chemical properties. This curve is determined by the design of the unit and the rating of its components. The operator of the generator can only affect this curve in the specification of the unit. Generally speaking, moving this curve to the right will make the unit more expensive.
4. The protection requirements dictated by the distribution system operator to ensure that the generator units will not interfere with the protection of the distribution system. This is a maximum requirement: The unit should trip for every disturbance below and to the right of this curve.

Three coordination requirements follow for these curves:

- Curve 2 should be to the right of curve 1.
- Curve 2 should be to the left of curve 3.
- Curve 2 should be to the left of curve 4.

The condition that curves 3 and 4 should be to the right of curve 1 follows from these requirements. There is no requirement on coordination between curves 3 and 4.

## 1.8 CONCLUSIONS

Power quality has been introduced as part of the modern, customer-based view on power systems. Power quality shares this view with deregulation and embedded generation. Deregulation and embedded generation are two important reasons for the recent interest in power quality. Other important reasons are the increased emission of disturbances by equipment and the increased susceptibility of equipment, production processes, and manufacturing industry to voltage disturbances.

Power quality has been defined as a combination of voltage and current quality. Voltage and current quality concern all deviations from the ideal voltage and current waveforms, respectively, the ideal waveform being a completely sinusoidal waveform of constant frequency and amplitude. A distinction is made between two types of power quality disturbances: (voltage and current) variations are (quasi-) steady-state disturbances that require or allow permanent measurements or measurements at predetermined instants; (voltage and current) events are disturbances with a beginning and an end, and a triggering mechanism is required to measure them. The difference in processing between variations and events is the basis for the structure of this book.

Signal processing forms an important part in power quality monitoring: the analysis of voltage and current measurements from sampled waveforms to system indices. Signal-processing techniques are needed for the characterization (feature extraction) of variations and events, for the triggering mechanism needed to detect events, and to extract additional information from the measurements.

The IEC EMC standards are based on the coordination of emission and immunity levels by defining a suitable compatibility level. In power system studies, the emission level and the compatibility level are determined by the existing level of disturbances.

A large number of papers have been written on power quality and related subjects. A database search resulted in several thousand hits (see Fig. 1.4). Also several books have been published on this subject. We will come across several papers and most of the books in the remaining chapters of this book. For an excellent overview of the various power quality phenomena and other issues, the reader is referred to the book by Dugan et al. [99]. Other overview books are the ones by Heydt [141], Sankaran [263], and Schlabbach et al. [271]. The latter one was originally published in German. The Swedish readers are referred to the overview book by Gustavsson [134]. Two books directed very much toward practical aspects of power quality monitoring are the *Handbook of Power Signatures* [95] and the *Field Handbook for Power Quality Analysis* [336]. Well-readable overview texts on power quality are also found in a number of IEEE standards—IEEE 1100 [164], IEEE 1159 [165], and IEEE 1250 [167]—and in some general power system books—*The Electric Power Engineering Handbook* [101, Chapter 15]—as well in some of the books from the IEEE Color Book Series.

## 1.9 ABOUT THIS BOOK

This book aims to introduce the various power quality disturbances and the way in which signal-processing tools can be used to analyze them. The various chapters and their relation are shown graphically in Figure 1.15. The figure shows horizontal as well as vertical lines that group the chapters into related ones. The vertical subdivision corresponds to the subdivision of power quality disturbances in variations and events. Chapters 2, 3, 4, and 5 discuss the most common power quality variations and their processing; Chapters 6, 7, 8, 9, and 10 discuss power quality events and their processing. The horizontal subdivision is based on the tools used and background needed. Chapters 2 and 6 are typical power system or power quality chapters in which the disturbances are described that will be subject to processing in the other chapters. Signal-processing tools and their application to power quality disturbances are described in detail in Chapters 3, 4, 7, 8, and 9. Chapter 5 and 10 use statistical methods for presenting the data resulting from the signal processing. These two

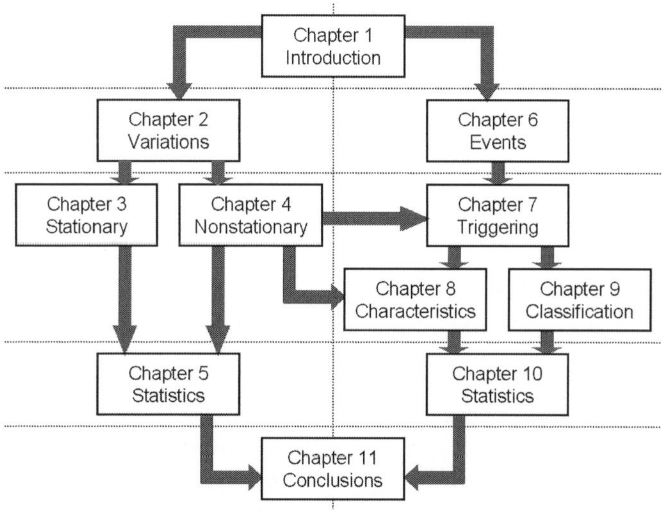


**Figure 1.15** Relation between different chapters in this book.

chapters are based on standard methods for statistical processing, extended with some methods under development and proposed in the literature.

Chapter 2 introduces the most common power quality variations: frequency variations, voltage variations, three-phase unbalance, voltage fluctuations leading to light flicker, and waveform distortion (harmonics). Chapter 3 introduces the basic features that are used to quantify the stationary signal waveforms and discusses signal-processing tools for extracting these features that are statistical time invariant. The two most commonly methods, rms and DFT, are discussed next along with some more advanced methods (such as MUSIC, ESPRIT, and Kalman filters) for estimating the spectral contents of a signal. Chapter 3 also introduces power quality indices: quantifiers that indicate the severity of a power quality disturbance. Both features, commonly referred to as characteristics in the power quality field, and indices will play an important part in several chapters. Chapter 4 discusses methods for processing signals that are not stationary, where the statistical characteristics or attributes of the signal are time varying, for example, the mean and variance of the signal are time dependent. The STFT, dyadic structured discrete wavelet transform, and a variety of Kalman filters will be discussed in detail. Block-based processing will be introduced where the signal is divided into blocks (time windows) during which it is assumed to be stationary. The term *quasi-stationary* is often used for this. Some signal-processing methods in Chapter 3 will be modified to form, for example, the sliding-window MUSIC/ESPRIT method and block-based AR modeling. Events are examples of signals that are nonstationary, so that the methods introduced in Chapter 4 will be used later when discussing the processing of power quality events, notably in Chapters 7 and 8. As we will see in Chapter 5 and in some of the examples in Chapters 3 and 4, the measurement of a power quality variation results in a large amount of data: features and indices as a function of time and at different locations. Chapter 5 will discuss methods for obtaining statistics from these data. Methods to be discussed include time aggregation (where the features obtained over a longer window of time are merged into one representative value for the whole window), site indices (quantifying the disturbance level at a certain location over several days, weeks, or even years), and system indices (quantifying the performance of a number of sites or even of a complete system). Chapter 5 will also discuss a number of relevant power quality standards, including IEC 61000-4-30 (power quality monitoring), EN 50160 (voltage characteristics), and IEEE 519 (harmonic limits).

Power quality events, sudden severe disturbances typically of short duration, are the subject of Chapters 6 through 10. The origin of the most common events (interruptions, dips, and transients) is discussed in Chapter 6. The three-phase character of voltage dips plays an important role in this chapter. Many event recordings will be shown to illustrate the various origins of the events. Some early triggering and characterization concepts, to be discussed in more detail in later chapters, will be used in Chapter 6. In Chapter 7 event triggering and segmentation are discussed. Both triggering and segmentation are methods that detect a sudden change in waveform characteristics. The term *triggering* is used to distinguish between events and variations or just to detect the beginning and ending points of events. The term *segmentation* refers to the further subdivision of event recordings. But, as we will see in Chapter 7, both are methods to detect instants at which the signal is nonstationary. The chapter starts with an overview of existing methods followed by a discussion of advanced methods for triggering and segmentations. Among others, wavelet filters and Kalman filters will be treated. Chapters 8 and 9 treat the further processing of event recordings resulting in parameters that quantify individual events. These parameters are generally referred to as *event characteristics* or *single-event indices*. Chapter 8 concentrates on methods to quantify the severity of an event, with magnitude and duration being the most commonly used characteristics for dips, interruptions, and transients. Reference is made to methods prescribed by IEC 61000-4-30 and to methods discussed in IEEE task forces P1159.2 and P1564 and in the International Council on Large Electric Systems (CIGRE) working group C4.07. Chapter 8 further contains a discussion on extracting three-phase characteristics for dips and on methods to characterize voltage and current transients. Chapter 9 treats a special type of event characterization: directed toward finding additional information about the origin of the event. The term *event classification* is used and the chapter concentrates on automatic methods for event classification, including the simplest linear discriminants, to somewhat more sophisticated Bayesian classifiers, Neyman–Pearson approaches, artificial neutral networks, support vector machines, and expert systems. Support vector machines are a relatively new method for power engineering which is based on the statistical learning theory and can be considered as the solution of a constrained optimization problem. Support vector machines provide a great potential for event classification in terms of both their generalization performance (i.e., classification performance on the test set) and their affordable complexity in machine implementation. Chapter 9 also contains an overview of machine learning and pattern classification techniques. Chapter 10 is equivalent to Chapter 5; it discusses methods of presenting the results from monitoring surveys by means of statistical indices. The differences in processing between events and variations make it appropriate to have separate chapters, where Chapter 10 concerns the statistical processing of events. The IEEE standard 1366 is discussed in detail as a method for presenting statistics on supply interruptions, better known as *reliability indices*. The statistical processing of voltage dips is presented as a three-step process: time aggregation, site indices, and system indices. Each chapter contains the main conclusions belonging to the subjects discussed in that chapter. Heavy emphasis is placed in the conclusion sections on gaps in the knowledge that may be filled by further research and development. Chapter 11 presents general conclusions on signal processing of power quality events.

CHAPTER 2

# ORIGIN OF POWER QUALITY VARIATIONS

This chapter describes the origin and some of the basic analysis tools of power quality variations. The consecutive sections of the chapter discuss (voltage) frequency variations, voltage (magnitude) variations, voltage unbalance, voltage fluctuations (and the resulting light flicker), and waveform distortion. A summary and conclusions for each of the sections will be given at the end of this chapter.

## 2.1 VOLTAGE FREQUENCY VARIATIONS

Variations in the frequency of the voltage are the first power quality disturbance to be discussed here. After a discussion on the origin of frequency variations (the power balance) the method for limiting the frequency variations (power–frequency control) is discussed. The section closes with an overview of consequences of frequency variations and measurements of frequency variations in a number of interconnected systems.

### 2.1.1 Power Balance

Storage of electrical energy in large amounts for long periods of time is not possible, therefore the generation and consumption of electrical energy should be in balance. Any unbalance in generation and production results in a change in the amount of energy present in the system. The energy in the system is dominated by the rotating

*Signal Processing of Power Quality Disturbances.* By Math H. J. Bollen and Irene Yu-Hua Gu

energy $E_{\text{rot}}$ of all generators and motors:

$$E_{\text{rot}} = \tfrac{1}{2} J \omega^2 \tag{2.1}$$

with $J$ the total moment of inertia of all rotating machines and $\omega$ the angular velocity at which these machines are rotating. An unbalance between generated power $P_g$ and the total consumption and losses $P_c$ causes a change in the amount of rotational energy and thus in angular velocity:

$$\frac{d\omega}{dt} = \frac{P_g - P_c}{J\omega} \tag{2.2}$$

The amount of inertia is normally quantified through the *inertia constant* $H$, which is defined as the ratio of the rotational energy at nominal speed $\omega_0$ and a base power $S_b$:

$$H = \frac{\tfrac{1}{2} J \omega_0^2}{S_b} \tag{2.3}$$

The base power is normally taken as the sum of the (apparent) rated powers of all generators connected to the system, but the mathematics that will follow is independent of the choice of base power. Typical values for the inertia constant of large systems are between 4 and 6 s.

Inserting (2.3) in (2.2), assuming that the frequency remains close to the nominal frequency, and replacing angular velocity by frequency give the following expression:

$$\frac{df}{dt} = \frac{f_0}{2H}(P_g - P_c) \tag{2.4}$$

where $P_g$ and $P_c$ are per-unit (pu) values on the same base as the inertia constant $H$.

Consider a 0.01-pu unbalance between generation and production in a system with an inertia constant of 5 s. This leads to a change in frequency equal to 0.05 Hz/s. If there would be a 0.01-pu surplus of generation, the frequency would rise to 51 Hz in 20 s; for a 0.01-pu deficit in generation the frequency would drop to 49 Hz in 20 s. It is very difficult to predict the load with a 1% accuracy. To keep the frequency constant some kind of control is needed.

The sudden loss of a large power station of 0.15 pu will result in a frequency drop of 1 Hz/s. In 1 s the frequency has dropped to 49 Hz. As the sudden unexpected loss of a large generator unit cannot be ruled out, there is obviously the need for an automatic control of the frequency and of the balance between generation and consumption.

For comparison we calculate the amount of electrical and magnetic energy present in 500 km of a 400-kV three-phase overhead line when transporting 1000 MW of active power at unity power factor. Assuming 1 mH/km and

12 nF/km as inductance and capacitance, respectively, gives for the electrical energy $\frac{1}{2}Cu^2 = 320\,\mathrm{kJ}$ and for the magnetic energy $\frac{1}{2}Li^2 = 1040\,\mathrm{kJ}$. For unity power factor the peaks in magnetic and electrical energy (current and voltage) occur at the same time, so that the maximum total electromagnetic energy equals 1360 kJ. As before we can express this as a time constant by dividing with the rated power. For a 1000-MVA base, we find a time constant of 1.4 ms. This is significantly less than the 4- to 6-s time constant for the rotational energy. This example confirms the statement at the start of this section that the energy present in a power system is dominated by the rotational energy of generators and motors.

### 2.1.2 Power–Frequency Control

To maintain the balance between generation and consumption of electrical energy most large generator units are equipped with power–frequency control. Maintaining the frequency close to its nominal value is a natural consequence of maintaining the balance between generation and consumption. The principle of power–frequency control is rather simple. The measured frequency is compared with a frequency setting (the nominal frequency, 50 or 60 Hz, in almost all cases). When the measured frequency is higher than the nominal frequency, this indicates a surplus of rotational energy in the system. To mitigate this, the generator reduces its active power output. More correctly, the mechanical input to the turbine generator is reduced. This leads after a transient to a new steady state with a lower amount of electrical energy supplied to the system.

The principle of power–frequency control is shown in Figure 2.1. The input to the speed governor is a corrected power setting (corrected for the deviation of the frequency from its setting). We will come back to the choice of the power setting below. The speed governor is a control system that delivers a signal to the steam valves with a thermal power station to regulate the amount of steam that reaches the turbine. The turbine reacts to this, with a certain delay, by changing the amount of mechanical power. Much more detailed models can be found in the literature [e.g, 6, Chapter 10; 26, Chapter 3]. For the purpose of this chapter, it is sufficient to know that there is a time delay of several seconds (10 s and more for large units) between a change in the power signal at the input of the governor and an increase in the mechanical power produced by the turbine. Also note that the speed governor is a

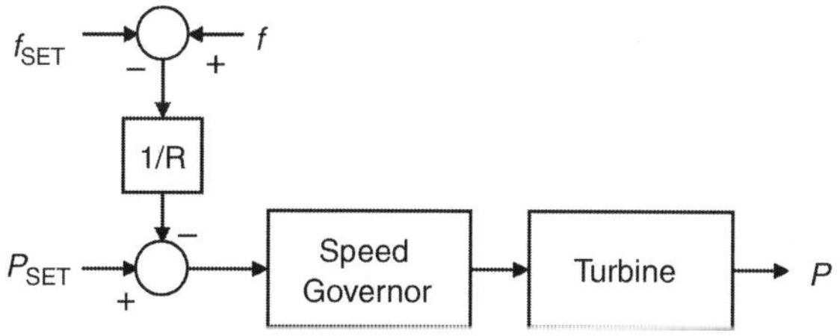


**Figure 2.1** Power–frequency control.

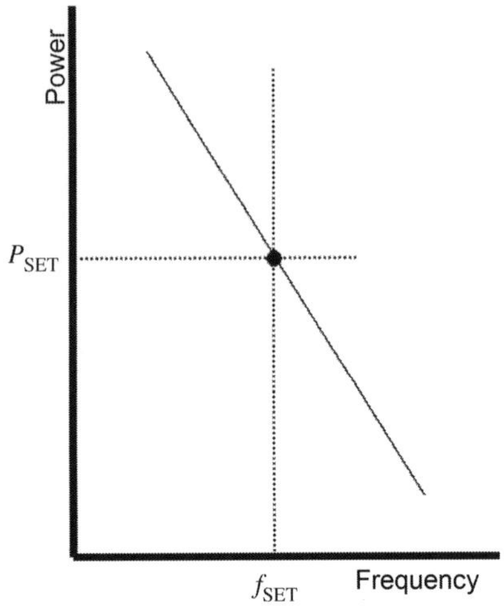


**Figure 2.2** Relation between system frequency and amount of power generated by one unit.

controller (its parameters are chosen during the design of the control system) whereas the turbine is a physical system (its parameters cannot be affected).

If we consider the system in steady state, where the production of the generator equals the input signal to the speed governor, the mechanical power is as follows:

$$P = P_{SET} - \frac{1}{R}(f - f_{SET}) \tag{2.5}$$

where $R$ is referred to as the *droop setting*. This relation is shown in Figure 2.2. When the system frequency drops, the power production increases. This compensates for the cause of the frequency drop (a deficit of generation). The frequency setting is equal to the nominal frequency of the system and the same for all generators connected to the system. In the Nordic system the frequency should not only be within a certain band but also on average be equal to 50 Hz to ensure that clocks indicate a correct time. When the integrated frequency error (the time difference for a clock) exceeds a certain value, the frequency setting of the generators is slightly changed. But the setting remains the same for all generators.

***2.1.2.1 Spinning Reserve*** To allow for an increase in generated power when there is a deficit of generation, for example, because a large unit has been disconnected from the system, the power produced by a generator should be less than its maximum capacity. The amount of extra power which can be produced in a matter of seconds is called *spinning reserve*. The total spinning reserve in an

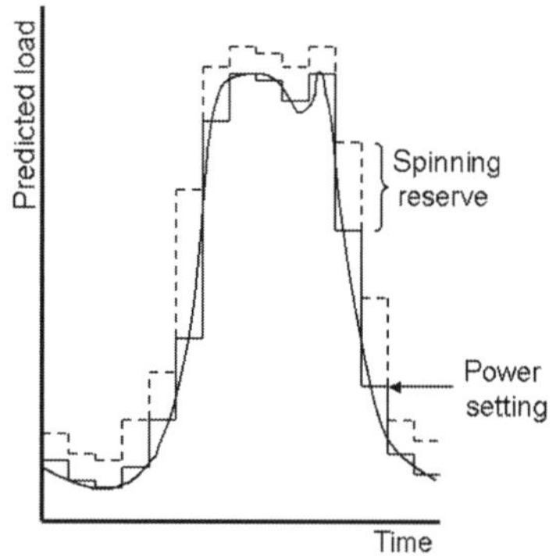


**Figure 2.3** Daily load curve (thick solid curve) for a power system, with the sum of generator power settings (thin solid lines) and the spinning reserve (dashed lines).

interconnected system should at least be equal to the largest unit connected to the system. For smaller systems during low load, the spinning reserve should be relatively high. However, in large interconnected systems like the Nordic system or the European interconnected system, a spinning reserve of a few percent is acceptable.

***2.1.2.2 Choice of Power Set Point*** Figure 2.3 shows a hypothetical daily load curve for a system. Such a curve is used to schedule the amount of generation capacity needed. The day is divided into a number of time intervals, typically of 15 to 30 min duration. For each interval the expected load is determined. This required generation is then spread over a number of generator stations. For each time interval the sum of the power settings is chosen equal to the expected load. The actual scheduling is in most countries done by a free-market principle where each generator can place bids. When such a bid is accepted for a certain time block, it will become the power setting of the generator for that block. Even the load is in principle market based, but the distribution companies typically place bids based on the expected consumption of their customers. For example, see [25] for a description of the various market principles.

Also for each time interval a spinning reserve can be decided, but this is typically kept at a fixed percentage of the total power. Even for the spinning reserve and the droop setting market principles can be applied (as discussed, e.g., in [335]).

***2.1.2.3 Sharing of Load*** A change in load, or a change in generation setting, results in a change in generated power for all generator units equipped with power–frequency control. Consider a system with $n$ generators with power setting $P_{\mathrm{i,SET}}$, $i = 1, \ldots, n$; droop setting $R_i$; and frequency setting $f_{\mathrm{SET}}$. Note that the power setting and the droop setting are different for each unit whereas the frequency setting is the same. The produced power for each generator at a system frequency $f$ is

$$P_i = P_{i,\mathrm{SET}} - \frac{1}{R_i}(f - f_{\mathrm{SET}}) \tag{2.6}$$

The sum of all power settings is equal to the total predicted load:

$$\sum_{i=1}^{n} P_{i,\mathrm{SET}} = P_C \tag{2.7}$$

Assume that the actual load deviates from the predicted load by an amount $\Delta P_c$, so that in steady state

$$P_g = P_c + \Delta P_c \tag{2.8}$$

Combining (2.6), (2.7), and (2.8) gives

$$P_g = \sum_{i=1}^{n} P_i = \sum_{i=1}^{n} P_{i,\mathrm{SET}} + \Delta P_c \tag{2.9}$$

which gives for the steady-state frequency

$$f = f_{\mathrm{SET}} - \frac{\Delta P_c}{\sum_{i=1}^{n}(1/R_i)} \tag{2.10}$$

The increase in consumption causes the system frequency to drop by an amount determined by the power–frequency control settings of all the generators that contribute. Each generator contributes to the increase in consumption by ratio of the inverse of its droop setting:

$$P_k = P_{k,\mathrm{SET}} + \frac{1/R_k}{\sum_{i=1}^{n}(1/R_i)}\Delta P_c \tag{2.11}$$

The droop setting is normally a constant value in per unit with the generator rating as a base. For a generator of rated power $S$ and per-unit droop setting $R_{\mathrm{pu}}$, the droop setting in hertz per megawatt is

$$R_k = R_{\mathrm{pu}}\frac{f_{\mathrm{SET}}}{S} \tag{2.12}$$

with typically $f_{\mathrm{SET}} = f_0$, the nominal frequency. The new steady-state frequency is obtained from inserting (2.12) in (2.10) under the assumption that the per-unit droop setting is the same for all units:

$$f = f_{\text{SET}} - \frac{\Delta P_c}{\sum_{i=1}^{n} S_k} R_{\text{pu}} f_{\text{SET}} \tag{2.13}$$

The relative drop in frequency is equal to the relative deficit in generation times the per-unit droop setting:

$$\frac{\Delta f}{f_{\text{SET}}} = -\frac{\Delta P_c}{\sum_{i=1}^{n} S_i} R_{\text{pu}} \tag{2.14}$$

Each generator contributes by the ratio of its rated power to any deficit in generation:

$$P_k = P_{k,\text{SET}} + \frac{S_k}{\sum_{i=1}^{n} S_i} \Delta P_c \tag{2.15}$$

Thus large generators contribute more than small generators. This calls for a spinning reserve which is a fixed percentage of the rated power of the generator unit.

### 2.1.3 Consequences of Frequency Variations

As far as the authors are aware, no equipment problems are being reported due to frequency variations. Still some of the consequences and potential problems are mentioned below.

***2.1.3.1 Time Deviation of Clocks*** Clocks still often synchronize to the voltage frequency (typically by counting zero crossings). A consequence of frequency variations is therefore that clocks will show an erroneous time. The size of the error depends on the deviation of the frequency from the nominal frequency.

Consider a system with nominal frequency $f_0$ and actual frequency $f(t)$. The number of zero crossings in a small time $\Delta t$ is

$$\Delta n_{\text{zc}} = 2f(t)\,\Delta t \tag{2.16}$$

Note that there are two zero crossings per cycle. In a long period $T$ (e.g., one day), the number of zero crossings is

$$N_{\text{zc}} = \int_0^T 2f(t)\,dt \tag{2.17}$$

Because the clock assumes a frequency $f_0$, the apparent elapsed time is

$$T + \Delta T = \frac{N_{\text{zc}}}{2f_0} \tag{2.18}$$

From (2.17) and (2.18) the time error $\Delta T$ is obtained as

$$\Delta T = \int_0^T \frac{f(t) - f_0}{f_0}\, dt \tag{2.19}$$

A frequency of 50.05 Hz instead of 50 Hz will cause clocks to run 0.1% faster. This may not appear much, but after one day the deviation in clocks is $0.001 \times 3600 \times 24 = 86.4$ s. Thus 0.05 Hz frequency deviation causes clocks to have an error of over 1 min per day. A frequency equal to 50.5 Hz (1% deviation) would cause clocks to be 15 min ahead after one day.

***2.1.3.2 Variations in Motor Speed*** Also the speed of induction motors and synchronous motors is affected when the voltage frequency changes. But as the frequency does not vary more than a few percent, these speed variations are hardly ever a problem. Very fast fluctuations in frequency could cause mechanical problems, but in large interconnected systems the rate of change of frequency remains moderate even for large disturbances. Variations in voltage magnitude will have a bigger influence.

***2.1.3.3 Variations in Flux*** Lower frequency implies a higher flux for rotating machines and transformers. This leads to higher magnetizing currents. The design of rotating machines and transformers is such that the transition from the linear to the nonlinear behavior (the "knee" in the magnetic flux-field, B-H, curve) is near the maximum normal operating flux. An increase of a few percent in flux may lead to 10% or more increase in magnetizing current. But the frequency variation is very rarely more than 1%, whereas several percent variations in voltage magnitude occur daily at almost any location in the system. One percent drop in frequency will give the same increase in flux as 1% rise in voltage magnitude. Where saturation due to flux increase is a concern, voltage variations are more likely to be the limiting factor than frequency variations.

***2.1.3.4 Risk of Underfrequency Tripping*** Larger frequency deviations increase the risk of load or generator tripping on overfrequency or on underfrequency. Overfrequency and underfrequency relays are set at a fixed frequency to save the power system during very severe disturbances like the loss of a number of large generating units. The most sensitive settings are normally for the underfrequency relays. In some systems the first level of underfrequency load shedding occurs already for 49.5 Hz, although 49 Hz is a more common setting. The loss of a large generator unit causes a fast drop in frequency due to the sudden deficit in generation followed by a recovery when the power–frequency control increases the production of the remaining units. The maximum frequency deviation during such an event is rather independent of the preevent frequency. Thus when the preevent frequency is lower, the risk of an unnecessary underfrequency trip increases.

Generally speaking large frequency variations point to unbalance between generation and (predicted) consumption. As long as this unbalance is equally spread over the system, it is of limited concern for the operation of the system. As shown in Section 2.1.2 the daily load variations are spread equally between all generator units that contribute to power–frequency control. However, fast fluctuations in frequency (time scales below 1 min) point to a shorter term unbalance that is associated with large power flows through the system. These fluctuating power flows cause a higher risk of a large-scale blackout.

Distributed generation has also become a concern with frequency variations. Most units are equipped with underfrequency and overfrequency protection. The underfrequency protection is the main concern as this is due to lack of generation. Tripping of distributed generation units will aggravate the situation even more.

***2.1.3.5 Rate of Change of Frequency*** Some equipment may be affected more by the rate of change in frequency than by the actual deviation from the nominal frequency. Any equipment using a phase-locked loop (PLL) to synchronize to the power system frequency will observe a phase shift in the voltage during a fast change in frequency. The design of a PLL is a tradeoff between speed (maximum rate of change of frequency) and accuracy (frequency deviation in the steady state).

Distributed generation is often equipped with a ROCOF (rate-of-chance-of-frequency) relay to detect islanding. These relays are reportedly also sensitive to the frequency drop caused by the tripping of a large generator unit.

No statistical measurement data are available on the rate of change of frequency, but it is possible to estimate expected values based on a knowledge of the underlying phenomenon. A large rate of change of frequency occurs during the loss of a large generator unit. The resulting unbalance between generation and consumption causes a drop in frequency in accordance with (2.4):

$$\frac{df}{dt} = 0.05p \qquad \text{Hz/s} \tag{2.20}$$

with $p$ the size of the largest unit as a fraction of the system size during low load. This does not mean that larger values of the rate of change of frequency are not possible, but they will only occur during the loss of more than one large unit at the same time. Such a situation has a much lower probability than the loss of one large unit. Also such a situation will severely endanger the survival of the system so that power quality disturbances become of minor importance. Some examples of fast changes in frequency will be shown in Section 5.3.2.

### 2.1.4 Measurement Examples

Examples of measured frequency variations are shown in Figures 2.4 and 2.5. As was explained in the earlier parts of this section, frequency variations are the same throughout an interconnected system and are related to the relative unbalance between generation and load and to power–frequency control. Generally speaking,

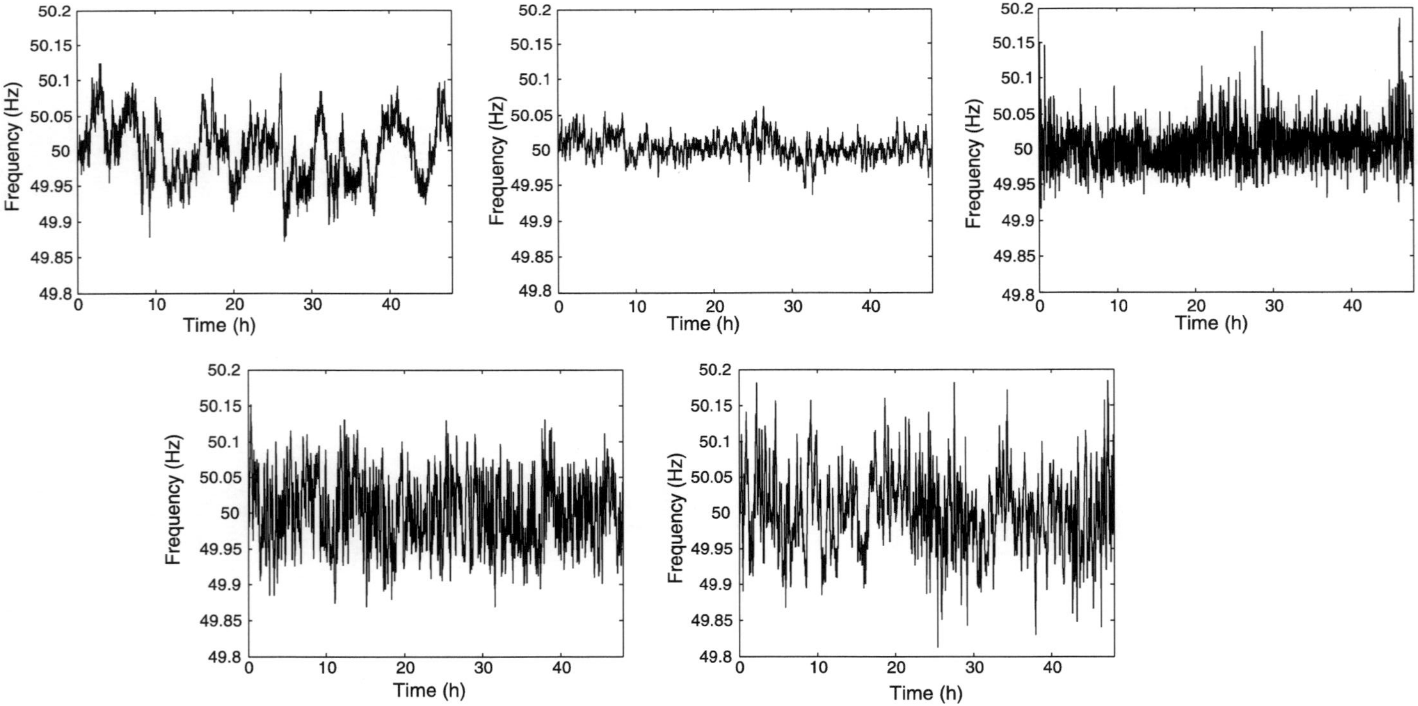


**Figure 2.4** Frequency variations measured in Sweden (top left), in Spain (top center), on the Chinese east coast (top right), in Singapore (bottom left), and in Great Britain (bottom right).

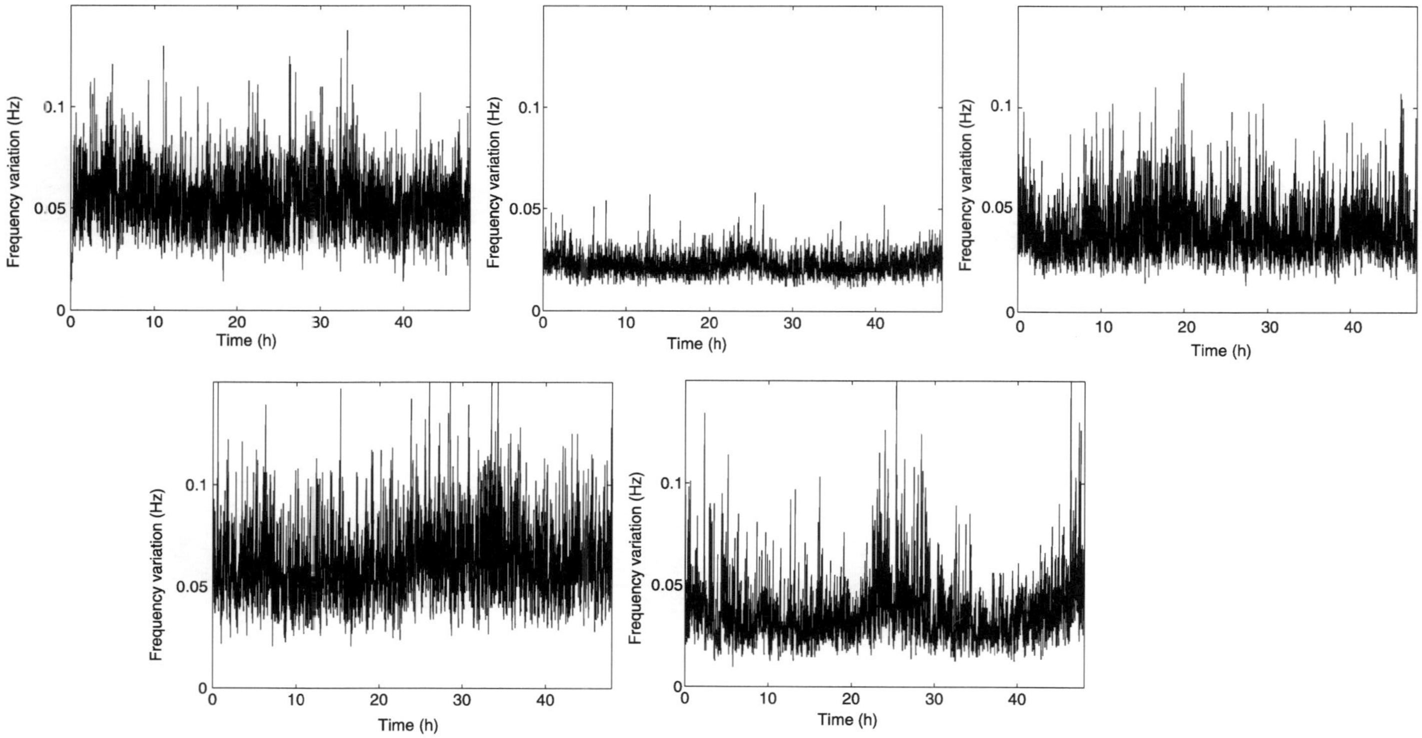


**Figure 2.5** Range in frequency during 1 min measured in Sweden (top left), in Spain (top center), on the Chinese east coast (top right), in Singapore (bottom left), and in Great Britain (bottom right).

the larger the system, the less the frequency variations. The data presented here were collected at five different locations in five different interconnected systems. Figure 2.4 gives the 1-min average frequency during a two-day (48-h) period, whereas Figure 2.5 gives the spread in frequency during each 1-min period. One may say that the first figure shows slow variations in frequency and the second figure the fast variations. Spain is part of the European interconnected system, one of the largest in the world. As expected, the range in frequency is small in this system and so are the fast variations. The Chinese system is smaller but still larger than the Nordic system (consisting of Norway, Sweden, Finland, and part of Denmark). Singapore and Great Britain are relatively small systems, which is visible from the relatively large variations in frequency.

The different systems show different patterns in variations, both at longer and at shorter time scales. These differences are related to the size of the system and to the control methods used. It should be noted, however, that in none of the systems presented here are the frequency variations of any concern.

## 2.2 VOLTAGE MAGNITUDE VARIATIONS

This section will discuss slow variations in the magnitude of the voltage. The section will start with an overview of the impact of voltage variations on end-user equipment followed by a discussion of several expressions to estimate voltage drops in the system. Expressions will be derived for a concentrated load and for load distributed along a feeder. The impact of distributed generation on voltage variations will also be discussed. The section concludes with an overview of voltage control methods, with transformer tap changers and capacitor banks being discussed in more detail.

### 2.2.1 Effect of Voltage Variations on Equipment

Voltage variations can effect the performance and the lifetime of the equipment. Some examples are as follows:

1. Any overvoltage will increase the risk of insulation failure. This holds for system components such as transformers and cables as well as for end-user equipment such as motors. This is obviously a long-term effect and in most cases not significant. Note, however, that a higher voltage during normal operation increases the base from which transient overvoltages start. This increases the peak voltage and thus the risk of insulation failure. Again this is probably an insignificant effect.
2. Induction motors:
   - Undervoltages will lead to reduced starting torque and increased full-load temperature rise. The reduced starting torque may significantly increase the time needed to accelerate the motor. In some cases the motor may not accelerate at all: It will "stall." The stalled motor will take a high current

but will not rotate (it becomes a short-circuited transformer). If a stalled motor is not removed by the protection, it will overheat very quickly. The further reduced voltage due to the high current taken by the stalled motor may lead to stalling of neighboring motors. Stalling normally does not occur until the voltage has dropped to about 70% of nominal.

- Overvoltages will lead to increased torque, increased starting current, and decreased power factor. The increased starting torque will increase the accelerating forces on couplings and driven equipment. Increased starting current also causes greater voltage drop in the supply system and increases the voltage dip seen by the loads close to the motor. Although the motor will start quicker, its effect on other load may be more severe.

3. Incandescent lamps: The light output and life of such lamps are critically affected by the voltage. The expected life length of an incandescent lamp is significantly reduced by only a few percent increase in the voltage magnitude. The lifetime somewhat increases for lower-than-nominal voltages, but this cannot compensate for the decrease in lifetime due to higher-than-nominal voltage. The result is that a large variation in voltage leads to a reduction in lifetime compared to a constant voltage.
4. Fluorescent lamps: The light output varies approximately in direct proportion to the applied voltage. The lifetime of fluorescent lamps is affected less by voltage variation than that of incandescent lamps.
5. Resistance heating devices: The energy input and therefore the heat output of resistance heaters vary approximately as the square of the voltage. Thus a 10% drop in voltage will cause a drop of approximately 20% in heat output.
6. An undervoltage will lead to an increased duty cycle for any equipment that uses a thermostat (heating, refrigerating, air conditioning). The result is that the total current for a group of such devices will increase. Even though individual heaters behave as a constant-resistance load, a group of them behave as constant-power loads. This phenomenon is one of the contributing factors to voltage collapse.
7. Electronic equipment may perform less efficient due to an undervoltage. The equipment will also be more sensitive to voltage dips. A higher-than-nominal voltage will make the equipment more sensitive to overvoltages. As the internal voltage control maintains the application voltage at a constant level (typically much lower than the 110 through 230 V mains voltage), a reduction in terminal voltage will lead to an increase in current which gives higher losses and reduced lifetime.
8. Transformers: A higher-than-nominal voltage over the transformer terminals will increase the magnetizing current of a transformer. As the magnetizing current is heavily distorted, an increase in voltage magnitude will increase the waveform distortion.

### 2.2.2 Calculation of Voltage Magnitude

The voltage as considered in this section is the rms value of a sinusoidal voltage waveform. We will neglect all distortion, so that the voltage waveform is described as

$$u(t) = \sqrt{2}u \cos(2\pi f_0 t) \tag{2.21}$$

where $u$ is the rms voltage and $f_0$ the voltage frequency. The time axis is chosen such that the phase angle of the voltage is zero. In any system it is possible to set the phase angle to zero for one of the voltages or currents without loss of generality. Voltage magnitude variations, or simply *voltage variations*, are variations in the value of $U$.

Note that in (2.21) the phase angle of the voltage is defined relative to the voltage maximum. More typically in power engineering, the phase angle is defined relative to the upward zero crossing. The zero crossing is easier to detect than the maximum. The *power system definition* would imply a sine function instead of cosine. The cosine function however fits better with the complex notation used for the calculations. The choice of reference does not affect the calculations as only phase *differences* have any physical meaning.

For calculations of the voltage magnitude, the complex notation is most often used. The voltage is written as the real part of a complex function of time:

$$u(t) = \mathrm{Re}\{\underline{U}\, e^{j2\pi f_0 t}\} \tag{2.22}$$

where $\underline{U} = Ue^{j\theta}$ is referred to as the *complex voltage* or simply the voltage where confusion is unlikely. In the same way a complex current can be calculated. A complex impedance is defined as the ratio between complex voltage and complex current. Complex power is defined as the product of complex voltage and the complex conjugate of the current. See any textbook on electric circuit theory for more details.

#### 2.2.2.1 Thevenin Source Model

To model the effect of a certain load on the voltage, the power system is represented through a Thevenin source: an ideal voltage source behind a constant impedance, as shown in Figure 2.6, with $\underline{E}$ the no-load voltage and $\underline{Z}$ the source impedance.

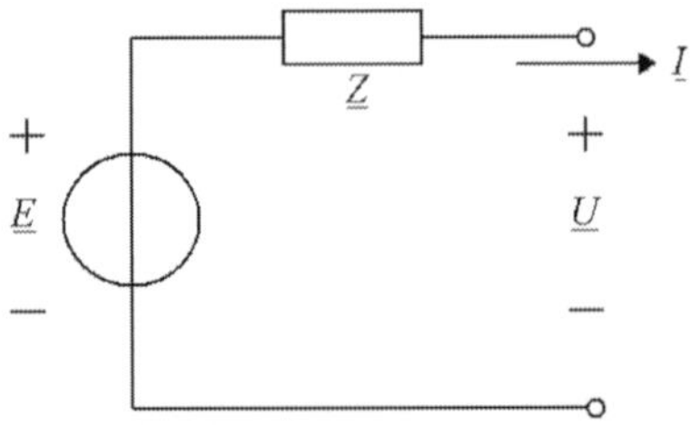


**Figure 2.6** Power system with load.

Thevenin's theorem states that any linear circuit, at its terminals, can be modeled as a voltage source behind an impedance. This holds for any location in the power system. The term *no-load voltage* does not imply that the load of the power system is neglected. It only refers to the loading of the Thevenin equivalent at the location under consideration. All other load is incorporated in the source model, that is, in the source voltage and the source impedance. Note also that the model in Figure 2.6 is a mathematical model. The source impedance $\underline{Z}$ and the source voltage $\underline{E}$ are not physical quantities. However, the values can often be approximated by physical quantities, for example, the impedance of a transformer and the voltage on primary side of the transformer.

**2.2.2.2 Changes in Voltage Due to Load** Consider the Thevenin model in Figure 2.6 for the source at the load terminals. The following relation holds between the load voltage and the no-load voltage:

$$\underline{U} = \underline{E} - \underline{Z}\underline{I} \tag{2.23}$$

The complex power delivered to the load is

$$\underline{S} = \underline{U}\underline{I}^* = P + jQ \tag{2.24}$$

with $P$ the active power and $Q$ the reactive power. Taking the load voltage along the positive real axis (so that $\underline{U} = U$) gives the following expression for the current as a function of the active and reactive power:

$$\underline{I} = \frac{P - jQ}{U} \tag{2.25}$$

This results in the following expression for the *complex voltage drop*, $\underline{U}_\Delta = \underline{E} - \underline{U}$:

$$U\underline{U}_\Delta = RP + XQ + j(XP - RQ) \tag{2.26}$$

The *scalar voltage drop* or simply the voltage drop is defined as the difference in absolute value between the no-load and the load voltage:

$$\Delta U = |\underline{E}| - |\underline{U}| = |U + \underline{U}_\Delta| - U \tag{2.27}$$

Inserting (2.26) for the complex voltage drop gives the following expression for the (scalar) voltage drop due to active and reactive power:

$$\frac{\Delta U}{U} = \sqrt{\left(1 + \frac{RP + XQ}{U^2}\right)^2 + \left(\frac{XP - RQ}{U^2}\right)^2} - 1 \tag{2.28}$$

Note that this expression cannot be used to calculate the voltage $U$, as this variable appears on both sides of the equal sign. It is possible to derive a closed expression for $U$, but this is outside of the scope of this book. Expression (2.28) can however be used to calculate the magnitude of the no-load voltage $E$ in case the voltage at the load terminals $U$ is known. The expression can thus be used to calculate the rise in voltage due to a reduction in load. If we linearize the relation between power and voltage, the voltage drop due to a load increase will be the same as the voltage rise due to a load decrease. The expression (2.28) could be used as an approximation for calculating the voltage drop due to a load increase. We will however obtain more practical approximated expressions for this in the next section.

**2.2.2.3 *Approximated Expressions*** In the previous section an exact expression has been derived for the voltage drop due to a load $P + jQ$. Such an exact expression will however not be used often. This is partly due to the complexity of the expression, even for such a simple case. Additionally, there are rarely any situations where active and reactive power are exactly known. Both active and reactive power are typically a function of the applied voltage, so that the "exact expression" remains an approximation after all. Therefore simplified but practical expressions are used to estimate the voltage drop.

The first-order approximation of (2.28) is obtained by replacing the square and the square root as their first-order approximations:

$$(1 + x)^2 \approx 1 + 2x \tag{2.29}$$

$$\sqrt{1 + x} \approx 1 + \tfrac{1}{2}x \tag{2.30}$$

The result is the following simple expression for the voltage drop due to active and reactive power flow:

$$\Delta U = \frac{RP + XQ}{U} \tag{2.31}$$

With $\theta$ the angle between voltage and current, we get

$$\Delta U = RI \cos\theta + XI \sin\theta \tag{2.32}$$

Expressions (2.31) and (2.32) are commonly used expressions for the voltage drop due to a load. As in most cases the voltage drop is limited to a few percent, this approximation is acceptable. Note that the same expression can be obtained by neglecting the imaginary part in (2.26), which is the normal derivation [193, p. 39].

Including second-order terms, we can make the following approximations:

$$(1 + x)^2 = 1 + 2x + x^2 \tag{2.33}$$

and

$$\sqrt{1+x} \approx 1 + \tfrac{1}{2}x - \tfrac{1}{8}x^2 \tag{2.34}$$

This results in the following second-order approximation for the voltage drop due to a load:

$$\Delta U = \frac{RP + XQ}{U} + \frac{3}{8}\frac{(RP + XQ)^2}{U^3} + \frac{1}{2}\frac{(XP - RQ)^2}{U^3} \tag{2.35}$$

For a small voltage drop we can also find an approximated expression for the change in phase angle. From (2.26) we find for the change in phase angle

$$\Delta\phi = \arctan\left(\frac{\mathrm{Im}(\underline{U}_\Delta)}{U + \mathrm{Re}(\underline{U}_\Delta)}\right) = \arctan\left(\frac{XP - RQ}{U^2 + RP + XQ}\right) \tag{2.36}$$

Using $\arctan x \approx x$, $U \approx 1$ pu, and $RP + XQ \ll 1$, we get the following approximation:

$$\Delta\phi \approx XP - RQ \tag{2.37}$$

Note that these expressions only give the change in voltage at a certain location due to the current at this location. Two possible applications are the daily voltage variation due to the daily load variation and the step in voltage due to a step in load current.

#### 2.2.2.4 Three Phases; Per Unit

The calculations before were done for a single-phase system. For a three-phase system we will consider balanced operation: Voltages and currents form a three-phase balanced set of phasors. Unbalanced voltages will be discussed in Section 2.3. For balanced operation the three-phase voltages can be written as follows in the time domain:

$$\begin{aligned} u_a(t) &= \sqrt{2}\,U\cos(2\pi f_0 t) \\ u_b(t) &= \sqrt{2}U\cos\left(2\pi f_0 t - \frac{2\pi}{3}\right) \\ u_c(t) &= \sqrt{2}U\cos\left(2\pi f_0 t + \frac{2\pi}{3}\right) \end{aligned} \tag{2.38}$$

where the voltage in phase $a$ has been used as a reference, resulting in a zero phase angle. The earlier expressions give the drop in phase voltage. The drop in line voltage is obtained by multiplying with $\sqrt{3}$. One should note, however, that $P$ and $Q$ in the earlier expressions are active and reactive power per phase. Let $P_3$

and $Q_3$ be the total active and reactive power, respectively. This results in the following approximated expression for the drop in line voltage:

$$\Delta U_{\text{line}} = \frac{RP_3 + XQ_3}{U_{\text{line}}} \tag{2.39}$$

Expressing all quantities in per-unit with a base equal to the nominal voltage results in the following well-known expression for the voltage drop in a three-phase system:

$$\Delta U = RP + XQ \tag{2.40}$$

where it has been assumed that the voltage is close to the nominal voltage.

**2.2.2.5 *Voltage Drop Along a Feeder*** Consider a low-voltage feeder with distributed load, as shown in Figure 2.7. The active and reactive load density at any location $s$ along the feeder is denoted by $p(s)$ and $q(s)$, respectively. The total active and reactive power downstream of location $s$ is denoted by $P(s)$ and $Q(s)$, respectively. These latter powers determine the current and thus the voltage drop. From Figure 2.7 the following difference equations are obtained:

$$P(s + \Delta s) = P(s) + p(s)\,\Delta s \tag{2.41}$$

$$Q(s + \Delta s) = Q(s) + q(s)\,\Delta s \tag{2.42}$$

$$U(s + \Delta s) = U(s) + r\frac{P(s)}{U_0}\Delta s + x\frac{Q(s)}{U_0}\Delta s \tag{2.43}$$

where $r + jx$ is the feeder impedance per unit length and all quantities are given in per unit. The approximation in (2.43) holds for small variations in voltage around $U_0$. If $U(s)$ is used instead of $U_0$, a nonlinear differential equation results, which

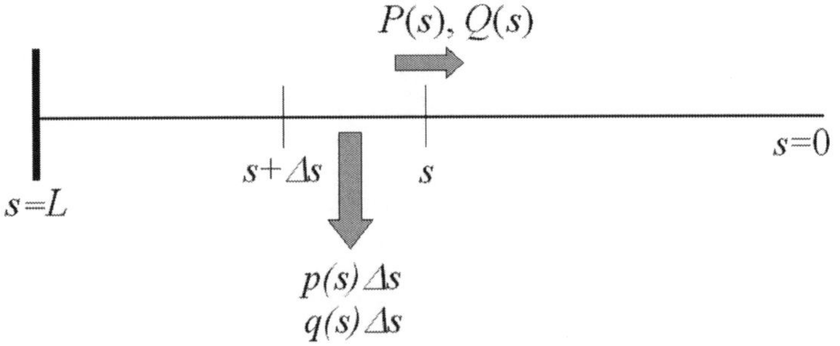


**Figure 2.7** Feeder with distributed load or generation.

is difficult to solve analytically. Alternatively, (2.43) can be obtained by considering constant-current load instead of constant-power load.

Taking the limit transition $\Delta s \to 0$ in (2.41) through (2.43) gives a set of three differential equations:

$$\frac{dP}{ds} = p(s) \tag{2.44}$$

$$\frac{dQ}{ds} = q(s) \tag{2.45}$$

$$\frac{dU}{ds} = \frac{1}{U_0}[rP(s) + xQ(s)] \tag{2.46}$$

Differentiation (2.46) and inserting (2.44) and (2.45) result in a second-order differential equation (if $r + jx$ is constant along the distance $s$):

$$\frac{d^2U}{ds^2} = \frac{1}{U_0}[rp(s) + xq(s)] \tag{2.47}$$

with boundary conditions

$$\frac{dU(0)}{ds} = 0 \tag{2.48}$$

$$U(L) = U_0 \tag{2.49}$$

The first boundary condition results from the fact that there is no load beyond the end of the feeder; the second one states that the voltage at the start of the feeder is known.

For a given load distribution $p(s)$, $q(s)$ the voltage profile along the feeder can be obtained. The case commonly studied in power system textbooks is uniformly distributed load along the feeder [e.g., 322]:

$$p(s) = p_0 \tag{2.50}$$

$$q(s) = q_0 \tag{2.51}$$

Combining (2.50) and (2.51) with (2.47) through (2.49) results in an expression for the voltage profile along a uniformly loaded feeder:

$$U(s) = U_0 - \frac{rp_0 + xq_0}{2U_0}(L^2 - s^2) \tag{2.52}$$

The voltage at the end of the feeder (i.e., the lowest voltage in case $p_0$ and $q_0$ are both positive) is equal to

$$U(L) = U_0 - \frac{rp_0 + xq_0}{2U_0} L^2 \tag{2.53}$$

Note that $s$ decreases from $s = L$ to $s = 0$ when going downstream along the feeder. From (2.53) an expression can be derived for the maximum feeder length under a minimum-voltage constraint:

$$L_{\max} = U_0 \sqrt{\frac{2}{rp_0 + xq_0} \times \frac{\Delta U_{\max}}{U_0}} \tag{2.54}$$

with $\Delta U_{\max}$ the maximum voltage drop along the feeder.

### 2.2.3 Voltage Control Methods

The voltage in the transmission network is controlled in a number of ways:

- The generator units control the voltage at their terminals through the field voltage.
- Shunt capacitor banks at strategic places in the transmission and subtransmission network compensate for the reactive power taken by the loads. In this way the reactive power in the transmission network is kept low. As the reactance of transmission lines dominates, the voltage drop is mainly due to the reactive power. Shunt capacitor banks will be discussed in Section 2.2.3.2.
- Series capacitor banks compensate for the reactance of long transmission lines. This limits the voltage drop due to the reactive power. Series capacitor banks also improve the stability of the system.
- Shunt reactors are used to compensate for the voltage rise due to long unloaded transmission lines or cables.

The voltage in the distribution network is controlled in a number of ways:

- By limiting the length of feeders (cables and lines). Note that the customer location cannot be affected by the design of the distribution network, so a given feeder length immediately implies a given amount of load. The relations between voltage drop limits and feeder length are discussed in Section 2.2.2.
- At a low voltage level the cross section of the feeder can be increased to limit the voltage drop.
- By installing transformer tap changers. Here one should distinguish between on-load tap changers and off-load tap changers. Both will be discussed in Section 2.2.3.1.

- Long distribution lines are sometimes equipped with series capacitor banks.
- Shunt capacitor banks are in use with large industrial customers, mainly to compensate for reactive power taken by the load. This also limits the voltage drop due to the load.
- For fast-fluctuating loads highly controllable sources of reactive power are used to keep the voltage constant. Examples are synchronous machines running at no load and static var compensators (SVCs).

***2.2.3.1 Transformer Tap Changers*** The transformation steps from the transmission network to the load typically consist of an high-voltage/medium-voltage (HV/MV) transformer with a relatively large impedance (15 to 30%) and an MV/low-voltage (LV) transformer with a small impedance (4 to 5%). Without any countermeasures the load variation would cause voltage variations of 10% and more over the HV/MV transformer. Together with the voltage variations due to cables or lines, due to the MV/LV transformer, plus the voltage variations in the transmission network and on the premises of the customer, the final load variation would be unacceptable.

The most common way of limiting the voltage variations is by installing on-load tap changers on the HV/MV transformers. The transformation from HV to MV sometimes takes place in two or more steps. In that case typically all these steps are equipped with on-load tap changers. For varying primary voltage and varying load, the voltage on the secondary side can be controlled by changing the turns ratio of the transformer. This enables compensation of the voltage variations in the transmission system and the voltage drop over the transformer. A typical range is $\pm 16\%$ of the nominal voltage for a total of $2 \times 16$ stages of 1% each [1].

To understand the method for voltage control, consider the voltage profile along the distribution system, as shown in Figure 2.8 for low load (top) and high load

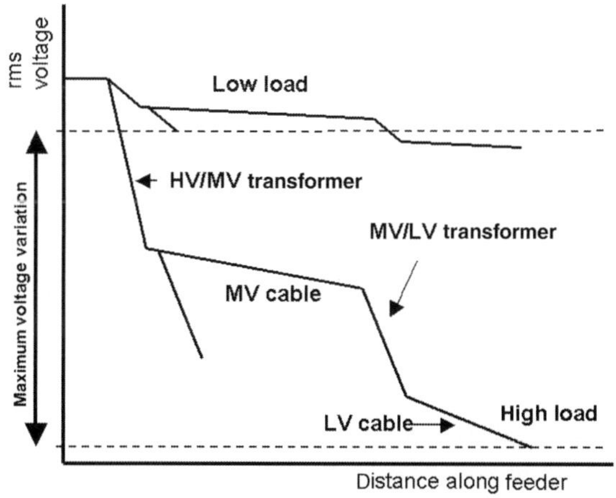


**Figure 2.8** Voltage profile in distribution system without voltage control.

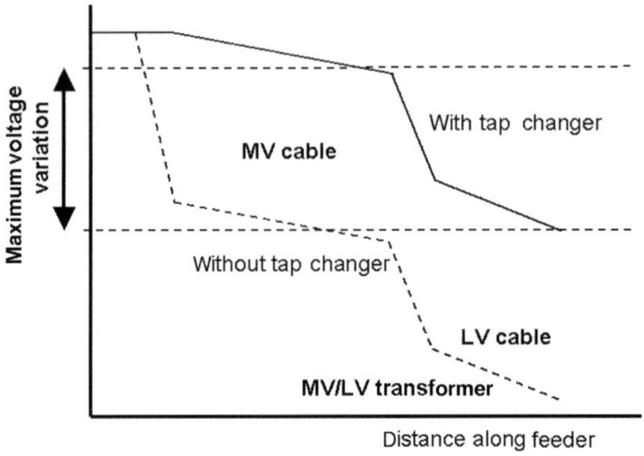


**Figure 2.9** Effect of on-load tap changers on voltage profile in distribution system.

(bottom). This is based on the distribution network as described above. The upper horizontal dashed line indicates the highest voltage for a customer close to the main substation. The lower dashed line gives the lowest voltage for a customer far away from the main substation. The difference between the two lines should be less than the permissible voltage range.

Assume that the permissible voltage range during the design stage is from 215 to 245 V. The design and control of the system should be such that the highest voltage is less than 245 V for the customer nearest the main substation. Also the lowest voltage should be above 215 V for the most remote customer. This will be very difficult without seriously restricting the maximum feeder length and the loading of the transformers.

The result of using an on-load tap changer is a constant voltage on the secondary side of the transformer. The effect on the voltage profile is shown in Figure 2.9. The voltage variation has decreased significantly. It becomes easier to keep the voltage variation within the permissible range. Alternatively, cable length can be longer so that more customers can be supplied from the same transformer. This limits the number of transformers needed. Note that the number of transformers needed is inversely proportional to the square of the feeder length.

The transformer tap changers are equipped with a delay to prevent them from reacting too fast. This delay varies between a few seconds and several minutes depending on the application. The resulting voltage variation due to a step in load current is shown in Figure 2.10. Transformer tap changers are not able to mitigate fast changes in voltage. They do however result in a constant voltage at a time scale of minutes.

For large transformers it is worth using *on-load tap changers*. But for smaller transformers, the costs become too high, simply because there are too many of them. Distribution transformers (10 kV/400 V) are typically equipped with *off-load tap changers*. As the relative impedance is only a few percent, there is also

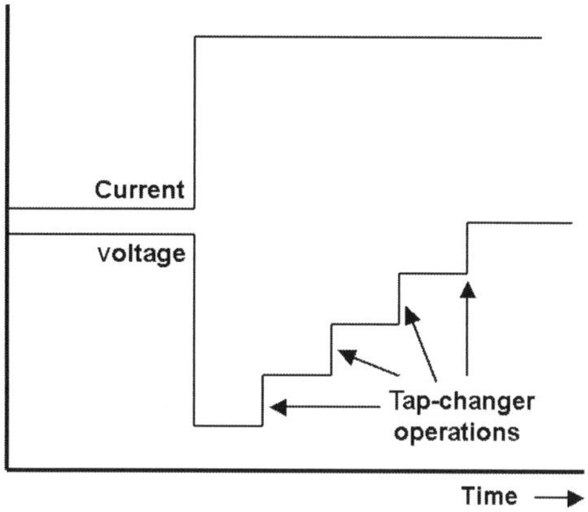


**Figure 2.10** Voltage variation due to a load step.

less need for voltage control. With off-load tap changers the transformer has a number of settings for the turns ratio. But the setting can only be changed when the transformer is not energized. The tap changer typically covers a band of $\pm 5\%$ around the nominal voltage. The taps are changed off load (i.e., when the transformer is disconnected from the system) in $2 \times 2$ stages of 2.5% each [1]. For example, a 10-kV/400-V transformer has tap settings of 10.5, 10.25, 10, 9.75, and 9.5 kV on the primary side. The secondary side nominal voltage is 400 V in all cases.

A smaller turns ratio (larger secondary voltage, smaller primary voltage) can be used for transformers near the end of a long feeder. The resulting voltage profile is shown in Figure 2.11, where the dotted line indicates the voltage profile with off-load tap changers. The use of off-load tap changers leads to a further decrease

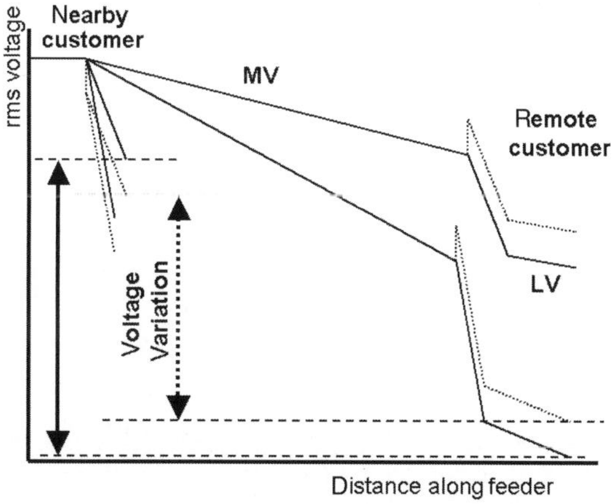


**Figure 2.11** Effect of off-load tap changers on voltage profile in distribution system.

in the voltage variation. Alternatively, longer feeder lengths are possible without exceeding the voltage limits.

#### 2.2.3.2 Shunt Capacitor Banks

Another common way of controlling the voltage is by installing shunt capacitor banks. By reducing the reactive power, the voltage drop is reduced. Capacitor banks can be installed at different voltage levels. In industrial systems they are often installed at low voltage, close to the equipment to limit the voltage drops in the distribution system. Industrial customers are often given financial incentives to limit their reactive power consumption, for example, by charging for kilovar-hours next to kilowatt-hours. In public supply networks capacitor banks are used in distribution and subtransmission substations. Installing a capacitor bank at a distribution substation limits the voltage drop over the HV/MV transformer and improves the voltage profile of the transmission system. In Sweden capacitor banks in the distribution and subtransmission networks are used to prevent voltage collapse in the transmission system. The distribution networks on average export reactive power to the transmission network.

A disadvantage of shunt capacitor banks is the voltage steps that arise when switching them. Energizing the bank may also lead to a transient overvoltage and the bank may create a resonance for harmonic currents produced by the load. Harmonic resonance and energizing transients will be discussed in Sections 2.5 and 6.3, respectively.

The voltage step when the capacitor is connected to the system can be calculated from the capacitor size $Q$ (in Mvar) and the fault level of the source $S_k$ (in MVA). The reactive power consumption of the capacitor bank is $-Q$; thus the capacitor bank generates reactive power. The resulting per-unit voltage step is

$$\frac{\Delta U}{E} = -\frac{Q}{S_k} \tag{2.55}$$

Note that a negative voltage drop indicates a rise in voltage. When the capacitor is removed from the system, the voltage shows a sudden decrease in rms value. The voltage steps should not exceed certain limits to prevent problems with end-user equipment. Typical steps in voltage magnitude due to switching of capacitor banks are between 1 and 3% of the nominal voltage.

In distribution systems the capacitor bank is typically connected to the secondary side of a transformer with on-load tap changers. The voltage step due to capacitor bank switching will be detected by the tap-changer control, which will bring back the voltage to within the normal range. The result is a (large) step followed by one or more (smaller) steps in the opposite direction, as in Figure 2.10.

#### 2.2.3.3 Distributed Generation and Voltage Variations

An increased penetration of distributed generation will impact the voltage in the distribution system. A single unit of relatively large size, 1 to 10 MW, will change the power flow and thus the voltage drop in the system. The generated power in many cases is not related to the consumed power so that the total load may become negative, leading to a voltage rise in the distribution system. The impact of distributed generation on the voltage control in transmission systems will occur for higher penetration levels and is more complex as it is also related to the closing of large generator stations as the generation is taken over by small units.

Power production by a distributed generator will in most cases lead to an increase in the voltage for all customers connected to the same feeder. During periods of high load, and thus low voltage, this improves the voltage quality. Generator units connected close to the load may be used as a way of controlling the voltage. This assumes, however, that the generator will be available during periods of high load. The operation of a distributed generator depends often on circumstances unrelated to the total load on the feeder. With combined heat and power the generated power is related to the heat demand, which has some relation to the electricity demand. The production of wind or solar power is completely independent of the electricity demand. The voltage rise due to distributed generation is discussed in several publications [e.g., 80, 99, 111, 178, 298].

The practical situation is that power production is steered by other factors than the need for voltage support. The extreme situations are high load without power production and low load with power production. The latter case may result in a voltage rise along the feeder, with the highest voltage occurring for the most remote customer. This will put completely new requirements on the design of the distribution system. Note that the off-line tap changers employed for distribution transformers may make the situation worse (see Figs. 2.9 and 2.11). For generation sources with a rather constant output (e.g., combined heat and power), the main impact is the voltage rise at low load, which simply calls for a change in distribution system design. Once the appropriate mitigation measures are in place in the system, the customer will have the same voltage quality as before. For generation with fluctuating sources (sun and wind) a statistical approach may be used. Standard requirements often allow voltage to be outside the normal range for a certain percentage of time. The introduction of such types of distributed generation may not lead to standard requirements being exceeded and thus does not require any mitigation measures in the system. The customer may however experience a deterioration of the voltage quality.

When a distributed generation unit is connected to a grid with line drop compensation on the transformer tap changers, the voltage along a distribution feeder may be lower or higher than without the generator. Connection of the unit close to the transformer will give a reduced voltage, whereas connection far away will lead to an increase in voltage [21, 188]. The impact also depends on the control mode of the line drop compensation relay [188]. In [143, 199] a method is proposed to determine the optimal setting of the tap changer (the so-called target voltage) from the voltages measured at several locations in the distribution network.

The use of distributed generation units with voltage source converter-based interfaces for controlling the voltage (and for providing other "ancillary services") is proposed in several publications. This remains a sensitive issue, however, as network operators often do not allow active power control by distributed generation units.

But according to [186] some utilities have somewhat loosened their requirement that distributed generation units should operate in a constant-power-factor mode and allow constant-voltage-mode operation. This statement however seems to refer to rather large units (of several megawatt), not to small units. A method for controlling the voltage is proposed in [42]. Under the proposed method the voltage source converter injects an amount of reactive power proportional to the difference between the actual voltage and the nominal voltage:

$$Q = \alpha(U - U_{\text{nom}}) \tag{2.56}$$

It is shown that this significantly reduces the voltage variations along a distribution feeder. By using proportional control only (as with power frequency control of generators), the risk of controller fighting is much reduced. The choice of the control parameter remains a point of discussion. A microgrid with converter-interfaced generation is discussed in [234]. A droop line in used for the voltage control as in (2.56). The droop setting is chosen as 4% of rated power. The method has however only been applied to one of the converters.

In [307] a control strategy is proposed for the connection of distributed generation units to weak distribution networks in South Africa. The unit operates at unity power factor for voltage below 105% of nominal. For higher terminal voltage the power factor may vary between unity and 0.95 inductive. Capacitive operation is blocked, as it would imply that the unit gives voltage support to the grid.

Some types of distributed generation show a strongly fluctuating power output (noticeably solar and wind power). If the unit significantly affects the voltage (in this case more than about 1%), it will pose an excessive duty on voltage-regulating equipment (tap changers and capacitor banks) resulting in premature failure of this equipment [99].

Regulations in distribution networks often require distributed generation (DG) units to disconnect during a fault and to come back after a fixed time (e.g., 5 min). In a system with a large penetration of distributed generation this would create a complex voltage variation: a voltage dip due to the fault followed by a sustained lower voltage corrected in one or more steps by the transformer tap changers. When the units reconnect, the voltage goes up, followed again by the operation of the tap changers [99]. The resulting voltage profiles along the feeder are shown in Figure 2.12, where the initiating event may be a fault at the distribution or transmission level but also the tripping of a large generator unit elsewhere in the system. Before the event the voltage along the feeder is higher than at the substation bus due to the power flow from the generator units back to the transformer (solid curve, normal operation). After the event, the generator units have tripped, leading to a drop in voltage for the loads connected to this feeder (dashed curve, tripping of DG). If other feeders are also equipped with distributed generation, the current through the transformer will also increase significantly, leading to a further drop in voltage at the terminals of the transformer. This drop will be more severe for generation connected to MV than for generation connected to LV.

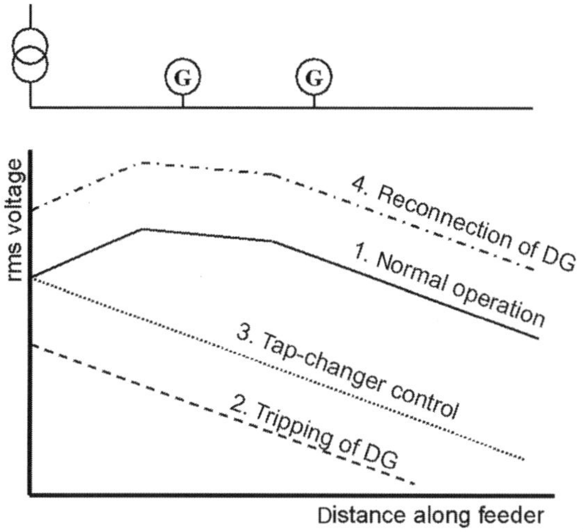


**Figure 2.12** Impact of tripping of local DG units due to external event on voltage profile along a feeder.

The impedance of an HV/MV transformer is higher (20 to 30%) than that of an MV/LV transformer (4 to 5%). The transformer tap changer on the HV/MV transformer will bring back the voltage to its normal level within seconds to minutes (depending on the local practice), but only for the secondary side of the HV/MV, not for the whole feeder (dotted curve, tap-changer control). If the units come back automatically after several minutes, this will lead to an overvoltage (dash-dot curve, reconnection of DG). The tap changers will alleviate the overvoltage and return the voltage profile to its normal shape.

The voltage variations due to variations in power output may occur in a time scale between the flicker range (1 s and faster) and the 10-min average as in EN 50160 and other standard documents. In [325, 326] the fluctuations in solar power generation are studied on time scales of 1 s and longer. Passing clouds may cause fast changes in power output that are on the borderline of the flicker spectrum. Further studies are needed on the severity of the problem. It may also be necessary to propose regulations for the permissible voltage variations between the flicker spectrum (time scale of 1 s and faster) and the limits in EN 50160 (10-min averages). We will introduce such a method in Section 5.2.4.

## 2.3 VOLTAGE UNBALANCE

This section will discuss the difference in voltage between the three phases. The method of symmetrical components is introduced to analyze and quantify the voltage unbalance. The origin of unbalance and the consequences of unbalance are also discussed.

### 2.3.1 Symmetrical Components

The method of symmetrical components is a way to describe unbalance in voltage and current in a three-phase system. Consider at first a balanced three-phase set of voltages of rms value $E$ written in the form of complex-voltage phasors:

$$\underline{U}_a = E \qquad \underline{U}_b = a^2 E \qquad \underline{U}_c = aE \tag{2.57}$$

with $a = -\frac{1}{2} + \frac{1}{2} j\sqrt{3} = 1e^{j120^\circ}$ a rotation over 120°.

Even in normal operation of the system, the voltages are not exactly balanced. To quantify the amount of unbalance, the actual complex voltages are written as the sum of three components:

$$\begin{aligned} \underline{U}_a &= \underline{U}^0 + \underline{U}^+ + \underline{U}^- \\ \underline{U}_b &= \underline{U}^0 + a^2\underline{U}^+ + a\underline{U}^- \\ \underline{U}_c &= \underline{U}^0 + a\underline{U}^+ + a^2\underline{U}^- \end{aligned} \tag{2.58}$$

The three complex voltages $\underline{U}^0$, $\underline{U}^+$, and $\underline{U}^-$ are called zero-sequence voltage, positive-sequence voltage, and negative-sequence voltage, respectively. These three complex numbers contain the same amount of information as the three complex phase voltages. The other way around, the component voltages are obtained from the phase voltages through the inverse transformation of (2.58):

$$\begin{aligned} \underline{U}^0 &= \tfrac{1}{3}(\underline{U}_a + \underline{U}_b + \underline{U}_c) \\ \underline{U}^+ &= \tfrac{1}{3}(\underline{U}_a + a\underline{U}_b + a^2\underline{U}_c) \\ \underline{U}^- &= \tfrac{1}{3}(\underline{U}_a + a^2\underline{U}_b + a\underline{U}_c) \end{aligned} \tag{2.59}$$

Similar expressions hold for the translation from component currents to phase currents and for the translation from phase currents to component currents.

A 3% drop in the voltage in phase $a$ with the voltages in the other two phases remaining at 1 pu,

$$\underline{U}_a = 0.97 \qquad \underline{U}_b = a^2 \qquad \underline{U}_c = a \tag{2.60}$$

results in the following values for the component voltages:

$$\underline{U}^0 = -0.01 \qquad \underline{U}^+ = 0.99 \qquad \underline{U}^- = -0.01 \tag{2.61}$$

In the same way a change of 0.03 rad in the phase angle of the phase $a$ voltage,

$$\underline{U}_a = \cos(0.03) + j\sin(0.03) \approx 1 + j0.03 \qquad \underline{U}_b = a^2 \qquad \underline{U}_c = a \tag{2.62}$$

results in

$$\underline{U}^0 = j0.01 \qquad \underline{U}^+ = 1 + j0.01 \qquad U^- = j0.01 \tag{2.63}$$

In both cases the magnitude of the negative-sequence voltage is 0.01 pu. The difference between a "magnitude unbalance" and a "phase unbalance" is only in the argument (direction) of the negative-sequence voltage.

To quantify the effect of voltage unbalance on rotating machines, consider the effect of a 3% drop in voltage in phase $a$. As shown in the earlier example, the symmetrical-component voltages for this case are

$$\underline{U}^+ = 1.0 \qquad \underline{U}^- = \underline{U}^0 = -0.01 \tag{2.64}$$

Assume further that the machine is operating at full load with 0.9 power factor:

$$\underline{I}^+ = 1.0 \angle{-25^\circ} \tag{2.65}$$

and that the negative-sequence impedance is one-tenth of the rated positive-sequence impedance. Negative-sequence losses are neglected so that the impedance is purely reactive:

$$\underline{Z}^- = 0.1 \angle 90^\circ \tag{2.66}$$

From this information the negative-sequence current is obtained as

$$\underline{I}^- = \frac{\underline{U}^-}{\underline{Z}^-} = 0.1 \angle 90^\circ \tag{2.67}$$

The phase currents are obtained from the addition of the contributions of the positive-sequence and negative-sequence components, resulting in

$$\underline{I}_a = 0.96 \angle{-20^\circ} \qquad \underline{I}_b = 1.10 \angle{-145^\circ} \qquad \underline{I}_c = 0.95 \angle{+90^\circ} \tag{2.68}$$

The current in phase $b$ increases by 10% whereas the currents in phases $a$ and $c$ decrease by about 5%. Note also the large change in phase angle for the currents, up to 30° in phase $c$.

### 2.3.2 Interpretation of Symmetrical Components

A balanced set of three-phase voltages only contains a positive-sequence component. Substituting (2.57) gives $\underline{U}^0 = 0$, $\underline{U}^+ = 1$, $\underline{U}^- = 0$. The *positive-sequence component* may be interpreted as the amount of balanced voltage in an unbalanced set of voltages. The torque produced by an induction motor is, for example, determined by the positive-sequence component of the voltage. The negative-sequence and zero-sequence component do not lead to any constant-torque production for an induction motor.

The *negative-sequence component* also forms a balanced set of three-phase voltages, but with opposite ("wrong") phase order. For $\underline{U}^0 = 0$, $\underline{U}^+ = 0$, $\underline{U}^- = 1$, (2.58) becomes

$$\underline{U}_a = 1 \qquad \underline{U}_b = a \qquad \underline{U}_c = a^2 \tag{2.69}$$

If the voltages at any point in the system would only contain a negative-sequence component, the induction motors would still operate "as normal," but they would rotate in the opposite direction. The difference between positive and negative sequence is simply a matter of definition.

The *zero-sequence component* forms a set of three in-phase voltages:

$$\underline{U}_a = 1 \qquad \underline{U}_b = 1 \qquad \underline{U}_c = 1 \tag{2.70}$$

The back transformation, from phase voltages to component voltages, can be physically interpreted as follows. The *zero-sequence component* is the average of the three phase voltages:

$$\underline{U}^0 = \tfrac{1}{3}(\underline{U}_a + \underline{U}_b + \underline{U}_c) \tag{2.71}$$

The *positive-sequence voltage* is the average of the three-phase voltages as experienced in the forward synchronously rotating frame. An induction motor or synchronous motor has phase order $a$, $b$, $c$. This implies that the phase $a$ voltage has a maximum, followed by a maximum in the phase $b$ voltage, followed by a maximum in the phase $c$ voltage. For the motor shown in Figure 2.13 the resulting rotation will be $a$–$b$–$c$.

The observer in the synchronously rotating frame (e.g., on the rotor winding) experiences $u_a$ now, $u_b$ 120 electrical degrees later, and $u_c$ 240 electrical degrees later. (An angle of 360 electrical degrees is defined as the rotation from one pole pair to the next pole pair.). The factor $a$ corresponds to a rotation of 120°. Thus $u_b$ 120 degrees later is, in complex notation, $a\underline{U}_b$. The average voltage experienced by the observer on the synchronously rotating frame is

$$\underline{U}^+ = \tfrac{1}{3}(\underline{U}_a + a\underline{U}_b + a^2\underline{U}_c) \tag{2.72}$$

Which is exactly the positive-sequence voltage. In the same way the *negative-sequence voltage* is the average voltage in the backwardly rotating

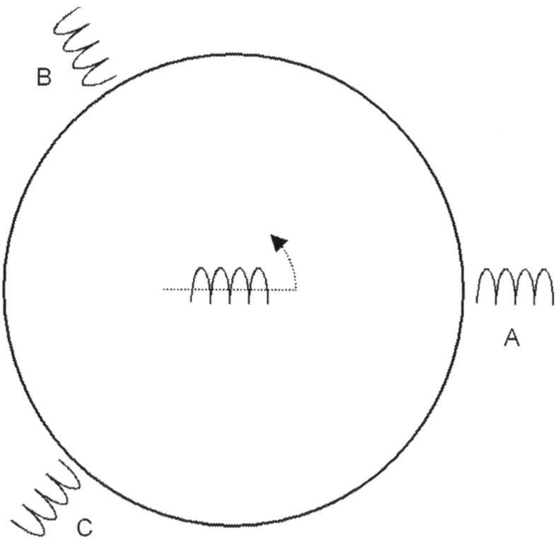


**Figure 2.13** Principle of rotation of synchronous motor.

synchronous frame:

$$\underline{U}^{-} = \tfrac{1}{3}(\underline{U}_a + a^2\underline{U}_b + a\underline{U}_c) \tag{2.73}$$

Summarizing:

- The positive-sequence voltage is the amount of voltage contributing to the power flow from generators to motors. One may state, with very little approximation, that the power flow from generation to load takes place in the positive sequence only.
- The negative-sequence voltage is an indication of the amount of unbalance in the system. As we will see later, unbalance indicates an inefficiency in the use of the three-phase system.
- The presence of zero-sequence voltage indicates a connection to earth. The zero-sequence current is a measure for the amount of current not returning through the phase conductors.

### 2.3.3 Power Definitions in Symmetrical Components: Basic Expressions

In the preceding sections equations have been given to transform phase voltages to component voltages and phase currents to component currents. The next step is to consider complex power flows in the three components (zero, positive, and negative sequence). Here we use the definitions as given in IEEE standard 1459 [171]. We consider voltages and currents to be sinusoidal. For nonsinusoidal voltages and currents additional definitions are needed, to be discussed in Section 2.5.

The total complex power in the three-phase system is the sum of the complex powers in the three phases:

$$\underline{S} = \underline{U}_a \underline{I}_a^* + \underline{U}_b \underline{I}_b^* + \underline{U}_c \underline{I}_c^* \tag{2.74}$$

Translating this into component voltages and currents gives

$$\underline{S} = 3\underline{U}^0 \underline{I}^{0*} + 3\underline{U}^+ \underline{I}^{+*} + 3\underline{U}^- \underline{I}^{-*} \tag{2.75}$$

Component power flows are defined as follows:

$$\underline{S}^0 = 3\underline{U}^0 \underline{I}^{0*} \qquad \underline{S}^+ = 3\underline{U}^+ \underline{I}^{+*} \qquad \underline{S}^- = 3\underline{U}^- \underline{I}^{-*} \tag{2.76}$$

Thus the total complex power $\underline{S}$ in a three-phase system is the sum of the powers in the three phases but also the sum of the powers in the three sequence components.

From the above expressions one can easily derive the definitions for active and reactive power for the symmetrical components. The active powers are defined as follows:

$$\begin{aligned} P^+ &= 3V^+ I^+ \cos(\theta^+) \\ P^- &= 3V^- I^- \cos(\theta^-) \\ P^0 &= 3V^0 I^0 \cos(\theta^0) \end{aligned} \tag{2.77}$$

with $\theta^+$ the angle between positive-sequence voltage and current, and so on. The total active power is equal to the sum of the active powers in the components but also equal to the sum of the active powers in the three phases:

$$P = P^+ + P^- + P^0 = P_a + P_b + P_c \tag{2.78}$$

The positive-, negative-, and zero-sequence reactive powers are defined in a similar way:

$$\begin{aligned} Q^+ &= 3V^+ I^+ \sin(\theta^+) \\ Q^- &= 3V^- I^- \sin(\theta^-) \\ Q^0 &= 3V^0 I^0 \sin(\theta^0) \end{aligned} \tag{2.79}$$

The total reactive power is equal to the sum of the reactive power values in the components:

$$Q = Q^{+} + Q^{-} + Q^{0} = Q_a + Q_b + Q_c \tag{2.80}$$

For the total apparent power, IEEE 1459 [171] gives two definitions. The *vector apparent power* is defined from the total active power $P$ and the total reactive power $Q$:

$$S_V = \sqrt{P^2 + Q^2} \tag{2.81}$$

The vector apparent power is the absolute value of the total complex power $\underline{S}$ as defined above. It can also be obtained from the complex apparent powers in the symmetrical components. Next to the vector apparent power, the so-called *arithmetic apparent power* is defined:

$$S_A = S_a + S_b + S_c \tag{2.82}$$

where $S_a = V_a I_a$ is the apparent power in phase $a$, and so on. The arithmetic apparent power may have some applications, but it is not related to the total active and reactive powers or to the symmetrical-component powers. Therefore it should be used with care and only when necessary. The interpretation of total reactive and apparent power under unbalanced operation remains a point of discussion.

It should be noted here that the negative- and zero-sequence power flow during normal operation is much smaller than the positive-sequence power. The power system is designed as a voltage source which results in small negative- and zero-sequence voltages and hence in small negative- and zero-sequence power flows. During severe disturbances in the system, like voltage dips, the situation may be different. Also note that the negative- and zero-sequence power flow is from the unbalanced load back to the system. The positive-sequence power is the sum of the actual energy consumption by the load and the negative- and zero-sequence power that is reinjected into the system.

### 2.3.4 The *dq*-Transform

The physical interpretation of the positive-sequence voltage can be used to explain the so-called *dq*-transform. The *dq*-transform (also known as Park's transform) is commonly used when modeling induction and synchronous motors and especially for the control of motor drives. The *dq*-transform makes it possible to control an alternating current (ac) motor as if it were a dc motor.

From the three-phase voltages in the time domain, two complex voltages are obtained as follows:

$$v_{\alpha\beta}(t) = \tfrac{1}{3}\sqrt{2}(v_a(t)\mathbf{e}_a + v_b(t)\mathbf{e}_b + v_c(t)\mathbf{e}_c) \tag{2.83}$$

where $\mathbf{e}_a = 1$, $\mathbf{e}_b = a = -\frac{1}{2} + \frac{1}{2}j\sqrt{3}$, and $\mathbf{e}_c = a^2 = -\frac{1}{2} - \frac{1}{2}j\sqrt{3}$ are the base vectors in the direction of the $a$-, $b$-, $c$-phase windings, respectively. The voltage $v_{\alpha\beta}$ is a complex voltage in the time domain. It can be interpreted as the rotating voltage that drives the air gap flux, as explained before. The factor $\frac{1}{3}\sqrt{2}$ in (2.83) can be arbitrarily chosen. A value equal to $\sqrt{\frac{2}{3}}$ is popular as it gives power invariance between the two systems. Here we have chosen the factor such that the positive-sequence voltage will be invariant.

Consider the three sinusoidal phase voltages

$$\begin{aligned} v_a(t) &= \mathrm{Re}\left\{\sqrt{2}\underline{U}_a e^{j2\pi f_0 t}\right\} \\ v_b(t) &= \mathrm{Re}\left\{\sqrt{2}\underline{U}_b e^{j2\pi f_0 t}\right\} \\ v_c(t) &= \mathrm{Re}\left\{\sqrt{2}\underline{U}_c e^{j2\pi f_0 t}\right\} \end{aligned} \tag{2.84}$$

Using $\mathrm{Re}\{\underline{z}\} = \frac{1}{2}(\underline{z} + \underline{z}^*)$ and the expressions for positive- and negative-sequence voltage derived earlier, we obtain for the $\alpha\beta$-voltage

$$v_{\alpha\beta}(t) = \underline{U}^+ e^{j2\pi f_0 t} + (\underline{U}^-)^* e^{-j2\pi f_0 t} \tag{2.85}$$

From the $\alpha\beta$-voltage (a rotating complex phasor), the $dq$-voltage (a stationary complex phasor) is calculated as follows:

$$v_{dq}(t) = v_{\alpha\beta}(t) e^{-j2\pi f_0 t} \tag{2.86}$$

The $dq$-voltage is again a complex phasor, but in this case a stationary one when only positive-sequence voltage is present. Using (2.85) and (2.86) the $dq$-voltage can be written as follows:

$$v_{dq}(t) = \underline{U}^+ + (\underline{U}^-)^* e^{-j4\pi f_0 t} \tag{2.87}$$

The average value of the $dq$-voltage is equal to the positive-sequence voltage. The value oscillates around the average with double the power system frequency (100 or 120 Hz) and an amplitude equal to the negative-sequence voltage.

### 2.3.5 Origin of Unbalance

Voltage unbalance (i.e., negative- and zero-sequence voltage) is due to unbalance in the load currents and to unbalance in the supplying network. The load unbalance is partly due to the natural variation between the single-phase loads in the three phases and partly due to large individual single-phase loads. Even if the loads are equally distributed over the three phases, the variation over time of the individual loads means there is never perfect balance between the load currents. This is mainly an issue in low-voltage networks. In medium-voltage networks and higher, the loads are in most cases three-phase loads. Of course the load unbalance from lower voltage levels propagates up to higher voltage levels, but as the unbalance is randomly distributed, the various contributions will cancel each other.

Unbalance due to large single-phase loads is mainly an issue at higher voltage levels. Examples of large single-phase loads are railway-traction supplies and arc furnaces [302].

Next to the voltage unbalance due to unbalanced load currents, voltage unbalance results from a balanced current flowing through unbalanced impedances. Transformers as well as transmission lines are not fully identical in the three phases. The center leg of a three-phase transformer takes a different magnetizing current than the outer legs. The three phases of an overhead transmission line have slightly different inductances and capacitances resulting in a coupling between positive- and negative-sequence voltages and currents.

Unbalance may also be due to differences between the phases in three-phase equipment. The unbalance in three-phase transmission is already mentioned, but even equipment that is supposed to be balanced, such as induction motors, may take some unbalanced current due to design limitations or erroneous design [306].

#### 2.3.5.1 Single-Phase Loads

Consider a single-phase load connected phase to neutral. Such loads are almost exclusively used in low-voltage systems. Let $\underline{I}$ be the current taken by this load. The phase currents are thus

$$\underline{I}_a = \underline{I} \qquad \underline{I}_b = 0 \qquad \underline{I}_c = 0 \tag{2.88}$$

In terms of symmetrical components this reads as

$$\underline{I}^+ = \underline{I}^- = \underline{I}^0 = \tfrac{1}{3}\underline{I} \tag{2.89}$$

In a balanced system there is no coupling between the symmetrical components, so that each of the component currents leads to a change in voltage in one component only:

$$\underline{U}^+ = \underline{E}^+ - \tfrac{1}{3}\underline{Z}^+\underline{I} \qquad \underline{U}^- = -\tfrac{1}{3}\underline{Z}^-\underline{I} \qquad \underline{U}^0 = -\tfrac{1}{3}\underline{Z}^0\underline{I} \tag{2.90}$$

with $\underline{E}$ the (balanced) source voltage before the connection of the single-phase load and $\underline{Z}^+, \underline{Z}^-$, and $\underline{Z}^0$ positive-, negative-, and zero-sequence source impedances, respectively.

The drop in positive-sequence voltage can be calculated in the same way as discussed in Section 2.2. The various approximations derived in that section hold for the positive-sequence voltage. The negative-sequence voltage due to a single-phase load is obtained from the second expression of (2.90). Using absolute values instead of complex numbers, this results in

$$U^- = \tfrac{1}{3} Z^- I \tag{2.91}$$

This can be written in terms of the fault level and the load power. The fault level $S_F$ for a system with source impedance $Z^+$ and phase voltage $E^+$ is

$$S_F = \frac{3(E^+)^2}{Z^+} \tag{2.92}$$

Here it is assumed that the drop in phase voltages is small, so that the absolute value of each phase voltage is about equal to the positive-sequence voltage. Note also that the fault level is determined by the positive-sequence source impedance.

The apparent power taken by the load is

$$S_{\text{load}} = E^+ I \tag{2.93}$$

Combining (2.91) with (2.92) and (2.93) results in the following expression for the negative-sequence unbalance:

$$\frac{U^-}{U^+} = \frac{S_{\text{load}}}{S_F} \tag{2.94}$$

where it has been assumed that positive- and negative-sequence source impedances are equal, $Z^+ = Z^-$, and that the drop in positive-sequence voltage is small, $U^+ \approx E^+$.

The zero-sequence voltage due to a single-phase load is calculated in a similar way. The only difference is that the zero-sequence source impedance is different from the positive- and negative-sequence source impedances. The zero-sequence source impedance is obtained from (2.92):

$$Z^0 = \frac{Z^0}{Z^+} \frac{3E^{+2}}{S_F} \tag{2.95}$$

Combining (2.95) with (2.93) and assuming $U^0 = \frac{1}{3} Z^0 I$ result in an expression for the zero-sequence unbalance:

$$\frac{U^0}{U^+} = \frac{Z^0}{Z^+} \frac{S_{\text{load}}}{S_F} \tag{2.96}$$

In low-voltage networks the zero-sequence source impedance is often similar to the positive-sequence source impedance, so that negative- and zero-sequence unbalances will be about equal.

**2.3.5.2 Phase-to-Phase Load** Consider next a load connected phase to phase. This is the case for most single-phase loads at distribution and transmission voltage levels. The current taken by the load is $\underline{I}$. The phase currents are in this case

$$\underline{I}_a = 0 \qquad \underline{I}_b = \underline{I}_c = \underline{I} \tag{2.97}$$

In terms of symmetrical components this reads as

$$\underline{I}^0 = 0 \qquad \underline{I}^+ = \tfrac{1}{3}j\sqrt{3}\underline{I} \qquad \underline{I}^- = -\tfrac{1}{3}j\sqrt{3}\underline{I} \tag{2.98}$$

The absolute value of the resulting negative-sequence voltage is

$$U^- = \tfrac{1}{3}j\sqrt{3}IZ^- \tag{2.99}$$

The fault level at the point of connection of the load is again obtained from (2.92). The apparent power of the load is

$$S_{\text{load}} = \sqrt{3}E^+I \tag{2.100}$$

Note that the load voltage equals the line voltage for phase-to-phase connected load. Combining (2.99) with (2.92) and (2.100) results in an expression for the (negative-sequence) unbalance due to a phase-to-phase connected load with apparent power $S_{\text{load}}$:

$$\frac{U^-}{U^+} = \frac{S_{\text{load}}}{S_F} \tag{2.101}$$

Note that this is the same expression as (2.94) for phase-to-neutral connected load. The difference is in the fact that phase-to-phase connected load does not cause any zero-sequence unbalance.

**2.3.5.3 Diversity of Single-Phase Loads** Consider a three-phase low-voltage feeder with $N_c$ customers equally spread over the three phases. Each customer takes a current $I$ which is normally distributed with expected value $\mu_I$ and standard deviation $\sigma_I$. In mathematical terms we can write this as

$$I = \mathcal{N}(\mu_I, \sigma_I) \tag{2.102}$$

We assume further that the currents are all in phase. The diversity between the loads is only in the magnitude of the currents. The total current for phase $a$ is also normally distributed but with a different expected value and standard deviation:

$$I_a = \mathcal{N}\left(\tfrac{1}{3}N_c\mu_I,\ \sqrt{\tfrac{1}{3}N_c}\sigma_I\right) \tag{2.103}$$

The currents in the other two phases have the same distribution. The symmetrical-component currents can be calculated from the complex currents:

$$\underline{I}_a = I_a \qquad \underline{I}_b = a^2 I_b \qquad \underline{I}_c = aI_c \tag{2.104}$$

The positive-sequence current is obtained from the standard expression

$$\underline{I}^+ = \tfrac{1}{3}\left(\underline{I}_a + a\underline{I}_b + a^2\underline{I}_c\right) = \tfrac{1}{3}\left(I_a + I_b + I_c\right) \tag{2.105}$$

The positive-sequence current, being the average of three normally distributed variables, is itself also normally distributed:

$$I^+ = \mathcal{N}\left(\tfrac{1}{3}N_c\mu_I,\ \tfrac{1}{3}\sqrt{N_c}\sigma_I\right) \tag{2.106}$$

The diversity of the load decreases with the square root of the number of customers connected to the feeder:

$$\frac{\sigma_+}{\mu_+} = \frac{1}{\sqrt{N_c}}\frac{\sigma_I}{\mu_I} \tag{2.107}$$

The negative-sequence current is obtained in a similar way:

$$\underline{I}^- = \tfrac{1}{3}(\underline{I}_a + a^2\underline{I}_b + a\underline{I}_c) = \tfrac{1}{3}(I_a + aI_b + a^2 I_c) \tag{2.108}$$

Using $a = -\frac{1}{2} + \frac{1}{2}j\sqrt{3}$ and $a^2 = -\frac{1}{2} - \frac{1}{2}j\sqrt{3}$ gives the following expression for the negative-sequence voltage:

$$\underline{I}^- = \left(\tfrac{1}{3}I_a - \tfrac{1}{6}I_b - \tfrac{1}{6}I_c\right) + j\left(\tfrac{1}{6}\sqrt{3}I_b - \tfrac{1}{6}\sqrt{3}I_c\right) \tag{2.109}$$

Both real and imaginary parts of the negative-sequence voltage are normally distributed with expected value zero and standard deviation $\frac{1}{6}\sigma_I\sqrt{2Nc}$. We can further prove that the real and imaginary parts are stochastically independent (the expected value of their product equals the product of their expected values). This allows us to calculate the expected value of the *square* of the absolute value of the negative-sequence current:

$$\mathcal{E}(|\underline{I}^-|^2) = \mathcal{E}(\mathrm{Re}\{\underline{I}^-)^2\} + \mathrm{Im}\{(\underline{I}^-)^2\}) = \tfrac{1}{9}N_c\sigma_I^2 \tag{2.110}$$

Calculating the expected value of the negative-sequence magnitude from the expected value of its square is not straightforward. But for small standard deviation, that is, for a large number of customers, we can approximately treat the stochastic variables as deterministic variables. This results in the following approximation:

$$I^- \approx \tfrac{1}{3}\sigma_I\sqrt{N_C} \tag{2.111}$$

Consider, for example, a feeder with 36 customers with expected current 10 A and standard deviation 5 A. The positive-sequence feeder current has an expected value of 120 A and a standard deviation of 10 A. The negative-sequence current is about 10 A.

### 2.3.6 Consequences of Unbalance

Voltage unbalance leads to unbalanced currents through three-phase equipment. Equipment that is especially prone to voltage unbalance includes rotating machines and three-phase diode rectifiers. As this equipment is normally connected phase to phase (or phase to neutral without a connection to ground in case of motor start), only the negative-sequence voltage affects the equipment. The zero-sequence voltage does not lead to any change in load currents. Therefore the zero-sequence voltage is normally not considered in characterizing the voltage unbalance.

Current unbalance leads to voltage unbalance and to an uneven heating of cables and lines. It also leads to an increase in losses in the cables and lines.

***2.3.6.1 Voltage Unbalance and Rotating Machines*** Rotating machines have a small negative-sequence impedance, typically one-fifth to one-tenth of the rated (positive-sequence) impedance. Therefore a negative-sequence voltage of only 2% (0.02 pu) will result in a negative-sequence current of 0.1 to 0.2 pu (10 to 20% of the rated current of the machine). This additional current will lead to increased losses and thus to increased heating of the machine. When the machine is used at its rated power for longer periods, some derating is needed to prevent overheating.

The total losses in the machine due to the positive- and negative-sequence currents are

$$P_{\text{loss}} = 3I^{+2}R^+ + 3I^{-2}R^- \tag{2.112}$$

with $R^+$ and $R^-$ positive- and negative-sequence resistances respectively. The equivalent (positive-sequence) current $I_{\text{eq}}$ that would lead to the same amount of losses is obtained from the following expression:

$$I_{\text{eq}} = I^+\sqrt{1 + \left(\frac{I^-}{I^+}\right)^2 \frac{R^-}{R^+}} \tag{2.113}$$

The thermal loading of the machine is only of interest for high mechanical loading and thus for positive-sequence currents close to the rated value. This is where the heating is highest and thus where there is a risk of overheating. Using $I^+ = 1$ in (2.113) and assuming $R^+ = R^-$ gives

$$I_{\text{eq}} = \sqrt{1 + \left(\frac{U^-}{Z^-}\right)^2} \tag{2.114}$$

where $I^- = U^-/Z^-$ has been used. Assuming a severe case where $Z^- = 0.1$ pu leads to the equivalent current values in Table 2.1 The increase in heating, as quantified by the equivalent current, requires derating of the machine to prevent overheating. A derating of 5% means that a machine with a 10-MW rating in case of balanced voltage has a rating of only 9.5 MW for unbalanced voltages.

From Table 2.1 one can conclude that 2% unbalance requires 2% derating and that 4% unbalance requires 8% derating. This effect is moderate if we consider that the unbalance in the public supply rarely exceeds 2%. We will see in section 5.6 that 2% is used as a de facto limit for the unbalance in many countries.

A more severe problem than the actual heating is the unequal distribution of the heating over the three phases. Depending on the phase angle between positive- and negative-sequence currents, the currents may add or subtract in some of the phases. Thus a 0.1-pu negative-sequence current (1% unbalance in Table 2.1) may result in 1.1-pu current in one of the phases. The losses in this phase are 121% of the normal (i.e., balanced) full-load losses. This is obviously compensated by lower losses in the other phases as the total increase in losses is only 0.5% according to Table 2.1. A severe case ($Z^- = 0.2$ pu, full addition of currents, no heat transfer between phases) has been the base for the required derating in Table 2.2. The actual derating needed is somewhere in between Table 2.1 and Table 2.2, depending on the construction of the machine. The recommended derating according to a report by the International Union for Electricity Applications (UIE) [302] is given in the final column of Table 2.2. It is assumed that for the latter the machine is designed to withstand 1% voltage unbalance.

**TABLE 2.1 Effect of Unbalance on Heating of Rotating Machines Assuming Uniform Heating**

| Unbalance (%) | Equivalent Current (pu) | Derating (%) |
|---|---|---|
| 1 | 1.005 | 0.5 |
| 2 | 1.020 | 2 |
| 3 | 1.044 | 4.5 |
| 4 | 1.077 | 8 |
| 5 | 1.118 | 12 |

**TABLE 2.2 Effect of Unbalance on Heating Rotating Machines Assuming Nonuniform Heating**

| Unbalance (%) | Maximum Phase Current (pu) | Maximum Phase Losses (%) | Derating (%) | UIE Recommendation (%) |
|---|---|---|---|---|
| 1 | 1.05 | 110 | 9 | 0 |
| 2 | 1.10 | 121 | 17 | 4 |
| 3 | 1.15 | 132 | 24 | 10 |
| 4 | 1.20 | 144 | 31 | 16 |
| 5 | 1.25 | 156 | 36 | 24 |

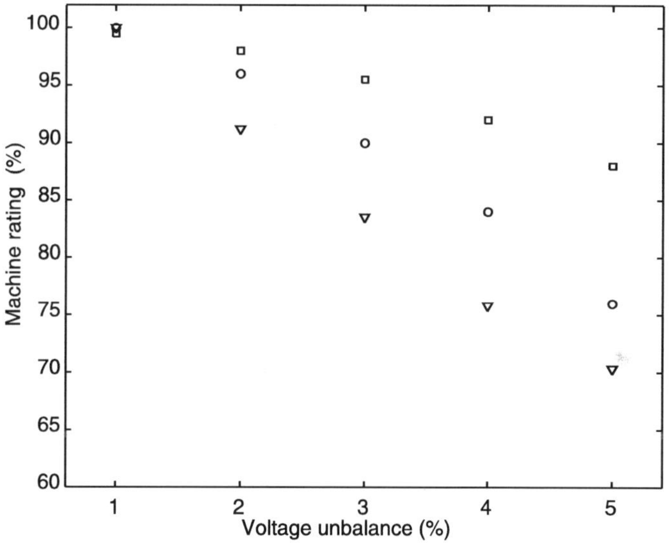


**Figure 2.14** Theoretical and practical derating of rotating machines due to unbalanced voltages: squares, based on total rms current; triangles, based on highest phase current; circles, UIE recommendation.

The ratings are also shown in Figure 2.14, where it has been assumed that the machine can tolerate a 1% unbalance without derating. We see that the UIE recommendations are somewhere in between the two extremes discussed here.

***2.3.6.2 Voltage Unbalance and Three-Phase Rectifiers*** Three-phase rectifiers are also sometimes heavily affected by voltage unbalance. But contrary to rotating machines there is no simple relation between the negative-sequence voltage and the negative-sequence current. This is very much related to the non linear nature of these devices. The concept of impedance is very hard to apply to power-electronic converters. Therefore this section will mainly present some qualitative discussion on the relation between voltage unbalance and the current taken by three-phase rectifiers.

With uncontrolled (diode) rectifiers we can distinguish between *dc current sources*, where the current on the dc side of the rectifier is more or less constant, and *dc voltage sources*, where the voltage on the dc side is more or less constant. These two types will be discussed in detail in Section 2.5. For a dc current source the effect of voltage unbalance is small. The difference in commutation instants will somewhat chance the duration of the pulses. This effect is discussed in [302, pp. 23–26; 319]. For ac voltage sources the effect is much bigger. As we will see in Section 2.5 the current pulse through the diodes is determined by the difference between the peak ac voltage and the dc voltage; see (2.243). For a three-phase rectifier the current shows six pulses, two for each phase-to-phase voltage. An unbalance in voltage thus causes a significant unbalance in current. Some examples of this phenomenon are shown in [33, p. 287; 202, pp. 37–42].

**2.3.6.3 *Current Unbalance and Losses*** Current unbalance leads to additional losses in the supply network. In the extreme case where all power is transported through only one of the three phases (i.e., the current is zero in two phases and three times its normal value in the third phase), the losses are three times as high as when the power transport is equally distributed over the three phases.

As shown in the examples before, an unbalanced load leads to a flow of negative- and zero-sequence power back to the system, that is, opposite to the flow of positive-sequence power. If we define the "useful power" as the net power flow to the load, we see that the positive-sequence power is higher than the useful power. As a result the positive-sequence current is higher than in a balanced situation with the same useful power being delivered. This leads to extra losses in the system, in addition to the losses due to the negative- and zero-sequence currents.

## 2.4 VOLTAGE FLUCTUATIONS AND LIGHT FLICKER

This section discusses fast changes in voltage magnitude (so-called voltage fluctuations) and the light flicker that is a consequence of it. The section starts with an overview of sources of voltage fluctuations followed by a mathematical model to describe voltage fluctuations. The model is used to relate current fluctuations and voltage fluctuations. The impact of voltage fluctuations on the light intensity of incandescent lamps is discussed as well as the perception of the light intensity fluctuations by human observers. The IEC flickermeter standard is discussed in detail followed by a discussion on flicker due to other types of lighting and other consequences of voltage fluctuations.

### 2.4.1 Sources of Voltage Fluctuations

Voltage fluctuations are generally speaking due to load variations. Any change in load current will obviously lead to a change in voltage, but those are generally not considered as voltage fluctuations.

There are two types of load that lead to light flicker, according to [303]: *loads that provoke separate voltage changes* and *loads that provoke voltage fluctuations*. The first group includes many heating and cooling loads. These loads often have a very short duty cycle, unless extreme (high or low) temperatures occur. Loads with an electrical motor as the main power consumer are the worst because the motor takes a high inrush current every time it is started. Examples are air conditioners and refrigerators. Also large photocopiers belong to this group of loads. In terms of the terminology introduced in Section 1.4 the resulting light flicker can be referred to as *light flicker due to repetitive events*.

The second group comprises those loads for which the current changes continuously. Examples are the arc furnace, arc and resistance welding, traction load, and wind turbines. The resulting light flicker can be referred to as *light flicker due to fast current variations*. This distinction is merely theoretical. The modern standards on flicker are such that both phenomena are included automatically.

An important source of voltage fluctuations is the arc furnace—a large electric oven in which metal is melted. The currents taken by an arc furnace vary at many different time scales. An example of a measured current with the resulting voltage fluctuations is shown in Figure 2.15. Both current and voltage magnitudes were obtained by taking the one-cycle rms value of the waveform and updating this calculation once every cycle. Arc furnaces take large amounts of power, are typically connected directly to the transmission system, and cause light flicker over large areas.

An example of a low-voltage load that leads to voltage fluctuations is the copy machine. Larger copy machines as are used in offices maintain the drum at a high

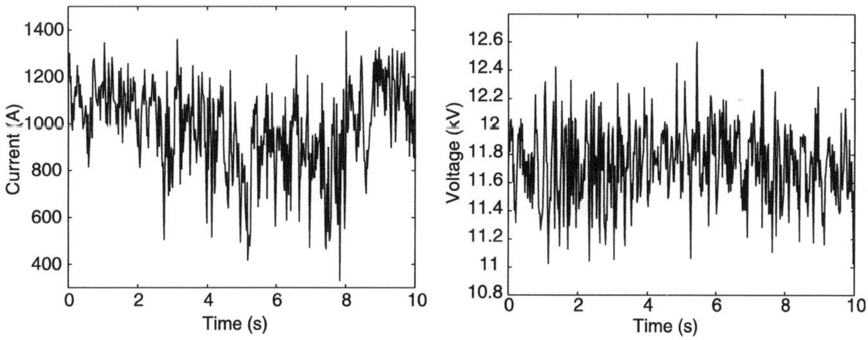


**Figure 2.15** Current fluctuations (left) and voltage fluctuations (right) due to large arc furnace: one-cycle rms voltage.

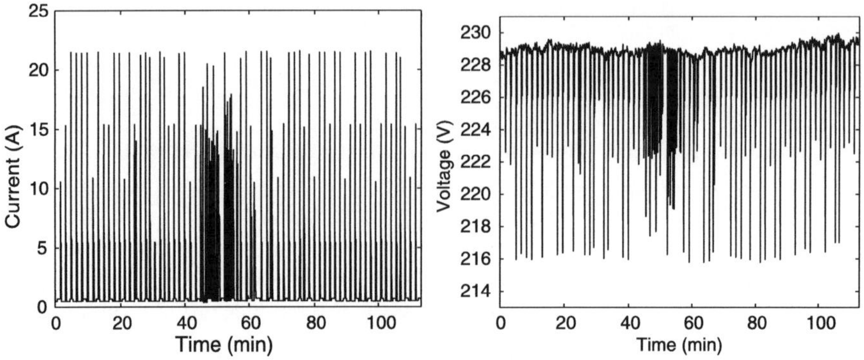


**Figure 2.16** Current taken by copy machine (left) and resulting voltage fluctuations (right).

temperature. Even when the copier is not in use, it regularly takes a high current to maintain this high drum temperature. An example of the current to a copier and the resulting voltage fluctuations are shown in Figure 2.16. These figures were obtained by sampling the voltage and current rms value every 5 s.

A domestic example of a load that leads to voltage fluctuations is the refrigerator. Starting the pump needed for circulating the cooling liquid leads to a voltage drop. Figure 2.17 shows the voltage measured during a four-day period in a small apartment. The plot shows maximum and minimum voltages obtained over 1-min

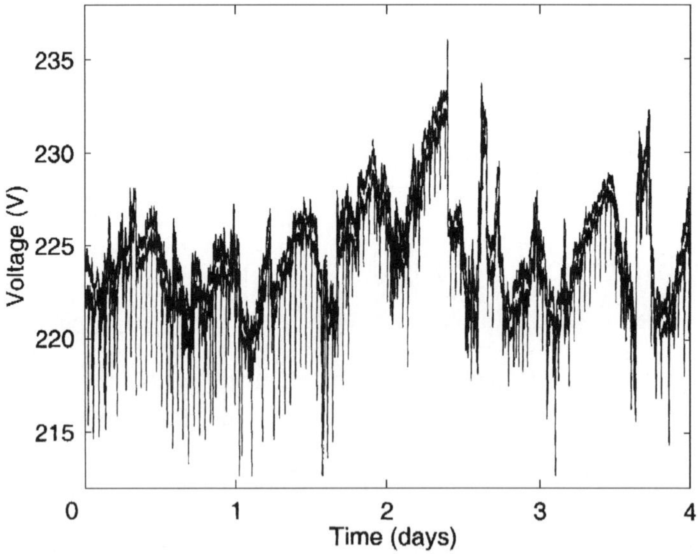


**Figure 2.17** Voltage fluctuations due to repetitive starting of a refrigerator.

**Figure 2.18** Refrigerator leading to the voltage fluctuations shown in Figure 2.17.

windows. The reduced minimum voltages are due to the starting of the refrigerator. Halfway during the measurement the refrigerator was moved to another wall outlet leading to an obvious reduction in the severity of the voltage fluctuations. The refrigerator that caused the voltage fluctuations is shown in Figure 2.18.

The introduction of distributed generation will generally lead to an increase of the voltage magnitude experienced by the customers. With highly variable sources of energy (like wind and sun) the voltage magnitude will also show a higher level of changes over a range of time scales. For slow voltage variations the indices defined in EN 50160 (10-min rms values) should be used. For the fastest fluctuations the flicker indices are an appropriate tool. A study of the different fluctuations in the power generated by a fixed-speed wind turbine is presented in [292]. Measurements were performed after the frequency components in the power fluctuations, between 0.1 and 10 Hz. The following components were found, next to a continuous spectrum:

- A 1.1-Hz fluctuation corresponding to the tower resonance.
- A 2.5-Hz fluctuation corresponding to the rotation speed of the gearbox.
- Four different components related to the rotation of the blades: $1p$, $2p$, $3p$, and $6p$. The $1p$ fluctuations are due to unbalance in the rotor and/or small differences between the blades. The $3p$ oscillations are due to the passing of

the blades in front of the tower. The $2p$ and $6p$ components are probably harmonics of the $1p$ and $3p$ fluctuations, respectively.

For low wind turbulence (wind from sea in this case), these discrete components dominate the spectrum. For high wind turbulence the fluctuations form a continuous spectrum. In [291] additional components are found at $4p/3$, $4p$, $14p/3$, $5p$, $9p$, $12p$, and $18p$.

There is some evidence that the turbines in a wind park may reach a state of "synchronized operation," thus amplifying the power pulsations due to the tower. The cause of this synchronous operation is not fully clear but it is thought to be due to interactions between the turbines through the network voltages [178]. Synchronous operation can only be expected for sites with a rather constant wind speed not affected by turbulence due to the terrain. Voltage fluctuations are especially a concern for wind turbines and a significant amount of literature is available on this subject. A survey of different studies [195, 233, 257, 285, 293, 294, 310] showed flicker levels due to wind power between $P_{st} = 0.04$ and $P_{st} = 0.5$ (where $P_{st} = 1$ is the acceptable limit; see Section 2.4.7).

The voltage variations due to variations in generated power may occur in a time scale between the flicker range (1 s and faster) and the 10-min average as in EN 50160 and other standard documents. In [325, 326] the fluctuations in solar power generation are studied on time scales of 1 s and longer. Passing clouds may cause fast changes in power output that are on the borderline of the flicker spectrum. Further studies are needed on the severity of the problem. It may also be necessary to propose regulations for the permissible voltage variations between the flicker spectrum (time scale of 1 s and faster) and the limits in EN 50160 (10-min averages). The "very-short variations" introduced in Section 5.2.4 are an appropriate tool for quantifying these variations.

The following list of sources of voltage fluctuations, taken from different publications, shows the variety of loads that may lead to flicker problems. However, most sources only cause local problems, that is, voltage fluctuations at the voltage level where they are connected. When this is in an industrial environment, the effect on other customers is small. Exceptions are large industrial installations such as arc furnaces and steel mills where the total current taken from the public supply shows large fluctuations. As the connection to the public supply is typically at the transmission level, customers over a large area will be affected. These are the loads that cause the main flicker concerns.

Possible sources of continuous voltage fluctuations include the following:

- Resistance welding machines [51, 271, 303]
- Rolling mills [303]
- Large industrial motors with variable loads [303]
- Arc furnaces [51, 118, 263, 271, 303]
- Arc welders [118, 263, 303]

- Saw mills [303]
- Railway traction [126]

Sources of separate voltage changes include the following:

- Switching of power factor correction capacitors [303]
- Large-capacity electric boilers [303]
- X-ray machines [303]
- Lasers [303]
- Large-capacity photocopying machines [303]
- Air conditioners [51, 271, 303]
- Refrigerators [271, 303]
- Startup of drives and steep load changes of drives [271]
- Connection and disconnection of lines [271]
- Elevators [51, 263]
- Particle accelerators [180]

### 2.4.2 Description of Voltage Fluctuations

Voltage fluctuations are described as an amplitude modulation of the fundamental-frequency voltage:

$$v(t) = \sqrt{2}V[1 + m(t)] \cos(2\pi f_0 t) \tag{2.115}$$

with $V$ the rms value of the undisturbed voltage waveform (the "carrier wave"), $f_0$ the fundamental frequency, and $m(t)$ the modulation. Expression (2.115) theoretically describes any voltage disturbance by an appropriate choice of $m(t)$. By giving $m(t)$ a rectangular shape with a duration between 20 ms and a few seconds, a voltage dip results. Even voltage transients can be described by (2.115). This is more than a theoretical observation because all voltage disturbances do in principle lead to light flicker. Repetitive voltage dips are equally disturbing as a continuous small fluctuation in the voltage magnitude.

Consider a fundamental-frequency signal modulated with a sinusoidal voltage fluctuation:

$$m(t) = M \cos(2\pi f_M t + \phi_M) \tag{2.116}$$

resulting in the following fluctuating voltage:

$$v(t) = \sqrt{2}V[1 + M \cos(2\pi f_M t + \phi_M)] \cos(2\pi f_0 t) \tag{2.117}$$

This can be written as the sum of three sine waves:

$$\begin{aligned} v(t) &= \sqrt{2}V\cos(2\pi f_0 t) \\ &\quad +\tfrac{1}{2}\sqrt{2}MV\cos[2\pi(f_0+f_M)t+\phi_M] \\ &\quad +\tfrac{1}{2}\sqrt{2}MV\cos[2\pi(f_0-f_M)t+\phi_M] \end{aligned} \tag{2.118}$$

The first term in (2.118) is the *carrier wave* and the second and the third terms are the *side lobes*: spectral components on opposite sides of the carrier wave. The voltage fluctuations can thus be described in the frequency domain as side lobes on opposite sides of the fundamental frequency. Note, however, that the modulation frequency cannot be directly found back in the spectrum. For example, a 7.3-Hz modulation on a 59.9-Hz fundamental component results in spectral bands at 52.6, 59.9, and 67.2 Hz, but there is no 7.3-Hz component.

Consider next a waveform with a pure phase modulation:

$$v(t) = \sqrt{2}V\cos[2\pi f_0 t + \phi(t)] \tag{2.119}$$

For $|\phi(t)| \ll 1$, that is, small changes in phase, we obtain from (2.119)

$$\begin{aligned} v(t) &= \sqrt{2}V\cos(2\pi f_0 t)\cos[\phi(t)] - \sqrt{2}V\sin(2\pi f_0 t)\sin[\phi(t)] \\ &\approx \sqrt{2}V\cos(2\pi f_0 t) - \sqrt{2}V\phi(t)\sin(2\pi f_0 t) \end{aligned} \tag{2.120}$$

Consider again a sinusoidal modulation signal:

$$\phi(t) = \Phi\cos(2\pi f_M t + \phi_M) \tag{2.121}$$

so that

$$\begin{aligned} v(t) &= \sqrt{2}V\cos(2\pi f_0 t) \\ &\quad -\tfrac{1}{2}\sqrt{2}V\Phi\sin[2\pi(f_0+f_M)t-\phi_M] \\ &\quad +\tfrac{1}{2}\sqrt{2}V\Phi\sin[2\pi(f_0-f_M)t-\phi_M] \end{aligned} \tag{2.122}$$

The result is again a carrier wave (the first term) and two side lobes (second and third terms). The difference with amplitude modulation is in the sign of the two side-lobe terms. The difference can be made visible by considering the phasor diagram for the fundamental frequency. At fundamental frequency, the carrier wave results in a constant vector. The side lobes result in vectors that rotate with the modulation frequency in opposite direction [55, Section 5.1, Section 6.2]. The result is shown in Figure 2.19.

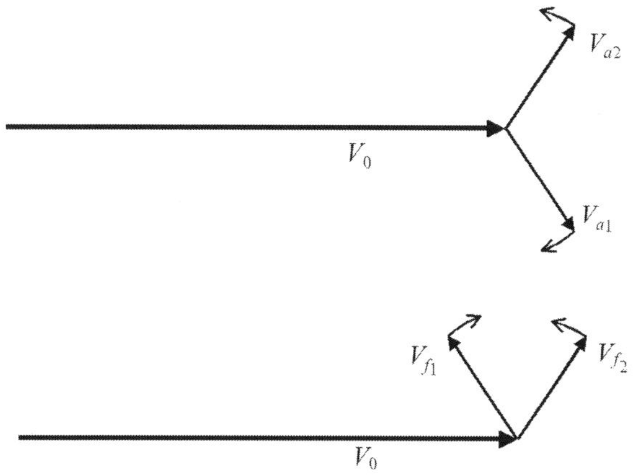


**Figure 2.19** Fluctuations in voltage magnitude (top) and voltage phase angle (bottom) shown as sum of three fundamental-frequency phasors: $V_0$ is the nonfluctuating (constant) part of the complex voltage; $V_{a1}$ and $V_{a2}$ are the two components, rotating in the complex plane, that lead to amplitude modulation; $V_{f1}$ and $V_{f2}$ lead to frequency or phase modulation.

***2.4.2.1 Relation Between Current and Voltage Fluctuations*** As mentioned before, voltage fluctuations are due to fluctuations in the load current. But there is no direct relation between the size of the current fluctuations and the size of the resulting voltage fluctuations. From Section 2.2.2, (2.37) and (2.40), we can obtain the following relations between fluctuations in power flow and fluctuations in voltage:

$$\begin{aligned} m_V(t) &= -R\,\Delta P(t) - X\,\Delta Q(t) \\ \phi_V(t) &= R\,\Delta Q(t) - X\,\Delta P(t) \end{aligned} \tag{2.123}$$

where $R + jX$ is the fundamental-frequency source impedance at the load terminals, $m_V(t)$ the per-unit amplitude modulation of the voltage, and $\phi_V(t)$ the phase modulation due to a change in complex power equal to $\Delta P(t) + j\Delta Q(t)$. The conclusion from (2.123) is that a fluctuation in the active and/or reactive part of the current leads to an amplitude modulation as well as a phase modulation. Note that an increase in power consumption gives a drop in voltage, hence the minus sign in the first line of (2.123).

Consider next a current with amplitude modulation $m(t)$ and phase modulation $\phi(t)$:

$$\underline{I}(t) = \underline{I}_0[1 + m(t)]e^{j\phi(t)} \tag{2.124}$$

which can be approximated as

$$\underline{I}(t) \approx \underline{I}_0[1 + m(t) + j\phi(t)] \tag{2.125}$$

Let $\underline{U}_0$ be the source voltage and $\underline{S}_0 = P_0 + jQ_0 = \underline{U}_0\underline{I}_0^*$ the nonfluctuating (average) power. The total complex power is obtained from

$$\underline{S} = \underline{U} \times \underline{I}^* = \underline{S}_0 + \Delta P(t) + j\,\Delta Q(t) \tag{2.126}$$

with $\Delta P(t) = P_0 m(t) + Q_0\phi(t)$ and $\Delta Q(t) = Q_0 m(t) - P_0\phi(t)$. The resulting amplitude and phase modulation in voltage can next be obtained from (2.123). The relation between phase and amplitude modulation in voltage and current is as follows:

$$m_V(t) = -(RP_0 + XQ_0)m(t) + (XP_0 - RQ_0)\phi(t) \tag{2.127}$$

$$\phi_V(t) = (RQ_0 - XP_0)m(t) - (RP_0 + XQ_0)\phi(t) \tag{2.128}$$

Even pure amplitude modulation in current, $\phi(t) = 0$, will lead to amplitude and phase modulation in the voltage. The ratio between the amount of phase and amplitude modulation depends on the power factor of the load and on the source impedance.

For fast fluctuations the above quasi-static approximation no longer holds. The voltage drop should be calculated by using differential equations in the time domain. Consider a Thevenin equivalent with a source voltage of constant amplitude:

$$e(t) = \sqrt{2}E\cos(2\pi f_0 t) \tag{2.129}$$

The voltage drop due to the load current $i(t)$ is obtained by assuming an *RL* series connection as source impedance:

$$u(t) = e(t) - L\frac{di}{dt} - Ri \tag{2.130}$$

Consider again a current waveform with amplitude modulation $m(t)$ and phase modulation $\phi(t)$:

$$i(t) = \sqrt{2}I_0[1 + m(t)]\cos[2\pi f_0 t - \psi + \phi(t)] \tag{2.131}$$

Using (2.120) this can be written, for small phase modulation, as the sum of three components:

$$\begin{aligned} i(t) &= \sqrt{2}I_0 \cos(2\pi f_0 t - \psi) \\ &+ \sqrt{2}I_0 m(t) \cos(2\pi f_0 t - \psi) \\ &- \sqrt{2}I_0 \phi(t) \sin(2\pi f_0 t - \psi) \end{aligned} \tag{2.132}$$

The first term is the nonfluctuating (average) component, which causes a steady-state voltage drop. There is no need to further consider this component. The second term in (2.132) is the amplitude modulation term. The resulting voltage fluctuation is found by substituting into (2.130):

$$\Delta u_1(t) = Xm(t)\sqrt{2}I_0 \sin(2\pi f_0 t - \psi) \tag{2.133}$$

$$- \left(L\frac{dm}{dt} + Rm(t)\right)\sqrt{2}I_0 \cos(2\pi f_0 t - \psi) \tag{2.134}$$

The third term in (2.132) is the phase modulation term, with the following expression for the resulting voltage fluctuations:

$$\Delta u_2(t) = X\phi(t)\sqrt{2}I_0 \cos(2\pi f_0 t - \psi) \tag{2.135}$$

$$+ \left(L\frac{d\phi}{dt} + R\phi(t)\right)\sqrt{2}I_0 \sin(2\pi f_0 t - \psi) \tag{2.136}$$

For slow current fluctuations $L(dm/dt)$ and $L(d\phi/dt)$ can be neglected compared to $Rm(t)$ and $R\phi(t)$, respectively. For fast current fluctuations these contributions have to be considered. Using a sinusoidal amplitude modulation, $m(t) = M \cos(2\pi f_M t + \phi_M)$, gives the following conditions for the quasi-stationary approximation in (2.127) and (2.128) to be valid:

$$f_M \ll \frac{f_0}{X/R} \tag{2.137}$$

To make the quasi-stationary and the time-domain expressions compatible, we use per-unit notation and assume that the voltage is close to 1 pu, so that we can substitute $P_0 = \sqrt{2}I_0 \cos(\psi)$ and $Q_0 = \sqrt{2}I_0 \sin(\psi)$. Combining this with the goniometric expressions for $\sin(a+b)$ and $\cos(a+b)$ results in

$$\begin{aligned} \sqrt{2}I_0 \sin(2\pi f_0 t - \psi) &= P_0 \sin(2\pi f_0 t) - Q_0 \cos(2\pi f_0 t) \\ \sqrt{2}I_0 \cos(2\pi f_0 t - \psi) &= P_0 \cos(2\pi f_0 t) + Q_0 \sin(2\pi f_0 t) \end{aligned} \tag{2.138}$$

This results in the following expression for voltage fluctuations due to amplitude-modulated current:

$$\Delta u_1(t) = Xm(t)[P_0 \sin(2\pi f_0 t) - Q_0 \cos(2\pi f_0 t)] - \left(Rm(t) + L\frac{dm}{dt}\right)[P_0 \cos(2\pi f_0 t) + Q_0 \sin(2\pi f_0 t)] \quad (2.139)$$

In the same way the voltage fluctuations due to phase-modulated currents are

$$\Delta u_2(t) = X\phi(t)[P_0 \cos(2\pi f_0 t) + Q_0 \sin(2\pi f_0 t)] + \left(R\phi(t) + L\frac{d\phi}{dt}\right)[P_0 \sin(2\pi f_0 t) - Q_0 \cos(2\pi f_0 t)] \quad (2.140)$$

These voltage waveforms are superimposed on the no-load voltage $e(t)$: The cosine terms are in phase with the steady-state voltage; they constitute phase modulation. The sine terms are in quadrature with the steady-state voltage, thus constituting a phase modulation. Note the minus sign in (2.120). The resulting amplitude modulation of the voltage is

$$m_V(t) = -(RP_0 + XQ_0)m(t) + (XP_0 - RQ_0)\phi(t) - LP_0\frac{dm}{dt} - LQ_0\frac{d\phi}{dt} \quad (2.141)$$

The phase modulation of the voltage is

$$\phi_V(t) = (RQ_0 - XP_0)m(t) - (XQ_0 + RP_0)\phi(t) - LQ_0\frac{dm}{dt} + LP_0\frac{d\phi}{dt} \quad (2.142)$$

### 2.4.3 Light Flicker

As mentioned above, the main interest in voltage fluctuations is due to their ability to cause light intensity fluctuations that are perceived by our brain as light flicker. Already very small voltage fluctuations are observable, somewhat larger ones are irritable, and at a certain level of voltage fluctuations they literary cause headache. For many years the severity of the voltage fluctuations at a certain location was obtained by comparing the fluctuations with the so-called *flickercurve*. For each repetition frequency a maximum permissible amplitude was defined for the voltage variation. The basic curve was valid for rectangular variations, but correction factors were available for nonrectangular (e.g., sinusoidal) variations. Such a curve was a useful tool in the design of systems, but it was not possible to uniquely quantify the severity of voltage fluctuations from measurements.

The more recent flickermeter standard, IEC 61000-4-15, addresses the issue in a more systematic way. The flickermeter standard is one of the most interesting power quality standards that have been issued over the years. It shows that it is possible to

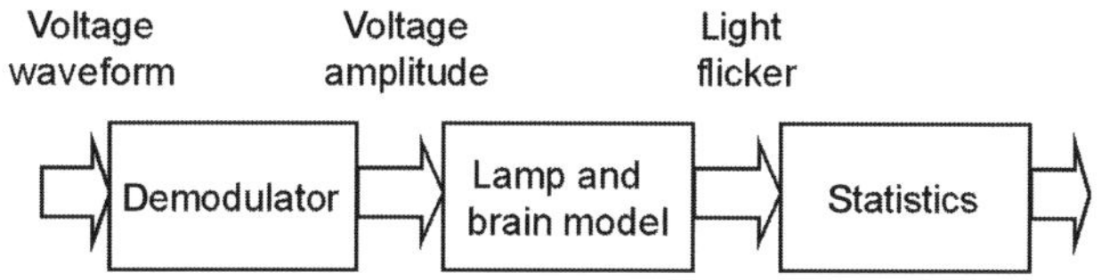


**Figure 2.20** From voltage fluctuations to light flicker.

use advanced scientific and engineering knowledge and make it into a workable standard document. The approach in the flickermeter standard is summarized in Figure 2.20.

From the measured voltage waveform, the fluctuations in magnitude are determined. This is done by means of a demodulator. The lamp model determines the fluctuations in light intensity due to the fluctuations in voltage amplitude. The second block models not only the lamp but also the way in which our brain observes the fluctuations. Very fast and very slow variations are not noticed by our brain. The response of this block more or less corresponds to the above-mentioned flickercurve. Finally, a rather complicated statistical block represents the way in which our brain interprets the severity of light intensity fluctuations.

The latest revision of the flickermeter standard IEC 61000-4-15 uses two lamp models: a 60-W, 120-V incandescent lamp and a 60-W, 230-V incandescent lamp. Models for other types of lamps could in principle be included, but no significant work towards their development has been done yet. A strong limitation with the development of such models is the large variety in lamp types. We will give some examples of the flicker response of nonincandescent lamps in Section 2.4.8.

### 2.4.4 Incandescent Lamps

An incandescent lamp consists of a coiled tungsten filament surrounded by a bulb filled with a mixture of nitrogen and argon in proportions dependent on the wattage of the lamp. The bulb is frosted on the inside with hydrofluoric acid to produce a diffused light instead of the glaring brightness of the unconcealed filament [49].

A voltage at the terminals of an incandescent lamp leads to a current through the filament of the lamp. The current heats up the filament, and when the filament reaches a high enough temperature, it will start to emit light. The steady-state temperature of the filament is around 3500 K. The higher the voltage, the higher the current, the higher the temperature, and the higher the light intensity. A fluctuation in voltage will thus lead to a fluctuation in light intensity.

Consider a voltage $v(t)$ over the terminals of an incandescent lamp. The lamp has a resistance $R$ so that the voltage leads to losses of the amount:

$$\mathcal{E}_{\text{in}} = \frac{v^2(t)}{R} \tag{2.143}$$

These losses will heat up the filament, whereas the energy loss to the environment (at a much lower temperature than the filament) will cool down the filament:

$$m_f c_1 \frac{dT_f}{dt} = \mathscr{E}_{\text{in}} - \mathscr{E}_{\text{out}} \tag{2.144}$$

with $T_f$ the temperature of the filament, $c_1$ the specific heat of tungsten, $m_f$ the mass of the filament, and $\mathcal{E}_{\text{out}}$ the heat transfer to the environment. The transfer of heat from the filament to the environment is, as with any case of heat transfer, a combination of conduction, convection, and radiation. As the bulb is filled with gas, conduction will be only a minor contribution. The ratio between convection and radiation depends on a lot of factors outside of the scope of this book. A detailed study of the physics of heat transfer would be needed. The heat transfer due to radiation is proportional to the fourth power of the temperature of the filament:

$$\mathscr{E}_{\text{rad}} = \sigma T_f^4 \tag{2.145}$$

The amount of heat transfer due to convection depends in a complicated way on the shape and the size of the light bulb. The commonly used linear relation between heat transfer and temperature difference only holds in exceptional cases, for example, with forced convection [49]. However, here we will assume that the total amount of heat transfer to the environment is linearly proportional to the temperature difference. We will see below that this assumption corresponds to the commonly used model of the lamp as a first-order low-pass filter. Using further that the environmental temperature is much lower than the temperature of the filament, we obtain for the total heat transfer

$$\mathscr{E}_{\text{out}} = c_2 T_f \tag{2.146}$$

The resistance of the metal in the filament is proportional to the temperature:

$$R = c_3 T_f \tag{2.147}$$

Combining (2.143), (2.144), (2.146), and (2.147) gives the following differential equation for the temperature of the filament:

$$m_f c_1 \frac{dT_f}{dt} = \frac{v^2(t)}{c_3 T_f} - c_2 T_f \tag{2.148}$$

Assume that the voltage varies sinusoidally:

$$v(t) = \sqrt{2} V \cos(\omega_0 t) \tag{2.149}$$

Substituting (2.149) in (2.148) gives

$$\tfrac{1}{2}m_f c_1 c_3 \frac{dT_f^2}{dt} + c_2 c_3 T_f^2 = V^2 + V^2 \cos^2(2\omega t) \tag{2.150}$$

which is a linear differential equation in $T_f^2$. In steady state, $T_f^2$ consists of a constant term and a term varying sinusoidally with twice the voltage frequency:

$$T_f^2 = T_{f0}^2 + \Delta T_f^2 \cos(2\omega_0 t + \xi) \tag{2.151}$$

Substituting (2.151) in (2.150) gives the following expression for the average temperature of the filament under steady state:

$$T_{f0} = \frac{V}{\sqrt{c_2 c_3}} \tag{2.152}$$

The average steady-state temperature depends on the efficiency of the heat transfer to the environment and the resistance of the filament. The amplitude of the temperature variation is

$$\Delta T_f^2 = \frac{V^2}{\sqrt{(m_f c_1 c_3 \omega_0)^2 + (c_2 c_3)^2}} \tag{2.153}$$

Note that the temperature does not vary sinusoidally but varies according to the expression

$$T_f(t) = \sqrt{T_{f0}^2 + \Delta T_f^2 \cos(2\omega_0 t + \xi)} \tag{2.154}$$

Alternatively, the lamp may be modeled as a first-order low-pass filter from $v^2(t)$ to $T_f^2(t)$. For this we rewrite (2.148) in the form

$$\frac{dT_f^2}{dt} + \frac{T_f^2}{\tau_f} = \frac{v^2(t)}{\tau_f c_2 c_3} \tag{2.155}$$

with $\tau_f = (m_f c_1 / 2c_2)$ the thermal time constant of the lamp.The thermal time constant depends on the mass of the filament and the efficiency of the heat transfer to the environment.

According to [303] typical time constants are 19 ms for a 230-V, 60-W lamp and 28 ms for a 120-V, 60-W lamp.

The differential equation (2.155) can be developed somewhat further by using knowledge of average filament temperature $T_{f0}$ for nominal rms voltage $V_0$.

From (2.152) we obtain:

$$T_{f0} = \frac{V_0}{\sqrt{c_2 c_3}} \tag{2.156}$$

which results in

$$c_2 c_3 = \left(\frac{V_0}{T_{f0}}\right)^2 \tag{2.157}$$

Substituting this in (2.155) gives

$$\frac{1}{T_{f0}^2}\frac{dT_f^2}{dt} + \frac{1}{\tau_f}\frac{T_f^2}{T_{f0}^2} = \frac{1}{\tau}\frac{v^2(t)}{V_0^2} \tag{2.158}$$

Before we continue we should again emphasize that the model as described in (2.158) holds under the assumption that the heat loss with the environment is linearly proportional to the temperature of the filament. For more accurate convection models and for including the heat loss due to radiation, the resulting differential equation will be nonlinear and cannot be solved without the use of numerical methods.

The first-order filter model only gives the temperature of the filament, not the amount of light emitted. If we assume that the filament behaves as a blackbody radiator, we can use Planck's radiation law to determine the amount of light emitted. According to Planck's law the amount of energy per unit of volume in a wavelength interval $[\lambda, \lambda + d\lambda]$ is found from

$$dW = \frac{8\pi hc}{\lambda^5}\frac{d\lambda}{e^{(hc/\lambda kT)} - 1} \tag{2.159}$$

with $h$ Planck's constant, $k$ Boltzmann's constant, and $c$ the speed of light. This relation is plotted in Figure 2.21 for five different values of the temperature $T$. These five values are chosen at 90, 95, 100, 105, and 110% of the normal temperature of the filament (3500 K). The figure clearly shows that even a relatively small variation in temperature (and thus in voltage) already gives a very large change in the amount of emitted radiation. The visible part of the electromagnetic spectrum (400 through 800 nm) is indicated by the dotted vertical lines.

The change in emitted energy with relatively small variations in temperature is even better visible in Figure 2.22. The emitted radiation as shown in Figure 2.21 has been integrated over the visible part of the spectrum from 400 through 800 nm. This can be used as a measure of the light intensity of the lamp, even though our eyes are not equally sensitive to this whole range of wavelengths. The range in temperature in Figure 2.22 is only $\pm 10\%$ around 3500 K, but the variation in light intensity is almost a factor of 4. Linearizing the curve around $T = 3500$ K

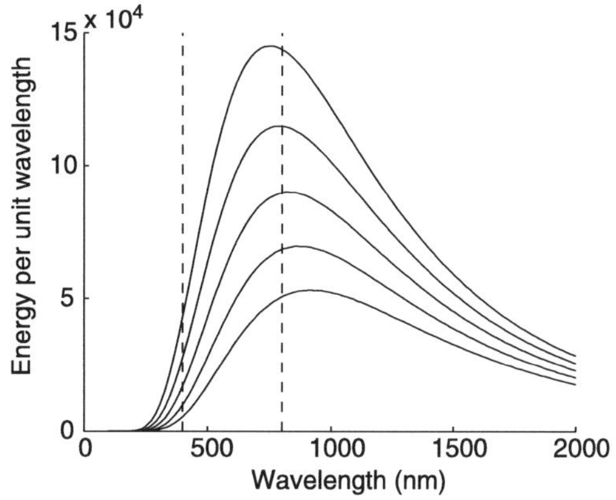


**Figure 2.21** Radiation spectra for blackbodies with temperatures of (top to bottom) 3850, 3675, 3500, 3325, and 3150 K. The vertical dotted lines indicate the visible part of the spectrum.

shows that a 1% change in temperature gives a 6.5% change in light intensity. This amplification effect, together with the high sensitivity of our eyes for fast fluctuations in light intensity, results in even very small voltage fluctuations already leading to irritable fluctuations in light intensity.

The reaction of a lamp to voltage fluctuations is described by means of a so-called gain factor. The gain factor is the ratio between the relative fluctuation in light intensity and the relative fluctuation in voltage:

$$G = \frac{(\Delta R)/R}{(\Delta V)/V} \tag{2.160}$$

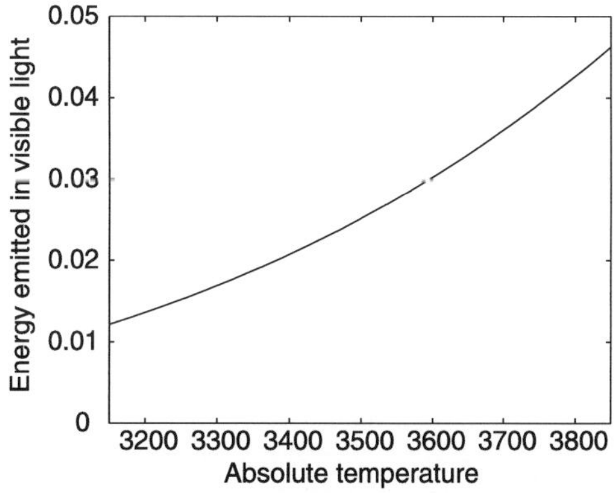


**Figure 2.22** Total energy emitted by blackbody in visible part of spectrum for temperatures ranging ±10% around 3500 K.

with $\Delta R$ the fluctuation in light intensity, $R$ the average light intensity, $\Delta V$ the fluctuation in voltage amplitude, and $V$ the average voltage amplitude. This gain factor $G$ is a function of the frequency of the fluctuation. Each lamp has its own gain factor as a function of frequency. In [271] and [303] the following relation is given:

$$G(f_M) = \frac{K}{\sqrt{1 + (2\pi f_M \tau)^2}} \tag{2.161}$$

with $K$ the gain factor and $\tau$ the time constant of the lamps. The behavior according to (2.161) is that of a first-order low-pass filter. Some measured examples of this function are presented in [303]. These examples were used as a basis for Figure 2.23. The gain factor is higher for a 230-V lamp than for a 120-V lamp of the same wattage. Thus for the same voltage fluctuations, the 230-V lamp will show larger light intensity fluctuations. The fluorescent lamp shows much less light intensity fluctuations than the incandescent lamps. Note, however, that typical incandescent lamps were chosen, but a "well-behaved, practically flicker-free electronic fluorescent lamp" was chosen instead of a typical one [303].

Figure 2.23 also shows the theoretical curves based on (2.161). For the 230-V lamp a gain factor $K = 3.8$ and a time constant $\tau = 21$ ms have been used. For the 120-V lamp we used $K = 3.5$ and $\tau = 29$ ms. This gain factor and time constant

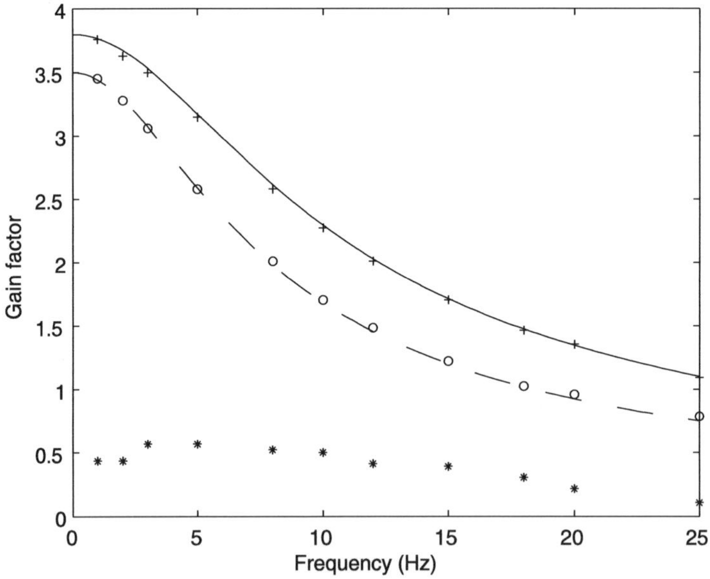


**Figure 2.23** Measured gain factor for 230-V, 60-W incandescent lamp (+), 120-V, 60-W incandescent lamp (O), and electronic fluorescent lamp ($*$) together with theoretical curves for 230-V, 60-W incandescent lamp (solid line) and 120-V, 60-W incandescent lamp (dashed line).

were manually chosen to get a good fit to the curves. According to [303], typical time constants are $\tau = 19$ ms for a 230-V, 60 W incandescent lamp and $\tau = 28$ ms for a 120-V lamp.

### 2.4.5 Perception of Light Fluctuations

In the previous section is was shown how voltage fluctuations lead to light intensity fluctuations. The presence of light intensity fluctuations is however not directly a problem. Only when voltage fluctuations lead to observable light flicker is there reason for concern. Slow changes in light intensity and steps far away in time are observed simply as changes and steps. With increasing frequency of the changes the sensation becomes one of flicker. The observer notices an unsteadiness in the light intensity without actually being able to observe the changes. After a while the sensation of flicker becomes uncomfortable. If the frequency of the changes is increased even further, the sensation becomes continuous and the observer is no longer aware of the fluctuations. The crossover from a sensation of flicker to a sensation of continuous light is called the *fusion frequency*. The fusion frequency depends on the average illumination. At high levels of illumination, during so-called *cone vision*, the fusion frequency may be as high as 60 Hz. Under *scotopic vision* (i.e., in the dark) the fusion frequency may drop to as low as 4 Hz [48, Vol. 27, p.183].

The fusion frequency, also referred to as *critical flicker frequency* depends also on the size of the fluctuations in light intensity. Experiments on the relation between the fusion frequency and the size of the fluctuations were performed by de Lange in the 1950s. Some of the results are summarized in [303, p. 20]. The main lasting result of these experiments was the description of the eye–brain behavior by means of a filter characteristic. This characteristic is still used and is referred to as the *de Lange filter*. Later experiments by Rashbass, Koenderink, and van Doorn resulted in a model for the relation between light intensity fluctuations and the sensation of flicker which became the basis for the IEC flickermeter standard. The model, referred to as the *Rashbass model*, is shown in Figure 2.24.

The Rashbass model consists of three blocks: a linear bandpass filter (the de Lange filter), a squaring circuit to obtain the amplitude of the observed fluctuations, and a first-order low-pass filter to model the memory function of the brain. For instantaneous flicker sensation below a certain threshold what is observed is not

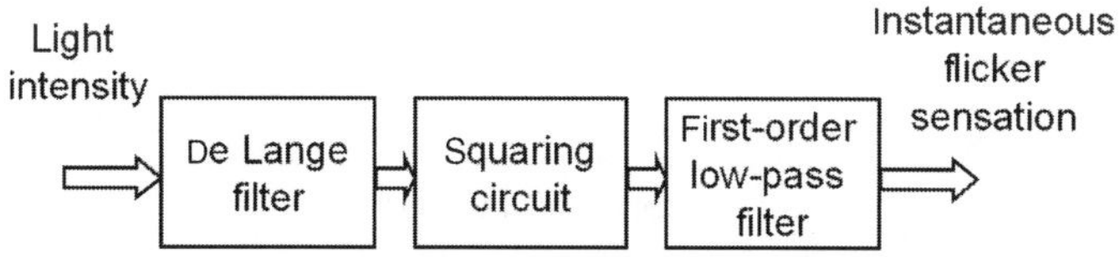


**Figure 2.24** Rashbass model for relation between light intensity fluctuations and flicker sensation.

flicker but merely a constant illumination. Above this threshold the observer notices a sensation of flicker. For even higher levels the flicker becomes annoying.

### 2.4.6 Flickercurve

Two curves were mentioned before: the lamp gain factor as a function of the frequency of the fluctuation and the smallest observable light intensity fluctuation as a function of the frequency of the fluctuation. Combining these two curves results in a so-called flickercurve: the smallest voltage fluctuation that leads to observable flicker as a function of the frequency of the fluctuation. As the lamp gain factor depends on the type of lamp, the flickercurve is different for each lamp type. However a number of national and international standards on flicker give flickercurves for standardized lamps—a 230-V, 60 W incandescent lamp being most commonly used as the standardized lamp in Europe. In the United States the 120-V, 60-W lamp is used as the standard. The latest revision of IEC 61000-4-15 includes standard models for both lamps. The same models are included in the current draft of IEEE standard 1453 [170].

Traditionally the assessment of flicker problems was based on rectangular voltage fluctuations. In the same way as for sinusoidal voltage fluctuations, a flickercurve can be obtained for rectangular voltage fluctuations. The standard curves according to IEC 61000-4-15 are shown in Figure 2.25. Note that both curves are obtained as the relative input voltage fluctuation that leads to one unit of perceptibility at the output. See the section in the flickermeter standard for more details.

We see from the figure that at 8.8 Hz a rectangular voltage fluctuation of only 0.2% will lead to visible flicker. The lower threshold level for rectangular

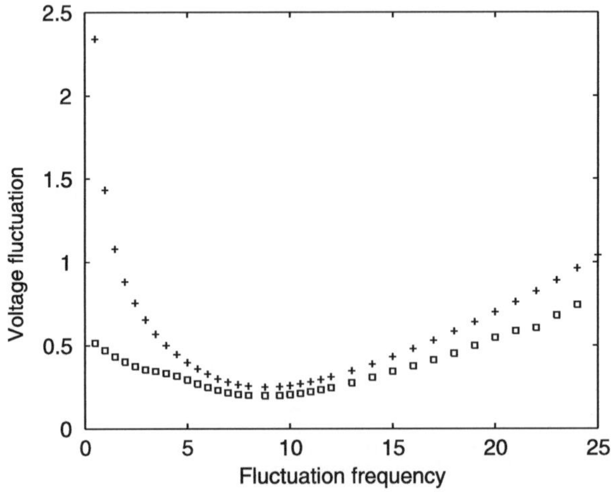


**Figure 2.25** Flickercurve for sinusoidal (plusses) and rectangular (squares) voltage fluctuations.

fluctuations is because the waveform contains multiple frequencies each of which contribute to the instantaneous flicker sensation.

### 2.4.7 Flickermeter Standard

The flickercurve as introduced in the previous section has been used for many years to assess the severity of voltage fluctuations. However, the curve only holds for one specific shape of the voltage fluctuations. Typically the curve was given for a rectangular shape and correction factors were given for other shapes of the voltage fluctuations (e.g., sinusoidal and triangular). For irregular voltage fluctuations and for measurements the curve was of limited use.

To determine the flicker due to any arbitrary voltage fluctuation the flickermeter concept was developed and implemented in an IEC standard [157, 303]. The flickermeter concept is based on the Rashbass model for flicker sensation, as shown in Figure 2.24. The difference is in the input, which is light intensity in the Rashbass model and voltage waveform in the flickermeter. The standard flickermeter consists of five blocks, as shown in Figure 2.26. The functionality of the different blocks will be discussed in some detail below.

***2.4.7.1 Voltage Adaptation*** The voltage adaptation block creates an output voltage with a constant long-term average. A 1% voltage fluctuation on a 242-V average rms will have the same effect as a 1% fluctuation around 200 V. The averaging period should be long enough to not affect the lowest fluctuation frequencies that are of concern. A 1-min averaging period is suggested in the flickermeter standard [157].

***2.4.7.2 Demodulation*** The aim of the second block of the flickermeter standard is to extract the voltage fluctuation from the voltage waveform. Let the voltage waveform be given by

$$v(t) = A[1 + m(t)]\cos(2\pi f_0 t) \tag{2.162}$$

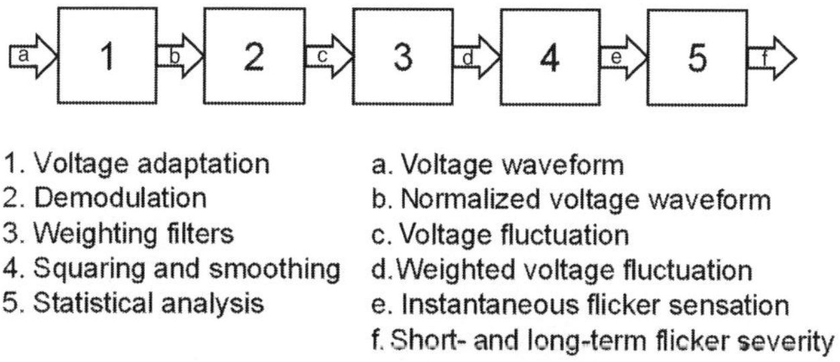


**Figure 2.26** Standard flickermeter.

Then the aim of the demodulation block is to extract the amplitude modulation $m(t)$. A number of methods are available for this in telecommunication [55]:

- *Heterodyning or Mixing* Multiplying the modulated signal with a signal of the same frequency as the carrier frequency, $y(t) = B\cos(2\pi f_0 t + \xi)$, results in the sum-and-difference signals:

$$v(t) \times y(t) = \tfrac{1}{2}AB\cos\xi + \tfrac{1}{2}ABm(t)\cos\xi + \tfrac{1}{2}AB[1 + m(t)]\cos(4\pi f_0 t + \xi) \tag{2.163}$$

  Ensuring that the two carrier frequency signals are in phase (so that $\cos\xi = 1$), typically, by means of a phase-locked loop, removing the double-frequency component by using a low-pass filter and removing the dc component ($\frac{1}{2}AB$) by using a high-pass filter result in a signal proportional to the fluctuation $m(t)$. Heterodyning is the most accurate method, but it is difficult to implement because of the need for a phase-locked loop and a narrow-band pass filter.
- *Envelope Detection* A simple circuit with one diode, two capacitors, and two resistors, will result in a signal that is proportional to the fluctuation $m(t)$. This method is commonly used in amplitude-modulated (AM) receivers because it is easy to implement in an analog circuit. Application of peak detection on digital signals is much less straightforward. Another disadvantage is that the peak level is not a good indication for the modulation of a heavily distorted waveform. As some sources of voltage fluctuation are also sources of distortion, envelope detection is not an appropriate method for demodulation of voltage fluctuation waveforms.
- *Square Demodulation* Instead of multiplying with a clean sine wave as in (2.163), the input is multiplied by itself, that is, squared. The result is similar to the output for the heterodyne demodulator for small fluctuations:

$$v(t)^2 = A^2\left[1 + 2m(t) + m(t)^2\right]\left[\tfrac{1}{2} + \tfrac{1}{2}\cos(4\pi f_0 t)\right] \tag{2.164}$$

  After removing the dc component and the double-frequency term, the following signal results:

$$v_2(t) = A^2 m(t) + A^2 m(t)^2 \tag{2.165}$$

  For small voltage fluctuations, $m(t) \ll 1$, the second term can be neglected. As voltage fluctuations rarely exceed a few percent, this is viewed as an acceptable approximation. When using amplitude modulation in telecommunication much higher modulation depths are used, ruling out square demodulation.

The flickermeter standard prescribes the use of a square demodulator. However, from the above reasoning one may conclude that a heterodyne demodulator should be consider acceptable as well.

**2.4.7.3 Weighting Filters** The third and fourth blocks of the flickermeter represent the behavior of the chain lamp–eye–brain. This part is similar to the Rashbass model, as shown in Figure 2.24, but with voltage fluctuations as input. The third block of the flickermeter contains the bandpass filter (i.e., the first block in Figure 2.24 plus the lamp model). Strictly according to the text of the flickermeter standard [157] the third block also contains the low-pass filter needed to remove the double-frequency component resulting from the squaring demodulator. In this way all filters are placed in one block, but as far as functionality is concerned, this low-pass filter is part of the second block. We will not further discuss this filter and the reader is referred to the flickermeter standard for more details [157, 303].

The weighting filters are defined in the flickermeter standard by its transfer function in the Laplace domain ($s$-domain):

$$F(s) = \frac{k\omega_1 s}{s^2 + 2\lambda s + \omega_1^2} \times \frac{1 + s/\omega_2}{(1 + s/\omega_3)(1 + s/\omega_4)} \tag{2.166}$$

where the various constants and frequencies are defined as in Table 2.3 [157, 303]. The flickermeter standard gives only values for 230- and 120-V, 60-W incandescent lamps.

The transfer functions of the resulting filters are shown in Figures 2.27 and 2.28. Figure 2.27 shows the transfer function of the filter for a 230-V, 60-W lamp as defined in the flickermeter standard [157]. The solid curve is the overall transfer function, that is, the absolute value of the expression in (2.166) with $s = j2\pi f$. The dotted curve is the contribution of the first factor in (2.166): This part of the filter has a bandpass characteristic with a maximum slightly above 9 Hz. The dashed curve is the contribution of the second factor, which has a low-pass characteristic with unity transfer for low frequencies. The gain decays to about half its initial value at 2 Hz and decays slowly after that. The total transfer has a bandpass characteristic with a maximum around 8.8 Hz.

**TABLE 2.3 Weighting Filter Constants for Two Types of Incandescent Lamps**

| | 230 V, 60 W | 120 V, 60 W |
|---|---|---|
| $k$ | 1.74802 | 1.6357 |
| $\lambda$ | $2\pi 4.05981$ Hz | $2\pi 4.167375$ Hz |
| $\omega_1$ | $2\pi 9.15494$ Hz | $2\pi 9.077169$ Hz |
| $\omega_2$ | $2\pi 2.27979$ Hz | $2\pi 2.939902$ Hz |
| $\omega_3$ | $2\pi 1.22535$ Hz | $2\pi 1.394468$ Hz |
| $\omega_4$ | $2\pi 21.9$ Hz | $2\pi 17.31512$ Hz |

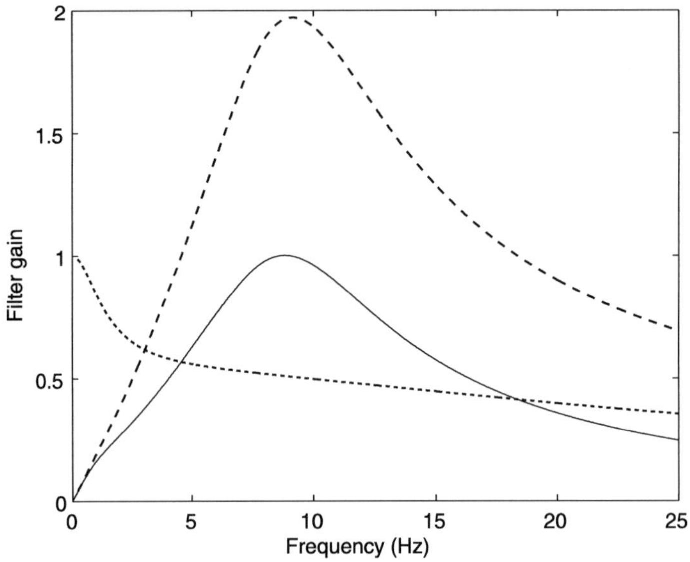


**Figure 2.27** Transfer function of weighting filter in flickermeter standard for 230-V, 60-W incandescent lamps (solid curve). The dashed and dotted curves give the transfer function of the two factors in the filter characteristic.

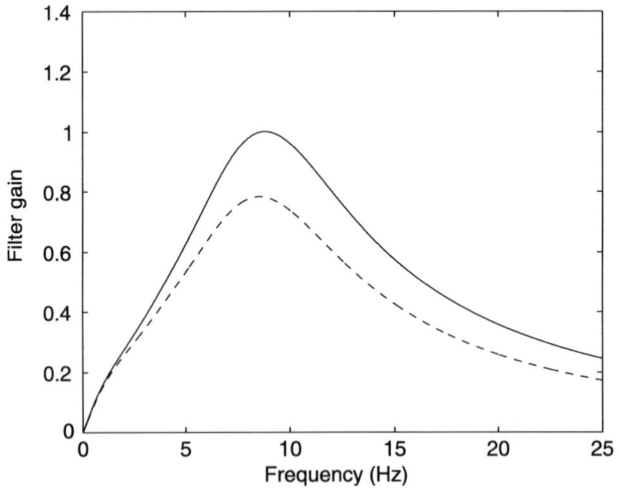


**Figure 2.28** Transfer function of weighting filter in flickermeter standard for 230-V (solid line) and 120-V (dashed line), 60-W incandescent lamps.

The weighting filters for 230-V and 120-V lamps are compared in Figure 2.28. The filter gain for the 120-V lamps is less because the light intensity fluctuations are less than for 230-V lamps, as was already shown in Figure 2.23.

#### 2.4.7.4 Squaring and Smoothing

The fourth block of the flickermeter consists of a squaring multiplier and a low-pass filter, as introduced in the Rashbass model in Figure 2.24. The low-pass filter is a simple first-order filter with a time constant of 300 ms (cutoff frequency of 0.5 Hz). The aim of this filter is to "simulate the storage effect in the brain" [157, page 25].

To understand the operation of this block, consider a voltage with a sinusoidal fluctuation:

$$v(t) = \sqrt{2}V\left[1 + M\cos(2\pi f_M t + \phi_M)\right]\cos(2\pi f_0 t) \tag{2.167}$$

The output of the demodulator (the input to the weighting filter) is the modulation or voltage fluctuation:

$$m(t) = M\cos(2\pi f_M t + \phi_M) \tag{2.168}$$

The effect of the weighting filters is scaling and phase shifts of the modulation signal. Both scaling and phase shifts are a function of the modulation frequency:

$$m'(t) = MF(f_M)\cos(2\pi f_M t + \phi'_M) \tag{2.169}$$

The result of the squaring is

$$[m'(t)]^2 = \tfrac{1}{2}[MF(f_M)]^2 + \tfrac{1}{2}[MF(f_M)]^2\cos(4\pi f_M t + 2\phi'_M) \tag{2.170}$$

The low-pass filter passes the first term without damping, whereas it damps the oscillating term. The damping of the oscillating term is equal to

$$F_{\text{LPF}} = \frac{1}{\sqrt{1+\omega^2\tau^2}} = \frac{1}{\sqrt{1+16\pi^2 f_M^2\tau^2}} \tag{2.171}$$

with $\tau = 300$ ms the time constant of the filter. Note that the frequency at the input of the filter is $2f_M$ so that $\omega = 4\pi f_M$.

The output of block 4 of the flickermeter is the instantaneous flicker sensation from the Rashbass model. For stationary voltage fluctuation, the instantaneous flicker sensation $P$ is given by the first term in (2.170):

$$P = \tfrac{1}{2}[MF(f_M)]^2 \tag{2.172}$$

with $F(f_M)$ the absolute value of the weighting filter transfer (2.166) for a voltage fluctuation of frequency $f_M$.

***2.4.7.5 Statistical Analysis*** The instantaneous flicker sensation, being the output of block 4 of the flickermeter, is statistically processed in the fifth and final block. The result is a 10-min or short-term flicker severity $P_{ST}$ and a 2-h or long-term flicker severity $P_{LT}$. At this stage it is very important to emphasize a difference in interpretation of the value of the instantaneous flicker sensation and the short-term flicker severity.

A unity value of the instantaneous flicker sensation corresponds to the "perceptibility threshold for 50% of observers viewing a 60-W, 230-V incandescent lamp" [303, page 29]. Thus when the instantaneous flicker sensation exceeds 1, more than half of the observers will notice a flickering of the light. The instantaneous flicker sensation is an interesting physical quantity but not of much use to characterize the severity of the voltage fluctuation. The severity of the voltage fluctuation should be related to the amount of annoyance caused by the resulting light flicker.

A unity value of the short-term flicker severity corresponds to a level which the majority of viewers find annoying. The short-term flicker severity is calculated from the probability distribution function of the instantaneous flicker sensation over a 10-min interval. The flickermeter standard prescribes that at least 50 samples per second shall be taken, resulting in at least 30 000 values over a 10-min interval. The flickermeter standard gives the following expression to calculate the short-term flicker severity from the instantaneous flicker sensation:

$$P_{ST} = \sqrt{0.0314P_{99.9} + 0.0525P_{99} + 0.0657P_{97} + 0.28P_{90} + 0.08P_{50}} \qquad (2.173)$$

where $P_{99}$ is the value not exceeded by 99% of the samples, and so on. Note that 0.1% of a 10-min interval corresponds to 600 ms. The 99 percentile is the value exceeded during 6 s, the 97 percentile during 18 s, and the 90 percentile during 1 min.

In the practical implementation according to the flickermeter standard, the percentile values are obtained as an average of a number of neighboring percentiles:

$$P_{50} = \tfrac{1}{3}(P_{70} + P_{50} + P_{20}) \qquad (2.174)$$

$$P_{90} = \tfrac{1}{5}(P_{94} + P_{92} + P_{90} + P_{87} + P_{83}) \qquad (2.175)$$

$$P_{97} = \tfrac{1}{3}(P_{97.8} + P_{97} + P_{96}) \qquad (2.176)$$

$$P_{99} = \tfrac{1}{3}(P_{99.3} + P_{99} + P_{98.5}) \qquad (2.177)$$

The square root in the expression for the short-term flicker severity, (2.173), cancels out the square in the relation between the voltage fluctuation and the instantaneous flicker severity, (2.172). As a result the short-term flicker severity is directly proportional to the amplitude of the voltage fluctuation.

For the interpretation of the statistical processing we will consider the basic expression (2.173) and neglect the effect of smoothing according to (2.174) through (2.177). Suppose that an instantaneous flicker sensation of 4 is present during at least half of a 10-min interval. Note that this corresponds to a voltage fluctuation equal to twice the perceptibility threshold. During the remainder of the interval the instantaneous flicker sensation is zero; thus there is no voltage fluctuation. All the percentiles in (2.173) are now equal to 4, resulting in a short-term flicker severity of

$$P_{ST} = \sqrt{0.0314 \times 4 + 0.0525 \times 4 + 0.657 \times 4 + 0.28 \times 4 + 0.08 \times 4} = 1.43$$

An instantaneous flicker severity $P = 1.96$ during 50% of the interval will lead to $P_{ST} = 1$. These calculations have been repeated for different durations of the high-flicker sensation during the interval, resulting in Table 2.4. In all cases, the instantaneous flicker sensation is considered zero during the remainder of the interval. The last column of Table 2.4 gives the amplitude of the voltage fluctuation in units of the perceptibility threshold.

From 12 consecutive values of the short-term flicker severity $P_{ST}$, the long-term flicker severity $P_{LT}$ is calculated by using the following expression:

$$P_{LT} = \sqrt[3]{\frac{1}{12}\sum_{i=1}^{12} P_{ST}(i)^3} \tag{2.178}$$

One value $P_{ST}$ equal to 4 during a 2-h interval and other values equal to zero during that interval will result in a long-term flicker severity of

$$P_{LT} = \sqrt[3]{\frac{1}{12}(4^3 + 0 + \cdots + 0)} = 1.75 \tag{2.179}$$

A value $P_{LT} = 1$ can be obtained from one $P_{ST} = 2.29$ value. Table 2.5 gives the resulting long-term flicker severity for different numbers of high $P_{ST}$ values

**TABLE 2.4 Bursts of Voltage Fluctuations of Different Duration and Resulting Short-Term Flicker Severity**

| Flicker Duration | $P_{ST}$ for $P = 4$ | $P$ for $P_{ST} = 1$ | Voltage Fluctuation (Units of threshold) |
|---|---|---|---|
| 50%, 5 min | 1.43 | 1.96 | 1.40 |
| 10%, 1 min | 1.31 | 2.33 | 1.53 |
| 3%, 18 s | 0.77 | 6.68 | 2.58 |
| 1%, 6 s | 0.58 | 11.92 | 3.45 |
| 0.1%, 0.6 s | 0.35 | 31.85 | 5.64 |

**TABLE 2.5 Long-Term Flicker Severity for Different Durations of High Short-Term Flicker Severity**

| Number of High $P_{ST}$ Values | $P_{LT}$ for $P_{ST} = 4$ | $P_{ST}$ Leading to $P_{LT} = 1$ |
|---|---|---|
| 1 | 1.75 | 2.29 |
| 2 | 2.20 | 1.82 |
| 3 | 2.52 | 1.59 |
| 4 | 2.77 | 1.44 |
| 5 | 2.99 | 1.34 |
| 6 | 3.17 | 1.26 |
| 7 | 3.34 | 1.20 |
| 8 | 3.49 | 1.14 |
| 9 | 3.63 | 1.10 |
| 10 | 3.76 | 1.06 |
| 11 | 3.88 | 1.03 |
| 12 | 4.00 | 1.00 |

during a 2-h interval. As before, it is assumed that there are no voltage fluctuations for the remainder of the 2-h interval.

Table 2.5 can be interpreted as follows: If during 10 min in a 2-h interval the voltage fluctuation is 2.29 times the threshold value (i.e., the threshold of annoyance), a unity long-term flicker severity will result. The same holds when during 20 min in a 2-h interval the voltage fluctuation is 1.82 times the threshold value, and so on.

Considering the whole transfer from voltage waveform to short-term flicker severity, we obtain, with reference to the notation in Figure 2.26, the following signals. At the input of the flickermeter we consider a stationary sinusoidal voltage fluctuation of amplitude $A \times M$ and fluctuation frequency $f_M$:

$$v_a(t) = A[1 + M\cos(2\pi f_M t)]\cos(2\pi f_0 t) \tag{2.180}$$

The effect of normalization is to bring the long-term average value of the voltage waveform to a reference value $V_R$.

$$v_b(t) = \sqrt{2}V_R[1 + M\cos(2\pi f_M t)]\cos(2\pi f_0 t) \tag{2.181}$$

The output of the squaring demodulator is proportional to the voltage fluctuation:

$$v_c(t) = 2V_R^2 M\cos(2\pi f_M t) \tag{2.182}$$

The effect of the weighting filters is a multiplication by the transfer function for the modulation frequency:

$$v_d(t) = 2V_R^2 M|F(f_M)|\cos(2\pi f_M t) \tag{2.183}$$

There is also a shift in phase angle due to the weighting filters, but that is compensated here by making an opposite shift of the time axis. The effect of smoothing and squaring is a small oscillation around a constant value. If we neglect the oscillation, the instantaneous flicker sensation is

$$P = \tfrac{1}{2} V_R^4 M^2 |F(f_M)|^2 \qquad (2.184)$$

Finally, the short-term flicker severity is

$$P_{\mathrm{ST}} = \sqrt{0.5096P} = 1.01 V_R^2 M |F(f_M)| \qquad (2.185)$$

### 2.4.8 Flicker with Other Types of Lighting

The flickermeter standards in their current form only apply to 120- and 230-V, 60-W incandescent lamps. Incandescent lamps of higher power than 60 W have a thicker filament to allow for the higher current without melting the filament. This gives a longer thermal time constant and makes the lamp less sensitive to voltage fluctuations. In the same way the 120-V, 60-W lamp (with a current of 0.5 A) is less sensitive than the 230-V, 60-W lamp (with a current of 0.26 A). Following this reasoning, a 230-V, 115-W lamp would be equally sensitive to voltage fluctuations as the standard 120-V, 60-W lamp. The other way around, smaller wattage incandescent lamps are more sensitive to voltage fluctuations. Fortunately low-wattage lamps are rarely used for applications where a constant illumination is important.

However, fluorescent lamps are very commonly used for applications where a constant illumination is important. Fluorescent lamps belong to the class of so-called luminescent lamps where light production is not linked to a high temperature as with incandescent lamps. Luminescent lamps produce less heat and are thus more efficient than incandescent lamps. The most commonly used luminescent lamps are [49]

- fluorescent lamps, which give a neutral white light;
- sodium-vapor lamps, which produce a yellow-orange light; and
- mercury-vapor lamps, producing a whitish blue-green light.

Fluorescent lamps are the most commonly used luminescent lamps for in-house applications. The basic light source of fluorescent lamps is emitted by an ionized mixture of argon and mercury vapor. The ionization is due to an electric current flowing through the gas. The ionized gas emits light in the ultraviolet part of the spectrum. This light is absorbed by the coating of the tube containing the gas and is reemitted as visual light by a phenomenon known as fluorescence. Hence the name "fluorescent lamp." To start and maintain the current through the gas an electronic circuit is needed: the so-called ballast. Traditionally this circuit mainly consisted of a large inductor—the magnetic ballast—but this is more and more being

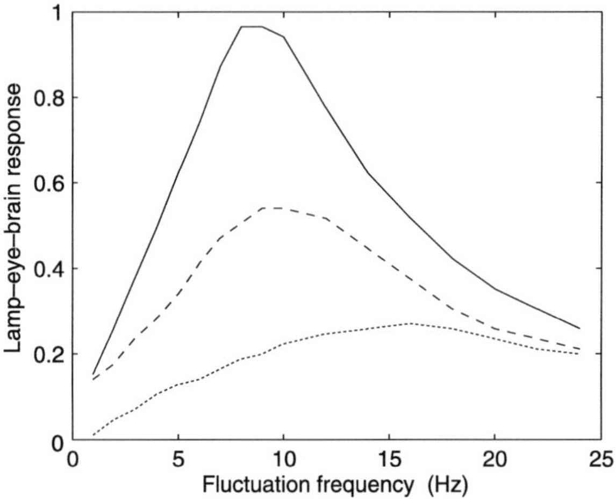


**Figure 2.29** Lamp–eye–brain model of incandescent (solid line) and fluorescent lamps with magnetic (dashed) and electronic (dotted) ballasts. (Data from [303, p. 80].)

replaced by an electronic circuit providing constant voltage instead of constant current—the electronic ballast.

The flickermeter standard can be adjusted to other types of lighting by changing the model for the lamp–eye–brain response. To make the flickermeter standard more adaptable, the lamp response should be separated from the eye–brain response. In that way the user of the standard can easily include the effect of new lamp types on the flicker severity.

In Figure 2.29 three types of 230-V lamps are compared. The vertical scale is a relative scale: The curves indicate the relative sensitivity to voltage fluctuations of the three lamp types. The original publication [303] also included the response of a 120-V magnetic-ballast fluorescent lamp. This one was similar to that of the 230-V magnetic-ballast fluorescent lamp shown in Figure 2.29.

The conclusion from Figure 2.29 is that fluorescent lamps with electronic ballast are much less sensitive to voltage fluctuations than incandescent lamps. However, lamps from different manufacturers may show completely different behaviors. Figure 2.30 compares the resulting light flicker due to an 8-Hz voltage fluctuation for 23 different compact fluorescent lamps (with electronic ballast). The unity value along the vertical axis corresponds to the light flicker of the standard incandescent lamp. The majority of the tested lamps show less than half the flicker of the incandescent lamp, but some have an even higher flicker than the incandescent lamp.

Another effect not considered in the flickermeter standard is that lamps may be powered through an electronic dimmer. The effect of an electronic dimmer is an increase of the light intensity fluctuations for the same voltage fluctuations. When the lamp is dimmed to 25% of its luminosity, the light intensity fluctuations almost double [303, p. 12].

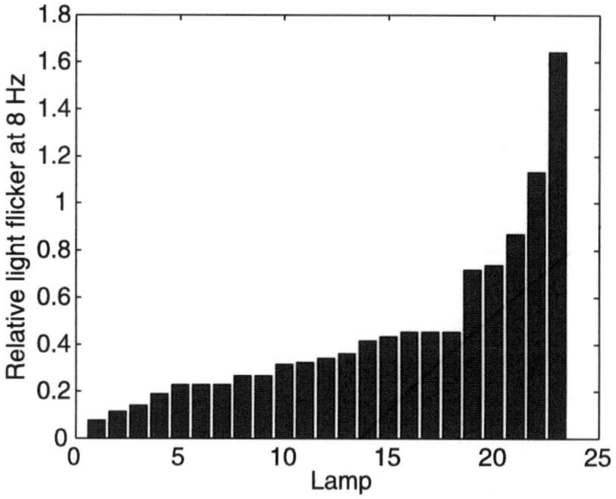


**Figure 2.30** Flicker due to different compact fluorescent lamps. (Data from [303, p. 17].)

Fluorescent lamps and dimmers are electronic equipment that may show non-linear behavior which is very difficult to combine with the flickermeter concept. Whereas the light intensity of incandescent lamps is a function of the rms of the voltage waveform, fluorescent lamps may also be affected by the distortion of the voltage, by changes in voltage phase angle, and by interharmonics. The latter issue has received some attention in the literature, but the other two issues are, as far as the authors are aware, not addressed. To include new lamp types a completely new lamp model may be needed, not just a change in the weighting functions of the flickermeter standard.

### 2.4.9 Other Effects of Voltage Fluctuations

Not only do voltage fluctuations lead to light flicker but also some other loads are adversely affected by fast variations/fluctuations in the voltage amplitude at their terminals. Some examples are [271]

- control action for control systems acting on the voltage angle,
- braking or accelerating moments for motors, and
- impairment of electronic equipment where the fluctuation of the supply voltage passes through to the electronic parts, for example, computers, printers, copiers, and components for telecommunication.

Two real-life examples are given in [303] where small voltage fluctuations lead to problems other than light flicker:

- Voltage fluctuations led to small speed variations in the motor driving a weaving machine. The results were small color variations in the final cloth. In that specific case this was a serious concern.

- Similar speed variations in a plastic extrusion process led to small variations in the diameter of the final product. These diameter variations exceeded the requirements under the International Organization for Standardization (ISO) 9000 specifications.

## 2.5 WAVEFORM DISTORTION

This section will discuss various aspects of nonsinusoidal voltage and current waveforms (waveform distortion). After an overview of the consequences of waveform distortion, a mathematical model (the Fourier series) will be introduced. The emphasis will be on harmonic distortion: a nonsinusoidal but periodic waveform. Different sources of waveform distortion will be discussed, with emphasis on single- and three-phase rectifiers. The section closes with a brief description of basic models for harmonic studies and harmonic resonances.

### 2.5.1 Consequences of Waveform Distortion

When discussing the consequences of waveform distortion, a distinction must immediately be made between the consequences of voltage distortion and the consequences of current distortion. Here we again reach the important distinction between "voltage quality" and "current quality" as introduced in Section 1.2: Voltage quality is how the network affects the customer or the load; current quality is how the customer or load affects the network. A similar distinction holds for waveform distortion: Distorted voltages affect the customer equipment; distorted currents affect the network components. However, voltage distortion in some cases also affects network components, especially shunt-connected equipment such as capacitor banks. A list of consequences of waveform distortion is given below. For a more complete discussion on the consequences of waveform distortion the reader is referred to the general literature on waveform distortion [e.g., 10, 11, 141, 162, 312] and to the indicated references.

In a survey of 80 utilities in the United States [255], the following problems were reported:

- Inadvertent trip of circuit breaker or fuse
- Transformer overheating
- Capacitor problems; mal-trip of capacitor fuse
- Malfunctioning of electronic equipment
- Digital clocks running fast
- Overheating of neutral conductors

*Transformers* Distortion of transformer voltages and currents may lead to an increase in audible noise [162], but the main effect is additional heating. Both voltage and current distortion lead to additional losses, but the effect is more pronounced for the current distortion because this is typically higher than the voltage distortion. When a current contains nonfundamental components, the rms current is higher than needed, as only the fundamental component transports useful power. The higher rms current leads to higher losses and thus to more heat development in series components such as transformers. The heating effect becomes more severe because the losses increase with increasing frequency. This phenomenon is present with all series components, but it is best documented for transformers [173].

The main phenomenon leading to overheating of transformers is due to the so-called stray flux: the magnetic flux in conducting parts of the transformer due to the current through the windings. The losses due to the stray flux (the "stray losses") are proportional to the square of the frequency so that especially higher order harmonics contribute. One should also note that the stray losses are not uniformly spread through the transformer but give rise to "hot spots" [332]. This makes the effect even more severe. Transformers supplying a heavy-distorted current will have to be derated due to this effect. Recommendations for the derating are given in IEEE Standard C57-110 [173]. Converter transformers for high-voltage dc (HVDC) links have shown unexpected hot spots in the tank due to heavy current distortion [10].

*Cables and Lines* Similar heating effects as for transformers appear in cables and lines. The effect is however not as pronounced due to the absence of stray losses. Cable losses show an increase with frequency due to skin effects and proximity effects. The derating is however not significant. A number of cases have been compared in [251] with a maximum derating of 6%. In [263, Section 4.8] a method is given to quantify the required derating of cables due to harmonic current distortion. The numerical example results in a required derating of 13% for a 300-kcmil conductor with a current of about 30% total harmonic distortion (THD).

For frequencies of 1 kHz and higher a pronounced local temperature rise may occur in cables due to the skin effect [271].

*Neutral Conductors* The neutral conductor in a three-phase system normally does not conduct any significant amount of current. However, even in a balanced condition the triple harmonics (three, nine, etc.) of the individual phases add in the neutral conductor. When the load contains large amounts of computers or energy-saving lamps the neutral current may exceed the phase current. This may result in overheating of the neutral conductor without tripping of the overload protection as the latter only protects the phase conductors. Potential overheating of the neutral conductor is the most dangerous consequence of harmonic distortion.

*Electronic Equipment* The direct effect of waveform distortion on electronic equipment is very difficult to quantify, if it is present at all. Equipment using the voltage zero crossing for obtaining phase-angle information will obviously be affected by the distortion when this causes shifts in voltage zero crossing or multiple zero crossings. The best example is the dc drive using the thyristor firing angle for speed control. More advanced equipment using a PLL is not affected by the waveform distortion.

There is an indirect effect of high-frequency distortion, as, for example, during notching [162]. The high-frequency voltage couples to the electronic or logic circuit, leading to malfunction. In the case of large voltage steps (notching) this may even lead to damage of the (low-power) electronic circuits. Notching is often associated with a high-frequency oscillation when the voltage recovers. This oscillation may involve a substantial part of the supply network, thus having the potential for more interference [328]. This form of interference is at the border area between conducted and radiated disturbances.

According to [162] malfunction of electronic equipment ("erratic, sometimes subtle, malfunctions of electronic equipment") appears for voltage waveforms with THD above 5% and when individual harmonics exceed 3%. Measurement instruments may give erroneous data. Malfunctions may occur in medical equipment with serious consequences. The high harmonic contents may also affect the performance of computer and television screens as well as audio- and video-recording equipment.

The high-frequency ripple due to modern rectifiers with active front end may interfere with sensitive electronic equipment. A real-life case is presented in [142] in which an active-front-end motor drive injects a 5-kHz ripple with an amplitude of about 9%. This ripple in turn leads to tripping of the electronic circuit of gas burners. It is interesting to notice that the electronic circuit trips because it detects an undervoltage condition.

The third-harmonic current in the neutral (see below) will lead to magnetic fields and neutral-to-ground voltages which in turn affect sensitive electronic equipment. According to [315] the neutral-to-ground voltage may also cause distortion to computer or television screens. Also the magnetic field due to the third-harmonic may interfere with screens. Several other examples are presented in [315] of the effect on sensitive equipment of third-harmonic currents in the grounding circuits.

Some types of compact fluorescent lamps no longer function when the voltage distortion becomes too high [14]. High-frequency voltage distortion leads to significant high-frequency current distortion, due to the capacitive character of the lamp for higher frequencies.

Most modern electronic equipment takes current from the supply only around a voltage maximum. The result is a flattening of the voltage waveform: a reduction of the "crest factor." The result of this flattening is that the dc voltage internally in the equipment becomes less. The internal dc voltage is more or less proportional to the maximum of the ac voltage waveform. The reduced dc voltage makes the equipment operate less efficiently; it becomes more sensitive to disturbances such as voltage dips, and the loading on the power supply becomes higher, leading to a reduced lifetime.

Changes in the peak voltage can cause changes in picture size and brightness for television screens [10] and also for computer screens.

*Signaling* Ripple control equipment and other types of signaling may experience difficulties due to waveform distortion [10]. High waveform distortion may be confused with the control signal. An indirect consequence of harmonic distortion is the installation of harmonic filters, which in turn also damps the control signals using the power grid as a communication channel. High-frequency distortion and the presence of filters will likely limit the use of the power grid for high-speed Internet communication and applications such as remote metering.

*Telephone Interference* A special case of radiated interference due to waveform distortion is the coupling of high-frequency currents from power lines to telecommunication lines. If the distortion frequencies are in the audio part of the spectrum, they may lead to what is called telephone interference. The problem occurs especially when single-phase overhead lines (e.g., in rural areas) use the same poles as the telephone lines. For three-phase lines the effect is less and only the zero-sequence component of the current has a significant influence on the telephone interference. Telephone interference is especially due to higher frequencies because the inductive coupling between conductors becomes more efficient with higher frequencies. The IEEE harmonic standard 519 defines so-called telephone interference factors to quantify the telephone interference due to different harmonic frequencies [162].

The high-frequency components in voltage or current may also couple to other equipment. In [218] a case is presented in which the high-frequency distortion is due to energy-efficient lighting coupling to the electronics of a hearing aid.

*Magnetic Fields* Any current through a wire causes a magnetic field. Harmonic current components cause magnetic fields at harmonic frequencies. For three-phase overhead lines, the main magnetic field at some distance from the line is due to the zero-sequence current. The power system frequency component of the zero-sequence current is small, but a line feeding a large amount of nonlinear load may contain a substantial zero-sequence current in the form of the third harmonic (150 or 180 Hz). In most cases the third harmonic is blocked by the distribution transformers, so that the concern is mainly for overhead low-voltage lines. For single-phase overhead lines the fundamental component will in all cases dominate the magnetic field.

*Capacitors* Capacitors are one of the main victims of waveform distortion. The displacement current through a capacitor increases linearly with frequency (for the same voltage) so that especially high frequency harmonics may lead to overheating and damage of capacitor banks. Harmonic distortion leads to higher thermal as well as dielectric stress. The thermal stress increases with the square of the frequency (for constant harmonic voltage). The dielectric stress is related to the voltage peak and thus to the rms voltage and the voltage crest factor. A discussion of partial discharges in relation to harmonic voltage distortion is presented in [97]. Capacitors are also network elements that often lead to amplification of the harmonic voltage and current levels by means of series or parallel resonances. This makes capacitors especially vulnerable to waveform distortion: not just capacitor banks in the system but also the small capacitors that are present in most energy-saving lamps, computer power supplies, and consumer electronics equipment.

Other examples of capacitors that may be affected are the ones in static power converters, in snubber circuits, and in electromagnetic interference (EMI) filters [162].

According to [312] problems start to appear when the harmonic load is more than 30% of the transformer rating. In such a case the capacitor banks should be applied as shunt filters.

*Rotating Machines* The impedance of rotating machines is mainly inductive so that the current through the machine reduces with frequency for constant voltage distortion. However at low harmonic orders, the impedance of a rotating machine is still rather low: the subtransient (or leakage) reactance times the harmonic order. Voltage waveform distortion has the same effect on rotating machines as voltage unbalance: It leads to additional losses and, worst of all, creates hot spots that may damage the machine. The local flux density and thus the heating may locally be twice the average [10]. When operating with a heavily distorted voltage, the machine will have to be derated. However, in most cases the voltage unbalance is the dominating factor in determining the derating of the machine.

The main effects of harmonic voltage distortion on rotating machines are oscillations in the torque which could lead to damage, especially when they occur near resonance frequencies [10]. Harmonic currents will also lead to additional noise being generated by the machines. A method for determining the torque oscillations from measured voltage waveforms is presented in [318].

For single-phase induction motors the losses could be more severe due to the presence of capacitors. Resonance between the capacitor and the motor leakage inductance may amplify the harmonic distortion, resulting in higher losses. A detailed analysis of this issue is presented in [201].

*Insulation* Whereas a low-voltage crest factor leads to reduced performance of electronic equipment, a high-voltage crest factor places additional stress on insulation. A high crest factor can be due to notching or due to resonances leading to amplification of certain frequencies. Also the corona starting level depends on the peak voltage and thus on the crest factor [10]. These phenomena are a combination of waveform distortion and voltage variations. The absolute value of the peak voltage (in volts or in percent of its nominal value) would be a more suitable way of description in this case.

*Protection* The effect of harmonic distortion on protection equipment remains unclear. The waveform distortion of the fault current may affect the slope at the zero crossing and thus the performance of circuit breakers [271, 312]. However, there is not likely any relation between the distortion of the fault current and the distortion of the normal operating current.

Fuses do show additional heating due to waveform distortion in the same way as cables, lines, and transformers. However, the thin wire in a fuse is not really prone to skin effects so that the change in performance characteristics is limited. According to [271, 312] the effect of harmonic distortion on the operation of fuses is small. The effects reported elsewhere may be due to measurement errors in which rms, peak, and fundamental currents were confused. Tests by a fuse manufacturer for frequencies up to 415 Hz showed no change in operating characteristics [312].

High voltage and current distortion (above 20%) may affect the operation of some types of relays and low-voltage circuit breakers [103, 115, 162, 312]. Older generation solid-state relays react to the peak current instead of to the rms or fundamental current. They may inadvertently trip on high-crest-factor currents.

*Reactance-Earthed Networks* In reactance-earthed networks, the presence of high harmonic distortion may lead to incorrect tuning of Pederson coils. It will also increase the earth-fault current, which further increases the risk of permanent earth faults and cross-country faults [271].

*Lighting* The main effect discussed in the literature is failure of the power factor correction capacitors with fluorescent lamps due to high harmonic distortion. According to [271] the presence of distortion reduces the lifetime of incandescent lamps. The phenomenon behind this is not mentioned: The skin effect could lead to additional heating of the filament; the harmonic currents could give rise to forces between the windings of the filament that could cause mechanical resonances. In [312] the increase in rms voltage due to the distortion is mentioned as the main contributing factor. This increase is however very small for practical voltage distortion levels. Disturbing noise levels could result from high voltage distortion with fluorescent and gas discharge lamps [271, 312].

### 2.5.2 Overview of Waveform Distortion

Waveform distortion includes all deviations of the voltage or current waveform from the ideal sine wave. This definition comes very close to the definition of "voltage quality" and "current quality" as given in Section 1.2. But variations in magnitude and frequency are not considered as waveform distortion, although a complete distinction between the different types of variations is not possible.

A number of different forms of waveform distortion can be distinguished: harmonic, interharmonic, and nonperiodic distortion. These three forms of distortion will be explained briefly below. In most studies, only the harmonic distortion is considered. Nonharmonic distortion (interharmonics and nonperiodic distortion) is much harder to quantify through suitable parameters and it is regularly neglected. Another reason for neglecting nonharmonic distortion is that harmonic distortion dominates in most cases. In other words, the waveform is close to periodic with a one-cycle window. The most recent IEC standards on measurements of harmonic distortion (IEC 61000-4-7 and IEC 61000-4-30) do include methods for quantifying nonharmonic distortion. These standards will be discussed in Section 3.2.3.

**2.5.2.1 *Harmonic Distortion*** When the waveform is nonsinusoidal but periodic with a period of one cycle (of the power system frequency, about 50 or 60 Hz), current and voltage waveforms can be decomposed into a sum of harmonic components. For the current this reads as

$$i(t) = I_0 + \sum_{h=1}^{H} I_h\sqrt{2}\cos\left(h\omega t - \beta_h\right) \tag{2.186}$$

With $\omega = 2\pi f_0$ and $f_0$ the *fundamental frequency* or *power system frequency*: $f_0 = 1/T$ with $T$ the (fundamental) period of the signal. In IEEE Standard 1459 [171] the dc component is included in the summation for $h = 0$ and with a phase angle of 45°. This may be mathematically correct but from an interpretation viewpoint the above notation is more appropriate.

In the same way, we write for the voltage waveform

$$v(t) = V_0 + \sum_{h=1}^{H} V_h\sqrt{2}\cos(h\omega t - \alpha_h) \tag{2.187}$$

The phase angle of the fundamental component of the voltage ($\alpha_1$) can be set to zero without loss of generality.

For 50-Hz systems we have $T = 20\,\text{ms}$ and $f_0 = 50\,\text{Hz}$. In most power system applications, the fundamental frequency ($h = 1$) dominates, especially for the voltage. The value of $H$ is infinite for a continuous signal, but for a discrete signal it is determined by the sample frequency. The highest frequency in a discrete signal is half the sample frequency. A commonly used sample frequency is 128 samples per (50 Hz) cycle (6.4 kHz), resulting in $H = 64$ in (2.186).

Within harmonic distortion, a further distinction can be made into *dc components*, *even-harmonic distortion*, and *odd-harmonic distortion*. The latter one is dominating is most cases.

In case the average value of the voltage or current (over an integer number of cycles) deviates from zero, a dc component is said to be present. The dc component has been treated as harmonic zero in (2.186), resulting in the additional term $I_0$. The dc components are often treated separately because their consequences are different from those of (other) harmonics. Also dc components require different measurement techniques. In the common use of the language, the term *harmonic* does not include the dc component or the fundamental component.

Odd-harmonic distortion may be defined as harmonic distortion in which the symmetry between the positive and negative half-cycle of the waveform is not broken. Even harmonic distortion affects the positive and negative half-cycle in a different way, as we will show in Section 2.5.3.

**2.5.2.2 Interharmonic Distortion** Often the voltage or current waveform contains components that are not a multiple integer of the power system frequency. To measure these so-called interharmonics it is needed to measure over a longer period than one cycle. For a voltage with only one interharmonic component, at frequency $\xi f_0$, present we can write

$$v(t) = V_0(t) + \sum_{h=1}^{H} V_h\sqrt{2}\cos(h\omega t - \alpha_h) + V_\xi\sqrt{2}\cos\left(\xi\omega t + \alpha_\xi\right) \tag{2.188}$$

The presence of interharmonics can be interpreted in the time domain as the signal being periodic but with a period of more than one cycle. Consider, for example, a 50-Hz signal distorted with a 155-Hz interharmonic. After 200 ms, 10 cycles of the signal have passed and 31 cycles of the interharmonic. The wave shape is thus periodic with a period of 200 ms (10 cycles of the power system frequency). But note that this signal does not contain any 5-Hz component.

Subharmonics are treated as a special case of interharmonic components, with frequencies less than 50 Hz, thus $\xi < 1$ in (2.188). Subharmonics are often treated separately as they can cause specific problems.

In [140] a further distinction is made between rational interharmonics and irrational interharmonics. The former lead to periodic signals whereas the latter signals are aperiodic. Mathematically such a distinction is correct, but in practice there is no way of distinguishing within a finite measurement window. A distinction could be made into signals with a short period (e.g., up to 10 cycles of the power system frequency) and those with a longer period, but such would require an arbitrary limit.

**2.5.2.3 Nonperiodic Distortion** Some signals contain no periodicity at all. Examples are the voltage during ferroresonance and the current taken by an arc furnace. In [140] a further distinction is made between noise and chaotic behavior, however without indicating how they can be distinguished by using measurements. This may be an issue justifying further investigation. The presence of chaotic behavior could lead to conflicts with the algorithms used in calculating characteristics and indices. Those algorithms are all based on the assumption that averaging will reduce the spread in values, which is not necessarily the case for chaotic behavior.

In terms of the spectrum of the signal, harmonic distortion corresponds to frequency components at integer multiples of the power system frequency, interharmonic distortion to noninteger multiples, and noise to a continuous spectrum in between the harmonic and interharmonic spectral lines (see Fig. 2.31).

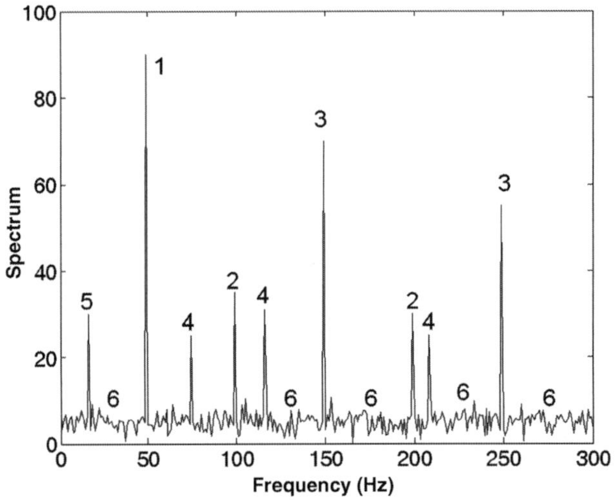


**Figure 2.31** Stylized spectrum of distorted signal with power system frequency (1), even-harmonic components (2), odd-harmonic components (3), interharmonic components (4), subharmonic components (5), and noise (6).

### 2.5.3 Harmonic Distortion

***2.5.3.1 Fourier Series*** A nonsinusoidal but periodic voltage waveform can be written as an infinite sum of harmonics according to

$$v(t) = V_0 + \sum_{h=1}^{\infty} \sqrt{2}V_h \cos(h\omega t - \alpha_h) \tag{2.189}$$

where

$$v_1(t) = \sqrt{2}V_1 \cos(\omega t) \tag{2.190}$$

is referred to as the *fundamental component of the voltage* or simply the *fundamental voltage* with rms voltage $V_1$. The phase angle of the fundamental voltage is taken as zero without any loss of generality. The term

$$v_h(t) = \sqrt{2}V_h \cos(h\omega t - \alpha_h) \tag{2.191}$$

is referred to as the *harmonic* *h* of the *hth harmonic component of the voltage*; $V_h$ is the rms value of harmonic $h$; $\alpha_h$ is its phase angle with reference to the fundamental voltage. Note that there is no unique way to define the phase-angle difference between sine waves of different frequency. By using a cosine function in (2.189), the phase angle is implicitly defined from the maximum of the sine wave. Using sine functions in (2.189) would lead to another value of $\alpha_h$. Instead of (2.189) we can write for a periodic signal

$$v(t) = V_0 + \sum_{h=1}^{\infty} \sqrt{2} V_h \sin(h\omega t' - \alpha'_h) \tag{2.192}$$

In this case we use as a reference angle $\alpha'_1 = 0$. From the fundamental component, we obtain a relation for the shift in time axis between the two formulations:

$$\cos(\omega t) = \sin(\omega t') \tag{2.193}$$

with as one of its solutions

$$t' = t + \frac{\pi}{2\omega} = t + \frac{T}{4} \tag{2.194}$$

with $T = 2\pi/\omega$ one cycle of the fundamental frequency. Substituting (2.194) in (2.192) gives for harmonic $h$

$$\alpha'_h = \alpha_h - (h-1)\frac{\pi}{2} \tag{2.195}$$

Note that both (2.194) and (2.195) are just one of an infinite number of solutions. Expression (2.194) is the one leading to the smallest shift in time axis, whereas (2.195) should be taken modulus $2\pi$ to get the smallest angle shift.

Splitting a signal in components according to (2.189) is strictly speaking only valid for continuous ("analog") signals. For sampled ("digital" or "discrete") signals, the decomposition in harmonic components reads as follows:

$$v[t_i] = V_0 + \sum_{h=1}^{H} \sqrt{2} V_h \cos(h\omega t_i - \alpha_h) \tag{2.196}$$

where $v[t_i]$ is the voltage sample at the sample instant $t_i$. The Fourier series is no longer an infinite series; instead the highest frequency $H\omega/2\pi$ is determined by the sample frequency $f_s$ according to

$$\frac{H\omega}{2\pi} \le \frac{f_s}{2} \tag{2.197}$$

where $H$ is the highest integer that fulfills (2.197). For a sample frequency that is an even multiple of the fundamental frequency, we obtain the commonly used form

$$H = \frac{1}{2}\frac{f_s}{f_0} \tag{2.198}$$

**2.5.3.2 *Total Harmonic Distortion*** It was shown above how a periodic signal can be decomposed into a number of harmonics. The signal can be totally characterized by the magnitude and phase of these harmonics. In power system applications the fundamental (50- or 60-Hz) component will normally dominate. This holds especially for the voltage. Often it is handy to characterize the deviation from the (ideal) sine wave through one quantity. This quantity should indicate how distorted the voltage or current is. For this the so-called total harmonic distortion, or THD, is most commonly used. The THD gives the relative amount of signal energy not in the fundamental component:

$$\mathrm{THD} = \frac{\sqrt{\sum_{h=2}^{H} V_h^2}}{V_1} \tag{2.199}$$

The THD is typically expressed as a percentage value (thus 7% instead of 0.07). In the mathematical analysis of a continuous signal, an upper limit $H = \infty$ should be chosen. Otherwise the upper limit is determined by the sample frequency or by a standard document. More discussion on the definition of THD can be found in Section 3.3.1.

**2.5.3.3 *Crest Factor*** The crest factor is a time-domain property indicating how much the top of the sine wave is distorted. It is defined as the ratio between the amplitude (maximum value) of a signal and its rms value:

$$C_r = \frac{V_{\max}}{V_{\mathrm{rms}}} \tag{2.200}$$

with $V_{\mathrm{rms}} = \sqrt{1/T \int_0^T v^2(t)\,dt}$. For a perfect sine wave the crest factor is equal to $\sqrt{2}$ so that it makes sense to introduce a *relative crest factor*, which is unity for a perfect sine wave:

$$c_r = \frac{1}{\sqrt{2}}\frac{V_{\max}}{V_{\mathrm{rms}}} \tag{2.201}$$

The crest factor indicates how much a signal deviates from a dc signal, whereas the relative crest factor indicates how much a signal deviates from a sine wave.

**2.5.3.4 *Odd Harmonics*** Odd-harmonic distortion is typically dominant in supply voltage and load current. To visualize the effect of odd harmonics on the

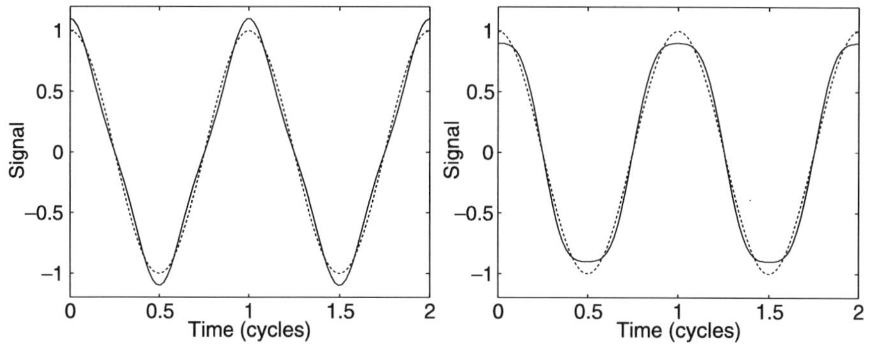


**Figure 2.32** Odd-harmonic distortion (solid curve) in phase with the fundamental (left) and in opposite phase (right). The dotted curve indicates the nondistorted (fundamental) waveform.

waveform, consider a 50-Hz signal distorted with a 150-Hz (third-harmonic) signal. Figure 2.32 shows a nondistorted sine wave (dashed curve) and a signal with 10% third-harmonic distortion (solid curve). The left-hand picture shows the waveform for a third-harmonic component in phase with the fundamental.

$$v(t) = \cos(\omega t) + 0.1\cos(3\omega t) \tag{2.202}$$

The right-hand picture shows the resulting waveform when the third harmonic is $180^\circ$, shifted compared to the fundamental:

$$v(t) = \cos(\omega t) + 0.1\cos(3\omega t + \pi) \tag{2.203}$$

The effect of the odd harmonic is an increase (left) or decrease (right) of the amplitude of the signal with 10%. The rms value increases only very little (0.5%) so that the crest factor increases or decreases by about 10%.

Generally speaking a third-harmonic component leads to a change in the crest factor. The effect of the distortion is the same for the positive and for the negative half of the sine wave. This holds for all odd harmonics, which can be shown by considering the following function:

$$f(t) = \sin(\omega t) + a_h\cos(h\omega t - \alpha_h) \tag{2.204}$$

with $h$ odd. Shifting the function over one half-cycle $1/2f_0$ gives

$$f\left(t + \frac{1}{2f_0}\right) = \sin(\omega t + \pi) + a_h\cos(h\omega t - \alpha_h + \pi) \tag{2.205}$$

because a shift over $h\pi$ equals a shift over $\pi$ for $h$ odd. Using $\sin(x+\pi) = -\sin(\pi)$ gives

$$f\left(t + \frac{1}{2f_0}\right) = -f(t) \tag{2.206}$$

The positive cycle is the same as the negative cycle of the voltage wave as long as only odd harmonics are present in the voltage.

**2.5.3.5 Even Harmonics** Even-harmonic distortion of voltage or current is normally rather small. Even harmonics are generated by some large converters, but modern rules on harmonic distortion state that equipment should not generate any even harmonics. In fact, a measurement of the supply voltage shows that the amount of even harmonics is indeed very small. Even harmonics are generated by transformer energizing. This event leads to a temporary increase in even-harmonic distortion and will be discussed in Section 6.2.

Figure 2.33 show the distortion due to a second harmonic of 10% magnitude in phase with the fundamental:

$$v(t) = \cos(\omega t) + 0.1\cos(2\omega t) \tag{2.207}$$

The result of even-harmonic distortion is that positive and negative half-cycles of the signal are no longer symmetrical. In this example, the positive half-cycle becomes narrower but higher and the negative half-cycle wider but lower. The zero crossings no longer arrive at 10-ms intervals.

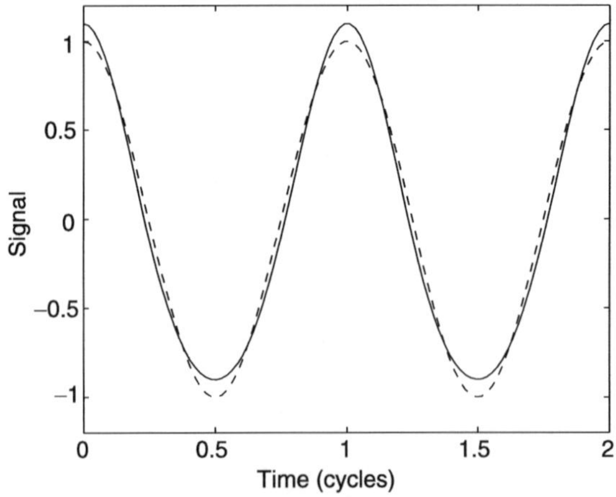


**Figure 2.33** Even-harmonic distortion (solid curve) in phase with fundamental (dashed curve).

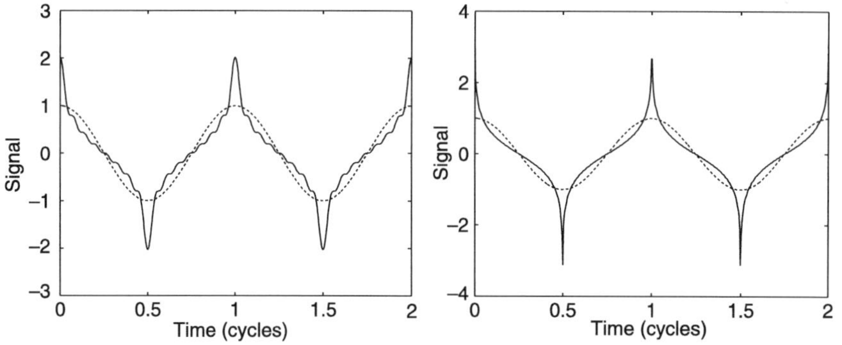


**Figure 2.34** Odd harmonics up to order 15 (left) and up to order 201 (right) in phase with fundamental.

***2.5.3.6 Many Harmonics*** Actual wave shapes always contain a whole spectrum of harmonics. As an example, Figure 2.34 shows the effect of a number of odd harmonics all in phase with the fundamental:

$$i(t) = \cos(\omega t) + \sum_{h=1}^{H} \frac{1}{2h+1} \cos[(2h+1)\omega t] \tag{2.208}$$

where $H = 7$ and $H = 100$ for the left- and right-hand pictures, respectively.

The oscillation in Figure 2.34 is a typical result of abruptly cutting off the harmonic spectrum (the phenomenon is called the *Gibb's effect* in signal processing), in this case after the 15th harmonic. Counting the oscillations in the signal would give an oscillation frequency of 800 Hz, corresponding to the 16th harmonic. This frequency is however not present in the signal as it is generated in accordance with (2.208). The 800-Hz oscillation is due to the absence of components of 800 Hz and higher, not due to the presence of an 800-Hz component. The result of including harmonics up to order 201 is a smooth waveform.

Figure 2.35 shows the wave shape that results from odd harmonics being in alternate phases compared to the fundamental. The mathematical expression for the signal is as follows:

$$i(t) = \cos(\omega t) + \sum_{h=1}^{H} \frac{(-1)^h}{2h+1} \cos[(2h+1)\omega t] \tag{2.209}$$

where $H = 7$ and $H = 100$. In words, the third harmonic is of amplitude $\frac{1}{3}$ in opposite phase compared to the fundamental, the fifth harmonic of amplitude $\frac{1}{5}$ in phase, the seventh of amplitude $\frac{1}{7}$ in opposite phase, and so on. The result is a rectangular wave shape for the signal. Note that the amplitude spectrum for the

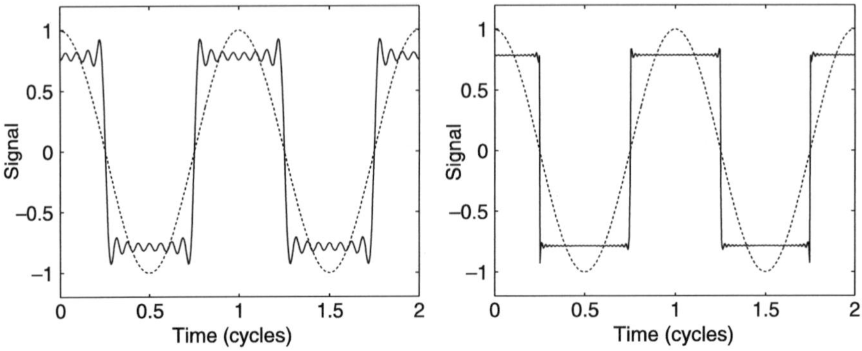


**Figure 2.35** Even harmonics up to order 15 (left) and up to order 201 (right) in alternating phase compared to fundamental.

signals in Figures 2.34 and 2.35 is exactly the same, so that the THD is exactly the same. The difference in shape, and thus in crest factor, is only due to the difference in phase angle of the individual harmonics. It shows that the THD alone cannot give any information about a waveform in the time domain.

Note also that even for $H = 100$ the oscillations remain present. This is related to the step in the time-domain signal. For a continuous signal the oscillations never disappear, no matter how many harmonic components are included. But for discrete (sampled) signals the oscillations disappear when the harmonic frequency becomes higher than the highest frequency present in the sampled signal.

***2.5.3.7 Three-Phase Balanced Systems*** Consider a balanced system with balanced but nonlinear load, so that the current wave shapes are identical in shape but shifted one third of a cycle of the fundamental frequency compared to each other:

$$i_a(t) = i(t) \qquad i_b(t) = i\left(t - \tfrac{1}{3}T\right) \qquad i_c(t) = i\left(t + \tfrac{1}{3}T\right) \tag{2.210}$$

Splitting the signals in harmonic components gives

$$\begin{aligned} i_a(t) &= \sum_{h=0}^{H} \sqrt{2} I_h \cos(h\omega t - \beta_h) \\ i_b(t) &= \sum_{h=0}^{H} \sqrt{2} I_h \cos\left(h\omega t - h\frac{2\pi}{3} - \alpha_h\right) \\ i_c(t) &= \sum_{h=0}^{H} \sqrt{2} I_h \cos\left(h\omega t + h\frac{2\pi}{3} - \alpha_h\right) \end{aligned} \tag{2.211}$$

**TABLE 2.6 Sequence Harmonics in Balanced System**

| | 1 | 2 | 3 | 4 | 5 | 6 | 7 | 8 | 9 | 10 | 11 | 12 | 13 | 14 | 15 | 16 | 17 |
|---|---|---|---|---|---|---|---|---|---|---|---|---|---|---|---|---|---|
| Positive sequence | + | | | + | | | + | | | + | | | + | | | + | |
| Negative sequence | | – | | | – | | | – | | | – | | | – | | | – |
| Zero sequence | | | 0 | | | 0 | | | 0 | | | 0 | | | 0 | | |

The fundamental currents form a positive-sequence set. For a harmonic of order $h$ we get the following expressions:

$$\begin{aligned} i_{ah}(t) &= \sqrt{2}I_h \cos\left(h\omega t - \beta_h\right) \\ i_{bh}(t) &= \sqrt{2}I_h \cos\left(h\omega t - h\frac{2\pi}{3} - \beta_h\right) \\ i_{ch}(t) &= \sqrt{2}I_h \cos\left(h\omega t + h\frac{2\pi}{3} - \beta_h\right) \end{aligned} \tag{2.212}$$

Note that a shift of one-third cycle at the fundamental frequency corresponds to a shift of $k/3$ cycles at harmonic $k$. The result can be positive, negative, or zero sequence. For the third harmonic, $h = 3$ in (2.212), we obtain a zero-sequence set:

$$\begin{aligned} i_{a3}(t) &= \sqrt{2}I_3 \cos\left(3\omega t - \beta_3\right) \\ i_{b3}(t) &= \sqrt{2}I_3 \cos\left(3\omega t - \beta_3\right) \\ i_{c3}(t) &= \sqrt{2}I_3 \cos\left(3\omega t - \beta_3\right) \end{aligned} \tag{2.213}$$

For a balanced system with balanced load, the third-harmonic currents are in phase in the three phases, just as with any zero-sequence fundamental component. The third harmonic is therefore called a *zero-sequence harmonic*. Along the same line of reasoning, the fifth harmonic is a *negative-sequence harmonic* and the seventh harmonic a *positive-sequence harmonic*. Repeating this for other harmonic numbers results in Table 2.6.

***2.5.3.8 Three-Phase Unbalanced Systems*** The calculations in the previous section only hold for a balanced load against a balanced system. Small unbalances in supply voltage or in the load can already lead to large deviations from this scheme. To obtain positive-, negative-, and zero-sequence components in an unbalanced system, a sequence transformation needs to be done for each harmonic component. Consider a harmonic $h$ with complex phase currents $\underline{I}_{ah}$, $\underline{I}_{bh}$, and $\underline{I}_{ch}$ as defined before. The sequence components for this harmonic are defined in the same way as for fundamental-frequency signals (see Section 2.3.1):

$$\begin{aligned} \underline{I}_h^0 &= \tfrac{1}{3}\left(\underline{I}_{ah} + \underline{I}_{bh} + \underline{I}_{ch}\right) \\ \underline{I}_h^+ &= \tfrac{1}{3}\left(\underline{I}_{ah} + a\underline{I}_{bh} + a^2\underline{I}_{ch}\right) \\ \underline{I}_h^- &= \tfrac{1}{3}\left(\underline{I}_{ah} + a^2\underline{I}_{bh} + a\underline{I}_{ch}\right) \end{aligned} \tag{2.214}$$

with $a = -\frac{1}{2} + \frac{1}{2}j\sqrt{3}$. Note that the superscripts $+$, $-$, and 0 are used for zero, positive, and negative sequences, respectively, instead of 0, 1, and 2. This is to prevent confusion with harmonic numbers.

As with the fundamental-frequency symmetrical components, this decomposition can be used to study the propagation of harmonic distortion through the system. However, for harmonic frequencies the coupling between the symmetrical components is much bigger than for the fundamental. Therefore one has to be careful in using symmetrical components to study harmonic propagation. This issue is discussed in detail in [11, 321].

**2.5.3.9 Transfer Through Transformers** The transfer of symmetrical components through three-phase transformers is explained in any power system analysis book treating symmetrical components. The main conclusions can be summarized as follows:

- The zero-sequence voltages and currents do not propagate through most transformers. The exception are star–star connected transformers grounded on both sides. For this transformer type the zero-sequence component is transferred without phase shift or with 180° phase shift.
- The amplitude ratio of negative-sequence voltages and currents is the same as for the positive sequence. The phase shift between primary- and secondary-side negative-sequence voltages is opposite to the phase shift for positive-sequence voltages. The same holds for the currents.

For the derivation of these rules, complex notation is used. As the complex notation is independent of frequency, the same rules hold for harmonics.

The most important application of this is with the propagation of harmonics through the system. Zero-sequence harmonics cannot propagate through most distribution transformers. The result is that the third-harmonic distortion at a medium voltage level is much lower than at a low voltage level. The fifth- and seventh-harmonic distortions on the other hand are similar at different voltage levels.

Note that this only holds for a balanced three-phase system. Small unbalances in the supply voltage already lead to large deviations from this scheme. Voltage unbalance leads to a third-harmonic current component which is not zero sequence and thus propagates through distribution transformers to medium voltage.

Another consequence of the third harmonic being a zero-sequence harmonic is that the third-harmonic currents add in the neutral conductor. Low-voltage load with a large third-harmonic component can lead to overloading of the neutral conductor. This is especially dangerous as the neutral conductor is normally not equipped with overload protection.

### 2.5.4 Sources of Waveform Distortion

Waveform distortion is due to the presence of nonlinear elements in the power system. A nonlinear element takes a nonsinusoidal current for a sinusoidal voltage. Thus even for a nondistorted voltage waveform the current through a nonlinear element is distorted. This distorted current waveform in turn leads to a distorted voltage waveform. Most elements of the power network are linear. The exceptions are transformers, especially during sustained overvoltages, and power-electronic components such as HVDC links and Flexible ac Transmission System (FACTS) devices. But the main distortion at most sites is due to nonlinear load, again mainly power-electronic converters.

In the forthcoming sections an overview is given of the way in which power-electronic converters lead to distorted current waveforms. A distinction is made between single- and three-phase sources and between so-called dc current sources (where the direct current is about constant) and dc voltage sources, with a constant dc voltage. Traditionally the emphasis has been on dc sources, where the harmonic spectrum can be obtained rather easily. However, most modern nonlinear equipment is a dc voltage source. Calculating the spectrum for this equipment is not that straightforward.

Some other sources of waveform distortion will be discussed as well, including sources of interharmonics.

***2.5.4.1 Single-Phase dc Current Source*** Figure 2.36 shows the well-known circuit diagram of a single-phase noncontrolled rectifier (or diode rectifier). If the dc load is mainly inductive (e.g., motor load), the current on the dc side can be assumed constant. Current and voltage on the ac side and current on the dc side are shown in Figure 2.37. The voltage is assumed to be sinusoidal.

The current is in phase with the voltage, so that the fundamental power factor is 1. The total power factor is less than 1 because the current is nonsinusoidal. As

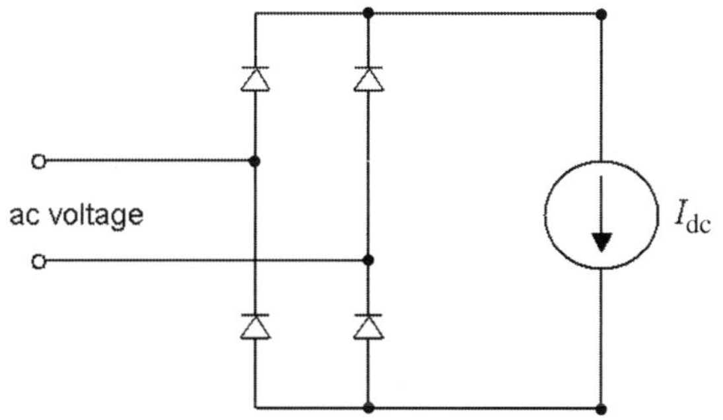


**Figure 2.36** Single-phase dc current source.

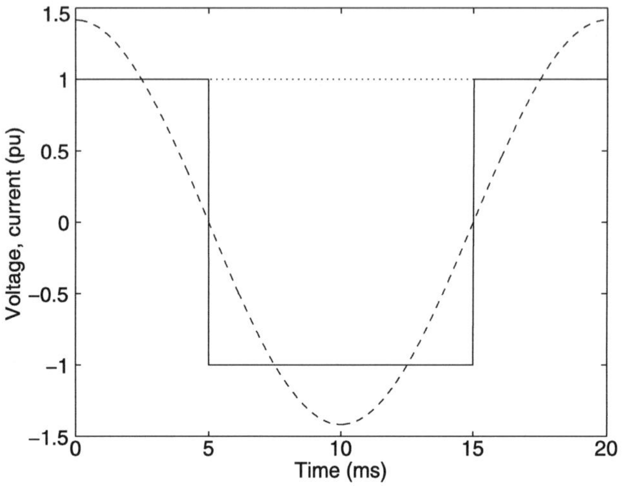


**Figure 2.37** Single-phase dc current source: ac-side voltage (dashed); ac-side current (solid); dc-side current (dotted).

we have a mathematical expression for voltage and current, we can apply the equations for continuous wave shapes given before. Note that this is only possible in idealized cases like this. The idealization lies here in the assumption that the voltage is sinusoidal and the current at the ac side changes direction instantaneously. Assume the voltage to be sinusoidal:

$$v(t) = \sqrt{2}V\cos(\omega t) \tag{2.215}$$

so that the current can be written as (see also Fig. 2.37)

$$i(t) = \begin{cases} -I_{\mathrm{dc}} & \frac{T}{4} < t < \frac{3T}{4} \\ +I_{\mathrm{dc}} & t < \frac{T}{4} \qquad t > \frac{3T}{4} \end{cases} \tag{2.216}$$

The active power taken from the supply is found as the average instantaneous power:

$$P = \frac{2\sqrt{2}}{T} V I_{\mathrm{dc}} \tag{2.217}$$

The rms voltage is equal to $V$, the rms current is equal to $I_{\mathrm{dc}}$, so that the total power factor equals

$$\Pi_{\mathrm{total}} = \frac{P}{VI_{\mathrm{dc}}} = \frac{2\sqrt{2}}{\pi} \approx 0.90 \tag{2.218}$$

Fundamental voltage and current are in phase, which implies a unity displacement power factor. The distortion power factor is equal to the total power factor. Using the relation between distortion power factor and THD enables us to calculate the THD:

$$\text{THD} = \sqrt{\tfrac{1}{8}\pi^2 - 1} \approx 48\% \tag{2.219}$$

To calculate the spectrum of the current, use the expression for the Fourier series:

$$I_h = \frac{\sqrt{2}}{T}\int_0^T i_a(t)\cos(h\omega t)\,dt \tag{2.220}$$

After some tedious mathematics this results in

$$I_h = \begin{cases} +\dfrac{2\sqrt{2}}{h} I_{\text{dc}} & h = 1, 5, 9, 13, \ldots \\ -\dfrac{2\sqrt{2}}{h} I_{\text{dc}} & h = 3, 7, 11, 15, \ldots \\ 0 & h = 0, 2, 4, 6, \ldots \end{cases} \tag{2.221}$$

Note that in practice there is hardly ever the need to apply expression (2.220) to a waveform. This only gives a result when a simple mathematical expression is known for the waveform, which is rarely ever the case. Even in this case, the mathematical expression is an approximation and the results are only used to give an insight into the kind of spectrum that can be expected. In reality the measured (sampled) waveform is applied to a DFT algorithm that gives the spectrum. The DFT algorithm will be discussed in Section 3.4.

The spectrum obtained in (2.221) is shown in Figure 2.38: The current only contains odd harmonics; the amplitude of the odd harmonics decays with the inverse of

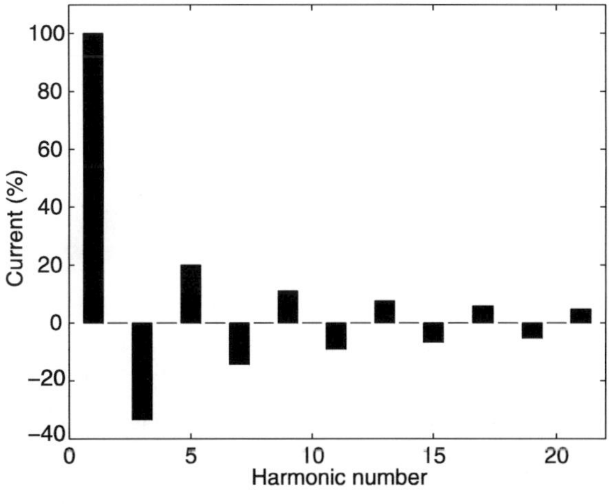


**Figure 2.38** Current spectrum of single-phase dc current source.

the frequency; and the harmonics are alternately in phase and in opposite phase with the fundamental current.

#### 2.5.4.2 Three-Phase dc Current Source

The basic three-phase noncontrolled rectifier is shown in Figure 2.39. In the "ideal" case (dc current source, no source impedance) a block-shaped current flows in each phase, as shown in Figure 2.40. The current can be described by the following mathematical expression:

$$i_a(t) = \begin{cases} I_{dc} & 0 < t < \frac{T}{6} \quad \frac{5T}{6} < t < T \\ 0 & \frac{T}{6} < t < \frac{2T}{6} \quad \frac{4T}{6} < t < \frac{5T}{6} \\ -I_{dc} & \frac{2T}{6} < t < \frac{4T}{6} \end{cases} \tag{2.222}$$

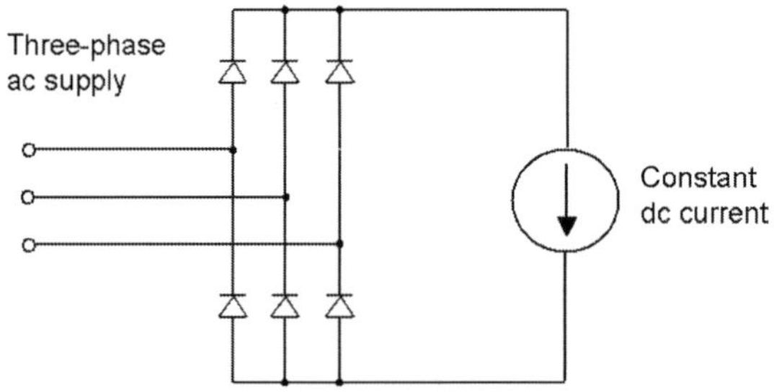


**Figure 2.39** Three-phase rectifier: dc current source.

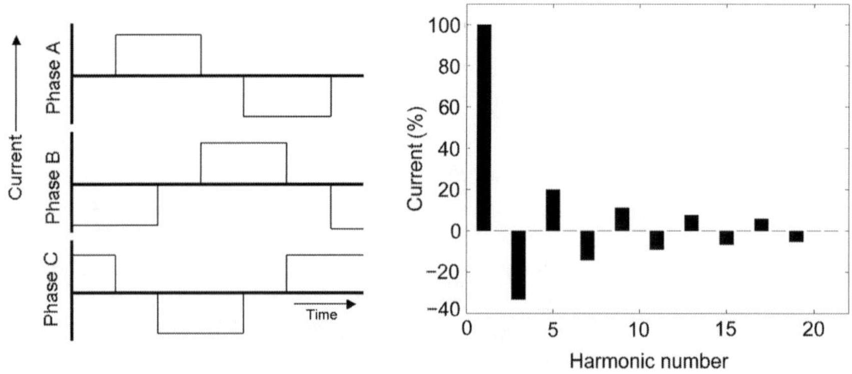


**Figure 2.40** Simplified wave shape of currents (left) taken by three-phase rectifier as shown in Figure 2.39 and its spectrum (right).

The spectrum of the current can again be calculated from the mathematical expression for the Fourier series (2.220):

$$I_h = \frac{\sqrt{2}I_{dc}}{2\pi h}\left[\sin\left(h\frac{\pi}{3}\right) - \sin\left(h\frac{5\pi}{3}\right) - \sin\left(h\frac{4\pi}{3}\right) + \sin\left(h\frac{2\pi}{3}\right)\right] \tag{2.223}$$

Working out the right-hand side of (2.223) results in the following harmonic spectrum:

$$I_h = \begin{cases} \frac{\sqrt{6}}{h\pi}I_{dc} & h = 1, 7, 13, 19, \ldots \\ -\frac{\sqrt{6}}{h\pi}I_{dc} & h = 5, 11, 17, 21, \ldots \\ 0 & h = 2, 4, 6, 8, \ldots \\ 0 & h = 3, 9, 15, 21, \ldots \end{cases} \tag{2.224}$$

The spectrum is shown in Figure 2.40. As this is a balanced system, the symmetrical components are as shown in Table 2.6. Thus

$$\begin{aligned} I_h^+ &= \frac{\sqrt{6}}{h\pi}I_{dc} \qquad h = 2n+1 \qquad n = 0, 1, 2, \ldots \\ I_h^- &= -\frac{\sqrt{6}}{h\pi}I_{dc} \quad h = 2n-1 \qquad n = 1, 2, \ldots \end{aligned} \tag{2.225}$$

All other contributions are zero for a balanced set of sinusoidal voltages. The harmonic components according to (2.225) are called *characteristic harmonics*. Unbalanced voltages and voltage distortion lead to additional harmonic and symmetrical components. These are referred to as *noncharacteristic harmonics*.

Three-phase rectifiers with a constant dc current source occur in practice as those with a large dc link inductance. They are, for example, used to power dc machines where any ripple in armature current causes a ripple in torque. As the latter should be small, the dc current will vary very little during the cycle.

***2.5.4.3 Twelve-Pulse Rectifier*** A 12-pulse rectifier consists of two 6-pulse rectifiers. A 6-pulse rectifier is the "normal" three-phase rectifier as shown in Figure 2.39. The two 6-pulse rectifiers are connected in series on the dc side and in parallel on the ac side. The result is that the dc bus voltage and thus the power are twice that for a 6-pulse rectifier, as shown in Figure 2.41. By connecting the two rectifiers in parallel on the dc side, a higher dc current can be obtained. But this type of configuration can also be used to mitigate some of the harmonic distortion created by the rectifier. To obtain a lower harmonic distortion one of the 6-pulse rectifiers is supplied through a Dy transformer, the other through a Yy transformer.

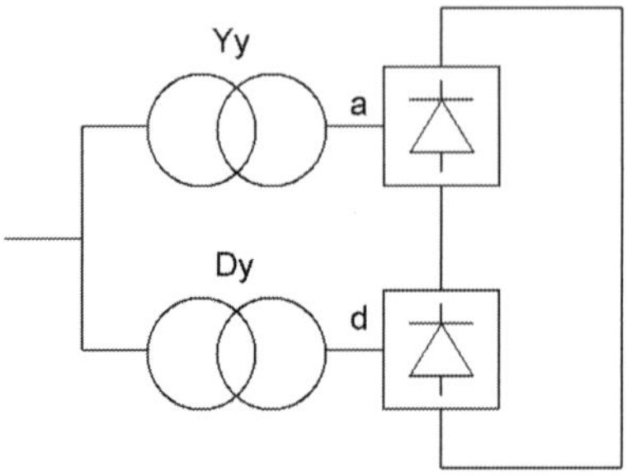


**Figure 2.41** Configuration of 12-pulse rectifier.

Considering Yy0 and Dy1 transformers, the (positive-sequence) voltages at the terminals of rectifier $\delta$ are shifted over $T/12$ ($30^\circ$ at 50 Hz) compared to rectifier $\alpha$. As the voltages are shifted, so are the current waveforms:

$$i_a^{(\delta)} = i_a^{(\alpha)}\left(t - \frac{T}{12}\right) \tag{2.226}$$

A shift over $T/12$ is $30^\circ$ at fundamental frequency but $h \times 30^\circ$ at harmonic $h$, so that

$$i_{ah}^{(\delta)} = i_{ah}^{(\alpha)} e^{-jh30^\circ} \tag{2.227}$$

It is easy to prove that positive- and negative-sequence voltages are shifted over this same angle:

$$I_h^{(\delta)+} = I_h^{(\alpha)+} e^{-jh30^\circ} \tag{2.228}$$

$$I_h^{(\delta)-} = I_h^{(\alpha)-} e^{-jh30^\circ} \tag{2.229}$$

From the secondary to the primary side of the Dy transformer, positive-sequence currents are shifted over $+30^\circ$, negative-sequence currents over $-30^\circ$. On the primary side of the transformers we get the following relations between the two rectifier currents:

$$I_h^{(\delta)+} = I_h^{(\alpha)+} e^{-jh30^\circ} e^{+j30^\circ} \tag{2.230}$$

$$I_h^{(\delta)-} = I_h^{(\alpha)-} e^{-jh30^\circ} e^{-j30^\circ} \tag{2.231}$$

The 5th harmonic is negative sequence so that

$$I_5^{(\delta)-} = I_5^{(\alpha)-} e^{-j180^\circ} \tag{2.232}$$

The currents through the two transformers cancel each other. The same holds for the (positive-sequence) 7th harmonic:

$$I_7^{(\delta)+} = I_7^{(\alpha)+} e^{-j180^\circ} \tag{2.233}$$

Repeating this for the 11th and 13th harmonics shows that they add:

$$I_{11}^{(\delta)-} = I_{11}^{(\alpha)-} e^{-j360^\circ} \tag{2.234}$$

$$I_{13}^{(\delta)+} = I_{13}^{(\alpha)+} e^{-j360^\circ} \tag{2.235}$$

In the same way it can be shown that the 17th and 19th harmonics will cancel but the 23rd and 25th will not.

Combining single- and three-phase loads has the same effect as a 12-pulse rectifier. Thus adding a three-phase load to a system with mainly single-phase load will reduce the harmonic distortion. Currently the voltage distortion in public supply networks is mainly due to single-phase loads. A growth in three-phase loads (air-conditioning equipment, motor drives) will initially lead to a reduction in harmonic distortion. This holds when single- and three-phase loads are connected at the same voltage level, as in most of Europe. When they are connected at different voltage levels separated by a delta–star-connected transformer, the harmonics will add, leading to higher levels of distortion [208].

A detailed discussion of harmonic mitigation through multipulse converters in given in [232]. By a clever choice of transformer windings a range of phase rotations can be obtained, resulting in converters with up to 48 pulses taking a current that is close to sinusoidal. Such converters are used in several FACTS devices that have been recently introduced in transmission systems.

***2.5.4.4 Notching*** Notching is due to the temporary shorting of two phases of a rectifier during the commutation of current from one phase to another. Consider the situation shown in Figure 2.42. The current commutates from phase 1 to phase 2. At time zero the two phase voltages are equal and the diode in phase 2 starts to conduct. Before time zero, diode 1 conducts and diode 2 not:

$$v_1(t) = V\sqrt{2}\cos\left(\omega t + \tfrac{1}{3}\pi\right) \qquad v_2(t) = V\sqrt{2}\cos\left(\omega t - \tfrac{1}{3}\pi\right) \tag{2.236}$$

The currents during commutation are described by the following differential equation (note that both diodes conduct, thus the two phases are shorted):

$$v_1(t) - L\frac{di_1}{dt} + L\frac{di_2}{dt} = v_2(t) \tag{2.237}$$

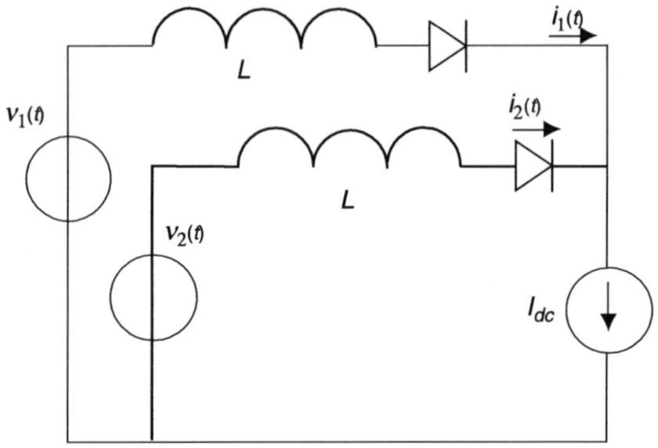


**Figure 2.42** Circuit explaining commutation from one phase to another.

We further know that $i_1(t) + i_2(t) = I_{\mathrm{dc}}$, so that $di_1/dt + di_2/dt = 0$, and

$$\frac{di_2}{dt} = \frac{\sqrt{6}V\sin(\omega t)}{2L} \tag{2.238}$$

with solution

$$i_2(t) = i_2(0) + \int_0^t \frac{di_2}{d\tau}d\tau = \frac{\sqrt{6}V}{2\omega L}[1 - \cos(\omega t)] \tag{2.239}$$

The commutation time is normally short compared to one cycle of the fundamental voltage, so that the cosine function can be approximated by $\cos(x) \approx 1 - \frac{1}{2}x^2$, leading to the following expression for the current during commutation:

$$i_2(t) = \frac{\pi\sqrt{6}f_0V}{2L}t^2 \tag{2.240}$$

The shape of the current during commutation is shown in Figure 2.43. The commutation is complete when the current through diode 2 becomes equal to the dc current. The current through diode 1 becomes zero at that moment and it extinguishes.

The duration of the commutation $T_C$ can be found from equating (2.240) to $I_{\mathrm{dc}}$. Using the approximate expression (2.240) gives

$$T_C = \sqrt{\frac{2LI_{\mathrm{dc}}}{\pi\sqrt{6}f_0V}} \tag{2.241}$$

During commutation the voltage in both phases is equal. There is a temporary phase-to-phase short circuit. The result is a serious distortion in the voltage.

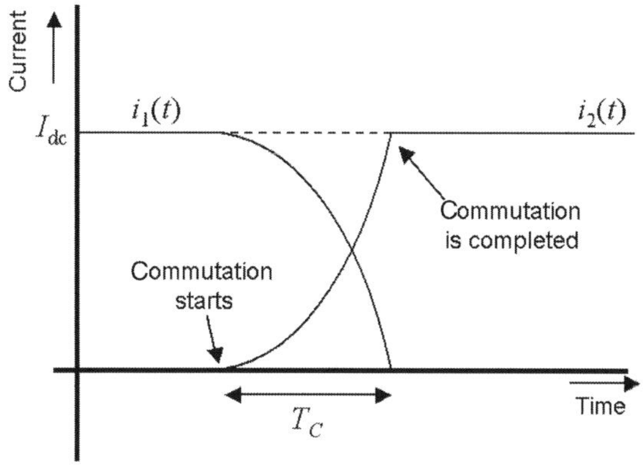


**Figure 2.43** Shape of current during commutation.

***2.5.4.5 Single-Phase dc Voltage Source*** An important source of harmonic distortion is formed by small electronic equipment, with personal computers and televisions the most typical (or, better, notorious) contributors. Early televisions contained a simple rectifier consisting of one diode and a capacitor to form the dc voltage needed for the screen. This caused a high amount of even-harmonic distortion in the current taken by the television. In [10] a measurement from 1970 is shown in which the current through a medium-voltage-distribution cable contains 5% second-harmonic distortion and 2% fourth-harmonic distortion. Televisions were special in that they needed a rather high dc voltage (about 300 V) whereas other electronic equipment would suffice with a few volts. Most other equipment therefore contained a transformer down to 10 through 50 V. The rectification took place with a bridge of four diodes at this lower voltage. The transformer impedance significantly mitigated the harmonic distortion.

Modern consumer electronics devices, personal computers, battery chargers, uninterrupted power supplies (UPSs), and so on, no longer contain such a transformer. Instead the supply voltage is directly rectified by means of a single-phase diode rectifier with a large capacitor on the dc side, as shown in Figure 2.44. This capacitor reduces the ripple in the dc voltage. The resulting dc voltage is rather constant, hence

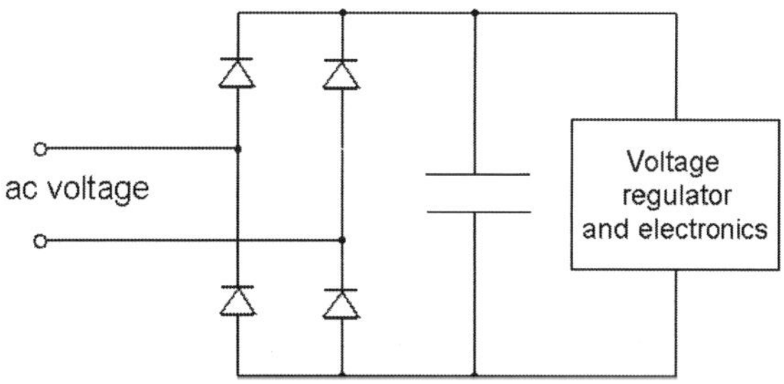


**Figure 2.44** Single-phase electronics load: four-pulse noncontrolled rectifier.

the term *dc voltage source*. For most calculations, the dc voltage can be assumed to be constant during the fundamental-frequency cycle.

The current cannot start to flow until the voltage on the dc side is lower than the voltage on the ac side. This is shown in Figure 2.45. The dc voltage is normally close to the peak of the ac voltage, and the current only flows during a short fraction of the cycle. There is an identical but opposite current peak half a cycle later. For a resistive source (Fig. 2.45) the current only flows when the source voltage is higher than the dc side voltage:

$$i(t) = \begin{cases} \dfrac{V_1\sqrt{2}\sin(\omega t) - V_{\mathrm{dc}}}{R} & V_1\sqrt{2}\sin(\omega t) > V_{\mathrm{dc}} \\ 0 & V_1\sqrt{2}\sin(\omega t) \le V_{\mathrm{dc}} \end{cases}$$

The duration of the current peak depends on the dc side voltage, its amplitude on the resistance of the source.

For an inductive source the situation is slightly more complicated. The current flow is determined by the following differential equation as long as one of the diodes is conducting:

$$L\frac{di}{dt} = V_1\sqrt{2}\sin(\omega t) - V_{\mathrm{dc}} \qquad (2.242)$$

The current will increase as long as the source voltage exceeds the dc side voltage. When the source voltage becomes lower than the dc side voltage, the current decreases until it reaches zero. At that moment the diode blocks further current flow. Like before, increasing the dc voltage decreases the duration of the current pulse, and increasing the inductance decreases its magnitude.

Obtaining a mathematical expression for the spectrum of the current is more difficult, but some conclusions can be drawn here already:

- The spectrum contains all odd harmonics, with the third harmonic dominating.
- There are no even harmonics present in the current.

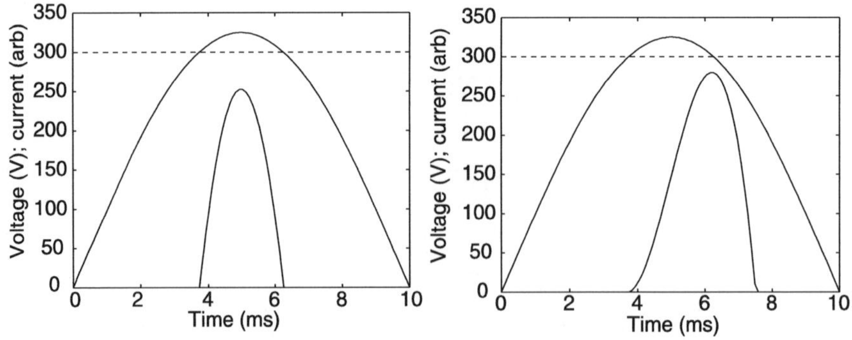


**Figure 2.45** Current taken by single-phase rectifier for resistive source (left) and inductive source (right).

- For a resistive source, all current harmonics are in phase with the voltage.
- The source inductance leads to a shift of the current pulse with respect to the fundamental voltage. This translates into a phase shift in harmonic current. The phase shift in degrees increases with increasing harmonic number.

To obtain some quantification of the current waveform, we return to the basic circuit. Assuming a resistive source with resistance $R$, the current during conduction is

$$i_{ac}(t) = \frac{|u(t)| - U_{dc}}{R} \tag{2.243}$$

The conduction period is calculated by equating the ac and dc voltages. For a resistive source the conduction period starts and ends when ac and dc voltages are (in absolute value) equal. Considering a clean (nondistorted) background voltage, we obtain the start of conduction $t_1$ from

$$\sqrt{2}U_{ac} \sin \omega t_1 = U_{dc} \tag{2.244}$$

resulting in

$$t_1 = \frac{1}{\omega} \arcsin\left(\frac{U_{dc}}{\sqrt{2}U_{ac}}\right) \tag{2.245}$$

The end of conduction occurs at instant $\pi/\omega - t_1$ (half a cycle minus the start of conduction).

Expression (2.245) does not allow us to calculate the conduction period because the dc voltage is unknown. Even though this type of rectifier is referred to as a dc voltage source, its dc voltage is not fully constant. Instead it depends on the dc load as well as on the ac voltage and the source impedance. To calculate the dc voltage in steady state, we use conservation of charge. The voltage over the dc capacitor is the same after one cycle, so that as much charge will have entered the capacitor as will have left it. In terms of current (the derivative of the charge), this reads as

$$\int_0^{2\pi/\omega} |I_{ac}(t)| \, dt = \int_0^{2\pi/\omega} I_{dc}(t) \, dt \tag{2.246}$$

If we assume that the dc is constant and determined from the load power $P$ and the dc voltage, we get after splitting the fundamental-frequency cycle into four equal parts

$$\int_{t_1}^{\pi/2\omega - t_1} \frac{\sqrt{2}U_{ac} \sin(\omega t)}{R} dt = \frac{\pi}{2\omega} \frac{P}{U_{dc}} \tag{2.247}$$

Performing the integration and using $\cos[\arcsin(x)] = \sqrt{1 - x^2}$ result in the expression

$$\frac{\sqrt{2}U_{ac}}{R} \sqrt{1 - \left(\frac{U_{dc}}{\sqrt{2}U_{ac}}\right)^2} - \frac{U_{dc}}{R} \left[\frac{\pi}{2} - \arcsin\left(\frac{U_{dc}}{\sqrt{2}U_{ac}}\right)\right] = \frac{\pi}{2} \frac{P}{U_{dc}} \tag{2.248}$$

Through an appropriate choice of base values, the system voltage amplitude $\sqrt{2}U_{ac}$ and the source impedance $R$ can both be set to unity. This results in the following expression for the dc voltage:

$$U_{dc}\sqrt{1-U_{dc}^2} - U_{dc}^2\left[\tfrac{1}{2}\pi - \arcsin(U_{dc})\right] = \tfrac{1}{2}\pi P \tag{2.249}$$

with $P$ the load power rated to the short-circuit power at the load terminals and $U_{dc}$ the dc voltage rated to the peak voltage of the ac source voltage. It is not possible to analytically obtain an expression for the dc voltage as a function of the load power from (2.249). However, a relation can easily be obtained numerically. The result is shown in Figure 2.46.

The duration of the positive pulse is from $t_1$ through $(\pi/\omega) - t_1$, which is as a fraction of the half-cycle (the "relative pulse duration" in [10]):

$$\alpha_c = 1 - \frac{2}{\pi}\arcsin\left(\frac{U_{dc}}{\sqrt{2}U_{ac}}\right) \tag{2.250}$$

Using the same base values as before, the expression becomes, in per-unit,

$$\alpha_c = 1 - \frac{2}{\pi}\arcsin(U_{dc}) \tag{2.251}$$

The results are shown in Figure 2.47. In [10] a range between 0.16 and 0.36 is mentioned for the relative pulse duration of televisions. (Note that Arrillaga [10] refers the duration of the positive pulse to the whole cycle, whereas we refer it to only one half-cycle.) Note that the size of a typical device is less than 1 A whereas the fault current may be 1 kA or more. Therefore the low-power part of Figure 2.47 has been enlarged and plotted on a logarithmic scale in the right-hand picture.

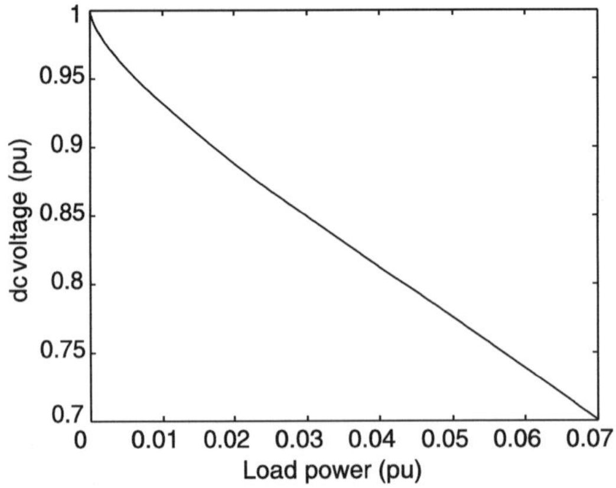


**Figure 2.46** The voltage at the dc bus for a single-phase dc voltage source as a function of the load power.

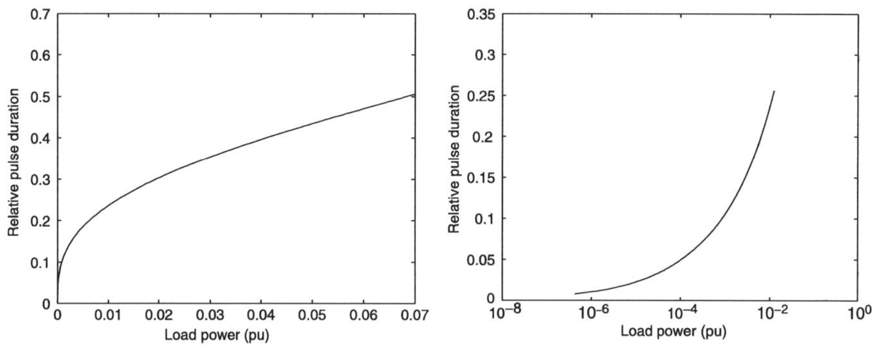


**Figure 2.47** Relative pulse duration versus load size for dc voltage sources.

For a 250-kVA source (corresponding to about 20 m of 2-mm$^2$ copper cable at 230 V), a 100-W load gives $P = 4 \times 10^{-4}$ in the figure. Such a load would have a relative pulse duration of 0.08 according to the above analysis. The same device supplied from a 100-kVA source would have a relative pulse duration of about 0.11.

In reality there is rarely just one device connected to the supply. If a number of identical rectifiers are connected to the same supply point, they can be modeled as one rectifier with a load equal to the sum of the loads of all the individual devices. In that case the load, the horizontal axis in Figures 2.46 and 2.47, can become significantly larger. In a modern office building there can easily be several hundred computers connected to the supply. The result is a reduced dc voltage and a wider pulse, which leads to a reduction in the harmonic distortion per device. The calculated waveform is shown in Figure 2.48 for two different load sizes. The solid line is for a dc voltage of 0.90 pu, the dashed line for a dc voltage of 0.99 pu. Using (2.249) this corresponds to a load size of 0.017 and 0.00059 pu, respectively.

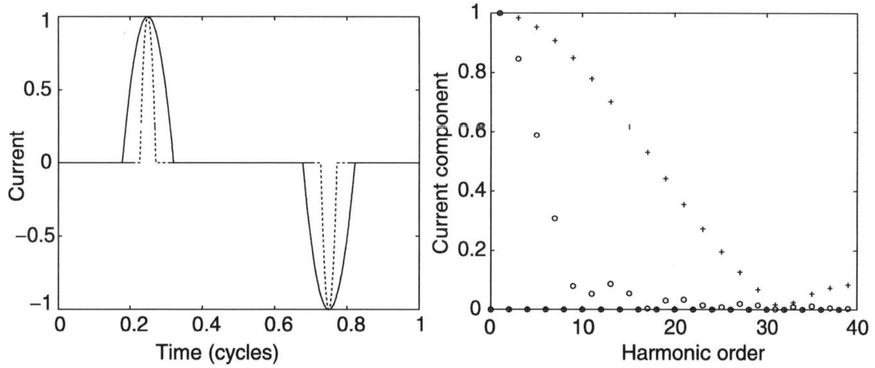


**Figure 2.48** Left: current taken by dc voltage source for two different load sizes: small load (dashed) and large load (solid line). Right: Spectrum for small load (+) and large load (O).

Figure 2.48 also shows the spectrum of the two waveforms as calculated by using the discrete Fourier transform routine in MATLAB. The small load takes a very narrow pulse from the supply which contains a large amount of higher harmonics. The large load takes a wider pulse with less high-order harmonics. Note, however, that even for the large load the third-harmonic component is still about 80% of the fundamental current.

The effect of adding more dc voltage sources (typically more computers) to the same node is a widening of the pulse (the total load increases, leading to a drop in dc voltage and thus a widening of the pulse) resulting in a reduction of the amount of higher order harmonics, however without much effect on the lower order harmonics.

With dc voltage sources, the effect of the background distortion is much larger than with dc current sources. As the rectifier only draws current around a voltage maximum, is it especially the crest factor that affects the current spectrum. These kinds of rectifiers do affect the crest factor a lot so that the background distortion from other rectifiers will affect the operation of a dc voltage source.

The current to an ordinary personal computer was measured when supplied from a clean voltage and when supplied from a distorted voltage. The two spectra are compared in Figure 2.49. The rectifier connected to the clean supply takes a more distorted current (134% THD) than the one connected to the dirty supply (106% THD).

Measurements of the spectra of a number of identical televisions are presented in [10]. A comparison is made of the current spectrum for 1 television, 10 televisions, and 80 televisions. The results are reproduced in Figure 2.50. These results confirm the measurements shown in Figure 2.49 that especially the higher harmonics become less when the number of rectifiers increases.

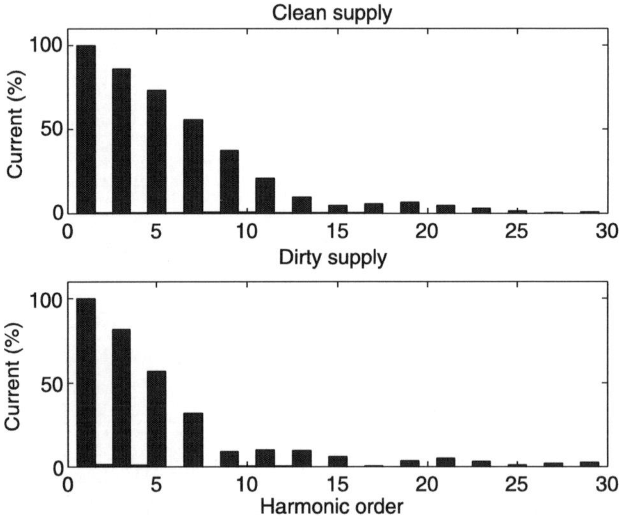


**Figure 2.49** Current spectrum of dc voltage source: clean supply and dirty supply.

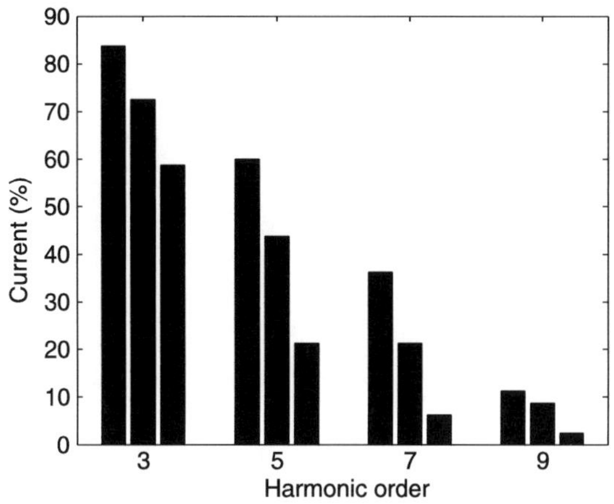


**Figure 2.50** Spectrum of current taken by (left to right) 1, 10, and 80 televisions.

Reference [191] presents the diversity in harmonic spectrum between computer workstations at different locations in the same distribution networks. The range of distortion was as follows:

- THD: 106 through 117%
- Third harmonic: 83 through 87%
- Fifth harmonic: 54 through 64%
- Seventh harmonic: 26 through 38%

Again the main variation is in the fifth and seventh harmonics. The third harmonic is rather independent of the location (thus of the background distortion).

The voltage distortion due to large numbers of dc voltage sources is treated in significant detail in [209, 210]. Analytical expressions are given for the current taken by a diode rectifier with a finite capacitor with a resistive load on the dc side. (Note that the expressions given in this chapter only hold for infinite capacitor sizes.) The expressions have been applied to a load consisting of typical desktop computers and typical single-phase adjustable-speed drives. Increase in nonlinear load shows a reduction in the relative harmonic distortion, especially for higher harmonic orders (five, seven, nine, etc.). The third-harmonic component shows only a slight reduction with increasing nonlinear load. The results from [210] are summarized in Figure 2.51. For harmonic orders up to 15, the resulting relative current distortion is given for increasing amounts of computer loads. Especially the fifth, seventh, and ninth harmonics show a significant attenuation with increasing electronic load.

We saw above how background distortion affects the current distortion of dc voltage sources. The main effect of the interaction between different sources through the background distortion is an overall reduction of the distortion. However, when even harmonic distortion is present in the background voltage

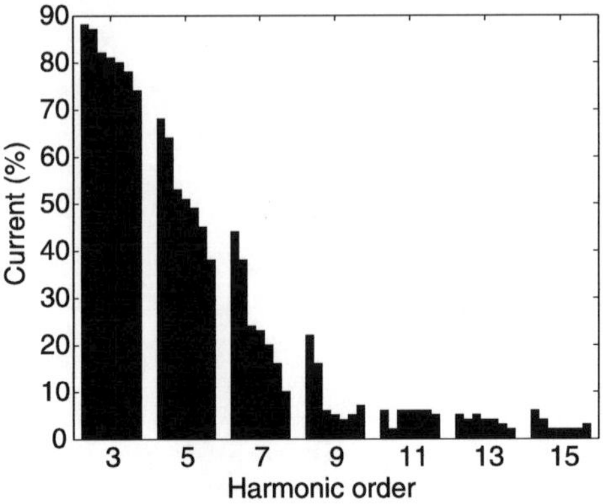


**Figure 2.51** Harmonic current distortion with increasing numbers of computers connected to the same supply.

this may result in an amplification of the overall distortion. Even harmonics in the voltage lead to a dc component and even harmonics in the current to a single-phase rectifier [231]. The explanation for this phenomenon is rather simple: Even harmonics affect the positive half wave in a different way than the negative half wave. The result is that the positive and negative half waves of the current are different, resulting in a dc component at the ac side of the rectifier. The problem is most severe when the peak voltage is different for the positive and negative half wave. As the current is determined by the difference between the ac and dc voltage around the peak of the ac voltage, a minor difference in ac peak voltage already leads to a significant change in input current.

A runaway effect may occur because the even harmonics in the current cause further even harmonics in the voltage. Also the dc component in the current leads to transformer saturation with more even harmonics. This additional increase in even-harmonic distortion closes the positive-feedback loop. Measurements presented in [271] show even harmonic distortion up to 1% for dc voltage sources, with the fourth and sixth harmonics dominating. It is however not clear how typical this measurement is.

The phenomenon was observed earlier with HVDC connections, where large second-harmonic distortion has been observed in some actual installations due to the amplification of the even-harmonic distortion [11]. The phenomenon also occurs for SVCs [212].

***2.5.4.6 Three-Phase dc Voltage Source*** Three-phase diode rectifiers with a capacitance on the dc side are commonly used to power ac adjustable-speed drives. The harmonic currents taken by a three-phase dc voltage source are similar to those taken by a single-phase dc voltage source. The main difference is that the supply current contains no triplen harmonics as long as the supply voltages are balanced. The harmonic components in the supply current under normal (balanced) conditions are called *characteristic harmonics*. In this case the harmonics are 5, 7, 11, 13, 17, 19, and so on. All other harmonics, when present, are called *noncharacteristic harmonics*.

The performance of a three-phase dc voltage source is best understood by considering the phase-to-phase voltages. Whenever one of the phase-to-phase voltages (in absolute value) exceeds the dc voltage, a current pulse is created. The equations describing the duration and shape of the current pulse are the same as for the single-phase dc source. Note that the source impedance should include the return path as well, being twice the impedance per phase. Note also that the ac voltage which drives the current is the phase-to-phase voltage. The result is that the dc voltage is about $\sqrt{3}$ times as high as for a single-phase dc voltage source. Expression (2.248) still holds but with $P$ the power per phase. Expression (2.249) holds for a three-phase dc voltage source but with another choice of base power voltage. The resulting spectrum of the currents through the diodes is thus the same as for a single-phase dc voltage source. The currents taken from the source are different in that the triplen harmonics are no longer present. The other harmonic components are not affected in magnitude but do experience a phase shift compared to the single-phase rectifier. This phase shift means that fifth- and seventh-harmonic components due to single- and three-phase rectifiers are in opposite phase [208].

The effect of voltage unbalance on three-phase dc voltage sources is the generation of so-called noncharacteristic harmonics. The main effect is the appearance of triplen harmonics. The current pulses through the diodes can be explained as due to three single-phase rectifiers with the same dc voltage. For a balanced voltage the pulses are identical, resulting in a cancellation of the third-harmonic components. More precisely, the third-harmonic component is the same in the three phases and thus of zero-sequence character only. Voltage unbalance causes the pulses through the diodes to no longer be identical. The third-harmonic component will no longer cancel in the line currents: It contains not only a zero-sequence component but also negative- and/or positive-sequence components.

The effect of background distortion on three-phase dc voltage sources is similar to the effect on their single-phase counterparts. A thorough study on the effect of second-harmonic distortion in the supply voltage is presented in [105]. It is shown that already a small amount of second-harmonic voltage will lead to a significant dc component which in turn will cause transformer saturation: 2% second-harmonic voltage gives up to 37% dc. The runaway effect that can theoretically occur is the same as described for HVDC links in [11]. It is also shown in [105] that the injected dc component becomes smaller when the dc voltage drops, and thus when the loading of the drive increases. Especially lightly loaded drives are prone to dc injection due to second-harmonic voltage. Only the positive-sequence component of the second-harmonic voltage causes a dc component; the negative-sequence component only gives (non-dc) even-harmonic currents: 2% second-harmonic voltage gives up to 48% second-harmonic current.

**2.5.4.7 Transformers** The transformer current in normal operation is dominated by the load current. The magnetizing current is less than 1% of the rated current of the transformer. The load current may obviously be distorted, but this cannot be attributed to the transformer. However, when the system voltage exceeds the rated voltage of the transformer, the magnetizing current may become significant. Some transmission operators have increased the normal operating voltage in their networks to allow for more power to be transmitted. This will however lead to an increase in (especially) third- and fifth-harmonic distortion. The increase in third-harmonic distortion is most noticeable, as this is normally small at transmission levels. An example of third-harmonic distortion at transmission level is shown in [202, Section 4.3]. A transformer with 30% overexcitation takes 0.18-pu third-harmonic current and 0.11-pu fifth-harmonic current (with the transformer rating as base) [10]. This is often a higher harmonic content than the load current.

***2.5.4.8 Synchronous Machines*** Although rotating machines are generally classified as linear load, their current is not fully sinusoidal due to the finite number of winding slots. Expressions for the resulting spectra are derived in [10, Section 4.3; 314]. Also the voltage generated by a synchronous generator is not fully sinusoidal due to the same phenomenon. The spectrum may contain third-harmonic components around 10% of nominal. As this is a zero-sequence harmonic, it will be blocked by the delta–star-connected generator transformer. With distributed generation, where generators are connected directly to medium- or low-voltage networks, the third-harmonic distortion could become a concern.

***2.5.4.9 Fluorescent Lighting*** Incandescent lamps consist of a thin metal wire which has a pure resistive behavior over a rather wide range of frequencies. Therefore the incandescent lamp is one of the most linear loads around. However, most other lamp types are nonlinear, with a wide range of current waveforms. Lamps with magnetic ballasts produce in general less distortion than those with electronic ballasts. The latter can be further divided into those with a waveform similar to that of a dc voltage source and those with a high-frequency ripple superimposed on a rather sinusoidal current waveform. Some measured spectra of different types of fluorescent lamps with different types of ballasts [315] are shown as circles in Figure 2.52. The current distortion varies significantly between the different lamps. Measurements from [122] are included in the same figure as plus signs. The low-distortion lamps were all of "compact fluorescent lamps with external ballast" whereas the high-distortion lamps were "self-ballasted electronic lamps."

An interesting conclusion from the measurements presented in [122] is that the total power factor is between 0.40 and 0.48 for all lamps. Lamps with a low distortion take a large capacitive fundamental current. A similar observation can be drawn from the measurements presented in [309]. The measurements presented in [109] show that the total power factor ranges between 0.4 and 0.55 for fluorescent lamps. Halogen lamps and incandescent lamps have total power factors of 0.95 and higher. The measurements from [122] (plusses) and [309] (squares) are summarized in Figure 2.53.

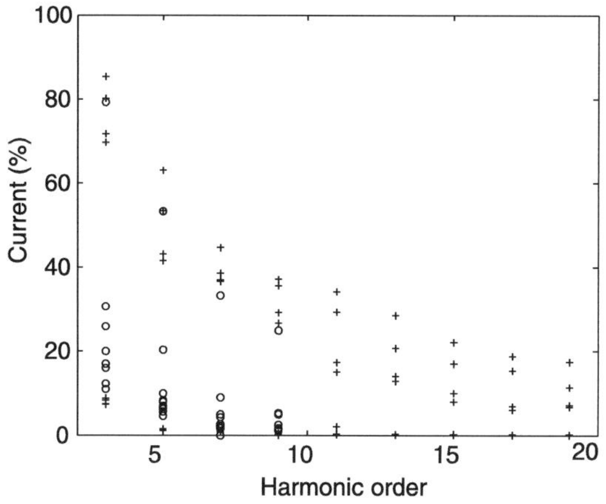


**Figure 2.52** Spectrum of current taken by different types of fluorescent lamps.

***2.5.4.10 Arc Furnaces*** Arc furnaces are a serious source of waveform distortion as well as flicker. During the operation of an arc furnace two different spectra can be observed. During the melting process the distortion is very high and highly fluctuating. This is also the period during which light flicker is most severe. The refining period is associated with a lower and more constant distortion [10]. Modeling of arc furnaces is a complicated issue that still has not been solved satisfactorily. The stochastic nature not only makes modeling very difficult but also hinders the comparison with actual measurements. Contributions to the development of arc-furnace models are [23, 54, 58, 69, 222].

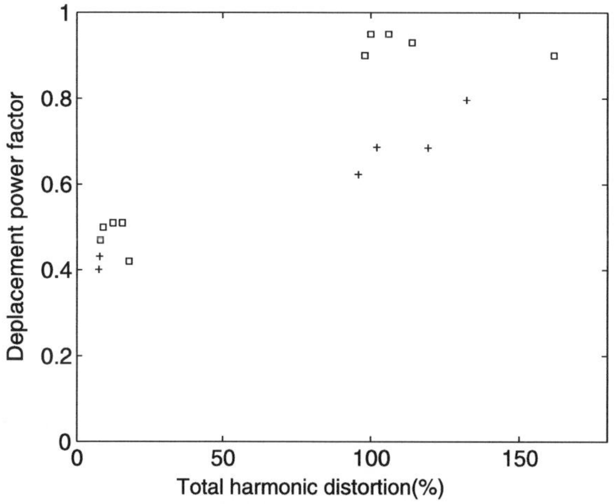


**Figure 2.53** Harmonic current distortion and fundamental reactive power taken by fluorescent lamps.

An example of the harmonic spectra of arc-furnace currents is given in [162]. The data are reproduced in graphical form in Figure 2.54. The distortion is not particularly high, not even during melting, but the even-harmonic distortion is much higher than for most other sources. During the refining stage the harmonic distortion is no longer of concern. A similar typical spectrum is given for the melting period in [63].

Next to the harmonic frequencies, the spectrum of an arc furnace contains interharmonic frequencies in the form of a continuous spectrum in between the harmonics, especially during the melting stage.

***2.5.4.11 Cycloconverters*** Cycloconverters change one frequency into another frequency without an intermediate dc link. The spectrum of the input current of a cycloconverter contains harmonics of both the input and the output frequencies. The same holds for HVDC links and for variable-frequency drives with a dc intermediate stage. However, in those latter cases, the waveform distortion is dominated by the (integer) harmonics of the power frequency. Significant interharmonics are rarely present.

Considering a six-pulse input as well as output stage (the most typical configuration) results in a spectrum for the input current with the following frequencies being present [10, 329]:

$$f = (6m \pm 1)f_{\text{in}} \pm 6nf_{\text{out}} \tag{2.252}$$

with $f_{\text{in}}$ the frequency of the input voltage (typically the fundamental power frequency), $f_{\text{out}}$ the output frequency, $n = 0, 1, 2 \ldots$, and $m = 1, 2, \ldots$.

The input current spectrum contains the characteristic harmonics of the power frequency (5th, 7th, 11th, 13th, etc.) with side bands due to the characteristic

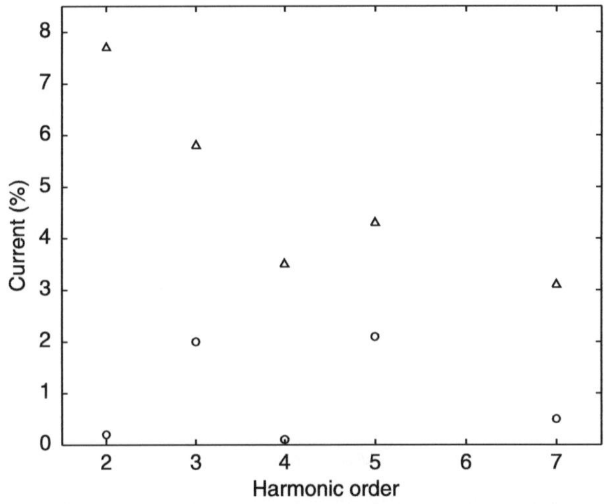


**Figure 2.54** Spectra of arc-furnace currents during melting (triangles) and refining (circles).

frequencies (6th, 12th, etc.) of the output frequency. An interesting measurement example of the spectrum of cycloconverter current is shown in [329]. Measurements are also presented in [60].

Consider, for example, a cycloconverter with 60-Hz input voltage feeding a large motor operating at a frequency of 1 Hz. The resulting spectrum of the input current contains the following frequencies: $(6m \pm 1) \times 60$ Hz $\pm\, 6n \times 1$ Hz: 282, 288, 294, 300, 306, 312, 318, 402, 408, 414, 420, 426, 432, 428, ....

Interharmonics are also generated when two grids with the same nominal frequencies are connected via an HVDC link, as the frequencies are rarely exactly the same.

For cycloconverters with single-phase output, expression (2.252) changes somewhat:

$$f = (6m \pm 1)f_{\text{in}} \pm 2nf_{\text{out}} \tag{2.253}$$

**2.5.4.12 *Integral Cycle Control*** Some equipment controls the average output power by switching between the on and off states with periods of several cycles of the power frequency. Examples are resistance welders and ovens. An example of the current taken by such a device is shown in Figure 2.55, where the on state lasts for $N$ cycles and the off state for $M - N$ cycles. The current has a period of $M \times T$, with $T = 1/f_0$ one cycle of the power frequency (50 or 60 Hz). The resulting spectrum thus consists of harmonics of $f_0/M$. The resulting spectrum is, according to [10],

$$I_h = \frac{2}{N\pi} \times \frac{\sin(Nh\pi)}{|1 - h^2|} \tag{2.254}$$

with $I_h$ the relative current for harmonic order $h$ of the fundamental power frequency ($I_1 = 1$). For integer values of $h$ the expression is zero; thus there are no harmonics present in the spectrum, only interharmonics. The spectrum for $M = 5$, $N = 1$ is shown in Figure 2.56. The resulting spectrum contains a significant amount of subharmonics and interharmonics around the fundamental power frequency. Increasing the duty cycle from 5 to 20 cycles but maintaining the same average current

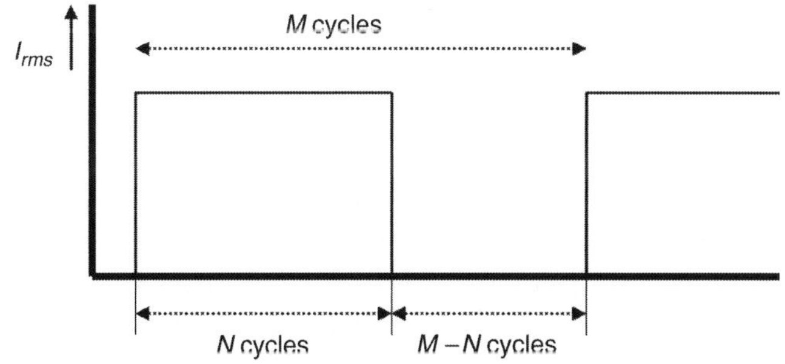


**Figure 2.55** Current taken by device with integral cycle control.

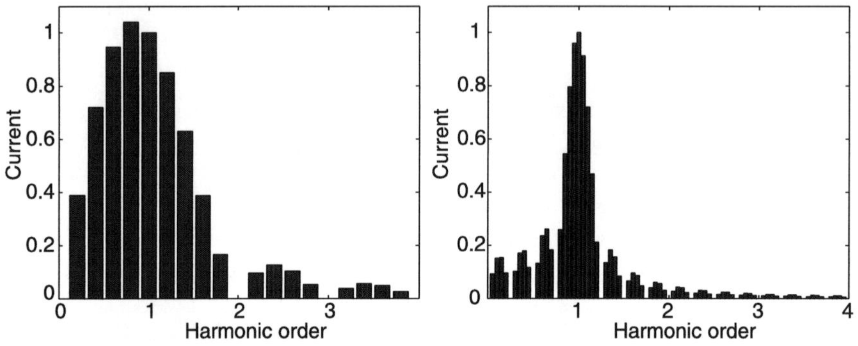


**Figure 2.56** Spectrum of current with integral cycle control. Left: 1 cycle on, 4 cycles off. Right: 4 cycles on, 16 cycles off.

($M = 20$, $N = 4$) results in a spectrum with less interharmonics. Observe that in both cases there are no (integer) harmonics present.

#### 2.5.4.13 Voltage Source Converters

Voltage source converters (VSCs) are known as a source of high-frequency harmonics. The switching frequency and multiples of the switching frequency (1 kHz and up) can be found back in the spectrum of the current. A systematic approach to determine the amplitude of these high-frequency harmonics is presented in [84], where it is also shown mathematically that pulse-width modulation leads to groups of frequency components around the switching frequency and its integer multiples.

Hysteresis control, used in smaller converters, leads to a noise like frequency spectrum around an *average switching frequency* determined by the design of the converter. If the switching frequency is close to a system resonance, it causes a large high-frequency ripple on the voltage. An increasing penetration of distributed generation with power-electronic interfaces will lead to an increasing level of high-frequency harmonics. The full consequences of this remain unclear. Standardized methods for the measurement and characterization of high-frequency current and voltage harmonics are not yet available.

The main application of voltage source converters remains on the load side of adjustable-speed drives and static UPSs. In both cases the high-frequency harmonics generated by the voltage source converters spread only over the local load. Several applications of voltage source converters in the general power system are becoming available: VSC-based HVDC and active-front end drives. Voltage source converters are also an important part of many types of distributed generation: for example, wind turbines with double-fed induction generators, microturbines, and photovoltaic converters. All these devices may inject high-frequency harmonics into the power system. The impact of this on the system is as yet unclear but certainly requires further study.

### 2.5.5 Harmonic Propagation and Resonance

In the previous section the various sources of harmonic distortion have been introduced. Each of these sources leads to a current that is nonsinusoidal even for a sinusoidal voltage at its terminals. These nonsinusoidal currents in turn cause nonsinusoidal voltages. The worst voltage distortion normally occurs with the terminals of the polluting equipment, and the distortion normally becomes less when moving away from its source. However, in some resonance cases the distortion may be more severe at a location some distance away from the source. In this section the basic methods are introduced for estimating the voltage distortion in the power system due to equipment taking nonsinusoidal current. We will only introduce some rather crude approximations. Several excellent books have been written about modeling of the power system and its load for calculating harmonic distortion [e.g., 10, 11] as well as some good overview papers [e.g., 19, 63, 247, 248, 327]. The reader is referred to these as well as to the wealth of literature about power system modeling.

***2.5.5.1 Current Source Model*** To determine the harmonic voltage distortion, the nonlinear load can be modeled as a harmonic current source and the system as an impedance. The value of the current $I_k$ is different for each harmonic number $k$. The current spectrum is determined for a nondistorted voltage wave or for a typical voltage waveform. The system impedance $Z_k$ is calculated for each harmonic frequency and the resulting harmonic voltage $V_k$ is calculated by using Ohm's law:

$$V_k = Z_k \times I_k \tag{2.255}$$

If higher accuracy is required, the distorted voltage can be used to calculate a new current spectrum which can in turn be used to calculate a new voltage distortion. But this requires more detailed load models, information on background distortion, detailed data on system inductances, capacitances and resistances, and detailed models on other load present in the system. In such a case it is best to use a power system analysis program. Most commercially available packages come with a module for harmonic calculations.

The current source model has its limitations in that the load current is in reality not independent of the voltage. It depends both on the fundamental component of the voltage and on the voltage wave shape. This is most obvious with dc voltage sources, where the crest factor of the voltage has a strong influence on the wave shape of the current and thus on the current spectrum. Also the absolute value of the voltage amplitude is known to affect the current spectrum. Some attempts have been made to linearize the voltage dependence around an operational point by representing the load by its Norton equivalent (current source in parallel with a source impedance). This neglects a lot of the dependencies and may not give any better results than the constant-current model.

When connecting a polluting load to a supply point, a detailed harmonic study can be done if the appropriate software is available and if there is time to obtain the data and do the simulation. If that is not the case, a first assessment should be made to decide if more accurate calculations are required.

For this first assessment the source impedance can be modeled as an $R$–$X$ series connection with a constant resistance or with a constant $X/R$ ratio. The inductance can be assumed independent of the frequency, so that the reactance increases linearly with frequency. If the capacitance plays a role, for example, during resonances, its value can again be assumed independent of the frequency. Thus the reactance decreases with frequency.

Two approximated expressions given by [313, pp. 146–147] are worth repeating here. For transformers the resistance is proportional with the frequency:

$$Z(h) = hR + jhX \tag{2.256}$$

The same relation was found by the authors from measurements on a 150/10-kV transformer [45]. However, it was not clear from those measurements if such a relation would be valid generally.

For generators the resistance increases with the square root of the frequency:

$$Z(h) = \sqrt{h}R + jhX \tag{2.257}$$

***2.5.5.2 Voltage Source Models*** The traditional way of representing a nonlinear load in a harmonic penetration study is as a harmonic current source. The underlying assumption is that the harmonic current spectrum is not too much affected by the system voltage (or by its fundamental or by its distortion). For the traditional sources of waveform distortion, large dc drives and HVDC links, this was a very acceptable model. The dc would be very constant and determined by the dc load. The effect of the ac voltage on the dc, and thus on the ac, would be small.

The modern sources of harmonic distortion—computers, televisions, lighting, and adjustable-speed drives—no longer fit under this category. The most severe harmonic polluters are the dc voltage sources. As we saw before, the harmonic spectrum is significantly affected by the supply voltage. This has led to a discussion of the appropriateness of the harmonic current source for calculating the harmonic voltage distortion in the system due to dc voltage sources. There have been suggestions that the harmonic voltage source would be a more appropriate model for the dc voltage source than the harmonic current source [235, 236]. (Watch out for the confusion in terminology: "current source" in "harmonic current source" is not the same as in "dc current source." The same holds for "voltage source".)

Let us consider the basic model of the diode rectifier with a large capacitor (the "dc voltage source") as seen from the ac side. This model is shown in Figure 2.57. The two switches each connect a dc voltage source to the ac supply. The resulting

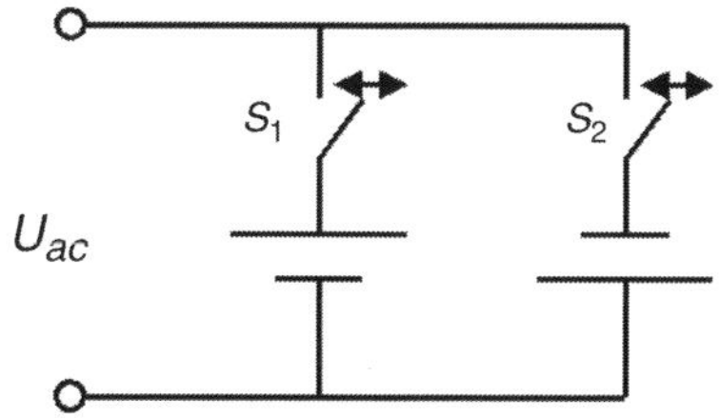


**Figure 2.57** Diode rectifier as voltage source.

current is due to the difference between the ac source voltage (typically distorted) and the dc voltage. The two dc voltage sources are of the same magnitude but opposite polarity. The switching instants can be found from the reasoning presented above when discussing the spectrum of the dc voltage source.

Thus during the conduction period the diode rectifier can be presented as a voltage source. However, this does not yet justify the use of harmonic voltage sources for harmonic penetration studies. The first argument against is that the dc voltage as well as the conduction period depend on the ac voltage. A more serious argument against is that the voltage outside the conduction period is not defined by the source. Outside the conduction period the current is zero, so that a current source model (with zero current) would be more appropriate. In the time domain modeling a voltage source model as in Figure 2.57 would be possible, but not in frequency-domain studies. As the voltage is not defined during the whole cycle, it is not possible to determine the spectrum of the voltage waveform, and thus it is not possible to determine the harmonic voltage sources needed for the penetration studies.

***2.5.5.3 Harmonic Resonances: Parallel Resonance*** The current source model for the distorting load can also be used to explain a phenomenon called *harmonic resonance*. Due to a combination of the source reactance and shunt capacitance at a certain location, the impedance seen by the current source becomes very large. The effect of this is a large voltage distortion, even for moderate current distortion.

Consider again a harmonic current source that injects a current into the system. Figure 2.58 shows a typical system, with $L$ the source impedance at the load bus and $C$ the capacitance connected to the load bus. Neglecting the resistance gives for the impedance seen by the harmonic current source, with $\omega = 2\pi f$,

$$Z(\omega) = \frac{j\omega L}{1 - \omega^2 LC} \tag{2.258}$$

This impedance becomes infinite at the resonance frequency:

$$f_{\text{res}} = \frac{1}{2\pi\sqrt{LC}} \tag{2.259}$$

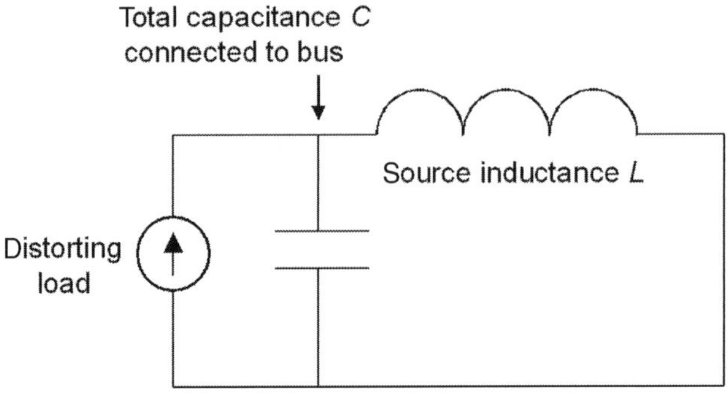


**Figure 2.58** Typical system structure leading to harmonic resonance.

Consider a system with a fault level $S_{\text{fault}}$ and a capacitor bank of size $Q_{\text{cap}}$. Resonance occurs in that case for harmonic number

$$n = \sqrt{\frac{S_{\text{fault}}}{Q_{\text{cap}}}} \tag{2.260}$$

The ideal current source will lead to an infinite harmonic voltage at the load bus and an infinite harmonic current through the capacitor and the inductor. The harmonic distortion will in practice be limited by two effects:

- The resistance present in the system will determine the impedance at the resonance frequency.
- The current source model is no longer valid for high-voltage distortion.

The resonance phenomenon is especially common with medium-voltage capacitor banks. With the commonly used ways of dimensioning these banks, resonance frequencies turn out typically between 250 and 500 Hz. Also long cables can lead to a resonance but normally at higher frequencies where the amount of current distortion is less and the amount of damping is higher.

At the resonance frequency, large currents flow through the source impedance and the capacitor bank, even though the load current is small. A heavily distorted current through, for example, a 132/11-kV transformer may indicate a resonance phenomenon at the low-voltage side, especially when the fifth or seventh harmonic is the dominant one.

To estimate the range of resonance frequencies that can be expected, consider a system with a source fault level $S$. The amount of apparent power connected to this source will be in the range $S/25 < S < S/12$. With a power factor between 0.7 and 0.9, the amount of reactive power taken will be $S/36 < Q < S/13$. If the power factor correction will compensate between 90 and 100% of this, the capacitor size will be $S/40 < Q_{\text{cap}} < S/13$. Resonance will occur for harmonic numbers between $\sqrt{13} < N_{\text{res}} < \sqrt{40}$, thus for $3.6 < N_{\text{res}} < 6.3$.

We see that especially the fifth harmonic is very susceptible to resonance, which is the major harmonic generated by three-phase rectifiers. A large six-pulse rectifier would thus very often require filters for the fifth harmonic, because even if the system study shows that no resonance will occur, changes in the system will change the resonance frequency. This effect is one of the reasons that the fifth-harmonic distortion in the voltage is often the dominant one.

The fifth harmonic can also along another route be found as the one most prone to resonance. Consider a capacitor bank used for voltage control, as in medium- and low-voltage-distribution systems. Connecting a bank with size $Q_{cap}$ to a source with short-circuit capacity $S_{fault}$ gives a voltage step equal to

$$\Delta V = \frac{Q_{cap}}{S_{fault}} \tag{2.261}$$

Combining this with (2.260) gives the following relation between the voltage step and the resonance frequency:

$$n_{res} = \sqrt{\frac{1}{\Delta V}} \tag{2.262}$$

Voltage steps less than 1 or 2% do not justify the connection of a capacitor bank. Voltage steps above 8 to 10% would lead to too large effects on the load when connecting or disconnecting the bank. Varying $\Delta V$ between 0.02 and 0.08 gives resonance frequencies between harmonic order 3.5 and harmonic order 7. Again the fifth harmonic is somewhere in the middle of this range.

Next to capacitor banks, underground cables are the main contribution to the capacitance in the system. The shunt capacitance of cables is around 0.35 μF/km. The reactive power produced by a capacitance $C$ at a (line) voltage $U$ is

$$Q = \omega C U^2 \tag{2.263}$$

For a 400-V network this results in 18 var/km, for 10 kV in 11 kvar/km, and for 130 kV in 1.9 Mvar/km. Knowing the total cable length $l_{cable}$ connected to a bus and the fault level $S_{fault}$, the harmonic order leading to resonance can be calculated from

$$n_{res} = \sqrt{\frac{S_{fault}}{l_{cable} Q}} \tag{2.264}$$

Consider, for example, a 10-kV bus with 180-MVA fault level. From this bus 14 underground cables originate with an average length of 2.4 km. The total cable length $l_{cable} = 14 \times 2.4 = 33.6$ km. Using $Q = 11$ kvar/km as found before results in

$$n_{\text{res}} = \sqrt{\frac{180\ \text{MVA}}{33.6 \times 11\ \text{kvar}}} = 22$$

Motor load affects the resonance frequency of the parallel resonance [162]. For harmonic studies motor load should be represented by its leakage reactance, which is about one-fifth to one-sixth of the nominal impedance of the motor. Thus even a relatively small motor load can already have a significant effect on the resonance frequency. What matters is the rated power of all motors connected to the system, not the actual loading in active power. The effect of motor load is automatically included when considering the fault level including the motor contribution in (2.260).

Consider the same system as before (10 kV, 180 MVA, etc.). The maximum motor load is estimated as 5 MVA. The resulting fault level is $180 + 6 \times 5 = 210$ MVA, resulting in parallel resonance for $n = \sqrt{210\ \text{MVA}/370\ \text{kvar}} = 23.8$. The effect of the motor load is that the resonance frequency shifts from harmonic order 22 to 24.

***2.5.5.4 Harmonic Resonance: Series Resonance*** Another case of harmonic resonance occurs when a significant amount of capacitance is present at a lower voltage level where there is a high background distortion at a higher level. The resonance between the transformer inductance and the capacitance may lead to high harmonic distortion at the secondary side of the transformer. This in turn may lead to capacitor failure. Modern electronic equipment often contains a capacitor over the ac terminals, which may lead to substantial level of harmonic voltage distortion, even if the equipment takes a fully sinusoidal current.

The network configuration leading to series resonance is shown in Figure 2.59 together with the equivalent circuit used to calculate the resulting voltage at the

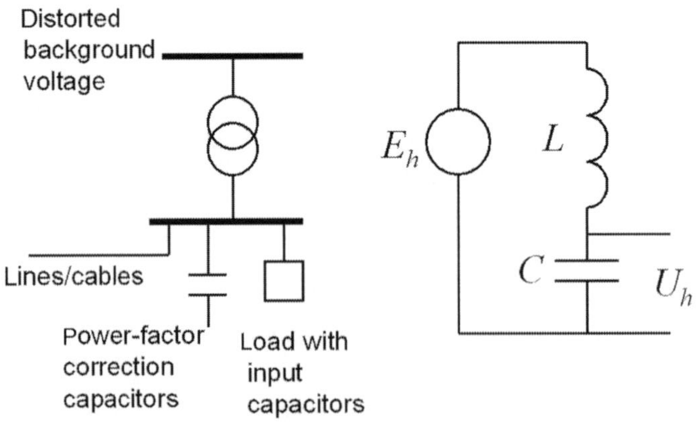


**Figure 2.59** Network configuration leading to series resonance (left) and equivalent circuit (right).

secondary side of the transformer. The voltage at the secondary side is obtained from the expression

$$U_h = \frac{1}{1 - h^2\omega^2 LC} E_h \tag{2.265}$$

with $C$ the total capacitance connected to the secondary side of the transformer, $L$ the transformer inductance, $h$ the harmonic order, and $\omega$ the (angular) fundamental power frequency. When $h^2\omega^2 LC \approx 1$ the secondary side voltage can be much higher than the primary side voltage.

Using $\omega C = Q_{\mathrm{cap}}/U^2$ and $\omega L = x_{\mathrm{tr}}(U^2/S_{\mathrm{tr}})$ results in the expression

$$U_h = \frac{1}{1 - h^2 x_{\mathrm{tr}}(Q_{\mathrm{cap}}/S_{\mathrm{tr}})} E_h \tag{2.266}$$

with $S_{\mathrm{tr}}$ the transformer rating, $x_{\mathrm{tr}}$ the (per-unit) transformer impedance, and $Q_{\mathrm{cap}}$ the capacitor size. Resonance occurs for

$$h = \sqrt{\frac{S_{\mathrm{tr}}}{x_{\mathrm{tr}} Q_{\mathrm{cap}}}} \tag{2.267}$$

Note that seen from the primary side of the transformer, the series connection has a low impedance. This will actually reduce the distortion on the primary side. In [202] an example is shown of a series resonance between the low-voltage capacitors and the combined impedance of the 130/10- and 10/0.4-kV transformers. The result is that the industrial installation cleans up the transmission voltage at the expense of an increased distortion at the low-voltage terminals.

In [47] a series resonance problem due to public lighting is discussed. The presence of public lighting led to heavy fifth-harmonic distortion, with fifth-harmonic levels up to 8% at 400 V and up to 4% at 22 kV. The fifth-harmonic voltage increased by 0.025 through 0.04 pu when switching on the lamps. The problem was due not to the current distortion injected by the lamps but to a harmonic series resonance between the distribution transformer and the power factor correction capacitors of the lamps. The current distortion of the lamps was very moderate with third-harmonic currents up to 12% and fifth-harmonic currents less than 4%.

A potential with the increased penetration of distributed generation is the occurrence of new resonances due to the increased amount of capacitance connected to the distribution grid. This capacitance may be involved in series or parallel resonances that cause amplification of harmonic distortion produced elsewhere. This problem occurs for voltage source converter-based interfaces and for induction machines but not for synchronous machines.

Various harmonic resonance issues are studied in [108] for a housing estate with a large number of photovoltaic inverters. The housing estate contains about 200

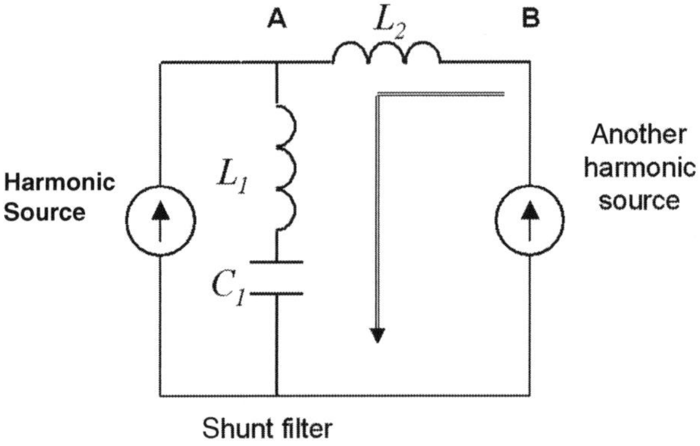


**Figure 2.60** Overloading of harmonic filter by remote source.

houses with photovoltaic installation supplied from an underground cable network. Measurements and simulation show that both parallel and series resonances occur, with resonance frequencies between 250 and 2000 Hz. The resulting high voltage distortion at the terminals of the inverters causes tripping of the inverter or high current distortion. The authors of [108] give the following values for the capacitance that should be considered for resonance studies: 0.6 to 6 μF per household, 0.5 to 10 μF per inverter. As the range is rather large, it will be difficult to determine resonance frequencies without measurements or more precise knowledge of the capacitance values.

Another resonance problem can occur with shunt passive filters. Consider a system as shown in Figure 2.60. A harmonic source at location A is equipped with a harmonic filter tuned to harmonic $k_1$. Seen from point B, the resonance frequency is

$$\omega_B = \frac{1}{\sqrt{(L_1 + L_2)C_1}} \tag{2.268}$$

If this corresponds to the frequency of a harmonic component $k_2$ produced by the load at B, the filter at A could become overloaded. Note that $k_2 < k_1$. Also note that this phenomenon will only occur if there is no filter for harmonic $k_2$ present at location A.

## 2.6 SUMMARY AND CONCLUSIONS

This section will summarize the material on power quality variations presented in this chapter and emphasize some of the conclusions. We will also give some references to further reading on these subjects.

### 2.6.1 Voltage Frequency Variations

Variations in voltage frequency are due to unbalance between generated and consumed electrical power. To limit power unbalance most large generators are equipped with power–frequency control. In large interconnected systems the power–frequency control leads to a stable frequency with deviations of more than 1% from the nominal frequency extremely unlikely. As the frequency variations are very small, their impact on end-user equipment is almost nonexistent. Therefore frequency variations are not widely covered in the power–frequency literature. Instead frequency variations and frequency control are part of the general power system literature. Further knowledge on the subject can be obtained from most books on power systems [e.g., 203, Chapter 8; 261, Chapter 3; 322, Chapter 4]

### 2.6.2 Voltage Magnitude Variations

Variations in the magnitude of the voltage are due to variations in active and reactive power flow. Exact and accurate expressions have been derived for the voltage drop $\Delta U$ over the source impedance $R + jX$ due to a load $P + jQ$. Expressed in per-unit, the following approximation has been derived:

$$\Delta U = RP + XQ \tag{2.269}$$

When the load is distributed uniformly over a line or cable, the voltage drop is only half that for a concentrated load.

Voltage control (i.e., mitigating voltage variations) is an important part of power system design for transmission as well as for distribution systems. Voltage control at the transmission level is important to guarantee the security of the system. Voltage control at distribution systems is a power quality issue. It mainly affects the performance of end-user equipment. Important devices for maintaining the distribution voltage within limits are transformer tap changers and capacitor banks. However, their switching leads to voltage steps up to a few percent that may in turn become a power quality issue.

Voltage variations reduce the lifetime of equipment, but the effect is small with the exception of incandescent lamps. The only voltage variation issue that is regularly discussed in connection with power quality is the appearance of overvoltages due to distributed generation. This may require a new approach to voltage control in distribution systems. The issue of voltage variations is also an important part of the literature on automatic distribution system design.

Further information on voltage variations can be found in the literature on distribution system design [e.g., 53; 160, Chapter 3; 193, Chapter 13; 261, Chapter 6; 322, Chapter 5; 324] and in the literature on distributed generation [e.g., Chapter 9; 178].

Voltage control for transmission systems is discussed in the literature on transmission system design. The issue is often referred to as "voltage stability." For further reading, try [203, Chapter 7; 289].

### 2.6.3 Voltage Unbalance

Unbalance in voltage and current in a three-phase system can be analyzed and quantified with the method of symmetrical components. Three complex voltages or currents are decomposed in a balanced component (positive sequence), an unbalanced component (negative sequence), and a common-mode component (zero sequence). Transfer of energy from generation to load takes place in the positive-sequence component only.

Voltage unbalance is due to unbalanced current and to unbalances in the system. The latter contribution is mainly a concern for large power transports over long distances. Unbalanced loads are mainly found in low-voltage networks, with the exception of traction and arc furnaces, which are connected to higher voltage levels. The negative-sequence unbalance due to a single-phase load is equal to the ratio between the apparent power of the load and the fault level of the supply. Current unbalance in low-voltage networks is due to the load diversity. Even when a feeder supplies identical loads equally spread over the phases, a negative-sequence current will result. More detailed stochastic models are needed to quantify the impact of load diversity on the unbalance. The main interest should be in calculating the probability that the negative-sequence unbalance exceeds a certain level.

Voltage unbalance leads to overcurrents for rotating machines and three-phase rectifiers. This could become a concern for the connection of induction and synchronous generators to low- and medium-voltage networks. Further work may be needed on quantifying the temperature rise due to voltage unbalance in distributed generation units.

The literature on voltage unbalance is small. A recent overview paper has been written by von Joanne [306]. Of the books on power quality, only the one by Schlabbach [271, Chapter 4] spends a separate chapter on unbalance. The other books discuss it together with other disturbances or not at all. One of the UIE guides on power quality is devoted completely to unbalance [302].

### 2.6.4 Voltage Fluctuations and Flicker

Voltage fluctuations are due to fluctuations in load current. Especially fluctuations in the frequency range 1 to 20 Hz are of importance as they cause light flicker. Examples of loads that lead to voltage fluctuations are arc furnaces, copy machines, and refrigerators. Renewable sources of energy that show a fast fluctuation in output power (e.g., wind power and solar power) are also a potential source of voltage fluctuations.

The IEC flickermeter standard gives a detailed, perception-based method for quantifying the severity of voltage fluctuations. The method results in flicker severity (instantaneous and average over 10-min and 2-h intervals) for 60-W, 120-V and 230-V incandescent lamps. The method will most likely become part of an IEEE standard in the near future as well. Nonincandescent lamps are generally less susceptible to voltage fluctuations. However, some lamps, show more light flicker than incandescent lamps. There are also strong indications that certain lamp types show light intensity fluctuations due to interharmonics. Further research is needed toward the development of models for nonincandescent lamps that can be included in the flickermeter standard.

Another issue for further research is the propagation of voltage fluctuations through the system and the addition of different flicker sources. Further development is also needed for methods to locate individual flicker sources.

Most books on power quality discuss flicker in at least some detail. A good overview is given in [271, Chapter 3]. An excellent explanation of the development and interpretation of the flickermeter standard is found in the UIE guide on flicker [303]. An overview on various aspects of flicker can also be found in [7, Section 7.3].

### 2.6.5 Waveform Distortion

The impact of waveform distortion is excess heating of end-user equipment and equipment in the system. The most dangerous consequence is the overheating of the neutral wire for low-voltage feeders with high penetration of single-phase distorting loads. Other victims are distribution transformers, capacitor banks, and capacitors in end-user equipment. Equipment malfunction may occur with higher levels of harmonic distortion.

Three types of waveform distortion can be distinguished: harmonic distortion (the waveform is periodic with the power system frequency), interharmonic distortion (the waveform is periodic with a longer period), and nonperiodic distortion or noise. Harmonic distortion is characterized by the harmonic spectrum of the voltage or current signal obtained by applying the Fourier transform. In three-phase systems the symmetrical-component transform can be combined with the Fourier transform.

Different sources of waveform distortion are discussed. For balanced load a distinction is made between dc sources (where the dc is assumed constant) and dc voltage sources (where the dc voltage is assumed constant). Among single-phase loads the latter are most common. A simple mathematical model is introduced for the dc voltage source.

A subject that remains underexposed in the literature is the origin and quantification of interharmonics and of nonperiodic distortion. A subject that has received significant attention, but without coming to a conclusion, is the development of methods for locating harmonic sources. More work is also needed on the development of aggregate load models and on the origin and impact of distortion due to active converters.

Waveform distortion has been a popular subject in the power quality literature. For a while the terms *power quality* and *harmonics* were regularly used as synonyms. There are hundreds of good papers written on this subject at conferences and in different journals. The classical book on power system harmonics (by Jos Arrillaga) was recently published in a revised edition [12]. Other books with a substantial contents on harmonics are [2; 11; 9; 99, Chapters 5 and 6; 141; 232]. Also the IEEE harmonics standard (519) contains a good overview text on waveform distortion. Interesting overview papers on waveform distortion are, among many others, [312, 19, 63, 247, 248, 327].

# CHAPTER 3

# PROCESSING OF STATIONARY SIGNALS

This chapter discusses a number of signal-processing methods for the analysis of the variations introduced in Chapter 2. What all those methods have in common is that they result in one or more parameters (features or characteristics) that quantify the deviation of the voltage or current waveform from the ideal. As shown in Section 1.2, power quality can be defined in a number of ways. One way is to define it as a set of parameters that can be obtained in a unique way. Such a definition set may be selected from the collection of parameters to be discussed in this chapter.

In the first part of the chapter the standard methods will be introduced. Regular reference will be made to the relevant IEC standards (IEC 61000-4-7 and IEC 61000-4-30). In modern power quality documents the reproducibility of characterization methods is very much emphasized and is therefore also an important part of this chapter. The second part of the chapter goes into further detail of a number of advanced methods to extract frequency components from a voltage or current spectrum.

## 3.1 OVERVIEW OF METHODS

In Section 1.2.3 we introduced the two types of power quality disturbances: variations and events. Variations are small deviations from the normal or desired voltage or current sine wave that can be measured at predefined instances (or more precisely over predefined windows of time), whereas events are larger disturbances that trigger a recording or further processing. In Chapter 2 we discussed the origin of the most common variations: (voltage) frequency variations: voltage

*Signal Processing of Power Quality Disturbances.* By Math H. J. Bollen and Irene Yu-Hua Gu

(magnitude) variations; unbalance; voltage fluctuations (flicker) and (waveform) distortion. For each of these variations there is a characteristic, or a set of characteristics, of the voltage or current that very much defines the disturbance.

Instead of the term *power quality variations* as used in this book, other authors use terms such as *normal operation*, *steady-state operation*, and *stationary signals* with a similar but not identical meaning. The latter term is commonly used in signal processing: A signal is stationary when it is *statistical time invariant* (or the statistics of the signal are independent of time). For example, the mean and the variance of a stationary signal do not change with time. Another way is to examine the underlying model (or system) associated with the signal. If a signal is stationary, then the underlying model of the signal is time invariant. Stationarity does not have to imply that the signal is periodic; small changes in the signal may occur as long as they are statistically the same at any instant of time. Two examples of stationary signals are shown in Figure 3.1. The upper curve is measured near the terminals of a wind turbine.

In the remainder of this chapter we will refer to this signal as the *normal case*. The lower curve is measured at the terminals of a large arc furnace; this signal will be referred to as the *arc-furnace case*. Note that the voltage is not of constant magnitude but still stationary. These two signals will be used in the forthcoming sections and in Chapter 5 to illustrate some of the methods for characterizing power quality variations. In Figure 3.1 only the voltage in one phase is shown for a 500-ms window. In later examples we will use the voltages and currents in the three phases over a time window up to 1 min.

Contrary to the stationarity, if a signal is statistical time varying, then it is a nonstationary signal. Two examples of nonstationary signals are shown in Figure 3.2.

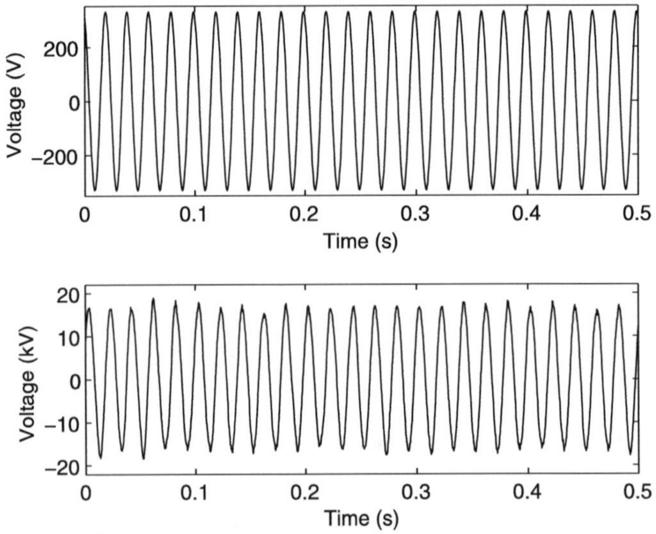


**Figure 3.1** Two examples of stationary signals: normal case (top) and arc-furnace case (bottom).

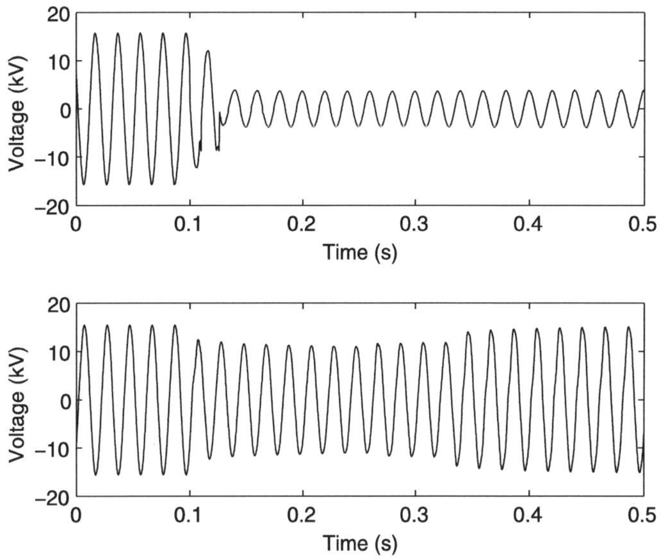


**Figure 3.2** Two examples of nonstationary signals.

In both cases the magnitude of the signals shows a sudden drop (implying that the mean value of the signal is a function of time). When considering the upper curve in Figure 3.2, we see that the signal can be considered as stationary before $t = 0.1$ and again after $t = 0.15$. The signal is however nonstationary when considered over the whole 500-ms window. In Chapters 4 and 8 we will discuss methods in which the signal is considered stationary over a short window (typically one cycle). For the voltages in the two other phases and a further discussion of these signals, refer to Figures 6.12 and 6.15. Many more examples of nonstationary signals (or "events") will be shown in Chapter 6 and further.

Note that it can be difficult in some occasions to judge whether a signal is stationary or nonstationary. To mathematically prove the stationarity requires the knowledge of the probability density function of the signal and is therefore not a straightforward task. Just as with the distinction between events and variations some kind of criterion is needed corresponding to our triggering criterion. Stationarity (or a more broad type, wide-sense stationarity, where a signal is statistically invariant to the time difference) of the signal is however a property that is implicitly assumed with all the signal-processing tools discussed in this chapter. Under this assumption the statistical properties of a signal are the same over any window. For example, a spectrum obtained over the period 0 to 100 ms in Figure 3.1 is expected to give the same results as the spectrum over the period 200 to 300 ms. The other way around is also true: The spectrum obtained over the whole 500-ms window is representative for any shorter window within the 500-ms window. Note again that this does not imply that each short window will result in exactly the same spectrum. The signal properties include the cycle-by-cycle variations, which are rather large for the arc-furnace case.

We will see in Chapter 5 that in many cases 3-s or 10-min values are considered to quantify a variation. In all methods, to obtain those values, it is implicitly assumed that the voltage or current signal is stationary. The 3-s or 10-min value is thus representative for the whole window. The so-called flagging concept (see Section 5.2.5) is introduced to detect nonstationarity (or, using power quality terminology, to detect the presence of an event during the window).

The deviation of the voltage or current signal from the ideal sine wave is characterized through a number of parameters (or *features*):

*Magnitude* The most important characteristic of the voltage as experienced by end-user equipment is its magnitude. As we will see in Section 3.6.1, the magnitude of the voltage can be estimated in a number of ways, but the rms value is by far the most commonly used method. The voltage magnitude is a way of quantifying voltage variations as discussed in Section 2.2.

*Frequency* Methods for estimating the frequency of the voltage are discussed in Section 3.2.1. The most commonly used method involves counting of zero crossings. The voltage frequency quantifies frequency variations as discussed in Section 2.1.

*Distortion* Distortion is normally treated as a multidimensional disturbance. The waveform is split into spectral components and the magnitude (and sometimes the frequency) of each component is used as a characteristic. Both commonly used and more advanced methods for estimating the spectrum of a voltage or current signal will be discussed in Sections 3.2.3, 3.4, 3.5, and 3.6. The origin of distortion of current and voltage signals has been discussed in Section 2.5.

*Unbalance* Three-phase unbalance is an issue that only concerns three-phase systems. The unbalance concerns the difference between the voltages and/or currents in the three phases. Characteristics for quantifying unbalance will be discussed in Section 3.2.4. The origin of the phenomenon has been discussed in Section 2.3.

*Flicker Severity* The flicker severity is a way of quantifying fast changes in the voltage magnitude at time scales of 1 s or shorter. Contrary to the first four parameters, the flicker severity cannot be obtained from one measurement window. Instead it requires information from a number of windows. The flicker severity has been introduced in Section 2.4 and will not be further discussed in this chapter.

*Very Short Variations* The term *very short variations* will be introduced in Section 5.2.4 to quantify changes in the voltage magnitude at a time scale between 3 s and 10 min. This characteristic also requires information from more than one measurement window. It will not be discussed in this chapter.

In the remainder of this chapter a number of signal-processing techniques will be introduced, ranging from very basic methods (e.g., the rms value) to very advanced methods (e.g., the Kalman filter). An important distinction is between so-called model-based methods and non-model-based methods.

Non-model-based methods (or nonparametric methods) simply decompose the signal into components or transform the signal into a different domain where the signal characteristics are easy to be extracted. These methods do not require any pre-knowledge about the signal and will always result in a value, even if the value has no physical meaning. Calculating the rms value to estimate the magnitude of the voltage or current is a non-model-based method. Even though the method is based on the assumption that the distortion of the signal is small, it will give a result in all cases. As another example, the Fourier series or discrete Fourier transform maps a signal in the time domain into the frequency domain. One of the main disadvantages of these methods is a relatively low frequency resolution, which is dependent on the length of the signal being processed.

Model-based methods (or parametric methods) form another important group of signal-processing methods for power system data analysis. Depending on the prior knowledge, one may assume that the signals are generated from certain models, for example, a sinusoid (harmonic) model or an autoregressive model. If the model is correctly chosen, the methods can achieve very high frequency resolution as compared with those in non-model-based methods. However, if an incorrect model is selected, it may lead to misleading results and very poor performance. Often there are some unknown parameters in the model that have to be tuned according to the given signal. One interesting model for the steady-state distortion in a power system is the *harmonic model*. Many different methods can be used for harmonic modeling, for example, the MUSIC method, ESPRIT method, and Kalman filters. Several model-based methods will be discussed in further detail in Section 3.5.

## 3.2 PARAMETERS THAT CHARACTERIZE VARIATIONS

Parameters are features or attributes of a voltage or current waveform that describe or quantify a certain disturbance. When using the word *parameter* here, we refer to the result of a calculation that is defined in a very accurate way such that two independent observers will obtain the same result. One of the communication problems within power quality remains that different persons use different definitions for the same parameter. Hypothetical examples of parameters are as follows:

- The rms voltage obtained over a one-cycle window starting at an upward zero crossing
- The third-harmonic component obtained by applying the DFT algorithm to a rectangular window with 200-ms duration
- The frequency obtained from the number of zero crossing of the voltage during a 3-s period

Several more examples will be discussed in the remainder of this section. These parameters may next be combined into so-called power quality indices. An example of an index is the total harmonic distortion, or THD. Note that some indices contain just one parameter: for example, the rms voltage as an index for voltage variations.

These indices will be used for a statistical description of the supply performance, to be discussed in Chapter 5.

In the remainder of this chapter and in Chapter 5 we will regularly refer to the IEC power quality measurement standard IEC 61000-4-30. This standard defines the methods to be used when quantifying a number of power quality variations. The standard distinguishes between two classes of instruments: instruments with class A performance and instruments with class B performance. Class A instruments are used for precise measurements, class B for less precise measurements. The standard document gives a number of examples:

- Of precise measurements
  Contractual applications
  Verifying compliance with standards
  Resolving disputes
- Of less precise measurements
  Statistical surveys
  Troubleshooting applications

The standard was officially published in March 2003 and there is not much published yet on the interpretation and use of this document. But already within the power quality community compliance with IEC 61000-4-30 has come to be interpreted as compliance for class A performance. But a monitoring instrument that fulfills the (much less severe) class B requirements is equally compliant with the standard, although for several disturbances the requirement for class B compliance is only that the manufacturer describe the method used to obtain the parameter.

In this book we will mainly discuss the class A requirements and refer to them most of the time simply as the IEC 61000-4-30 requirements. The class B requirements will only be discussed in some cases.

### 3.2.1 Voltage Frequency Variations

***3.2.1.1 What Is Frequency?*** With (voltage) frequency variations the parameter to be estimated is the frequency of the voltage or the frequency of the system. There is a fundamental difference between these two frequencies even though in practice their values are very close. The *frequency of the system* is a measure of the speed with which the electrical machines rotate. In a synchronously interconnected system in steady state, all electrical machines rotate with the exactly same speed. But a system is never completely in steady state and in practice there are small differences in speed between the different machines. The system frequency would have to be defined as the weighted average of the frequencies for the individual machines. Such a definition is used in simulation studies [e.g., 261, Chapter 8], but it is not a very practical measurement definition.

The *frequency of the voltage* is the repetition rate of the voltage waveform at a specific location. Most of the time the frequency of the voltage is very close to the frequency of the system. In other words, the frequency can be assumed to be the same anywhere in the system. Only during system instabilities could the frequency vary notably between different locations. An example will be given below.

In Figure 3.3 a comparison is made between the frequency variation as measured at two different locations during a large disturbance in the transmission grid. The solid curve contains 100-ms measurements performed by the transmission operator. The dotted and dashed curves are 1-min average, maximum and minimum values obtained from a 230-V wall outlet. The correspondence between the two measurements is very good, even though they were measured at two different locations during a large disturbance in the system.

#### 3.2.1.2 Standard Estimation Methods

There are a number of different methods to estimate the frequency:

- Counting of zero crossings
- Phase-locked loops
- For three-phase systems, $\alpha\beta$-transform and $dq$-transform methods
- Advanced signal-processing methods

The most commonly used method is simply counting of zero crossings. The IEC power quality measurement standard IEC 61000-4-30 [158, page 25] defines the frequency as follows: "the ratio of the number of integral cycles counted during the 10-s time clock interval, divided by the cumulative duration of the integer cycles." This definition clearly defines not only which feature to use but also very exactly how it should be measured.

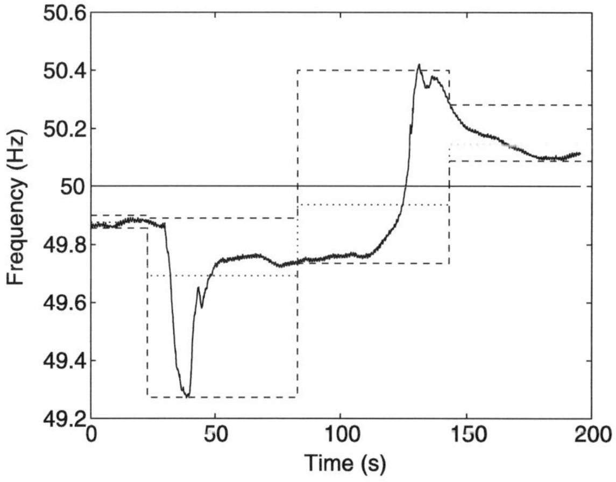


**Figure 3.3** Frequency variation measured during a large disturbance.

For measurement of the *power frequency*, IEC 61000-4-30 prescribes the following method (for class A performance):

- The frequency should be obtained every 10 s over a 10-s interval.
- Harmonics and interharmonics should be attenuated.
- Each measurement interval should begin on an absolute 10-s time instant. The error in time instant is limited to one cycle of the power system frequency (20 ms in a 50-Hz system, 16.7 ms in a 60-Hz system).
- The error in the frequency estimation is limited to 10 mHz.

For a frequency of exactly 50 Hz there are 500 cycles during a 10-s interval. If the frequency drops to 49.99 Hz, there are only 499 integral cycles during a 10-s interval, with a total duration of 9.982 s. (500 cycles would last 10.002 s, which would exceed the length of the interval). The frequency would then be obtained from the ratio

$$f = \frac{499}{9.982\ \text{s}} = 49.99\ \text{Hz}$$

For measurements involving more than one channel (e.g., three-phase measurements) a reference channel should be designated. The frequency should be estimated from the sampled voltage waveform obtained in the reference channel.

The estimation of the voltage frequency is in most monitors based on the measurement of the time elapsed between a known number of zero crossings. As time measurement can be done very accurately, frequency estimation reaches a very high accuracy for stationary signals. For nonstationary signals the frequency estimation sometimes shows large errors. This is mainly due to the phase–frequency dilemma. It is not possible to consider changes in frequency and phase independent from each other. In telecommunication, phase modulation and frequency modulation are the same. An assumption has to be made to estimate frequency and/or phase angle. For estimating frequency variations, the voltage phase angle is assumed constant. When analyzing power quality events (e.g., estimating the phase-angle jump with voltage dips), the frequency is assumed constant (see Section 8.2). The result is that a voltage dip or another event causing a change in phase angle shows up as a large frequency variation. The flagging concept, introduced in Section 5.2.5, has been introduced to remove such frequency values from the statistics. The other way around, a frequency variation gives an apparent drift in phase angle with voltage dips.

#### 3.2.1.3 Use of *dq*-Transform in Three-Phase System

With most power quality measurements, the three phase or line voltages are available for processing. Despite this, the frequency is typically only obtained from one channel, the so-called reference channel. It is however possible to consider all three phases in the calculation of the frequency by using the *dq*-transform as introduced in

Section 2.3.4. In (2.83) and (2.86) the *dq*-transform is defined as follows:

$$v_{dq}(t) = e^{-j2\pi f_0 t} \tfrac{1}{3}\sqrt{2}(v_a(t)\mathbf{e}_a + v_b(t)\mathbf{e}_b + v_c(t)\mathbf{e}_c) \tag{3.1}$$

where $\mathbf{e}_a = 1, \mathbf{e}_b = a = -\frac{1}{2} + \frac{1}{2}j\sqrt{3}$, and $\mathbf{e}_c = a^2 = -\frac{1}{2} - \frac{1}{2}j\sqrt{3}$.

For the method presented in Section 2.3.4 it was stated that the power system frequency, $\omega_0/2\pi$ in (3.1) needs to be known exactly. If an incorrect value of the frequency is used, this results in a slow rotation of the *dq*-voltage in the complex plane.

This phenomenon can be used to determine the frequency as well. The value of $\omega_0$ which gives a stable *dq*-voltage is the actual power system frequency. This method can be used in a software PLL [17]. The method can also be used to obtain the *differential phase* with a synchronous system operating exactly at the nominal frequency. Let the frequency be equal to

$$f_0 = f_{\text{nom}} + \Delta f \tag{3.2}$$

When estimating the frequency, instead of using the exact frequency to determine the *dq*-voltage, we use the nominal frequency:

$$v_{dq}(t) = v_{\alpha\beta}(t)e^{-j2\pi f_{\text{nom}} t} \tag{3.3}$$

Instead of (2.87) we obtain the following relation between the *dq*-voltage and the symmetrical-component voltages:

$$v_{dq}(t) = \underline{U}^{+} e^{-j2\pi\Delta f t} + (\underline{U}^{-})^{*} e^{-j2\pi(2f_{\text{nom}}t + \Delta f)} \tag{3.4}$$

The effect on the second (negative-sequence) term is small, as the frequency deviation is rarely more than 1%. Furthermore the magnitude of the negative-sequence term is much smaller than the magnitude of the positive-sequence term. The effect on the first (positive-sequence) term is a slow rotation in phase angle. The rate of chance of the phase angle is a measure of the relative frequency compared to the nominal frequency:

$$\frac{d \arg\{v_{dq}\}}{dt} = 2\pi\,\Delta f \tag{3.5}$$

This expression cannot be used immediately. The unbalance and distortion that are always present will cause small oscillations in the *dq*-voltage. These have to be removed before differentiating the *dq*-voltage. Differentiation should only be applied to a signal after removal of all unwanted high-frequency components.

This method has been applied to a 1-min recording of the three phase voltages. The phase of the *dq*-voltage is shown as the top-left curve in Figure 3.4: the left-hand picture shows the nonfiltered phase. A fourth-order Butterworth filter with a cutoff frequency of 10 Hz has been used to remove the oscillations.

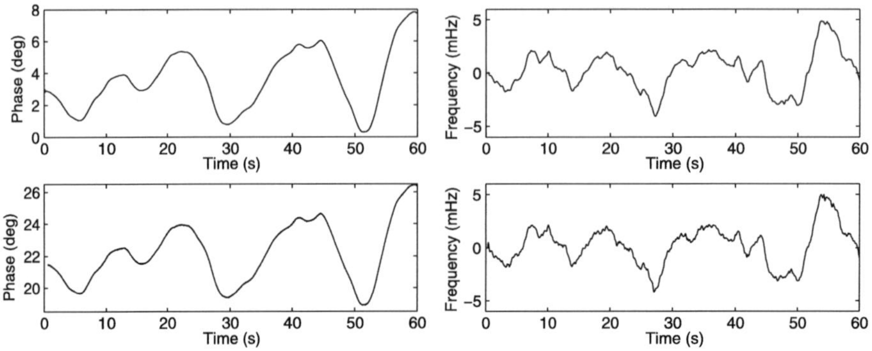


**Figure 3.4** Phase (left) and frequency (right) obtained from *dq*-voltage (top) and from *dq*-current (bottom).

The frequency is calculated from the filtered phase by applying (3.5) to the output of the low-pass filter. The result is shown as the top-right curve in Figure 3.4. An additional fourth-order Butterworth filter (with a cutoff frequency of 2.5 Hz) has been applied to the result to remove further oscillations. The figure shows that the frequency oscillates with a period of a few seconds. Note the huge difference (more than 100 times) in vertical scale compared with Figure 3.3. In Figure 3.4 frequency oscillations during normal operation are shown, whereas Figure 3.3 shows a frequency swing due to a large disturbance in the system.

For comparison also the *current frequency* has been obtained by applying (3.5) to the *dq*-current. The result is shown in the two lower curves in Figure 3.4. The overall pattern of the current frequency is the same as that of the voltage frequency. This indicates that the underlying phase-angle variations are due to frequency oscillations in the system. At a shorter time scale the voltage and current do deviate however. This is better visible in Figure 3.5 where a 10-s interval is shown in more detail. The current frequency oscillates around the voltage frequency with a period of about 1 s.

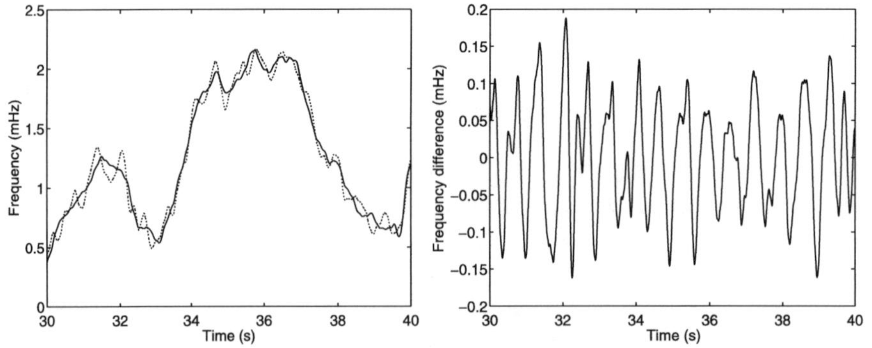


**Figure 3.5** Frequency as obtained from voltage (left, solid line) and current (left, dashed line); difference between voltage frequency and current frequency (right).

### 3.2.2 Voltage Magnitude Variations

***3.2.2.1 What Is Voltage Magnitude?*** The magnitude of the voltage is different for each location in the system. There is no longer any confusion between the local value and the system value, as there is with frequency variations. However, each location is different now, which in principle requires a separate measurement for every location in the system.

When considering voltage magnitude variations, a balanced three-phase model is assumed for the three phase voltages:

$$
\begin{aligned}
v_a(t) &= \sqrt{2}V\cos(\omega_0 t + \phi) \\
v_b(t) &= \sqrt{2}V\cos(\omega_0 t + \phi - 120^\circ) \\
v_c(t) &= \sqrt{2}V\cos(\omega_0 t + \phi + 120^\circ)
\end{aligned}
\tag{3.6}
$$

The challenge for a signal-processing method is to estimate or extract the value of the voltage magnitude $V$. When the voltage is completely balanced and nondistorted, most methods give the same result. Unbalance and distortion, which are always present, cause the different methods to give different results, as will be discussed below.

We will distinguish between single-phase measurements and three-phase measurements. With single-phase measurements the voltage magnitude is extracted by using the waveform of only one phase. This may be a phase voltage or a line voltage. Below we will refer only to phase voltages, but the calculations for line voltages are exactly the same. Single-phase measurements may be performed because only one voltage is available or because the voltage magnitude needs to be estimated for each phase separately. But before the general discussion on single- and three-phase measurements we will discuss the standard method according to IEC 61000-4-30 [158].

***3.2.2.2 Standard Estimation Methods: IEC 61000-4-30*** The IEC power quality measurement standard IEC 61000-4-30 prescribes the use of the rms voltage over a certain period for all instruments. For instruments with class B performance, the period is to be specified by the manufacturer. For instruments with class A performance, a 10-cycle interval should be used in 50-Hz systems and a 12-cycle interval in 60-Hz systems. In both cases the length of the interval is about 200 ms. The length of the interval is determined not by the clock time but by the fundamental frequency. The length of the interval should be exactly 10 or 12 cycles of the fundamental frequency. However, the standard document does not state how the length of one cycle has to be determined. A possible method would be to use the last 10-s frequency value obtained in accordance with the standard. This is in fact suggested in the standard document for the rms calculation with voltage dips. We will come back to this in Chapter 7. Another method would be to use a PLL to synchronize the sampling frequency with the power system frequency.

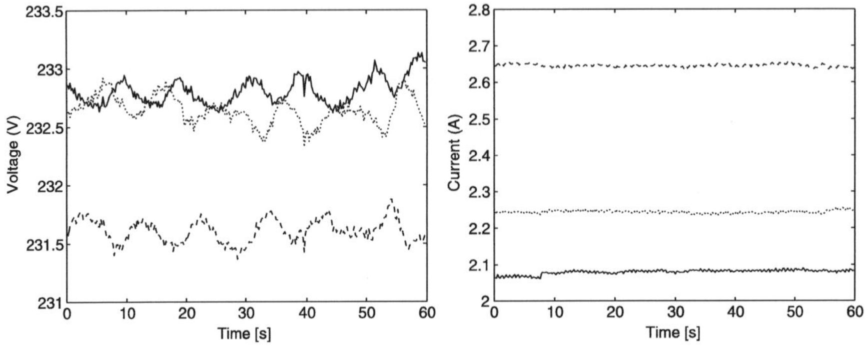


**Figure 3.6** The rms voltages (left) and currents (right) for normal case obtained using IEC 61000-4-30 method; 200-ms window.

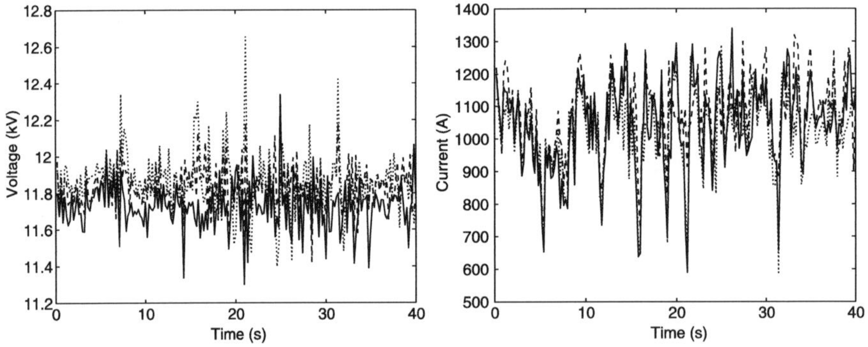


**Figure 3.7** The rms voltages (left) and currents (right) for arc-furnace case obtained using IEC 61000-4-30 method; 200-ms window.

The method as defined in IEC 61000-4-30 has been applied to the normal case and to the arc-furnace case (see Fig. 3.1). The rms value has been calculated every 200-ms window for both voltage and current. The results are shown in Figures 3.6 and 3.7. The window length has been chosen as exactly 200 ms, assuming the frequency to be exactly 50 Hz. For both voltages and currents the values in the three phases are shown separately.

For the normal case both voltage and current are rather constant. Note that the vertical range in current is significantly larger than for the voltage (33% vs. 0.8%) so that the amplitude fluctuations in voltage appear to be larger even though they are in fact much smaller. For the arc-furnace case both voltage and current show severe fluctuations over a range of time scales. Again the fluctuations are more severe in current than in voltage.

***3.2.2.3 Single-Phase Measurements*** The voltage magnitude $V$ can be obtained in a number of ways from the sampled waveform of one of the phase voltages:

- As the rms value of the voltage waveform:

$$V_{\mathrm{I}} = \sqrt{\frac{1}{N}\sum_{i=1}^{N} v_i^2} \tag{3.7}$$

with $N$ an integer multiple of the number of samples during one half-cycle of the power system frequency. This method is used in IEC 61000-4-30 with a window length of 10 or 12 cycles for 50- or 60-Hz systems, respectively. Some examples and a discussion on the influence of the window length follow in the next section.

- From the amplitude of the waveform:

$$V_{\mathrm{II}} = \frac{\max(|v_0|, |v_1|, \ldots, |v_N|)}{\sqrt{2}} \tag{3.8}$$

with $N$ again an integer multiple of the number of samples during one half-cycle of the power system frequency. The implementation of this method is not straightforward. At first a low-pass filter is needed to remove high-frequency components that may inflate the voltage magnitude. When taking the peak value over a number of cycles, a kind of averaging of the peak values of the individual cycles should be taken. Otherwise a longer window will always result in a higher peak value, thus again inflating the result. Another solution that tackles this problem is to determine an *average half-cycle* over the measurement window. If, for example, the measurement window has a length of 10 cycles of the fundamental frequency, the average is taken of the absolute value of the voltage over 20 half-cycles. The amplitude over this averaged half-cycle is next determined as the highest value over this average half-cycle. This method also limits the impact of high-frequency components on the peak value.

The difference between the rms value and the amplitude is mainly due to lower order harmonics. In most low-voltage supplies, the presence of low-voltage equipment causes a flattening of the top of the sine wave, resulting in a reduction of the amplitude compared to the rms voltage.

- As the absolute value of the fundamental component:

$$V_{\mathrm{III}} = |V_1| \tag{3.9}$$

with $V_1$ the complex fundamental component obtained from any of the spectrum estimation techniques to be discussed later. The calculation of the fundamental component is straightforward in those cases where the spectrum also has to be obtained. The difference between the rms value and the amplitude of the fundamental is small in most cases. With THD the total harmonic distortion we obtain the following relation:

$$V_{\mathrm{rms}} = V_{\mathrm{fund}}\sqrt{1 + \mathrm{THD}^2} \tag{3.10}$$

For a THD of 8% the error is only 0.3%. To quantify the voltage magnitude, the difference between the rms value and the fundamental is thus in all cases small. For the current to individual equipment the current distortion can be more than 100%. The resulting error when using the fundamental instead of the rms will be more than 40% in that case. For groups of loads (e.g., at the low-voltage side of a distribution transformer) the distortion is in most cases below 30%, leading to errors up to 4%, which are in many cases acceptable.

The most commonly used method is the rms voltage (method I), but there is no obvious reason for this. The rms voltage determines the performance for any type of equipment that is a pure resistance (e.g., incandescent lamps, heating and tea cookers). The fundamental voltage (method III) is the component of the voltage that is involved in the energy transfer to most rotating loads, all other components only lead to fluctuating torques and losses in the system. The voltage amplitude (method II) is a good measure of the performance of many electronic devices, where the level of the internal dc voltage is related to the voltage amplitude, not to the rms value. But the difference between the methods is not significant as long as voltages are concerned.

For current measurements the difference between the estimation results is much larger because the waveform distortion is much higher. Again, which method is more appropriate depends on the application. The rms current is an appropriate measure for the loading of cables, lines, and transformers as long as the distortion is not too high. For high distortion a correction has to be made, especially for transformers. This so-called $K$-factor model will be discussed in Section 3.3. The fundamental current is again a good measure for the energy transfer from generators to the load and for the average torque produced by rotating machines. The amplitude is a measure for the loading of electronic equipment (e.g., at the output of a UPS).

***3.2.2.4 rms Voltage*** The most commonly used method is the one calculating the rms voltage over a window of several cycles. The effect of changing the length of the window is shown in Figure 3.8 for a 1-min window of the phase $a$ voltage of the normal case (as shown in Fig. 3.6). In all four cases one value per window has been calculated. For a one- or two-cycle window the rms voltage has a rather noisy appearance. Applying a longer window gives a more smooth function of time. Note that the longest window length used in Figure 3.8 corresponds to the basic window prescribed by IEC 61000-4-30.

In all examples in Figure 3.8 it was assumed that the power system frequency, and thus the cycle length, is known. Even as the frequency may be known, one cycle is not always exactly an integer number of time steps. Interpolation would be an option, but a very time-consuming one.

Suppose we have a window over which we want to calculate the rms value. However, the window length is not equal to an integer number of sampling time steps. The result is an error in the estimation. The mathematics for this are solved

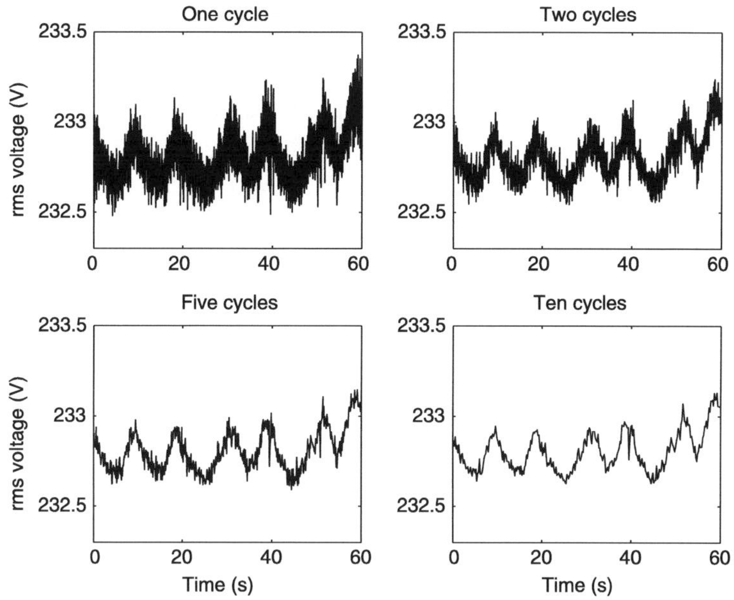


**Figure 3.8** Effect of window length on rms voltage versus time.

in [316, Chapter 7; 317]. Assume a sinusoidal signal of unity rms value:

$$y(k\,\Delta t) = \sqrt{2}\cos\left(2\pi f\,k\Delta t + \alpha_t\right) \tag{3.11}$$

The rms value is calculated over a window $(0, (K-1)\,\Delta t)$ where the first and the last values are included so that the window length is equal to $K\,\Delta t$. The initial phase $\alpha_t$ is the phase angle of the signal at the start of the window. With the sliding window moving through the signal, the initial phase increases with time:

$$\alpha_t = \alpha_0 + 2\pi f t \tag{3.12}$$

with $t$ the time stamp at the start of the window. Without loss of generality we can choose $\alpha_0 = 0$. The resulting estimated rms value can be written as

$$Y_{\text{rms}} = \sqrt{1 + \varepsilon_{\text{MS}}} \tag{3.13}$$

with the mean-square error given by the following rather complicated expression:

$$\varepsilon_{\text{MS}} = \frac{\sin\left(2f\,T_w\pi\right)}{T_W f_s \sin[(2f/f_s)\pi]} \times \cos\left[2\left(T_w - \frac{1}{f_s}\right)f\pi + 4\pi f t\right] \tag{3.14}$$

with $T_W = K\,\Delta t$ the window length and $f_s = 1/\Delta t$ the sampling frequency. The second factor in (3.14) is an oscillating term with twice the frequency as the original signal. The first factor determines the amplitude of this oscillation. If the window length is exactly equal to one cycle, $T_w = 1/f$, the error is zero. The same holds whenever the window length is an integer multiple of one half-cycle.

For small deviations $T_w = 1/f + \delta_T$ we can approximate

$$\sin\left[2f\left(\frac{1}{f} + \delta_T\right)\pi\right] = \sin(2f\,\delta_T\,\pi) \approx 2f\,\delta_T\,\pi \tag{3.15}$$

In general we may assume that the sampling frequency is significantly higher than the power system frequency, $f_s \gg f$, so that $\sin\,[(2f/f_s)\pi] \approx (2f/f_s)\pi$. If we further assume that $T_w \approx 1/f$, we get the following approximated expression for the mean-square error:

$$\varepsilon_{\mathrm{MS}} \approx f\,\delta_T \cos(4\pi f t) \tag{3.16}$$

This results in the following approximated expression for the estimated rms value:

$$Y_{\mathrm{rms}} \approx 1 + \tfrac{1}{2}\,\delta_T f \cos(4\pi f t) \tag{3.17}$$

For a sampling frequency of 2 kHz, the maximum error is half the sampling time step, $\delta_T = 250\ \mu\mathrm{s}$. The amplitude of the oscillation becomes, for a 50-Hz system, $\frac{1}{2} \times 250\ \mu\mathrm{s} \times 50\ \mathrm{Hz} = 0.625\%$.

Assume, as another example, that the sampling frequency is synchronized to the nominal frequency, exactly 2000 Hz, in a 50-Hz system. Assume that the actual frequency is 49.5 Hz, resulting in a cycle length of 20.2 ms. The error is 200 μs in that case, resulting in an oscillation with an amplitude of 0.5%. This may appear a small error, but one should consider that the daily variation in rms voltage is of the order of 5 to 10%. Thus a 0.5% error is a substantial fraction of the daily variation.

***3.2.2.5 Three-Phase Measurements*** When the three sampled waveforms in a three-phase system are available, a few more estimation methods are possible:

- The average of the magnitude values from the three phases:

$$V_{\mathrm{IV}} = \mathrm{mean}(V_a, V_b, V_c) \tag{3.18}$$

  where $V_a$, $V_b$, and $V_c$ are the magnitude estimations obtained from one of the methods for single-phase measurements given above. Typically the rms voltage (method I) is used.
- The rms value of the amplitudes from three phases:

$$V_{\mathrm{V}} = \sqrt{\tfrac{1}{3}\left(V_a^2 + V_b^2 + V_c^2\right)} \tag{3.19}$$

- The absolute value of the positive-sequence voltage:

$$V_{\mathrm{VI}} = |\underline{V}_p| \tag{3.20}$$

with $\underline{V}_p$ the complex positive-sequence voltage. Methods to obtain the positive-sequence voltage will be discussed in Section 3.2.4.

- The instantaneous three-phase rms:

$$V_{\mathrm{VII}} = \sqrt{\tfrac{1}{3}[v_a^2(t) + v_b^2(t) + v_c^2(t)]} \tag{3.21}$$

This results in an instantaneous value. For a balanced three-phase system with sinusoidal voltages, the instantaneous three-phase rms is constant. However, the presence of waveform distortion (the fifth- and seventh-harmonic components) and unbalance causes oscillations with amplitudes similar to the negative-sequence and harmonic components. This means that also here time averaging is needed to obtain a value that can be used for further processing.

- The highest overdeviation and underdeviation for the three phases. The overdeviation is defined as

$$\Delta_{\mathrm{over}} = \begin{cases} 0 & \text{if} \quad V_{\mathrm{rms}} < V_{\mathrm{ref}} \\ \dfrac{V_{\mathrm{rms}} - V_{\mathrm{ref}}}{V_{\mathrm{ref}}} & \text{if} \quad V_{\mathrm{rms}} \geq V_{\mathrm{ref}} \end{cases} \tag{3.22}$$

where $V_{\mathrm{ref}}$ is typically the nominal voltage. But any other reference value can be decided upon. The overdeviation indicates how much higher the rms voltage is than the reference voltage, if it is higher. The underdeviation is defined in a similar way as

$$\Delta_{\mathrm{under}} = \begin{cases} 0 & \text{if} \quad V_{\mathrm{rms}} > V_{\mathrm{ref}} \\ \dfrac{V_{\mathrm{ref}} - V_{\mathrm{rms}}}{V_{\mathrm{ref}}} & \text{if} \quad V_{\mathrm{rms}} \leq V_{\mathrm{ref}} \end{cases} \tag{3.23}$$

Note that both overdeviation and underdeviation are always nonnegative. For a three-phase measurement the three-phase overdeviation is the highest of the overdeviation values for the three phases and the underdeviation is the highest of the three underdeviation values.

The presence of three-phase unbalance leads to a difference in results between the different methods. When presenting the results of voltage variation measurements in a three-phase system, it is important to clearly indicate which method is used. In the opinion of the authors, the positive-sequence voltage is the most appropriate choice for a three-phase system.

***3.2.2.6 Phase-Angle Information*** Sometimes it is needed to estimate not only the magnitude of voltage or current but also the phase angle. It is not possible to obtain an absolute phase angle. One may determine the phase angle of a voltage or current compared to a synchronous reference, as we used earlier to estimate the frequency from the chance in phase angle. But in most cases one is only interested in the phase-angle difference between two signals. For the calculation of symmetrical components the angular difference between the phase voltages or currents is needed. The angle between voltage and current is needed to calculate the active and reactive power.

A straightforward way of calculating the angle between two signals is by obtaining the fundamental component, for example, by means of the discrete Fourier transform algorithm. This will give the angle for each of the signals compared to a synchronous reference. This method has been used to calculate the angle between voltage and current for the arc-furnace measurement. The results are shown in Figure 3.9, where a 200-ms window has been used to calculate the complex voltage and current.

An alternative method is to use the basic relations between voltage, current, and power. The active power can be calculated from the voltage and current magnitudes and the angle between them:

$$P = VI \cos \psi \tag{3.24}$$

But it can equally be calculated from the average instantaneous power, which in turn is equal to the product of (instantaneous) voltage and current:

$$P = \overline{v(t) \times i(t)} \tag{3.25}$$

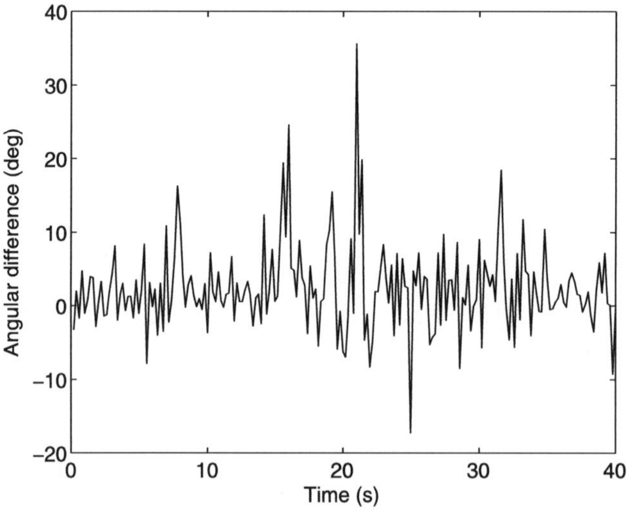


**Figure 3.9** Angular difference between voltage and current calculated from complex voltage, arc-furnace case.

Combining (3.24) and (3.25) gives a method for calculating the angle directly from the waveforms. This expression holds not only for voltage and current but for any set of waveforms:

$$\cos\phi = \frac{\overline{v(t) \times i(t)}}{V_{\text{rms}} I_{\text{rms}}} \tag{3.26}$$

where the rms value is used to estimate the signal magnitude. The disadvantage of this method is that the sign of the angle cannot be determined. However, in many cases the sign of the angle is known, for example, when determining the angle between the phase voltages. This method has been used to determine the angle between the phase $a$ and phase $b$ voltages for the arc-furnace case. The results are shown in Figure 3.10. The calculations have been repeated by using the fundamental voltages (the "correct method" in this case). The difference between the two methods was on average 0.06° with a highest value of 0.3°.

### 3.2.3 Waveform Distortion

We have seen in Section 2.5 that a periodic waveform can be split into a sum of harmonic components with harmonic magnitudes (rms values) $V_k$ and phase angles $\phi_k$. Each harmonic magnitude, phase angle, and complex voltage $V_k e^{j\phi_k}$ can be used as a parameter to quantify the distortion. A number of harmonic standards set limits on the magnitudes of individual voltage and current harmonics. Therefore, estimation of these magnitudes is a common part of power quality monitoring. The presence of different frequencies also provides information on the origin of the disturbance and is thus an essential part in trouble shooting. Estimation of the harmonic phase

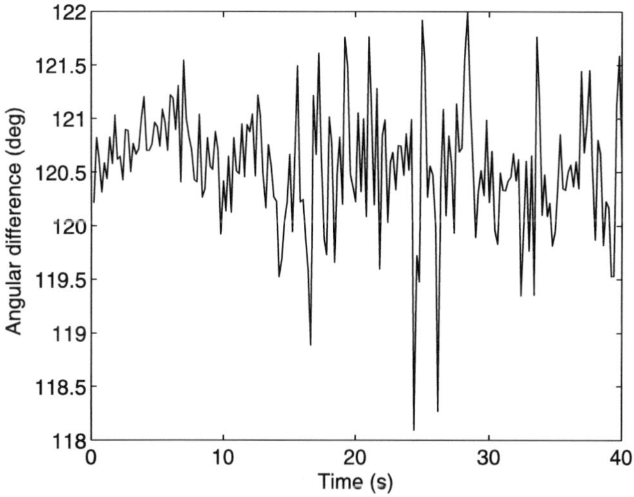


**Figure 3.10** Angular difference between phase $a$ and phase $b$ voltage, arc-furnace case.

angle or complex voltage is less common. Methods for estimating the harmonic magnitudes, for example, will be discussed later in this chapter. In this section we will discuss the standard methods for extracting the harmonic spectrum as defined in IEC 61000-4-7 [155] and in IEC 61000-4-30 [158]. The DFT and other methods for extracting the spectrum will be discussed in more detail in Section 3.4, where we will also discuss some combinations of harmonic voltages and currents into *harmonic indices*. The most commonly used index is the THD, but other combinations may be more appropriate for specific applications.

**3.2.3.1 Anti-Aliasing Filters** The first step with any harmonic measurement is to remove unwanted frequency components. When an analog signal is sampled with a sample frequency $f_s$, the highest frequency component that is present in the digital signal, the so-called Nyquist frequency, is equal to half the sample frequency:

$$f_N = \tfrac{1}{2} f_s \tag{3.27}$$

If the analog signal contains frequency components higher than the Nyquist frequency, these will appear as lower frequency components in the digital signal. A component at frequency $f_x$ with $\frac{1}{2}f_s < f_x < f_s$ will show up as a component with frequency $f_s - f_x$ after sampling with a sample frequency $f_s$. For example, with a sample frequency of 6400 Hz, a 6-kHz component will show up as a 400-Hz components, thus affecting the estimated value of the eighth harmonic (in a 50-Hz system). This phenomenon will be explained further when discussing the discrete Fourier transform in Section 3.4.

To prevent high-frequency components from affecting the measured spectrum, an anti-aliasing filter is used. The anti-aliasing filter is an analog low-pass filter that is placed before the analog–digital conversion, as shown in Figure 3.11. The

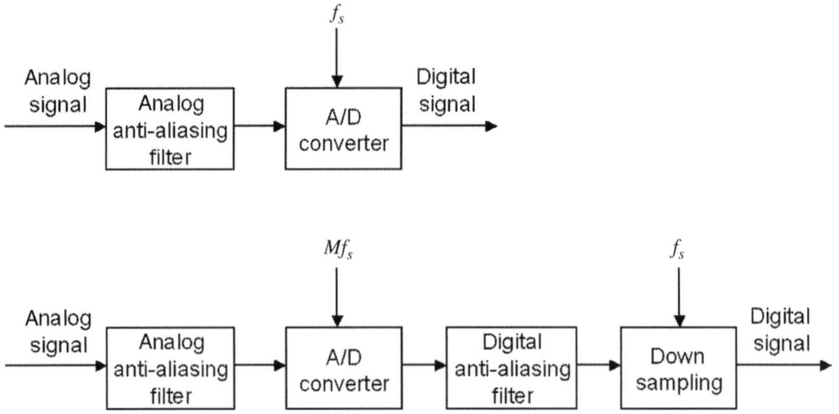


**Figure 3.11** Anti-aliasing filter without (top) and with (bottom) oversampling.

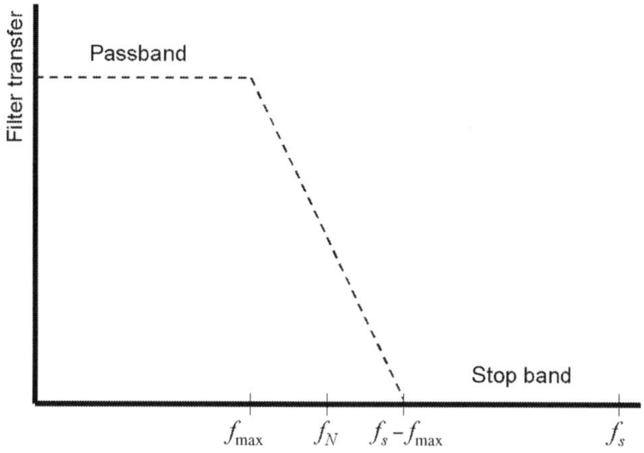


**Figure 3.12** Frequency characteristic of anti-aliasing filter with $f_{\max}$ the highest frequency of interest, $f_s$ the sample frequency, and $f_N = \frac{1}{2}f_s'$ the Nyquist frequency.

requirements for the configuration without oversampling are shown in Figure 3.12. The sampling frequency should be chosen such that there is some margin between the highest frequency of interest $f_{\max}$ and the Nyquist frequency $f_N$. Additional frequency components that appear in the frequency band between the highest frequency of interest and the Nyquist frequency are not of concern. Therefore the lowest frequency that should be removed equals $f_s - f_{\max}$. The highest frequency of interest is typically determined by the highest harmonic order. If harmonic orders up to 40 have to be included in the spectrum, the highest frequencies of interest are 2000 Hz in a 50-Hz system and 2400 Hz in a 60-Hz system. The anti-aliasing filter should be designed such that its passband reaches up to $f_{\max}$ and its stop band starts at $f_s - f_{\max}$.

When designing a power quality monitor, it is important to allow sufficient margin between the highest frequency of interest and the Nyquist frequency. The smaller this margin, the higher the order of the anti-aliasing filter that is needed. A high-order analog filter is not only more expensive, it also leads to a higher phase error in the passband. An alternative configuration [13, Section 5.4.2] is shown in the bottom part of Figure 3.11. The sampling rate of concern for the analog filter equals $Mf_s$ in this case, so that the stop band only has to start at $Mf_s - f_{\max}$, allowing for a much simpler analog filter. A second (digital) anti-aliasing filter reduces the start of the stop band to $f_s - f_{max}$. The DFT is applied to a down-sampled version of the digital signal. Applying the DFT to the original digital signal would place too high computational demands on the monitor. The output of the digital–analog converter may however be used for the detection of high-frequency transients.

One of the results of using a high-order filter is the introduction of time delay and phase errors. A decreasing phase angle is not a problem by itself. What matters is the difference in time delay for different frequencies. A phase angle shift $\phi(f)$ at a frequency $f$ corresponds to a time delay of

$$\Delta T(f) = \frac{\phi(f)}{2\pi f} \tag{3.28}$$

Note that a time delay equal to $[\phi(f) + 2\pi]/2\pi f$ gives the same phase-angle shift between the input and output of the filter. This additional delay is proportional with frequency. We will later see that any linear phase shift does not affect the performance of the filter. Therefore (3.28) can be used for further analysis. As long as all frequency components have the same time delay, the output waveform will be the same as the input waveform; the filter does not introduce any distortion.

A time delay of 1 μs is more severe for a frequency of 2 kHz than for a frequency of 100 Hz, so that it is appropriate to express the difference in time delay in degrees. We will refer to this as the *phase error*. Note that a phase shift between output and input of a filter is not the same as a phase error. The phase error is the deviation from the linear phase. If we use the phase shift (or time delay) at $f_0 = 50$ Hz as a reference, we get for the phase error

$$\delta_\phi(f) = [\Delta T(f) - \Delta T(f_0)] \times 2\pi f \tag{3.29}$$

Using (3.28) we obtain

$$\delta_\phi(f) = \phi(f) - \phi(f_0)\frac{f}{f_0} \tag{3.30}$$

The IEC standard for harmonic measurements (IEC 61000-4-7) sets the requirement that the stop-band attenuation is at least 50 dB and that the −3-dB frequency should be above the frequency range of interest. We have used these requirements to design anti-aliasing Butterworth filters for two sampling rates: 128 and 256 samples per cycle in a 50-Hz system. The highest frequency of interest is 2000 Hz (harmonic 40). The resulting filter characteristics are presented in Table 3.1.

The filter transfer and the phase error as a function of frequency are shown in Figure 3.13. The three curves correspond to the three filters in Table 3.1. The solid line holds for a sample frequency of 6.4 kHz (128 samples per cycle). The

**TABLE 3.1 Properties and Performance of Anti-Aliasing Filters**

| | Filter Requirements (kHz) | | Filter Properties | | Attenuation (%) | | Phase Error (deg) | |
|---|---|---|---|---|---|---|---|---|
| $f_s$ (kHz) | Passband Edge | Stop-Band Edge | $N$ | $f_N$(Hz) | $h = 37$ | $h = 40$ | $h = 37$ | $h = 40$ |
| 6.4 | 2.1 | 4.3 | 9 | 2268 | 1 | 5 | 31 | 43 |
| 12.8 | 2.1 | 10.7 | 4 | 2537 | 4 | 7 | 12 | 15 |
| 12.8 | 3.0 | 10.7 | 5 | 3384 | 0.1 | 0.3 | 5 | 6 |

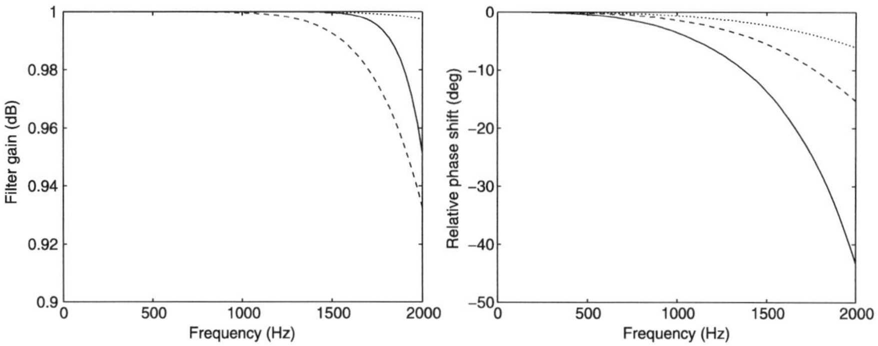


**Figure 3.13** Frequency response (left) and phase-angle error (right) for three anti-aliasing filters.

dashed and dotted lines are both for a sample frequency of 12.8 kHz. Both filters fulfill the IEC requirements. By slightly raising the requirements (moving the end of the passband from 2.1 to 3 kHz), a much better performance is obtained.

***3.2.3.2 Standard Methods: IEC 61000-4-30, IEC 61000-4-7*** The measurement of harmonic distortion is defined in two IEC standards: IEC 61000-4-7 [155] and IEC-61000-4-30 [158]. The first one defines the way in which harmonic current distortion shall be measured when comparing equipment emission with the emission limit. The latter document defines the way in which voltage quality shall be measured. In this section we will give an overview of the general measurement method as defined in these two documents. For details and when applying the standards, the reader is referred to the original documents.

According to the IEC standards, the Fourier series shall be obtained over a rectangular window with a length equal to 10 cycles in a 50-Hz system and equal to 12 cycles in a 60-Hz system. The length of this basic window is about 200 ms for each system frequency. The window length is not exactly 200 ms as the frequency always differs somewhat from the nominal frequency. We will, somewhat incorrectly, refer to this as the 200-ms window and to the resulting spectrum as the 200-ms values. The more correct term used in the IEC documents is *10/12-cycle window*.

The use of a rectangular window requires that the measurement window is synchronized with the actual power system frequency, hence the use of a 10/12-cycle window instead of a window of exactly 200 ms. The IEC standard requires that 10/12 cycles corresponds with an integer number of samples within 0.03%. To ensure synchronism between the measurement window and the power system frequency, most instruments use a PLL generating a sampling frequency that is an integer multiple of the actual power system frequency. A synchronization error leads to cross-talk between the different harmonic frequencies. The 50-Hz component is by far the dominating component in most cases so that the main concern is the cross-talk from the 50-Hz component to higher order components. This will be discussed in further detail in Section 3.4.

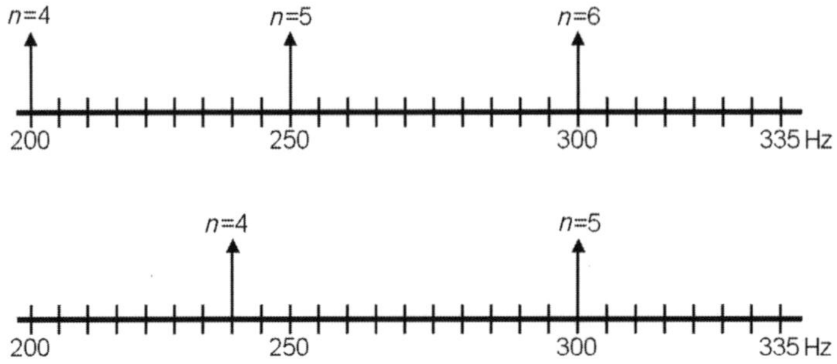


**Figure 3.14** Spectrum from applying DFT to 10/12-cycle window with harmonic frequencies indicated for 50-Hz system (top) and 60-Hz system (bottom).

The result of applying the DFT to the basic window is a spectrum with 5-Hz spacing between frequency components. The spectrum thus contains both harmonics and interharmonics. Figure 3.14 shows the frequency band between 200 and 335 Hz with the harmonic frequencies indicated. Note that the frequency resolution is the same as the window length is the same. However, the number of frequency components between harmonic frequencies is higher in a 60-Hz system.

The spectrum has been calculated over a 200-ms window for both the arc-furnace case and the normal case. The 10-cycle (200-ms) window for the arc-furnace case is shown in Figure 3.15 together with the result of applying a DFT algorithm. The voltage is measured on the secondary (20-kV) side of the 135/20-kV transformer supplying the arc furnace. The current taken by an arc furnace is highly distorted and highly fluctuating and as a consequence so is the voltage. Only the absolute values of the spectral components are shown. The 50-Hz component is much larger than any of the other components and falls outside the scale. The spectra of both voltage and current contain frequency components up to a 100 Hz.

The waveform and spectrum for the normal case are shown in Figure 3.16. The sampling rate was only 2048 Hz so that it was not possible to obtain the spectral components above 1 kHz. The distortion of the waveform is very small and its

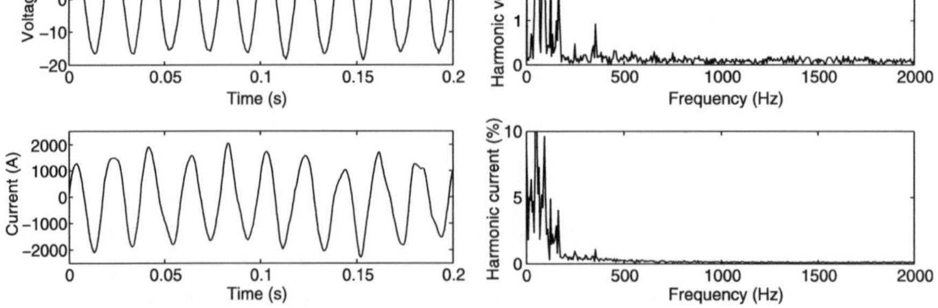


**Figure 3.15** Voltage (top) and current (bottom) waveform (left) and spectrum (right) measured over 200-ms window, arc-furnace case.

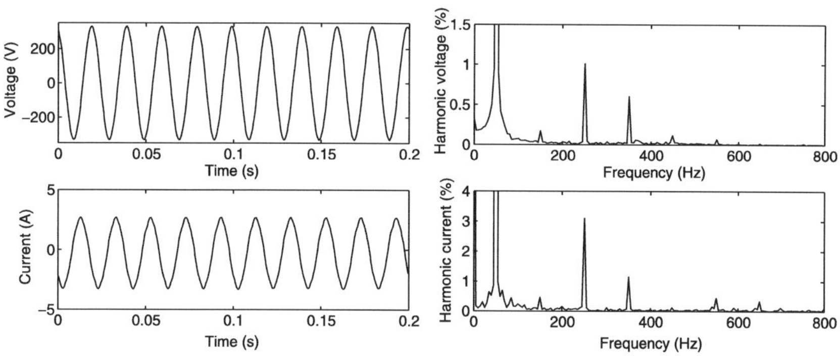


**Figure 3.16** Voltage (top) and current (bottom) waveform (left) and spectrum (right) measured over 200-ms window, normal case.

spectrum contains mainly odd harmonics, with the fifth and seventh harmonics being dominant. This voltage spectrum is typical for a network with three-phase diode or thyristor rectifiers. The lack of single-phase equipment means that the third-harmonic component is small. The absence of frequency components above 600 Hz is due to the anti-aliasing filter. Note the difference in character with the spectra in Figure 3.15. Whereas Figure 3.15 shows a continuous spectrum, Figure 3.16 shows a line spectrum. The latter is much more common. Continuous spectra are only obtained for loads that show a very fast cycle-to-cycle change in current amplitude and/or waveform. The arc furnace is a typical example. Most loads, however, show a line spectrum. Note the frequency components, in both voltage and current, around 50 Hz. These are due to the small variations in amplitude (amplitude modulation) resulting in side bands around the fundamental frequency, as discussed in Section 2.4.2. Measurement of a dc component requires special care, so no conclusions should be drawn from the value for zero frequency. The apparent dc component in the voltage may be due to a small offset in the measurement circuit.

A third example is shown in Figure 3.17. This recording was obtained in an office building and triggered by the interharmonic contents. The main interharmonic

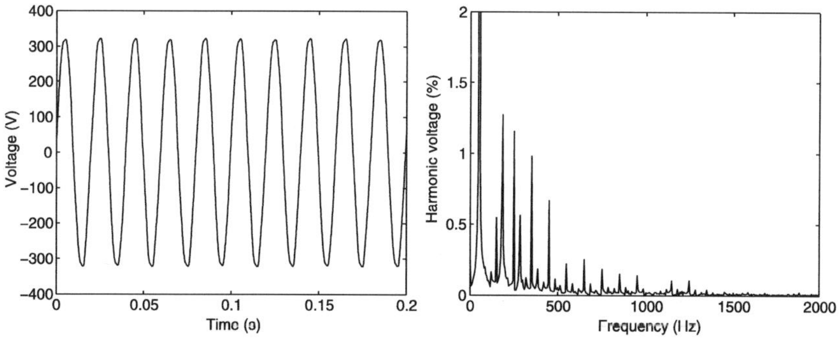


**Figure 3.17** Voltage waveform (left) and spectrum (right) measured over 200-ms window, interharmonic case.

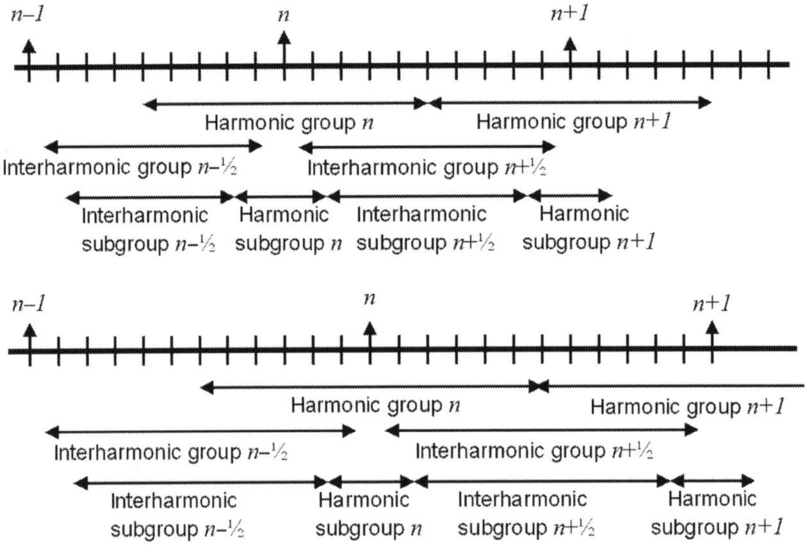


**Figure 3.18** Combining frequency components into harmonic and interharmonic groups and subgroups: for 50-Hz system (top) and for 60-Hz system (bottom).

(near 185 Hz) is due to signaling used by the local utility. The 285-Hz component is produced by single-phase rectifiers exposed to the 185-Hz component. Harmonic resonance in the system amplifies the 285-Hz component in the voltage. The spectrum in Figure 3.17 further shows the standard harmonic components present in office buildings, basically all odd harmonics starting at the third harmonic.

The output of the discrete Fourier transform consists of a number of frequency components that are indicated through their harmonic order: $c_n$ is the absolute value for harmonic order $n$ (frequency $50 \times n$ in a 50-Hz system, $60 \times n$ in a 60-Hz system). The frequency components (absolute value) between order $n$ and $n + 1$ are referred to as

$$c_{n+1/10}, c_{n+2/10}, c_{n+3/10}, \ldots, c_{n+9/10}$$

in a 50-Hz system and as

$$c_{n+1/12}, c_{n+2/12}, c_{n+3/12}, \ldots, c_{n+11/12} \tag{3.31}$$

in a 60-Hz system. These frequency components are combined into groups and subgroups according to the definitions given below. The various combinations are visualized in Figure 3.18:

- *Harmonic Group* In a 50-Hz system there are 9 frequency components between two harmonic orders $n$ and $n + 1$. The lowest four are added to group $n$, the highest four to group $n + 1$, and the middle one is divided equally between the two groups. The harmonic group $C_{ng}$ is defined in a

50-Hz system through the following equation:

$$C_{ng}^2 = \tfrac{1}{2}c_{n-5/10}^2 + \sum_{i=-4}^{4} c_{n-i/10}^2 + \tfrac{1}{2}c_{n+5/10}^2 \tag{3.32}$$

In a 60-Hz system there are 11 frequency components between two harmonic orders. The harmonic group $C_{ng}$ is defined in a 60-Hz system as

$$C_{ng}^2 = \tfrac{1}{2}c_{n-6/12}^2 \sum_{i=-5}^{4} c_{n+i/12}^2 + \tfrac{1}{2}c_{n+6/12}^2 \tag{3.33}$$

- *Interharmonic Group* All frequency components between harmonic orders $n$ and $n+1$ are combined in interharmonic group $n+\frac{1}{2}$. In a 50-Hz system

$$C_{(n+1/2)g}^2 = \sum_{i=1}^{9} c_{n+i/10}^2 \tag{3.34}$$

and in a 60-Hz system

$$C_{(n+1/2)g}^2 = \sum_{i=1}^{11} c_{n+i/12}^2 \tag{3.35}$$

- *Harmonic Subgroup* The harmonic subgroup is formed by the harmonic order plus the adjacent frequency components:

$$C_{ns}^2 = c_{n-1/10}^2 + c_n^2 + c_{n+1/10}^2 \tag{3.36}$$

in a 50-Hz system and

$$C_{ns}^2 = c_{n-1/12}^2 + c_n^2 + c_{n+1/12}^2 \tag{3.37}$$

in a 60-Hz system. The adjacent frequency components are due to fast fluctuations of the level of the harmonic component. This phenomenon is the same as the occurrence of side bands of the power system component in case of voltage fluctuations, as discussed in Section 2.4.

- *Interharmonic Subgroup* The interharmonic subgroups are formed by the frequency components in between the harmonic subgroups. In a 50 Hz-system

$$C_{(n+1/2)s}^2 = \sum_{i=2}^{8} c_{n+i/10}^2 \tag{3.38}$$

and in a 60-Hz system

$$C_{(n+1/2)s}^2 = \sum_{i=2}^{10} c_{n+i/12}^2 \tag{3.39}$$

The grouping into groups and subgroups has been applied to the spectral components for the arc-furnace case in Figure 3.15. The resulting harmonic and interharmonic groups and subgroups are shown in Figure 3.19. The results

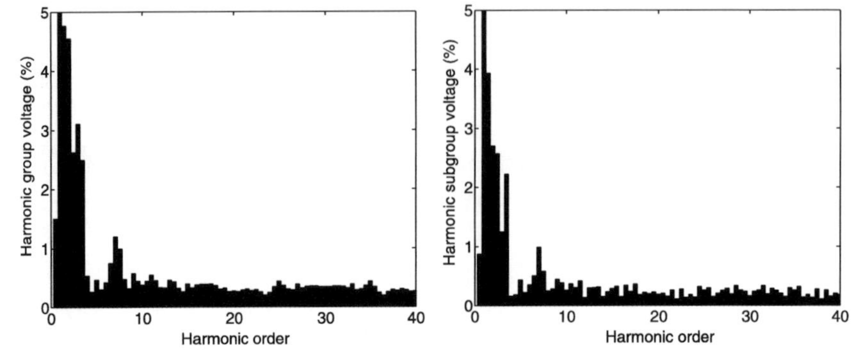


**Figure 3.19** Harmonic and interharmonic groups (left) and subgroups (right) obtained over 200-ms window for arc-furnace case.

for the normal case are shown in Figure 3.20. The values are given in percentage of the fundamental components (group or subgroup 1). The harmonic groups contain all frequency components, whereas the harmonic subgroups only contain frequency components close to the integer multiple of the power system frequency. The result is that the harmonic groups have a higher value that the harmonic subgroups. Also the interharmonic groups contain more components than the interharmonic subgroups and thus have higher values. The difference between harmonic groups and subgroups is especially large for the arc-furnace case where a significant amount of signal energy is present between the harmonics frequencies. For the normal case the amount of energy between the harmonics is small, with the exception of interharmonic components 0.5 and 1.5. These are significantly higher when using interharmonic groups instead of subgroups. The high values of subgroups 0.5 and 1.5 indicate significant cycle-to-cycle variations in the voltage magnitude, possibly with a high flicker severity as a result.

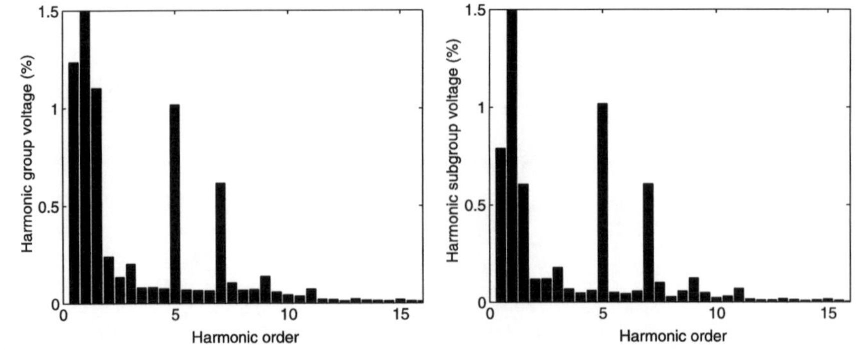


**Figure 3.20** Harmonic and interharmonic groups (left) and subgroups (right) obtained over 200-ms window for normal case. Note the difference in horizontal scale compared with Figure 3.19.

**3.2.3.3 Disturbances in Frequency Range 2 to 9 kHz** The characterization of harmonic distortion as defined in IEC standards includes harmonics up to order 40, that is, 2 kHz in a 50-Hz system and 2.4 kHz in a 60-Hz system. Also other documents do not consider harmonics above order 40. An informative annex with IEC 61000-4-7 discussed a characterization method for disturbances in the frequency range 2 to 9 kHz. The lower limit of this frequency range is determined by the upper limit of the harmonic spectrum in a 50-Hz system. The upper limit is determined by the lower limit of radiated disturbances as defined in CISPR 16-1 [65].

The standard document proposes using a 100-ms window that no longer needs to be synchronized to the power system frequency. The power system frequency component should be removed through an analog high-pass filter, so that leakage from the power system frequency to higher frequencies no longer is a concern. Further, the disturbances in this higher frequency range are typically not linked to the power system frequency but are due to active controllers that operate at a certain switching frequency. For the same reason there is no longer a need for separate protocols in 50- and 60-Hz systems.

The frequency range of interest for these measurements is up to 9 kHz, so that a higher sampling frequency should be used than for the measurement of harmonic distortion. As mentioned before (see Fig. 3.12), a certain margin is needed between the highest frequency of interest and the Nyquist frequency. The sampling frequency for measurements in this frequency range should therefore be in the range 30 to 50 kHz. A switching frequency of 40.96 kHz is a convenient choice as it results in $4096 = 2^{12}$ samples per 100-ms window. This allows for the use of the computationally efficient fast Fourier transform algorithm.

Measurements in this frequency range require analog filtering on both low- and high-frequency sides. A high-pass filter is needed to remove the fundamental frequency and some of the lower order harmonics. A suitable stop-band edge is slightly above the seventh of ninth harmonic. Analog filtering is needed here so that the whole dynamic range of the analog–digital converter can be used for the high-frequency component. A low-pass filter is needed to prevent aliasing. The passband edge should be above 9 kHz. For 40 kHz sampling frequency, the stop-band edge should be less than 31 kHz. The filter requirements are summarized in Figure 3.21.

Applying a DFT algorithm to the 100-ms window results in frequency values with 10 Hz separation. The resulting values are grouped into 200-Hz bands using the following grouping algorithm:

$$G_b = \sqrt{\sum_{f=b-90\,\text{Hz}}^{b+100\,\text{Hz}} C_f^2} \tag{3.40}$$

with $C_f$ the rms value of the frequency component at frequency $f$ and $b$ the center frequency of the band. This grouping algorithm results in 35 frequency bands centered around 2100, 2300, ..., 8700 and 8900 Hz. The choice of frequency bands is based on CISPR 16-1, where the same 200-Hz bandwidth is used for measurement

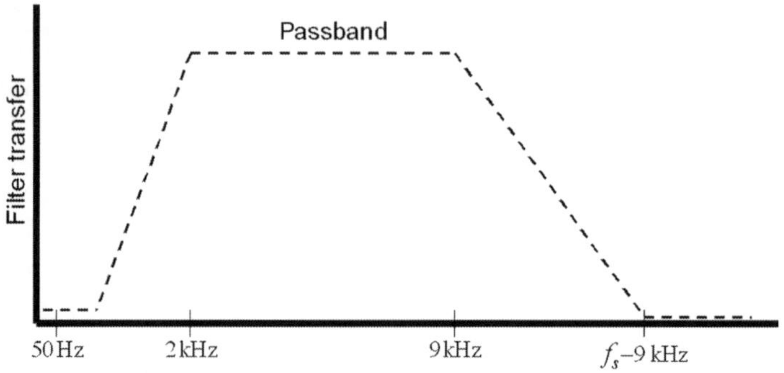


**Figure 3.21** Analog filter requirements for measurements in frequency range 2 to 9 kHz.

of radiated disturbances. The characterization method is however not compatible with the method used for harmonic and interharmonic distortion. The grouping algorithms for harmonics and interharmonics, as defined in (3.32) and further, adds the power of all frequency components within each group or subgroup. The grouping algorithm used in the frequency band 2 to 9 kHz, as in (3.40), also adds the power over all components within each frequency band but over a different frequency band. For a single-frequency signal (a line spectrum) the result will be consistent for both methods. Thus a 1-V signal at 2010 Hz will result in a 1-V value for the 40th harmonic as well as a 1-V value for the 2100-Hz band.

The situation is different for broadband signals, as are more common for the higher frequency ranges. To compare between different frequency bands, it is important to consider Parseval's theorem. Parseval's theorem relates the energy of the signal in the time domain and the frequency domain. For digital (sampled) signals Parseval's theorem reads as

$$\frac{1}{N}\sum_{i=1}^{N}[g(t_i)]^2 = \sum_{k=0}^{N/2}|C_k|^2 \tag{3.41}$$

The left hand-side is the square of the rms value of the signal, which is independent of sampling rate or measurement window. Doubling the measurement window will double the number of frequency components. For a broadband spectrum the energy will be spread over the frequency components, so that the amplitude of each component will be reduced by a factor of $\sqrt{2}$. The basic frequency components below 2 kHz are obtained from a 200-ms window, the ones above 2 kHz from a 100-ms window. Let the amplitude of a 5-Hz component (obtained from a 200-ms window) be equal to 1. The amplitude of a 10-Hz component (obtained from a 100-ms window) is then equal to $\sqrt{2} \approx 1.41$ and the amplitude of a 200-Hz band is equal to $\sqrt{40} \approx 6.32$. All this assumes a flat continuous spectrum within the 200-Hz frequency band. Table 3.2 gives the resulting amplitudes for the harmonic and interharmonic groups and subgroups in 50- and 60-Hz systems when the 200-Hz band results in a value of 1 V.

**TABLE 3.2 Relation Between Amplitude of Groups and Subgroups for Flat Continuous Spectrum When 200-Hz Band Results in 1 V.**

| | 50 Hz | 60 Hz |
|---|---|---|
| Harmonic group | 0.5 V | 0.55 V |
| Harmonic subgroup | 0.27 V | 0.27 V |
| Interharmonic group | 0.47 V | 0.52 V |
| Interharmonic subgroup | 0.42 V | 0.47 V |

### 3.2.4 Three-Phase Unbalance

***3.2.4.1 Standard Methods*** Three-phase unbalance is normally characterized by means of the negative-sequence voltage. The standard parameter is the ratio between the negative- and positive-sequence voltage: the so-called negative-sequence unbalance or simply unbalance:

$$u_n = \frac{U^-}{U^+} \tag{3.42}$$

Normally, only the absolute value of the unbalance is considered. But one may in the same way define a phase angle of the unbalance as the phase angle between the negative- and positive-sequence voltages. The phase angle of the negative-sequence voltage indicates the kind of unbalance: phase to ground (difference in voltage magnitude for the phase-to-ground voltages) or phase to phase (difference in phase angle for the phase-to-ground voltages) and which phases are involved. The angle of the negative-sequence voltage will be used in Section 6.2.3 to determine the character of the unbalance during a voltage dip. A similar approach could be used to determine the character of voltage unbalance during normal operation.

In Figure 3.22 the magnitude of the phase voltages is shown for 2% unbalance in the voltage. Although the negative-sequence voltage is the same in all cases,

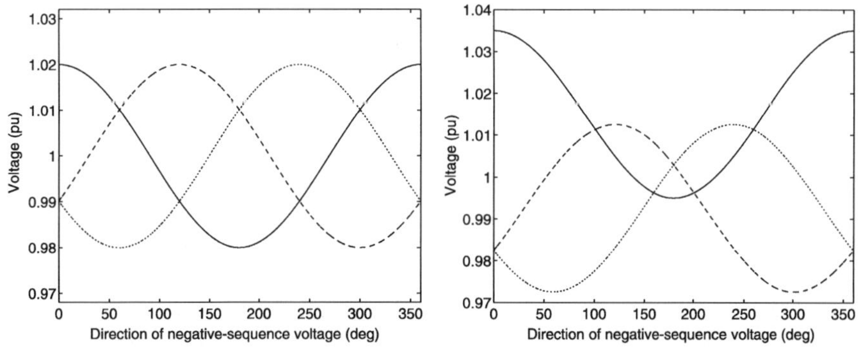


**Figure 3.22** Magnitude of phase voltages for 2% negative-sequence voltage with different phase angle: no zero-sequence voltage (left); 1.5% zero-sequence voltage.

the phase voltage magnitudes are different. The angle of the negative-sequence voltage determines which phase voltage is increased or decreased in amplitude. The presence of a zero-sequence component causes a shift in the curves. Note that it is possible for the phase voltages to be almost identical despite 2% negative sequence and 1.5% zero sequence. The magnitude of the phase voltages is not a good indicator for the negative-sequence unbalance. We will come back to this later.

The IEC standard 61000-4-30 and several other documents mention an alternative method of calculating the unbalance. When the fundamental components of the phase-to-phase voltages $U_{ab}$, $U_{bc}$, and $U_{ca}$ are known, the unbalance can be calculated from the expression

$$u_n = \sqrt{\frac{1 - \sqrt{3 - 6Q}}{1 + \sqrt{1 - 3Q}}} \tag{3.43}$$

with

$$Q = \frac{U_{ab}^4 + U_{bc}^4 + U_{ca}^4}{\left(U_{ab}^2 + U_{bc}^2 + U_{ca}^2\right)^2} \tag{3.44}$$

Note that this is an exact expression, not an approximation. A proof of this expression will be given in Section 3.2.4.3. We saw above that the difference between fundamental voltage and rms voltage is small, so that it is also appropriate to use the rms voltages.

Under IEC 61000-4-30, the basic measurement window used to calculate the unbalance is the same as for voltage magnitude and harmonic distortion: 10 cycles in a 50-Hz system, 12 cycles in a 60-Hz system, about 200 ms in each case.

Next to the negative-sequence unbalance a "zero-sequence unbalance" may be calculated as the ratio between zero- and positive-sequence voltage.

***3.2.4.2 Comparison Between Different Definitions*** Four different parameters are defined within National Electrical Manufacturers Association (NEMA) and IEEE standard documents; unfortunately they are all referred to as "unbalance." A comparison of the different definitions is given in [34, 238]. The different definitions may lead to significantly different results, thus making it very difficult to interpret and compare results:

- The maximum deviation from the average phase voltage, referred to the average of the phase voltage. This definition originates from IEEE standard 112 and was take over by the more recent IEEE 1159. We will refer to this one as $u_{112}$.

- The difference between the highest and the lowest rms voltage, referred to the average of the three voltages. This definition can be found in IEEE 936 and will be referred to as $u_{936}$.
- The ratio of negative- and positive-sequence voltage. This definition is given in IEEE 1159 and is the same definition as used in IEC documents. We will refer to this definition as the *true value* and use the notation $u_n$ as before.
- The maximum deviation from the average line voltage, referred to the average of the line voltages. This is the NEMA definition, which we will refer to as $u_{\text{NEMA}}$. Note that line voltages are used instead of the phase voltages used in IEEE 112 and IEEE 1159.

As an example consider a system with 3% drop in the phase $a$ voltage compared to an otherwise balanced set of voltage phasors:

$$\underline{V}_a = 0.97 \qquad \underline{V}_b = a^2 \qquad \underline{V}_c = a \tag{3.45}$$

with $a = 1\angle 120^\circ$. The positive-, negative-, and zero-sequence voltages are obtained from the standard expressions as given, for example, in Section 2.3.1. This results for this example in

$$\underline{V}_1 = 0.99 \qquad \underline{V}_2 = -0.01 \qquad \underline{V}_0 = -0.01 \tag{3.46}$$

The values of the three above-mentioned indices are for this example

$$\begin{aligned} u_n &= 1.01\% \\ u_{936} &= 2.02\% \\ u_{112} &= 3.03\% \\ u_{\text{NEMA}} &= 1.01\% \end{aligned}$$

For this example, the NEMA definition gives exactly the same value as the IEC definition. The other definitions deviate by a factor of 2 or 3. The immediate conclusion from this example is that both IEEE definitions deviate significantly from the IEC value and from the NEMA definition of voltage unbalance.

To further assess the difference between the different definitions, a 2% negative-sequence voltage is added to a balanced voltage (100% positive-sequence voltage). The direction of the negative-sequence voltage varies over the whole range from zero to 360°. For each direction the resulting unbalance is calculated according to the different definitions. The results are shown in Figure 3.23: Both $u_{112}$ and $u_{\text{NEMA}}$ give a value close to the true value of 2%. However, the value of $u_{936}$ is between 3 and 3.5%, which deviates significantly from the values obtained from other definitions.

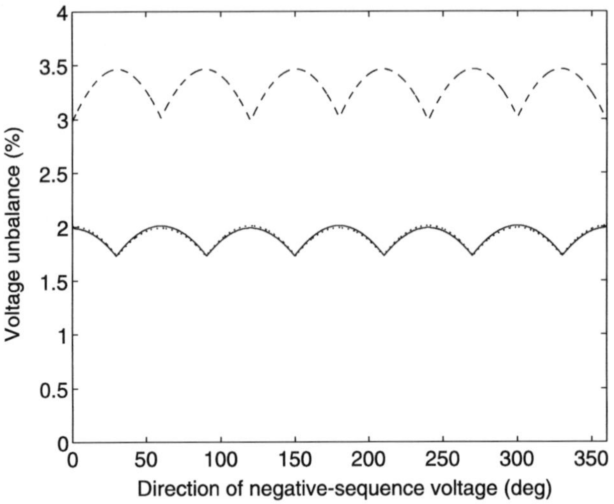


**Figure 3.23** Voltage unbalance according to different definitions for 2% negative-sequence voltage: $u_{NEMA}$ (dotted line), $u_{936}$ (dashed line), and $u_{112}$ (solid line).

The situation becomes worse with the presence of zero-sequence voltage. Obviously, the value of $u_{NEMA}$ is not affected by the presence of any zero-sequence voltage as the line voltages are used. But both IEEE definitions are affected significantly, as shown in Figure 3.24. Next to 2% negative-sequence voltage (of random

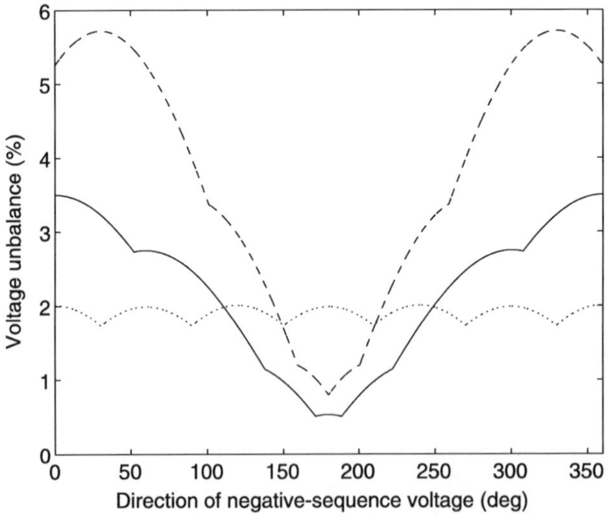


**Figure 3.24** Voltage unbalance according to different definitions for 2% negative-sequence and 1.5% zero-sequence: $u_{NEMA}$ (dotted line), $u_{936}$ (dashed line), and $u_{112}$ (solid line).

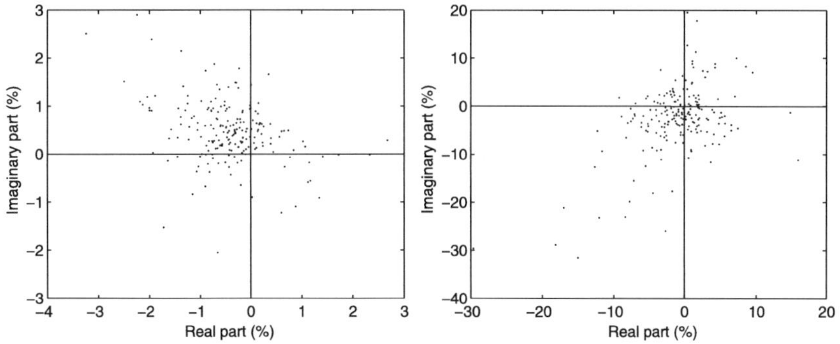


**Figure 3.25** Complex unbalance for arc-furnace case: voltage (left) and current (right).

direction, as before), 1.5% zero-sequence voltage is present. The direction of the zero-sequence voltage is along the direction of the positive-sequence voltage so that the magnitude of the phase $a$ voltage increases by 1.5%. The value of $u_{936}$ varies now between 1 and 5.5%, $u_{112}$ between 0.5 and 3.5%. For both definitions the difference with the IEC value of 2% is significant. Note again that this does not mean that these parameters are less correct than the others. The choice of parameter depends on the application. Any well-defined parameter can be used for this. What is a concern, however, is that different parameters with the same name give significantly different results.

As mentioned before, voltage or current unbalance are by nature complex values, even though in almost all cases only their absolute value is considered. The complex positive- and negative-sequence voltages and currents have been calculated and are shown in Figure 3.25 for the arc-furnace case and in Figure 3.26 for the normal case. Each dot represents one value of the *complex negative-sequence unbalance*, that is, the ratio between the complex values of

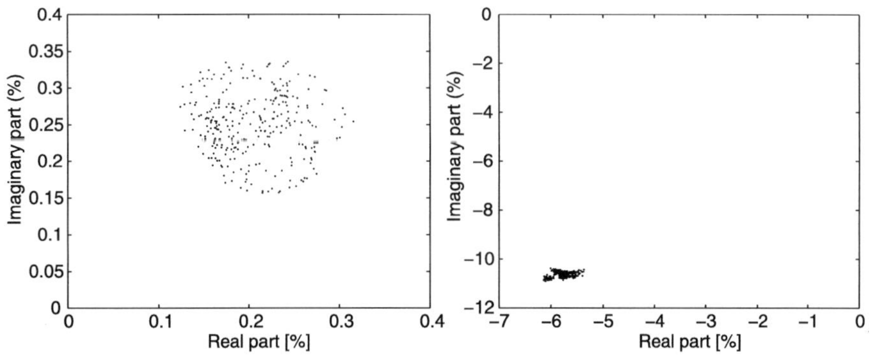


**Figure 3.26** Complex unbalance for normal-measurement case: voltage (left) and current (right). The origin is located lower left for the voltage and upper right for the current.

negative- and positive-sequence voltages. The axes have been chosen for each figure in such a way that the origin is part of the plot. For the arc-furnace case the complex unbalance fluctuates around the origin of the complex plane, leading to large fluctuations in the phase angle. For the normal-measurement case the fluctuations are around a point away from the origin. The fluctuations in phase angle are as a result much less.

**3.2.4.3 Calculating Unbalance from Voltage Magnitudes** The negative-sequence unbalance is, according to the normal definition of the symmetrical-component voltages, obtained from the complex phase voltages. However, it is also possible to calculate the absolute value of the negative-sequence unbalance from the fundamental components of the phase-to-phase voltages. Instead of the fundamental components the rms voltages can be used when the harmonic distortion is limited, an approximation that holds in almost all cases.

Consider the voltage phasors shown in Figure 3.27*a*. As only the phase-to-phase voltages are known, we create the triangle as shown in Figure 3.27*b*, where

$$A = |\underline{U}_{bc}| \qquad B = |\underline{U}_{ca}| \qquad C = |\underline{U}_{ab}| \tag{3.47}$$

We can use the cosine rule to calculate the (unknown) angles from the (known) lengths:

$$\begin{aligned} A^2 &= B^2 + C^2 - 2BC\cos\alpha \\ B^2 &= A^2 + C^2 - 2AC\cos\beta \\ C^2 &= A^2 + B^2 - 2AB\cos\gamma \end{aligned} \tag{3.48}$$

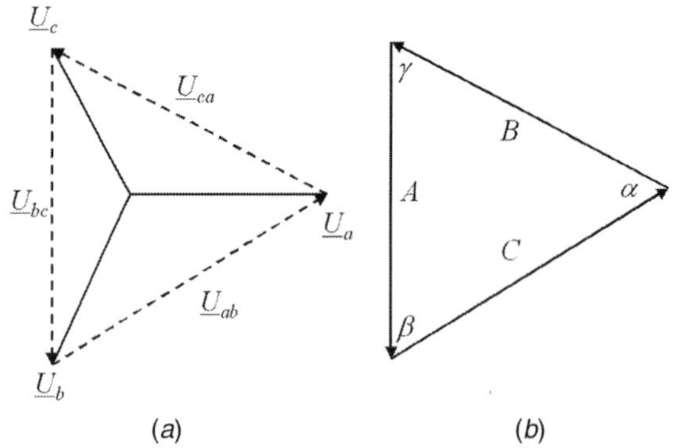


**Figure 3.27** (*a*) Unbalanced voltage phasors and (*b*) triangle used to calculate negative-sequence unbalance.

This results in the following expressions for the angles:

$$\begin{aligned}\cos\,\alpha &= \frac{B^2 + C^2 - A^2}{2BC} \\ \cos\,\beta &= \frac{A^2 + C^2 - B^2}{2AC} \\ \cos\,\gamma &= \frac{A^2 + B^2 - C^2}{2AB}\end{aligned} \tag{3.49}$$

Knowing the angles we can determine the complex line voltages. Using $\underline{C}$ as a reference we obtain the following expressions, with reference to the phasor diagram in Figure 3.28:

$$\begin{aligned}\underline{C} &= C \\ \underline{B} &= -B\cos\,\alpha + jB\,\sin\,\alpha \\ \underline{A} &= -A\,\cos\,\beta - jA\,\sin\,\beta\end{aligned} \tag{3.50}$$

Positive- and negative-sequence voltages can be obtained from the standard expressions

$$\begin{aligned}\underline{U}^{+} &= \underline{C} + a^2\underline{B} + a\underline{A} \\ \underline{U}^{-} &= \underline{C} + a\underline{B} + a^2\underline{A}\end{aligned} \tag{3.51}$$

Note that the zero-sequence voltage obtained from the line voltage is always zero. Note also that the positive- and negative-sequence voltages as obtained from the line voltages are 30° rotated as compared to the values obtained from the phase voltages.

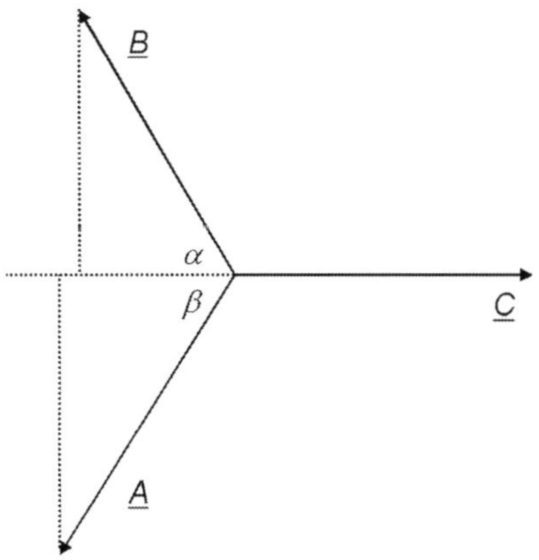


**Figure 3.28** Phasor diagram for phase-to-phase voltages in Figure 3.27.

This is not a concern here as we are only interested in the absolute value of the unbalance.

Combining (3.50) and (3.51) results in the following expression for the positive-sequence voltage,

$$\underline{U}^{+} = C + \tfrac{1}{2}B \cos \alpha + \tfrac{1}{2}jB\sqrt{3}\cos \alpha - \tfrac{1}{2}jB \sin \alpha + \tfrac{1}{2}B\sqrt{3}\sin \alpha$$
$$+ \tfrac{1}{2}A \cos \beta - \tfrac{1}{2}jA\sqrt{3}\cos \beta + \tfrac{1}{2}jA \sin \beta + \tfrac{1}{2}A\sqrt{3} \sin \beta \qquad (3.52)$$

and an equivalent expression for the negative-sequence voltage,

$$\underline{U}^{-} = C + \tfrac{1}{2}B\cos \alpha - \tfrac{1}{2}jB\sqrt{3}\cos \alpha - \tfrac{1}{2}jB\sin \alpha - \tfrac{1}{2}B\sqrt{3} \sin \alpha$$
$$+ \tfrac{1}{2}A\cos \beta + \tfrac{1}{2} jA\sqrt{3} \cos \beta + \tfrac{1}{2}jA \sin \beta - \tfrac{1}{2}A\sqrt{3} \sin \beta \qquad (3.53)$$

Before continuing we derive some expressions to simplify (3.52) and (3.53).

Applying the sine rule to the triangle in Figure 3.27 results in

$$\frac{\sin \alpha}{A} = \frac{\sin \beta}{B} \qquad (3.54)$$

which can be rewritten as

$$B \sin \alpha = A \sin \beta \qquad (3.55)$$

From the earlier results of applying the cosine rule, (3.49), we get the relations

$$B \cos \alpha + A \cos \beta = C$$
$$B \cos \alpha - A \cos \beta = \frac{B^2 - A^2}{C} \qquad (3.57)$$

Substituting (3.55) through (3.57) into (3.52) and (3.53) results in the expressions

$$\underline{U}^{+} = \frac{3}{2}C + B\sqrt{3} \sin \alpha + \frac{1}{2}j\sqrt{3}\frac{B^2 - A^2}{C} \qquad (3.58)$$

for the positive-sequence voltage and

$$\underline{U}^{-} = \frac{3}{2}C - B\sqrt{3} \sin \alpha - \frac{1}{2}j\sqrt{3}\frac{B^2 - A^2}{C} \qquad (3.59)$$

for the negative-sequence voltage. The last remaining unknown in these expressions is $\sin \alpha$, which can be obtained from (3.49) and $\sin^2 \alpha + \cos^2 \alpha = 1$:

$$\sin \alpha = \sqrt{1 - \frac{(B^2 + C^2 - A^2)^2}{4B^2C^2}} = \frac{\sqrt{(A^2 + B^2 + C^2)^2 - 2(A^4 + B^4 + C^4)}}{2BC} \tag{3.60}$$

The resulting expression for the positive-sequence voltage as a function of the magnitude of the line voltages is

$$\underline{U}^+ = \frac{3}{2}C + \frac{1}{2C}\sqrt{(A^2 + B^2 + C^2)^2 - 2(A^4 + B^4 + C^4)} + j\frac{\sqrt{3}(B^2 - A^2)}{2C} \tag{3.61}$$

The equivalent expression for the negative-sequence voltage is

$$\underline{U}^- = \frac{3}{2}C - \frac{1}{2C}\sqrt{(A^2 + B^2 + C^2)^2 - 2(A^4 + B^4 + C^4)} - j\frac{\sqrt{3}(B^2 - A^2)}{2C} \tag{3.62}$$

These expressions can be used to calculate the complex unbalance, but normally one is only interested in the absolute value of the unbalance. The absolute values of the positive- and negative-sequence voltages read as

$$|\underline{U}^+| = \tfrac{3}{2}(A^2 + B^2 + C^2) + \tfrac{3}{2}\sqrt{(A^2 + B^2 + C^2)^2 - 2(A^4 + B^4 + C^4)} \tag{3.63}$$

$$|\underline{U}^-| = \tfrac{3}{2}(A^2 + B^2 + C^2) - \tfrac{3}{2}\sqrt{(A^2 + B^2 + C^2)^2 - 2(A^4 + B^4 + C^4)} \tag{3.64}$$

Note that these expressions are symmetrical in the three line voltages $A$, $B$, and $C$. This is to be expected as none of these three values has any different influence from the others. The fact that the expressions for the complex voltages are not symmetrical is due to the choice of $\underline{C}$ as the reference for the positive real axis. Finally the unbalance is obtained from the ratio of (3.64) and (3.63) and by substituting $A = U_{bc}, B - U_{ac}$, and $C - U_{ab}$:

$$u^- = \frac{|\underline{U}^-|}{|\underline{U}^+|} = \frac{\sqrt{1 - \sqrt{3 - 6Q}}}{\sqrt{1 - \sqrt{3 + 6Q}}} \tag{3.65}$$

with

$$Q = \frac{U_{ab}^4 + U_{bc}^4 + U_{ac}^4}{(U_{ab}^2 + U_{bc}^2 + U_{ac}^2)^2} \tag{3.66}$$

***3.2.4.4 The dq-Transform*** We saw in Section 2.4 that the *dq*-voltage can be written as a function of positive- and negative-sequence voltages:

$$\underline{v}_{dq} = \underline{U}^{+} + (\underline{U}^{-})^{*} e^{-j4\pi f_0 t} \tag{3.67}$$

This relation can be used to estimate the positive-sequence voltage $\underline{U}^{+}$ and the negative-sequence voltage $\underline{U}^{-}$. In the complex plane the *dq*-voltage will describe a circle. The center of the circle is the positive-sequence voltage; the radius gives the negative-sequence voltage. The positive-sequence voltage can be extracted by averaging or by applying a suitable low-pass filter. The negative-sequence voltage can be extracted from a bandpass filter or by applying a model-based technique. A Kalman filter is used in [112, Chapter 4; 113] to estimate positive- and negative-sequence voltages from the *dq*-voltage.

The negative-sequence voltage can also be obtained from the *backward dq-transform*, defined as

$$\underline{v}_{dq}^{-} = v_{\alpha\beta} e^{+j\omega_0 t} \tag{3.68}$$

The standard definition as in (3.67) would be referred to as the *forward dq-transform*. The backward *dq*-transform is again related to the symmetrical components. Combining (3.68) with (2.85) gives the relation

$$\underline{v}_{dq}^{-} = (\underline{U}^{-})^{*} + \underline{U}^{+} e^{+j4\pi f_0 t} \tag{3.69}$$

The average value of the backward *dq*-transform is equal to the (complex conjugate of the) negative-sequence voltage.

In power-electronic controllers for mitigating voltage disturbances, the use of the *dq*-transform is rather common. (It is often referred to as *vector control*). The method originated with the control of adjustable-speed drives. The advantage of vector control is that it allows an ac machine to be controlled as a dc machine; in normal operation the *dq*-voltage is a dc value which makes control much easier. When using the *dq*-transform for power quality mitigation, the voltages cannot always be assumed balanced. This requires a high switching frequency (with the associated high losses) or a separation of positive- and negative-sequence components. Some publications discuss the use of the *dq*-voltage to obtain a so-called instantaneous negative-sequence voltage [16, 83, 96]. Note that it is essential for control algorithms to obtain instantaneous values as the output of the controller is also an instantaneous value.

From the normal $\alpha\beta$-voltage a *negative-sequence $\alpha\beta$-voltage* is calculated:

$$v_{\alpha\beta}^{-}(t) = \tfrac{1}{2} v_{\alpha\beta}(t) - \tfrac{1}{2} j v_{\alpha\beta}\left(t - \tfrac{1}{4}T\right) \tag{3.70}$$

with $T$ one cycle of the power system frequency. Substituting (2.85) in (3.70) shows that the negative-sequence $\alpha\beta$-voltage is only a function of the negative-sequence voltage:

$$v_{\alpha\beta}^{-}(t) = (\underline{U}^{-})^{*} e^{-j2\pi f_0 t} \tag{3.71}$$

The negative-sequence voltage can be obtained by multiplying with a power system frequency signal, for example, obtained from a PLL:

$$\underline{U}^{-} = \{v_{\alpha\beta}^{-} e^{j2\pi f_0 t}\}^{*} \tag{3.72}$$

The difference between the normal $\alpha\beta$-voltage and the negative-sequence $\alpha\beta$-voltage can be used to obtain an instantaneous positive-sequence voltage:

$$\underline{U}^{+} = \{v_{\alpha\beta} - v_{\alpha\beta}^{-}\} e^{-j2\pi f_0 t} \tag{3.73}$$

This method may be very suitable for control algorithms; it has some disadvantages when used to extract features from signals. The implementation is rather complicated as an accurate PLL is needed. Although the output is referred to as *instantaneous negative-sequence voltage*, is does not follow changes in voltage immediately. The quarter-cycle delay in (3.70) results in an error in the estimated negative-sequence voltage during one quarter-cycle after any chance. This could be a problem for fast-fluctuating signals. In developing the algorithms it has been assumed that the signal is distortion free. The presence of distortion gives high-frequency oscillations on the $\alpha\beta$-voltages. Second-harmonic distortion may interfere with the quarter-cycle delay.

**3.2.4.5 Symmetrical-Component Waveforms** Most of the above-mentioned methods use complex arithmetic to obtain positive- and negative-sequence components. There is, however, a method to obtain *symmetrical-component waveforms* without the need for complex arithmetic. To understand the method we have to go back to the original definition introduced in (2.59). The parameter $a = -\frac{1}{2} + \frac{1}{2} j\sqrt{3}$ is a rotation of 120° in the complex plane. In the time domain this corresponds to a shift over one-third of a cycle. This results in the following definition of the symmetrical-component waveforms:

$$\begin{aligned} u^{+}(t) &= \tfrac{1}{3}\left\{u_a(t) + u_b\left(t - \tfrac{2}{3}T\right) + U_c\left(t - \tfrac{1}{3}T\right)\right\} \\ u^{-}(t) &= \tfrac{1}{3}\left\{u_a(t) + u_b\left(t - \tfrac{1}{3}T\right) + U_c\left(t - \tfrac{2}{3}T\right)\right\} \\ u^{0}(t) &= \tfrac{1}{3}\left\{u_a(t) + u_b(t) + U_c(t)\right\} \end{aligned} \tag{3.74}$$

The magnitude of the symmetrical components can be obtained from any of the magnitude estimation methods discussed before. This results in a method with limited computation effort, especially as one is in most cases only interested in the magnitudes. The disadvantage is that there is almost one cycle delay in calculation. This makes the method less suitable for fast-fluctuating voltages or currents. The method is used in [204] as part of a digital-signal-processor (DSP)-based controller mitigating unbalanced load currents.

## 3.3 POWER QUALITY INDICES

In the first section of this chapter we introduced a number of parameters to quantify different voltage and current variations, for example, the voltage magnitude, the negative-sequence voltage, and the magnitude of each harmonic component. To describe the quality of the voltage or current these parameters are not always sufficient. Instead certain combinations of parameters are needed. In the forthcoming sections we will discuss some combinations of parameters that are or can be used to quantify the voltage or current quality. Note that each basic parameter can in principle also be used as an index. This is, for example, the case when one is only interested in voltage or frequency variations. The unbalance as defined in Section 3.2.4.1 is strictly speaking not a parameter but a combination of two parameters: negative-sequence voltage and positive-sequence voltage. The power quality indices discussed below will mainly be directed toward waveform distortion.

### 3.3.1 Total Harmonic Distortion

The THD is a commonly used power quality index to quantify the distortion of a waveform. The THD is defined as the relative signal energy present at nonfundamental frequencies, normally written as

$$\mathrm{THD} = \frac{\sqrt{\sum_{h=2}^{H} V_h^2}}{V_1} \tag{3.75}$$

Note that the dc component is not considered in our definition of THD, although some other publications include it. The dc component is normally small enough for it not to affect the result of the THD. Special care should further be taken to measure the dc component, so that the result may not be sufficiently reliable in all cases.

An alternative to the THD is to quantify the distortion through the *total waveform distortion* (TWD):

$$\mathrm{TWD} = \frac{\sqrt{V_{\mathrm{rms}}^2 - V_1^2}}{V_1} \tag{3.76}$$

In IEEE standard 519 expression (3.75) is used to quantify the distortion. In the more recent IEEE 1459 expression (3.76) is used instead. But neither document mentions the measurement window required to obtain the spectrum. For a one-cycle window THD and TWD are the same. In IEC 61000-4-7 the THD is defined as in (3.75) with $H = 40$. The harmonic components shall be obtained over a 10- or 12-cycle window (50- and 60-Hz systems, respectively). The IEC document also defines group total harmonic distortion (THDG) and subgroup total harmonic distortion (THDs) where $V_h$ in (3.75) are the harmonic groups and subgroups, respectively, as defined in Section 3.2.3, (3.32) and further.

For pure harmonic distortion (period equal to one cycle, no dc component) the two definitions lead to the same result. For signals with interharmonic components and noise, expressions (3.75) and (3.76) give different results. The difference between the two is the signal energy at nonharmonic frequencies. Referring to this as the *total nonharmonic distortion* (TnHD), we obtain the expression

$$\mathrm{TnHD} = \frac{\sqrt{V_{\mathrm{rms}}^2 - \sum_{h=0}^{H} V_h^2}}{V_1} \tag{3.77}$$

The TnHD is related to the THD and the TWD by the relation

$$\mathrm{TWD}^2 = \mathrm{THD}^2 + \mathrm{TnHD}^2 + \frac{V_0^2}{V_1^2} \tag{3.78}$$

With $V_0$ the dc component of the voltage. The distinction between harmonic and nonharmonic distortion is only relevant for analysis windows longer than one cycle of the fundamental frequency. The TnHD is also referred to as *total interharmonic distortion* due to the difficulty in distinguishing between interharmonics and noise.

The harmonic spectrum can further be subdivided into even harmonics and odd harmonics with their corresponding THD definitions. The *total even-harmonic distortion* may be defined as

$$\mathrm{THD}_{\mathrm{even}} = \frac{\sqrt{\sum_{n=1}^{H/2} V_{2n}^2}}{V_1} \tag{3.79}$$

and the total odd-harmonic distortion as

$$\mathrm{THD}_{\mathrm{odd}} = \frac{\sqrt{\sum_{n=2}^{H/2} V_{2n-1}^2}}{V_1} \tag{3.80}$$

From (3.75), (3.79), and (3.80) we obtain

$$\mathrm{THD}^2 = \mathrm{THD}_{\mathrm{even}}^2 + \mathrm{THD}_{\mathrm{odd}}^2 \tag{3.81}$$

In most cases the total odd-harmonic distortion dominates so that

$$\mathrm{TWD} \approx \mathrm{THD}_{\mathrm{odd}} \tag{3.82}$$

The THD values for the other contributions to the distortion are useful for diagnostics purposes. High values for even or nonharmonic distortion often indicate an abnormal state for the system or for a piece of equipment.

Expression (2.199) gives the THD of the voltage, often indicated by the notation $THD_V$. The THD of the current is defined in a similar way:

$$\mathrm{THD}_I = \frac{\sqrt{\sum_{h=2}^{H} I_h^2}}{I_1} \tag{3.83}$$

In case there in no confusion, the general notation THD may be used. The American harmonics standard IEEE 519 uses the terms *harmonic factor* and *distortion factor* as synonyms for THD. This standard also introduces the *total demand distortion* (TDD) for the current taken by a customer:

$$\mathrm{TDD} = \frac{\sqrt{\sum_{h=2}^{H} I_h^2}}{I_{1,\,\max}} \tag{3.84}$$

When the maximum current is difficult to determine, the rated or subscribed current may be used instead. The advantage of using maximum or rated current is that it gives a better indication of the effect of the distorted current on the system.

For cases with a large variation in fundamental voltage, for example, during voltage dips, it may be suitable to use a similar definition to characterize voltage distortion. The nominal voltage could be used as a reference:

$$\mathrm{THD}_{\mathrm{nom}} = \frac{\sqrt{\sum_{h=2}^{H} V_h^2}}{V_{\mathrm{nom}}} \tag{3.85}$$

There remains some confusion about the definition of THD. In (3.75) it is defined as the ratio between the rms of all the harmonics and the fundamental. This is the standard definition, and one is recommended to use this definition unless clearly indicated otherwise. But some older standards use the ratio between the rms of all the harmonics and the total rms. The definition of THD would then become

$$\mathrm{THD}_{\mathrm{rms}} = \frac{\sqrt{\sum_{h=2}^{H} I_h^2}}{I_{\mathrm{rms}}} \tag{3.86}$$

Power quality monitors and measurement analysis software typically allow the user to choose the base value for the THD calculation.

The two definitions (2.199) and (3.86) can be related through

$$\mathrm{THD} = \mathrm{THD}_{\mathrm{rms}} \times \frac{I_{\mathrm{rms}}}{I_1} \tag{3.87}$$

Using

$$I_{\mathrm{rms}}^2 = I_1^2 + \sum_{h=2}^{H} I_h^2 = I_1^2(1 + \mathrm{THD}^2) \tag{3.88}$$

results in the relations

$$\mathrm{THD}_{\mathrm{rms}} = \frac{\mathrm{THD}}{\sqrt{1+\mathrm{THD}^2}} \tag{3.89}$$

$$\mathrm{THD} = \frac{\mathrm{THD}_{\mathrm{rms}}}{\sqrt{1-\mathrm{THD}_{\mathrm{rms}}^2}} \tag{3.90}$$

For a THD of 25% the rms-based definition would give a value of 24%. The difference is thus negligible for the voltage distortion, which is very rarely above 10%. For the current distortion much higher values occur. For a THD of 100% according to the standard definition, the rms-based definition would give a value of 71%. The difference could thus be significant for current measurements.

Note that the THD is a characteristic of the waveform. It does not directly quantify the impact of the disturbance, as its value is rated to the fundamental component. For voltage waveforms the variations in fundamental component are small, so that the THD can be easily interpreted. With current distortion the situation becomes different; a low THD during high load can have a bigger impact on the system than a high THD during low load. For this reason the TDD is introduced in IEEE 519. The total distortion refers the signal energy present at nonharmonic frequencies to the rated or maximum current; see also Section 5.6.6.

### 3.3.2 Crest Factor

The crest factor as introduced in Section 2.5 can also be used as a parameter to quantify the harmonic distortion. The crest factor cannot be related to the commonly used magnitude spectrum; instead a time-domain approach is needed.

One may distinguish between a *low-frequency crest factor* and a *high-frequency crest factor*. The low-frequency crest factor of the voltage is a measure of the operation of electronic equipment. The high-frequency crest factor is a measure of the effect of the waveform on insulation aging. The low-frequency crest factor of the current is a measure of the effective loading of electronic series components.

Note again that also the crest factor does not directly quantify the impact of the waveform distortion. For this it needs to be multiplied by the rms value of the voltage or current. If one is interested in the effect on equipment, it could be more appropriate to use the voltage or current magnitude estimated from the peak value as an index.

### 3.3.3 Transformers: *K*-factor

The $K$-factor of a current waveform is used to quantify the effect of a distorted current on the loading of transformers. The $K$-factor is defined as follows [164]:

$$K = \frac{\sum_{h=1}^{H} h^2 I_h^2}{\sum_{h=1}^{H} I_h^2} \tag{3.91}$$

Note that the denominator is equal to the rms current. The $K$-factor gives a multiplication factor for the rms current to obtain the actual value of the so-called winding eddy current losses. The various loss terms due to harmonic currents in transformers were discussed in Section 2.5. To quantify the severity of a waveform in the light of transformer losses, the actual value of the rms current should also be considered. A high $K$-factor during low load may be less severe than a lower $K$-factor during high load. The result of this is again an alternative for the current magnitude. The following index can be used to quantify the actual loading of a transformer:

$$I_K = \sqrt{\sum_{h=1}^{H} h^2 I_h^2} \tag{3.92}$$

The literature gives conflicting information on the upper limit of $H$ that should be used. According to the IEEE Emerald Book [164, p. 103] the weighing factor is proportional to $h^2$ up to about harmonic order 15 and proportional to $h$ or even less for higher frequencies. Even though the harmonic components are of low magnitude for harmonic orders above 15, the difference between a factor $h$ and $h^2$ may still be significant. Expression (3.92) only considers harmonic frequencies. If nonharmonic frequencies are present, the summation should include the nonharmonic components.

### 3.3.4 Capacitor Banks

Harmonic distortion in the voltage leads to additional heating of capacitor banks. As capacitor banks are more affected by high-frequency harmonics than by low-frequency harmonics, the standard THD is not a suitable parameter. A harmonic voltage $V_k$ leads to a current through a capacitor bank $C$ equal to

$$I_k = k\omega_0 C V_k \tag{3.93}$$

The current is thus proportional to the harmonic order. Weighting the voltage harmonics by their harmonic order results in a parameter that quantifies the effect of the voltage distortion on a capacitor bank:

$$\mathrm{THD}_{\mathrm{cap}} = \frac{\sqrt{\sum_{h=2}^{H} (hV_h)^2}}{V_1} \tag{3.94}$$

But the value of this index alone does not determine the losses in the capacitor. Also the fundamental component contributes and is in many cases still the dominant contribution. An equivalent voltage magnitude can be determined as follows:

$$V_C = \sqrt{\sum_{h=1}^{H} h^2 V_h^2} \tag{3.95}$$

Again the summation should include any nonharmonic components. Note the similarity with (3.92). Like with the $K$-factor definition before, an upper limit should be posed on the harmonic order to be considered. This upper limit will strongly depend on the configuration of the capacitor. Some capacitor banks include a small inductor to prevent their overheating by high-frequency voltage components. Capacitors present at the terminals of low-voltage equipment are rarely equipped with such an inductor.

### 3.3.5 Motors and Generators

A harmonic voltage at the terminals of a rotating machine (a motor or a generator) leads to a harmonic current limited by the leakage inductance:

$$I_h = \frac{V_h}{h\omega_0 L} \tag{3.96}$$

with $L$ the leakage inductance of the machine. A THD for rotating machines may be defined as

$$\mathrm{THD}_{\mathrm{mach}} = \frac{\sqrt{\sum_{h=2}^{H} (V_h/h)^2}}{V_1} \tag{3.97}$$

where only positive- and negative-sequence components should be included (thus normally no triple harmonics). Zero-sequence components do not cause any current through a rotating machine.

The inductance limiting the harmonic currents is the same as the one limiting the negative-sequence current. It therefore makes sense to combine harmonics and unbalance into one index. The total rms value of the non–effective current is equal to

$$I_{\text{non-eff}} = \sqrt{(V^-)^2 + \sum_{h=2}^{H} \left(\frac{V_h}{h}\right)^2} \tag{3.98}$$

Rating this to the nominal (positive-sequence) voltage gives the following *effective unbalance*:

$$u_{n,\mathrm{eff}} = \frac{V_1}{V_{\mathrm{nom}}} \sqrt{u_n^2 + \sum_{h=2}^{H} \left(\frac{1}{h}\frac{V_h}{V_1}\right)} \tag{3.99}$$

This index is a combination of unbalance, harmonic distortion, and voltage variation. It quantifies the effect of voltage quality on the heating of the machine. However, this index does not serve as a tool to estimate the additional aging (or the required derating) due to harmonics and unbalance. The model used is much too simple for that. As we discussed in the section on unbalance, the main consequence of distortion and unbalance is the forming of hot spots, which is not considered in this index. Despite this the effective unbalance as defined in (3.99) may be used as an indicator to trigger the need for machine derating.

### 3.3.6 Telephone Interference Factor

One of the potential impacts of harmonic currents is audible noise on telephone lines that run in parallel with distribution feeders. To quantify the severity of a distorted current for a telephone line, the so-called *telephone interference factor* (TIF) has been introduced. The TIF is a dimensionless quantity that is indicative of the waveform, not of the actual impact (in the same way as the THD of the current). The telephone interference factor is defined as

$$\mathrm{TIF} = \frac{\sqrt{\sum_{h=1}^{H} (W_h I_h)^2}}{I_{\mathrm{rms}}} \tag{3.100}$$

where $I_h$ is the rms value of harmonic component $h$ and $W_h$ a weighting function as defined in a look-up table or curve. The weighting factor accounts for the coupling between the power line and the telephone line, the response of a standard telephone set, and the sensitivity of our ear to different frequencies. The weighting factor is plotted as a function of the harmonic order in Figure 3.29. As the transfer function only depends on the frequency of the disturbance, the weighting factor for the same harmonic order is different in a 50-Hz and in a 60-Hz system.

As mentioned before, the TIF is a characteristic of the waveform. To quantify the impact of the distortion the TIF has to be multiplied by the rms current, resulting in the so-called *It* product:

$$It = \sqrt{\sum_{h=1}^{H} (W_h I_h)^2} \tag{3.101}$$

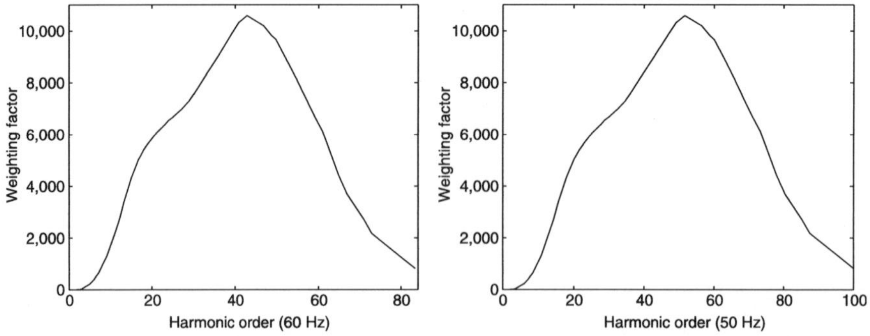


**Figure 3.29** Weighting factor for calculation of TIF 60-Hz system (left) and 50-Hz system (right).

A unity value of the *It* product corresponds to the telephone interference due to a 1-kHz current with an rms value of 0.2 mA.

Instead of the TIF the impact of harmonic current distortion is sometimes also quantified by means of the so-called *equivalent psophometric current*:

$$I_{\text{pe}} = \frac{1}{16{,}000}\sqrt{\sum_{h=1}^{H}(hP_h I_h)^2} \tag{3.102}$$

where $P_h$ is a weighting factor which equals 1000 for a frequency of 800 Hz. The equivalent psophometric current is the amount of 800-Hz current that would give the same level of telephone interference as the current being characterized.

### 3.3.7 Three-Phase Harmonic Measurements

***3.3.7.1 Differences Between Three Phases*** All the spectra shown before are obtained for the current or voltage in phase *a* only. In a three-phase system three different values will typically result for the three phases. As an example, the second-, third- and fifth-harmonic groups are plotted for the arc-furnace case during a 3-s period in Figure 3.30. Each value is obtained by taking the spectrum

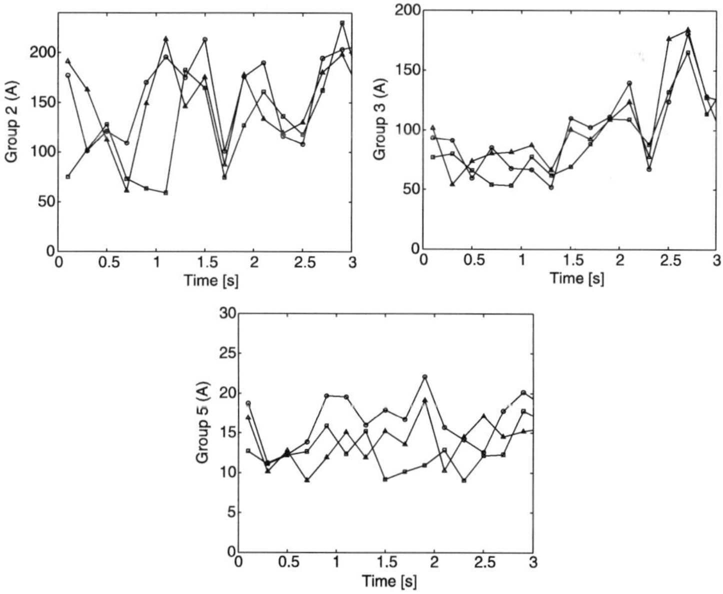


**Figure 3.30** Time variation of second- (left), third- (middle), and fifth- (right) harmonic groups for arc-furnace case during 3-s period: circle, phase *a*; square, phase *b*; triangle, phase *c*.

over a 200-ms window and applying the grouping method introduced before. We see that the spectra of the individual phases appear to have a *correlated* and a *noncorrelated* component. These two components need to be better quantified however. In this section we will introduce a number of methods to distinguish between balanced and nonbalanced harmonics.

For three-phase measurements, three magnitude values result. These values are often combined into one value by a processed referred to as *three-phase aggregation*. A number of simple aggregation methods are being used, neither of which includes the three-phase character of the system. Some examples of simple aggregation methods are as follows:

- Highest of the three harmonic magnitudes
- Average of the harmonic magnitudes in the three phases
- Middle of the three harmonic magnitudes
- The rms of the harmonic magnitudes in the three phases

As with single-phase measurements it is again possible to calculate a THD value or any of the aggregated harmonic parameters introduced before. When using the highest of the three harmonic magnitudes, the resulting THD value depends on the order in which the calculations are made. Calculating the THD from the highest value per harmonic,

$$\mathrm{THD} = \frac{\sqrt{\sum_{h=2}^{H} \max\left(V_{ha}, V_{hb}, V_{hc}\right)}}{\max\left(V_{1a}, V_{1b}, V_{1c}\right)} \tag{3.103}$$

will yield a different value than taking the highest THD per phase,

$$\mathrm{THD} = \max\left(\mathrm{THD}_a, \mathrm{THD}_b, \mathrm{THD}_c\right) \tag{3.104}$$

with $\mathrm{THD}_a$ the THD in phase $a$, calculated from (3.75), and so on.

In (3.103) the highest of the three fundamental voltages was used as a reference. As the unbalance in the fundamental voltages is normally rather small, using the average of the three voltages or the positive-sequence voltage does not significantly affect the result.

In the forthcoming sections we will introduce a number of aggregation methods that better incorporate the three-phase character of the power system. All methods are based on the calculation of symmetrical components so that it is essential that the complex harmonic components are available. When only the harmonic magnitudes are available, these methods cannot be used.

#### 3.3.7.2 Symmetrical-Component THD Values

When calculating positive-, negative-, and zero-sequence harmonics, one may decide to calculate THD values for each component separately, resulting in the *total positive-sequence harmonic distortion*,

$$\mathrm{THD}_{\mathrm{pos}} = \frac{\sqrt{\sum_{h=2}^{H} (V_h^+)^2}}{V_1^+} \tag{3.105}$$

the *total negative-sequence harmonic distortion*,

$$\mathrm{THD}_{\mathrm{neg}} = \frac{\sqrt{\sum_{h=2}^{H} (V_h^-)^2}}{V_1^+} \tag{3.106}$$

and the *total zero-sequence harmonic distortion*,

$$\mathrm{THD}_{\mathrm{zero}} = \frac{\sqrt{\sum_{h=2}^{H} (V_h^0)^2}}{V_1^+} \tag{3.107}$$

where the positive-sequence fundamental voltage has been used as a reference in all three definitions. Note, however, that at harmonic frequencies the positive- and negative-sequence components are equivalent as neither of them is synchronous with the fundamental-frequency positive-sequence component. The zero-sequence component has a physical meaning, as it is related to the current through the neutral conductor and/or to stray currents. The positive- and negative-sequence components can be combined into one *non-zero-sequence component*, with a THD defined as

$$\mathrm{THD}_{\mathrm{non\text{-}zero}} = \frac{\sqrt{\sum_{h=2}^{H} (V_h^+)^2 + (V_h^-)^2}}{V_1^+} \tag{3.108}$$

The zero-sequence part of the distortion is blocked by Dy-connected transformers whereas the non-zero-sequence part freely propagates through the system. A high value for the zero-sequence THD points to a local source of the distortion. For harmonic studies different models are needed for zero- and non-zero-sequence components.

***3.3.7.3 Balanced and Unbalanced THD*** The three symmetrical components have a different meaning for harmonic components than for the fundamental component. The positive-sequence component for a harmonic is no longer in all cases the dominant one for an almost balanced system. Therefore it is more appropriate to add all the *balanced harmonic components* into one THD value. As we saw in Section 2.5 the balanced component is the positive-sequence component for harmonics 1, 4, 7, ..., (i.e., $3n - 2$ for $n \in \mathbf{N}$); the negative-sequence component for harmonics 2, 5, 8, ..., $3n - 1$; and the zero-sequence component for harmonics 3, 6, 9, ..., $3n$. The balanced THD (*total balanced harmonic distortion*) is defined as the square root of the signal energy in all balanced components:

$$\mathrm{THD_{bal}} = \frac{\sqrt{\sum_{n=2}^{H/3} (V^{+}_{(3n-2)})^2 + \sum_{n=1}^{H/3} (V^{-}_{(3n-1)})^2 + \sum_{n=1}^{H/3} (V^{0}_{(3n)})^2}}{V_1^{+}} \tag{3.109}$$

Using the expressions for positive-, negative-, and zero-sequence components, we obtain the following direct way of calculating the balanced component for harmonic $h$:

$$V_{h,\mathrm{bal}} = \tfrac{1}{3}(\underline{V}_{ha} + a^h \underline{V}_{hb} + a^{2h} \underline{V}_{hc}) \tag{3.110}$$

With $V_{h,\mathrm{bal}}$ the magnitude of $V_{h,\mathrm{bal}}$, we get for the balanced THD

$$\mathrm{THD_{bal}} = \frac{\sqrt{\sum_{h=2}^{H} V^2_{h,\,\mathrm{bal}}}}{V_1^{+}} \tag{3.111}$$

The unbalanced THD can be obtained by summation over all other components:

$$\mathrm{THD_{ubal}} = \frac{\sqrt{\sum_{n=1}^{H/3} A_n}}{V_1^{+}} \tag{3.112}$$

with

$$A_1^2 = (V_2^0)^2 + (V_2^{+})^2 + (V_3^{+})^2 + (V_3^{-})^2 \tag{3.113}$$

and

$$\begin{aligned} A_n^2 = {} & (V^{-}_{(3n-2)})^2 + (V^{0}_{(3n-2)})^2 + (V^{0}_{(3n-1)})^2 \\ & + (V^{+}_{(3n-1)})^2 + (V^{p}_{(3n)})^2 + (V^{-}_{(3n)})^2 \qquad n > 1 \end{aligned} \tag{3.114}$$

or by using one of the following relations:

$$\mathrm{THD^2_{bal}} + \mathrm{THD^2_{ubal}} = \mathrm{THD^2_{pos}} + \mathrm{THD^2_{neg}} + \mathrm{THD^2_{zero}} \tag{3.115}$$

$$\mathrm{THD^2_{bal}} + \mathrm{THD^2_{ubal}} \approx \tfrac{1}{3}(\mathrm{THD}_a^2 + \mathrm{THD}_b^2 + \mathrm{THD}_c^2) \tag{3.116}$$

Along the line of (3.110) we may define two unbalanced harmonic components:

$$\underline{V}_{h,\,\text{ubal1}} = \tfrac{1}{3}(\underline{V}_{ha} + a^{h+1}\underline{V}_{hb} + a^{2h-1}\underline{V}_{hc}) \tag{3.117}$$

and

$$\underline{V}_{h,\,\text{ubal2}} = \tfrac{1}{3}(\underline{V}_{ha} + a^{h-1}\underline{V}_{hb} + a^{2h+1}\underline{V}_{hc}) \tag{3.118}$$

Expressions (3.117) and (3.118) can be used to define two unbalanced THD values, the sum of the squares of which is equal to the square of the unbalanced THD according to (3.112). There is no physical reason for the existence of two unbalanced components, but they cannot be simply added as they affect the different phase voltages in a different way. It is however allowed to add the signal energies of the two unbalanced components to obtain a *total unbalanced harmonic distortion* or unbalanced THD.

#### 3.3.7.4 Time-Domain Approach

The expression for the balanced component, (3.110), can be used as the basis for a time-domain approach to distinguish between balanced and unbalanced distortion. The second term of the right-hand side of (3.110) corresponds to a shift of one-third of the fundamental cycle forward in time. The third term corresponds to a similar shift backward in time. For a distorted current, different in the three phases $i_a(t)$, $i_b(t)$, and $i_c(t)$, the balanced component of the distorted waveforms is obtained from the time-domain equivalent of (3.110):

$$i_{\text{bal}}(t) = \tfrac{1}{3}\left[i_a(t) + i_b\left(t + \tfrac{1}{3}T\right) + i_c\left(t - \tfrac{1}{3}T\right)\right] \tag{3.119}$$

From the waveform of $i_{\text{bal}}(t)$ the spectrum and the THD can be calculated. The latter is identical to the result of (3.109). Note that expression (3.119) is the same as the definition of positive-sequence waveform as introduced by (3.74) in Section 3.2.4.5.

The remainder of the phase current, after subtraction of the balanced component, forms the unbalanced component of the distorted waveforms:

$$\begin{aligned} i_{a,\,\text{ub}}(t) &= i_a(t) - i_{\text{bal}}(t) \\ i_{b,\,\text{ub}}(t) &= i_b(t) - i_{\text{bal}}\left(t - \tfrac{1}{3}T\right) \\ i_{c,\,\text{ub}}(t) &= i_c(t) - i_{\text{bal}}\left(t + \tfrac{1}{3}T\right) \end{aligned} \tag{3.120}$$

where $T$ is the duration of one (power frequency) cycle. The unbalanced THD can be calculated from these current waveforms or from the balanced THD and the THD values for the phase currents by using (3.116).

The collection of all harmonic and interharmonic groups or subgroups obtained from the balanced current is referred to as the *spectrum of the balanced component*. The groups or subgroups obtained from the unbalanced currents are merged into one *spectrum of the unbalanced current* by taking the rms:

$$C_{\text{ubal}} = \sqrt{\tfrac{1}{3}(C_{a,\text{ub}}^2 + C_{b,\text{ub}}^2 + C_{c,\text{ub}}^2)} \qquad (3.121)$$

with $C_{a,\text{ub}}$ the spectrum obtained from $i_{a,\text{ub}}(t)$, and so on.

Figure 3.31 shows the balanced and unbalanced currents for the arc-furnace case during two 200-ms windows. On the left the balanced current is of significantly higher amplitude than the unbalanced current. This is the case during most of the measurement period. However, on the right the balanced and unbalanced currents are of similar amplitude. This corresponds with a sharp drop in positive-sequence current.

The second- and third-harmonic currents as a function of time are presented in Figures 3.32 and 3.33. In both cases the unbalanced harmonic component is higher than the balanced harmonic component when the harmonic groups are used. In an arc furnace the waveform distortion occurs independent of the phase of the fundamental component. The distortion is more or less random and not (as for most three-phase converters) shifted one-third of a (power system frequency) cycle in the three phases. Using harmonic subgroups the balanced and unbalanced harmonic components are about the same for the second harmonic. For the third harmonic we see that the balanced subgroup is significantly smaller than the unbalanced subgroup. The third harmonic is a *zero-sequence harmonic*: The zero-sequence component is the balanced one. As there is no zero-sequence path for the current in this case, the balanced third harmonic is small. The harmonic subgroup only includes the third harmonic proper and two side bands, whereas the harmonic group also contains interharmonic frequencies.

The calculations have been repeated for the normal case, resulting in Figure 3.34. Only the harmonic groups are shown; the results for the harmonic subgroups are the same as this waveform does not contain any significant interharmonic components.

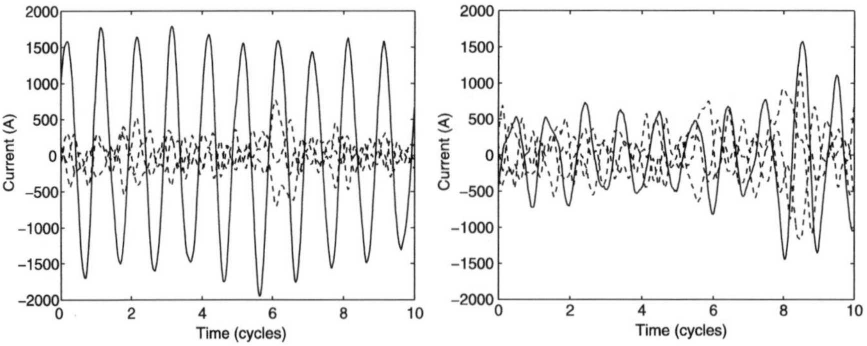


**Figure 3.31** Balanced (solid) and unbalanced (dashed) currents during two different 200-ms windows, arc-furnace case.

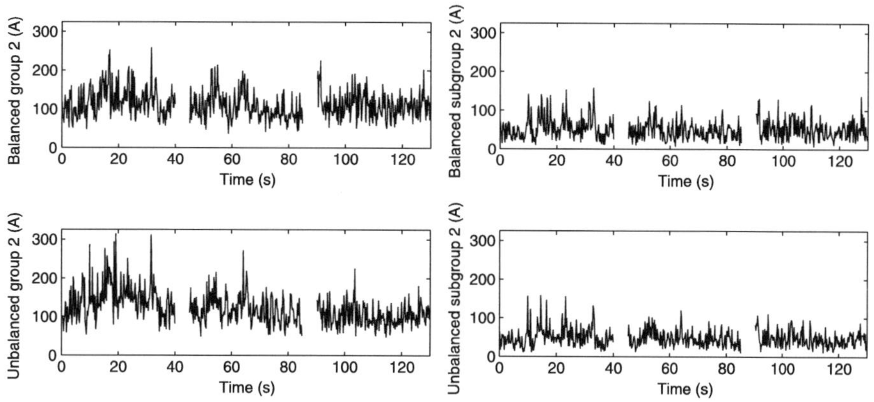


**Figure 3.32** The 200-ms balanced (top) and unbalanced (bottom) second-harmonic component of the current to an arc-furnace installation. The figure on the left shows harmonic groups, the one on the right harmonic subgroups.

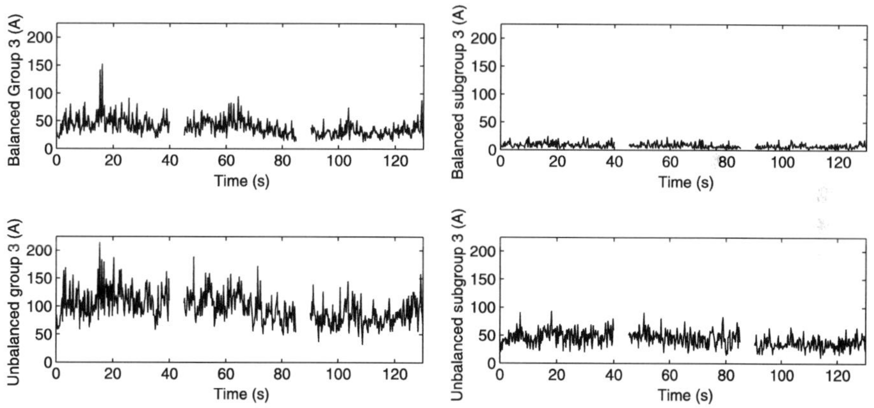


**Figure 3.33** The 200-ms balanced (top) and unbalanced (bottom) third-harmonic components of the current to an arc-furnace installation. The figure on the left shows harmonic groups, the one on the right harmonic subgroups.

### 3.3.8 Power and Power Factor

A large number of definitions exist for the reactive power, apparent power, and power factor in the case of distorted voltages and current. We will not repeat the discussions that have resulted in all these different definitions. Instead we will only summarize the method as proposed in IEEE standard 1459 [171]. The standard document distinguishes between balanced sinusoidal, balanced nonsinusoidal, nonbalanced sinusoidal, and nonbalanced nonsinusoidal three-phase systems. Here we

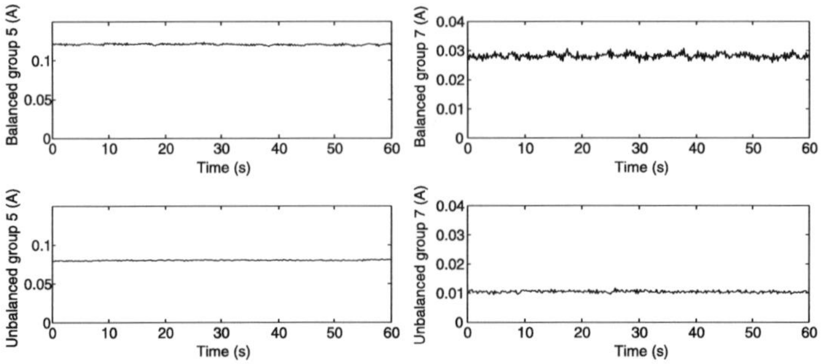


**Figure 3.34** The 200-ms balanced (top) and unbalanced (bottom) fifth- (left) and seventh- (right) harmonic groups for the normal case.

will only describe the latter case as the other cases do not occur in practice and merely have educational value.

The method uses as input the three line-to-neutral voltages (phase voltages) $v_a$, $v_b$, $v_c$ and the three line currents $i_a$, $i_b$, $i_c$. The definition of instantaneous power is simple: It is the flow of energy at any moment in time:

$$p(t) = v_a i_a + v_b i_b + v_c i_c \tag{3.122}$$

The active power $P$ is the average instantaneous power over an integer number of cycles. No window length is defined in the standard, but the same basic window length and time aggregation may be used as for other parameters in the IEC power quality measurement standard IEC 61000-4-30.

The definitions of reactive and apparent power are less obvious. For single-phase systems there is general agreement that the apparent power should be defined as the product of rms voltage and rms current. For three-phase systems there are two alternatives: *arithmetic apparent power* and *vector apparent power*. The arithmetic apparent power is obtained as the sum of the apparent power in each phase:

$$S = S_a + S_b + S_c \tag{3.123}$$

where $S_a$ is the product of rms voltage and rms current in phase $a$, and so on. The vector apparent power for the three-phase system is obtained from the active, reactive, and distortion power for the three-phase system:

$$S_V = \sqrt{P^2 + Q_B^2 + D_B^2} \tag{3.124}$$

with $Q_B = Q_{Ba} + Q_{Bb} + Q_{Bc}$ the Budeanu reactive power for the three-phase system and $D_B = D_{Ba} + D_{Bb} + D_{Bc}$ the Budeanu distortion power for the three-phase system. Budeanu's reactive power is calculated for each phase as the sum of the reactive power in each harmonic component; for phase $a$ this reads as

$$Q_{Ba} = \sum_{h=2}^{H} V_{ha} I_{ha} \sin(\theta_{ha}) \tag{3.125}$$

Budeanu's distortion power is the term that makes up the power triangle with the active and apparent powers in each phase; for phase $a$ this reads as

$$D_{Ba} = \sqrt{S_a^2 - P_a^2 - Q_{Ba}^2} \tag{3.126}$$

With two apparent power definitions there are also two power factor definitions: *Arithmetic power factor* $P_{FA}$ is the ratio of the active power and the arithmetic apparent power and *vector power factor* $P_{FV}$ is the ratio of the active power and the vector apparent power.

For the fundamental component the power factor of the positive-sequence component should be used to calculate the so-called *positive-sequence power factor.* This index replaces the fundamental or displacement power factor used for single-phase systems.

However, neither the vector apparent power nor the arithmetic apparent power gives a correct indication of the actual loading of the system. Therefore IEEE 1459 introduces so-called *effective voltage* and *effective current.* These are positive-sequence voltages and currents that are assumed to transport energy to a purely resistive balanced three-phase load adjusted to permit maximum transfer of power [104]. The virtual balanced circuit is chosen such that it has exactly the same power losses as the actual circuit. As part of the losses occur in the neutral wire in an unbalanced current, the expressions are different for three- and four-wire circuits. In both cases the equivalent voltages and current are defined from the rms phase and line voltages and the rms line and neutral currents. The equivalent current is obtained by assuming the series losses to be invariant. The equivalent voltage is obtained by assuming the shunt losses to be invariant.

For a four-wire system the following definitions are given:

$$I_e = \sqrt{\tfrac{1}{3}(I_a^2 + I_b^2 + I_c^2 + I_n^2)} \tag{3.127}$$

$$V_e = \sqrt{\tfrac{1}{6}(V_a^2 + V_b^2 + V_c^2) + \tfrac{1}{18}(V_{ab}^2 + V_{bc}^2 + V_{ca}^2)} \tag{3.128}$$

For a three-wire system the expressions are

$$I_e = \sqrt{\tfrac{1}{3}(I_a^2 + I_b^2 + I_c^2)} \tag{3.129}$$

$$V_e = \sqrt{\tfrac{1}{9}(V_{ab}^2 + V_{bc}^2 + V_{ca}^2)} \tag{3.130}$$

The effective apparent power is defined from the product of effective voltage and effective current,

$$S_e = 3V_e I_e \tag{3.131}$$

and the effective power factor as the ratio between active power and effective apparent power,

$$P_{Fe} = \frac{P}{S_e} \tag{3.132}$$

In a sinusoidal three-phase system effective voltage and current can be written in terms of symmetrical components:

$$V_e = \sqrt{(V^+)^2 + (V^-)^2 + \tfrac{1}{2}(V^0)^2} \tag{3.133}$$

$$I_e = \sqrt{(I^+)^2 + (I^-)^2 + 4(I^0)^2} \tag{3.134}$$

where $I_0$ and $V_0$ are zero in a three-wire system.

Despite the thoroughness of the definitions in IEEE 1459, the discussion concerning apparent power and power factor definitions is not over yet. The discussion will likely shift to the definition of effective voltage and current. Some assumptions have to be made, among others on the resistance of the return path. In [104] a distinction is made between a "European School" and a "North American School." The definition in IEEE 1459 is the one according to the North American School.

## 3.4 FREQUENCY-DOMAIN ANALYSIS AND SIGNAL TRANSFORMATION

### 3.4.1 Continuous and Discrete Fourier Series

***3.4.1.1 Continuous-Time Periodic Signals*** A continuous-time real-valued signal $v(t)$ with period $T$ can be decomposed into an infinite sum of harmonics as

$$v(t) = \sqrt{2}V_0 + \sum_{k=1}^{\infty} v_k(t) \tag{3.135}$$

where

$$v_k(t) = \sqrt{2}V_k \cos(2\pi k f_0 t + \phi_k) \tag{3.136}$$

The term $V_0$ in (3.135) is the dc component of the signal, and $v_k(t)$ is the $k$th harmonic with rms, or effective magnitude value $V_k$ and phase angle $\phi_k$. Using the Euler equation $\cos(x) = (e^{jx} + e^{-jx})/2$, it follows that (3.135) is equivalent to the following exponential form:

$$v(t) = \sqrt{2}V_0 + \sum_{k=-\infty,\, k\neq 0}^{\infty} \frac{\sqrt{2}}{2} \underline{V}_k e^{j2\pi k f_0 t} \tag{3.137}$$

where $\underline{V}_k = V_k e^{j\phi_k}$ is called the *harmonic phasor* for harmonic $k$ in power system terminology.

Comparing (3.137) to the analysis equation of Fourier series (FS),

$$v(t) = \sum_{k=-\infty}^{\infty} c_k e^{j2\pi k f_0 t} \tag{3.138}$$

it follows that the complex coefficients of FS are related to the harmonic phasor by

$$c_0 = \sqrt{2}V_0 \qquad c_k = \frac{\sqrt{2}}{2} \underline{V}_k \qquad k = \pm 1, \pm 2, \ldots \tag{3.139}$$

where (3.139) implies that the rms magnitude and phase angle of the $k$th harmonic, $k = 1, 2, \ldots$, are related to the complex coefficient of FS by

$$V_k = \sqrt{2}|c_k| = \sqrt{2}\sqrt{[\mathrm{Re}(c_k)]^2 + [\mathrm{Im}(c_k)]^2}, \quad \phi_k = \arg(c_k) = \tan^{-1}\frac{\mathrm{Im}(c_k)}{\mathrm{Re}(c_k)} \tag{3.140}$$

From the FS theory, the coefficients $c_k$ can be obtained from

$$c_0 = \frac{1}{T}\int_{t_0}^{t_0+T} v(t)\, dt \quad \underline{c}_k = \frac{1}{T}\int_{t_0}^{t_0+T} v(t) e^{-jk2\pi f_0 t}\, dt \tag{3.141}$$

Substituting (3.139) into (3.141), it follows that the dc element and the phasor of the $k$th harmonic from a periodic waveform $v(t)$ can be computed from

$$V_0 = \frac{1}{T}\int_{t_0}^{t_0+T} v(t)\, dt \quad \underline{V}_k = \frac{\sqrt{2}}{T}\int_{t_0}^{t_0+T} v(t) e^{-jk2\pi f_0 t}\, dt \tag{3.142}$$

It is also interesting to notice that the kth harmonic is related to the complex FS coefficients by

$$v_k(t) = \tfrac{1}{2}(c_k e^{j2\pi k f_0 t} + c_{-k} e^{-j2\pi k f_0 t}) \tag{3.143}$$

***3.4.1.2 Discrete-Time Periodic Signals*** When the continuous-time signal $v(t)$ in (3.135) is properly sampled (i.e., there is an integer number of samples $N$ in each signal period, $N = f_s/f_0$, where $f_0$ is the power system frequency, in addition to satisfying the Nyquist sampling rate $f_s$), the corresponding discrete-time signal $v(n)$ remains periodic. As in most cases we shall deal with such a discrete-time instead of continuous-time signal. For a discrete-time periodic signal $v(n)$, discrete Fourier series (DFS) should be used to replace the continuous-time Fourier series. A discrete-time real-valued periodic signal $v(n)$ with a bandwidth $f_s/2$ (assuming Nyquist sampling frequency is used) and a period $T = 1/f_0$ can be decomposed into an finite sum of harmonics using

$$v(n) = \sum_{k=0}^{N-1} c_k e^{j2\pi kn/N} \tag{3.144}$$

and the complex coefficients can be obtained by

$$c_k = \frac{1}{N}\sum_{n=0}^{N-1} v(n) e^{-j2\pi kn/N} \qquad 0 \le k < N \tag{3.145}$$

To yield DFS from FS, note that the discrete time $t_n = n\,\Delta t$ and $\Delta t = T/N$ (i.e., each signal period contains an integer number of discrete samples). The relations (3.145) and (3.144) are referred to as DFS and IDFS (inverse DFS), respectively. Similar to that in the continuous-time FS, the dc element and the phasor of harmonics of $v(n)$ can be expressed as

$$V_0 = \frac{1}{N}\sum_{n=0}^{N-1} v(n) \qquad \underline{V}_k = \frac{\sqrt{2}}{N}\sum_{n=0}^{N-1} v(n) e^{-j2\pi kn/N} \qquad k = 1, 2, \ldots, N-1 \tag{3.146}$$

### 3.4.2 Discrete Fourier Transform

Fourier series (or discrete Fourier series) may be very useful from the mathematical viewpoint but their practical application is rather limited. Since the signal is assumed to be periodic, the length of the signal is infinite, $n \in (-\infty, \infty)$. However, in practical situations we have a finite length of measurement data. Discrete Fourier transform allows us to use a finite length of data (which is aperiodic). However, it is assumed that this finite-length of data can be periodically extended such that the finite length signal is a part of the corresponding periodic signal.

Let the finite-length signal consist of $N$ samples from sampling a continuous-time voltage or current signal in an equal interval $t_n$:

$$t_n = n\Delta t = \frac{nT}{N} \qquad n = 0, 1, \ldots, N-1 \tag{3.147}$$

where $N$ is the number of samples within the measurement window $T$. Applying a Fourier transform to a windowed signal (of size $T$) implies that the signal can be periodically extended with the period $T$. Whether this assumption holds or not is not relevant.

The dc element and harmonic phasors of a DFT can be computed by

$$V_0 = \frac{1}{N}\sum_{n=0}^{N-1} v(n) \qquad \underline{V}_k = \frac{\sqrt{2}}{N}\sum_{n=0}^{N-1} v(n)\exp\left(-j\frac{2\pi kn}{N}\right) \qquad k = 1,\ldots,N-1 \tag{3.148}$$

Let $x_n = v(n)$ be the data samples and $X_k$ be the DFT coefficients. The DFT of $x_n$ is defined as

$$X_k = \mathrm{DFT}\{x_n\} = \sum_{n=0}^{N-1} x_n W_N^{kn} \qquad k = 0, 1, \ldots, N-1 \tag{3.149}$$

where $W_N = \exp[-j(2\pi/N)]$. It is worth noticing the factor $N/\sqrt{2}$ difference between (3.148) and (3.149) when computing the harmonics of voltage or current from the DFT spectrum. This difference is due to the normalization used in deriving the expression of DFT (i.e., the normalization factor 1/N is applied in the IDFT: $x_n = (1/N)\sum_{k=0}^{N-1} X_k W_N^{-kn}$). Comparing (3.148) and (3.149), it follows that the dc component and the harmonic phasors are related to the DFT coefficients by

$$V_0 = \frac{1}{N}X_0 \qquad \underline{V}_k = \frac{\sqrt{2}}{N}X_k \qquad k = 1, 2, \ldots, N-1 \tag{3.150}$$

Applying the DFT to $N$ voltage or current samples results in $N$ complex coefficients in the DFT domain. The first coefficient $X_0$ is the dc component scaled by $1/N$. The coefficients $X_k$, $k = 1, \ldots, N/2$, are associated with the complex harmonics for positive frequencies. The remaining coefficients $X_k$, $k = N/2 + 1, \ldots, (N-1)$, are associated with the complex harmonics for negative frequencies. For real signals, the positive and negative frequencies contain the same information so that we only need to consider the coefficients $X_k$, $k = 0, 1, \ldots, N/2$. The highest harmonic in the DFT coefficient is associated with the frequency $N/2T$, which is equivalent to the Nyquist frequency $f_s/2$.

**3.4.2.1 *Discrete Fourier Transform with Rectangular Window*** Since the discrete-time Fourier transform (DTFT) is a linear transform and the DFT spectrum of a discrete-time signal can be obtained by sampling the corresponding continuous DTFT spectrum, we shall first analyze the DTFT signal spectrum.

Applying the DTFT to a finite-length signal can be interpreted as multiplying an infinitely long signal with a finite-length window function that is zero outside the window. If we limit ourselves to a rectangular window function,

$$w(n) = \begin{cases} 1 & |n| \leq \frac{1}{2}T_w \\ 0 & \text{otherwise} \end{cases} \tag{3.151}$$

This is equivalent to applying the DTFT to the windowed signal

$$v_w(n) = v(n)w(n) \tag{3.152}$$

Since a multiplication in the time domain is equivalent to a convolution in the frequency domain, the resulting DTFT spectrum is the convolution of the "actual spectrum" $V(f)$ and the spectrum of the window function $W(f)$,

$$W(f) = \frac{\sin(\pi T_w f)}{\pi f} = T_w \operatorname{sinc}(\pi T_w f) \tag{3.153}$$

Since the DTFT is a linear transform, the DTFT spectrum of a signal is equal to the sum of the DTFT spectrum of individual signal components. As a result we can treat each single-frequency signal component separately. Consider a single-frequency signal $v_k(n)$ with the frequency $f_k$ and magnitude $1/T_w$. Using a rectangular window of size $T_w$ to $v_k(n)$ and applying the DTFT, the resulting DTFT spectrum is

$$V_k(f) = \frac{\sin[\pi T_w(f - f_k)]}{\pi T_w(f - f_k)} \tag{3.154}$$

The spectrum has a maximum value equal to unity for $f = f_k$ and is zero for $f = f_k - m/T_w$, $m \neq 0$.

The DFT is known to result in a discrete line spectrum. This can be obtained by sampling the continuous DTFT spectrum at frequencies

$$f_k = \frac{k}{T_w} \qquad k = 0, 1, \ldots, \tfrac{1}{2}N \tag{3.155}$$

according to the relation between the DTFT and the DFT.

**Example 3.1** Influences of Window Size to DFT Spectra. Four cases are shown in Figure 3.35. The first one (upper left) shows the DTFT spectrum of a 100-Hz periodic signal where a 20-ms rectangular window is applied (i.e., containing two cycles of signal within the window). The magnitude spectrum is equal to unity at $f = 100$ Hz and equal to zero for all integer multiples of 50 Hz ($1/T_w = 50$). In

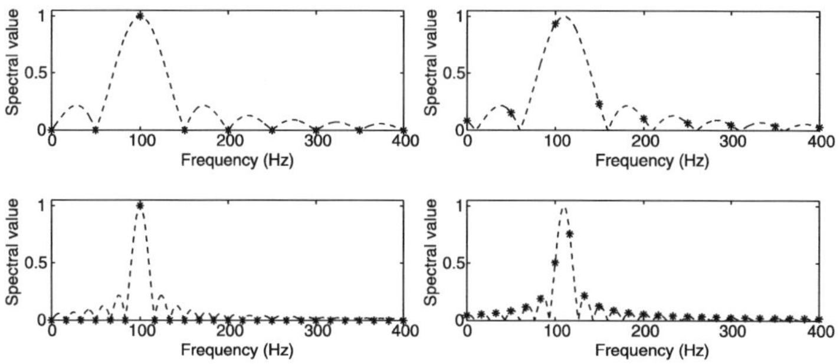


**Figure 3.35** Magnitude spectrum for 100-Hz signal (left) and 110-Hz signal (right). First row corresponds to the magnitude spectra from a 20-ms windowed signal. Second row corresponds to the magnitude spectra from a 60-ms windowed signal. The dashed lines in the figure correspond to the DTFT spectra, while the stars correspond to the DFT spectra.

the second case (lower left) the window length is increased to 60 ms (i.e., six cycles of signal within the window). This leads to a narrower bandwidth in the window function in (3.154) and hence a higher frequency resolution of the spectrum. The magnitude spectrum is equal to unity at $f = 100$ Hz and equal to zero for all other integer multiples of $1/T_w = 16\frac{2}{3}$ Hz. The spectral peaks are closer ("compressed") with much narrower main lobe and side lobe, implying a higher frequency resolution. As shown in Figure 3.35, the DFT line spectrum that resulted from sampling the DTFT spectrum is indicated by stars. For the DFT spectra in the first column of Figure 3.35, we see that all frequency values except the 100-Hz value are equal to zero. The DFT thus results in the correct spectrum. This corresponds to the cases that all frequency components are an integer multiple of $1/T_w$.

In the second column of Figure 3.35, we consider a single-frequency periodic signal with frequency 110 Hz and the DFT spectra corresponding to using 20- and 60-ms windows are shown. It is worth noting that when applying a 20-ms (or 60-ms) window, the windowed signal contains 2.2 (or 6.6) cycles. Recall the assumption in the DFT theory that applying a DFT to a block of data implies that the actual signal is an infinite-length periodic signal that can be obtained by periodically extending the block data on both sides. Noting that if we periodically extend this 20-ms (or 60-ms) data block, we will obtain another periodic signal that is not the same as the original 110-Hz periodic signal! In these two cases, the spectral lines obtained from the DFT do not coincide with the zeros in the corresponding continuous spectrum. Instead of obtaining a single spectral line at 110 Hz from the main lobe, we obtain a spectral line at 100 Hz plus many other spectral lines leaked from the side lobes of the window frequency response due to window "spectral leakage."

The results from Example 3.1 have shown that the size of a windowed signal must contain one or several complete cycles in order to obtain the correct DFT line spectrum of the signal. A windowed signal with incomplete cycle(s) will lead to incorrect line spectrum due to the window spectral leakage and will not satisfy the periodic signal assumption.

*This implies that when analyzing power system measurement data, it is important that the window length be synchronized to the power system frequency* (which means a complete cycle of the fundamental signal component as well as an integer number of cycles of the harmonic signal components). In such a way, the DFT will result in the spectrum lines exactly at all harmonic frequencies of the signal. There will be no spectral lines at the interharmonics.

#### 3.4.2.2 Interpolation of Discrete Spectral Lines

A more densely spaced spectral line can be obtained by padding zeros at the end of original signal samples, that is for $N$ signal samples, padding $\hat{N} - N$ zeros behind the signal, $[x_0, x_1, \ldots, x_{N-1}, 0, \ldots, 0]$, before applying the DFT. The DFT coefficients $X_k$, $k = 0, 1, \ldots, \hat{N}/2$, are then extracted as the spectrum. The frequency difference between two adjacent DFT spectral lines becomes $\Delta f = f_s/\hat{N}$ instead of the previous $f_s/N$.

It should be noted that zero padding does not increase the frequency resolution of the DFT spectrum; rather it only generates the interpolated spectrum. The frequency resolution is defined as the narrowest frequency distance between two closely located frequency components of a signal that the system is able to distinguish:

$$(\Delta\omega)_{3\text{dB}} = 2\pi|f_1 - f_2| \tag{3.156}$$

This is determined by the length and the type of the window that is used in the DFT, noting that using the DFT implies multiplication of data with a rectangular window of length $T_w = N\,\Delta t$ ($N$ is the actual number of data samples). This rectangular window of length $N\,\Delta t$ implies that the DFT spectrum has a frequency resolution of

$$(\Delta\omega)_{3\text{dB}} = 0.89\frac{2\pi}{N\,\Delta t} \tag{3.157}$$

despite the length $\hat{N}$ after the zero padding. Whether one would be able to obtain nearby frequency components in the DFT spectrum after zero padding depends on this exact frequency resolution. If this frequency resolution is too low, then the spectrum lines between two harmonic frequencies cannot be explained as due to individual frequency components.

#### 3.4.2.3 Synchronization

When using a rectangular window, it is important to synchronize the measurement window accurately with the power system frequency. The reason for this can easily be understood from the reasoning on the previous DFT results. For example, when the power system frequency is 50.2 Hz whereas the window size is 20 ms, the spectral lines of the DFT no longer correspond exactly to the zero crossings of the continuous frequency response for the power signal frequency component. The DFT spectrum of a single power system frequency signal becomes an interharmonic with spectral leakage as a consequence. Fortunately the power system frequency rarely deviates much from its nominal value so that the effect is small in most circumstances. However, one needs to take care in some cases, for example during measurements in small (islanded) system or during island operations of industrial systems.

**Example 3.2** Window Synchronization. An example of spectral leakage is shown in Figure 3.36. The first example (on the left) shows the line spectrum resulting from the DFT for a 51-Hz signal over a 20-ms window. The vertical scale is enlarged compared with Figure 3.35. The spectrum obtained contains about 2% dc and second-harmonic component, about 1% third-harmonic component, and so on. As the dc and second-harmonic components are normally much lower, it could be contributed by the nonsynchronized window. This example shows that nonsynchronization may have a serious impact on the overall spectral results. When a spectrum cannot be explained well, one may examine other possible artifacts which might be introduced from, for example a nonsynchronized window and discrete periodicity.

When the power system frequency equals 51 Hz, its fifth-harmonic component is 255 Hz. The spectral leakage due to this is shown in the second example of Figure 3.36 (on the right). The nonsynchronization gives about 10% leakage to the neighboring spectral lines. Note that the fifth-harmonic component is typically 1 to 10% of the fundamental power system frequency component, while the fourth- and sixth-even-harmonic components are up to 1% of the fundamental. In the figure, these even-harmonic values are again rather high. Care should be taken in the interpretation of these results.

#### *3.4.2.4 Interharmonics*

Interharmonics are signal components at frequencies that are not integer multiples of the power system frequency, for example, a 190-Hz component in a 60-Hz system. To correctly measure an interharmonic component requires a measurement window which is also an integer multiple of

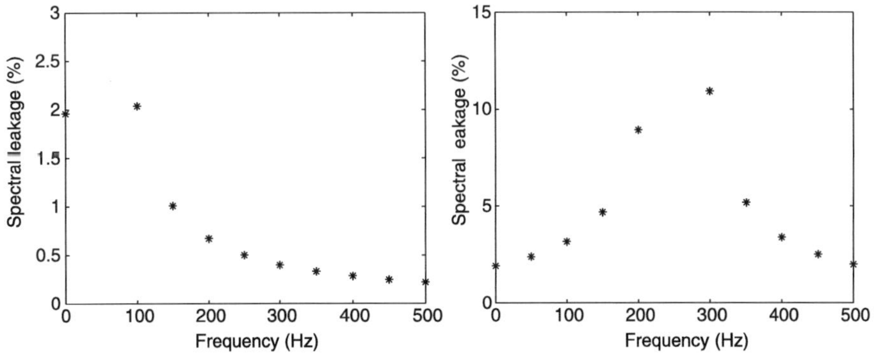


**Figure 3.36** Spectral leakage due to nonsynchronization between measurement window and power system frequency. Left: 51-Hz signal over a 20-ms windows. Right: 255-Hz signal over a 20-ms window.

the interharmonic cycle length. For the above example the minimum window size is 100 ms (6 cycles of 60 Hz and 19 cycles of 190 Hz). However, if the interharmonic frequency is 187 Hz [287], a 1-s window is needed. As the interharmonic frequency is rarely known beforehand, a very long window length would be needed, especially where more interharmonic components may be present.

From this reasoning one may conclude that there is no complete solution to the interharmonic measurement problem. The approach in the IEC standards (IEC 61000-4-7 and IEC 61000-4-30) is a pragmatic one: The 200-ms window gives 5 Hz separation between spectral lines, which is considered sufficient for characterization purposes. The use of a longer window would place higher computational demands on the monitor and also reduce the time resolution. In [290] it is shown that the use of a Hanning window instead of a rectangular window may lead to better results.

#### 3.4.2.5 More Examples

**Example 3.3** DFT of a Sampled Voltage Signal. The following (hypothetically) sampled voltage signal is obtained over a 20-ms window:

$$v_n = \begin{bmatrix} 0 & 1 & 1 & 0 & 0 & -1 & -1 & 0 \end{bmatrix}$$

Applying the DFT to the above sampled data results in the following complex numbers:

$$\text{DFT}\{v_n\} = \begin{bmatrix} 0 & 1.41 - j3.41 & 0 & -1.41 + j0.59 & 0 & -1.41 - j0.59 & 0 & 1.41 + j3.41 \end{bmatrix}$$

To obtain the actual spectrum of voltage harmonics, we have to multiply the first value by $1/N = 0.125$ and the other values by $\sqrt{2}/N = 0.177$. Only the first five values are needed for the spectrum, which are

$$\begin{aligned} V_0 &= 0 \\ V_1 &= 0.25 - j0.6036 \\ V_2 &= 0 \\ V_3 &= -0.25 + j0.1036 \\ V_4 &= 0 \end{aligned}$$

The DFT magnitude spectrum can then be obtained as {0, 0.6533, 0, 0.2706, 0}.

**Example 3.4** DFT Spectrum of a One-Cycle (20-ms) Measured Current Signal. Figure 3.37*a* shows the waveforms of a measured current signal. It is clear from the figure that the signal is (or is close to) periodic with a period of 20 ms. Applying the DFT to the first 20 ms of sampled signal results in the line spectrum at discrete frequencies 0, 50, 100, 150 Hz, and so on. Figure 3.37*b* shows the magnitude values of the first few harmonics obtained from using (3.150). The figure shows that the 50-Hz component (or the first harmonic) dominates, which is already

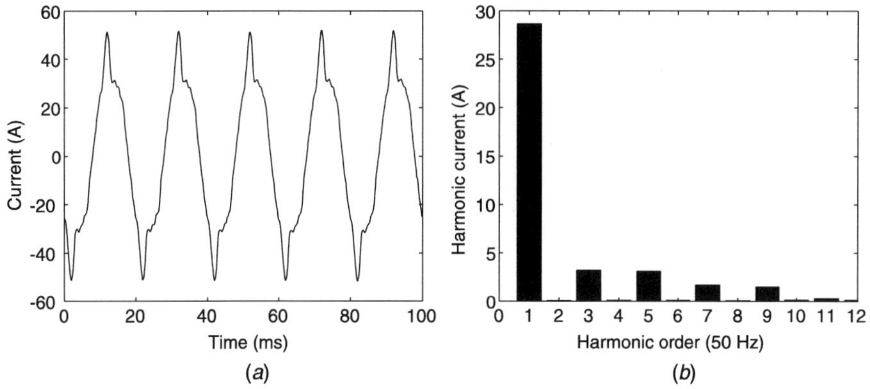


**Figure 3.37** The DFT spectrum obtained from a one-cycle window. (*a*) The original discrete-current waveforms: five cycles of the fundamental frequency; (*b*) The DFT spectrum using a one-cycle window.

visible from the original signal in Figure 3.37*a*. The current contains the significant components at odd harmonics 150, 250, 350, and 450 Hz while the even harmonics are small. This can be confirmed from Figure 3.37*a* by noticing the nearly symmetric wave shape for the two half-cycles (i.e., the positive and negative part of each cycle) in the original signal waveform.

**Example 3.5** DFT Spectrum of a Five-Cycle (100-ms) Measured Current Signal. Figure 3.38 contains the results obtained by applying the DFT to a

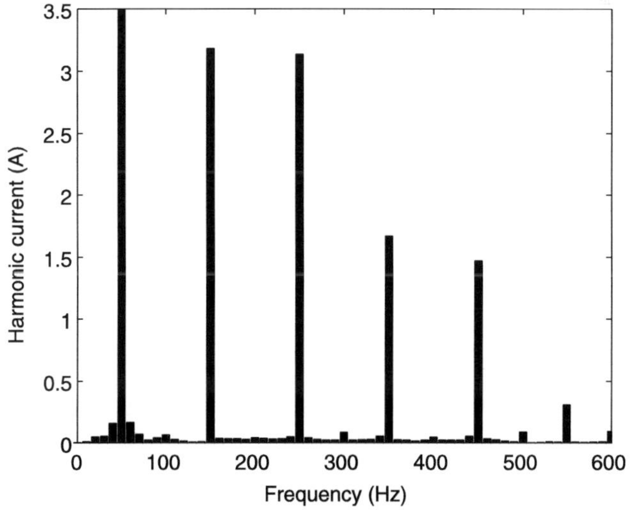


**Figure 3.38** The DFT spectrum obtained from a five-cycle window. Note the difference in vertical scale with Figure 3.37*b*.

five-cycle windowed signal shown in Figure 3.37*a*. This results in the line spectrum located at frequencies 0, 10, 20, 30, 40 Hz, and so on. The figure shows that the main contributions are still found at the multiples of 50 Hz. As the actual data length has increased five times, a higher frequency resolution is clearly visible in this case as compared with that in the Example 3.4. When the current shows a cycle-to-cycle variation, the spectrum lines also contain components at nonharmonic frequencies.

**Example 3.6** Observing Interharmonics in the DFT Spectrum by Using a Large Data Window. An example of a voltage magnitude spectrum with interharmonic components is shown in Figure 3.39. The upper curve shows the DFT spectrum obtained from a signal containing 10 cycle periods, the bottom curve the spectrum of a signal containing 12 cycles. A clear line spectrum can be seen: Interharmonic frequencies are present around 185 and 285 Hz. (Note that the same signal was used as in Figure 3.17.) The upper curve shows a spread in the spectrum because the window length was not an integer multiple of the cycle length of the interharmonics. Recalculating the spectrum with a 600-ms window shows that the interharmonic components are around 183 and 283 Hz. Three cycles of the fundamental frequency (60 ms) correspond to close to 11 cycles of the 183-Hz component and close to 17 cycles of the 283-Hz component.

**Example 3.7** Interharmonics in Arc-Furnace Measurement Data. In Figure 3.15 the spectrum was shown of the voltage at the terminals of a large arc-furnace

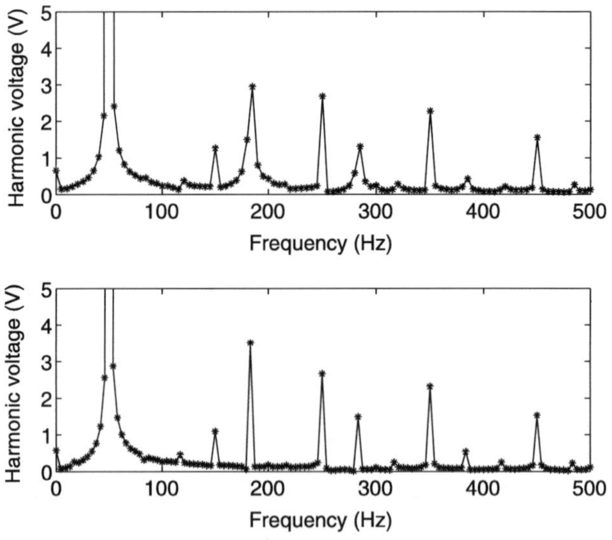


**Figure 3.39** The DFT spectrum of voltage signal with discrete interharmonics: corresponding to using a 10-cycle window (top) and a 12-cycle window (bottom). The voltage signal is the same as that used in Figure 3.17 but with different lengths.

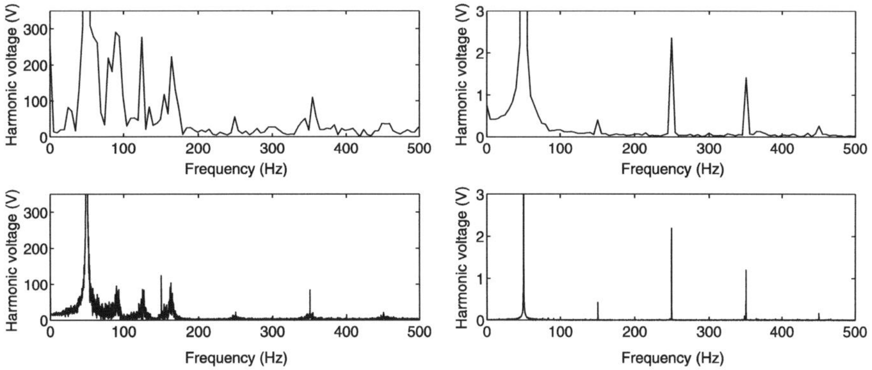


**Figure 3.40** The DFT spectrum of voltage for the arc-furnace case (left) and the normal case (right) obtained using a 200-ms window (top) and a 1-s window (bottom).

installation. That spectrum was obtained by using the signal with a 200-ms-long window. The spectrum is repeated in the upper curve of Figure 3.40. The bottom curve gives the DFT spectrum of the signal with a 1-s window. As clearly visible from the figure, using a 1-s window leads to a significantly higher frequency resolution (in fact, five times higher frequency resolution as compared with that using the 200-ms window). There are no discrete interharmonics visible; instead the spectrum shows continuous noise.

For comparison Figure 3.40 also shows the spectrum of the voltage for the normal case. As mentioned before, the spectrum is dominated by the fifth and seventh harmonics. Comparing the 200-ms and 1-s spectra shows another interesting phenomenon. The noise present between the harmonic component appears to be much smaller for the 1-s spectrum than for the 200-ms spectrum. The discrete spectral components (at harmonic frequencies) are however not affected by the window length. The amount of signal energy in a certain frequency window (e.g., between 250 and 350 Hz) is not affected, but it is spread over a larger frequency range. To compare two broadband spectra obtained over different windows, a scaling of the values is needed, either according to Parseval's theorem in (3.41) or noting the scaling factor in (3.153). However, for most harmonic studies the scaling is done such that the amplitudes of discrete components are correctly reproduced.

## 3.5 ESTIMATION OF HARMONICS AND INTERHARMONICS

### 3.5.1 Sinusoidal Models and High-Resolution Line Spectral Analysis

In many cases waveform distortion is related to disturbances in some narrow bands or disturbances at discrete frequencies (at either harmonic or interharmonic frequencies of a power system). To characterize harmonic and interharmonic disturbances, sinusoidal models are the most suitable choice.

**3.5.1.1 *Sinusoidal Models*** Let the discrete-time signal $v(n)$ of finite length $L$ be represented by a sinusoidal model with $K$ sinusoidal components in noise,

$$v(n) = \sum_{k=1}^{K} a_k \cos(n\omega_k + \phi_k) + \omega(n) \tag{3.158}$$

where $a_k$ is the magnitude, $\phi_k$ is the initial phase angle, $\omega_k = 2\pi f_k$ is the harmonic (or interharmonic) frequency in radius, and $K$ is the total number of sinusoids. To prevent model ambiguity, $a_k \geq 0$ is assumed. In the model, $a_k$ and $\omega_k = 2\pi f_k$ are assumed to be deterministic and unknown and $\phi_k$ is unknown and assumed to be random and uniformly distributed in $[-\pi, \pi]$. Alternatively, the model in (3.158) can be equivalently expressed in the form of complex exponentials in noise (which is frequently referred to as the *harmonic model* in signal-processing terminology),

$$v(n) = \sum_{k=1}^{K} \underline{A}_k e^{jn\omega_k} + \omega(n) \tag{3.159}$$

where $A_k = |\underline{A}_k| e^{j\phi_k}$ is the complex magnitude of the $k$th-harmonic signal component,

$$s_k(n) = \underline{A}_k e^{jn\omega k} \qquad k = 1, \ldots, K \tag{3.160}$$

The magnitudes in the complex exponential model (3.159) and in the sinusoidal model (3.158) are related by

$$a_k = 2|\underline{A}_k| \tag{3.161}$$

Using the conventional expression of power system voltage or current waveforms in noise,

$$v(n) = \sum_{k=1}^{k} \sqrt{2} V_k \cos(\omega_k n + \varphi_k) + \omega(n) \tag{3.162}$$

(3.159) and (3.162) are related by

$$\begin{aligned} \sqrt{2} V_k &= 2|\underline{A}_k| = 2\sqrt{[\mathrm{Re}(\underline{A}_k)]^2 + [\mathrm{Im}(\underline{A}_k)]^2} \\ \varphi_k &= \phi_k = \tan^{-1} \frac{\mathrm{Im}(\underline{A}_k)}{\mathrm{Re}(\underline{A}_k)} \end{aligned} \tag{3.163}$$

*Analysis of Harmonic and Interharmonic Disturbances in Power Systems*

- When using sinusoidal models for power system voltage or current waveforms, it is often assumed that the power system frequency is $f_0$ and $f_k = \tilde{k}f_0$ is the $\tilde{k}$th-harmonic frequency of the power system, $k = 2,3,\ldots$. If $K - 1$ significant harmonics are located in the first several odd-harmonic frequencies (including the fundamental), $f_k = (2\tilde{k} + 1)f_0, \tilde{k} = 1, 2, \ldots, K - 1$, then only the fundamental and the first $(K - 1)$ odd harmonics will be included in the model.

  If $K$ significant harmonics are located in the fundamental and even-harmonic frequencies $f_k = 2\tilde{k}f_0, \tilde{k} = 1, 2, \ldots, 2(K - 1)$, then only the fundamental (or first harmonic) and the $K - 1$ even harmonics will be included in the model. In such a case, the model does not need to contain any odd harmonics.
- It is worth noticing that in the sinusoidal (or harmonic) model the range of $\omega_k = 2\pi f_k$ [where $f_k = \tilde{f}_k(\mathrm{Hz})/f_s$ is the normalized frequency] in (3.158) or (3.159) is $[-\pi, \pi]$; therefore $f_k$ can be located at any nonharmonic frequency. There is no restriction in the mathematical model that $f_k$ should be located at the harmonics of the power system frequency $f_0$.

Since the frequency resolution in the model-based approach is high (provided that the model is correct), the estimation of interharmonics becomes a realistic issue. However, in the harmonic model it is assumed that the model order $K$ (the number of the harmonics) is known. This is usually not true for practical applications.

### 3.5.2 Multiple Signal Classification

The MUSIC method employs a harmonic model and estimates the frequencies and powers of the harmonics in the signal. The MUSIC algorithm is a noise subspace-based method.

For a giving data sequence $v(n)$ of length $L = N + M - 1$ in (3.159), its autocorrelation matrix $\mathbf{R}_v$ of size $M \times M$, [where $M$ is the dimension of space spanned by $v(n)$ and $K$ is the dimension of the signal subspace, $M > K$] can be estimated from the data samples by using

$$\hat{\mathbf{R}}_v = \frac{1}{N}\mathbf{V}^H\mathbf{V} \tag{3.164}$$

where the data matrix $\mathbf{V}$ is of size $N \times M$ and is described by

$$\mathbf{V} = \begin{bmatrix} \mathbf{v}^{\mathrm{T}}(0) \\ \mathbf{v}^{\mathrm{T}}(1) \\ \vdots \\ \mathbf{v}^{\mathrm{T}}(N-1) \end{bmatrix}^{\mathrm{T}} = \begin{bmatrix} v(0) & v(1) & \cdots & v(M-1) \\ v(1) & v(2) & \cdots & v(M) \\ \vdots & \vdots & \ddots & \vdots \\ v(N-1) & v(N) & \cdots & v(N+M-2) \end{bmatrix} \tag{3.165}$$

where $M$ is the time window length of the (row) data vector and the superscript $(\cdot)^{\mathrm{H}}$ is the Hermitian operator [223] (for the complex conjugate and transpose of the matrix). Substituting (3.159) in (3.164) yields

$$\hat{\mathbf{R}}_v = \mathbf{R}_s + \mathbf{R}_w = \mathbf{EPE}^{\mathrm{H}} + \sigma_w^2\mathbf{I} \tag{3.166}$$

where the matrices $\mathbf{E}$ and $\mathbf{P}$ are defined as

$$\mathbf{E} = \begin{bmatrix} \mathbf{e}_1 & \mathbf{e}_2 & \cdots & \mathbf{e}_K \end{bmatrix} \qquad \mathbf{P} = \mathrm{diag}\{|\underline{A}_1|^2 |\underline{A}_2|^2 \cdots |\underline{A}_K|^2\} \tag{3.167}$$

and

$$\mathbf{e}_l = \begin{bmatrix} 1 & e^{j\omega_l} & e^{j2\omega_l} & \cdots & e^{j(M-1)\omega_l} \end{bmatrix}^{\mathrm{T}} \qquad l = 1, 2, \ldots, K \tag{3.168}$$

are the eigenvectors of $\mathbf{R}_s$.

*Signal and Noise Subspaces* Assuming $\hat{\mathbf{R}}_v$ is of full rank $M$, the eigenvalues of $\hat{\mathbf{R}}_v$ are arranged in decreasing order (i.e., $\lambda_1 \geq \lambda_2 \geq \cdots \geq \lambda_M$), and the corresponding eigenvectors are $\mathrm{s}_1, \mathrm{s}_2, \ldots, \mathrm{s}_M$, it follows that

$$\hat{\mathbf{R}}_v \mathbf{s}_i = \lambda_i \mathbf{s}_i \qquad i = 1, 2, \ldots, M \tag{3.169}$$

These eigenvectors can be divided into two groups, $K$ eigenvectors corresponding to the $K$ largest eigenvalues belong to the signal subspace and the remaining $M - K$ eigenvectors belong to the noise subspace. In the MUSIC method, noise subspace is used to estimate the unknown harmonic frequencies $\omega_k$. First, the following *pseudospectrum* is computed,

$$P_{\mathrm{music}}(e^{j\omega}) = \frac{1}{\sum_{i=K+1}^{M} |\mathbf{e}^{\mathrm{H}}\mathbf{s}_i|^2} \tag{3.170}$$

where $\mathrm{s}_i$, $i = K + 1, \ldots, M$, are the eigenvectors associated with the noise subspace that are orthogonal to the signal eigenvector $\mathrm{e} = [1\ e^{j\omega}\ e^{j2\omega} \cdots e^{j(M-1)\omega}]^{\mathrm{T}}$, and $\mathbf{e}^{\mathrm{H}}$ denotes the complex-conjugate transpose. This implies that the denominator has zero values at the frequencies related to the signal eigenvectors. *It is worth emphasizing that* $\mathrm{P}_{music}(e^{j\omega})$ *in (3.170) does not relate to any real power spectrum; rather the only purpose of this pseudospectrum is to generate peaks whose frequencies correspond to those of the dominant frequency components.*

The Z-domain equivalent of (3.170) is

$$P_{\mathrm{music}}(z) = \left[\sum_{i=K+1}^{M} S_i(z) S_i^*\left(\frac{1}{z^*}\right)\right]^{-1} \tag{3.171}$$

where $S_i(z) = \sum_{m=0}^{M-1} s_i(m)z^{-m}$, $s_i(m)$ denoting the $m$th element in the $i$th eigenvector. Since signal subspace and noise subspace are orthogonal, the denominator in (3.171) should be zero at those harmonic frequencies.

*Estimation of Frequencies* The frequencies of the complex exponentials $\omega_k = 2\pi f_k, k = 1, 2, \ldots, K$, in (3.159) can be obtained either from the frequency locations corresponding to the $K$ highest peaks in the pseudospectrum $P_{\text{music}}(e^{j\omega})$ or from (3.171) as the angles of $K$ roots of $P_{\text{music}}(z)$ that are closest to the unit circle in the $z$-domain. Note that these frequencies are the same for $v(n)$ in (3.162).

*Estimation of Harmonic Magnitudes and Powers* Once the $\omega_k$ are estimated, $P_k = |\underline{A}_k|^2$ can be estimated by jointly solving the $K$ equations

$$\sum_{k=1}^{K} P_k |\mathbf{e}_k^{\mathrm{H}} \mathbf{s}_i|^2 = \lambda_i - \hat{\sigma}_w^2 \qquad i = 1, 2, \ldots, K \tag{3.172}$$

where the noise power is estimated by

$$\hat{\sigma}_w^2 = \frac{1}{M-K} \sum_{i=K+1}^{M} \lambda_i \tag{3.173}$$

Noting $\mathbf{e}_k^{\mathrm{H}} \mathbf{s}_i = S_i(e^{j\omega_k}) = \sum_{m=0}^{M-1} s_i(m) e^{-jm\omega_k}$, (3.172) is equivalent to

$$\begin{bmatrix} |S_1(e^{j\omega_1})|^2 & |S_1(e^{j\omega_2})|^2 & \cdots & |S_1(e^{j\omega_K})|^2 \\ |S_2(e^{j\omega_1})|^2 & |S_2(e^{j\omega_2})|^2 & \cdots & |S_2(e^{j\omega_K})|^2 \\ \vdots & \vdots & \ddots & \vdots \\ |S_K(e^{j\omega_1})|^2 & |S_K(e^{j\omega_2})|^2 & \cdots & |S_K(e^{j\omega_K})|^2 \end{bmatrix} \begin{bmatrix} P_1 \\ P_2 \\ \vdots \\ P_K \end{bmatrix} = \begin{bmatrix} \lambda_1 - \hat{\sigma}_w^2 \\ \lambda_2 - \hat{\sigma}_w^2 \\ \vdots \\ \lambda_K - \hat{\sigma}_w^2 \end{bmatrix} \tag{3.174}$$

The solution of (3.174) yields the harmonic power $P_k, k = 1, 2, \ldots, K$. Finally, using (3.161) and (3.162), it follows that the magnitude of the $k$th harmonic of $v(n)$ in (3.162) is

$$V_k = \sqrt{2P_k} \tag{3.175}$$

**Example 3.8** Analysis of Synthetic Signal Consisting of Only Odd Harmonics. In this example, the synthetic signal consists of the first six odd harmonics (i.e., odd harmonics from the 3rd to the 13th) of the power system frequency $f_0 = 50$ Hz in additive white noise $\omega(n)$, described by

$$x(n) = s(n) + w(n) = \sum_{k=2}^{7} a_{2k-1} \cos\left(2\pi n(2k-1)\frac{f_0}{f_s} + \phi_{2k-1}\right) + w(n) \tag{3.176}$$

**TABLE 3.3 Typical and High Levels of Odd-Harmonic Voltage Magnitude Levels in Europe**

| Harmonic No. (%) | 3 | 5 | 7 | 9 | 11 | 13 | 15 | 17 | 19 | 21 | 23 | 25 |
|---|---|---|---|---|---|---|---|---|---|---|---|---|
| Typical value | 1.5 | 4 | 4 | 0.8 | 2.5 | 2 | $\leq$0.3 | 1 | 0.8 | $\leq$0.3 | 0.8 | 0.8 |
| High value | 2.5 | 6 | 5 | 1.5 | 3.5 | 3 | $\leq$0.3 | 2 | 1.5 | $\leq$0.3 | 1.5 | 1.5 |

*Note*: The high levels are based on EN 50160 [106].

The magnitudes of odd harmonics $a_{2k-1}$ are set to be the typical values described in Table 3.3, the initial phases are set to $\{0^\circ\ 30^\circ\ 60^\circ\ 90^\circ\ 120^\circ\ 180^\circ\}$ (ordered according to the harmonic numbers), and the noise power is set to $\sigma_w^2 = 1.0$.

Figure 3.41*a* shows the signal waveforms and Figure 3.41*b* the pseudospectrum from the MUSIC method. As can be seen from Figure 3.41*b*, the peaks indeed

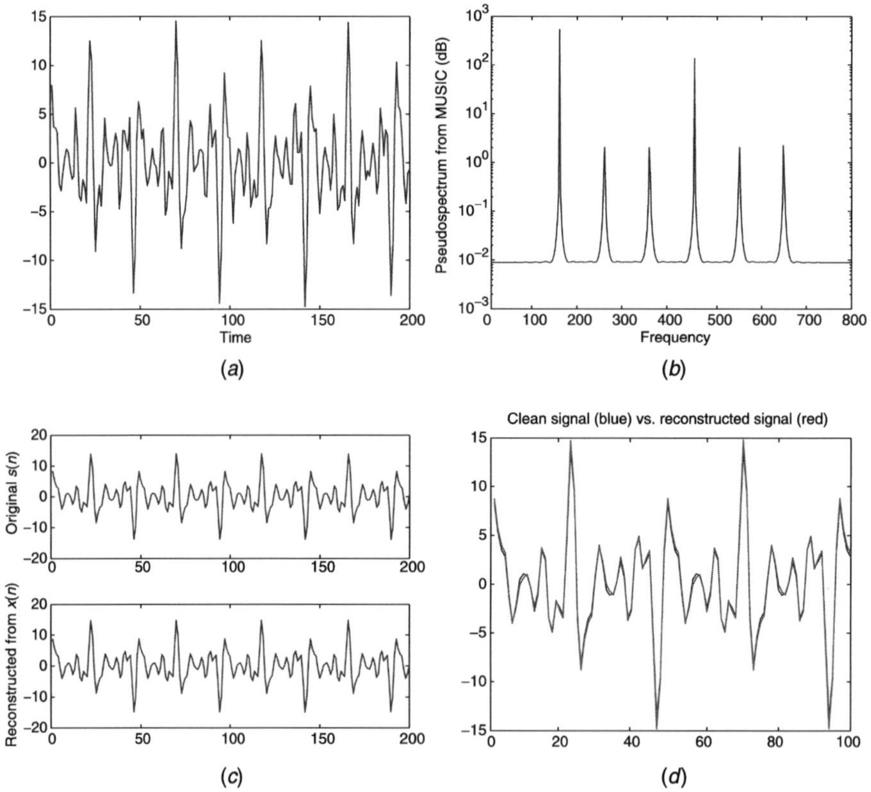


**Figure 3.41** Estimating odd-harmonic disturbances by MUSIC: (*a*) original synthesized noisy waveforms $x(n)$; (*b*) pseudospectrum obtained from MUSIC where peaks indicate frequencies of harmonics; (*c*) original clean signal $s(n)$ versus reconstructed signal using parameters estimated from the MUSIC; (*d*) overlapped plot of (*c*).

**TABLE 3.4 Typical and High Levels of Even-Harmonic Voltage Magnitude Levels in Europe**

| Harmonic No. (%) | 2 | 4 | 6,8, . . . , 24 |
|---|---|---|---|
| Typical value | 0.5 | 0.3 | 0.3 |
| High value | 2 | 1 | 0.5 |

*Note*: The high levels are based on EN 50160 [106].

correspond to the odd-harmonic frequencies. Since this is a pseudospectrum, only the frequencies of peaks are meaningful, while their magnitudes are irrelevant to the actual harmonic magnitudes! The resulting harmonic frequencies and magnitudes from the estimation are shown in Table 3.5. Figure 3.41*c* and *d* show the reconstructed waveforms using the estimated parameters as compared with the original clean signal.

**Example 3.9** Analysis of Synthetic Signal Consisting of Both Even and Odd Harmonics. In this example, the synthetic signal consists of the first 12 harmonics of $f_0 = 50$ Hz (i.e., even and odd harmonics from the 2nd to the 13th) in additive white noise $w(n)$,

$$x(n) = s(n) + w(n) = \sum_{k=2}^{13} a_k \cos\left(2\pi nk\frac{f_0}{f_s} + \phi_k\right) + w(n) \tag{3.177}$$

where the harmonic magnitudes are set to the typical values in Tables 3.3 and 3.4, the initial phases are {0° 20° 30° 40° 50° 60°} for the six odd harmonics and {70° 80° 90° 100° 110° 120°} for the six even harmonics, and the noise power $\sigma_w^2 = 1.0$.

Figure 3.42*a* shows the signal waveforms and Figure 3.42*b* the pseudospectrum from the MUSIC method. As can be seen from Figure 3.42*b*, the peaks indeed correspond to the first even- and odd-harmonic frequencies. The resulting harmonic frequencies and magnitudes from the estimation are shown in Table 3.6.

**TABLE 3.5 Estimated Results from MUSIC**

| | | Magnitude | |
|---|---|---|---|
| Harmonic No. | Frequency (Hz) | Estimated | True |
| 3 | 149.9649 | 1.6816 | 1.5 |
| 5 | 250.0043 | 4.2272 | 4.0 |
| 7 | 350.0021 | 4.1496 | 4.0 |
| 9 | 449.9321 | 0.9307 | 0.8 |
| 11 | 549.9942 | 2.6324 | 2.5 |
| 13 | 649.9668 | 2.1037 | 2.0 |

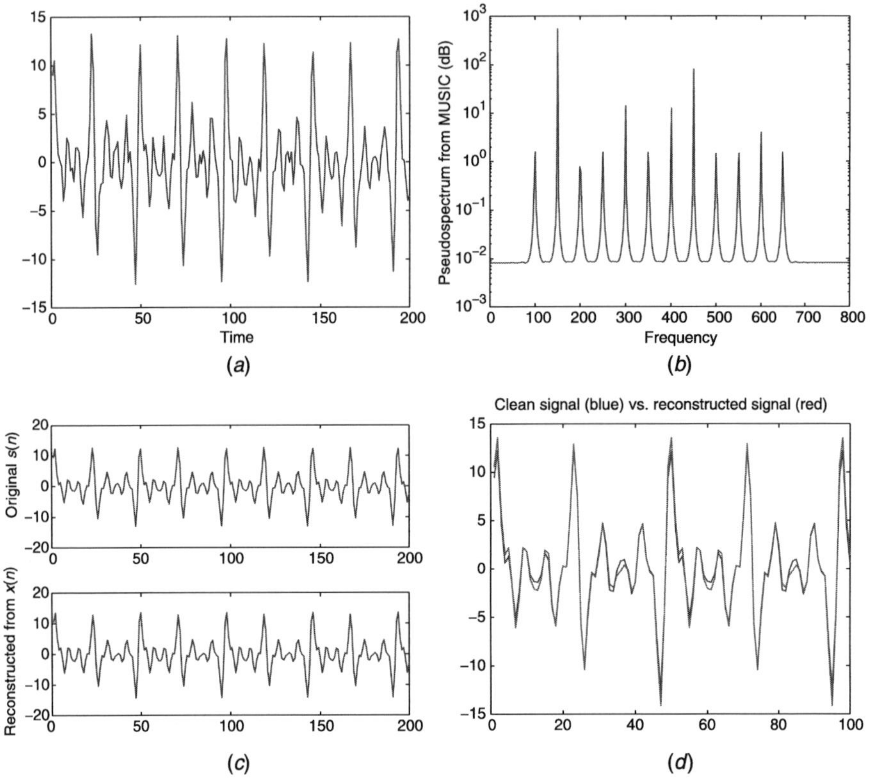


**Figure 3.42** Estimating even- and odd-harmonic disturbances by MUSIC: (*a*) original synthesized waveform $x(n)$; (*b*) pseudospectrum obtained from MUSIC where peaks indicate frequencies of odd and even harmonics; (*c*) original clean signal $s(n)$ versus reconstructed signal using parameters estimated from MUSIC; (*d*) overlapped plot of (*c*).

**Example 3.10** Analysis of Synthetic Signal Consisting of Interharmonics and Harmonics. In this example the synthetic signal consists of five harmonics (from the second to the fifth harmonics) and two interharmonics in $f_1 = 82$ Hz and $f_2 = 182$ Hz in additive white noise $w(n)$ synthesized according to

$$x(n) = s(n) + w(n) = \sum_{k=2,3,4,5,7} a_k \cos\left(2\pi nk\frac{f_0}{f_s} + \phi_l\right)$$

$$+ \sum_{l=1}^{2} b_l \cos\left(2\pi n\frac{f_l}{f_s} + \phi_l\right) + w(n) \qquad (3.178)$$

Again the harmonic magnitudes are set to the typical values in Tables 3.3 and 3.4, the magnitudes of interharmonics are set to $b_1 = 3.0$ and $b_2 = 5.0$ (these values are taken from one of the real case scenarios), the initial phases are {0° 30° 60° 120°

**TABLE 3.6 Estimated Results from MUSIC**

| Harmonic No. | Estimated Frequency (Hz) | Magnitude | |
|---|---|---|---|
| | | Estimated | True |
| 2 | 100.0246 | 0.7465 | 0.5 |
| 3 | 149.9710 | 1.6615 | 1.5 |
| 4 | 200.3171 | 0.4375 | 0.3 |
| 5 | 250.0051 | 4.2550 | 4.0 |
| 6 | 300.1083 | 0.5232 | 0.3 |
| 7 | 350.0066 | 4.1306 | 4.0 |
| 8 | 400.6756 | 0.2374 | 0.3 |
| 9 | 449.9169 | 0.9349 | 0.8 |
| 10 | 499.9248 | 0.4670 | 0.3 |
| 11 | 549.9893 | 2.6231 | 2.5 |
| 12 | 600.2224 | 0.3702 | 0.3 |
| 13 | 649.9953 | 2.1896 | 2.0 |

150° 180° 210°} (according to the increasing order in frequencies), and the noise power $\sigma_w^2 = 1.0$.

Figure 3.43 shows the original signal waveforms and the pseudospectrum as well as the reconstructed signal using the estimated parameters. Table 3.7 shows that all harmonic and interharmonic frequencies are correctly estimated.

From the results of Examples 3.8 to 3.10 one can observe that the estimated frequencies of harmonics and interharmonics are rather accurate. However, the magnitude estimates appear to be less accurate, especially for those harmonics with small magnitude values.

**Example 3.11** Analysis of Measurement Data Containing Harmonic and Interharmonic Components: Signaling. In this example, measurement data are used for the analysis. The sample rate of the data is $f_s = 7200$ Hz, the voltage fundamental frequency is $f_0 = 50$ Hz, and the data used for the analysis contains 24 cycles. To reduce the strong influence of the 50-Hz component, prefiltering is applied to remove the low-frequency components up to 90 Hz. Figure 3.44 shows the original measurement data, the data after the prefiltering, and the pseudospectrum obtained by MUSIC where the peak frequencies indicate the frequencies of the harmonics and interharmonics. Table 3.8 shows the estimated frequencies and magnitudes of these harmonics and interharmonics.

**Example 3.12** Analysis of Measurement Data: Arc-Furnace Case. In this example the measurement data obtained from a large arc-furnace installation are used for the analysis. The sample rate of the data is 3000 Hz, the voltage fundamental frequency is 50 Hz, and the data used for the MUSIC analysis contains 12 cycles.

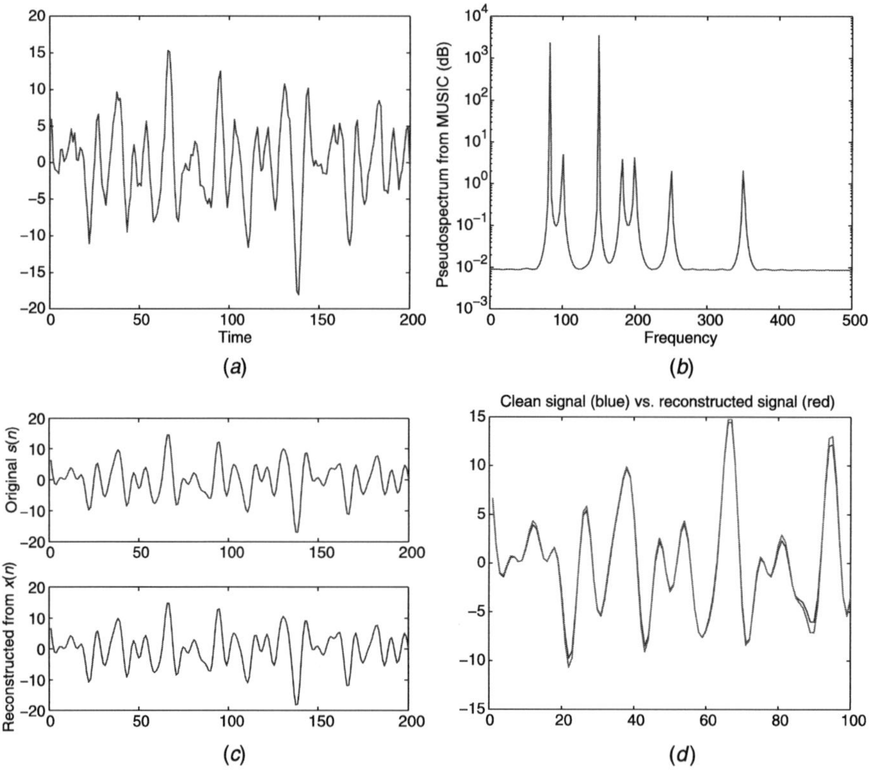


**Figure 3.43** Estimating harmonic and interharmonic disturbances by MUSIC (*a*) original synthesized waveform $x(n)$; (*b*) pseudospectrum obtained from MUSIC where peaks indicate frequencies of harmonics and interharmonics; (*c*) original clean signal $s(n)$ versus reconstructed signal using the parameters estimated from MUSIC; (*d*) overlapped plot of (*c*).

**TABLE 3.7 Estimation of Harmonics and Interharmonics from MUSIC**

| Frequency (Hz) | | Magnitude | |
|---|---|---|---|
| Estimated | True | Estimated | True |
| 82.0026 | 82 | 3.1132 | 3.0 |
| 100.1299 | 100 | 0.8107 | 0.5 |
| 149.9843 | 150 | 3.1344 | 1.5 |
| 182.0080 | 182 | 5.1710 | 5.0 |
| 200.0131 | 200 | 1.1796 | 0.3 |
| 249.9974 | 250 | 4.2145 | 4.0 |
| 349.9922 | 350 | 2.6820 | 4.0 |

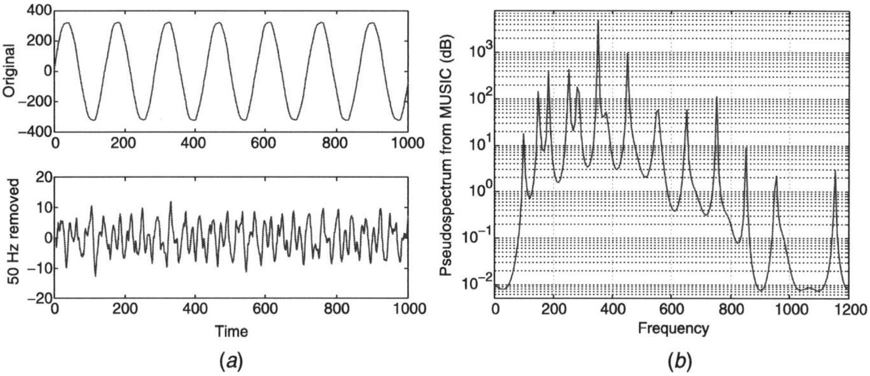


**Figure 3.44** Estimating harmonic and interharmonic disturbances by MUSIC: (*a*) measured voltage waveforms and waveforms after prefiltering; (*b*) pseudospectrum obtained from MUSIC where peaks indicate frequencies of harmonics and interharmonics. The power system frequency 50 Hz.

**TABLE 3.8 Estimated Results from MUSIC**

| Harmonic No. | Magnitude/$\sqrt{2}$ | Interharmonic Frequency (Hz) | Magnitude/$\sqrt{2}$ |
|---|---|---|---|
| 1.9808 | 0.2586 | 184.25 | 1.7503 |
| 2.9866 | 1.0810 | 284.71 | 1.8802 |
| 5.0227 | 1.2524 | 380.28 | 0.8633 |
| 7.0188 | 2.2996 | 477.27 | 1.0270 |
| 9.0257 | 2.0941 | 562.30 | 1.9540 |
| 11.0372 | 0.9877 | 686.65 | 0.2836 |
| 13.0352 | 0.9019 | 792.19 | 0.4241 |
| 15.0380 | 0.0909 | 982.18 | 0.5010 |
| 17.0387 | 0.3447 | | |
| 19.0637 | 0.2074 | | |
| 23.0511 | 0.3291 | | |
| 25.0793 | 0.3209 | | |

To reduce the strong influence of the 50-Hz component, prefiltering is applied to remove the low-frequency components up to 90 Hz.

Figure 3.45 shows the original data, the data after prefiltering, and the pseudo-spectrum obtained by MUSIC where the peak frequencies indicate the frequencies of the harmonics and interharmonics. Table 3.9 shows the estimated frequencies and magnitudes of harmonics and interharmonics.

**Example 3.13** Analysis of Measurement Data Containing Disturbances: "Clean" Supply. In this example, the measurement data obtained from a relatively clean supply are used. The sample rate of the data is 2048 Hz, the voltage fundamental

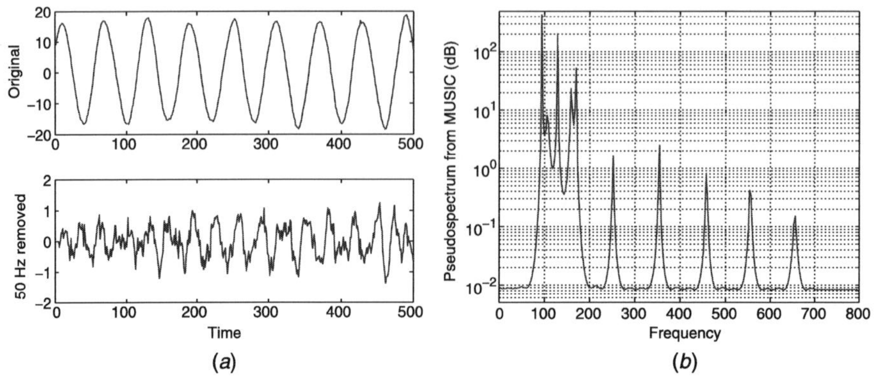


**Figure 3.45** Estimating major harmonics and interharmonics by MUSIC: (*a*) measured voltage waveforms and waveforms after prefiltering; (*b*) pseudospectrum obtained from MUSIC where peaks indicate main disturbance frequencies of harmonics and interharmonics. The power system frequency is 50 Hz.

**TABLE 3.9 Estimated Results from MUSIC**

| Frequency (Hz) | Magnitude/$\sqrt{2}$ |
|---|---|
| 93.9400 | 0.3346 |
| 106.5116 | 0.0614 |
| 128.8169 | 0.1064 |
| 158.5212 | 0.0431 |
| 169.3702 | 0.1938 |
| 251.6060 | 0.0668 |
| 353.9118 | 0.0971 |
| 458.0709 | 0.0733 |
| 555.1282 | 0.0563 |
| 655.5907 | 0.0576 |

frequency is 50 Hz, and the data used for the MUSIC analysis contains 10 cycles. To reduce the strong influence of the 50-Hz component, prefiltering is applied to remove the low-frequency components up to 90 Hz.

Figure 3.46 shows the original data, the data after prefiltering, and the pseudospectrum obtained by MUSIC. The peak frequencies indicate that the disturbances are mainly due to the 3rd, 5th, 7th, 9th, and 11th odd harmonics. It is worth noting that the component at frequency 91.1075 Hz is a remainder of the strong 50-Hz component, since the prefilter removes the frequencies up to 90 Hz. Table 3.10 shows the estimated frequencies and magnitudes of harmonics and interharmonics.

*Discussion: Preprocessing to Reduce Influence of Strong 50-Hz Component* In the first three examples, the synthetic signal does not include the power system fundamental-frequency (50-Hz) component. For real measurements

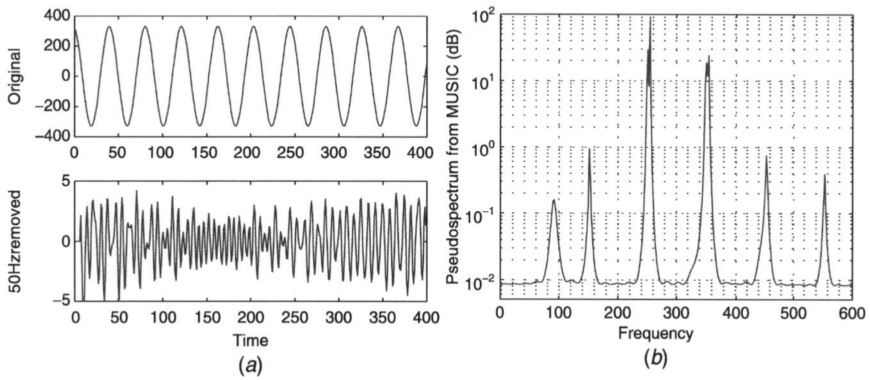


**Figure 3.46** Estimating major harmonics by MUSIC: (*a*) measured voltage waveforms and waveforms after prefiltering; (*b*) pseudospectrum obtained from MUSIC where peaks indicate main harmonic frequency components.

**TABLE 3.10 Estimated Results from MUSIC**

| Frequency (Hz) | Magnitude/$\sqrt{2}$ |
|---|---|
| 91.1075 | 0.8134 |
| 152.5058 | 0.2451 |
| 249.2435 | 1.4161 |
| 349.0544 | 0.9265 |
| 441.4898 | 0.3734 |
| 551.8829 | 0.1883 |

this will not be the case. Practically, the sharp contrast between the strong power at the power system fundamental frequency and the relative weak power at the harmonic or interharmonic frequencies will make the MUSIC algorithm less accurate in estimating the harmonics. Therefore, some preprocessing is needed before applying the MUSIC algorithm. A simple preprocessing used in the above examples (where the data were obtained from the measurements) consists of applying the DFT, removing several spectral lines around the 50-Hz component, followed by an inverse DFT. The main disadvantage is that this may result in a "pseudoharmonic" at the cutoff frequency (in the above examples, at around 90 Hz) in the MUSIC estimation. The reason is that such a simple preprocessing is equivalent to applying an ideal filter with a rectangular shaped transfer function. Alternatively, a notch filter centered around the power system frequency or a high-pass filter with a stop band including the power system frequency can be applied.

### 3.5.3 Estimation of Signal Parameters via Rotational Invariance Techniques

The ESPRIT method also employs sinusoidal (or harmonic) models and estimates the frequencies and powers of the harmonics. However, the difference to the

MUSIC method is that ESPRIT is a signal-subspace-based method rather than a noise-subspace-based method. ESPRIT considers the generalized eigenvalue problem in the signal subspace and exploits the rational property of the signal.

For a given data sequence $v(n)$ of length $L$ that is modeled by (3.158),

$$v(n) = s(n) + w(n) = \sum_{k=1}^{K} a_k \cos(n\omega_k + \phi_k) + w(n)$$

we use its complex exponential form in (3.159) for the sake of mathematical convenience. Let us consider a time-windowed data vector of size $M$,

$$\begin{bmatrix} v(n) \\ v(n+1) \\ \vdots \\ v(n+M-1) \end{bmatrix} = \sum_{k=1}^{K} \begin{bmatrix} s_k(n) \\ s_k(n+1) \\ \vdots \\ s_k(n+M-1) \end{bmatrix} + \begin{bmatrix} w(n) \\ w(n+1) \\ \vdots \\ w(n+M-1) \end{bmatrix} \tag{3.179}$$

Noting that the signal in (3.159) consists of $K$ harmonic components $s(n) = \sum_{k=1}^{K} s_k(n) = \sum_{k=1}^{K} \underline{A}_k e^{jn\omega_k}$ and the $k$th signal component vector can be written as

$$\mathbf{s}_k(n) = \begin{bmatrix} s_k(n) \\ s_k(n+1) \\ \vdots \\ s_k(n+M-1) \end{bmatrix} = \underline{A}_k e^{jn\omega_k} \begin{bmatrix} 1 \\ e^{j\omega_k} \\ \vdots \\ e^{j(M-1)\omega_k} \end{bmatrix} \tag{3.180}$$

and substituting (3.180) into (3.179), it follows that

$$\mathbf{v}(n) = \mathbf{E}\boldsymbol{\Phi}^n \underline{\mathbf{A}} + \mathbf{w}(n) \tag{3.181}$$

where $\boldsymbol{\Phi} = \text{diag}\{e^{j\omega_1} \quad e^{j\omega_2} \quad \cdots \quad e^{j\omega_K}\}$, $\underline{\mathbf{A}} = [\underline{A}_1 \quad \underline{A}_2 \quad \cdots \quad \underline{A}_K]^{\mathrm{T}}$, and $\mathbf{E}$ is defined as

$$\mathbf{E} = \begin{bmatrix} 1 & 1 & \cdots & 1 \\ e^{j\omega_1} & e^{j\omega_2} & \cdots & e^{j\omega_K} \\ \vdots & \vdots & \ddots & \vdots \\ e^{j(M-1)\omega_1} & e^{j(M-1)\omega_2} & \cdots & e^{j(M-1)\omega_K} \end{bmatrix} \tag{3.182}$$

Using the harmonic model in (3.159), it follows that the time shift of a signal sample equals its phase shift and is described by the rotation on the unit circle,

$$s(n+1) = \sum_{k=1}^{K} \underline{A}_k e^{j\omega_k(n+1)} = \sum_{k=1}^{K} s_k(n) e^{j\omega_k} \tag{3.183}$$

Considering the time-shifted data vector

$$\tilde{\mathbf{v}}(n) = \mathbf{v}(n+1) = [v(n+1) \quad v(n+2) \quad \cdots \quad v(n+M)]^{\mathrm{T}} \tag{3.184}$$

and noting (3.158), it follows that

$$\tilde{\mathbf{v}}(n) = \mathbf{E}\boldsymbol{\Phi}^{n+1}\underline{\mathbf{A}} + \mathbf{w}(n+1) \tag{3.185}$$

Using the definition of the autocorrelation matrix and cross-correlation matrix as well as the relations in (3.158) and (3.185), it follows that

$$\mathbf{R}_{\mathbf{v}} = E\{\mathbf{v}(n)\mathbf{v}^{\mathrm{H}}(n)\} \quad \mathbf{R}_{\mathbf{v}\tilde{\mathbf{v}}} = E\{\mathbf{v}(n)\tilde{\mathbf{v}}^{\mathrm{H}}(n)\} \tag{3.186}$$

where $E(\cdot)$ denotes the expectation. These expressions can be written as

$$\begin{aligned} \mathbf{R}_{\mathbf{v}} &= \mathbf{E}\underline{\mathbf{A}}\underline{\mathbf{A}}^{\mathrm{H}}\mathbf{E}^{\mathrm{H}} + \sigma_w^2\mathbf{I} = \mathbf{R}_{\mathbf{s}} + \sigma_w^2\mathbf{I} \\ \mathbf{R}_{\mathbf{v}}\tilde{\mathbf{v}} &= \mathbf{E}\underline{\mathbf{A}}\underline{\mathbf{A}}^{\mathrm{H}}\boldsymbol{\Phi}^{\mathrm{H}}\mathbf{E}^{\mathrm{H}} + \sigma_w^2\mathbf{Q} = \mathbf{R}_{\mathbf{s}\tilde{\mathbf{s}}} + \sigma_w^2\mathbf{Q} \end{aligned} \tag{3.187}$$

where $\tilde{\mathbf{s}}(n) = \mathbf{s}(n+1)$ and $\mathbf{Q}$ is defined as

$$\mathbf{Q} = \begin{bmatrix} 0 & 0 & \cdots & 0 & 0 \\ 1 & 0 & \cdots & 0 & 0 \\ 0 & 1 & \cdots & 0 & 0 \\ \vdots & \vdots & \ddots & \vdots & \vdots \\ 0 & 0 & \cdots & 1 & 0 \end{bmatrix} \tag{3.188}$$

Now, let us consider the generalized eigenvalue problem in the signal subspace,

$$\mathbf{R}_{\mathbf{s}}\mathbf{u}_i = \lambda_i\mathbf{R}_{\mathbf{s}\tilde{\mathbf{s}}}\mathbf{u}_i \qquad i = 1, \ldots, K \tag{3.189}$$

where $\mathbf{u}_i$ is the generalized eigenvectors of $(\mathbf{R}_{\mathbf{s}}, \mathbf{R}_{\tilde{\mathbf{s}}\mathbf{s}})$ or is referred to as the eigenvector of $\mathbf{R}_{\mathbf{s}} - \lambda_i\mathbf{R}_{\tilde{\mathbf{s}}\mathbf{s}}$. Substituting the relations in (3.187) into (3.189), it follows that

$$\mathbf{E}\underline{\mathbf{A}}\underline{\mathbf{A}}^{\mathrm{H}}\mathbf{E}^{\mathrm{H}}\mathbf{u}_i = \lambda_i\mathbf{E}\underline{\mathbf{A}}\underline{\mathbf{A}}^{\mathrm{H}}\boldsymbol{\Phi}^{\mathrm{H}}\mathbf{E}^{\mathrm{H}}\mathbf{u}_i \qquad i = 1, 2, \ldots, K \tag{3.190}$$

which is equivalent to

$$\mathbf{E}\underline{\mathbf{A}}\underline{\mathbf{A}}^{\mathrm{H}}(\mathbf{I} - \lambda_i\boldsymbol{\Phi}^{\mathrm{H}})\mathbf{E}^{\mathrm{H}}\mathbf{u}_i = 0 \qquad i = 1, 2, \ldots, K \tag{3.191}$$

From (3.191), the frequencies of harmonics $f_i$ can be found from

$$\lambda_i = e^{j\omega_i} \qquad i = 1, 2, \ldots, K \tag{3.192}$$

where $\omega_i = 2\pi f_i$, $i = 1, 2, \ldots, K$.

*Alternative Approach of Computing* $\mathbf{\Phi}$ Finding the rotation matrix $\mathbf{\Phi}$ is shown [207] to be equivalent to finding the eigenvalues of $\mathbf{\Psi}$, which can be estimated from

$$\mathbf{U}_2 = \mathbf{U}_1 \mathbf{\Psi} \tag{3.193}$$

This can be achieved as follows. First, the $K$ eigenvectors in the signal subspace are found from the eigenvectors of $\mathbf{R}_v$ (this can be obtained by applying singular value decomposition to the data matrix $\mathbf{V} = \mathbf{L\Sigma U}$, where $\mathbf{U}$ forms the orthogonal bases for the underlying $M$-dimensional vector space. The matrix $\mathbf{U}$ can be written as $\mathbf{U} = [\mathbf{U}_s | \mathbf{U}_n]$, where $\mathbf{U}_s$ is a submatrix of size $M \times K$ formed by extracting the first $K$ column vectors (eigenvectors) in $\mathbf{U}$ that correspond to the $K$ largest eigenvalues in $\mathbf{\Sigma}$. From $\mathbf{U}_s$, we further form two matrices of size $(M-1) \times K$ as follows: The matrix $\mathbf{U}_1$ contains the first $M-1$ rows of $\mathbf{U}_s$ and matrix $\mathbf{U}_2$ contains the last $M-1$ rows of $\mathbf{U}_s$. The least-squares (LS) ESPRIT can be implemented by first computing the LS estimate of $\mathbf{\Psi}$,

$$\hat{\mathbf{\Psi}}_{\text{LS}} = (\mathbf{U}_1^{\text{H}} \mathbf{U}_1)^{-1} \mathbf{U}_1^{\text{H}} \mathbf{U}_2 \tag{3.194}$$

Next, noting that the diagonal elements of $\mathbf{\Phi}$, $\phi_l = e^{j\omega_l}$, $l = 1, 2, \ldots, K$, are the eigenvalues of $\mathbf{\Psi}$, we can obtain the harmonic frequencies by computing the eigenvalues of $\mathbf{\Psi}$.

*Estimating Remaining Unknown Parameters*

*Magnitude and Power.* Once the frequencies of the harmonics are estimated, the magnitudes and powers of the harmonics can be estimated using the same method as described in MUSIC (Section 3.5.2). One should notice that the $A_k$ in ESPRIT are associated with the complex exponential model. The magnitude $a_k$ in the sinusoidal model can be obtained by using the relation $a_k = 2|\underline{A}_k|$ [see (3.161)].

*Damping Factor.* The ESPRIT method can also be applied to estimate the dominant components in a voltage or current transient (see Sections 6.3 and 8.10). The signal is in that case modeled as a number of damped sinusoids in noise, that is,

$$x(n) = \sum_{k=1}^{K} a_k e^{-\beta_k n} \cos(2\pi f_k n + \phi_k) + w(n)$$

where $f_k$ is a normalized frequency. In such a case, $\mathbf{\Phi}$ in (3.181) becomes

$$\mathbf{\Phi} = \begin{bmatrix} e^{-\beta_1 + j\omega_1} & 0 & \cdots & 0 \\ 0 & e^{-\beta_2 + j\omega_2} & \cdots & 0 \\ \vdots & \vdots & \ddots & \vdots \\ 0 & 0 & 0 & e^{-\beta_K + j\omega_K} \end{bmatrix} \tag{3.195}$$

and (3.192) becomes

$$\lambda_i = e^{-\beta_i + j\omega_i} \qquad i = 1, 2, \ldots, K \tag{3.196}$$

Hence, the damping factor and the frequency can be obtained by

$$\beta_i = -\text{Re}[\ln(\lambda_i)] \qquad f_i = \frac{\text{Im}[\ln(\lambda_i)]}{2\pi}, \quad i = 1, 2, \ldots, K \tag{3.197}$$

where Re(·) and Im(·) denote the real and the imaginary parts, respectively.

*Initial Phases* The initial phases are included in $\underline{\mathbf{A}} = [\underline{A}_1 e^{j\phi_1} \quad \underline{A}_2 e^{j\phi_2} \quad \cdots \quad \underline{A}_K e^{j\phi_K}]^T$ and can be computed in the following way. First, using (3.181) the LS estimate of $\mathbf{\Phi}^n \underline{\mathbf{A}}$ can be computed [using $\mathbf{E}\backslash\mathbf{v}(n)$ in MATLAB]. Since the frequencies and the damping factors in (3.195) are already estimated, these parameters can be treated as known parameters. Hence, the angles in the vector $\underline{\mathbf{A}}$ can be computed from the LS estimate ($\mathbf{\Phi}^n \underline{\mathbf{A}}$) and the initial phases $\phi_i$, $i = 1, \ldots, K$, can be obtained.

**Example 3.14** Analysis of Synthetic Signal Consisting of Damped Sinusoids at Odd-Harmonic Frequencies. Similar to the signal used in the Example 3.8, the synthetic data consist of the first six odd harmonics; however, a damping factor is introduced to each of the harmonics. It should be noted that for stationary signals there is no damping; however, adding small damping factors here is purely for the purpose of explaining a possible application of ESPRIT. The synthetic data with damped sinusoids in noise can be described by

$$x(n) = \sum_{k=2}^{7} a_{2k-1} e^{-\beta_{2k-1} n} \cos\left(2\pi(2k-1)n\frac{f_0}{f_s} + \phi_{2k-1}\right) + w(n) \tag{3.198}$$

It is worth mentioning that in (3.198) we use the actual frequency values $f_k = (2k-1)f_0$ (in hertz); therefore it should be normalized by dividing $f_s$. Correspondingly, some adjustment is required in ESPRIT to obtain the damping factor in a proper scale. This can be done as follows: If we replace $\omega_i = 2\pi f_i/f_s$ in (3.196), we obtain the following equivalent expression:

$$\lambda_i = e^{(-\beta_i + j\omega_i)} = e^{(-\beta_i f_s + j2\pi f_i)/f_s} \tag{3.199}$$

Therefore when using unnormalized frequencies we actually estimate scaled damping factors $f_s\beta_i$ in the ESPRIT algorithm instead of $\beta_i$, which are $f_s$ times larger than the actual $\beta_i$ values.

The magnitudes, the initial phases, and the noise power are used exactly the same as those in the Example 3.8, the scaled damping factors $\tilde{\beta}_{2k-1} = \beta_{2k-1} f_s$ are set to $\{0.8\ 0.1\ 0.4\ 0.1\ 0.5\ 0.3\}$, $f_s = 2400$ Hz, and $f_0 = 50$ Hz.

**TABLE 3.11 Estimation Results from LS ESPRIT**

| Estimated Frequency (Hz) | | Scaled Damping $f_s\beta$ | | Magnitude | | Initial Phase (deg) | |
|---|---|---|---|---|---|---|---|
| Estimated | True | Estimated | True | Estimated | True | Estimated | True |
| 149.9487 | 150.0 | 0.7746 | 0.8 | 1.5036 | 1.5 | 5.5238 | 0.0 |
| 249.9858 | 250.0 | 0.1152 | 0.1 | 4.0078 | 4.0 | 32.7393 | 30.0 |
| 349.9907 | 350.0 | 0.3931 | 0.4 | 3.9545 | 4.0 | 62.2265 | 60.0 |
| 449.9177 | 450.0 | 0.2910 | 0.1 | 0.6558 | 0.8 | 107.3496 | 90.0 |
| 549.9928 | 550.0 | 0.4750 | 0.5 | 2.4369 | 2.5 | 122.0686 | 120.0 |
| 649.9941 | 650.0 | 0.4845 | 0.3 | 2.1541 | 2.0 | 181.0159 | 180.0 |

Using LS ESPRIT, we estimate the frequencies, magnitudes, initial phases, and damping factors of the harmonics, and the results are given in Table 3.11. Figure 3.47 shows the original clean signal $s(n)$ and the reconstructed harmonic components as well as the entire signal.

**Example 3.15** Analysis of Synthetic Signal Consisting of Damped Sinusoids in Interharmonic and Harmonic Frequencies. In this example, the synthetic signal is similar to that used in the Example 3.10, which consists of harmonics 2, 3, 4, 5, and 7 and two interharmonics (82 and 182 Hz), except the newly introduced damping factors. The synthetic data with damped sinusoids in noise are described by

$$x(n) = \sum_{k=2,3,4,5,7} a_k e^{-\beta_k n} \cos\left(2\pi nk\frac{f_0}{f_s} + \phi_k\right) + \sum_{l=1}^{2} b_l e^{-\beta_l n} \cos\left(2\pi n\frac{f_l}{f_s} + \phi_l\right) + w(n) \tag{3.200}$$

and the true values of scaled damping factors $\tilde{\beta}_k$, initial phases, and magnitudes are given in Table 3.12. The sampling rate $f_s = 2400$ Hz and the fundamental frequency $f_0 = 50$ Hz. Using the LS ESPRIT method, the resulting estimated values of the frequencies, magnitudes, initial phases, and damping factors are given in Table 3.12. Figure 3.48 shows the original clean signal $s(n)$ and the reconstructed signal components as well as the entire signal.

**Example 3.16** Analysis of Measurement Data Containing Harmonic and Interharmonic Components: Signaling. In this example, we use the same measurement data as in the Example 3.11. The data block size used for the ESPRIT analysis is 24 cycles. Similarly, a prefilter is applied to remove the low-frequency components up to 90 Hz. Figure 3.49 shows the original data, the data after the prefiltering, and the resulting line spectrum (i.e., the magnitude spectrum) obtained by ESPRIT. Table 3.13 shows the estimated frequencies and magnitudes of these harmonics and interharmonics. Comparing the results obtained from MUSIC in the

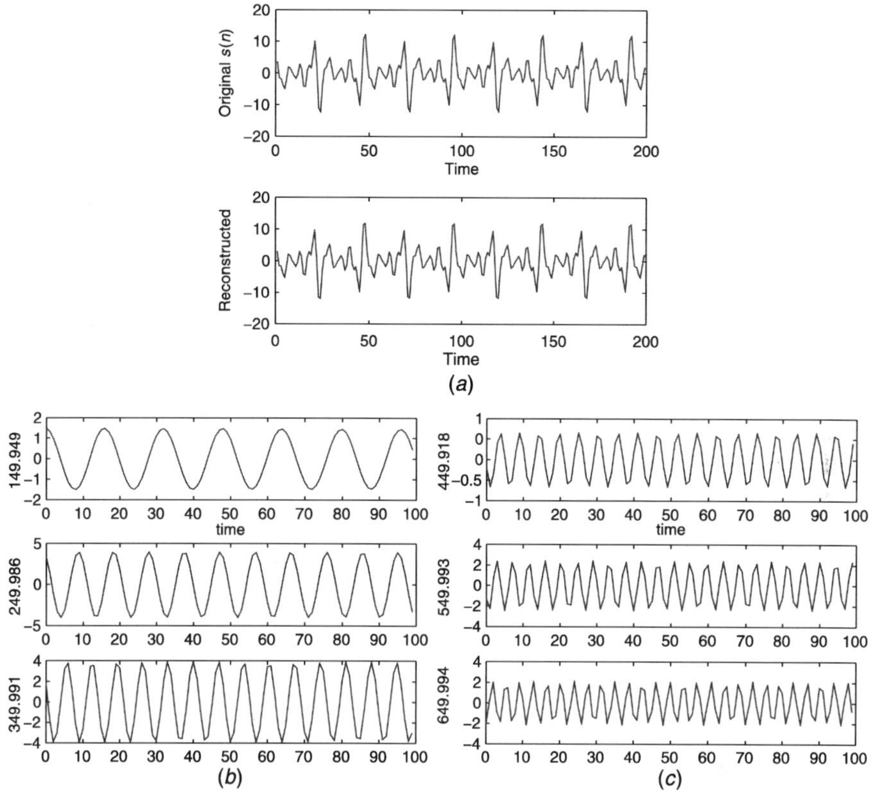


**Figure 3.47** Reconstructed odd harmonics from parameters estimated from LS ESPRIT: (*a*) original clean signal $s(n)$ versus reconstructed signal using estimated parameters; (*b*, *c*) reconstructed six odd-harmonic components.

**TABLE 3.12 Estimation Results from LS ESPRIT Method**

| Estimated Frequency (Hz) | | Scaled Damping $f_s\beta$ | | Magnitude | | Initial Phase (deg) | |
|---|---|---|---|---|---|---|---|
| Estimated | True | Estimated | True | Estimated | True | Estimated | True |
| 82.0386 | 82 | 2.1915 | 2.0 | 3.0538 | 3.0 | −2.3062 | 0 |
| 100.0522 | 100 | 0.3660 | 0.3 | 0.4267 | 0.5 | 18.9884 | 30 |
| 150.0003 | 150 | 0.5476 | 0.5 | 3.0408 | 1.5 | 59.1542 | 60 |
| 181.9968 | 182 | 0.1737 | 0.2 | 4.9290 | 5.0 | 119.6340 | 120 |
| 200.0262 | 200 | 0.4028 | 0.1 | 1.1351 | 0.3 | 143.0823 | 150 |
| 250.0131 | 250 | 2.1999 | 2.0 | 4.2572 | 4.0 | 178.3761 | 180 |
| 350.0087 | 350 | 0.8288 | 0.8 | 2.5988 | 4.0 | 208.0081 | 210 |

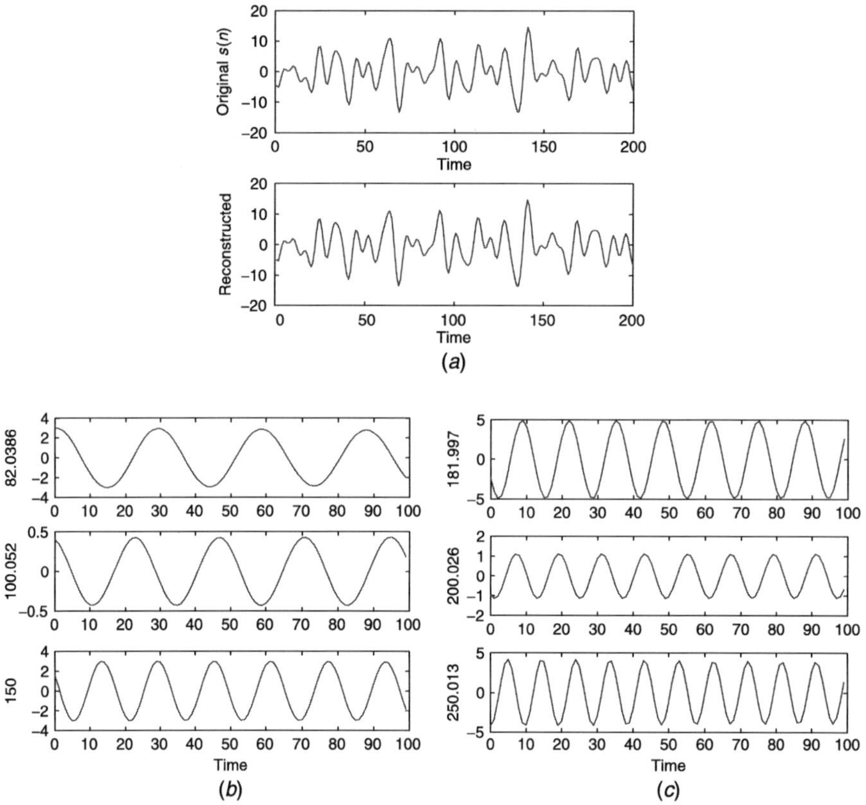


**Figure 3.48** Reconstructed signal and harmonics and interharmonics from LS ESPRIT estimated parameters: (*a*) original clean signal *s*(*n*) versus reconstructed signal using estimated parameters; (*b*, *c*) reconstructed first six harmonic and interharmonic components.

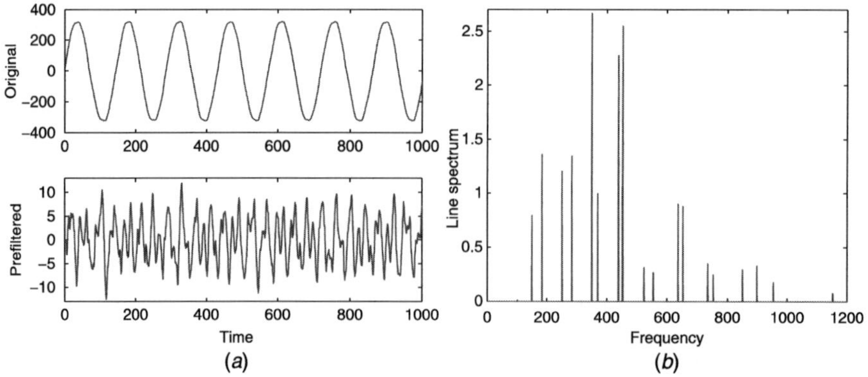


**Figure 3.49** Estimating harmonic and interharmonic disturbances using ESPRIT method: (*a*) measured voltage waveforms and waveforms after prefiltering; (*b*) estimated line spectrum (magnitude vs. frequency) from ESPRIT.

**TABLE 3.13 Estimated Results Associated with Figure 3.49 Using ESPRIT**

| Harmonic No. | Magnitude/$\sqrt{2}$ | Interharmonic Frequency (Hz) | Magnitude/$\sqrt{2}$ |
|---|---|---|---|
| 2.0150 | 0.0108 | 184.23 | 1.3615 |
| 3.0016 | 0.7975 | 284.21 | 1.3470 |
| 5.0203 | 1.2096 | 369.63 | 0.9980 |
| 7.0083 | 2.6668 | 438.73 | 2.2784 |
| 9.0546 | 2.5521 | 524.42 | 0.3150 |
| 11.1076 | 0.2673 | 636.72 | 0.9041 |
| 13.0634 | 0.8847 | 735.48 | 0.3495 |
| 15.0569 | 0.2513 | | |
| 17.0002 | 0.2932 | | |
| 17.9526 | 0.3348 | | |
| 19.0810 | 0.1786 | | |
| 23.0300 | 0.0814 | | |
| 25.0232 | 0.1586 | | |

Example 3.11, there are some differences, for example, an extra 18th harmonic is present from using ESPRIT and some terms of interharmonics are different at the higher frequencies.

**Example 3.17** Analysis of Measurement Data: Arc-Furnace Case. In this example, the measurement data were obtained from a large arc-furnace installation, similar to the one used in Example 3.12. The data block size used for the ESPRIT analysis contains 12 cycles, and a prefilter is applied to remove the low-frequency components up to 90 Hz.

Figure 3.50 shows the original data and the data after the prefiltering and the line (or magnitude) spectrum estimated from ESPRIT. Table 3.14 shows the estimated frequencies and magnitudes of harmonics and interharmonics. The presence of

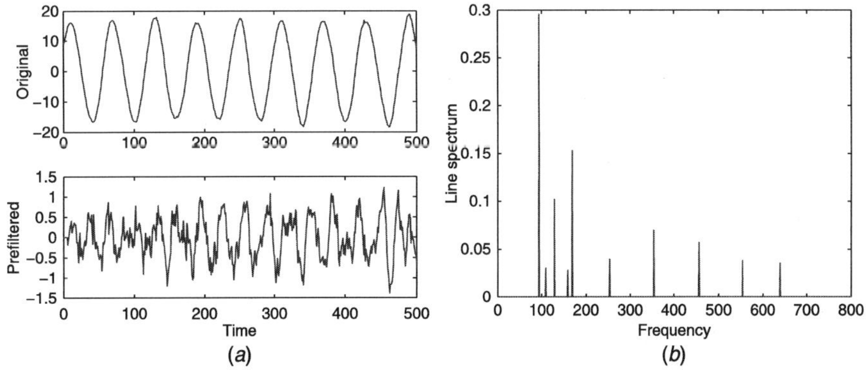


**Figure 3.50** Estimating major harmonics and interharmonics using ESPRIT: (*a*) measured voltage waveforms and waveforms after prefiltering; (*b*) estimated line spectrum (magnitude vs. frequency) from ESPRIT.

**TABLE 3.14 Estimated Results Associated with Figure 3.50 Using ESPRIT**

| Frequency (Hz) | Magnitude/$\sqrt{2}$ |
|---|---|
| 94.1153 | 0.2959 |
| 108.7025 | 0.0307 |
| 128.9766 | 0.1024 |
| 158.5671 | 0.0281 |
| 169.4523 | 0.1533 |
| 252.5995 | 0.0400 |
| 353.9231 | 0.0700 |
| 457.3694 | 0.0569 |
| 553.3138 | 0.0381 |
| 640.4327 | 0.0355 |

the 94-Hz component should again be treated with some suspicion as it is close to the 90-Hz cutoff frequency used to remove the power system frequency component. The spectrum of this signal, obtained by using a DFT, is shown in Figure 3.40. In this case a peak is present in the spectrum around 90 Hz.

**Example 3.18** Analysis of Measurement Data Containing Disturbances: Clean Supply. In this example, the measurement data are the same as in Example 3.13. The data block size used for the ESPRIT analysis contains 10 cycles, and the prefilter is applied to remove the low-frequency components up to 90 Hz.

Figure 3.51 shows the original data, the data after the prefiltering, and the line spectrum obtained by ESPRIT where the peak frequencies indicate the disturbances are mainly caused by odd harmonics (3rd, 5th, 7th, 9th, and 11th). It is noticed that

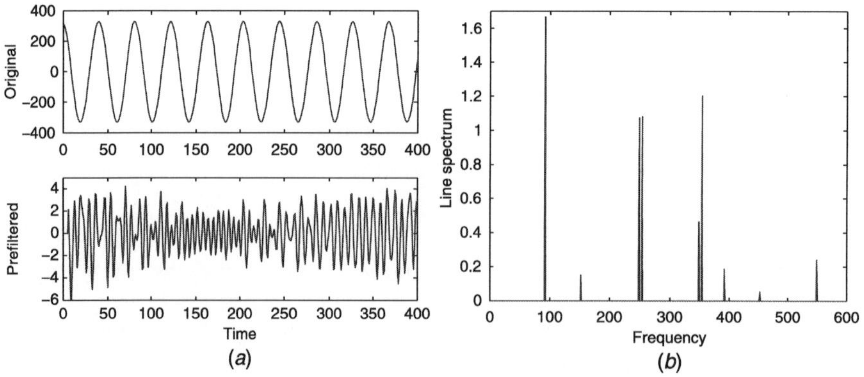


**Figure 3.51** Estimating major harmonics using ESPRIT: (*a*) measured voltage waveforms and waveforms after prefiltering; (*b*) line spectrum (magnitude vs. frequency) obtained from ESPRIT.

**TABLE 3.15 Estimated Results Associated with Figure 3.51 Using ESPRIT**

| Frequency (Hz) | Magnitude/$\sqrt{2}$ |
|---|---|
| 91.9385 | 1.6692 |
| 152.2721 | 0.1523 |
| 249.2084 | 1.0754 |
| 254.3887 | 1.0846 |
| 348.8702 | 0.4639 |
| 353.8094 | 1.2055 |
| 391.5418 | 0.1876 |
| 451.8858 | 0.0535 |
| 549.3013 | 0.2432 |

the line spectrum split appears for the 5th- and 7th-harmonic components. It is worth noting that the component at frequency 91.1075 Hz results from the remaining strong 50-Hz influence, since the prefilter removes the frequencies up to 90 Hz. Inspecting the spectrum obtained from applying a DFT shows that there is no significant frequency component present around 90 Hz. Table 3.15 shows the estimated frequencies and magnitudes.

*Discussion: ESPRIT Versus MUSIC* Both MUSIC and ESPRIT methods can estimate the parameters in the sinusoidal models. While the MUSIC method is based on the noise subspace, ESPRIT is based on the signal subspace.

The two methods both serve similar purposes and the results are also rather similar. The impact on the resulting spectrum based on using the noise subspace or the signal subspace is not immediately obvious; however, methods based on the signal subspace are often considered more reliable. Therefore, ESPRIT is often more preferred for line spectrum estimation. More experience on the use of these methods for analyzing power system harmonics and interharmonics is needed. From the limited number of experiments presented here we can however already conclude that both methods are appropriate tools for estimating the frequency of interharmonic components.

It is worth noting that the above frequency spectra resulting from the two methods are presented in different ways: The results of the MUSIC method are presented as continuous pseudospectra (where the peak positions indicate the harmonic/interharmonic frequencies but the magnitudes do not associate with any real frequency magnitudes!); the results of the ESPRIT method are presented as line spectra, where the estimated magnitudes and frequencies of harmonics/interharmonics are shown. One should notice that the way to present these results is however independent of the method: Obviously, the results from MUSIC can also be depicted as those shown for ESPRIT. The computation results in both cases contain a set of sinusoidal frequencies with their corresponding power (or magnitude) values.

### 3.5.4 Kalman Filters

***3.5.4.1 State-Space Modeling of Power System Harmonics*** Kalman filters are special types of filters. Their solutions are based on a set of state-space equations. Kalman filters are useful tools for many power system applications, for example, real-time tracking harmonics [262, 182], estimating voltage and current parameters in power system protection [22, 120], and estimating the parameters of transients [121]. Give the observation data $z(n)$, a Kalman filter is described by a set of state equations and a set of observation equations as follows:

$$\begin{aligned} &\text{State equations:} && \mathbf{x}(n) = \mathbf{A}(n-1)\mathbf{x}(n-1) + \mathbf{w}(n) \\ &\text{Observation equations:} && \mathbf{z}(n) = \mathbf{C}(n)\mathbf{x}(n) + \mathbf{v}(n) \end{aligned} \tag{3.201}$$

where $\mathbf{x}(n)$ is a vector of state variables, $\mathbf{A}(n-1)$ is the state transition matrix, $\mathbf{w}(n)$ is a vector of model noise assumed to be zero-mean white with a covariance matrix $E\{\mathbf{w}(n)\mathbf{w}^{\mathrm{T}}(n)\} = \mathbf{Q}_w$, $\mathbf{z}(n)$ is a vector of observations, and the matrix $\mathbf{C}(n)$ connects the measurement $\mathbf{z}(n)$ with the state vector $\mathbf{x}(n)$; $\mathbf{v}(n)$ is a vector of observation noise assumed to be zero-mean white with a covariance matrix $E\{\mathbf{v}(n)\mathbf{v}^{\mathrm{T}}(n)\} = \mathbf{Q}_v$ and is statistically independent of model noise $\mathbf{w}(n)$. For stationary data, $\mathbf{A}(n)$ and $\mathbf{C}(n)$ are time independent, that is, $\mathbf{A}(n) = \mathbf{A}$ and $\mathbf{C}(n) = \mathbf{C}$.

To apply a Kalman filter, one of the most essential issues is to define the state variables $\mathbf{x}(n)$ according to the given problem (e.g., the state variables are often chosen to be the unknown parameters or a function of the parameters that one wishes to estimate). Once the state variables are defined, the exact expressions of the state equations and the observation equations can be determined with fixed matrices $\mathbf{A}(n)$ and $\mathbf{C}(n)$. Further, one should carefully examine whether the white-noise assumption holds for $\mathbf{w}(n)$ and $\mathbf{v}(n)$ under the given state and observation equations. If the assumption does not hold, the performance of a Kalman filter can be poor, for example, with large estimation errors and less reliable estimates.

Although a Kalman filter can be used under many different signal models, we shall only consider a practical power system data analysis example where the sinusoidal models in (3.158) are used. Our purpose of using a Kalman filter here is to estimate harmonic-related power system distortions and to estimate the parameters associated with these harmonics (e.g., the magnitudes and phase angles).

For notational convenience, we use the complex exponential form (3.159) of the sinusoidal model and rewrite it as

$$z(n) = \sum_{k=1}^{K} s_k(n) + v(n) \tag{3.202}$$

where the harmonic component is $s_k(n) = \underline{A}_k e^{jn\omega_k}$, $k = 1, \ldots, K$. For a total of $K$ harmonics, we define a state vector $\mathbf{x}(n)$ that contains $2K$ elements:

$$\mathbf{x}(n) = [s_{1,r}(n) \quad s_{1,i}(n) \quad \cdots \quad s_{K,r}(n) \quad s_{K,i}(n)]^{\mathrm{T}} \tag{3.203}$$

where $s_{k,r}(n)$ and $s_{k,i}(n)$ are the real and imaginary parts of $s_k(n)$ at time $n$, respectively

$$\begin{aligned} s_{k,r}(n) &= \mathrm{Re}[s_k(n)] = a_k \cos(n\omega_k + \phi_k) \\ s_{k,i}(n) &= \mathrm{Im}[s_k(n)] = a_k \sin(n\omega_k + \phi_k) \end{aligned} \tag{3.204}$$

Consider the state variables $s_{k,r}(n+1)$ and $s_{k,i}(n+1)$ at time $n+1$. It follows that

$$\begin{aligned} s_{k,r}(n+1) &= a_k \cos[(n+1)\omega_k + \phi_k] = s_{k,r}(n)\cos(\omega_k) - s_{k,i}(n)\sin(\omega_k) \\ s_{k,i}(n+1) &= a_k \sin[(n+1)\omega_k + \phi_k] = s_{k,r}(n)\sin(\omega_k) + s_{k,i}(n)\cos(\omega_k) \end{aligned} \tag{3.205}$$

Note that $a_k$, $\omega_k$, and $\phi_k$ in (3.205) remain constants from $n$ to $n+1$. Combining (3.203) and (3.205), the state transition matrix $\mathbf{A}(n)$ in (3.201) can be written as

$$\mathbf{A}(n) = \begin{bmatrix} \cos(\omega_1) & -\sin(\omega_1) & 0 & \dots & 0 & 0 & 0 \\ \sin(\omega_1) & \cos(\omega_1) & 0 & \dots & 0 & 0 & 0 \\ \vdots & \vdots & \vdots & \ddots & \vdots & \vdots & \vdots \\ 0 & 0 & 0 & \dots & 0 & \cos(\omega_K) & -\sin(\omega_K) \\ 0 & 0 & 0 & \dots & 0 & \sin(\omega_K) & \cos(\omega_K) \end{bmatrix} \tag{3.206}$$

where $\omega_k$ can be chosen as the true power system harmonics or some interharmonics. Choosing power system harmonics implies that $\omega_k = k\omega_0$, $k = 1, 2, \dots, K$, $\omega_0 = 2\pi f_0/f_s$ is the fundamental frequency (in radius) of the power system, and $f_s$ is the sampling frequency (in hertz). It is worth mentioning that to implement the Kalman filter algorithm the matrix $\mathbf{A}(n)$ should be fully specified. *This also implies that the frequencies $\omega_k$, $k = 1, 2, \dots, K$, are fixed and prespecified, rather than the parameters to be estimated.*

By modeling the (scalar) measurement data $z(n)$ as sinusoids in white noise [as in (3.158)], $\mathbf{C}(n)$ in (3.201) becomes a vector of size $2K$ as follows:

$$\mathbf{C}(n) = [1 \quad 0 \quad \cdots \quad 1 \quad 0] \tag{3.207}$$

Further, the observation noise vector $\mathbf{v}(n) = [v(n)\ 0\ \cdots\ 0]^{\mathrm{T}}$ and the model noise vector $\mathbf{w}(n) = [w(n)\ 0\ \cdots\ 0]^{\mathrm{T}}$ in (3.201) are both of size $2K$.

With the definition of (3.203), (3.204), (3.206), and (3.207) for the state equations and observation equations in (3.201), the harmonic model in (3.159) is fully specified in the state space.

*Estimation of Time-Dependent Harmonic-Related Parameters* Based on the two sets of equations in (3.201), a Kalman filter algorithm estimates the state variables using the measurement data. Once the state variables are estimated, the magnitudes $A_k(n)$ and the initial phase angles $\phi_k$ of the harmonics $s_k(n)$, $k = 1, 2, \ldots, K$, can be computed at each time instant $n$ as follows:

$$A_k(n) = \sqrt{[s_{k,r}(n)]^2 + [s_{k,i}(n)]^2} \qquad \phi_k(n) = \tan^{-1}\left(\frac{s_{k,i}(n)}{s_{k,r}(n)}\right) - n\omega_k \tag{3.208}$$

Note that the $\omega_k$ are predetermined fixed parameters. In reality the power system frequency may deviate slightly from its nominal value (50 or 60 Hz), leading to the deviation of the true harmonic frequencies and hence affecting the estimated state variables (i.e., the magnitudes and phases of harmonics). However, if the frequency deviation is small, as is normally the case, one can assume that the estimation errors are small and the estimation results are sufficiently close to those in the true harmonic frequencies. However, for large frequency deviations, as in island operation of industrial systems, the estimated parameters of the preselected harmonics may actually relate to interharmonics. For example, assume one is interested in estimating odd harmonics. Assuming the fundamental frequency is 52 Hz for a nominal frequency of 50 Hz, then the 3rd-, 5th-, 7th-, 9th-, and 11-th harmonic frequencies {150, 250, 350, 450, 550} have become {156, 260, 364, 468, 572} Hz. Hence, what a Kalman filter actually estimates are the interharmonics rather than the originally chosen harmonics. One can see that the larger the frequency drift and the higher the harmonic number, the more error will be introduced to the estimated harmonics.

It is also worth mentioning that variables $\omega_k$ and $\phi_k$ are in general not independent, that is, variations in $n\omega_k + \phi_k$ interact with each other. Since $\omega_k$ are set as constants in the above Kalman filter model, all frequency variations in the system will therefore show up as changes in phase angle.

### *Discussion*

1. *Harmonic Frequencies and Phases.* It is important to notice that the matrices $\mathbf{A}(n)$ and $\mathbf{C}(n)$ in (3.201) must be fully specified before applying a Kalman filter algorithm. Using the definition of $\mathbf{A}(n)$ in (3.206) implies that the $\omega_k$, $k = 1, 2, \ldots, K$, are predetermined. Using the prior knowledge of power systems, a reasonable choice is to set $\omega_k$ as the actual or nominal power system frequency and its harmonics up to some order. The principle of choosing $K$ is that the number of harmonics should be sufficiently high so that the model noise $\mathbf{w}(n)$ in (3.202) can be reasonably assumed as white. If a reasonable guess can be made, for example, the signal only consists of odd harmonics, then the harmonic model may include the power system fundamental and its odd harmonics up to some order $K$. This can reduce the computation in a Kalman filter.

However, if we assume that $z(n)$ contains all harmonics while in reality some harmonics may not be there, one may find that the estimated magnitudes of the corresponding harmonics $A_k(n)$ are close to zero. Hence, the impact of including nonexisting harmonics to a Kalman filter algorithm is mainly the extra computational cost.

Often one does not know in advance which harmonics are included in the measurement data. In that sense, a Kalman filter under the above model cannot estimate the harmonic frequencies. Compared with the MUSIC and ESPRIT methods, this is clearly a disadvantage.

2. *Interharmonics.* Interharmonic frequencies are usually unknown in advance and can be any frequency between two harmonic frequencies. It is therefore not feasible for a Kalman filter to estimate interharmonics in practical situations. On the contrary, if the signal contains a strong interharmonic component which has not been included in the Kalman filter model, this component would effectively be included in the model noise $\mathbf{w}(n)$. In such a case, $\mathbf{w}(n)$ may deviate from the white-noise assumption, making Kalman filter estimation less reliable. Compared with MUSIC and ESPRIT methods, this is another disadvantage.

3. *Variations in Power System Frequency.* As mentioned above the frequency of a harmonic component will change as a consequence of a slight deviation in the power system frequency $f_0$. See Section 2.1 and some of the measurements presented earlier in this chapter for a discussion on frequency variations. The frequency deviation will affect the estimated $\phi_k$ since $\phi_k$ and $\omega_k$ are not independent variables. The real fundamental frequency may be estimated from the phase-angle variations in the estimated fundamental component. Implementation of this method would be similar to the method described in Section 3.2.1 for estimating the frequency from the *dq*-voltages. The Kalman-filter-based method would however be much more complicated and not give a higher accuracy.

4. *Time-Varying Parameters in Harmonics.* Often, the parameters in harmonics have slow variations over time, especially for nonstationary disturbance data (see Chapter 4). In this case the Kalman filter has a clear advantage in that it can estimate the time-varying harmonics (or the time-varying magnitudes and phases of harmonics). Since Kalman filters use iterative processing, one naturally obtains the time trajectories of the estimated parameters (e.g., harmonic magnitudes as a function of time). Compared with Kalman filters, both the MUSIC and ESPRIT methods employ batch processing (i.e., the whole block of data is processed once). Only one value is estimated for each parameter. To obtain time-dependent estimates of parameters, one requires to modify the MUSIC and ESPRIT methods, for example, use a (overlapped) sliding data window each time (see Chapter 4 for the sliding-window MUSIC and ESPRIT). However, this will bring a rather high computational cost to the estimation.

5. *Impact of Setting Model Noise Variance $\sigma_w^2$ to Estimated State Variables.* In the Kalman filter, the variance of model noise $\mathbf{w}(n)$ is assumed to be known in advance. However, this is usually not the case. For harmonic modeling of power system disturbances, we actually consider the signal in the low-frequency band with a bandwidth $B = f_0 K$ consisting of $f_0$ values of 50 or 60 Hz and its harmonics. Noise within this band will affect the estimated harmonics, while noise outside this band has little influence on the estimation results. Since we do not know the

actual model noise variance $\sigma_w^2$, it is usually set by the users before the Kalman filter algorithm starts. The principle of selection is that a small $\sigma_w^2$ should be chosen in order to obtain relatively small variances in the estimated state variables after the algorithm converges. However, this also implies that the algorithm will take a longer time to converge as compared with using a larger $\sigma_w^2$. If one anticipates a steady-state harmonic disturbance or slow variations in harmonic disturbances, setting a relatively small $\sigma_w^2$ will be a good choice.

6. *Numerical Stability and Convergence.* The above conventional Kalman filters (also called covariance Kalman filters) are prone to problems caused by numerical computation of ill-conditioned quantities in finite word length. Several reasons may cause the divergence of a Kalman filter algorithm. Some of the main causes include the following:

(a) The model for the physical problem is incorrect or too far away from reality. In such a case, proper model adjustment is required.

(b) Lack of second-order statistics of **w** and **v** leads to inappropriate estimation of $\mathbf{Q}_w$ and $\mathbf{Q}_v$.

(c) The error covariance matrix **P** can become nonpositive semidefinite due to the finite-length computation after some recursive steps (even though this is theoretically not possible).

The last case mainly occurs when the covariance matrix reduces rapidly by processing very accurate measurements or a linear combination of the state variables is of great precision while other combinations are unobservable. Obviously if **P** approaches zero very fast, then the Kalman filter gain **G** in (3.215) becomes very small. This implies that the newly incoming data samples play a less important role in estimating the state vector **x** [see (3.216)], leading to a phenomenon called *data saturation*. Consequently, the estimation may further drift away from the true model as the iterations continuous, leading to the divergence.

Some possible solutions to Kalman filter instability or divergence include enforcing a minimum possible gain $\mathbf{G}(n)$ after some iterations, employing square-root covariance Kalman filters to reduce the ill-conditioning and numerical instability in finite-length computation, applying exponential weighting or a sliding window to data samples in Kalman filters, and applying extended Kalman filters to model a nonlinear dynamic system. Some detailed solutions and algorithms will be included in the remaining part of this section.

#### 3.5.4.2 Conventional (Covariance) Kalman Filter Algorithm for Harmonic Estimation

Let $\hat{\mathbf{x}}(n|n)$ and $\hat{\mathbf{x}}(n|n-1)$ be the *mean-square linear estimates* of state vector $\mathbf{x}(n)$ at time $n$ using the observations $z(n)$ up to time $n$ and $n-1$, respectively,

$$\hat{\mathbf{x}}(n|n) = E(x_n|z_0, z_1, \ldots, z_n) \quad \hat{\mathbf{x}}(n|n-1) = E(x_n|z_0, z_1, \ldots, z_{n-1}) \qquad (3.209)$$

where $E(\cdot)$ denotes the expectation. Let the corresponding filtering error vector $\mathbf{e}(n|n)$ and prediction error vector $\mathbf{e}(n|n-1)$ be,

$$\mathbf{e}(n|n) = \mathbf{x}(n) - \hat{\mathbf{x}}(n|n) \quad \mathbf{e}(n|n-1) = \mathbf{x}(n) - \hat{\mathbf{x}}(n|n-1) \tag{3.210}$$

Further, let the corresponding error covariance matrices from the filtering and the prediction be

$$\mathbf{P}(n|n) = E[\,\mathbf{e}(n|n)\mathbf{e}^{\mathrm{T}}(n|n)] \qquad \mathbf{P}(n|n-1) = E[\mathbf{e}(n|n-1)\,\mathbf{e}^{\mathrm{T}}(n|n-1)] \tag{3.211}$$

The Kalman filter yields the best *linear* estimates that minimize the mean-square error (MSE)

$$\xi = \mathrm{tr}[\,\mathbf{P}(n|n)] = E[\,\mathbf{e}^{\mathrm{T}}(n|n)\,\mathbf{e}(n|n)] \tag{3.212}$$

where the ensemble average $E[\mathbf{e}^{\mathrm{T}}(n|n)\mathbf{e}(n|n)]$ in (3.212) can be replaced by the sample average $1/N\sum_n \mathbf{e}^{\mathrm{T}}(n|n)\mathbf{e}(n|n)$ if the process is ergodic. It is worth noting that a window is effectively introduced due to the summation index $n$.

To use a Kalman filter, one should remember that the following assumptions are made: The model noise $\mathbf{w}(n)$ and observation noise $\mathbf{v}(n)$ are zero-mean white and statistically independent and the observations $\mathbf{z}(n)$ and state variables $\mathbf{x}(n)$ are zero-mean stochastic processes. If the noise is nonwhite, the estimates may deviate from the optimal solutions. Further, if the measurement data have a non-zero-mean value, preprocessing should be added to remove the mean value.

The algorithm for implementing a Kalman filter uses an iterative process described in Table 3.16. Equations (3.213), (3.214), (3.215), (3.216), and (3.217) in Table 3.16 form a complete (covariance) Kalman filter algorithm.

In the algorithm, $K$ is a predefined number. If the MSE $\xi(n)$ in (3.217) appears large and does not converge, one may consider increasing the number of harmonics $K$ (so that the noise can be somewhat closer to white).

#### 3.5.4.3 Square-Root Covariance Kalman Filters

Potter [243] proposed a method for propagating the error covariance matrix in square-root form in the absence of model noise. The proposed method can successfully maintain the positive-semidefinite nature of the error covariance and is known to provide twice the effective precision of the conventional implementation in ill-conditioned problems. The conditional number of matrix $\mathbf{A}$ is related to eigenvalue spread and is defined as $K(\mathbf{A}) = \sqrt{\lambda_1}/\sqrt{\lambda_n}$, where $\lambda_1$ is the maximum eigenvalue and $\lambda_n$ is the minimum eigenvalue of $\mathbf{A}^{\mathrm{T}}\mathbf{A}$. It is shown that the conditional number of $\mathbf{P}$ in the conventional (covariance) Kalman filter and $\mathbf{S}$ in the square-root covariance Kalman filter are

**TABLE 3.16 (Covariance) Kalman Filter Algorithm**

(a) *Initial Step* ($n = 0$) Given the observations $\mathbf{z}(n)$, the state vector $\hat{\mathbf{x}}(0|0)$, error covariance $\mathbf{P}(0|0)$, estimated covariance matrices of model noise $\mathbf{Q}_w$ and observation noise $\mathbf{Q}_v$,

$$\begin{aligned} \hat{\mathbf{x}}(0|0) &= E[\mathbf{x}(0)] \qquad \mathbf{P}(0|0) = E\{[\mathbf{x}(0) - E(\mathbf{x}(0))][\mathbf{x}(0) - E(\mathbf{x}(0))]^{\mathrm{T}}\} \\ \mathbf{Q}_w &= E[\mathbf{w}(0)\mathbf{w}^{\mathrm{T}}(0)] \quad \mathbf{Q}_v = E[\mathbf{v}(0)\mathbf{v}^{\mathrm{T}}(0)] \end{aligned} \tag{3.213}$$

the above ensemble average $E(\cdot)$ should be replaced by the sample averages under the ergodic process assumption.

(b) Iteration Steps: for $n = 1, 2, \ldots,$ do:

1. Assuming the estimate of state vector $\hat{\mathbf{x}}(n-1|n-1)$, the error covariances $\mathbf{P}(n-1|n-1)$ are obtained from the $(n-1)$th iteration. The predicted state vector and prediction error covariance matrix at the $n$th iteration are then computed from

$$\begin{aligned} \hat{\mathbf{x}}(n|n-1) &= \mathbf{A}(n-1)\hat{\mathbf{x}}(n-1|n-1) \\ \mathbf{P}(n|n-1) &= \mathbf{A}(n-1)\mathbf{P}(n-1|n-1)\mathbf{A}^{\mathrm{H}}(n-1) + \mathbf{Q}_w \end{aligned} \tag{3.214}$$

2. The Kalman filter gain $\mathbf{G}(n)$ is then computed from

$$\mathbf{G}(n) = \mathbf{P}(n|n-1)\mathbf{C}^{\mathrm{H}}(n)[\mathbf{C}(n)\mathbf{P}(n|n-1)\mathbf{C}^{\mathrm{H}}(n) + \mathbf{Q}_v]^{-1} \tag{3.215}$$

3. The estimate of the state vector and the corresponding error covariance matrix are updated after obtaining a new measurement data $\mathbf{z}(n)$ at time $n$ using

$$\begin{aligned} \hat{\mathbf{x}}(n|n) &= \hat{\mathbf{x}}(n|n-1) + \mathbf{G}(n)[\mathbf{z}(n) - \mathbf{C}(n)\hat{\mathbf{x}}(n|n-1)] \\ \mathbf{P}(n|n) &= [\mathbf{I} - \mathbf{G}(n)\mathbf{C}(n)]\mathbf{P}(n|n-1)| \end{aligned} \tag{3.216}$$

The MSE of the filter at time $n$ can be obtained from

$$\xi(n) = \mathrm{tr}\{\mathbf{P}(n|n)\} \tag{3.217}$$

If $\xi(n)$ is small, for example, below a preselected threshold $\tilde{\epsilon}$ (a small constant) (or $|\xi(n) - \xi(n-1)| < \tilde{\epsilon}$), then the algorithm is considered as converged and the estimated variables are close to the optimal solutions.

4. The estimates of magnitude $A_k(n)$ and the phase angle $\phi_k(n)$ are then computed by using (3.208). Note that there is a factor of 2 difference between the magnitudes in the sinusoidal model and in the complex exponential model, that is, $a_k(n) = 2|A_k(n)|$.
5. (*Optional*) In some applications, the estimated signals $\hat{\mathbf{z}}(n)$ and the residuals $\boldsymbol{\varepsilon}(n)$ are required. They can be estimated as,

$$\hat{\mathbf{z}}(n|n) = \mathbf{C}(n)\hat{\mathbf{x}}(n|n) \qquad \boldsymbol{\varepsilon}(n) = \mathbf{z}(n) - \hat{\mathbf{z}}(n|n) \tag{3.218}$$

6. (*Optional*) To obtain the estimate of an individual component, for example, the $k$th sinusoidal signal component $\hat{s}_k(n)$, by noting (3.203), it follows that

$$\hat{s}_k(n) = x(n|n; 2k) \tag{3.219}$$

where $x(n|n; 2k)$ is the $(2k)$th component of the vector $\mathbf{x}(n|n)$.

square-root related by

$$K(\mathbf{S}) = K(\mathbf{P})^{1/2} \tag{3.220}$$

where $\mathbf{P} = \mathbf{S}\mathbf{S}^{\mathrm{T}}$.

To avoid the divergence in Kalman filters, square-root covariance Kalman filters are often employed to replace the conventional covariance Kalman filters to increase the numerical stability [181, 137]. The first square-root filter was developed by Potter [243] for the limited case of uncorrelated scalar observations where model noise is assumed to be nonexistent. Table 3.17 summarizes the algorithm. For other variations readers are referred to the [181, 137, 183, 223, 227].

#### 3.5.4.4 Exponentially Weighted Kalman Filters

In the conventional Kalman filter algorithms, all measurement samples play an equal weight in the estimation whether it is an old or new measurement sample (i.e., independent of time $n$). For a dynamic system it is desirable that the newly incoming measurements play a more important role than the old measurement data or a forgetting factor should be introduced to the old data samples. This can be achieved by using a so-called exponentially weighted Kalman filter which minimizes the following

**TABLE 3.17 Potter's Square-Root Covariance Kalman Filter**

| | |
|---|---|
| **State-space model:** | |
| State equations: | $\mathbf{x}(n) = \mathbf{A}(n-1)\mathbf{x}(n-1) + \mathbf{w}(n)$ |
| Observation equations: | $\mathbf{z}(n) = \mathbf{C}(n)\mathbf{x}(n) + \mathbf{v}(n)$ |
| **Define**: | $\mathbf{P}(n\|n-1) = \mathbf{S}(n\|n-1)\mathbf{S}^{\mathrm{T}}(n\|n-1)$, $\mathbf{P}(n\|n) = \mathbf{S}(n\|n)\mathbf{S}^{\mathrm{T}}(n\|n)$ |
| **Initialization ($n = 0$)**: | $\hat{\mathbf{x}}(0\|0) = E[\mathbf{x}(0)]$; $\mathbf{P}(0\|0) = E\{[\mathbf{x}(0) - E(\mathbf{x}(0))] \times [\mathbf{x}(0) - E(\mathbf{x}(0))]^{\mathrm{T}}\}$; $\mathbf{S}^{\mathrm{T}}(0\|0) = \mathrm{chol}[\mathbf{P}(0\|0)]$; $\mathbf{Q}_w = E[\mathbf{w}(0)\mathbf{w}^{\mathrm{T}}(0)]$; $\mathbf{Q}_v = E[\mathbf{v}(0)\mathbf{v}^{\mathrm{T}}(0)]$, where chol(**P**): the Cholesky decomposition; $\mathbf{S}^{\mathrm{T}}$: upper triangular matrix |
| **Compute**: | **for** $n$=1, 2,..., **do**: $\hat{\mathbf{x}}(n\|n-1)=\mathbf{A}(n-1)\hat{\mathbf{x}}(n-1\|n-1)$; $\mathbf{S}(n\|n-1)=\mathbf{A}(n-1)\mathbf{S}(n-1\|n-1)$; $\mathbf{F}_n=\mathbf{S}^{\mathrm{T}}(n\|n-1)\mathbf{C}^{\mathrm{T}}(n)$; $\alpha_n=(\mathbf{F}_n^{\mathrm{T}}\mathbf{F}_n+\mathbf{Q}_v)^{-1}$; $\gamma_n=(1+\sqrt{(\alpha_n\mathbf{Q}_v)})^{-1}$; $\mathbf{G}(n)=\alpha_n\mathbf{S}(n\|n-1)\mathbf{F}_n$; $\hat{\mathbf{x}}(n\|n)=\hat{\mathbf{x}}(n\|n-1)+\mathbf{G}(n)[z(n)-\mathbf{C}(n)\hat{\mathbf{x}}(n\|n-1)]$; $\mathbf{S}(n\|n)=\mathbf{S}(n\|n-1)-\gamma_n\mathbf{G}(n)\mathbf{F}_n^{\mathrm{T}}$ **end** |

weighted LS [227]:

$$J_n = [\mathbf{x}(0) - E(\mathbf{x}(0))]^{\mathrm{H}} \mathbf{W}^{-1}(0)[\mathbf{x}(0) - E(\mathbf{x}(0))] + \sum_{i=1}^{n} [\mathbf{z}(i) - \mathbf{C}(i)\hat{\mathbf{x}}(i|i-1)]^{\mathrm{H}} \mathbf{W}^{-1}(i)[\mathbf{z}(i) - \mathbf{C}(i)\hat{\mathbf{x}}(i|i-1)] \tag{3.221}$$

To decrease the influence of previous measurement samples, one needs to properly set a time-dependent weight $\mathbf{W}^{-1}(n)$. Assuming the current time is $n$, then the weight should be $\mathbf{W}^{-1}(n) > \mathbf{W}^{-1}(n-1) > \cdots > \mathbf{W}^{-1}(0)$, or equivalently

$$\mathbf{W}(n) < \mathbf{W}(n-1) < \cdots < \mathbf{W}(0) \tag{3.222}$$

This can be obtained by setting $\mathbf{W}(n-1) = \mathbf{W}(n)\exp(c_{n-1})$, leading to

$$\mathbf{W}(n) < \mathbf{W}(n)\exp(c_{n-1}) < \cdots < \mathbf{W}(n)\exp\left(\sum_{i=0}^{n-1} c_i\right) \tag{3.223}$$

where $c_i = c > 0$ is a constant determined empirically. This leads to the exponentially weighted Kalman filter algorithm summarized in Table 3.18.

One can observe that the only difference between the conventional Kalman filter algorithm in Table 3.16 and the exponentially weighted Kalman filter algorithm in Table 3.18 is that the latter has introduced an extra weighting factor $e^{c_{n-1}}$ when

**TABLE 3.18 Exponentially Weighted Kalman Filter Algorithm**

**State-space model**:

State equations: $\mathbf{x}(n) = \mathbf{A}(n-1)\mathbf{x}(n-1) + \mathbf{w}(n)$

Observation equations: $\mathbf{z}(n) = \mathbf{C}(n)\mathbf{x}(n) + \mathbf{v}(n)$

**Initialization**:

$\hat{\mathbf{x}}(0|0) = E[\mathbf{x}(0)]$, $\mathbf{P}(0|0) = E\{[\mathbf{x}(0) - E(\mathbf{x}(0)][\mathbf{x}(0) - E(\mathbf{x}(0)]^{\mathrm{T}}\}$

$\mathbf{Q}_w = E[\mathbf{w}(0)\mathbf{w}^T(0)]$, $\mathbf{Q}_v = E[\mathbf{v}(0)\mathbf{v}^{\mathrm{T}}(0)]$

**Compute**:

**for** $n=1, 2, \ldots,$ **do**:

$\hat{\mathbf{x}}(n|n-1) = \mathbf{A}(n-1)\hat{\mathbf{x}}(n-1|n-1)$

$$\mathbf{P}(n|n-1) = \mathbf{A}(n-1)\mathbf{P}(n-1|n-1)e^{c_{n-1}}\mathbf{A}^{\mathrm{H}}(n-1) + \mathbf{Q}_w \tag{3.224}$$

$\mathbf{G}(n) = \mathbf{P}(n|n-1)\mathbf{C}^{\mathrm{H}}(n)[\mathbf{C}(n)\mathbf{P}(n|n-1)\mathbf{C}^{\mathrm{H}}(n) + \mathbf{Q}_v]^{-1}$

$\hat{\mathbf{x}}(n|n) = \hat{\mathbf{x}}(n|n-1) + \mathbf{G}(n)[\mathbf{z}(n) - \mathbf{C}(n)\hat{\mathbf{x}}(n|n-1)]$

$\mathbf{P}(n|n) = [\mathbf{I} - \mathbf{G}(n)\mathbf{C}(n)]\mathbf{P}(n|n-1)$

**end** (for)

computing $\mathbf{P}(n|n-1)$ in (3.224). This leads to a larger $P(n|n-1)$ since $e^{c_{n-1}} > 1$ and subsequently an increase in $\mathbf{G}(n)$. Consequently, more weight will be added to the newly input measurement sample $\mathbf{z}(n)$ for estimating the state vector $\mathbf{x}(n|n)$ owing to a larger Kalman filter gain $\mathbf{G}(n)$.

Further analysis can show that such a Kalman filter is equivalent to employing the exponential weighted covariance matrices $\mathbf{P}(0|0)$, $\mathbf{Q}_w(k-1)$, and $\mathbf{Q}_v(k)$, $k = 1, 2, \ldots, n-1$.

#### 3.5.4.5 Sliding-Window Kalman Filters

Instead of using an exponential weight to gradually "forget" the old data, another commonly used method is to discard the old data samples after a certain time duration. This leads to a so-called sliding-window Kalman filter [227]. Let us assume that the model noise $\mathbf{w}$ is zero and the length of the data window is $N$; the sliding-window Kalman filter algorithm can be summarized as in Table 3.19. Note that notation of the state

**TABLE 3.19 Sliding-Window Kalman Filter Algorithm**

**State-space model:**
State equations: $\mathbf{x}(n) = \mathbf{A}(n, n-1)\mathbf{x}(n-1)$
Observation equations: $\mathbf{z}(n) = \mathbf{C}(n)\mathbf{x}(n) + \mathbf{v}(n)$

**Initialization:** (n = 0)
Set sliding window size = $N$
$\hat{\mathbf{x}}(0|0) = E[\mathbf{x}(0)]$, $\mathbf{P}(0|0) = E\{[\mathbf{x}(0) - E(\mathbf{x}(0))][\mathbf{x}(0) - E(\mathbf{x}(0))]^T\}$
$\mathbf{Q}_v = E[\mathbf{v}(0)\mathbf{v}^T(0)]$

**Compute**:

```
for n=1, 2,..., do:
  if n<N then set C(d) ≡ 0
    x̂(n|n-1)=A(n, n-1)x̂(n-1|n-1)
    x̂(d|n-1)=A(d, n-1)x̂(n-1|n-1)
    P(n|n-1)=A(n, n-1)P(n-1|n-1)A^H(n, n-1)
    G(n)=P(n|n)C^H(n)Q_v^-1
    G̃(n)=P(n|n)A^H(d, n)C^H(d)Q_v^-1
  if n ≠ N do:
    x̂(n|n)=x̂(n|n-1)+G(n)[z(n)-C(n)x̂(n|n-1)]
    -G̃(n)[z(d)-C(d)x̂(d|n-1)]
    P(n|n)=[P^-1(n|n-1)+C^H(n)Q_v^-1 C(n)-A^H(d, n)C^H(d)
    Q_v^-1 C(d)A(d, n)]^-1
  else if n=N, do:
    x̂(n|n)=P(N|N)[P^-1(N|N)x̂(N|N)-A^H(0, N)P^-1(0|0)x̂(0|0)]
    P(n|n)=[P^-1(N|N)-A^H(0, N)P^-1(0|0)A(0, N)]^-1
end

where: d=n-N;
  (n-d)-step state transition matrix:
  A(n, d)=A(n, n-1)A(n-1, n-2) ··· A(d+1, d)
```

transition matrix $\mathbf{A}(n-1)$ used previously is replaced by $\mathbf{A}(n, n-1)$ in Table 3.19 to indicate the one-step transition.

It is interesting to compare the sliding-window Kalman filter algorithm in Table 3.19 with the conventional Kalman filter algorithm in Table 3.16. First, notice the following equivalent expressions of $\mathbf{G}(n)$ in (3.215) and $\mathbf{P}(n|n)$ in (3.216):

$$\mathbf{P}(n|n) = [\mathbf{I} - \mathbf{G}(n)\mathbf{C}(n)]\,\mathbf{P}(n|n-1) = [\mathbf{P}^{-1}(n|n-1) + \mathbf{C}^{\mathrm{H}}(n)\,\mathbf{Q}_v^{-1}\,\mathbf{C}(n)]^{-1}$$

$$\mathbf{G}(n) = \mathbf{P}(n|n-1)\mathbf{C}^{\mathrm{H}}(n)[\,\mathbf{C}(n)\,\mathbf{P}(n|n-1)\,\mathbf{C}^{\mathrm{H}}(n) + \mathbf{Q}_v]^{-1} = \mathbf{P}(n|n)\,\mathbf{C}^{\mathrm{H}}(n)\,\mathbf{Q}_v^{-1} \tag{3.225}$$

Observing the algorithm in Table 3.19, since $\mathbf{C}(d) = \mathbf{0}$ for $n < N$, a sliding-window Kalman filter degenerates to a conventional Kalman filter, as it is supposed to be. For $n > N$, we can observe that an extra term $\tilde{\mathbf{G}}(n)[\mathbf{z}(d) - \mathbf{C}(d)\hat{\mathbf{x}}(d|n-1)]$ is introduced in computing $\hat{\mathbf{x}}(n|n)$. This is used to remove the influence of old (discarded) data outside the window. Similarly, for computing $\mathbf{P}(n|n)$ the term $\mathbf{A}^{\mathrm{H}}(d, n)\mathbf{C}^{\mathrm{H}}(d)\mathbf{Q}_v^{-1}\mathbf{C}(d)\mathbf{A}(d, n)$ is introduced to remove the influence of the old sample that is shifed outside the window. There is an additional auxiliary Kalman filter gain $\mathbf{G}(n)$ being computed. Further, it is worth noting that the state transition matrix can be multistep rather than the one step in the previously addressed Kalman filter algorithms; for example, $\mathbf{A}(n, d)$ denotes a $(n-d)$-step state transition matrix. Obviously, there is an increase in computational demand. Whether selecting a sliding-window or a conventional Kalman filter is often a tradeoff between the computational cost and the nonstationarity of data.

***3.5.4.6 Extended Kalman Filters*** Kalman filters addressed so far in this chapter are designed to estimate the state vector in *a linear system model.* However, if the underlying system model is nonlinear, we need to extend the Kalman filters through a linearization procedure. This leads to a so-called extended Kalman filter (EKF) [137, 183, 223]. Instead of state-space modeling of linear systems by (3.201), nonlinear systems are modeled by the following two sets of equations,

$$\begin{array}{lrl} \text{State equations:} & \mathbf{x}(n+1) = & \mathbf{f}(\mathbf{x}(n), n) + \mathbf{w}(n) \\ \text{Observation equations:} & \mathbf{z}(n) = & \mathbf{h}(\mathbf{x}(n), n) + \mathbf{v}(n) \end{array} \tag{3.226}$$

To obtain a linear system, we assume that there exists a (local/global) nominal state trajectory $\bar{\mathbf{x}}(n)$ and a (local/global) nominal output trajectory $\bar{\mathbf{z}}(n)$ and let the deviations be denoted by

$$\delta\mathbf{x}(n) = \mathbf{x}(n) - \bar{\mathbf{x}}(n) \qquad \delta\mathbf{z}(n) = \mathbf{z}(n) - \bar{\mathbf{z}}(n) \tag{3.227}$$

Recalling the Taylor series expansion of a function $f(y)$ in a neighborhood of $y = y_0$,

$$f(y) = f(y_0) + f'(y_0)(y - y_0) + \frac{1}{2}f''(y_0)(y - y_0)^2 + \cdots + \frac{1}{n!}f^{(n)}(y_0)(y - y_0)^n + e_n(y)$$

Assuming that $\mathbf{x}(n)$ and $\mathbf{z}(n)$ are close to the nominal state and output trajectories, the state and observation equations can be approximated by Taylor series expansion of $\mathbf{f}$ and $\mathbf{h}$ in a neighborhood about nominal values. Linear approximation can be obtained by retaining up to the first-order terms in the Taylor series expansion.

For the state and the observation equations in (3.226), this approximation yields the following state-space model,

$$\delta\mathbf{x}(n+1) = \mathbf{A}_n\delta\mathbf{x}(n) + \mathbf{w}(n) \qquad \delta\mathbf{z}(n) = \mathbf{C}_n\delta\mathbf{x}(n) + \mathbf{v}(n) \tag{3.228}$$

where $\mathbf{A}_n$ and $\mathbf{C}_n$ are the Jacobian matrices of $\mathbf{f}$ and $\mathbf{h}$, defined as

$$\begin{aligned} \mathbf{A}_n &= \left.\frac{\partial\mathbf{f}(\mathbf{x}(n),n)}{\partial\mathbf{x}}\right|_{\mathbf{x}=\bar{\mathbf{x}}} = \begin{bmatrix} \frac{\partial f_1}{\partial x_1} & \cdots & \frac{\partial f_1}{\partial x_N} \\ \vdots & \ddots & \vdots \\ \frac{\partial f_N}{\partial x_1} & \cdots & \frac{\partial f_N}{\partial x_N} \end{bmatrix} \\ \mathbf{C}_n &= \left.\frac{\partial\mathbf{h}(\mathbf{x}(n),n)}{\partial\mathbf{x}}\right|_{\mathbf{x}=\bar{\mathbf{x}}} = \begin{bmatrix} \frac{\partial h_1}{\partial x_1} & \cdots & \frac{\partial h_1}{\partial x_N} \\ \vdots & \ddots & \vdots \\ \frac{\partial h_N}{\partial x_1} & \cdots & \frac{\partial h_N}{\partial x_N} \end{bmatrix} \end{aligned} \tag{3.229}$$

In the EKFs, the local nominal trajectory is selected as $\bar{\mathbf{x}}_n = \hat{\mathbf{x}}(n|n-1)$ about which the state vector is to be linearized. Consequently, applying (3.227) we have $\delta\hat{\mathbf{x}}(n|n) = \hat{\mathbf{x}}(n|n) - \hat{\mathbf{x}}(n|n-1)$ and $\delta\hat{\mathbf{x}}(n|n-1) = 0$. After several manipulation steps, one can obtain the EKF algorithm [137], which is summarized in Table 3.20.

#### *3.5.4.7 Examples*

**Example 3.19** Analysis of Synthetic Signal Consisting of Even and Odd Harmonics. In this example, the synthetic signal $z(n)$ consists of seven harmonics (i.e., the harmonics from the second to the ninth excluding the fourth harmonic, i.e., set $a_4 = 0.0$) in the additive white noise $v(n)$ and can be described by

$$z(n) = \sum_{k=2}^{9} a_k \cos\left(2\pi nk\frac{f_0}{f_s} + \phi_k\right) + v(n) \tag{3.230}$$

The magnitudes of the harmonics are set to {0.5, 1.5, 0.0, 4.0, 0.3, 4.0, 0.3, 0.8}, with the typical ratio of values described in Table 3.3, except that the fourth harmonic is set to zero. The initial phases are set to $\{100°, 20°, 120°, 40°, 140°, 60°, 160°, 80°\}$ (in the ascent order of harmonics), and the power of measurement noise is set to $\sigma_v^2 = 0.5$.

In the Kalman filter algorithm, we set $2K = 18$ and $\upsilon_v^2 = 0.5$. Figure 3.52 shows the estimated harmonic magnitudes in time using the data generated by (3.230). The first column in Figure 3.52 shows the simulation results where a small model noise

**TABLE 3.20 Extended Kalman Filter Algorithm**

**State-space model**:

State equations: $\mathbf{x}(n+1) = \mathbf{f}(\mathbf{x}(n),\ n) + \mathbf{w}(n)$

Observation equations: $\mathbf{z}(n) = \mathbf{h}(\mathbf{x}(n),\ n) + \mathbf{v}(n)$

**Initialization**:

$\hat{\mathbf{x}}(0|0) = E[\mathbf{x}(0)];\ \mathbf{P}(0|0) = E\{[\mathbf{x}(0) - E(\mathbf{x}(0))][\mathbf{x}(0) - E(\mathbf{x}(0))]^{\mathrm{T}}\}$

$\mathbf{Q}_w = E[\mathbf{w}(0)\mathbf{w}^{\mathrm{T}}(0)];\ \mathbf{Q}_v = E[\mathbf{v}(0)\mathbf{v}^{T}(0)]$

**Compute**:

**for** $n=1,\ 2, \ldots,$ **do**:

$$\hat{\mathbf{x}}(n|n-1) = \mathbf{f}(\hat{\mathbf{x}}(n-1|n-1),\ n-1)$$

$$\mathbf{A}_{n-1} = \left.\frac{\partial \mathbf{f}(\mathbf{x}, n1)}{\partial \mathbf{x}}\right|_{\mathbf{x}(n-1)=\hat{\mathbf{x}}(n-1|n-1)}$$

$$\mathbf{P}(n|n-1) = \mathbf{A}_{n-1}\mathbf{P}(n-1|n-1)\mathbf{A}_{n-1}^{\mathrm{H}} + \mathbf{Q}_w$$

$$\mathbf{C}_n = \left.\frac{\partial \mathbf{h}(\mathbf{x}, n)}{\partial \mathbf{x}}\right|_{\mathbf{x}(n)=\hat{\mathbf{x}}(n|n-1)}$$

$$\mathbf{G}(n) = \mathbf{P}(n|n-1)\mathbf{C}_n^{\mathrm{H}}[\mathbf{C}_n\mathbf{P}(n|n-1)\mathbf{C}_n^{\mathrm{H}} + \mathbf{Q}_v]^{-1}$$

$$\hat{\mathbf{x}}(n|n) = \hat{\mathbf{x}}(n|n-1) + \mathbf{G}(n)\,[\mathbf{z}(n) - \mathbf{h}(\hat{\mathbf{x}}(n|n-1),\ n-1)]$$

$$\mathbf{P}(n|n) = [\mathbf{I} - \mathbf{G}(n)\mathbf{C}_n]\mathbf{P}(n|n-1)$$

**end**

variance $\sigma_w^2 = 0.0001$ was used, while the second column in Figure 3.52 shows the results where a larger value $\sigma_w^2 = 0.1$ was applied.

One may observe that the magnitudes of the fourth harmonic are very small or negligible, corresponding to the case where the fourth harmonic is excluded in the model. From the first column of Figure 3.52 one may observe that good results are obtained for the estimated harmonic magnitudes after the convergence of the algorithm (after about 50 samples). The estimated magnitudes fluctuated a bit with time, partly due to influence of noise within the signal bandwidth.

Comparing the estimated harmonic magnitudes in the first and second columns of Figure 3.52, one may observe that the estimated harmonic magnitudes have smaller fluctuations (or smaller variances in estimation) in the first column corresponding to using a smaller $\sigma_w^2 = 0.0001$, while the fluctuations are much larger in the second column corresponding to using a larger $\sigma_w^2 = 0.1$. Generally speaking, choosing a smaller $\sigma_w^2$ leads to smaller variance in the estimation but a slower convergence speed. In this case, the difference in convergence speed is not very obvious for the two different $\sigma_w^2$ settings.

Figure 3.53 shows the estimated harmonic signals in time by using the estimated magnitudes and phase angles from the Kalman filter. In this simulation, $\sigma_w^2 = 0.0001$ was used.

**Example 3.20** Analysis of Synthetic Signal Consisting of Fundamental and Its Odd Harmonics. In this example, the synthetic data consist of the power system fundamental component in 50 Hz and its first six odd harmonics (i.e., odd harmonics

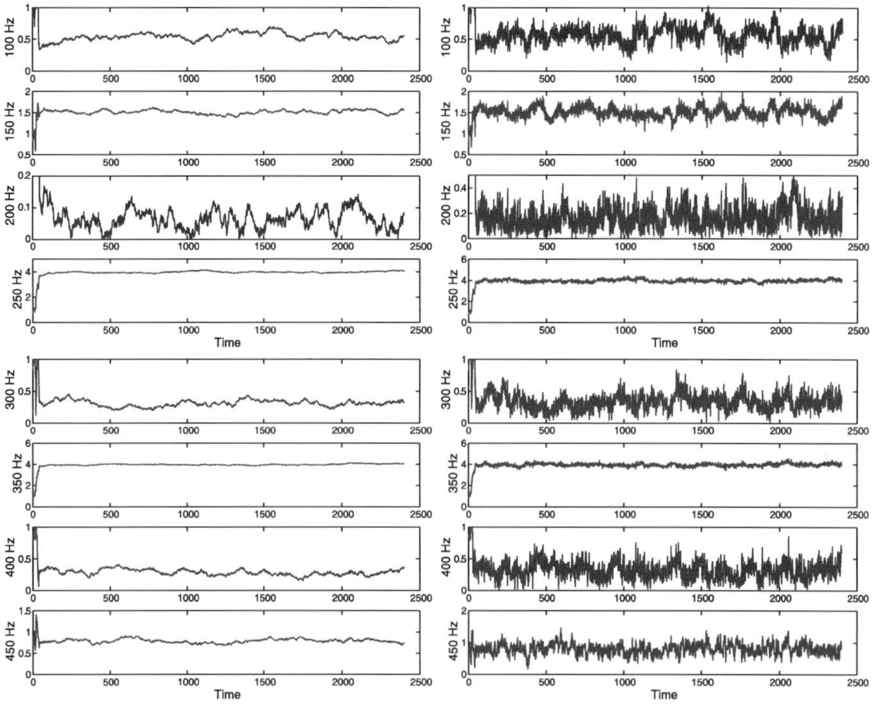


**Figure 3.52** Estimated harmonic magnitudes from Kalman filter. Left: The model noise variance was set to $\sigma_w^2 = 0.0001$. Right: The model noise variance was set to $\sigma_w^2 = 0.1$. The vertical axis indicates the estimated magnitude and the horizontal axis indicates the time.

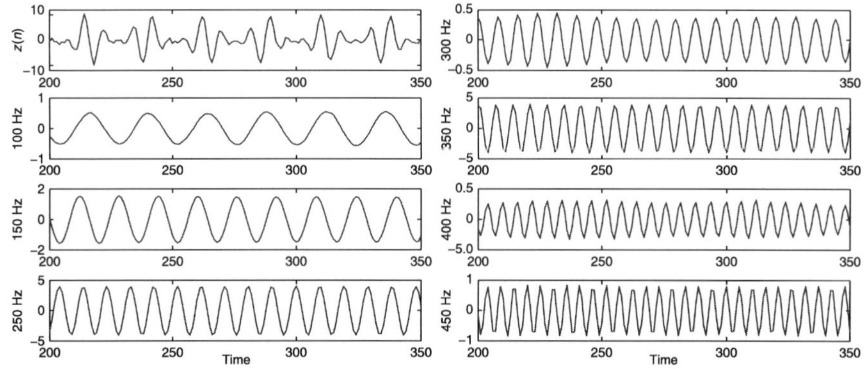


**Figure 3.53** Original signal and estimated harmonic components using parameters obtained from Kalman filter. In the Kalman filter the model noise variance $\sigma_w^2 = 0.0001$ was used. From the top to bottom and left to right: the original signal $z(n)$; the decomposed second, third, fifth, sixth, seventh, eighth, and ninth harmonics.

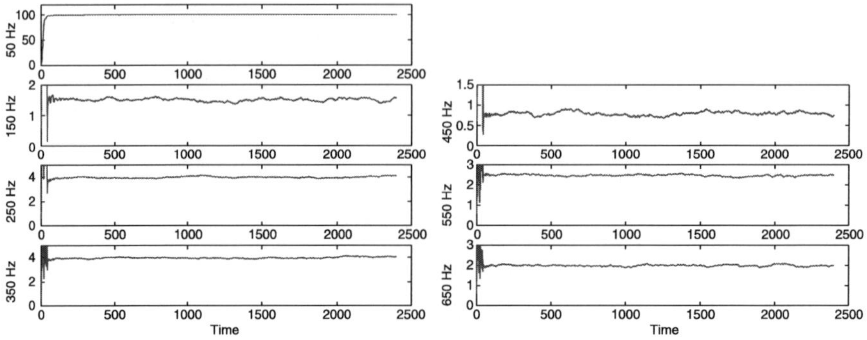


**Figure 3.54** Estimated harmonic magnitudes in time using Kalman filter. The vertical axis indicates the estimated magnitude and the horizontal axis indicates the time.

from the 3rd to the 13th) plus the additive white noise $v(n)$ and can be described by

$$z(n) = \sum_{k=1}^{7} a_{2k-1} \cos\left(2\pi n(2k-1)\frac{f_0}{f_s} + \phi_{2k-1}\right) + v(n) \tag{3.231}$$

The magnitude of the fundamental is set to 100.0, and the magnitudes of odd harmonics from 3 to 13 are set to {1.5, 4, 4, 0.8, 2.5, 2}, exactly as the typical values described in Table 3.3. The initial phases of the fundamental and the odd harmonics are set to $\{0°, 0°, 30°, 60°, 90°, 120°, 180°\}$, and the noise power is set to $\sigma_v^2 = 0.5$. Note that this synthetic signal is very close to the realistic (typical) measured data.

In the Kalman filter algorithm, we choose a small variance value for the model noise as $\sigma_w^2 = 0.0001$. The other parameters are set to $\sigma_v^2 = 0.5$, $2K = 26$ (with $w_k = 2\pi f_0 k$, $k = 1, \ldots, 13$). Figure 3.54 shows the estimated magnitudes of the fundamental and odd harmonics in time using the above synthetic data $z(n)$ as

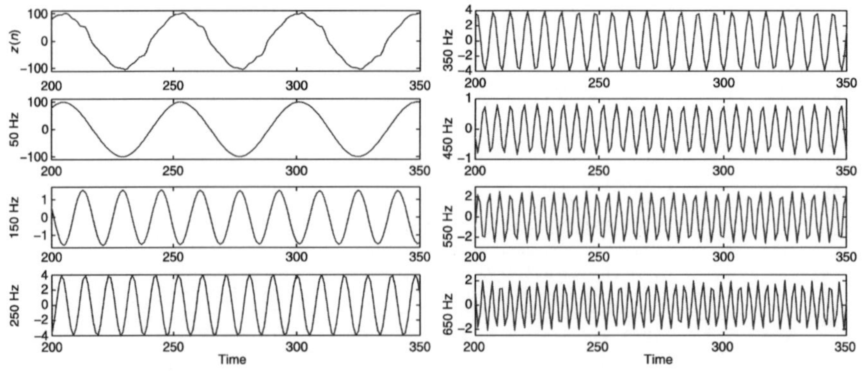


**Figure 3.55** Original signal $z(n)$ (synthetic) and estimated signal components including fundamental and odd harmonics.

the input. All the magnitudes of even harmonics are negligible and therefore are not included in the figure. One can see from the figure that the Kalman filter provides rather good estimates for the magnitudes of the fundamental and the harmonics, even though the fundamental-signal component is 25 to 100 times stronger than the harmonics. This is very encouraging and a clear advantage of the Kalman filter, as compared with the MUSIC or ESPRIT method, which usually requires a preprocessing to reduce the strong influence of the fundamental component. Figure 3.55 shows the original signal $z(n)$ and the estimated signal components (i.e., the fundamental and the odd harmonics), where the data are taken after the algorithm converges.

## 3.6 ESTIMATION OF BROADBAND SPECTRUM

In Section 3.5, we mainly consider the power system harmonics and interharmonics as the main cause of the disturbances, where line spectra can be used to characterize these disturbances. For other types of disturbances, broadband spectrum analysis may be required to quantify the disturbances. One of the main barriers for doing so is that, for power systems, very little is known about the disturbances in the high-frequency band and hence interpretation becomes difficult. Since there is limited knowledge on the causes of high-frequency elements in a power system, it is a common practice to simply ignore the high-frequency band. However, by doing so some useful and important information that could be used to quantify the disturbances in terms of their underlying causes can be lost. In this section, we shall briefly describe some useful and simple broadband spectrum analysis methods that can potentially be used for power system disturbance analysis.

To obtain the entire frequency spectrum of a disturbance data recording, some stochastic signal models can be used. This is because most real data contain measurement noise. A simple and the most frequently used model is the AR model. A more general rational model is the ARMA model.

### 3.6.1 AR Models

In an AR($N$) model, the data sample at time $n$ is assumed to be the weighted combination of the previous data samples $x(n-i)$, $i = 1, \ldots, N$, in white noise $w(n)$,

$$x(n) = -\sum_{i=1}^{N} a_i x(n-i) + b_0 w(n) \tag{3.232}$$

Where $N$ is the model order. This is equivalent to $x(n)$ being the output of an all-pole system $H(z)$ whose input is the white noise $w(n)$. The transfer function of the all-pole system is

$$H_{AR}(z) = \frac{b_0}{1 + \sum_{i=1}^{N} a_i z^{-i}} \tag{3.233}$$

Estimating the frequency spectrum of $x(n)$ thus becomes estimating the model parameters $a_i$ under a selected criterion. For example, using the MSE criterion, model parameters $\{a_i, i = 1, \ldots, N\}$ are estimated by solving a set of so-called normal equations

$$\sum_{m=1}^{N} a_m r_x(k, m) = -r_x(k, 0) \qquad k = 1, \ldots, N \tag{3.234}$$

where $r_x(k, m) = E(x(n-k)x(n-m))$ is the data autocorrelation sequence, $k$, $m \geq 0$ and $x(n)$ is assumed to be real valued, and $E(\cdot)$ denotes the expectation and is replaced by the summation under the ergodic data assumption. Once $\{a_i\}$ are estimated, $b_0$ can be computed by solving

$$|b_0|^2 = r_x(0, 0) + \sum_{m=1}^{N} a_m r_x(0, m) \tag{3.235}$$

The magnitude spectrum $A(e^{j\omega})$ and power spectrum $P(e^{j\omega})$ can be obtained by

$$A(e^{j\omega}) = |H_{\mathrm{AR}}(e^{j\omega})| = \left| \frac{b_0}{1 + \sum_{i=1}^{N} a_i e^{-j\omega}} \right| \qquad P(e^{j\omega}) = \left| H_{\mathrm{AR}}(e^{j\omega}) \right|^2 \tag{3.236}$$

### 3.6.2 ARMA Models

In an ARMA($N, M$) model, the current data sample is assumed to be the weighted combination of previous data samples $x(n-i)$, $i = 1, 2, \ldots, N$, and the white noise $w(n-i)$, $i = 0, 1, \ldots, M$, as follows:

$$x(n) = -\sum_{i=1}^{N} a_i x(n-i) + \sum_{i=0}^{M} b_i w(n-i) \tag{3.237}$$

where $N$ is the number of poles and $M$ is the number of zeros. This is equivalent to $x(n)$ being the output of an pole–zero system $H(z)$ whose input is the white noise $w(n)$. The transfer function of the system can be written as

$$H_{\mathrm{ARMA}}(z) = \frac{\sum_{i=0}^{M} b_i z^{-i}}{1 + \sum_{i=1}^{N} a_i z^{-i}} \tag{3.238}$$

Estimating the magnitude or power spectrum of data thus becomes optimally estimating the model parameters $\{a_i\}$ and $\{b_i\}$ under a selected criterion (e.g., the MSE criterion). These model parameters can be estimated by solving a set of Yule–Walker equations. In the AR case, it is a set of linear equations in terms of unknown parameters. However, in the ARMA case the equations are nonlinear in $b_i$'s, making the estimation more complicated. The parameter estimation is usually performed in two steps. In the first step, the solution for $\{a_i\}$ is obtained by solving a set of Yule–Walker equations for $k > M$,

$$r_x(k) + \sum_{l=1}^{N} a_l r_x(k-l) = 0 \qquad k > M \tag{3.239}$$

Depending on the available length of the autocorrelation sequence $r_x(k)$, one could use the so-called modified Yule–Walker equation method [set $k = M+1, +2, \ldots, M+N$ in (3.239)] or the least-squares modified Yule–Walker equation method [set $k = M+1, \ldots, M+L$, where $L > N$ in (3.239)]. The parameters $\{a_i\}$ are obtained by solving the set of equations in (3.239).

Once $\{a_i\}$ are obtained, the parameters $\{b_i\}$ are then estimated as follows. First, $\{c_k\}$ is obtained from the solution of the following set of Yule–Walker equations:

$$r_x(k) + \sum_{l=1}^{N} a_l r_x(k-l) = c_k \qquad k = 0, 1, \ldots, M \tag{3.240}$$

where $c_k = \sum_{l=1}^{M} b_{l+k} h^*(l)$. Since there are $M$ equations and $M$ unknowns in (3.240), the exact solution of $\{c_k\}$ can be obtained. Then the causal part of the power spectrum $[P_y(z)]_+$ is the calculated by

$$[P_y(z)]_+ = \left[ [c_M(z)]_+ A_N^*\left(\frac{1}{z^*}\right) \right]_+ \tag{3.241}$$

where + denotes the causal part, $*$ denotes the complex conjugate, $A_N(z) = \sum_{i=1}^{N} a_i z^{-i}$, and $c_M(z) = \sum_{i=1}^{M} c_i z^{-i}$.

Using the conjugate symmetry, $P_y(z)$ can then be determined. Finally, $\{b_l\}$ are obtained by performing the spectral factorization to $P_y(z) = B_M(z) B_M^*(1/z^*)$, where $B_M(z) = \sum_{i=0}^{M} b_i z^{-i}$.

The magnitude spectrum $A(e^{j\omega})$ and power spectrum $P(e^{j\omega})$ can be obtained by using

$$A(e^{j\omega}) = \left|H_{\text{ARMA}}(e^{j\omega})\right| = \left| \frac{\sum_{i=0}^{M} b_i e^{-j\omega}}{1 + \sum_{i=1}^{N} a_i e^{-j\omega}} \right| \qquad P(e^{j\omega}) = \left|H_{\text{ARMA}}(e^{j\omega})\right|^2 \tag{3.242}$$

From the magnitude or power spectrum of the data, one can analyze the elements that are part of the disturbance.

## 3.7 SUMMARY AND CONCLUSIONS

This chapter discusses methods to quantify voltage and current variations, that is, to extract features or characteristics from the voltage or current waveform. Most of the methods are based on the assumption that the signal is stationary. Under this assumption the (statistical) properties of the signal are constant; a value for a feature obtained over a longer period also characterizes the signal over a shorter period, and the other way around.

Several methods introduced here can also be applied for the study of power quality events. The difference is merely in the length of the measurement window: Much shorter measurement windows are needed for events than for variations. We will come back to some of the methods in Chapters 7 to 9 in association with events.

Most of the discussion in this chapter is concentrated on voltage variations, but the methods can also be applied to current variations in most cases. With current measurements one should however keep in mind that the magnitude variations are typically much larger than those of voltages. This means the fundamental component cannot be used as a reference value. Instead a rated value should be used. Also with current measurements the assumption that harmonic components are small compared with the fundamental component is not always valid.

### 3.7.1 Frequency Variations

The standard method to characterize frequency variations consists of counting the voltage zero crossings during a time period with an accurately known length (10 s according to IEC 61000-4-30). The high accuracy of this method is based on the very high accuracy for time measurements. The high accuracy that can be obtained by the standard method also limits the need for more advanced methods. Despite this, an alternative method is introduced, based on the *dq*-transform, that is better suited to present short-time changes in frequency.

Another issue with frequency measurements is the ambiguity between frequency and phase-angle variations. It is not possible to uniquely distinguish between the two (cf. frequency and phase modulation in communication theory). This issue is not of concern for slow variations as considered in most studies. However, during events such as short circuits in the transmission system the voltage may experience changes in both phase angle and frequency. No satisfactory solution to this dilemma has been proposed.

### 3.7.2 Voltage Magnitude Variations

The vast majority of voltage (magnitude) variation studies uses the rms value to quantify the voltage magnitude. In IEC 61000-4-30 a basic measurement window of 10 or 12 cycles (in 50- and 60-Hz systems, respectively) is prescribed: The measurement window shall be synchronized to the actual power system frequency. Nonsynchronization leads to a small error in the estimated rms value, where the relative error in rms value is half the relative error in frequency.

A number of alternative methods are introduced: among others, voltage amplitude, fundamental component, instantaneous three-phase rms, and positive-sequence voltage. The latter two values are suitable alternatives in a three-phase system. The characterization of voltage variations in three-phase systems is not part of any standard document yet. Some further development and standard-setting work is needed on this issue. The instantaneous three-phase rms may be an appropriate method, possibly even to characterize flicker in a three-phase system.

Several of the methods for quantifying waveform distortion also result in a value for the fundamental (power system frequency) component, which can be used to quantify the voltage variation. Although the difference between rms and fundamental voltage is small, IEC 61000-4-30 does not allow the use of fundamental voltage. A discussion on this may be needed.

### 3.7.3 Three-Phase Unbalance

The standard method for quantifying three-phase unbalance is as the ratio between negative- and positive-sequence voltage. The ratio between zero- and positive-sequence voltage is a second characteristic, but one rarely used. The standard measurement window, according to IEC 61000-4-30, is the same as for voltage magnitude variations: 10 or 12 cycles. Positive- and negative-sequence voltage can be calculated from the voltage phasors in the individual phases or from the *dq*-transform. From the fundamental component or rms of the three phase-to-phase voltages it is possible to exactly calculate the (negative-sequence) unbalance.

Alternative definitions for the three-phase unbalance are given in a number of IEEE standards. These methods should however be handled with care, as the results may deviate significantly from the IEC values, especially if a zero-sequence component is present.

### 3.7.4 Waveform Distortion

The vast majority is methods for quantifying waveform distortion is based on a knowledge of the frequency components present in the signal. The main exception is the crest factor, which is unfortunately rarely used in power quality studies. Most studies obtain the spectrum from a, DFT over an integer number of cycles of the power system frequency: in IEC 61000-4-7: 10 cycles in a 50-Hz system; 12-cycles in a 60-Hz system. The resulting values are next aggregated in harmonic and interharmonic groups and subgroups. It is important that the window is synchronized to the actual power system frequency to prevent "leakage" from dominant components to nearby harmonic and interharmonic components. The result of leakage may be that interharmonics or even harmonics appear to be of much higher amplitude that they actually are.

In a three-phase system, the application of the DFT or any other method will result in three spectra, one for each of the three phase-to-ground or phase-to-phase voltages. A method is presented to quantify the waveform distortion in a way that considers the three-phase character of the system. For this a distinction is made between balanced and unbalanced harmonics. For each harmonic frequency, one balanced and two unbalanced components exist. Alternatively, the time-domain waveforms can be separated in a balanced components (one for the three-phase system) and unbalanced components (one for each of the three phases). The DFT is next applied to the balanced and unbalanced components.

### 3.7.5 Methods for Spectral Analysis

In addition to the DFT a number of alternative methods have been discussed in this chapter to estimate the spectral contents of a signal. Each of these methods has specific advantages compared with the DFT. However, it should be noted that the DFT is the easiest to implement and is computationally efficient.

MUSIC and ESPRIT, as the candidate methods for high-resolution spectral line estimation, are introduced for analyzing harmonics and interharmonics from stationary voltage or current signals under sinusoidal models. Similar to the DFT, they are batch processing methods where a block of data is required for each estimation. They can be considered as offline processing methods or processing methods that require a long delay. The resulting estimates are the frequencies and the magnitudes (and damping factors) of the dominant components in the input data. A clear advantage over the DFT is high-frequency resolution, which enables one to distinguish two closely spaced spectral line components. Further, the noise is included in the signal modeling, which enables a good estimation from noisy data. One disadvantage is that for power quality measurement data a preprocessing step is often required. Due to the presence of the prominent power system frequency component (typically a scale of 100 time larger as compared with the remaining harmonics and interharmonics), prefiltering is required to significantly reduce the strong power system frequency component, so that a high estimation accuracy can be achieved for the remaining components. This prefilter may pose a limitation to the estimation of frequency components that are close to the power system frequency. The ESPRIT method, being a signal-subspace-based method, is generally viewed as more accurate than the noise-subspace-based-MUSIC method. More experience is needed in the use of these methods for analysis of power quality disturbances. Both methods have especially shown good potential in detecting interharmonic components. In addition, the methods can also be used to estimate the frequency components in transient voltages and currents (see Chapter 8).

Kalman filters are based on state-space modeling and offer recursive estimation for each new incoming signal sample. They are therefore more suitable for online processing. Kalman filters in this chapter are driven by their application to power quality analysis where a sinusoidal model is adopted, and the state variables are set to be the parameters of the power system harmonics and interharmonics. Under such a setting the frequencies of the sinusoids have to be prespecified, which is clearly a disadvantage. This mainly imposes a limitation for interharmonic estimation as there could be an unlimited number of possible interharmonic frequencies. The application of Kalman filters is also limited when there is a relatively large variation in the power system frequency.

Assuming a practically nonexisting (or minor) harmonic (or interharmonic) component in the model will only result in a nearly zero magnitude estimate in the Kalman filter at the expense of more computation. A main advantage of Kalman filters over ESPRIT and MUSIC is its recursive nature. As a result, a Kalman filter generates the time trajectories of the estimated parameters rather than the individual estimated values in MUSIC and ESPRIT. Another advantage is that Kalman filters are also known to be suitable for nonstationary signal analysis (see Chapter 4).

Stability and convergence are practical problems when applying Kalman filters to measured data rather than synthetic data. Here, nonaccurate models or model deviations, unknown second-order noise statistics, and numerical instability due to finite-length computation may arise. Some solutions to the above problems include using the square-root covariance Kalman filters to reduce the ill-conditioning and hence stabilize the algorithms, using exponentially weighted or sliding-window Kalman filter algorithms to emphasize the most recently measured data samples, and using extended Kalman filters to include some nonlinear system models. More experience is still needed for analysis of power quality disturbances using Kalman filters even though Kalman filters are generally considered a relatively mature signal-processing subject.

It should be emphasized that whether a signal or some of its parameters would be estimated by a Kalman filter is dependent on the selection of the state-space model. For example, a Kalman filter can also be used to estimate AR model parameters. The Kalman filter examples addressed in this chapter are only motivated by harmonic- and interharmonic-based power quality analysis.

Most power quality signals possess a line spectrum, that is, a limited number of frequency components at discrete frequencies. The above methods (MUSIC, ESPRIT, and Kalman filters) are very suitable for the analysis of such signals. However some power quality signals can better be described by broadband spectra. This especially holds when the interest is not in a few harmonic and interharmonic components but in the band-limited distortions. Such examples include the current taken by an arc furnace and the high-frequency part of active rectifiers (e.g., used in certain types of motor drives, in interfaces for distributed generation, and in most energy-saving lamps). Two methods are presented in this chapter that are directly aimed at quantifying a broadband spectrum: AR and ARMA models. Further development of these methods is needed, especially when the signals contain both narrow-band and broadband components.

### 3.7.6 General Issues

All the methods presented in this chapter have assumed that the measurement location is known. The measurement location may however have a significant influence on the resulting values of characteristics. For example, the harmonic distortion at the terminals of a piece of equipment may be different from the distortion at the service entrance, which may in turn be different from the distortion in the substation. The impact of the measurement location may be much bigger than the impact of any of the details for obtaining the characteristics. A discussion on this subject is needed. Such a discussion will provide some guidance in the placing of power quality monitoring equipment. It will also provide a background for setting accuracy requirements for the monitors.

A somewhat related discussion concerns the impact of the disturbances on equipment. In Section 3.3 we presented some combinations of parameters that may be used to quantify the impact of variations on different types of equipment. A further discussion on this is needed. Again the accuracy requirements for the monitors may be coordinated with the uncertainty of the impact of a disturbance on equipment. Both of these general issues require a combination of basic academic research, measurement campaigns by network operators, and standard development work. The second issue even involves a contribution from equipment manufacturers.

## 3.8 FURTHER READING

In the literature there exist rich references on the theories and applications of signal processing which are beyond the scope of this chapter. Readers are referred to [230, 207, 221, 223, 102, 183] for more details on the basic signal theories and [175, 282, 33] for further inspiration to power system applications.

## CHAPTER 4

# PROCESSING OF NONSTATIONARY SIGNALS

In Chapter 3 we discussed disturbance data that are stationary (or statistical time invariant) where the statistics of a data sequence is independent of time. The analysis of the disturbance data in Chapter 3 consisted of estimating waveform characteristics such as magnitude and frequency contents. For analyzing real power system disturbance data, however, strictly stationary signals rarely exist. Even during the normal operation the statistics of magnitude and frequency contents of voltages and currents may change slowly over time. As long as these changes are small, we use the term *variations* as introduced in Section 1.2.3. For larger sudden changes of statistics the term *events* is used.

In this chapter we will extend the methods introduced in Chapter 3 to nonstationary signals where the signals are statistical time varying. We will mainly concentrate on nonstationary variations, but the examples will also contain some disturbances that should be classified as events. The extraction of characteristics (or features and attributes) from events will be discussed in more detail in Chapters 7 and 8. The processing of nonstationary signals in most cases results in signal characteristics or attributes as a function of time, thus quantifying the dynamic nature of the disturbance. Such a function of time can next be used to extract information on the origin of the event. By correlating variations in the level of a certain component with a certain stage in a production process, a link between the production process and that component can be established. It is possible in the same way to correlate certain disturbances with consumption patterns of electricity, such as the early-evening television peak in the harmonic distortion.

*Signal Processing of Power Quality Disturbances.* By Math H. J. Bollen and Irene Yu-Hua Gu

At a short time scale the time variation of characteristics during an event can be used to draw conclusions on the origin of the event. Some examples will be shown in this chapter, but detailed discussions will be included in Chapter 6 and some later chapters. Standard methods for analyzing the time variation of voltage and current variations are presented in Chapter 5.

## 4.1 OVERVIEW OF SOME NONSTATIONARY POWER QUALITY DATA ANALYSIS METHODS

For nonstationary power quality disturbances, the primary interest is to quantify the existing components (e.g., harmonics, interharmonics, or some narrow-band components) with the corresponding magnitudes and phase angles, to detect the positions where the disturbances start and end, to estimate and track some dynamically changing parameters in time, and more advanced, to associate parameters that are estimated from three-phase voltage and current waveforms measured in different locations and different voltage levels. Since the statistical time-variant characteristics (or attributes) of nonstationary disturbance signals are often caused by the underlying change of the states in the power systems, such nonstationary disturbances in the power systems are sometimes referred to as *non-steady-state* disturbances. When the terminology *nonstationary signal* emphasizes the statistical time-varying nature of the disturbances and non-steady-state the underlying changes in the power system states (or attributes of the disturbances), they reflect and emphasize the different aspects of the same disturbance signal. Therefore, in the subsequent context these two terms are often used or interchanged without further clarification if they both hold for a disturbance signal.

For nonstationary signals, we will require some joint time–frequency (or time–scale, time–transform) domain analysis in order to track the time-evolving characteristics of signal components. Signal-processing techniques that find applications to analyzing such disturbances can again be roughly categorized into non-model-based (or, nonparametric) methods and model-based (or, parametric) methods.

### 4.1.1 Non-Model-Based Methods

This class of methods primarily decomposes the signal into time-dependent frequency (or frequency-related) components, for example, using the discrete STFT (often referred to as the sliding-window DFT), subband filters, or wavelets. In this way it is easier to extract the dynamic characteristics of a signal from the time-evolving signal components than from the original signal waveforms. These methods are very different from the non-model-based methods described in Chapter 3 where the signal characteristics are *static* extracted either from the time domain (e.g., rms) or from the frequency domain (such as the Fourier series or Fourier transform). For example, the frequency components provided by the DFT are time independent; in fact, for stationary signals these components remain the same over time. However, this is very different if the voltage or current waveforms are nonstationary. In such a case, applying DFT to a block of data yields signal components as the averaging effect of data within the analysis window. For different windowed data, the DFT results are different as the signal is nonstationary.

We shall discuss non-model-based methods, including the STFT in Section 4.2 and the discrete wavelet transform in Section 4.3, and both are explained under the framework of subband filters. These are time–frequency (or time–scale) signal decomposition methods that are more suitable for analyzing and characterizing non-steady-state power system disturbances.

### 4.1.2 Model-Based Methods

Another important class of methods for analyzing non-steady-state power system disturbances is formed by modeling the disturbances and estimating the parameters of interest in the model. Frequently used models include harmonic and damped sinusoidal models, AR and ARMA models, as described in Chapter 3. However, some modifications should be made in order to handle the nonstationarity of the signal. If we divide the data into blocks and can reasonably assume that the data within each block are stationary, then the methods described in Chapter 3 are still valid to each data block. As we shift the data window (block) forward by using a sliding window, we can estimate the nonstationary model parameters. A simplest way is to divide data into a fixed-size block (either overlap or nonoverlap). A more sophisticated way is to employ an automatic *segmentation* method where data are partitioned into disjoint blocks (usually of different length); each is due to one specific cause in the power system and hence can be described by one specific underlying model. The model-based methods can next be applied to each segmented data and the parameters of interest can be estimated.

We shall first introduce the block-based processing in Section 4.4. Nonstationary signal modeling and parameter estimation are then described by modifying the corresponding model-based methods in Chapter 3. This includes block-based AR models in Section 4.4.3 and block-based harmonic models (both MUSIC and ESPRIT) in Section 4.4.4. In all these methods, a fixed block size is used. However, the best suitable way is to use an adaptive block size where each block of data is associated with a stationary model caused by one underlying reason. To automatically determine the size of such blocks is nontrivial. This issue will be further addressed in Chapter 7 where an automatic data segmentation method is included.

In addition to modifying the existing methods by block-based processing, in Section 4.5 we will discuss Kalman filters that can be directly used to characterize non-steady-state power system disturbances.

## 4.2 DISCRETE STFT FOR ANALYZING TIME-EVOLVING SIGNAL COMPONENTS

One of the main interests for analyzing nonstationary voltage and current disturbance data is to estimate the frequency contents of data as a function of time. For example, one may be interested in the dominant frequency components of a capacitor switching transient or the changes in the third-harmonic voltage distortion during a voltage dip. For power system disturbances, accurately describing the amplitude and phase-angle jump as a function of time does require some nontrivial methods, especially for short-duration events.

A suitable way to extract such information is to apply time–frequency (or time–scale) signal decomposition where one obtains the time-evolved signal components in different frequency bands [129]. A discrete STFT [230, 246] is a time–frequency decomposition method among many others. A discrete STFT is utilized for the time–frequency decomposition of nonstationary signals, where the use of Fourier transform alone becomes inadequate.

For a given signal $x(n)$, the complex-signal component in the frequency band $k$ at time instant $n$ can be obtained from the discrete STFT, defined as

$$X_n(e^{j\omega_k}) = \sum_m x(m)w(n-m)e^{-j\omega_k m} \qquad k = 0, 1, \ldots, N-1 \tag{4.1}$$

where $\omega_k = 2\pi k/N$ is the frequency in radians, $N$ is the number of frequency bands, $w(m)$ is a selected symmetric window (e.g., a Hamming window) of size $L$, $L \leq N$ if perfect signal reconstruction is required by the synthesis filter bank. A STFT is usually implemented by employing a DFT with a sliding window as follows:

$$\begin{aligned}&[X_n(0) \quad X_n(1) \quad \cdots \quad X_n(N-1)] \\ &\quad = \mathrm{DFT}_N[x(n-L+1)w(0) \quad x(n-L+2)w(1) \quad \cdots \quad x(n)w(L-1)]\end{aligned} \tag{4.2}$$

where $k$ is an index associated with the frequency $\omega_k = 2\pi k/N$, $n$ denotes the time, and $\mathrm{DFT}_N$ denotes the $N$-point DFT, $L = N$ often used. However, if $L < N$ is selected, then the windowed data in the right-hand side of (4.2) should first pad zeros to the length $N$ (i.e., equal to the number of bands) before applying the DFT. The time-dependent output is obtained by sliding the data window forward. It is worth noting that $X_n(e^{j\omega_k})$ is usually a complex value. In fact, $X_n(e^{j\omega_k})$ in (4.1) can be viewed as the low-pass shifted version of the $k$th complex bandpass filter output $\widetilde{X}_n(\omega_k)$, obtained as,

$$\widetilde{X}_n(e^{j\omega_k}) = e^{j\omega_k n} X_n(e^{j\omega_k}) \tag{4.3}$$

where

$$\widetilde{X}_n(e^{j\omega_k}) = \sum_m x(n-m)w(m)e^{j\omega_k m} = x(n) * h_k(n) \tag{4.4}$$

and the impulse response of the $k$th complex bandpass filter is

$$h_k(n) = w(n)e^{j\omega_k n} \qquad k = 0, 1, \ldots, N-1 \tag{4.5}$$

**TABLE 4.1 Filter Bandwidth in STFT Related to Window Type and Size $L$**

| Window Type | Side-Lobe Attenuation (dB) | Approximate Width of Main Lobe, $\Delta_\omega$ |
|---|---|---|
| Rectangular | −13 | $4\pi/L$ |
| Bartlett | −25 | $8\pi/(L-1)$ |
| Hanning | −31 | $8\pi/(L-1)$ |
| Hamming | −41 | $8\pi/(L-1)$ |
| Blackman | −57 | $12\pi/(L-1)$ |

whose center frequency is

$$f_k = \frac{f_s k}{N} \qquad \text{Hz} \tag{4.6}$$

### 4.2.1 Interpretation of STFT as Bank of Subband Filters with Equal Bandwidth

A discrete STFT can be interpreted as the low-pass representation of the associated bandpass filter output. All these bandpass filters [with impulse response specified in (4.5)] have an equal bandwidth that is determined by the type and size of the selected window. For a Hamming window of length $L$, the corresponding 3-dB bandwidth of the bandpass filters is

$$2B = \frac{4f_s}{L} \qquad \text{Hz} \tag{4.7}$$

The bandwidth in (4.7) is defined as the main-lobe width of the frequency response of the window function [246]. Table 4.1 lists the corresponding bandwidths of the subband filters for other types of windows [230]. It is worth noting that the level of side-lobe attenuation in Table 4.1 is associated with the possible leakage of signal components from the neighboring frequency bands: The larger the attenuation, the less *leakage* of unwanted signal components from the neighboring bands.

### 4.2.2 Time Resolution and Frequency Resolution

It is worth noting that once the type and size of the data window are chosen for the STFT, there is a fixed frequency resolution in all bands. This is because the product of the time resolution $\Delta_t$ and the frequency resolution $\Delta_\omega$ is constrained by the uncertainty principle [67],

$$\Delta_t \Delta_\omega \geq \tfrac{1}{2} \tag{4.8}$$

where the rms bandwidth $\Delta_\omega$ of the window $w$ is defined as,

$$
\begin{aligned}
&\text{Center of window:} \qquad x^* = \frac{1}{\|w\|_2}\int_{\infty}^{\infty} x|w(x)|^2 dx \\
&\text{Radius of window:} \qquad \Delta = \frac{1}{\|w\|_2}\left\{\int_{\infty}^{\infty} (x - x^*)^2 |w(x)|^2 dx\right\}^{1/2}
\end{aligned}
\tag{4.9}
$$

and the rms time duration $\Delta_t$ is defined in a similar way.

Figure 4.1 shows the frequency responses of bandpass filters corresponding to the STFT with a Hamming window of size $L = 16$. The side-lobes of the frequency response are removed as they are about 40 dB below the main lobe. Increasing the window size leads to a reduced filter bandwidth, and vice versa. The bandwidth also determines whether each band contains one or more harmonics. If more than one harmonic is within the filter bandwidth, it is nonresolvable and will be recognized as the output of that bandpass filter due to the low frequency resolution of the filter.

To obtain different resolutions in different frequency bands, one needs to redo the STFT with a different window size and then extract the outputs from the corresponding bands. More efficiently, one can directly convolve the data $x(n)$ with the filter impulse response $h_k(n)$ in (4.5) [where the size of the window $w(n)$ is determined by the desired resolution] at the selected frequency band, as indicated by (4.4).

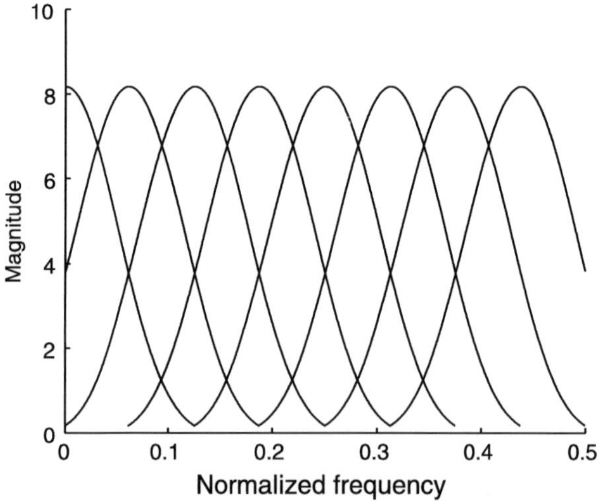


**Figure 4.1** Bandpass filter frequency responses from discrete STFT (Hamming window $L = 16$); actual frequency $= f_s \times$ normalized frequency (in hertz).

### 4.2.3 Selecting Center Frequencies of Bandpass Filters

In the discrete STFT the original signal (e.g., a data recording containing a voltage dip) is split into a set of bandpass components. From (4.6) and (4.7) we can see that both the center frequencies and the bandwidths of filters can be selected.

To analyze power system disturbances, it is desirable that the center frequencies of the bandpass filters are located at the harmonics of the power system frequency, so that one may obtain signal components related to the harmonics.

However, it should be noted that even though the center frequencies of bandpass filters are set in the harmonic frequencies, the outputs from bandpass filters do not correspond to the pure harmonic components! Rather, the output usually contains several harmonics and interharmonics depending on the frequency resolution of the filter. Therefore, one should consider the output from a bandpass filter as a *pseudoharmonic* rather than a pure harmonic. This can be explained as follows. For a selected window type and size (or size of the data block), the frequency resolution $\Delta_\omega$ of each bandpass filter is fixed. The 3-dB frequency resolution is equal to the main-lobe width of the window spectrum (See Table 4.1). If two harmonics (or interharmonics) are within the same main lobe, then these two components cannot be resolved by the bandpass filter and hence will only be recognized as one narrow-band component. For example, for a disturbance data sequence sampled at $f_s = 15{,}360$ Hz, that is, 256 samples per 60-Hz cycle, choosing a Hamming window of size $L = 256$ results in a bandwidth of 240 Hz for the bandpass filters, 120 Hz to each side of the filter center frequency $f_c$. This implies that within each band there are three harmonics (at frequencies $f_c - 60$, $f_c$, and $f_c + 60$ Hz). Since the window size is equal to the number of samples in one 60-Hz cycle, the other two harmonics are located at the frequencies $f_c - 120$ and $f_c + 120$ Hz where the window magnitude response has zero values and hence is fully attenuated (see the next section for more discussion). In this case, we consider the output from the bandpass filter not as a pure harmonic (in fact, three weighted harmonics) but as a pseudoharmonic.

### 4.2.4 Leakage and Selection of Windows

Table 4.1 indicates that the side-lobe attenuation varies depending on the type of window. For a rectangular window the attenuation is the smallest. This implies that the leakage of signal components from the neighboring frequency bands is large. If the signal component in the main lobe is weak as compared to the components in the neighboring bands (e.g., a voltage signal in the fundamental frequency is often significantly stronger than the harmonic components), this leakage can mask the main-lobe signal component. Therefore, when choosing the type of window, the side-lobe attenuation level of the window should be taken into account.

Once a window type is selected, choosing the size of the window can also play a role in preventing the leakage, for example, from a harmonically dominant disturbance signal. To reduce the leakage of harmonic components from the neighboring bands, it is essential to choose a proper window size $L$ such that the zeros in the spectrum of window function are located exactly at the harmonic frequencies. For example, a rectangular window of size $L$ has the magnitude spectrum

$$|W_{\text{rec}}(\omega)| = L\left|\frac{\sin(\omega L/2)}{\omega L/2}\right| \tag{4.10}$$

whose zeros are located at $\omega L/2 = k\pi$, or equivalently at the frequencies $f = 2\pi k f_s/L$, $k = 1, 2, \ldots$, where $f_s$ is the sample rate. Let us assume that the sample rate is chosen such that each cycle of voltage or current waveform contains an integer $m$ samples, that is, $f_0 = mfs$. By setting the harmonic frequencies at the zeros of $|W_{\text{rec}}(\omega)|$, the leakage effect from the neighboring harmonics can be minimized. This can be done by choosing the size of the window $L$ equal to the multiples of $m$, that is, $L = nm$, where $n$ is an integer $n = 1, 2, \ldots$. In such a way, $|W_{\text{rec}}(2\pi f_l/f_s)| = 0$ holds for harmonic frequencies $f_l = lf_0$, $l = 1, 2, \ldots$, and hence yields the highest attenuation to the harmonics from the neighboring bands. For example, consider a disturbance recording from a $f_0 = 50$ Hz power system with a sample rate of $f_s = 2400$ Hz (i.e., $m = 48$ samples for one cycle of signal). If we choose the window size $L = 48n$ (e.g., $n = 1$), then for the $k$th band, $(k \pm i)$th harmonics, $i = 1, 2, \ldots$, have the maximum attenuation as desired.

**Example 4.1** Analysis of Measured Data: Voltage Dip. Consider one of the test signals in IEEE Project Group 1159.2 [166] shown in Figure 4.2. The sample frequency used is $f_s = 15{,}360$ Hz, or 256 samples per 60-Hz cycle. The discrete STFT is applied to the signal. The center frequencies of bandpass filters are set at the power system harmonics by choosing the total number of bands as $N = 256$.

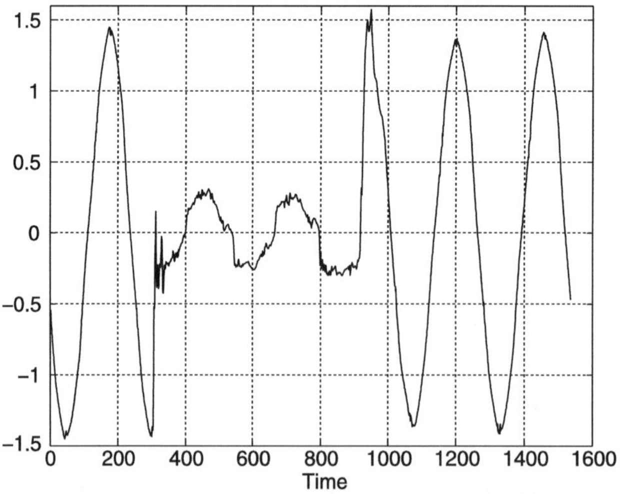


**Figure 4.2** Measured waveforms containing voltage dip. (From IEEE Project Group 1159.2.)

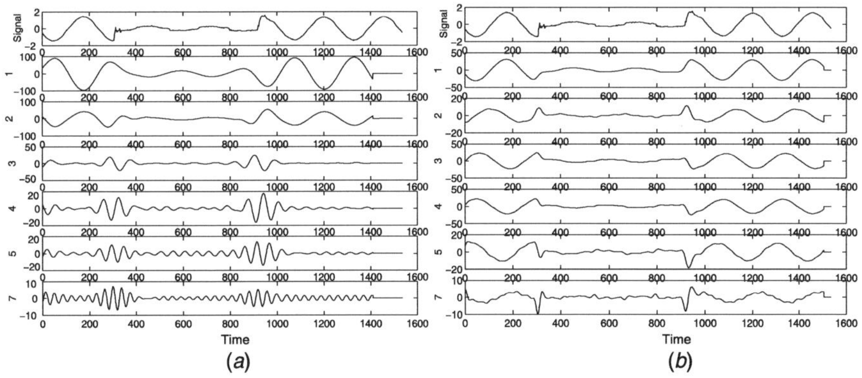


**Figure 4.3** Discrete STFT for analyzing voltage dip: (*a*) $L = 256$; (*b*) $L = 64$. From top to bottom: measurement data containing voltage dip; outputs of bandpass filters centered at harmonics 1, 2, 3, 4, 5, 7.

To obtain signal components from the real bandpass filters instead of complex bandpass filters, the real part of the complex output signal $\widetilde{X}_n(e^{j\omega_k})$ is used.

Figure 4.3 shows the outputs from the bandpass filters using the STFT with $N = 256$ bands and a Hamming window of sizes $L = 256$ and $L = 64$, respectively. The window size $L = 256$ corresponds to one cycle of the power system frequency. Changes in the time domain within one cycle are still visible due to the sliding Hamming window; however, they are spread due to the limited time resolution. This effect is most obvious for dip initialization and voltage recovery. Using (4.7) the bandwidth of the bandpass filters used in Figure 4.3*a* is 240 Hz, that is, 120 Hz to each side of the center frequency. There is a certain overlap between two neighboring bands; hence the strong 60-Hz component is visible in the second band (centered at 120 Hz). Using a smaller window (e.g., $L = 64$ as in Fig. 4.3*b*) gives a higher time resolution but a lower resolution (blurred curves) in the frequency domain (as the bandwidth is 480 Hz on each side of the center frequency). As one can see, there is still a clear 60-Hz component present in the passband centered at the seventh-harmonic frequency. To ensure that each band only contains one harmonic, a window size of two cycles should be used. However, this will further reduce the time resolution.

Using a rectangular window for the STFT gives a bandwidth $2B = 2f_S/L$ and thus 60 Hz on each side of the center frequency for a one-cycle window. However, compared to the Hamming window case there are more oscillations in the frequency domain due to the Gibbs effect and the lower attenuation in the side lobes of the window frequency response.

Figure 4.4 displays the absolute values of the complex filter outputs, which provides an alternative way of analysis. The figure plots the magnitudes of the harmonic signals as a function of time, where one should realize the limited time and frequency resolution as discussed before. The large peaks in bands 3 and higher are due to the high frequencies presented in the voltage steps at the dip initiation and

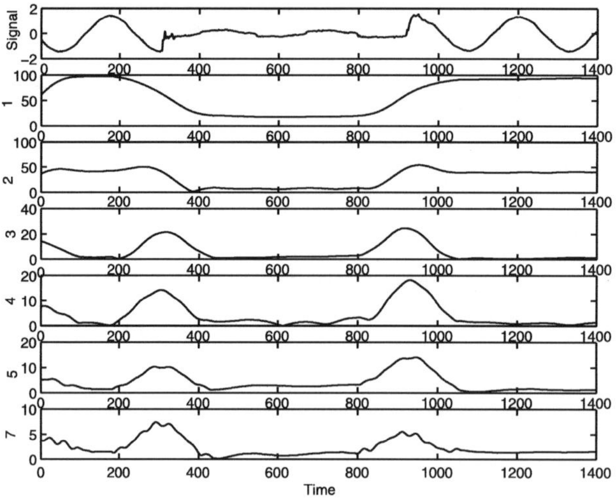


**Figure 4.4** Discrete STFT for analyzing a voltage-dip signal, $L = 256$. From top to bottom: measurement containing a voltage dip; output magnitudes from complex bandpass filters centered at harmonics 1, 2, 3, 4, 5, 7.

voltage recovery. Of more interest for the harmonic analysis are the signals before, between, and after the two peaks. We can see a steady increase in the third harmonic, probably related to the reaction of equipment to the dip, the fifth harmonic somewhat remained a constant during the event but at a higher level than that before the dip initiated, and the seventh harmonic seems to remain the same before and after the dip. We also see that the harmonic levels before and after the dip are about equal.

## 4.3 DISCRETE WAVELET TRANSFORMS FOR TIME–SCALE ANALYSIS OF DISTURBANCES

Contrary to the STFT where all the corresponding bandpass filters have the same frequency resolution, a dyadic DWT (discrete wavelet transform) corresponds to a set of bandpass filters with different frequency resolution if a dyadic structured DWT (having an octave bandwidth relation in any two neighboring bands) is applied [67, 281, 220]. The DWTs are frequently used in the analysis of power system disturbances, especially for detecting transitions such as the starting and ending positions of a dip.

For multiresolution decomposition of signals, a dyadic structured DWT is employed. This is done by first splitting the input signal into a low- and a high-frequency component using a low-pass and a high-pass filter associated with a scaling function and a wavelet function followed by a down-sampling of 2. Then, the low-frequency component is further split into a low- and a high-frequency one followed by a down-sampling of 2. The process is repeated a number of times (the number of scales), resulting in multiresolution signal components at different bands.

From the subband filter point of view, a dyadic DWT can be interpreted by a set of bandpass filters with an octave bandwidth (or the ratio of bandwidth to filter center frequency is a constant for all filters). Note that for a data sequence the length of the decomposed signal components in each band is only half of the length in its immediate lower band. It is worth mentioning that also interband *aliasing and imaging* are introduced by the subband filters. Choosing proper wavelets and their support (or length) may reduce the aliasing terms between different subbands. This is associated with an important property of wavelets: the number of vanishing moments (i.e., moments equal to zero). The number of vanishing moments is associated with the smoothness of wavelets and is a key factor in many applications, for example singularity detection and noise removal. For a large number ($N$) of vanishing moments, the wavelet has approximately $N/5$ continuous derivatives. Readers are referred to other references for further details [56, 281]. To generate equal-length signal components for different frequency bands, one may apply the analysis filters followed by the synthesis filters.

From the signal space $V$ and the basis function point of view, wavelet theory is described by the scaling functions $\phi(t)$ and wavelet functions $w(t)$. The dilations and translations of these functions form the basis functions: the scaling functions $\{\phi_{jk}(t) = 2^{j/2}\phi(2^j t - k)\}$ span the subspace $V_j$ and the wavelet functions $\{w_{jk}(t) = 2^{j/2}w(2^j t - k)\}$ span the subspace $W_j$, $W_j \oplus V_j = V_{j+1}$, $V_j \cap W_j = \{0\}$, and $j$ and $k$ are related to the scaling and the translation, respectively.

The scaling function $\phi(t)$ and wavelet function $w(t)$ are associated with the low-pass filter coefficients $h_0(k)$ and high-pass filter coefficients $h_1(k)$ by the dilation equation and wavelet equation as follows:

$$\begin{aligned} \phi(t) &= \sum_k 2h_0(k)\phi(2t - k) \\ w(t) &= \sum_k 2h_1(k)\phi(2t - k) \end{aligned} \tag{4.11}$$

They are often described by the two-channel filter bank. In fact, wavelets are special types of subband filters that satisfy certain conditions.

### 4.3.1 Structure of Multiscale Analysis and Synthesis Filter Banks

Let the low-pass filter and high-pass filter in a two-channel analysis filter bank be $H_0(z)$ and $H_1(z)$ and the low-pass filter and high-pass filter in the synthesis filter bank be $F_0(z)$ and $F_1(z)$. After applying analysis filters at each scale, a down sample of 2 is applied to the output of filters. Also, prior to applying synthesis

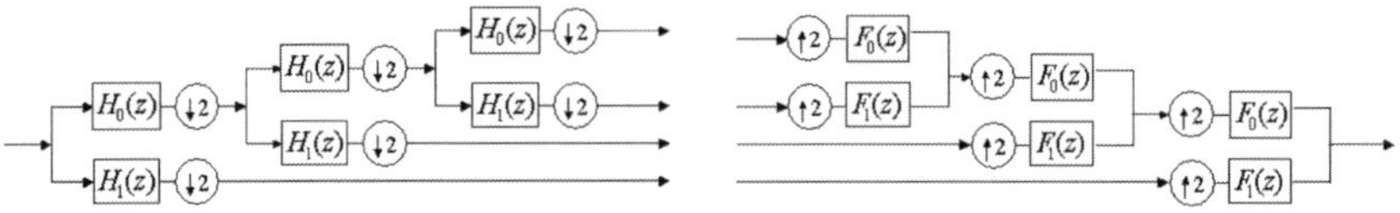


**Figure 4.5** Structure of analysis filter bank (left) and synthesis filter bank (right).

filters at each scale, a up sample of 2 is applied to the input signal components. Figure 4.5 shows the structure of a three-scale analysis and synthesis filter bank, where the subband filters have an octave bandwidth.

It is worth noting that this is equivalent to using four subband filters, with the following equivalent analysis filter transfer functions:

$$\widetilde{H}_0(z) = H_0(z)H_0(z^2)H_0(z^4) \qquad \widetilde{H}_1(z) = H_0(z)H_0(z^2)H_1(z^4)$$
$$\widetilde{H}_2(z) = H_0(z)H_1(z^2) \qquad \widetilde{H}_3(z) = H_1(z)$$

Similarly, we can obtain the four equivalent synthesis filter transfer functions as follows:

$$\widetilde{F}_0(z) = F_0(z^4)F_0(z^2)F_0(z) \qquad \widetilde{F}_1(z) = F_1(z^4)F_0(z^2)F_0(z)$$
$$\widetilde{F}_2(z) = F_1(z^2)F_0(z) \qquad \widetilde{F}_3(z) = F_1(z)$$

Since an analysis filter bank and synthesis filter bank are often combined to obtain the full-length bandpass signal components, we shall briefly describe the design of analysis filter banks and synthesis filter banks as well as the concept of perfect reconstruction (PR), including orthogonal PR and PR with linear phase.

### 4.3.2 Conditions for Perfect Reconstruction

If a signal passes through an analysis filter bank followed by a synthesis filter bank without any other processing, it is desirable that the signal remain the same except some time delay or possibly a signal amplification. To achieve this, a PR is required in designing the analysis and synthesis filter banks. The PR for a two-channel filter bank is achieved if the following two conditions are satisfied:

$$\begin{aligned} \text{Aliasing cancellation condition:} \quad & F_0(z)H_0(-z) + F_1(z)H_1(-z) = 0 \\ \text{No distortion condition:} \quad & F_0(z)H_0(z) + F_1(z)H_1(z) = 2z^{-l} \end{aligned} \tag{4.12}$$

where $l$ is the time delay. Using the aliasing cancellation condition in (4.12) it follows that

$$F_0(z) = H_1(-z) \qquad F_1(z) = -H_0(-z) \tag{4.13}$$

or, equivalently, $f_0(n) = (-1)^n h_1(n), f_1(n) = -(-1)^n h_0(n)$ in the time domain. Substituting the anti-aliasing condition in (4.13) to the no-distortion condition in (4.12) yields

$$P_0(z) - P_0(-z) = 2z^{-l} \tag{4.14}$$

where

$$P_0(z) = H_0(z)F_0(z) \tag{4.15}$$

is a *half-band filter*. The PR conditions naturally imply that the analysis and synthesis filter banks are biorthogonal. The PR filter banks satisfy the following biorthogonal conditions:

$$\begin{aligned} &\sum_n h_0(n)\tilde{f}_0(n-2k) = \delta(k) \qquad \sum_n h_0(n)\tilde{f}_1(n-2k) = 0 \\ &\sum_n h_1(n)\tilde{f}_1(n-2k) = \delta(k) \qquad \sum_n h_1(n)\tilde{f}_0(n-2k) = 0 \end{aligned} \tag{4.16}$$

where $\tilde{f}_i(n)$ is defined as the time reversal of $f_i(n)$, $i = 1, 2$.

### 4.3.3 Orthogonal Two-Channel PR Filter Banks

The PR conditions only imply the biorthogonality. A biorthogonal filter bank ensures that the synthetic filter bank is the inverse of the analysis filter bank. Often it is desirable that both the analysis filter bank and the synthesis filter bank are FIR (finite impulse response) filters. However, the inverse of an FIR filter often yields an IIR (infinite impulse response) filter.

If two filters are orthogonal, then the inverse of a filter is equal to its transpose. For an orthogonal filter bank, there is a nice relation in which the coefficients of an analysis filter are the time reversal of the corresponding synthesis filter coefficients (assuming real-valued filter coefficients). For example, the impulse responses of the corresponding pair of analysis and synthesis filters of order $N = 3$ that are orthogonal to one another are related by

$$\{h(n)\} = \{h(0),\ h(1),\ h(2),\ h(3)\} \qquad \{f(n)\} = \{h(3),\ h(2),\ h(1),\ h(0)\} \tag{4.17}$$

To obtain an orthogonal two-channel filter bank, the filter coefficients must satisfy the double-shift orthogonality conditions:

$$\begin{aligned} &\sum_n h_i(n)h_i(n-2k) = \delta(k) \qquad i = 1, 2 \\ &\sum_n h_0(n)h_1(n-2k) = 0 \end{aligned} \tag{4.18}$$

This implies that an extra condition must be imposed between $H_0(z)$ and $H_1(z)$ so that the two filters will be orthogonal. One choice by Smith-Barnwell [281] is to set the coefficients of the high-pass filter as the alternating flip of the low-pass filter. Assuming odd $N$, this is equivalent to

$$H_1(z) = -z^{-N} H_0(-z^{-1}) \tag{4.19}$$

Substituting (4.19) into the anti-aliasing condition in (4.13), it follows that for $N$ odd

$$\begin{aligned} F_0(z) &= H_1(-z) = z^{-N} H_0(z^{-1}) \\ F_1(z) &= -H_0(-z) = z^{-N} H_1(z^{-1}) \end{aligned} \tag{4.20}$$

For a given low-pass analysis filter $H_0(z)$, we can use the relations in (4.19) and (4.20) to form the remaining filters in the orthogonal PR FIR filter banks.

For two-channel orthogonal PR filter banks, the half-band filter in (4.15) becomes

$$P_0(z) = H_0(z)H_0(z^{-1}) \tag{4.21}$$

Once the coefficients of $H_0(z)$ are found, the coefficients of the remaining three filters can be determined. For convenience we assume the filter coefficients are real valued and $N$ is an odd number; then the relations of the filter coefficients can be summarized as follows:

$$\begin{array}{lll} H_0\text{: given coefficients:} & h_0(n) & \{h_0(0), h_0(1), \ldots, h_0(N)\} \\ H_1\text{: alternating flip:} & h_1(n) = (-1)^n h_0(N-n) & \{h_0(N), -h_0(N-1), \ldots, \\ & & \quad (-1)^N h_0(0)\} \\ F_0\text{: time reversal:} & f_0(n) = h_0(N-n) & \{h_0(N), h_0(N-1), \ldots, h_0(0)\} \\ F_1\text{: alternating signs:} & f_1(n) = (-1)^n h_0(n) & \{-h_0(0), h_0(1), \ldots, \\ & & \quad -h_0(N-1), h_0(N)\} \end{array} \tag{4.22}$$

### 4.3.4 Linear-Phase Two-Channel PR Filter Banks

It is often desirable that a filter bank have a linear phase. The impulse response of a linear-phase filter is either symmetric or antisymmetric, which can be equivalently expressed in the time domain or in the $z$-domain by

$$\begin{array}{llll} \text{Time domain:} & h(n) = h(N-n) & \text{or} & h_1(n) = -h_1(N-n) \\ z\text{-Domain:} & H(z) = z^{-N} H(z^{-1}) & \text{or} & H_1(z) = -z^{-N} H_1(z^{-1}) \end{array} \tag{4.23}$$

For a two-channel biorthogonal PR filter bank with a linear phase, it is known that the length and symmetry of the filters must satisfy one of the following conditions:

$$\begin{array}{l}\text{Even lengths for both filters: } h_0(n) \text{ is symmetric and } h_1(n) \text{ is antisymmetric} \\ \text{Odd lengths for both filters: } h_0(n) \text{ and } h_1(n) \text{ are both symmetric}\end{array} \qquad (4.24)$$

Condition (4.24) implies that if $z_i$ is a root of the filter, then $z_i^{-1}$ is also a root. This principle should be used when we factorize $P(z)$ in (4.15) to $F_0(z)$ and $H_0(z)$ so that linear-phase filters can be obtained.

Once the coefficients of $H_0(z)$ and $F_0(z)$ are found, the remaining two filters can be obtained by using the relations given in (4.13). For convenience we again assume that the filter coefficients are real valued and $N_0$ and $N_1$ are both odd numbers; the relations of filter coefficients can then be summarized as

$$\begin{array}{lll}\text{Given } H_0 \text{ and } F_0\text{:} & \{h_0(0), \ldots, h_0(N_0)\}, \{f_0(0), \ldots, f_0(N_1)\} & \\ H_1\text{:}\ h_1(n) = (-1)^n f_0(n) & \{h_1(n)\} = \{f_0(0), \ -f_0(1), \ldots, f_0(N_1-1), \ -f_0(N_1)\} & \\ F_1\text{:}\ f_1(n) = -(-1)^n h_0(n) & \{f_1(n)\} = \{-h_0(0), \ h_0(1), \ \ldots, \ -h_0(N_0-1), \ h_0(N_0)\} & \end{array} \qquad (4.25)$$

It is worth mentioning that although the quadrature mirror filter (QMF) selection by Croisier, Estaban, and Galand [281],

$$H_0(z) = H_1(-z) \qquad \text{or} \qquad H_0(e^{\omega}) = H_1(e^{\omega+\pi}) \qquad (4.26)$$

does yield linear-phase PR filter banks, this choice is only meaningful when IIR filters are allowed in the filter banks (otherwise, only Haar filters will satisfy the FIR filter constraint). Therefore, if two-channel linear-phase PR FIR filter banks are required, one should *not* choose (4.26).

### 4.3.5 Possibility for Two-Channel PR FIR Filter Banks with Both Linear-Phase and Orthogonality

For two-channel PR FIR filter banks simultaneous requirements on linear phase and orthogonality can only lead to a trivial solution where filters contain two coefficients (Haar filter). Linear-phase filters and orthogonality are not compatible for two-channel PR FIR filter banks. Therefore, one should design either a linear-phase biorthogonal filter bank or an orthogonal filter bank without the linear phase for the two-channel case. The selection should be dependent on the applications to see whether or not the linear phase (or orthogonality) is essential.

### 4.3.6 Steps for Designing Two-Channel PR FIR Filter Banks

The steps for designing two-channel PR FIR filter banks can be briefly summarized as follows:

1. Design a low-pass filter $P_0(z) = F_0(z)H_0(z)$ that satisfies the no-distortion condition in (4.12). One possibility is to use the Daubechies function for $P_0(z)$:

$$P_0(z) = \left[\tfrac{1}{2}(1 + z^{-1})\right]^{2p} Q_{2p-2}(z) \tag{4.27}$$

   where $Q(z)$ is a polynomial of degree $2p - 2$ chosen to satisfy the no-distortion condition in (4.12).
2. Factorize the half-band filter $P_0(z)$ into $H_0(z)$ and $F_0(z)$ that satisfies either the linear-phase condition in (4.23) or the orthogonality condition in (4.18).
3. Find the high-pass filter coefficients $h_1(n)$ and $f_1(n)$ by using the anti-aliasing condition in (4.13).

**Example 4.2** Design a Two-Channel PR FIR Filter Bank Using the Daubechies function in (4.27), selecting $p = 2$, and choosing $Q(z) = (-1 + 4z^{-1} - z^{-2})$ yield

$$P_0(z) = \left[\tfrac{1}{2}(1 + z^{-1})\right]^4 Q(z) = \tfrac{1}{16}(-1 + 9z^{-2} + 16z^{-3} + 9z^{-4} - z^{-6}) \tag{4.28}$$

There are six roots in $P(z)$:

$$z_1 = 2 - \sqrt{3} \qquad z_2 = z_1^{-1} = 2 + \sqrt{3} \qquad z_{3,4,5,6} = -1 \tag{4.29}$$

In the following, we show two cases to assign the factorized $P_0(z)$ to the filters $H_0(z)$ and $F_0(z)$. The first case leads to the orthogonal solution and the second case to the linear-phase solution.

**Case 1** Orthogonal PR Filter Banks. The search for orthogonal filter banks leads to a 4/4 pair, the so-called maxflat low-pass filters. This is done by assigning three roots $z_1 = 2 - \sqrt{3}$, $z_{3,4} = -1$ to the filter $H_0(z)$ and another three roots $z_2 = 2 + \sqrt{3}$, $z_{5,6} = -1$ to $F_0(z)$. For the orthogonality between $H_0(z)$ and $F_0(z)$, $z_1$ and $z_2$ must be assigned to two separate filters. This leads to the low-pass analysis and synthesis filters, each having two zeros at $z = -1$,

$$\begin{aligned} H_0(z) &= \alpha(1 + z^{-1})^2(1 - (2 - \sqrt{3})z^{-1}) \\ F_0(z) &= \beta(1 + z^{-1})^2(1 - (2 + \sqrt{3})z^{-1}) \end{aligned} \tag{4.30}$$

where $\alpha$ and $\beta$ are constants. After the normalization, we obtain the filters

$$H_0(z) = \sum_{i=0}^{3} h_0(i)z^{-i} \qquad F_0(z) = \sum_{i=0}^{3} f_0(i)z^{-i} \tag{4.31}$$

with the coefficients

$$\begin{aligned}\{h_0(0), h_0(1), h_0(2), h_0(3)\} &= \frac{1}{4\sqrt{2}}\{(1+\sqrt{3}),\ (3+\sqrt{3}),\ (3-\sqrt{3}),\ (1-\sqrt{3})\}\\ \{f_0(0), f_0(1), f_0(2), f_0(3)\} &= \frac{1}{4\sqrt{2}}\{(1-\sqrt{3}),\ (3-\sqrt{3}),\ (3+\sqrt{3}),\ (1+\sqrt{3})\}\end{aligned} \tag{4.32}$$

where $4\sqrt{2}$ is used for the normalization. (Hint: The above results can be obtained by setting $\alpha = 1 + \sqrt{3}$ and $\beta = 1 - \sqrt{3}$ and then applying normalization.) Using filter relations in (4.20) in the $z$-domain [or (4.22) in the time domain] yields the coefficients of the high-pass analysis filter,

$$\begin{aligned}\{h_1(0),\ h_1(1),\ h_1(2),\ h_1(3)\} = \frac{1}{4\sqrt{2}}\{1-\sqrt{3},\ -(3-\sqrt{3}),\ 3+\sqrt{3},\\ -(1+\sqrt{3})\}\end{aligned} \tag{4.33}$$

and the coefficients of the high-pass synthesis filter,

$$\begin{aligned}\{f_1(0),\ f_1(1),\ f_1(2),\ f_1(3)\} = \frac{1}{4\sqrt{2}}\{-(1+\sqrt{3}),\ (3+\sqrt{3}),\\ -(3-\sqrt{3}),\ (1-\sqrt{3})\}\end{aligned} \tag{4.34}$$

The above filter coefficients can be generated by the MATLAB program in Table 4.2, which requires some functions in the MATLAB Wavelet Toolbox.

**Case 2** Linear-Phase Biorthogonal PR Filter Banks. Since a linear-phase filter requires symmetry or antisymmetry, the closest even-length split of $P_0(z)$ leads to a 3/5 (or 5/3) analysis and synthesis filter pair. Since each filter has at least one root at $z = -1$ and for symmetry the roots $2 - \sqrt{3}$ and $2 + \sqrt{3}$ must remain in

**TABLE 4.2 MATLAB Code to Generate Four Filter Coefficients in Case 1**

| | |
|---|---|
| *f* = *dbwav f*('*db2*'); | % generate Daubechies wavelets |
| [*f*0, *f*1, *h*0, *h*1] = *orthfilt*(*f*) | % compute corresponding analysis and synthesis filters |

one filter, this leads to the following filters,

$$H_0(z) = \left[\tfrac{1}{2}(1+z^{-1})\right]^2 = \tfrac{1}{4}(1+2z^{-1}+z^{-2}) \tag{4.35}$$

$$\begin{aligned} F_0(z) &= \tfrac{1}{2}(-1+4z^{-1}-z^{-2})\left[\tfrac{1}{2}(1+z^{-1})\right]^2 \\ &= \tfrac{1}{8}(-1+2z^{-1}+6z^{-2}+2z^{-3}-z^{-4}) \end{aligned} \tag{4.36}$$

where $H_0(z)$ contains a double root at $z_{1,2} = -1$, $F_0(z)$ contains four roots $z_1 = 2-\sqrt{3}$, $z_2 = 2+\sqrt{3}$ and a double root $z_{3,4} = -1$. After the normalization, the coefficients of the two low-pass filters become

$$h_0(n) = \tfrac{1}{4}\sqrt{2}\{1,2,1\} \qquad f_0(n) = \tfrac{1}{8}\sqrt{2}\{-1,2,6,2,-1\} \tag{4.37}$$

where $\sqrt{2}$ is introduced for the normalization. This split corresponds to factorizing the Daubechies function $P_0(z) = H_0(z)F_0(z)$ in (4.27) by

$$H_0(z) = \left[\tfrac{1}{2}(1+z^{-1})\right]^N \quad F_0(z) = \left[\tfrac{1}{2}(1+z^{-1})\right]^{2p-N} Q(z) \tag{4.38}$$

Here $H_0(z)$ is a spline filter. In general, $H_0(z)$ and $F_0(z)$ are interchangeable. Here we choose a $3/5$ split, since experiments indicated that a better performance can be achieved when choosing a shorter analysis low-pass filter and a longer synthesis filter [66, 281].

For obtaining the remaining two filters, employing the anti-aliasing condition in (4.13) in the z-domain (or, (4.25) in the time-domain) yields the coefficients of the 2 high-pass filters

$$h_1(n) = \tfrac{1}{8}\sqrt{2}\{1,\ 2,\ -6,\ 2,\ 1\} \qquad f_1(n) = \tfrac{1}{4}\sqrt{2}\{1,\ -2, 1\} \tag{4.39}$$

The above filter coefficients can be generated by the MATLAB codes in Table 4.3 using the functions in the MATLAB Wavelet Toolbox.

**Example 4.3** Analyzing Voltage Dips by Using Wavelets. In this example, the same voltage dip signal as in Figure 4.2 is analyzed by the wavelet filters. Figure 4.6 shows the outputs of dyadic wavelet filters using a biorthogonal mother wavelet generated by the function bior4.4 and Daubechies wavelet by db4 in the MATLAB Wavelet Toolbox, respectively. In each case the number of scales is $M = 7$, leading to center frequencies of the first five bands equal to 30, 90, 180, 360, and 720 Hz. A first observation from Figure 4.6 is that the results from using different mother wavelets are rather different. The sharpness of the edge of the band-pass filter depends on the mother wavelet and its order. Daubechies wavelet db4 is

**TABLE 4.3 MATLAB Code to Generate Four Filter Coefficients in Case 2**

| | |
|---|---|
| [*df*, *rf*] = *biorwav f*('bior2.2') | % generate biorthogonal wavelets |
| [*f*0, *f*1, *h*0, *h*1] = *biorfilt*(*df*, *rf*) | % compute the analysis and synthesis filters |

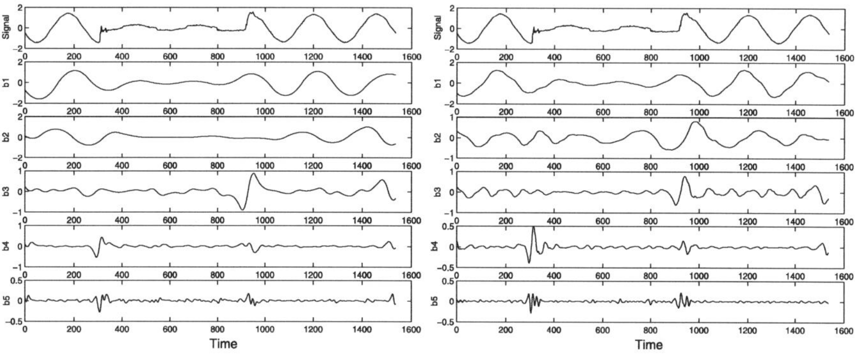


**Figure 4.6** Outputs from dyadic wavelet ($M = 7$): (*a*) wavelet = bior4.4; (*b*) wavelet = db4; From top to bottom: measured voltage dip; outputs from first five bandpass filters.

expected to be better than bior4.4 for detecting transient signal changes such as dip initiation and voltage recovery. However, biorthogonal wavelets provide filters with linear phase, which is required when the phase of the signal is essential for the analysis. Comparing (*a*) and (*b*) in Figure 4.6, both wavelets are able to clearly detect the dip initiation, which is very sharp. In recognizing the smoother transition at voltage recovery, db4 gives a slightly better result. Note that in the STFT with $L = 64$, both points are easily detected in band 7, as shown in Figure 4.3.

### 4.3.7 Discussion

***4.3.7.1 Time Resolution and Frequency Resolution*** As mentioned in Section 4.2, the product of the time resolution $\Delta_t$ and the frequency resolution $\Delta_\omega$ of a bandpass filter is constrained by the following uncertainty principle:

$$\Delta_t \Delta_\omega \geq \tfrac{1}{2} \tag{4.40}$$

Using a dyadic structure for the wavelet filters, the bandwidths of two adjacent bandpass filters have an octave relation. This corresponds to monotonically decreasing the frequency resolution of the bandpass filters, where the low-pass filter and its immediate bandpass filter have the highest resolution. Conversely, the bandpass filters have a monotonically increasing time resolution, with the highest time resolution in the high-pass filter.

The situation is different when using the STFT, where all bandpass filters have the same bandwidth and hence the same frequency resolution. Consequently, all bandpass filters also have the same time resolution when using the STFT.

***4.3.7.2 Center Frequencies and Bandwidths of Multiscale Wavelet Filters*** For multiscale wavelet filters, the center frequencies of filters are determined once the number of scales and the signal sample frequency $f_s$ are given. Let the total number of scales be $M$, the center frequencies of bandpass filters,

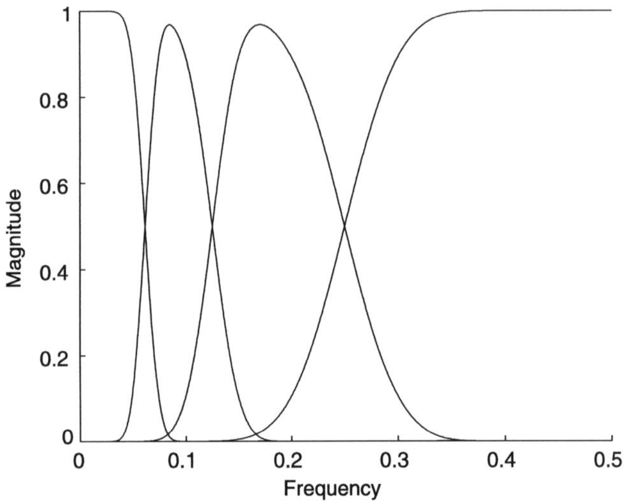


**Figure 4.7** Frequency response of bandpass filters using three-scale DWT.

starting from the lowest frequency, be ($f_A$, $f_M$, ..., $f_2$, $f_1$), and the corresponding filter bandwidths be ($B_A, B_M, \ldots, B_2, B_1$). Then, the center frequencies and the corresponding (ideal) bandwidths of the filters are

$$\begin{aligned} &\text{Bandwidth:} && B_k = \frac{f_s}{2^{k+1}} && B_A = B_M \quad k = 1, 2, \ldots, M \\ &\text{Center frequency:} && f_k = \frac{3f_s}{2^{k+2}} && f_A = f_M - B_M \quad k = 1, 2, \ldots, M \end{aligned} \tag{4.41}$$

where $f_A$ and $B_A$ are the center frequency and the bandwidth of the low-pass filter. From (4.41) one can see that these dyadic wavelet filters have an octave bandwidth relation (see Fig. 4.7 for a three-scale case), starting from the low-pass and the first bandpass band with the narrowest bandwidth. Hence, different frequency bands have different frequency resolution and hence different time resolution. Obviously, the lowest band has the highest frequency resolution corresponding to the lowest time resolution. It is worth mentioning that (4.41) also imposes an inflexible constraint, that is, the center frequencies of bandpass filters can be inconveniently located at nonharmonic frequencies!

### 4.3.8 Consideration in Power Quality Data Analysis: Choosing Wavelets or STFTs?

Whether wavelets or STFTs should be selected for analyzing the nonstationary power system disturbances, some issues need to be carefully considered [129]:

(a) *Which Basis Functions Are More Suitable: Fourier or Wavelets*? The STFTs offer a tool for time–frequency analysis while wavelets offer a tool for time–scale analysis where scales are related to logarithmic frequencies.

A STFT uses the Fourier bases $\{e^{j\omega_k n}\}$ or $\{\cos(\omega_k n), \sin(\omega_k n)\}$, which are global basis functions, while wavelet filters use wavelet bases that are local. In general, Fourier bases are more suitable for a smooth signal while wavelet bases are more suitable for a piecewise-smooth signal.

(b) *Detecting Singular Points and Sudden Changes.* Obviously, wavelets are more suitable for detecting singular points in the signals. Since dyadic wavelet filters naturally use different frequency resolution, wavelets can easily detect both the sudden and slow transition positions, for example, a sudden onset in the dip initiation and a slow voltage recovery from the dip. The discrete STFT requires proper tuning of suitable time resolution by selecting the window size in order to detect the changes at different rates.

(c) *Disturbance Analysis Based on Harmonic Distortions.* If time-dependent harmonic analysis is essential, that is, decomposing a signal into harmonic-related time sequences (although each sequence may contain one pure harmonic or several harmonics and interharmonics depending on the frequency resolution of the filter) by using subband filters centered at the exact frequencies of the selected harmonics, STFTs are more suitable. In wavelets, the center frequencies of bandpass filters are fixed once the number of scales is chosen. These frequencies are likely to be located at inconvenient frequencies other than those of the desired harmonics. Also, the number of harmonics within each band increases with the increase of filter center frequencies. The combination of these makes wavelets no longer a best choice for harmonic analysis.

## 4.4 BLOCK-BASED MODELING

### 4.4.1 Why Divide Data into Blocks?

The models described in Chapter 3 assume that the signal under consideration is stationary (or statistically time invariant). This often implies that the states of the underlying power system remain the same (or in a steady state) during the measurement. However, if the power system states change during the measurement, the recorded data are no longer stationary. As a result the model assumption used in Chapter 3 does not hold, and the stationary models described in Chapter 3 are no longer appropriate. In such a case the model parameters may change over time, that is, the models become time variant. For example, a voltage waveform sequence is captured while the system changes from one state to another due to some disturbances. Under the stationary assumption, the model is applied to the entire data sequence. However, due to the change of system states, the estimated model (or the model parameters) only represents some kind of averaging. It reflects neither the first state nor the second.

In such situations, we need to modify the methods described in Chapter 3 to handle the corresponding nonstationary signal models. This can be obtained by block-based signal modeling. Instead of using all available data samples from the recorded data sequence, we can divide the data into blocks such that within each block the data are approximately stationary. The blocks can be either *overlapped* or *nonoverlapped* depending on the application. Since the stationarity holds for each data block, we can estimate the parameters of the model associated with each data block. Obviously, the model parameters change from block to block (associated with the different time duration). As a result we obtain a series of stationary models whose parameters vary from block to block (hence vary with time).

### 4.4.2 Divide Data into Fixed-Size Blocks

Let the data samples $x(n)$, $n = 0, 1, \ldots,$ in a recording be divided into blocks of *fixed size L*. The size of blocks is usually determined empirically such that the data within each block can be considered as approximately stationary. Let the overlap of adjacent blocks be $K$, $K < L$, and the $m$th sample in the $j$th block be $x^{(j)}(m)$, $m = 0, 1, \ldots, L-1$, $j = 1, 2, \ldots$. Then the data sample in the $j$th block, $x^{(j)}(m)$, is related to the sample $x(n)$ in the original data sequence by

$$x^{(j)}(m) = x\big(m + (j-1)(L-K)\big) \tag{4.42}$$

or the index $m$ in the blocked data $x^{(j)}(m)$ is related to the actual time index $n$ in the original data sequence $x(n)$ by

$$n = m + (j-1)(L-K) \tag{4.43}$$

### 4.4.3 Block-Based AR Modeling

This section describes how to modify the AR models in Section 3.6.1 to the corresponding block-based AR models that are suitable for nonstationary data. Block-based AR modeling of data can be summarized by the following steps:

1. *Dividing Data into Blocks* Data samples $x(n')$, $n' = 0, 1, \ldots,$ are divided into blocks of size $L$, $x^{(j)}(n)$, $j = 1, 2, \ldots,$ where $n'$ denotes the time index in the data sequence and $n$ the index in the $j$th data block.

2. *Estimating Model Parameters in Each Block* For each block of data $x^{(j)}(n)$, $j = 1, 2, \ldots,$ an AR model of order $N$ is employed in a similar way to that in Section 3.6.1. For the sake of convenience we repeat (3.232) and (3.233), adding the superscript $j$ to indicate the parameters and the model in the $j$th block. The $j$th block of data is thus modeled by the following AR process:

$$x^{(j)}(n) = -\sum_{k=1}^{N} a_k^{(j)} x^{(j)}(n-k) + b_0^{(j)} w(n) \tag{4.44}$$

This corresponds to a block-related AR model with the transfer function

$$H_{\mathrm{AR}}^{(j)}(z) = \frac{b_0^{(j)}}{1 + \sum_{k=1}^{N} a_k^{(j)} z^{-k}} \tag{4.45}$$

whose parameters $\{a_k^{(j)}, k = 1, \ldots, N\}$ vary in different blocks $j$.

To estimate the model parameters, one solves a set of normal equations that are related to the $j$th-block data. Analogous to (3.234) and (3.235), we have the following set of equations for the block-based case,

$$\begin{aligned} &\sum_{m=1}^{N} a_m^{(j)} r_{x^{(j)}}(k,m) = -r_{x^{(j)}}(k,0) \qquad k = 1, \ldots, N \\ &|b_0^{(j)}|^2 = r_{x^{(j)}}(0,0) + \sum_{m=1}^{N} a_m^{(j)} r_{x^{(j)}}(0,m) \end{aligned} \tag{4.46}$$

where $r_x^{(j)}(k,m) = E[x^{(j)}(n-k)x^{(j)}(n-m)]$ is the data autocorrelation sequence (assuming real-valued data), and $k, m \geq 0$. The expectation $E[\cdot]$ is replaced by the summation if the data are assumed to be ergodic. Note that the summation index $n$ is dependent on the assumption made on the data outside the block. For example, in the autocorrelation method data samples are assumed to be zero valued outside the block (or the window). In the covariance method no assumption is made on the data values outside the block. In the prewindow (or postwindow) method data samples are assumed to be zeros prior to the block (or after the end of block). It is worth noting that using a different window assumption in estimating the autocorrelation sequence will lead to somewhat different estimation results, especially when the block size $L$ is relatively small. To reduce the window effect in the model, the covariance method (i.e., no window) is often chosen to compute the autocorrelation sequence $r_x^{(j)}(k,m)$ if the block size is small.

Once the parameters $\{b_0^{(j)}, a_k^{(j)}, k = 1, \ldots, N\}$ are estimated, the magnitude spectrum of the modeled signal can be obtained directly from (4.45) by substituting $z$ with $e^{j\omega}$, leading to

$$|H_{\mathrm{AR}}^{(j)}(e^{j\omega})| = \left| \frac{b_0^{(j)}}{1 + \sum_{k=1}^{N} a_k^{(j)} e^{-jk\omega}} \right| \tag{4.47}$$

3. *Finding Residuals of Models* (optional depending on application) If the model residuals are required, they can be computed by using the inverse filter $1/H^{(j)}(z)$, $j = 1, 2, \ldots$. The residual $e(n')$ at each time instant can be computed from the inverse filter

$$e(n') = \hat{x}(n') + \sum_{k=1}^{N} a_k^{(j)} \hat{x}(n'-k) \tag{4.48}$$

where $\hat{x}(n')$ is the predicted signal. It is worth mentioning that when using (4.48) one should use the $j$th-block model parameters to compute the residuals $e(n')$ only in the corresponding time interval $n' \in [(j-1)(L-K), j(L-K)-1]$ (where $K$ is the overlap between the two adjacent blocks). Further, computing $e(n')$ requires using $N$ previous estimated samples $\hat{x}(n'-1), \ldots, \hat{x}(n'-N)$. Therefore, the estimated samples $\hat{x}(n')$ from the previous block are used to calculate the residuals at the beginning of the current block. Using this boundary continuation condition is essential to preventing artificial errors introduced to the residuals.

If there is no overlap between the blocks, that is, $K = 0$, then the time interval for computing the residuals $e(n')$ in (4.48) is $n' \in [(j-1)L, jL-1]$. Repeating the process over all blocks $j = 1, 2, \ldots$ yields the entire residual sequence $e(n')$.

**4.4.3.1 Applications: Detecting and Interpreting Transitions** The residual sequence $e(n)$ from a nonstationary AR model can be used for detecting transition points and analyzing and characterizing the disturbances, for example, interpreting the phenomena and finding the underlying causes of the disturbances [128].

The basic idea is that large residuals from an AR model usually appear around the transition points of the data. This can be explained as follows:

1. At the transition points, a much higher model order is required to model the disturbance. Therefore, using a fixed model order $N$ throughout the whole data sequence, large residuals will appear around the transition points as a result of model mismatch.
2. Large residual errors are also caused by the discontinuity of models between adjacent blocks. Recall (4.48); at the left boundary position of block $j$, the computation of residuals uses $N$ data samples from the previous block $(j-1)$, which is best modeled by $H^{(j-1)}(z)$. Since transition causes large differences between the models in blocks $j$ and $j-1$, the residual values at block $j$, computed by using data samples best described by $H^{(j-1)}(z)$, can become large.

Consequently, transition points are detectable by allocating the time where the model residuals are prominent. Another advantage is that the phase-angle relation in the original data is maintained in the residual sequence, making it very useful for interpreting the disturbances.

**Example 4.4** Detecting Start and End Points of Voltage Dip from Prominent Residual Values. Figure 4.8*a* shows the original voltage waveform containing a voltage dip measured at 33 kV with the sampling rate $f_s = 2$ kHz. Figure 4.8*b* shows the results from applying block-based AR models to the voltage waveform. The AR models used were of order $N = 10$ with block size $L = 40$ samples (one cycle) and an overlap $K = 30$ samples. Figure 4.8*b* clearly shows that the start and end points of the dip are associated with the prominent peaks in the residual sequence. These points indicate the time instants of fault initiation and fault clearing. Their difference gives the fault-clearing time, an important diagnostic of the power system protection. Note that the voltage drop gives a large negative value of the residual, while voltage recovery gives a large positive value.

**Example 4.5** Interpreting Cause of Disturbance as Due to Synchronized Capacitor Switching. This example shows the use of AR model residuals for interpreting the

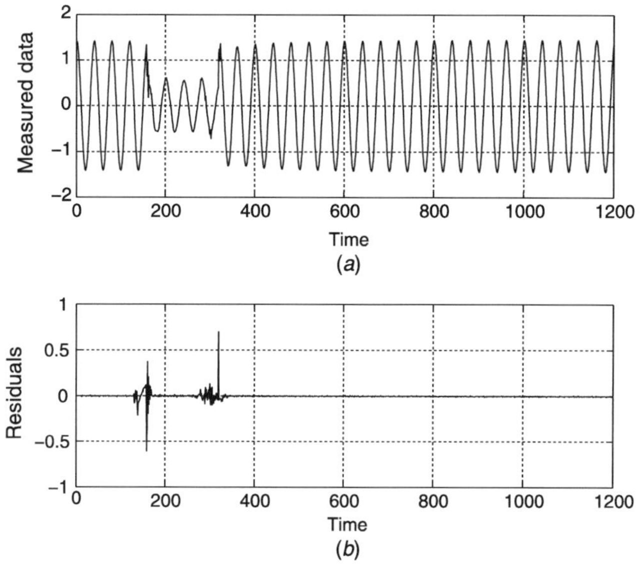


**Figure 4.8** Using residual sequence for detecting start and end points of voltage dip: (*a*) measurement data; (*b*) residual sequence.

cause (synchronized capacitor switching) of a disturbance. The measurements were recorded in a-10 kV system with a sampling rate $f_s = 6.4$ kHz. In the experiment, the AR model order was chosen as $N = 20$, block size $L = 128$ samples (equal to one cycle of 50 Hz), and overlap $K = 112$ samples. Figure 4.9 shows the three-phase voltage waveforms measured during the synchronized capacitor switching and the residuals obtained from AR modeling of these signals. The residuals in Figure 4.9 show a transition in phases *a* and *b* followed by a transition in all three phases. This can be brought back to synchronized switching of a nongrounded star-connected capacitor bank. Phases *a* and *b* were switched near the zero crossing of their voltage difference. Phase *c* was switched 5 ms later, that is, when the voltage in that phase was close to zero. Despite the small transient in the time-domain waveforms, the transition points are clearly visible in the residuals. In addition, it is noted that at the first transition instant the residuals are opposite and of about equal magnitude in the two phases involved. The first instant corresponds to the instant where the first two phases were switched. This is a phase-to-phase event with equal magnitude but opposite sign change in the affected phases. For the second transition instant, the residual is largest in one phase (the phase being switched) and about half the magnitude in the opposite direction in the other two phases. The second transition corresponds to the closing of the third phase. This is the same behavior as observed in the time-domain waveforms. The residual sequences indicate that the disturbances in the three phase voltages were caused by the synchronized capacitor switching. For further discussion on transients in three phases, readers are referred to Section 8.10.3.

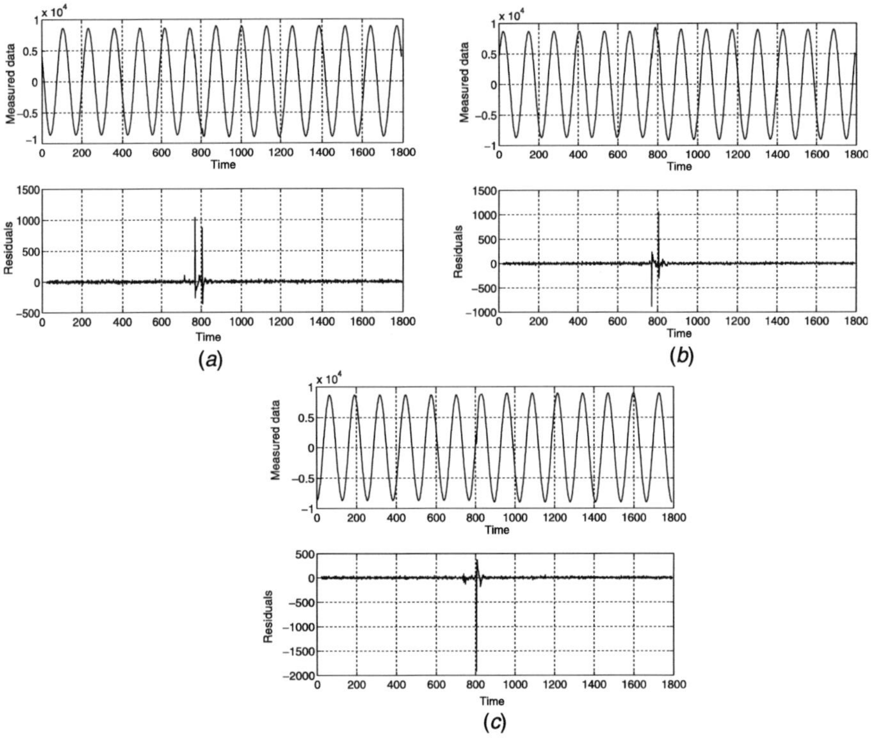


**Figure 4.9** Using residual sequence to detect transients from 5-ms synchronized capacitor switching. (*a*–*c*) original data (three phases) and corresponding residual sequences.

**Example 4.6** Interpreting the Cause of Disturbance as Due to Self-Extinguishing Fault in a Reactance-Earthed System. The measurements were recorded in a 10-kV system with a sampling rate $f_s = 6.4$ kHz. The AR model order was chosen as $N = 20$, block size $L = 128$ samples (equal to one cycle of 50 Hz), and overlap $K = 112$ samples.

Figure 4.10 shows the original measurement data, including one of the three phase voltages and the zero sequence of voltage (i.e., the average of the three phase voltages), and the corresponding residual sequences. The main transition occurs at the same time instant in all three phases, and the residuals are of equal sign and of similar amplitude (only one phase is shown). This points to a zero-sequence phenomenon, as shown on the bottom in Figure 4.10. The main transition corresponds to the initiation of a self-extinguishing fault. It also follows from the residuals that another transition took place in about 45 ms after the main transition. It is not known what the cause of this transition is. From the analysis, one can see that the disturbance is due to a self-extinguishing fault in a reactance-earthed system.

**Example 4.7** AR Models for Interpreting Synchronized Capacitor-Switching Event. To extract information from the time-varying AR model spectra, a data recording containing a capacitor-switching transient event is studied here.

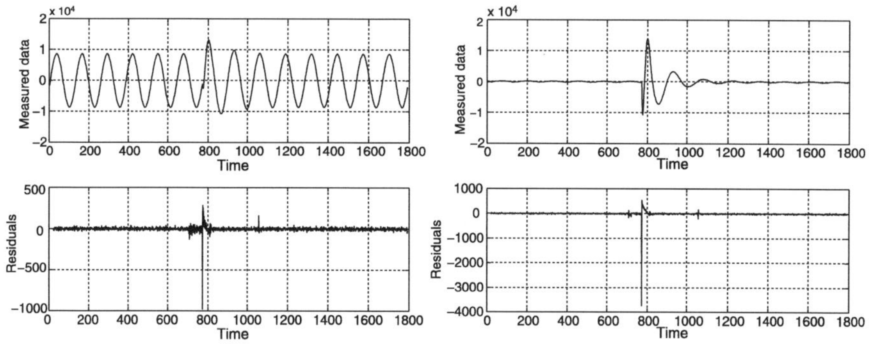


**Figure 4.10** Using residual sequence to detect transition positions of transient fault event. Top: original data of phase *a* and zero sequence. Bottom: corresponding residual sequences.

Figures 4.11*a* and *b* show the original voltage waveform and the resulting time-dependent power spectrum at the fundamental frequency (50 Hz) from using AR models of order $N = 20$. Figure 4.11*b* shows a small increase in the 50 Hz voltage due to reactive power compensation by the capacitor bank (see Section 2.2.3). To capture the detailed contribution of the disturbance other than 50 Hz, we use a two-level cascade AR models. In the first level, low order ($N_1 = 4$) AR models are applied. Since the model order is low, only the major contribution of the data (i.e., mainly the 50-Hz-component) is modeled. The residuals of the first-level AR models are then fed into the second-level AR models as the inputs. The AR model order in the second level is chosen sufficiently high so that all the detailed disturbances can be included ($N_2 = 16$ was chosen in this case). The first-level processing is essential to remove the strong influence of the 50 Hz-component so that the detailed disturbances with much smaller powers can be estimated by the

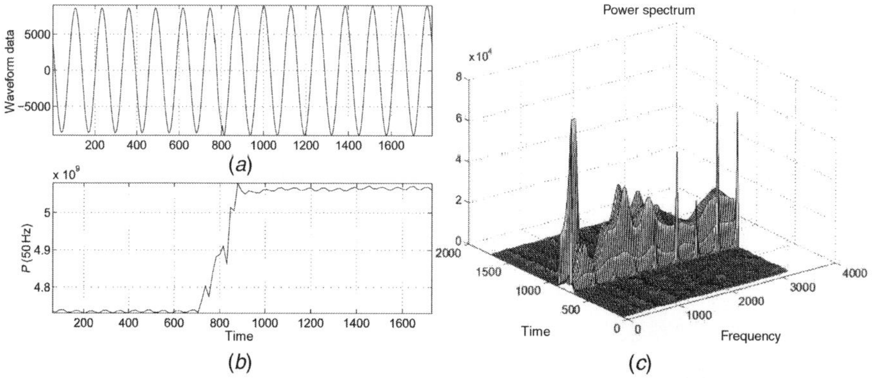


**Figure 4.11** Spectra of synchronized capacitor-switching event: (*a*) Original data $s(n)$ (phase *b*); (*b*) time-varying power spectrum at frequency $f = 50$ Hz obtained from AR models of order $N = 20$; (*c*) time–frequency power spectra of detailed disturbances from second level AR models (order $N_2 = 16$) (with peak clipping to $8 \times 10^4$).

second-level AR models. Figure 4.11*c* shows that the time–frequency spectra of the detailed disturbances are obtained from the second-level AR models. A large high-frequency content is present in the signal during the time window between the two switching events.

**Example 4.8** AR Models for Interpreting Transformer-Energizing Event. The time-varying AR model spectra have been applied to a recording of a transformer-energizing transient measured at 10 kV with 6.4-kHz sampling rate. Figures 4.12*a* and *b* show the original voltage waveform containing a transformer-energizing event and the resulting time-dependent power spectrum at the fundamental frequency (50 Hz) from using AR models of order $N = 20$. The effect of transformer energizing is a drop in the fundamental voltage due to the increased system loading plus the transformer inrush current. This is visible in Figure 4.12*b*. To capture the detailed contribution of the disturbance other than 50 Hz, we use a two-level cascade AR model similar to that in the Example 4.7; that is, low-order ($N_1 = 4$) AR models were applied in the first level to remove the main contribution from the 50-Hz component. The residuals from the first-level AR models are modeled by the second-level AR models where model order $N_2 = 16$ was chosen. Figure 4.12*c* shows the time–frequency spectra of the detailed disturbances from the second-level AR models. The high-frequency content during the transformer energizing is clearly visible.

Comparing Figure 4.11*c* and Figure 4.12*c*, one can see the spectral differences between the two events. Comparing with the first spectral peak, the capacitor-switching event contains more high frequencies than the transformer-energizing event (note the difference in vertical scale). There are also significant differences in spectral shape. A full interpretation of these differences lies in understanding the physics behind the two types of events. In both Figures 4.11 and 4.12, block

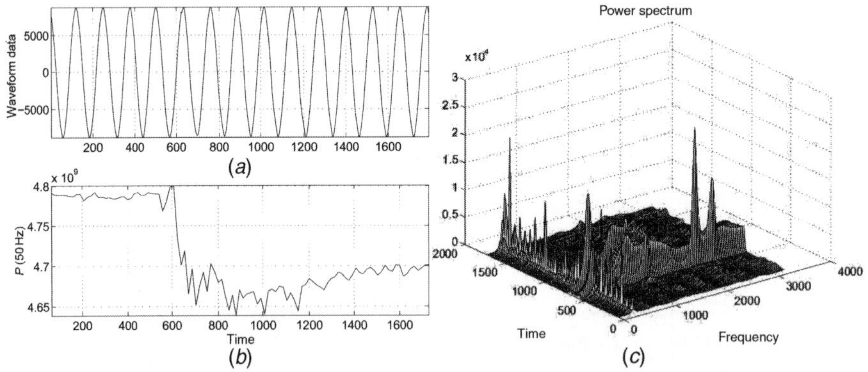


**Figure 4.12** Spectra of transformer-energizing event: (*a*) original data waveform (phase *a*); (*b*) time-varying power spectrum at frequency $f = 50$ Hz obtained from AR models of order $N = 20$; (*c*) time–frequency power spectra of detailed disturbances from second-level AR model (filter order in the second level, $N_2 = 16$).

size $M = 128$ samples (one cycle) and overlap $K = 112$ samples were used. The original data were sampled at $f_s = 6.4$ kHz.

### 4.4.4 Sliding-Window MUSIC and ESPRIT

The MUSIC and ESPRIT methods described in Section 3.5.2 are suitable for stationary signals, where an entire data sequence (or only one block of data) is used for the estimation of sinusoidal model parameters. However, when the signals are nonstationary, these methods should be modified as block-based (or sliding-window) processing according to the principles described in Section 4.4.1. The nonstationary sinusoidal model obtained by modifying (3.158) is

$$v^{(j)}(m) = \sum_{k=1}^{K} a_k^{(j)} \cos(m\omega_k^{(j)} + \phi_k^{(j)}) + w^{(j)}(m) \tag{4.49}$$

where the $j$th block of data $v^{(j)}(m)$ and the data sequence $v(n)$ are related by (4.42). The sliding-window MUSIC (or ESPRIT) algorithm is briefly described in Table 4.4.

It should be noted that for different blocks of data the number of harmonics $K^{(j)}$ (see Table 4.4) may vary. If the aim of analysis is to decompose the signal into the time-dependent components $s_k(n)$, then the problem may arise when the number of harmonics varies or when the frequencies of the harmonics have relatively large variations over different data blocks. The variations in harmonic magnitudes usually do not impose a problem in decomposing the signal. The performance of the final results are also dependent on correctly tracking the contours of the estimated harmonic frequencies over time.

**Example 4.9** Sliding-Window ESPRIT Analysis of Nonstationary Data: Measurement from Clean Supply. The sliding-window ESPRIT method is applied to analyze the data measured from a "clean"-supply case. The measurement data used in this example are the same as the clean-supply data used in Examples 3.13 and 3.18. The difference is that in Chapter 3 only one block of data was analyzed, whereas in this example we use a sliding-window ESPRIT in order to extract the time-dependent information of harmonics from the whole data sequence (containing 122,800 samples, or 60 s). The sample frequency of the data is 2048 Hz (or 41

**TABLE 4.4 Sliding-Window MUSIC (or ESPRIT) Algorithm for Sinusoidal Modeling of Nonstationary Signals**

```
For j=1,2,..., Total Number of Blocks, do:
    Applying the MUSIC (or ESPRIT) algorithm to the jth block of
    data v^(j)(m), m=0,1, ..., L-1
    Resulting in the time-dependent parameters {a_l^(j), ω_l^(j),
    φ_l^(j), l=1,...,K^(j)}.
End {for}
```

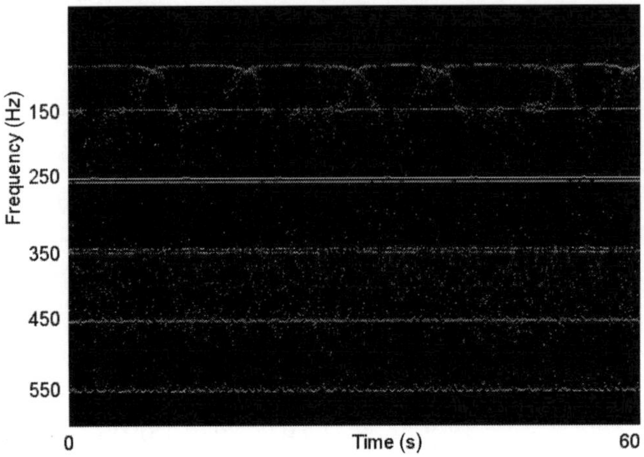


**Figure 4.13** Time-dependent line spectra from sliding-window ESPRIT analysis: clean supply. The white points indicate the detected harmonic/interharmonic frequencies (however, they are not proportional to the values of magnitude).

samples per 50-Hz cycle). As in Chapter 3 prefiltering is applied before the ESPRIT algorithm is performed to remove the low-frequency components up to 90 Hz. In the experiment, the sliding-window ESPRIT (where the number of sinusoids in the model is chosen to be $K = 10$) has a fixed window size of 10 cycles, or 410 data samples. Further, we use overlapped windows (with 80% overlap). Figure 4.13 shows the results obtained from the sliding-window ESPRIT analysis. In Figure 4.13, the top horizontal frequency "line" trajectory is about 100 Hz, probably due to the influence of the 90-Hz low-pass prefiltering. The remaining line trajectory locations correspond to the 3rd, 5th, 7th, 9th, and 11th harmonics. Note that the 5th harmonic is almost continuously present in the data sequence, while the 9th and 11th harmonics are mostly present, though less consistent. One can also notice that there are double lines around the 5th and 7th harmonics, probably due to the line spectrum splitting in ESPRIT (this sometimes happens and the phenomenon may be further studied). Further, one can also observe that there are many interharmonics, especially between the 7th- and 9th-harmonic frequency band.

**Example 4.10** Sliding-Window ESPRIT Analysis of Measurement Data: Signaling. In this example, we use the same measurement data "signaling" as in the Examples 3.11 and 3.16. The sample rate is 7200 Hz (or 144 samples per 50-Hz cycle). Similarly, a prefilter is applied to remove the low-frequency components up to 90 Hz. The sliding window used for the ESPRIT analysis contains 12 cycles, or 1728 data samples. Due to short data length from the measurement (4603 samples in total), we use highly overlapped sliding windows: For each new block, the sliding window is shifted forward by 10 samples. Figure 4.14 shows the results obtained from the sliding-window ESPRIT analysis. The top horizontal frequency line trajectory in Figure 4.14 is about 100 Hz, probably due to the

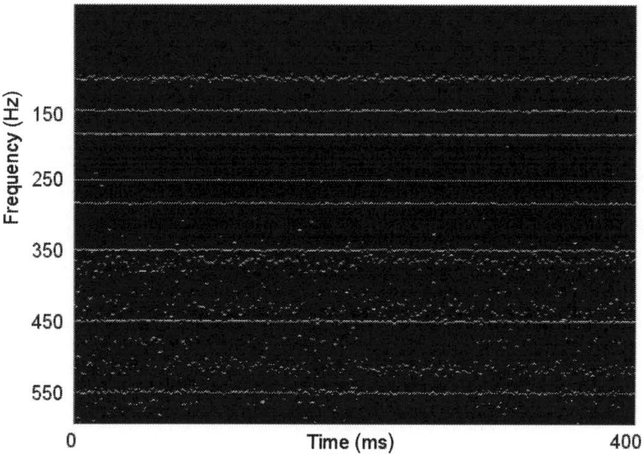


**Figure 4.14** Time-dependent line spectra from sliding-window ESPRIT analysis: signaling. The white points indicate the detected harmonic/interharmonic frequencies (however, they are not proportional to the values of magnitude).

influence of the 90-Hz low-pass prefiltering. The location of the second horizontal line trajectory in the top corresponds to the 3rd-harmonic frequency. It is also interesting that the horizontal lines correspond to the interharmonics at around 184 and 284 Hz, which are consistently present over most blocks, while the locations of the fourth, sixth, seventh, and eighth horizontal line trajectories correspond to the frequencies of the 5th, 7th, 9th, and 11th harmonics. Further, one can observe that about the 7th-harmonic frequency there are more "noisy" peaks. This indicates a relatively large amount of disturbance that is present in the high-frequency band, which can be a useful piece of information for quantifying the underlying cause of the disturbance.

**Example 4.11** Sliding-Window ESPRIT Analysis of Measurement Data: Arc-Furnace Case. In this example, we use the same measurement data "arc furnace" as in Examples 3.12 and 3.17. The sample rate is 3000 Hz (or 60 samples per 50-Hz cycle). Similarly, a prefilter is applied to remove the low-frequency components up to 90 Hz. The sliding window used for ESPRIT analysis contains 12 cycles, or 720 data samples. Since the measurement data are rather long (containing 120,330 samples, or about 40 s), we use a sliding window that is shifted forward by two cycles for each new block.

Figure 4.15 shows the results obtained from the sliding-window ESPRIT analysis. From the frequency scale the top horizontal frequency line in Figure 4.15 is about 90 Hz, probably due to the influence of the 90-Hz low-pass prefiltering. There are many interharmonics (or narrow-band disturbances) between 90 Hz to the third-harmonic frequency. Beyond this band of disturbances, there are four horizontal line trajectories whose locations correspond to the 5th-, 7th-, 9th-, and 11th-harmonic frequencies. It is also observed that the higher the frequencies, the less

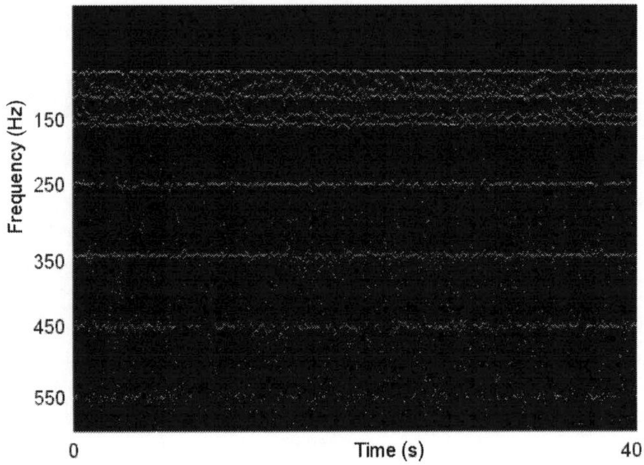


**Figure 4.15** Time-dependent line spectra from sliding-window ESPRIT analysis: arc furnace. The white points indicate the detected harmonic/interharmonic frequencies (however, they are not proportional to the values of magnitude).

consistent (or weaker) is the continuity of these line trajectories, probably because the reduced magnitudes of harmonics have led to selecting other sinusoids with relatively larger magnitudes.

**Example 4.12** Sliding-Window ESPRIT Analysis of Measurement Data: Soft-Starting of Wind Turbine. In this example, we use data samples measured during the soft starting of a wind turbine. The sample rate is 2048 Hz (or each 50-Hz cycle contains 41 samples). The measured data sequence contains 14,337 samples, or 7 s. The size of each data block is 10 cycles, or 410 samples per block. The total number of blocks is 347, and the overlap between two successive sliding windows is 90% (i.e., the sliding window moves forward one cycle for each new block). No prefiltering is applied in this case. Figures 4.16 and 4.17 show the three-dimensional (3D) time-dependent magnitude (line) spectra and the two-dimensional (2D) plot of the estimated time-dependent sinusoidal frequencies, respectively. From Figure 4.17 one can observe that the 3rd, 5th, 7th, 9th, and 11th harmonics are mostly present during the whole measurement time, although the presence of the 9th harmonic is less consistent over the blocks. It is also shown that there are many interharmonic disturbances between and around the 7th and 9th harmonics. From the 3D magnitude spectrum in Figure 4.16, it becomes more obvious that although the 3rd harmonic is present continuously, it is not as significant as the 5th and 7th harmonics. The 5th harmonic has the largest magnitudes, and the 7th harmonic has the second largest magnitudes. Further, the magnitudes of these harmonics significantly increase during the motor-starting dip, as shown from the peaks in the magnitude spectrum. From the analysis, one can also conclude that the frequencies of the harmonics stay rather constant despite their magnitudes

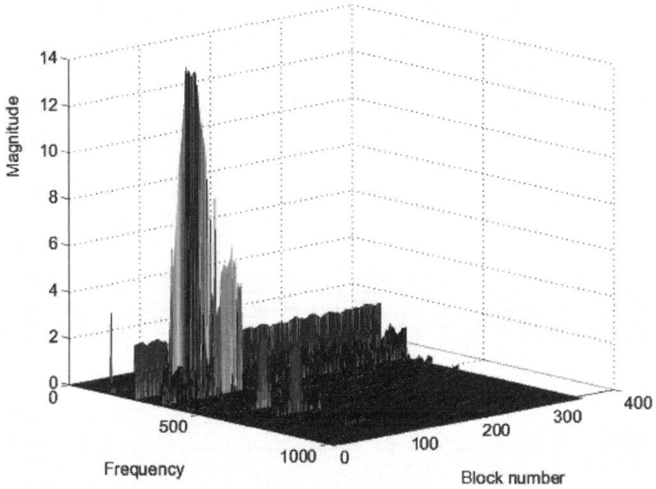


**Figure 4.16** Time-dependent line spectra from sliding-window ESPRIT analysis: soft starting of wind turbine. The 3D plot shows the magnitude line spectra versus time (i.e., the block number) and frequency; the $z$ axis is the magnitude of sinusoids.

changing significantly during the dip. It is also observed that the magnitudes of the harmonics remain at rather low and constant values outside the dip interval.

By observing the results in Examples 4.9 to 4.12 using the sliding-window ESPRIT, we can conclude that it is much more important to observe and extract

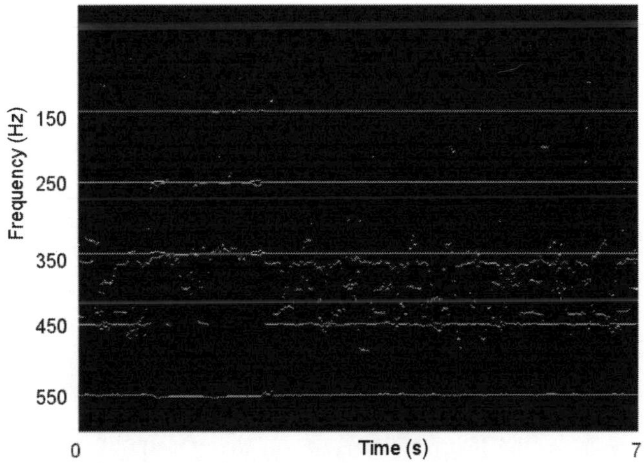


**Figure 4.17** Time-dependent line spectra from sliding-window ESPRIT analysis: soft starting of wind turbine. The white points correspond to the frequencies of the estimated time-dependent sinusoids. Note that the values of magnitude are not included. The frequency components below 100 Hz have been removed from the figure.

some consistent information of harmonics/interharmonics from the time-dependent (or block-dependent) frequency trajectories obtained from the sliding-window ESPRIT rather than to put emphasis on the estimated harmonics/interharmonics from individual blocks. By observing these harmonic/interharmonic disturbances as time-evolving processes, each being under some continuity constraint, we can obtain more reliable dynamic information on the harmonics/interharmonics.

## 4.5 MODELS DIRECTLY APPLICABLE TO NONSTATIONARY DATA

### 4.5.1 Kalman Filters

As mentioned in Section 3.5.4, Kalman filters [183, 102, 223] can be directly applied to nonstationary signals. This is a clear advantage of the Kalman filter. In the following we use a Kalman filter to analyze the nonstationary disturbance data.

**Example 4.13** Analyzing a Nonstationary Disturbance Signal by Directly Using Kalman Filter. The synthetic signal used in Example 3.20 is modified to obtain the corresponding nonstationary signal. The nonstationary signal is generated by using the time-varying harmonic magnitudes while the frequencies of the power system (50 Hz) and its harmonics remain unchanged.

The synthetic signal consists of the power system fundamental component in 50 Hz and its first six odd harmonics (i.e., odd harmonics from 3rd to 13th) in the additive white noise $v(n)$ and can be described by

$$z(n) = \sum_{k=1}^{7} a_{2k-1}(n) \cos\left(2\pi n(2k-1)\frac{f_0}{f_s} + \phi_{2k-1}\right) + v(n) \tag{4.50}$$

where $\{A_3, \ldots, A_{2k-1}\} = \{1.5, 4.0, 4.0, 0.8, 2.5, 2.0\}$ and

$$a_{2k+1}(n) = \begin{cases} (1 + 0.001n)A_{2k+1} & 0 < n \leq 500 \\ a_{2k+1}(500) & 500 < n \leq 1300 \\ [1 - 0.001(n - 1300)]a_{2k+1}(1300) & 1300 < n \leq 2400 \end{cases} \tag{4.51}$$

The magnitude of the power system fundamental is set to 100.0, and the initial phase of the fundamental and the odd harmonics are set to $\{0°, 0°, 30°, 60°, 90°, 120°, 180°\}$, which are the same as those in the Example 3.20. The variances of model noise and measurement noise in the Kalman filter are set to $\sigma_w^2 = 0.001$ and $\sigma_v^2 = 0.0005$, respectively. Figure 4.18 shows the original synthetic signal, the residuals of Kalman filters, and the estimated time-varying magnitudes of the odd harmonics.

From Figure 4.18, one can observe that the Kalman filter converges after about 50 samples. Also, the filter can well track the time-varying magnitudes of these odd harmonics.

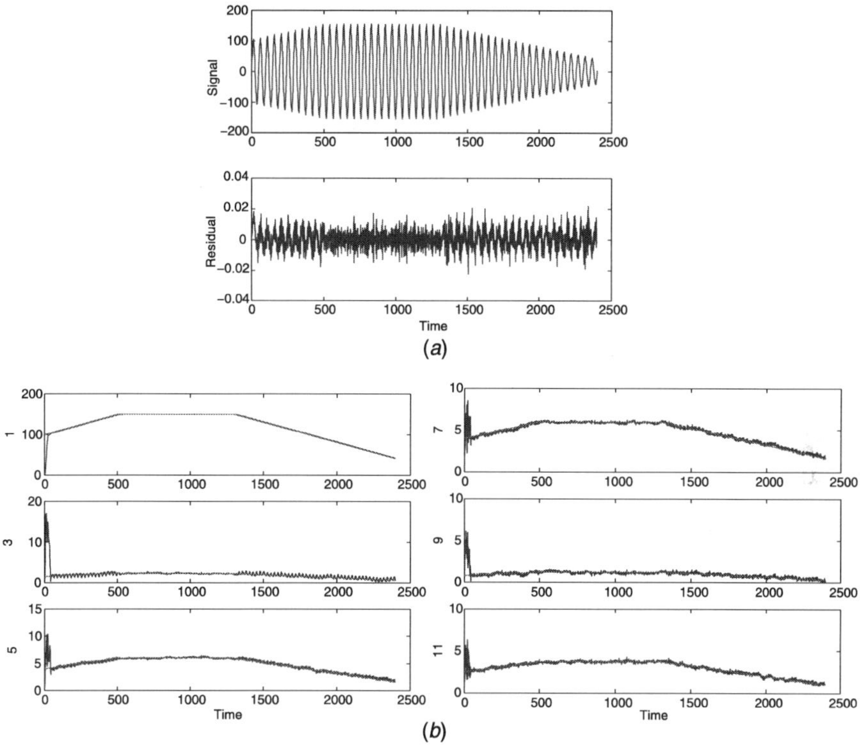


**Figure 4.18** Estimated harmonic magnitudes in time using Kalman filter: (*a*) synthetic signal and residual from Kalman filter; (*b*) estimated magnitudes of odd harmonics, where vertical axis shows magnitude value (odd-harmonic numbers are also indicated) and horizontal axis denotes time in samples.

**Example 4.14** Analyzing of Measurement Data Using Kalman Filter: Soft Starting of Wind Turbine. For the measurement data sequence obtained from the soft-starting case (same as the data used in Example 4.12), we apply the Kalman filter to estimate the harmonics and their magnitudes. From the analysis result in Example 4.12 we can conclude that the harmonic frequencies are nearly constant; only their magnitude values change with time. Such a case is very suitable for the harmonic model used in the Kalman filter. The sample rate is 2048 Hz, or about 41 samples for each 50-Hz cycle. The data sequence contains 14,337 samples, corresponding to 7 s of time. In the experiment, the model order used in the Kalman filter is chosen to be 13 (i.e., the number of sinusoids used in the model). The variance of measurement noise is set to $\sigma_v^2 = 0.0005$, and the variance of model noise is $\sigma_w^2 = 0.001$. Figure 4.19 shows the original data and the residuals from the Kalman filter. Figure 4.20 shows the estimated magnitudes of the voltage fundamental and the first five odd harmonics. As can be seen, the magnitudes for the 5th, 7th, and 11th harmonics are significantly larger during the dip. The magnitudes of odd harmonics remain rather constant before and after the

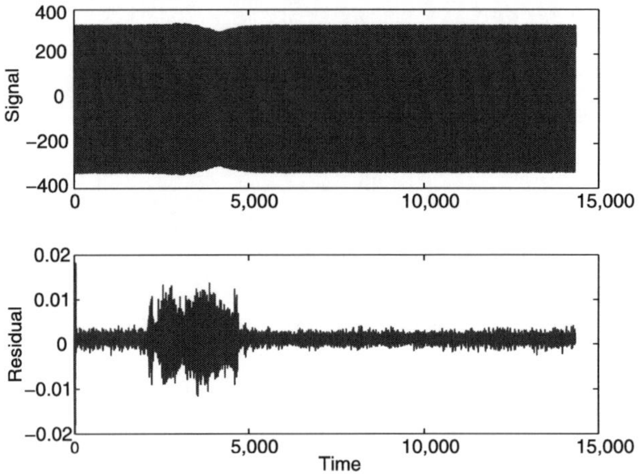


**Figure 4.19** Original measurement data from soft starting of wind turbine (top) and residuals from Kalman filter estimation (bottom).

voltage dip. Further, one can observe that the disturbances are mainly caused by the relatively strong 5th and 7th harmonics.

To explain the harmonic spectrum in time, notice that the wind turbine is a delta-connected three-phase load. During the balanced operation, which is normally the case, such a load does not produce any even or triple harmonics. What remains are the well-known characteristic harmonic pairs 5/7, 11/13, .... A very high level of voltage distortion is present at the terminals of the wind turbine during about 1 s.

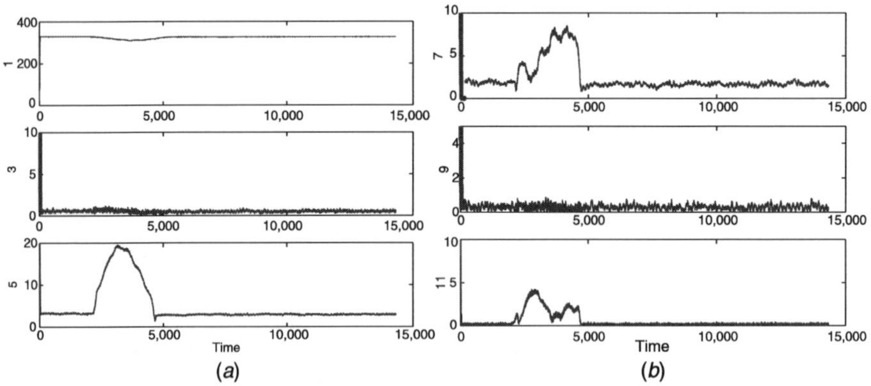


**Figure 4.20** Estimated time-dependent harmonic magnitudes from Kalman filter: (*a*) magnitudes in time for fundamental, 3rd, and 5th harmonics; (*b*) magnitudes in time for 7th, 9th, and 11th harmonics. The horizontal axis indicates the time in samples.

**Example 4.15** Analyzing of Measurement Data Using Kalman Filter: Clean-Supply Case. For the measurement of the data sequence obtained from the clean supply studied in Example 4.9, we shall apply the Kalman filter to estimate the harmonics and their magnitudes. As was concluded from the sliding-window ESPRIT result in the Example 4.9, the disturbances predominantly consist of odd harmonics whose frequencies are about constants but with changing magnitudes. Such a case is again very suitable for employing the Kalman filter. The data sequence contains 122,800 samples, with a sample frequency of 2048 Hz, or about 41 samples per 50-Hz cycle. In the experiment, the Kalman filter model order is chosen to be 13 (i.e., the number of sinusoids). The variance of measurement noise is set to $\sigma_v^2 = 0.0005$, and the variance of model noise is $\sigma_w^2 = 0.001$. Figure 4.21 shows the original measurement data, the residuals from the Kalman filter estimation, and the estimated magnitudes of the voltage fundamental and the first five odd harmonics. As can be seen in Figure 4.21*b*, the third, fifth, and seventh harmonics are the main contributors to the distortion. The magnitudes of the harmonics appear to remain rather stationary

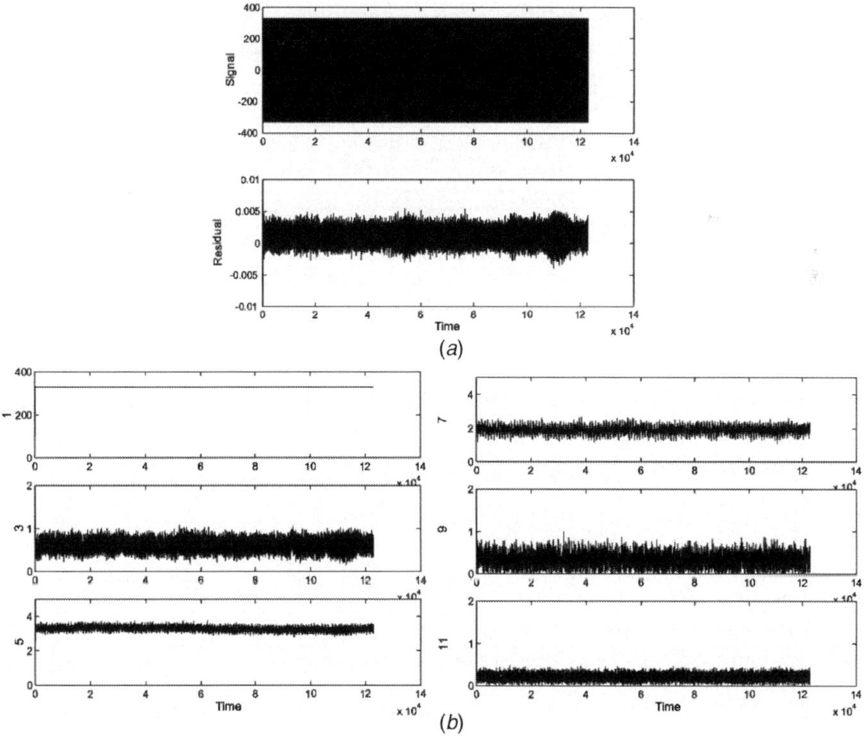


**Figure 4.21** Analysis of clean-supply measurement data using Kalman filter: (*a*) measurement data from clean supply (top) and residuals from Kalman filter estimation (bottom); (*b*) estimated magnitudes in time for fundamental, 3rd, 5th, 7th, 9th, and 11th harmonics. The horizontal axis indicates the time in samples.

over the entire data sequence. Since the residuals from the Kalman filter are small, one can conclude that the estimation is rather accurate.

### 4.5.2 Discussion: Sliding-Window ESPRIT/MUSIC Versus Kalman Filter

Sliding-window ESPRIT and MUSIC methods are described in Section 4.4.3 along with several examples for sliding-window ESPRIT. From these examples, one can see that sliding-window ESPRIT can be used to estimate nonstationary harmonics and interharmonics whose parameters (frequencies and/or magnitudes) are changing with time. However, postprocessing is required to trace (or link) the estimated harmonic frequencies into trajectories, for example, using the continuation (or smoothing) constraint, before the time trajectories of harmonic/interharmonic magnitudes are obtained. A main advantage of sliding-window ESPRIT remains its ability to estimate the time-varying closely spaced frequencies of harmonics/interharmonics (i.e., high frequency resolution). Kalman filters under the harmonic model described in Chapters 3 and 4 require a set of prespecified frequencies in advance.

Compared with a Kalman filter using a harmonic model, the Kalman filter requires that the frequencies of harmonics or interharmonics be specified and fixed in advance (to obtain the state transition matrix **A** before running the Kalman filter algorithm). This may severely limit a Kalman filter's ability to estimate interharmonic components since there are an unlimited number of possible interharmonics. However, if one can reasonably assume that the disturbance is predominantly caused by harmonics, that the power system fundamental frequency remains a constant in time (which implies constant frequencies of harmonics in time) though the magnitudes of harmonics can change with time, then a Kalman filter is a better choice (due to its simplicity and effectiveness) as compared with the sliding-window ESPRIT method.

## 4.6 SUMMARY AND CONCLUSION

Most real-world power quality signals are nonstationary, which implies that they are statistically time variant or their characteristics are evolving with time rather than remaining constant. Quantifying such power quality disturbances requires us to find time-dependent, or dynamic, attributes of the measurements.

Non-model-based methods described in Chapter 3 only analyze signals in each individual domain separately (e.g., time domain or frequency domain). For nonstationary signals joint domain analysis (e.g., time–frequency domain, time–scale domain) are required to quantify the dynamic nature of the measurements. We have addressed the discrete STFTs as a signal-processing tool for time–frequency analysis and multiresolution wavelets for time–scale analysis. For analyzing harmonic-related power quality disturbances STFTs are attractive since one may set the center frequencies of the corresponding subband filters at the harmonic frequencies of interest. The main disadvantage of STFTs is the uniform and single frequency resolution over the entire signal bandwidth once a sliding-window (of fixed-size) FFT is applied. In this respect, wavelets are particularly attractive as multiresolution is naturally introduced into subband filters associated with the wavelets. Wavelets are both effective and efficient for detecting the time positions of transitions or sudden changes in a power quality measurement. Different-scale signal discontinuities can be detected without the prior knowledge on the scale. A main disadvantage of wavelets is, however, that the center frequencies of the subband filters are difficult to be set in the harmonic frequencies, making them less attractive to harmonic-related disturbance analysis.

If the underlying model of a measurement sequence can be reasonably well guessed, then model-based analysis is a better choice than the non-model-based one. Model-based analysis may result in more accuracy and often higher frequency resolution. Again, the dynamic nature of signal characteristics should be taken into account in the analysis. This is done by either applying block-based models or directly applying nonstationary models.

For measurement signals that fit well the (time-varying) harmonic models, we have described the sliding-window MUSIC and ESPRIT algorithms for the estimation of time-dependent harmonic/interharmonic frequencies and their magnitudes.

For measurement signals that are suitable for broadband analysis, block-based AR models and analysis have been addressed. Here, dividing data into blocks is essential to guarantee that each individual block of data is nearly stationary and hence that stationary signal models and analysis can be applied. Examples that include analyzing both synthetic and measured power quality data have shown good potentials. Kalman filters described in Chapter 3 are also suitable for nonstationary data analysis. Examples of nonstationary power quality data (both synthetic and measurement data) analysis are given to demonstrate their potential applications.

For measurement data that cannot be anticipated as primarily consisting of harmonic and/or interharmonic distortions, block-based AR models provide broadband characterization and analysis of the distortions over the entire frequency band. However, much experience is still needed to accumulate in this respect, especially the interpretation of spectra in the high-frequency band and the analysis and understanding of power system behaviors therein.

## 4.7 FURTHER READING

There exists a rich literature on the theories and various applications of time–frequency and time–scale analysis. For more details, readers are referred to [246, 230, 221] for the basic theories on discrete STFTs, [281, 67, 66, 220] for wavelet filters, and [56] for the general theories on time–frequency and time–scale analysis. Reference [223] is useful for its summaries of the theories of signal processing and the library of programs provided. Several power system applications on wavelet filters are mentioned in Section 7.1.3.

# CHAPTER 5

# STATISTICS OF VARIATIONS

This chapter deals with the interpretation and use of the features and characteristics introduced in Chapters 3 and 4. A five-step procedure is introduced starting with the basic characteristics and resulting in performance indicators concerning the level of one or more power quality disturbances throughout the whole system. The steps involve calculating the basic characteristics, time aggregation (Section 5.2), presenting variations as a function of time (Section 5.3), calculating performance indicators for individual sites (so-called *site indices*, to be discussed in Section 5.4), and finally the calculation of performance indicators for the whole system (*system indices*, Section 5.5). The general procedure is introduced in Section 5.1 and an overview of relevant power quality standards and other limits and objectives is given in Section 5.6. This chapter is to a large extent based on standard methods, but new methods are introduced where needed. The chapter contains a large number of measurement results, ranging from a $3 \times 40$-s recording of arc-furnace voltages and currents up to a three-month recording of the harmonic distortion at the wall outlet in an apartment. The chapter concludes with a summary in which the conclusions and the need for further research and development are strongly emphasized.

The organization of this chapter is based on the systematic five-step approach. The steps are however not always that clearly visible in practical power quality measurements. This will also be clear from the various examples throughout the text. In fact, most examples discuss more than one step. However, it remains important to consider the general systematic approach, both for setting up a measurement or survey and for interpreting the results.

*Signal Processing of Power Quality Disturbances.* By Math H. J. Bollen and Irene Yu-Hua Gu

## 5.1 FROM FEATURES TO SYSTEM INDICES

From the analysis in the previous chapters we have obtained values to characterize the voltage and current quality over a certain short time window and at a certain location. An example of such a characterization of the voltage quality is the rms voltage, unbalance, and harmonic spectrum over a 10-cycle window. We saw that the IEC power quality monitoring standard IEC 61000-4-30 gives an accurate description of how this could be done [158]. There are however other methods as well, for example, taking a snapshot of the voltage waveform, typically one cycle. Standard methods for power quality measurements are important and should be followed whereever feasible. This does not mean that any measurement not following the standard is not of use. Any standard is a compromise that can never cover all cases. especially when power quality measurements are performed to get a better understanding of the causes of equipment failure or malfunctions, it is often necessary to deviate from the standard methods. The various examples in this chapter will address both standardized and nonstandardized methods.

Whatever the method used, monitoring rarely involves just one measurement. Instead it typically involves a large number of measurements obtained over a longer period and often at a number of locations as well. To get an impression of the power quality at a certain location, a monitoring duration of at least one or two weeks is often recommended [190]. Longer duration monitoring periods (permanent and semipermanent monitoring) are becoming more common as well. Recording and storing all possible measurement values for a longer duration and for a number of locations result in a very large amount of data. This not only leads to storage and handling problems, but also the interpretation of the data becomes rather difficult as well. Therefore significant data reduction is needed at a number of stages in the monitoring process. An overview of the data reduction process from sampled voltage or current waveforms to site indices is shown in Figure 5.1.

- The first data reduction already takes place in the monitor. Even though the monitor is connected to the supply continuously, it does not store all data. Some monitors take a snapshot of the voltage waveform at regular intervals. The interval between snapshots may vary between a few seconds to several minutes. The main reason for this data reduction is limited memory size of the monitor. A high time resolution is in many cases not needed because the overall trend changes rather slowly. Other monitors obtain values over a certain period. The standard monitoring period according to IEC 61000-4-30 is 10 cycles in a 50-Hz system and 12 cycles in a 60-Hz system (thus about 200 ms in both cases).
- The next step involves time aggregation of the data. Suppose that a snapshot of the voltage waveform is obtained every 10-s. A 1-min rms voltage may be obtained as the average of 6 rms values obtained with the 10-s interval. If the monitor records 200-ms values, a 1-min average can be obtained from 300 such values. Standard aggregation periods in IEC 61000-4-30 are 3 s, 10 min, and 2 h.

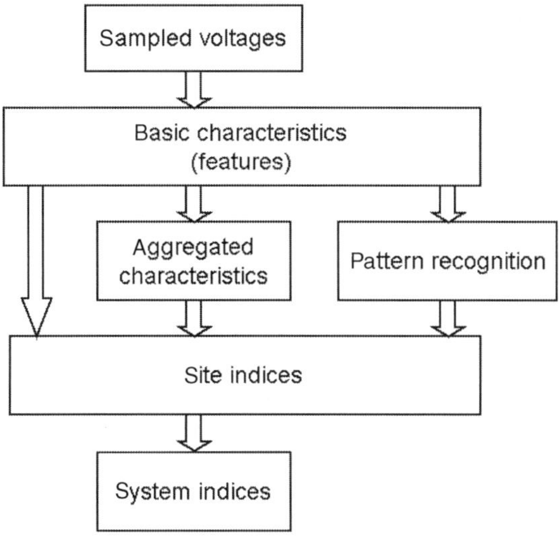


**Figure 5.1** Road to system indices for power quality variations.

- The monitoring together with the first two steps results in a number of indices (e.g., 10-min rms voltage) as a function of time during the whole monitoring period (e.g., two weeks). The next step of data reduction involves the description of the power quality during the whole period with a limited number of indices. Examples of such site indices are the average, median, maximum, and 95 percentile of the characteristics.
- An alternative approach would be to look for patterns in the characteristics as a function of time. In many cases clear daily patterns are visible, but other patterns may be present as well. Patterns could be compared to load patterns (industrial activity), weather (temperature), or system properties (capacitor banks, HVDC links) to find the cause or main contributing factor of the disturbance level.
- The final step involves the extraction of system indices from the site indices for all monitor locations. Examples of system indices are the average, median, maximum, and 95 percentile of the site indices.

## 5.2 TIME AGGREGATION

Time aggregation is the process by which characteristics obtained over a certain period of time are replaced by one representative characteristic. Averaging may be an aggregation method, but the term *aggregation* is more general. Strictly speaking the term aggregation could apply to the whole chain from sampled voltages to system indices, but it is normally only used for the step from basic characteristics

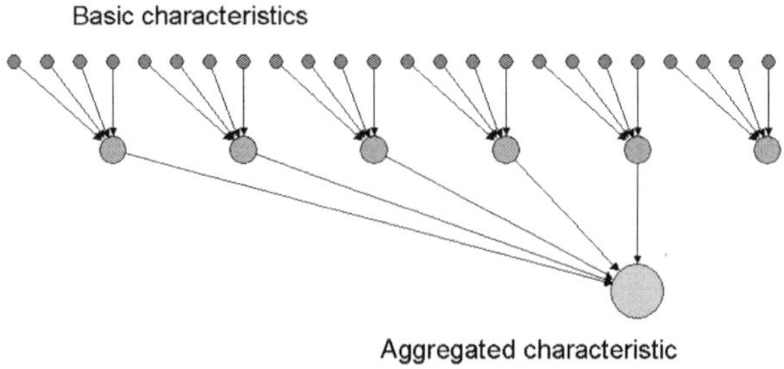


**Figure 5.2** From basic characteristics to aggregated characteristics.

(i.e., those obtained directly from the sampled voltages or currents) to the characteristics that are presented as a function of time or that form the basis for the calculation of site indices. In addition to time aggregation, phase aggregation and monitor aggregation are used. Phase aggregation concerns the combination of the characteristics for the three phases into a single three-phase characteristic. Phase aggregation is preferably done when calculating the basic characteristics (e.g., using symmetrical components), but instead in many cases the highest or average values in the three phases are used. Monitor aggregation concerns the combination of results from two nearby monitors, for example, two monitors located in the same substation. Phase aggregation and monitor aggregation may also take place when calculating site or system indices. In this section we will mainly discuss time aggregation.

The basic principle of the time aggregation process is shown in Figure 5.2. The figure shows a two-stage example, but the aggregation may also just take one stage or more than two stages. In this example a first stage of aggregation uses four values of the basic characteristics. During the second stage five values of the first-level aggregated value are used. Alternatively the second-level aggregation can be obtained directly from 20 basic characteristic values. There are two different reasons for using an intermediate stage. A two-stage process may be more efficient in calculations and memory use. In some cases the intermediate value itself is also presented as a function of time and/or used to calculate site indices.

The aggregation process typically involves a kind of averaging: The actual arithmetic average can be used, but also the rms of the basic values is often used. But even the median or the highest value may be an appropriate choice for certain applications. Even though a standard aggregation method is defined in the IEC power quality monitoring standard, there are sometimes good arguments to deviate from the standard method depending on the application. We will come back to this later.

### 5.2.1 Need for Aggregation

Applying the DFT to a 10/12-cycle window and grouping as described in Chapter 3 result in a number of harmonic and interharmonic groups and subgroups. The values are calculated (about) each 200 ms. Considering all frequencies up to the 40th harmonic, this will result in 40 harmonic groups and 40 interharmonic groups for each 200 ms, that is, 400 data points per second. A typical measurement period is one week, after which the total number of measurement data would add up to more than 240 million data points. This does not include the voltage variations, frequency, unbalance, and flicker characteristics. This amount of data requires some kind of averaging and/or statistical treatment. Even when storage size becomes less of a concern, there still remains the interpretation problem. A too large data set makes interpretation of the results all but impossible without further processing, that is, data reduction.

Another reason for further processing is that the 200-ms values are not of much interest for most loads. The consequences of harmonic distortion are mainly accelerated aging in various loads and system components. The 200-ms values are therefore not considered of much use for describing the electromagnetic environment as experienced by most equipment. Aggregation intervals of 1 min through 30 min duration are based on thermal time constants of end-user and power system equipment. What matters to the equipment are the rms values over periods corresponding to the thermal time constant. The 10-min rms as used in IEC 61000-4-30 (see Section 5.2.2) is based on this reasoning. Recently, however, electronic equipment with shorter thermal time constants has shown problems due to disturbance levels of shorter duration; the term *bursts of harmonics* is sometimes used for this when referring to waveform distortion. The 3-s values in IEC 61000-4-30 are partly aimed at quantifying the impact of disturbances on this kind of equipment.

Some equipment may suffer from erroneous operation almost instantaneously when certain disturbance levels are exceeded. Such should however not be covered by characteristics for power quality variations but instead requires an event-based approach.

A third argument for using aggregated values is that the disturbance levels are rather constant on time scales below a few minutes. Therefore there is no need to keep the 200-ms values. As we will see later, the 10-min values are normally used to calculate site and system indices. In the public supply, for which these methods have been mainly developed, the distortion is typically due to a large number of small polluting sources. The overall distortion is therefore expected to follow a smooth daily pattern. Some examples of daily distortion patterns will be given in Section 5.3.

In some industrial environments or close to industrial installations, the waveform distortion may vary on a much shorter time scale than normally in the public supply. Examples are steel industry and railway traction. Consecutive 200-ms values may differ significantly in such a case. But still the 200-ms values are normally not used for calculating site and system indices. If strongly fluctuating distortion is expected, site indices may be calculated from the 3-s values. A large deviation between the site indices based on 3-s values and those based on 10-min values points to strongly fluctuating distortion.

In Section 5.2.4 we will introduce an alternative method to quantify changes in disturbance levels on a short time scale without the need to store a large amount of data. The method quantifies the deviation between nonaggregated and aggregated values and applies a time aggregation method to this deviation.

### 5.2.2 IEC 61000-4-30

***5.2.2.1 Basic Characteristics*** The IEC power quality monitoring standard IEC 61000-4-30 defines the following basic characteristics (the term *power quality parameters* is used in the standard document). Note that the list below includes characteristics for voltage as well as for current. The characteristics for current measurements are part of an informative annex with the standard. The definitions of the various basic characteristics are give in detail in Chapter 3.

- The basic characteristic for (voltage) frequency variations is the 10 s frequency value.
- The three basic characteristics for voltage (magnitude) variations are defined: the 10/12-cycle rms value, the 10/12-cycle underdeviation value, and the 10/12-cycle overdeviation value. Note that the latter two can be calculated from the rms value. They do not provide any additional information at this stage. However, they need to be calculated at the basic time window, not after time aggregation.
- The basic characteristic for current (magnitude) variations is the 10/12-cycle rms current.
- For voltage fluctuations a reference is made to the IEC flickermeter standard (IEC 61000-4-15). The basic characteristic used in that standard is the *instantaneous flicker sensation.* No time window is associated with the instantaneous flicker sensation as there is with the other basic characteristics.
- Unbalance is characterized by two basic characteristics: the 10/12-cycle negative-sequence unbalance (ratio of negative- and positive-sequence voltages) and the 10/12-cycle zero-sequence unbalance (ratio of zero- and positive-sequence voltage.
- For voltage waveform distortion the 10/12-cycle harmonic and interharmonic subgroups are used as basic characteristics with power quality monitoring. The IEC standard for measurement of waveform distortion (IEC 61000-4-7) also defines so-called *harmonic groups* and *interharmonic groups.*
- The basic characteristics for current distortion are the same as for voltage distortion: 10/12-cycle harmonic and interharmonic subgroups.

***5.2.2.2 Frequency*** No aggregation method is defined for frequency values. The 10-s values are to be used directly to calculate site and system indices. As frequency variations are rarely a concern, such a calculation is normally not needed. In large interconnected systems, frequency variations are mainly a concern for the operational security of the transmission system. Large frequency variations often indicate a reduced security and increased power flows through the system. This will become a concern long before equipment is impacted by the frequency variations. This is no longer a power quality issue, but performance indices like the ones discussed in this chapter could also be used to quantify the security of an interconnected system.

***5.2.2.3 Voltage Fluctuations*** The aggregation for voltage fluctuations should be done in accordance with the IEC flickermeter standard (IEC 61000-4-15). From the probability distribution of the instantaneous flicker sensation values over a 10-min interval a short-term flicker severity is calculated. Twelve of the short-term flicker severity values are used to calculate one 2-h long-term flicker severity value:

$$P_{\mathrm{lt}} = \sqrt[3]{\frac{1}{12}\sum_{i=1}^{12} P_{\mathrm{st}}(i)^3} \tag{5.1}$$

where $P_{\mathrm{lt}}$ and $P_{\mathrm{st}}$ are long-term and short-term flicker severity, respectively. The flickermeter standard is described in more detail in Section 2.4.5.

***5.2.2.4 Sliding Reference Voltage*** The so-called *sliding reference voltage* is obtained from the 10/12-cycle rms voltages by means of a first-order filter with a 1-min time constant. The value of the sliding reference voltage is updated every 10/12 cycles using the expression

$$U_s(n) = 0.9967 U_s(n-1) + 0.0033 U_{10/12} \tag{5.2}$$

This value is not used to describe any variation but as a reference value to detect voltage dips, swells, and interruptions.

***5.2.2.5 Harmonic Emission*** The harmonic emission testing standard IEC 61000-4-7 [155] requires smoothing of the groups and subgroups by using a digital first-order filter with a time constant of 1.5 s:

$$C_{\mathrm{aver}}(n) = 0.882 C_{\mathrm{aver}}(n-1) + 0.118(n) \tag{5.3}$$

with $C_{\mathrm{aver}}$ the 1.5-s value and $C_{10/12}$ the 10/12-cycle value. This aggregated value is compared with the requirements defined in the relevant emission standard.

The discrepancy between IEC 61000-4-7 and IEC 61000-4-30 is due to the difference in aim of the two documents. IEC 61000-4-7 is a testing standard describing how to measure the current to equipment under test, whereas IEC 61000-4-30 is a voltage quality standard defining how to quantify the voltage distortion in an existing system. However, calculation of harmonic groups and subgroups is the same in both documents.

#### *5.2.2.6 Magnitude, Unbalance, and Distortion*

In IEC 61000-4-30 [158] the following time aggregation intervals are defined for voltage magnitude, unbalance, and distortion:

- A 150/180-cycle interval: 150 cycles in a 50-Hz system, 180 cycles in a 60-Hz system
- A 10-min interval: exactly 10 min, thus no longer synchronized to the power system frequency
- A 2-h interval

The 150-cycle interval is aggregated from fifteen 10-cycle intervals in a 50-Hz system. In a 60-Hz system the 180-cycle window is aggregated from fifteen 12-cycle intervals. The 150/180-cycle value is the rms value of fifteen 10/12-cycle values:

$$C_{150/180} = \sqrt{\frac{1}{15}\sum_{i=1}^{15} C_{10/12}[i]^2} \tag{5.4}$$

For distortion $C_{10/12}[i]$ is obtained from one of the expressions (3.32) through (3.39). For voltage magnitude the 10/12-cycle rms value is used, for unbalance the ratio between the 10/12-cycle negative- or zero-sequence and positive-sequence voltages. The resulting values from (5.4) are referred to as 3-s values even though the window length is not exactly 3 s. The term 150/180-cycle value is more correct but less often used.

The 10/12-cycle values also form the basis for the calculation of the 10-min values. If the system frequency is exactly 50 or 60 Hz the 10-min interval is formed by exactly 3000 intervals of 10/12-cycle duration. For a frequency of 49.9 Hz, only 2994 intervals of 10 cycles are needed to form a 10-min interval:

$$C_{10\,\mathrm{min}} = \sqrt{\frac{1}{N}\sum_{i=1}^{N} C_{10/12}^2(i)} \tag{5.5}$$

where $N$ is the number of 10/12-cycle intervals in the 10-min interval, $N \approx 3000$. As the 10/12-cycle intervals and the 10-min intervals are not synchronized with each other, the starting instant of a 10-min interval will fall somewhere in a 10/12-cycle interval. The aggregation rule is that each 10/12-cycle interval belongs to the 10-min interval in which it starts.

A final level of aggregation is formed by the 2-h intervals. The 2-h values are obtained as the rms of 12 consecutive 10-min values:

$$C_{2\mathrm{h}} = \sqrt{\frac{1}{12}\sum_{i=1}^{12} C_{10\,\mathrm{min}}^2(i)} \tag{5.6}$$

When trying to interpret the method used, the first observation that can be made is that the rms of the basic characteristics is used instead of the (arithmetic) average value. To quantify the difference, consider average $\bar{C}$ and rms value $C_{\mathrm{rms}}$ of a characteristic $C_i$, with

$$\bar{C} = \frac{1}{N}\sum_{i=1}^{N} C_i \tag{5.7}$$

and

$$C_{\mathrm{rms}} = \sqrt{\frac{1}{N}\sum_{i=1}^{N} C_i^2} \tag{5.8}$$

Defining further $D_i = C_i - \bar{C}$ gives after rewriting (5.8)

$$C_{\mathrm{rms}}^2 = \bar{C}^2 + \frac{2}{N}\bar{C}\sum_{i=1}^{N} D_i + \frac{1}{N}\sum_{i=1}^{N} D_i^2 \tag{5.9}$$

According to its definition, the average value of $D_i$ is zero, so that we obtain

$$C_{\mathrm{rms}}^2 = \bar{C}^2 + \frac{1}{N}\sum_{i=1}^{N} D_i^2 \tag{5.10}$$

As the second term only contains nonnegative value, we can conclude easily that the rms value is greater than or equal to the (arithmetic) average:

$$C_{\mathrm{rms}} \geq \bar{C} \tag{5.11}$$

Those familiar with statistics will recognize the standard deviation in the second term on the right-hand side of (5.10). The difference between the rms and the average depends on the standard deviation of the individual values. The larger the standard deviation, the more the rms value exceeds the average.

The next issue is the difference between calculating a characteristic immediately over a longer period and aggregating it from values taken over the shorter periods. If we consider a simple aggregation, the comparison is between the following two calculation methods:

1. Obtain a characteristic over two windows of length $T$ and calculate the aggregated value $C$ from the two individual values $C_1$ and $C_2$ using $C = \sqrt{\frac{1}{2}(C_1^2 + C_2^2)}$
2. Obtain the characteristic directly over the window $2T$.

For the rms voltage, the two methods will give exactly the same result:

$$C = \sqrt{\frac{1}{2}\left(\frac{1}{N}\sum_{i=1}^{N} v_i^2 + \frac{1}{N}\sum_{i=N+1}^{2N} v_i^2\right)} \tag{5.12}$$

$$= \sqrt{\frac{1}{2N}\sum_{i=1}^{2N} v_i^2} \tag{5.13}$$

The latter expression is exactly the rms value over the period $2T$.

To study the effect of the aggregation algorithm on the distortion characteristics, we again introduce a simplified case. Consider a waveform with 25% fifth-harmonic distortion. The fifth-harmonic component is constant in amplitude but changes phase after every cycle. The resulting waveform is shown in Figure 5.3.

Next the spectrum of this waveform is determined in two ways. The first method takes the spectrum over each individual cycle and averages the absolute values. The result is indicated by plus symbols in Figure 5.4. The spectrum contains a fundamental component of amplitude 1 and a fifth-harmonic component of amplitude 0.25. The second method takes immediately the spectrum over the whole four-cycle window, indicated by circles in Figure 5.4. The latter method does not give any fifth-harmonic component but instead would indicate that there is interharmonic distortion present in the signal.

Note that the one-cycle spectrum has been obtained by taking the rms over the absolute values of the four individual spectra. Taking the average of the complex spectra would result in zero for the fifth harmonic, because the individual values would cancel each other due to the different phase angles.

The two spectra in Figure 5.4 cannot be immediately compared because they have a different frequency resolution. Therefore the interharmonic values in the

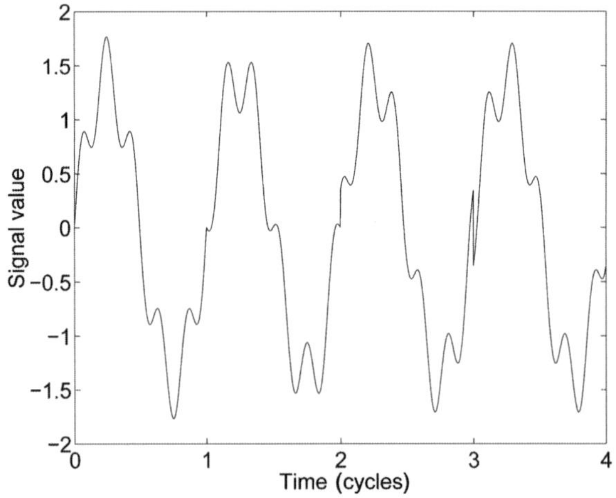


**Figure 5.3** Example waveform with 25% fifth-harmonic component with changing phase angle.

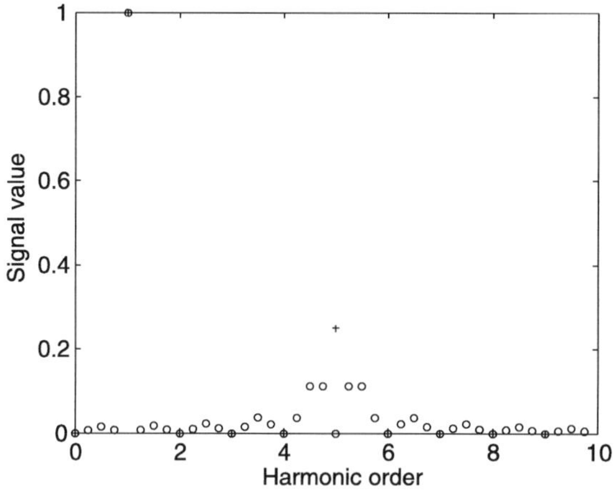


**Figure 5.4** Spectrum of waveform in Figure 5.3: one-cycle window (+) versus four-cycle window (○).

four-cycle spectrum have been merged with the harmonic values in the same way as with the grouping in IEC 61000-4-30. For example, the new fifth-harmonic component is calculated as follows:

$$C_5 = \sqrt{0.5C_{4.5}^2 + C_{4.75}^2 + C_5^2 + C_{5.25}^2 + 0.5C_{5.5}^2} \tag{5.14}$$

The result is shown in Figure 5.5, which shows how the spectral energy appears to shift from the fifth harmonic to other frequencies. This phenomenon is well known in telecommunication: Side bands appear due to phase modulation of the carrier

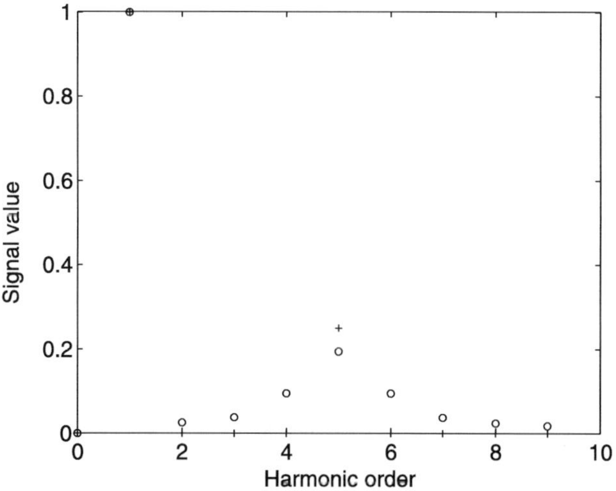


**Figure 5.5** Comparison of spectra from Figure 5.4 after merging interharmonic frequencies with harmonic frequencies.

wave. A longer window will give a higher frequency resolution but will also cause leakage to other frequencies because of time-varying harmonics. The leakage is limited by limiting the window length and aggregating.

### 5.2.3 Voltage and Current Steps

Part of the variations in voltage magnitude are in the form of sudden steps, for example, due to transformer tap-changer operations, capacitor switching, or switching or large loads. Such steps do not normally cause the rms voltage to reach a value outside its normal operating range. Typical voltage steps are up to 5 V, thus only 2% of the nominal voltage. Presenting voltage variations in the normal way will cause these steps to get lost in the overall daily variations. When information is needed on steps in voltage, another method is required.

When a method to detect voltage steps is to be implemented in a monitor, that is, when voltage waveforms are available, the steps can be detected in the time domain. A triggering method like the ones discussed in Section 7.3 and Chapter 8 can be used for that. It is however also possible to detect steps from the basic characteristics. A simple detection criterion is in that case that the difference between two consecutive values exceeds a certain threshold:

$$\Delta U = |U_k - U_{k-2}| > \delta \tag{5.15}$$

When a step occurs in the middle of a basic measurement window, the rms value for this window will be halfway between the prestep and poststep voltage. By taking the difference with the value two time steps back in time, the size of the total voltage step is obtained.

This method has been applied to the 3-s rms voltage recorded at a residential location. The results are shown in Figure 5.6 for two recordings. The first recording (left) was obtained around 10 in the morning. It shows three periods of about 30 s

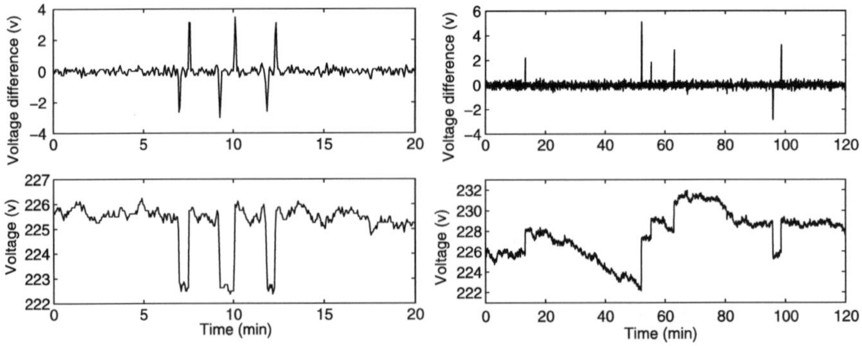


**Figure 5.6** A 3-s rms voltage (bottom) and difference between two consecutive 3-s rms voltages (top).

duration during which the rms voltage is about 3 V lower. The voltage drop is too low to be registered as a voltage dip, but still it may be worth knowing that such a disturbance has occurred. The function in the top figure, calculated as the difference between two consecutive 3-s values, clearly shows these events. A simple threshold is sufficient to detect the disturbance. The equal amplitude of the difference function indicates that the steps have a common origin. The size of the steps, about 1% of the nominal voltage, points to tap-changer operation as the origin.

The figures on the right-hand side of Figure 5.6 show a number of voltage steps at the same location, three days later between 8 and 10 in the morning. Also here the voltage difference shows easily distinguishable peaks. The difference in step size points to a different mechanism. The largest step is about 2.2% of the nominal voltage and is probably due to capacitor bank switching.

Results obtained at another location are shown in Figure 5.7. The recording on the left shows a number of voltage steps that can again be easily detected from the difference function. However, for the recording on the right this is much more difficult. The step in voltage is clearly visible in the plot of rms voltage versus time, but the voltage difference is dominated by a large number of changes between consecutive 3-s values. These fast fluctuations in voltage are due to the operation of an arc furnace in the neighborhood. A higher threshold value or an alternative detection method is needed to detect voltage steps in this recording.

The voltage difference as a function of time, as calculated by using (5.15) and plotted in Figures 5.6 and 5.7, can be used to count the number of voltage steps within a predefined interval. The use of a predefined interval means that the number of steps can be treated as a variation, even though a triggering method is used to detect them. The results of the counting process are shown in Figure 5.8 for two different sites. A threshold equal to 0.75% of the nominal voltage (1.725 V) has been used. When the voltage difference according to (5.15) is above the threshold for a number of consecutive samples, this is counted as one voltage step.

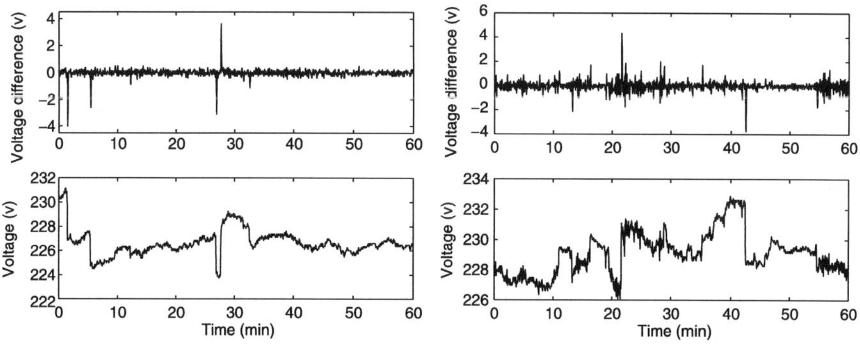


**Figure 5.7** A 3-s rms voltage (bottom) and difference between two consecutive 3-s rms voltages (top).

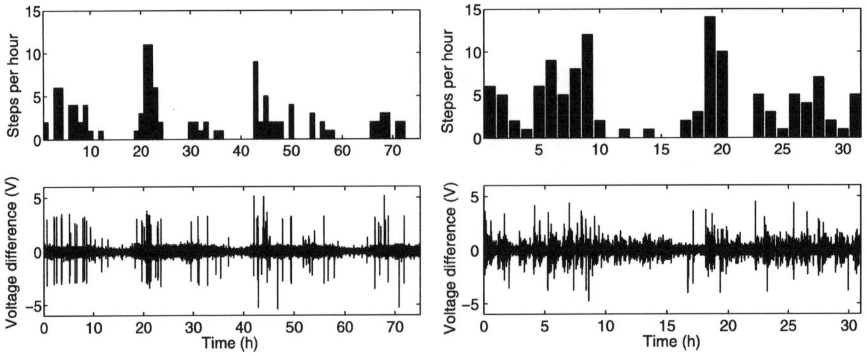


**Figure 5.8** Number of voltage steps per hour for two different locations; the bottom curves plot the voltage difference.

The left-hand plots were obtained for the same site as in Figure 5.6. The voltage difference has a low level and the voltage steps can be detected accurately. The right-hand plots were obtained for the site in Figure 5.7. The high background level due to the arc furnace means there are likely some false alarms in the counting of voltage steps. To resolve this, a discussion should be started about what constitutes a voltage step and what is merely a severe voltage fluctuation. The fact that the two sites in Figure 5.8 show a similar number of voltage steps per hour should be seen as a coincidence. Repeating this measurement for a total of seven sites in three towns in two countries gave maximum values for the number of steps per hour ranging from 2 to 14.

In this case the method has been applied to the 3-s rms voltages, but it may equally be applied to voltages obtained over a longer period. However, this would make it harder to detect a number of steps within a short period, as in the left-hand plots in Figure 5.6. The method could also be applied to the rms current, enabling the detection of switching of large local loads. Applying the method to the harmonic distortion could detect switching actions that lead to changes in resonance frequency. To process these disturbances it may be better to use techniques for power quality events.

### 5.2.4 Very Short Variations

The IEC standard for power quality measurements defines two aggregated time intervals for variations in the rms voltage: a 3-s (*very short time*) interval and a 10-min (*short-time*) interval. Only the 10-min values are used to quantify the performance of the system (e.g., in EN 50160). The fastest variations (*fluctuations*) in voltage magnitude are covered by the flickermeter standard. Variations with a time scale between 3 s and 10 min are not covered by any standard. This section proposes an approach to characterize these medium-scale variations. The term *very short variations* (VSVs) is suggested for variations at this time scale.

**5.2.4.1 *Definition of Very Short Variations*** The IEC power quality measurement standard prescribes the use of 3-s or very short time values and 10-min or short-time values of the rms voltage. The 10-min values $U_{sh}$ are obtained as the rms of the 3-s values $U_{vs}$ over the preceding 10 min:

$$U_{sh}(t_k) = \sqrt{\frac{1}{N} \sum_{i=k-N+1}^{k} U_{vs}^2(t_i)} \tag{5.16}$$

with $N$ the number of 3-s values in the 10-min window and $t_k$ a time sample corresponding to the end of a 10-min clock interval: A value is calculated at 12:10, at 12:20, at 12:30, and so on, over the preceding 10-min interval.

To characterize the voltage variations at time scales shorter than 10 min, the difference between the 3-s values and the 10-min values is used. Using the 10-min values as defined in IEC 61000-4-30 would result in a step at every 10-min time stamp. Therefore the 10-min value is updated every 3 s in the proposed method. The *3-s very short variation value* is defined as the difference between the 3-s rms voltage and the rms of the 3-s values over the preceding 10 min:

$$\Delta U_{vs}(t_k) = U_{vs}(t_k) - U_{sh}(t_k) \tag{5.17}$$

where the short-time voltage $U_{sh}$ is calculated as in (5.16), with the difference that the value is updated for every new 3-s value. This can be interpreted as a high-pass residual of the very short values after taking the 10-min averages. From the 3-s VSV values a *10-min very short variation value* is calculated for every 10-min time stamp:

$$\Delta U_{sh}(t_k) = \sqrt{\frac{1}{N} \sum_{i=k-N+1}^{k} \Delta U_{vs}^2(t_i)} \tag{5.18}$$

with $t_k$ a time sample corresponding to a 10-min time stamp, as in (5.16). The result is that a voltage measurement results in three values over every 10-min interval:

- The short-term flicker severity as defined in IEC 61000-4-15
- The 10-min VSV as in (5.18)
- The short-time (10-min) rms voltage as in (5.16) as defined in IEC 61000-4-30

This method allows the inclusion of 3-s values without the need to store excessive amounts of data.

**5.2.4.2 *Measurement Examples*** The characterization method described here has been applied to the very short time voltages $U_{vs}$ measured over a three-day period in a residential apartment. The results are shown in Figure 5.9. The

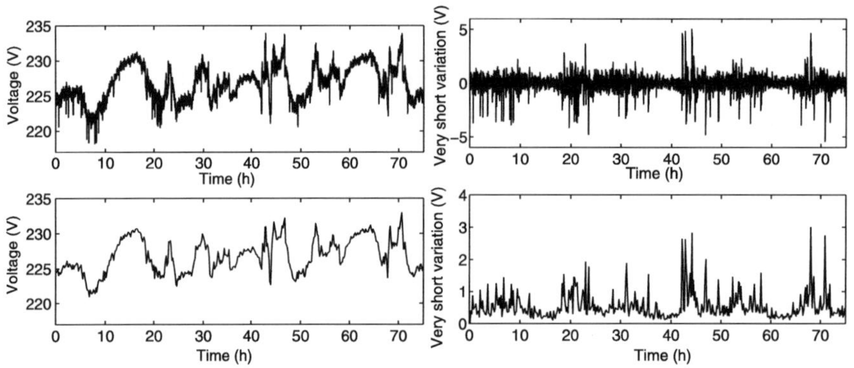


**Figure 5.9** The 3-s (top) and 10-min (bottom) rms voltages (left); 3-s (top) and 10-min (bottom) VSV voltages (right).

measurement data are the same as used in Figure 5.6 and the left-hand plot in Figure 5.8. The curves on the left show the 3-s and 10-min rms values. As expected, the 10-min values show a smoother pattern than the 3-s values. The difference between the two, the 3-s VSV, is shown in the top-right-hand curve. This curve shows a number of spikes in the positive and negative directions superimposed on a noisy background. The 10-min values aggregate both spikes and background noise. However, the rms used means that the presence of a spike in any 10-min interval will have a dominating influence.

The second example, with the results presented in Figure 5.10, uses the same set of data as in Figure 5.7 and the right-hand plot in Figure 5.8. As mentioned before, this measurement was done at a location with a severe flicker intensity. The VSV values however are not any different from those at the other site. Both figures show a more or less continuous level with superimposed spikes. These spikes

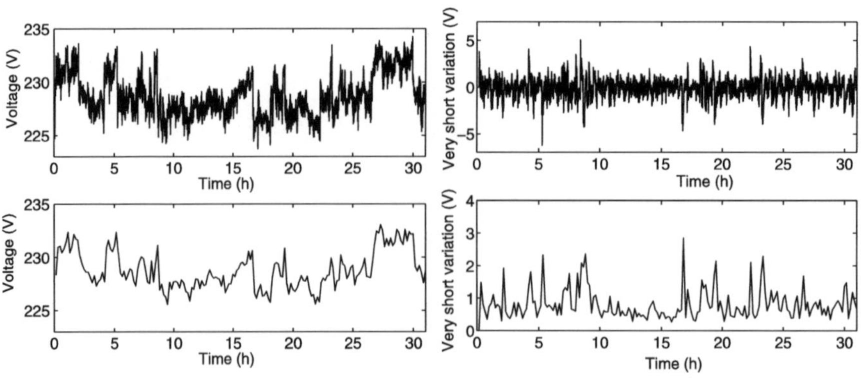


**Figure 5.10** The 3-s (top) and 10-min (bottom) rms voltages (left); 3-s (top) and 10-min (bottom) VSV voltages (right).

are due to sudden steps in rms voltage, the same ones that were considered in Section 5.2.3.

The impact of a voltage step on the VSV values can be understood from its impact on the 3-s and 10-min rms values. Consider a step in voltage magnitude from $U_1$ to $U_1 + \Delta U$. The 3-s rms values will be equal to $U_1$ before the step. One 3-s rms value will be between $U_1$ and $U_1 + \Delta U$, depending on where in the 3-s interval the step will take place. Note that the location of the 3-s interval is predetermined so that the voltage step can take place anywhere in that interval. The further 3-s values after the voltage step are equal to $U_1 + \Delta U$. The sliding-window 10-min values, on the other hand, will slowly increase from $U_1$ to $U_1 + \Delta U$ over a 10-min period starting at the instant of the step. The 3-s VSV is the difference between the 3-s and 10-min rms values. It will be zero before the step, increase to a value close to $\Delta U$ in two time steps (6 s), and next decay linearly to zero in a 10-min period.

The 10-min VSV value is obtained from the 3-s VSV over a 10-min interval. But this interval will rarely correspond to the 10-min interval over which the 3-s values decay. A single step in voltage magnitude will thus result in two nonzero values for the 10-min VSV. The algorithm used for Figures 5.9 and 5.10 has been applied to a synthetic step in voltage magnitude. The resulting rms voltages and VSVs are shown in Figure 5.11; a step from 227 to 232 V was simulated. The 3-s rms value reaches the higher value after only one sample with an intermediate value. The 3-s VSV, however, takes 10 min to come back to zero.

The calculations leading to Figure 5.11 have been repeated for different locations of the voltage step within the 10-min time aggregation interval. The resulting two nonzero 10-min VSV values are shown in Figure 5.12. The horizontal axis gives the time delay between the start of the 10-min time aggregation interval and the voltage step. For small values of this delay, the triangularly shaped 3-s VSV is almost completely inside the interval, resulting in a high 10-min value. With increasing delay, the first 10-min value decreases and the next one increases until it finally

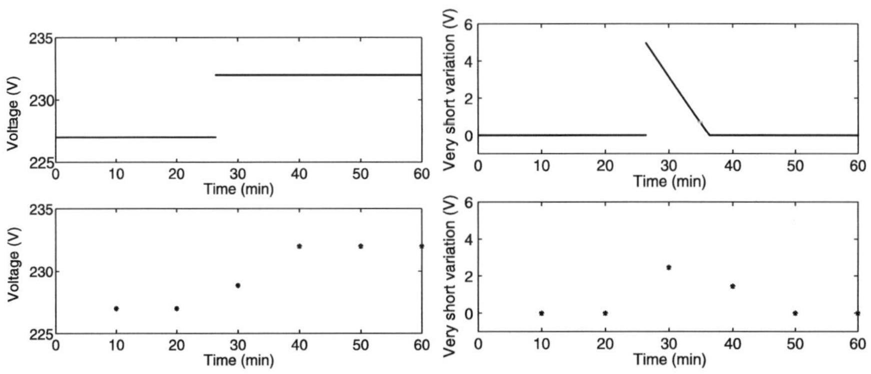


**Figure 5.11** The 3-s (top) and 10-min (bottom) rms voltages (left); 3-s (top) and 10-min (bottom) VSV voltages (right) for a synthetic step in voltage magnitude.

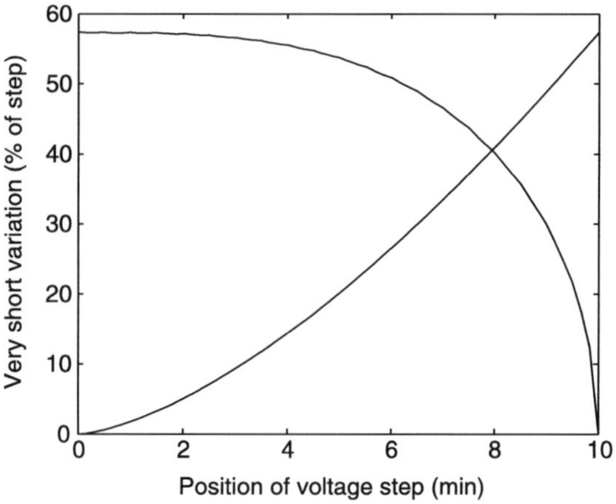


**Figure 5.12** A 10-min VSV due to a voltage step as function of location of step within 10-min aggregation interval.

becomes larger than the first value. An interesting observation is that the highest of the two values (the one that causes the peak in plots as in Fig. 5.6) is always around 50% of the voltage step. The peaks in the VSV are thus a good indication of the size of voltage steps.

The VSV concept is rather complex and is therefore worth considering if it offers anything new compared to existing or simpler methods. In Figures 5.13 and 5.14 the VSV for two different sites is compared to the voltage difference introduced in the previous paragraph. We can observe from Figure 5.13 that there is some correlation between the two characteristics. Especially in the left-hand figure, the spikes in the two curves show a strong correlation. The VSV is however not limited to a short band between the spikes. The VSV characteristics include information over

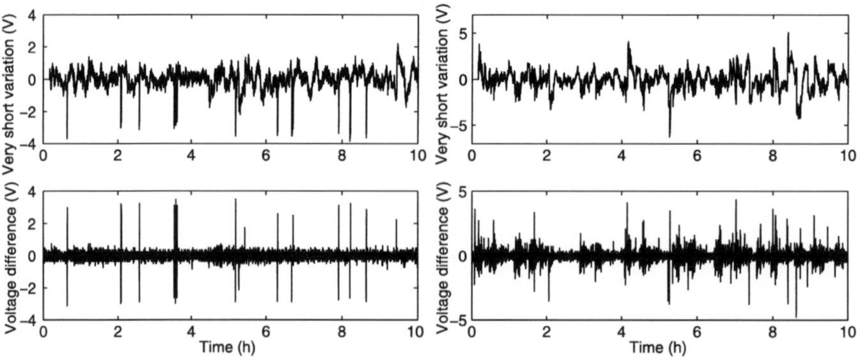


**Figure 5.13** Comparison of 3-s VSV (top) and voltage difference (bottom) for two sites: variations versus time.

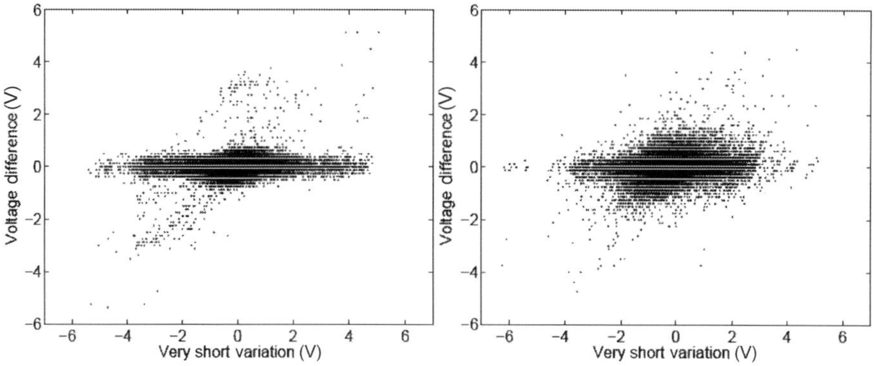


**Figure 5.14** Comparison of 3-s VSV (horizontal scale) and voltage difference (vertical scale) for two sites: scatter diagram.

a 10-min interval and thus over the whole range of time scales from 3 s to 10 min. The voltage difference, however, only includes information over a time window of three samples (9 s). The measurement on the left was obtained at a site characterized by a large number of short-duration drops in voltage. These drops were about 3 V and the voltage recovered within a few minutes. These voltage drops correspond to the negative excursions in VSV and the positive and negative excursions of the voltage step. This explains the strong correlation for this site. Note that a short-duration drop corresponds to two voltage steps in opposite direction. The site presented on the right shows a more typical behavior with a large number of small voltage steps.

The correlation can be better presented in a scatter diagram where two characteristics are plotted against each other, as in Figure 5.14. It is difficult to find any clear correlations, but generally speaking one may say that high values of the voltage difference are associated with high values of the VSV but not the other way around. These diagrams confirm that the short-time variation gives additional information.

As mentioned before, one of the measurement sites suffers from a high flicker severity. That makes it possible to verify if the VSV is independent of flicker severity. From the available measurements it was not possible to directly obtain the flicker severity. Instead the difference between the highest and the lowest rms voltage for every 3-s window was used as an indicator of the amount of voltage fluctuations. This *3-s voltage range* is shown in the top-left plot of Figure 5.15. The 3-s voltage range shows the familiar pattern of arc-furnace operation. The peaks correspond to the successive melts by the arc furnace [286, 303]. The aggregation method defined in IEC 61000-4-30 was applied to the 3-s voltage range to obtain a 10-min aggregated voltage range. This 10-min value is shown in the bottom left plot of Figure 5.15. The patterns of spikes and periods with low values are again reminiscent of the flicker produced by an arc furnace [303].

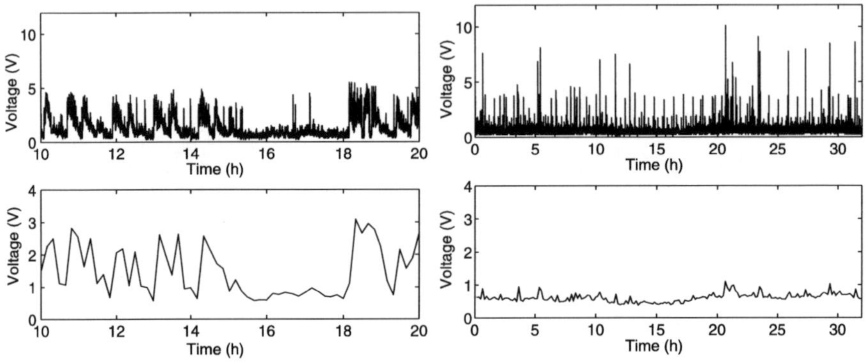


**Figure 5.15** A 3-s range in voltage and 10-min aggregated values. Left: measurement site with high flicker severity. Right: normal site.

The same calculations have been performed for recordings obtained at a site without a flicker problem. The results are shown on the right-hand side in Figure 5.15. The peak values in the 3-s voltage range are actually bigger for the site without flicker than for the site with flicker. However, the 10-min aggregated values are significantly lower. The spikes in the 3-s voltage range are probably due to the repeated starting of the refrigerator in the apartment.

To find any correlation between the VSV and the voltage range, Figure 5.16 shows a plot with the two characteristics along the two axes. The 10-min aggregated values were used to obtain this figure. The plot shows a random spread of points, indicating that the two parameters are uncorrelated. The correlation coefficient was in fact 0.2, indicating a very low, if any, correlation between the two. The correlation coefficient between the 3-s values is about the same. This confirms that our approach results in an index that is independent of the flicker indices.

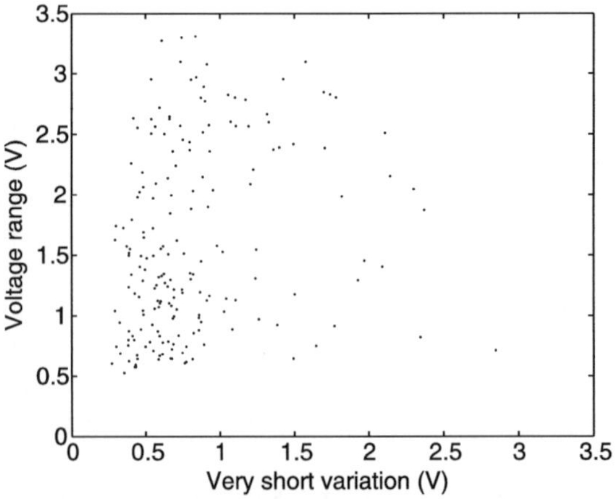


**Figure 5.16** Correlation between 10-min VSV and 10-min aggregation of 3-s voltage range.

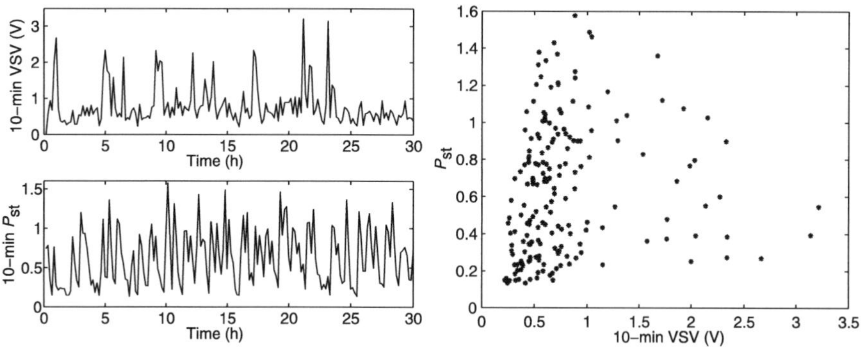


**Figure 5.17** Correlation between 10-min VSV and 10-min flicker severity.

The 10-min VSV is compared with the 10-min (short-term) flicker severity in Figure 5.17. The plots on the left show these two characteristics as a function of time. Some of the spikes in the VSV reappear in the flicker severity, but other do not. Most of the spikes in flicker severity do not reappear in the VSV. The scatter plot on the right clearly shows that there is very little correlation between the two characteristics.

We will give some more examples of VSV measurements in Sections 5.3.4 and 5.4.4. The concept of VSVs is new so that it is difficult to draw any hard conclusions. The preliminary conclusion from the measurements at seven locations is however that the 10-min VSVs consist of two "components":

- A continuous level up to about 1 V due to the combined effect of local load switching. The measurements at two of the domestic locations clearly showed a daily pattern.
- A number of spikes up to 3 or 4 V due to switching actions in the power system. The peak in VSV is about half the voltage step.

### 5.2.5 Flagging

The method of calculating characteristics (features) for variations is based on the assumption of quasi-stationarity. The signal is assumed to not deviate too much from the normal sine wave. However, sometimes there are large variations, which we referred to as events in Section 1.4, for which the basic assumption no longer holds. When, for example, a dip or a transient occurs during the measurement window, this will lead to a high apparent level of waveform distortion. The general consensus is that such a value does not represent a high level of distortion. The disturbance should be counted as an event in the event statistics, not as a high distortion level in the distortion (variation) statistics. A related problem is that some power quality monitors register a large frequency deviation during a voltage dip. This is mainly related to the phase-angle jump associated with the voltage dip causing a shift in voltage zero crossing. As the frequency estimation is based on the measurement of the time between zero crossings, errors may appear due to such a phase-angle jump.

Another reason for "flagging" a certain measurement window is to prevent double counting. When, for example, a voltage dip occurs during the measurement window, this will give a low rms value and in many cases also a high unbalance and a high flicker severity. Without flagging the same disturbances would show up in the voltage-dip statistics as well as in the statistics on voltage variations, unbalance, and flicker. To prevent this, the measurement window is flagged.

The concept of flagging is introduced in IEC 61000-4-30. When the one-cycle rms voltage drops below the voltage-dip threshold or rises above the voltage-swell threshold within the 200-ms basic measurement window, this window should be flagged. The 3-s, 10-min, and 2-h windows that contain the flagged basic measurement window should also be flagged. Note that the standard document only prescribes that certain measurement windows are flagged. It does not state that these measurements should not be used. That part of the interpretation of the measurements is not included in the standard. We will give some guidelines related to the processing of flagged data below.

Consider as an example harmonic distortion. Harmonic distortion is a variation: a steady-state phenomenon which is always present. The analysis and characterization of harmonic distortion take place through the DFT, which assumes a certain level of stationarity of the signal. A voltage dip is an example of an event: a short-duration reduction of the rms voltage. Calculating the Fourier series during a voltage dip will still result in a spectrum, but this spectrum no longer has any relation to the distortion of the voltage waveform. For this reason a 200-ms value should be flagged when during the window the rms voltage drops below the *voltage-dip threshold*. The voltage-dip threshold is typically set at a value around 90% of the nominal voltage.

An example of a basic measurement window including a voltage dip is shown in Figure 5.18: The predip and postdip rms voltage is 230 V; during the dip the

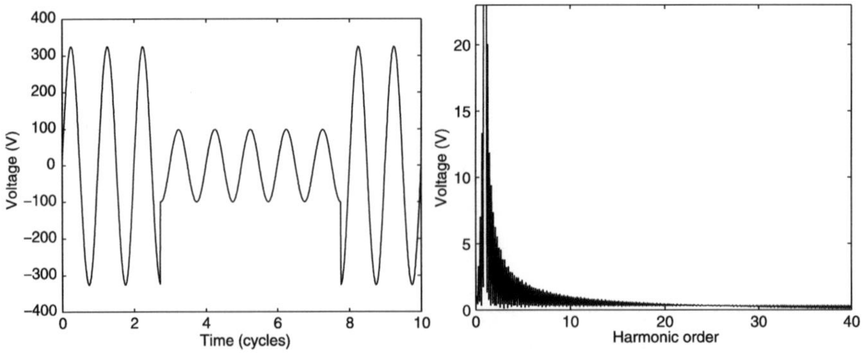


**Figure 5.18** Voltage dip of 5 cycles duration within 10-cycle window (left) and spectrum of this waveform (right).

rms voltage is equal to 70 V; the dip duration is five cycles. Both before and during the dip, the distortion of the waveform is zero. Note that this is a simulated waveform; measured voltages always contain distortion. The spectrum taken over the 10-cycle window is also shown in Figure 5.18. The vertical scale is adjusted to show the higher order components in the spectrum; the magnitude of the first-order (fundamental) component is about 150 V. Although the signal does not possess any distortion, the resulting spectrum shows a large amount of nonfundamental components. The reason for this is obvious: The DFT is valid under the assumption that the signal is periodic. A periodic continuation of the signal in Figure 5.18 contains a substantial amount of nonfundamental components.

Applying the grouping algorithm as defined in Section 3.2.3 results in the harmonic groups shown in Figure 5.19. The vertical axis is given in percent with respect to the nominal voltage of 230 V. Calculating the THD over the harmonic orders 2 through 40 gives

$$\mathrm{THD} = \frac{\sqrt{\sum_{n=2}^{40} C_{ng}^2}}{V_{\mathrm{nom}}} = 10.6\% \qquad (5.19)$$

with $C_{ng}$ the harmonic group for harmonic order $n$ calculated according to (3.32).

In Figure 5.19 we see that the second-harmonic subgroup has a value of 8% over the 200-ms window. If there is no second-harmonic distortion present in the signal, the value will be zero for all other windows within the 3-s aggregation time. This results in a 3-s value equal to

$$C_{150/180} = \sqrt{\tfrac{1}{15} \times 0.08^2} = 0.02$$

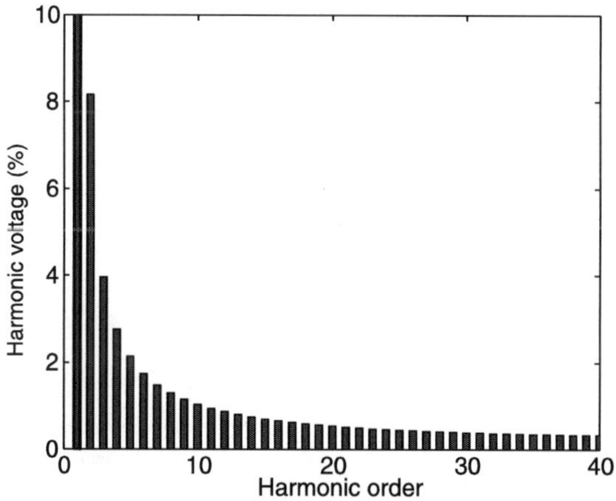


**Figure 5.19** Harmonic groups calculated from spectrum shown in Figure 5.18.

and a 10-min value equal to

$$C_{10\text{min}} = \sqrt{\tfrac{1}{3000} \times 0.02^2} = 0.0004$$

The flagging concept leads to both values being flagged. However, one may decide to remove the flagged 3-s values but not the flagged 10-min values. For harmonics of order 5 and higher it may not be needed to remove any flagged data. The 200-ms value for the fifth harmonic is 2%, which is not much higher than the typical distortion.

The flagging concept as introduced in IEC 61000-4-30 applies only to interruptions, voltage dips, and voltage swells. However, transients may also cause unrealistic harmonic values. An example is shown in Figure 5.20. The transient is obtained at 10 kV in an urban distribution network. Even though the values of the 200-ms harmonic groups are not as high as with the voltage dip in the previous example, they may still affect the resulting site indices. The eighth harmonic group is 3%, which results in a value of at least 0.8% for the 3-s, value which is much higher than can be expected under normal circumstances. Defining an appropriate flagging concept for transients is however rather complicated. The triggering for transients should be better understood before any standard flagging method can be defined. See Section 7.1.2 for a discussion of triggering methods for transients.

Another example is shown in Figure 5.21. Although the transient appears more severe in the time domain, its effect on the spectrum is less. The energy of the transient is more equally spread over the spectrum.

As mentioned before one of the aims of flagging is to prevent double counting. One case where double counting occurs is with voltage dips. When voltage-dip statistics are reported, the dips should not also be counted as a severe voltage variation.

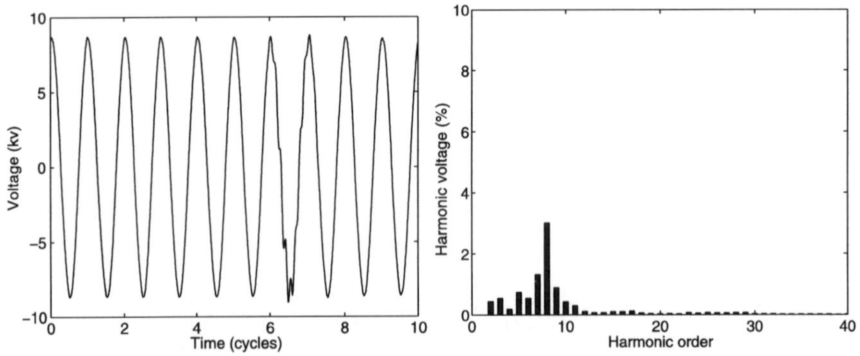


**Figure 5.20** Measured voltage transient (left) and harmonic groups obtained over 10-cycle window (right).

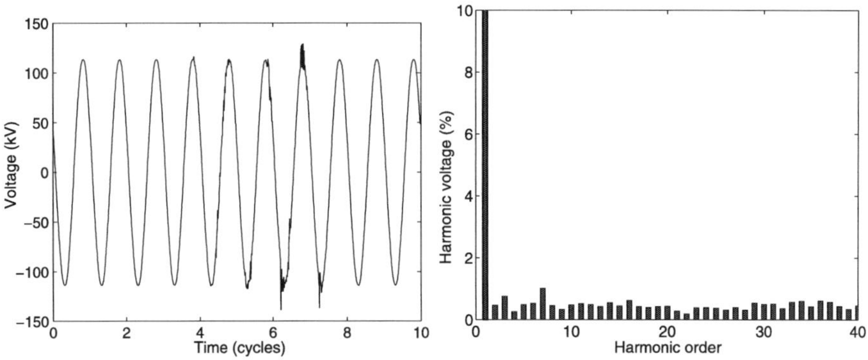


**Figure 5.21** Measured voltage transient (left) and harmonic groups obtained over 10-cycle window (right).

This should however be considered with care as the flagging, when used in the wrong way, may make the supply look better than it is.

As an example we have applied the flagging concept to three days of measurements in a hotel with a low average voltage and regular short drops in voltage due to the starting of refrigerators. During every 1-min interval, the lowest, average, and highest rms voltages were recorded. When the minimum rms voltage was less than 90% of nominal (lower than 207 V in this case), the interval was flagged. The results are shown in Figure 5.22 for a 24-h period. The continuous line connects the nonflagged data; the dots indicate the flagged data. As expected, it is especially the low rms voltages that are flagged. When these are removed from the data set, the supply voltage looks significantly better than it actually is. Of course, when voltage-dip statistics are also presented for this site, the dip frequency will be very high.

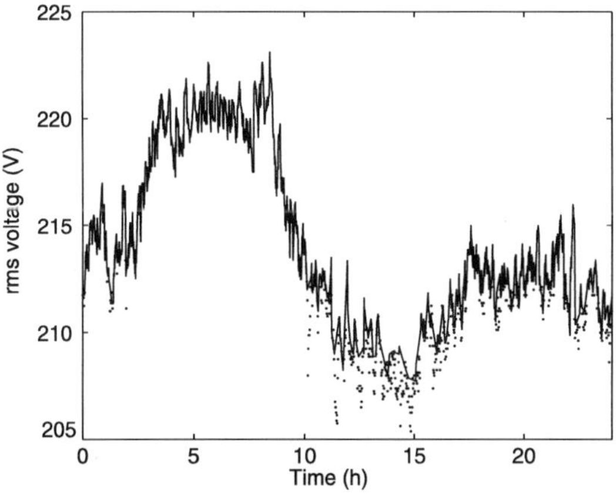


**Figure 5.22** Flagged data (dots) and nonflagged data (continuous line) for rms voltage.

### 5.2.6 Phase Aggregation

Many power quality measurements are performed in three-phase systems; that is, they involve three voltages and three currents. In many cases even a fourth voltage (neutral to ground) and a fourth current (the neutral current) are measured, but these are rarely used to calculate indices. Many methods for calculating power quality characteristics result in one value for each phase, for example, rms voltage, flicker severity, and waveform distortion. Somewhere along the path from characteristics to system indices, the values for the individual phases have to be merged into one three-phase value.

The most appropriate method is considering the three-phase character of the system when calculating the basic characteristic. Examples are the use of positive-sequence voltage for voltage magnitude and balanced and unbalanced harmonics for waveform distortion. Time aggregation, site indices, and system indices are next determined as discussed before. For unbalance and frequency no three-phase aggregation is needed; only one value results for a three-phase system.

However, in many cases, as when postprocessing historical records, only the three rms voltages are available. When this concerns phase-to-phase voltages, the positive-sequence voltage can still be calculated. This is not possible when only the rms values for the three phase-to-ground voltages are known.

A more commonly used method, unfortunately, is to use the worst or the average value over the three phases. For voltage magnitude there is no unique method to determine the worst phase (one phase may show an overvoltage and the other an undervoltage) and most standard documents are not clear about this.

When the characteristics and indices are directed toward single-phase equipment, one may decide to use another approach. A low level in one phase cannot be compensated for by a high level is another phase. There are three different approaches: Phase aggregation can take place together with time aggregation, phase aggregation can take place when calculating site indices, and phase aggregation can take place when calculating system indices:

- *Phase aggregation can take place at any of the time aggregation steps.* The 150/180-cycle value can be obtained as the rms over 45, instead of 15, 10/12-cycle values. Alternatively the 10-s value can be obtained over 600, instead of 200, 150/180-cycle values. Note that both methods use a kind of averaging over the three phases. When taking the average of the three 10/12-cycle values as a basic characteristic, similar results may be expected. The disadvantage of this approach is that the averaging is somewhat of a contradiction with the use of the 95 percentile for site and system indices. Extreme values that take place in just one phase are not found back in the indices.
- *Phase aggregation can take place when calculating site indices.* When, for example, the 95 percentile of the 10-min values is used as a site index, this is calculated not over 1008 but over 3024 samples. This method somewhat assumes that a load will experience the three phases consecutively in time. If the differences between the phases are random, that is, a rather high voltage

in one phase may occur next week in another phase, this approach is appropriate. However, this may not be the case.

- *The three phases are considered as three monitoring sites.* Phase aggregation only takes place with the calculation of the system indices. The disadvantage of this approach is that the three phases are typically not independent sites. This could be a problem, especially when other locations are only monitored in one or two phases.

Further discussion and study of these issues are needed. Appropriate three-phase indices need to be developed for voltage fluctuations as well.

## 5.3 CHARACTERISTICS VERSUS TIME

Most power quality measurements concern the recording of voltages and/or currents over a longer period: ranging from several minutes to several years. Plotting the basic or aggregated characteristics as a function of time is a very common way of presenting the results. This will give the user a very quick visual interpretation of the disturbance level. Some disturbances, noticeably harmonic distortion, show a very strong daily pattern, whereas others, such as the frequency, behave very random. In this section we will present some measurement results that have been obtained over a longer period of time: starting with arc-furnace data obtained during a period of several minutes to data recorded in a residential apartment over a period of several months.

### 5.3.1 Arc-Furnace Voltages and Currents

Measurements have been performed of voltages and currents with an electric arc furnace. The arc-furnace installation consists of a 40-MVA transformer supplying the arc furnace proper, a thyristor-controlled reactor (TCR) to mitigate voltage fluctuations, and second-, third-, and fourth-harmonic filters. This whole installation is supplied through a 63-MVA, 135/20-kV transformer with 10% impedance. The short-circuit capacity at 135 kV is 2300 MVA. The measurements shown below are voltage and current on the secondary side of the 135/20-kV transformer. Voltage and current were sampled with a sampling frequency of about 3 kHz during snapshots with durations between 20 and 40 s. The snapshots were taken at irregular intervals related to certain operational stages of the installation. The waveform data were available for the analysis presented below allowing for a high degree of flexibility in our analysis of the data.

It should be noted that the measurements were performed during the installation and testing phase, so no conclusions should be drawn about the performance of the TCR or the harmonic filters.

***5.3.1.1 Current Fluctuations as Function of Time*** Three consecutive snapshots, of 40 s length each, of the current to the arc-furnace installation have

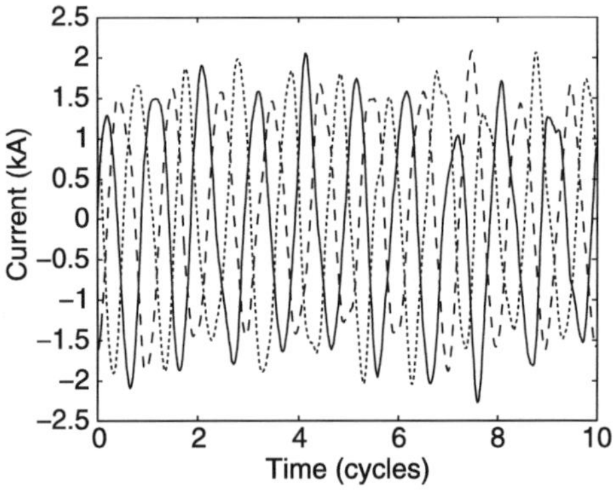


**Figure 5.23** Waveforms of three phase currents supplying arc-furnace installation.

been used for further analysis. The three phase currents during a 10-cycle window are shown in Figure 5.23. The currents show strong fluctuations, but the distortion is relatively small. We will come back to the distortion later; first we will concentrate on the fluctuations.

The current waveform for phase *a* during the whole $3 \times 40$-s recording is shown in Figure 5.24. The gaps between the measurements are shown as 5 s duration in the figure. In reality the gaps were 48 and 40 s. The currents in the other two phases show a similar behavior.

The basic measurement window according to IEC 61000-4-30 has a length of 10 cycles. In this study, we somewhat deviated from the standard by taking a 200-ms window. Synchronizing to the actual power system frequency would require interpolation and resampling of the waveform as the sampling frequency is not an integer multiple of the power system frequency. From the current waveforms in the three phases, the rms value has been calculated for every 200-ms window. The results

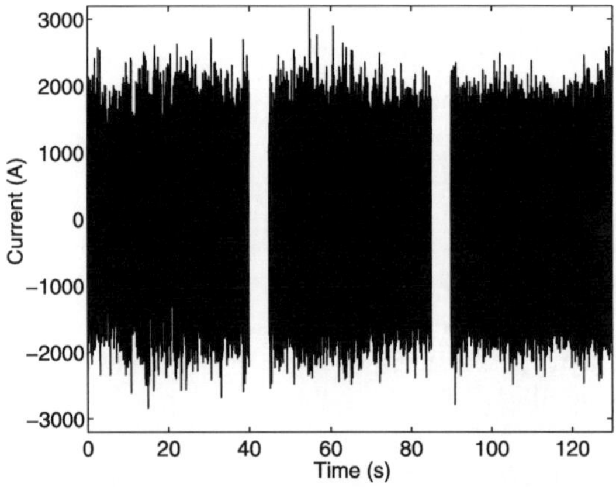


**Figure 5.24** Current waveform measured during three 40-s windows.

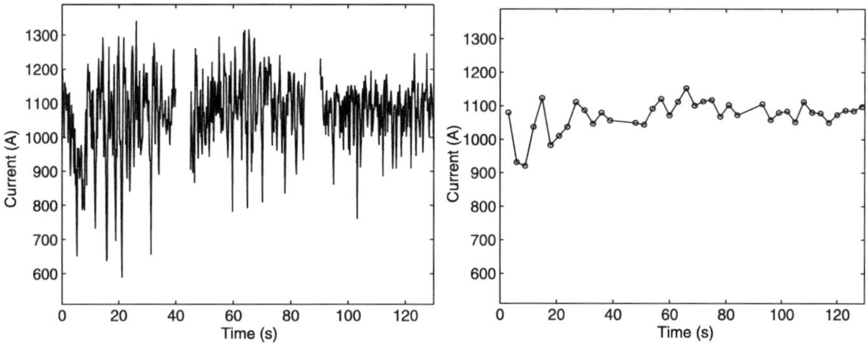


**Figure 5.25** The 200-ms (left) and 3-s (right) rms currents versus time.

are shown in Figure 5.25, again only for phase *a*. Where the original data ($3 \times 40$ s duration) contained $120 \times 3000 = 360{,}000$ samples per phase, the rms current only contains $120 \times 5 = 600$ samples per phase. Calculation of the rms current is thus a form of data reduction, but it also offers an easier way of interpreting the current fluctuations.

The 200-ms values of the rms current were next aggregated into 3-s values. The results are shown on the right-hand side of Figure 5.25 using the same vertical scale as the nonaggregated 200-ms values. As expected, the 3-s values show much less fluctuations than the 200-ms values. The average value is 1072 A for both the 200-ms and the 3-s values. The standard deviation, however, is 110 A for the 200-ms values but only 47 A for the 3-s values. The fact that the ratio between the standard deviations is less than $\sqrt{15}$ is due to the fact that there are fluctuations at any time scale; in other words, successive values are not stochastically independent.

The 200-ms rms current can also be calculated in the other two phases. An alternative is to calculate symmetrical-component currents. Positive- and negative-sequence currents are shown in Figure 5.26 again calculated over 200-ms and

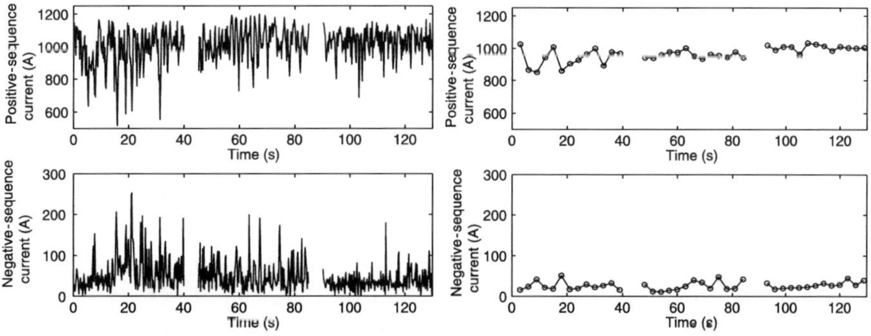


**Figure 5.26** The 200-ms (left) and 3-s (right) positive- and negative-sequence currents.

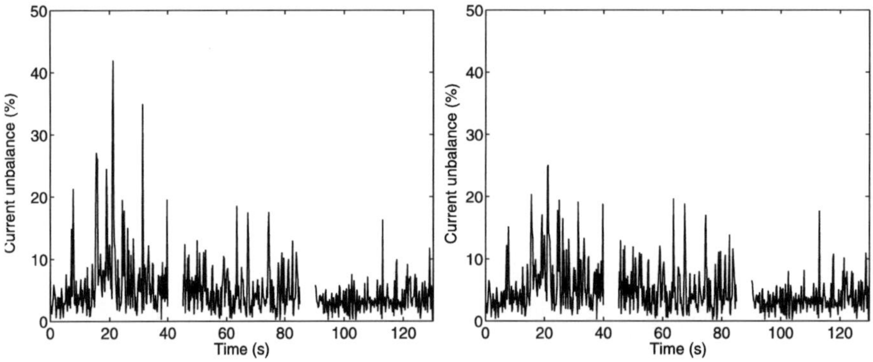


**Figure 5.27** The 200-ms (negative-sequence) current unbalance relative to actual positive-sequence current (left) and relative to average positive-sequence current (right).

3-s windows. The positive-sequence current is in all cases higher than the negative-sequence current. The zero-sequence current (not shown here) remains small during the whole measurement period. As before with the rms currents, we see that the 3-s values show substantially less fluctuations that the 200-ms values. The time aggregation from 200 ms to 3 s thus will not only lead to a data reduction but may also make the supply look better than it is. When considering voltage fluctuations, the time scale up to a few seconds is incorporated in the flicker standards. The choice of time scale to quantify current fluctuations depends among other things on the setting of overcurrent relays and the thermal time constant of series equipment (switchgear, cables, and transformers).

The ratio between negative- and positive-sequence currents is defined as the (negative-sequence) current unbalance. This ratio is shown on the left-hand side of Figure 5.27. The use of the ratio between positive- and negative-sequence current as a feature has one big disadvantage: Variations in the ratio may be due to variations in positive-sequence current as well as variations in negative-sequence current. From a system viewpoint, the per-unit value of the negative-sequence current would be a better indicator. As a base the rated power of the supply transformer (leading to a rated current of 1820 A in this case) or the short-circuit capacity (14.3 kA rated current) can be chosen.

To allow for a better comparison with the unbalance, the average positive-sequence current (1013 A during the 120-s recording) was used as a reference. This results in the average unbalance given in the standard definition. The results are shown on the right in Figure 5.27. The high values in the left-hand plot are no longer present in the right-hand plot.

#### *5.3.1.2 Current Fluctuations: Probability Distribution Functions*

Instead of plotting the characteristics versus time, the probability distribution function of a characteristic may be presented. For a characteristic that varies in a very random

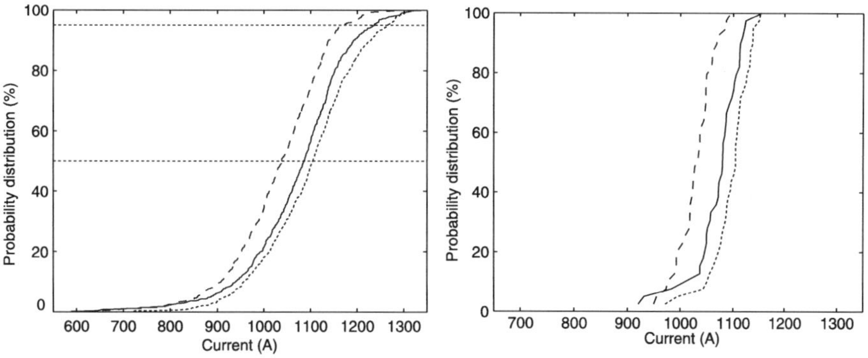


**Figure 5.28** Probability distribution of 200-ms (left) and 3-s (right) rms current for phase *a* (solid), phase *b* (dashed), and phase *c* (dotted).

**TABLE 5.1 Statistics for 200-ms rms Current**

| | 200-ms Values (A) | | | 3-s Values (A) | | |
|---|---|---|---|---|---|---|
| Index | Phase *a* | Phase *b* | Phase *c* | Phase *a* | Phase *b* | Phase *c* |
| Mean value | 1072 | 1030 | 1097 | 1071 | 1029 | 1097 |
| 50 percentile | 1085 | 1038 | 1105 | 1080 | 1034 | 1105 |
| 90 percentile | 1193 | 1139 | 1230 | 1114 | 1069 | 1137 |
| 95 percentile | 1237 | 1169 | 1265 | 1121 | 1087 | 1138 |
| 99 percentile | 1294 | 1227 | 1300 | | | |
| Maximum | 1340 | 1289 | 1321 | 1153 | 1095 | 1152 |

manner, like the arc-furnace currents in this example, the probability distribution function may actually contain more information than the time-domain plot.

Figure 5.28 plots the probability distribution function of the 200-ms and 3-s rms currents in the three phases. The current in phase *c* is higher than the phase *a* current, which in turn is higher than the phase *b* current. In a plot of the rms currents versus time such a trend is very difficult to extract. Comparing the two distributions it also becomes clear immediately that the spread in 3-s values is much smaller than the spread in 200-ms values. From the probability distributions some statistics of the currents have been obtained. These are summarized in Table 5.1. Note that the 99 percentile has not been calculated for the 3-s values. For the 99 percentile to be relevant at least 100 data points are needed, whereas only thirty-nine 3-s values are available from this measurement.

***5.3.1.3 Current Distortion Versus Time*** The arc-furnace current contains an interesting spectrum of harmonics, although the worst harmonics are already removed by the harmonic filters. As the currents in the three phases were available,

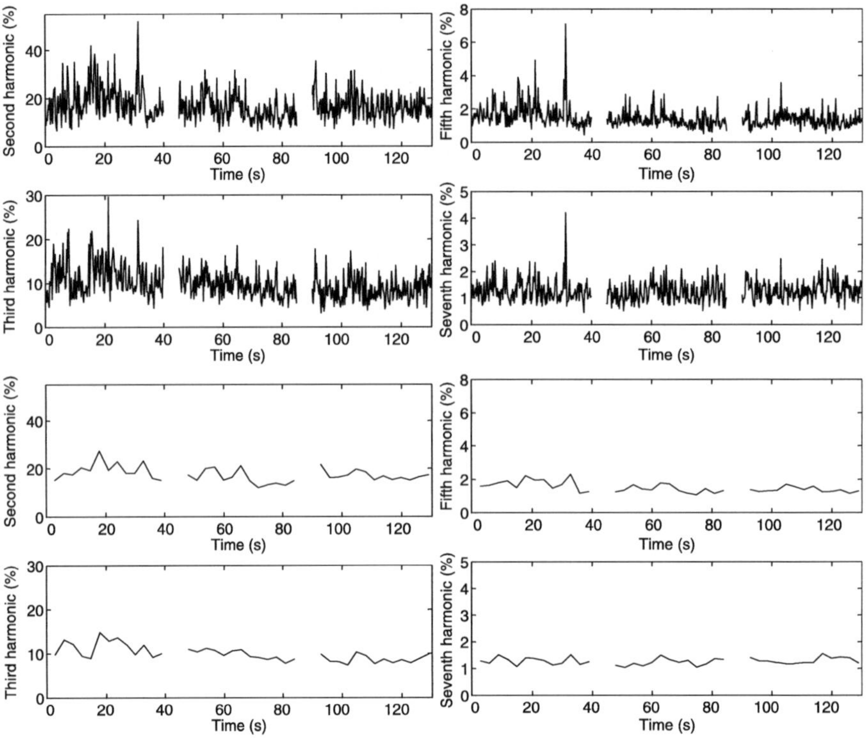


**Figure 5.29** The 200-ms second (top) and 3-s (bottom) third-, fifth-, and seventh-harmonic groups as function of time.

it was possible to calculate the balanced and unbalanced spectra as introduced in Section 3.3.7.

To characterize the distortion, we used the 200-ms harmonic groups in IEC 61000-4-30 as well as the 3-s aggregated values. The time variation of the second-, third-, fifth-, and seventh-harmonic groups over the monitoring period of $3 \times 40$ s has been plotted in Figure 5.29. The second- and third-harmonic currents are very high, but the fifth- and seventh-harmonic components are moderate. (Note the difference in vertical scale for the different harmonic orders.)

#### 5.3.1.4 Current Spectra

The calculations that resulted in Figure 5.29 have been repeated for all harmonic orders through 40. Reproducing the time variations of each individual harmonic would result in 40 plots similar to Figure 5.29. Although it may be worth studying the time variation of each harmonic separately, reproducing them is outside the scope of this book. They all show the same random behavior with high spikes as the harmonics plotted in Figure 5.30. Instead a representative spectrum has been calculated for the whole measurement period. There are a large number of options to obtain such a representative spectrum, falling all more or less in one of the three following categories:

- The average or median value can be taken for each harmonic group. This will result in an "average spectrum" over the measurement period. The advantage is that most engineers have a good feel for the average. However, short-duration very high distortion values could cause problems even though the average is well below any danger level.
- The maximum value could be used for each harmonic group. The worst case may however be reached for only one value during a whole measurement. High maximum values could falsely give the impression of very bad power quality. Another reason for not taking the maximum value is that events like transients or dips will give a very high distortion level even though the waveform may be close to sinusoidal. Also occasional bad data (for whatever reason) can give very high distortion values that are hard to recognize without looking at all the waveforms individually.
- A high percentile value like 90, 95, or 99% could be used. Incidental very high values do not affect the result whereas those values that occur often enough to potentially cause problems are included in the statistics.

From the 200-ms and 3-s harmonic groups spectra have been calculated by using the maximum values and the 50, 90, 95, and 99 percentile values, the latter one only for the 200-ms values. The resulting spectra are reproduced in Figure 5.30. The average fundamental component (harmonic group 1) has been used as a reference to calculate the percentage values. In a system where the harmonic distortion varies a lot at shorter time scales, it is very important to clearly define which spectrum is being presented. The average 3-s values are about the same as the 200-ms spectra, but the 95 and 90 percentiles are significantly lower.

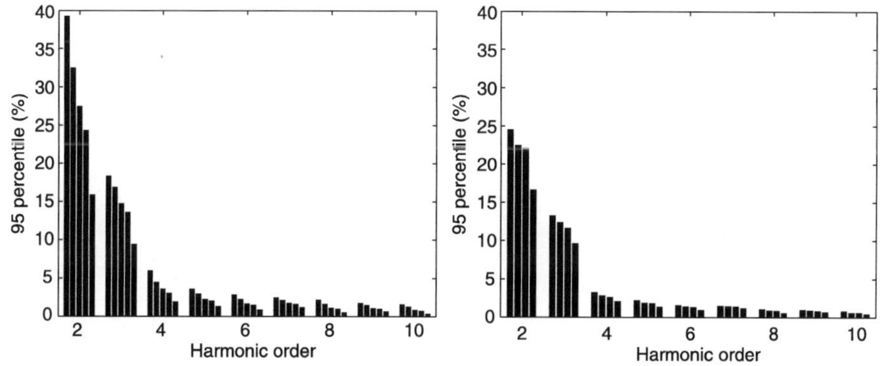


**Figure 5.30** Harmonic spectra obtained from 200-ms values (left) and 3-s values (right). Left to right: maximum, 99%, 95%, 90%, and 50%. The 99% spectrum is not shown for 3-s values.

### 5.3.2 Voltage Frequency

***5.3.2.1 Basic Measurement: 3-s Values*** A measurement of the frequency has been performed during an eight-day period in a residential apartment. The maximum, minimum, and average frequencies were obtained every 3 s. Figure 5.31 shows the measurement results as obtained during a 3-min period (60 measurement points). The highest frequency measured during these 3 min was 50.023 Hz, the lowest frequency 49.978 Hz, and the average frequency 50.0076 Hz (thus 7.6 mHz above nominal).

From the 3-s average frequency the so-called *overdeviation* and *underdeviation* have been calculated according to IEC 61000-4-30. The (frequency) overdeviation is defined as

$$f_{\text{over}} = \begin{cases} f - 50\ \text{Hz} & f > 50\ \text{Hz} \\ 0 & f \leq 50\ \text{Hz} \end{cases} \tag{5.20}$$

In the same way the (frequency) underdeviation is defined as

$$f_{\text{under}} = \begin{cases} f - 50\ \text{Hz} & f < 50\ \text{Hz} \\ 0 & f \geq 50\ \text{Hz} \end{cases} \tag{5.21}$$

Note that the overdeviation is always nonnegative, whereas the underdeviation is always nonpositive. Alternatively the underdeviation may be defined in such a way that it is also always nonnegative, by simply taking the opposite of (5.21).

The overdeviation and underdeviation are shown in Figure 5.32 over the same 3-min period in Figure 5.31. Note that the shape of the 3-s average frequency in

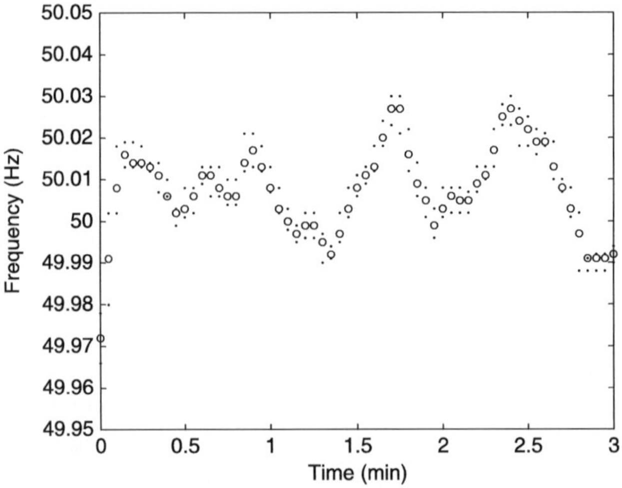


**Figure 5.31** Frequency measurements: For every 3-s window, both highest and lowest values are indicated by a small dot. The average over each 3-s window is shown as a circle.

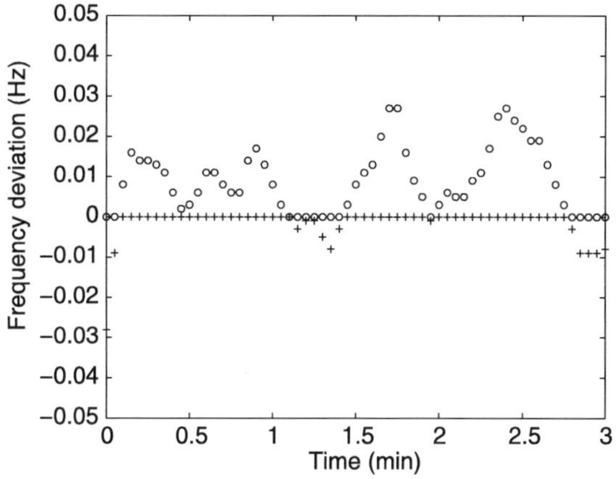


**Figure 5.32** Overdeviation (circle) and underdeviation (plus) over 3-min period.

Figure 5.31 is also found in Figure 5.32. The average overdeviation over this period is 9.1 mHz, the average underdeviation −1.5 mHz.

On average half of the underdeviation values are equal to zero. The same holds for the overdeviation. The average values are strongly influenced by these zero values. Therefore we also define the so-called *average nonzero overdeviation* and *average nonzero underdeviation*. These averages only include the nonzero values. For this 3-min period the average nonzero overdeviation is 11.8 mHz and the average nonzero underdeviation −6.8 mHz.

Figure 5.33 shows the 3-s averages over a 1-h period. The plot on the left shows the average frequency as a function of time. Instead of plotting the frequency as a function of time, the probability distribution function of the frequency can be shown, resulting in the plot on the right. The left-hand plot shows a step in frequency around the middle of the period. Before as well as after this step the frequency oscillates around an average value. The step is due to the loss of a large generator

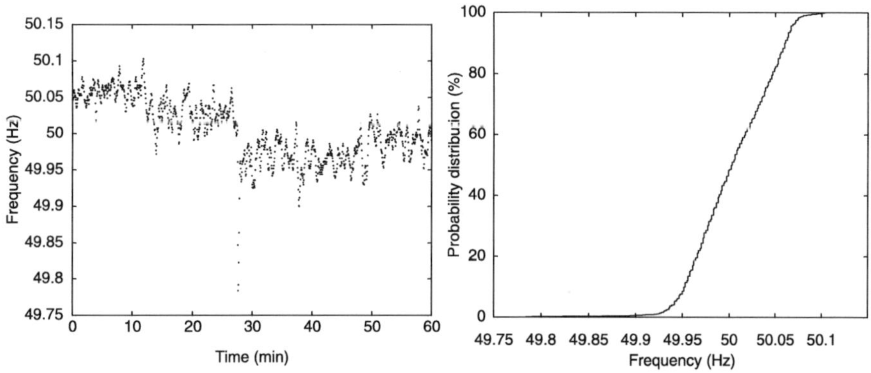


**Figure 5.33** Average frequency (3-s average) over 1-h period as function of time (left) and probability distribution function (right).

**TABLE 5.2 Frequency Statistics Obtained from 3-s Values Over 1-h Period Shown in Figure 5.33**

| Quantity | Value |
|---|---|
| Highest frequency | 50 Hz + 95 mHz |
| 97.5 percentile | 50 Hz + 72 mHz |
| Average nonzero overdeviation | +38.5 mHz |
| Average overdeviation | +19.8 mHz |
| Average frequency | 50 Hz + 0.5 mHz |
| Average underdeviation | −15.4 mHz |
| Average nonzero underdeviation | −32.1 mHz |
| 2.5 percentile | 50 Hz − 58 mHz |
| Lowest frequency | 50 Hz − 205 mHz |

unit. We will come back to this in the next section. Statistics can be obtained over the 1-h period in the same way as over the 3-min period above. The results for the 1-h period are presented in Table 5.2.

***5.3.2.2 Time Aggregation: 3-min Values*** To reduce the data set and obtain a better impression of the longer term stability of the frequency, the 3-s values were aggregated into 3-min values. Each 3-min value was calculated as the (arithmetic) average of sixty 3-s values. The time stamp of the 3-min value was taken as the time stamp of the last 3-s value. The minimum and maximum values over each 3-min period were calculated as the minimum and maximum of the 3-s values, respectively. When there were no nonzero overdeviation values, the average nonzero overdeviation was taken equal to zero. The same rule has been applied to calculate the average nonzero underdeviation. Thus a zero value for the average nonzero overdeviation means that the average frequency is below 50 Hz for the whole aggregation window. The result of the aggregation is shown in Figures 5.34 and 5.35 for a 24-h period.

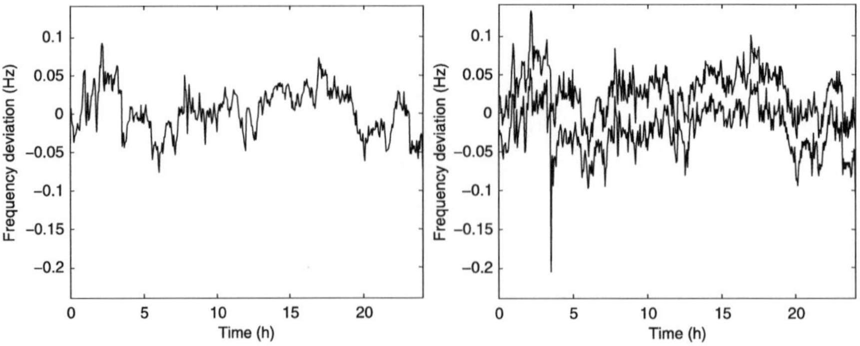


**Figure 5.34** Aggregated average frequency (left) and maximum/minimum frequency over each 3-min window (right) during 24-h period.

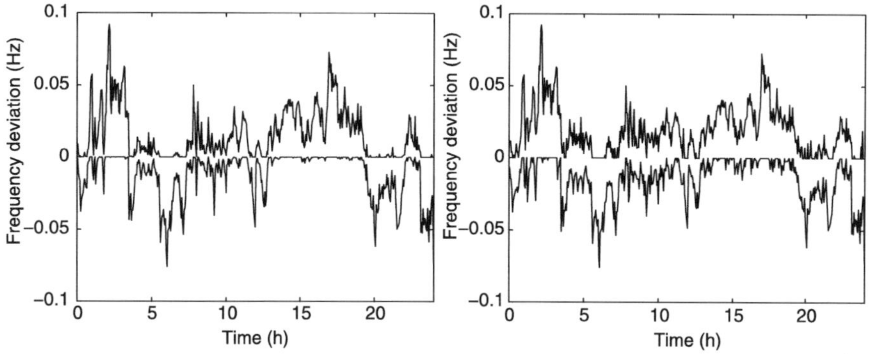


**Figure 5.35** Average overdeviation and underdeviation (left) and average nonzero overdeviation and underdeviation (right) over each 3-min window during 24-h period.

***5.3.2.3 Frequency Events*** In the frequency plot in Figure 5.33 a sudden negative excursion of the frequency is visible about halfway along the time axis. Such a sudden drop in frequency is due to the loss of a large generator somewhere in the interconnected system. One may refer to these as *frequency events*. Four such events are shown in Figure 5.36. The solid line indicates the average frequency (over

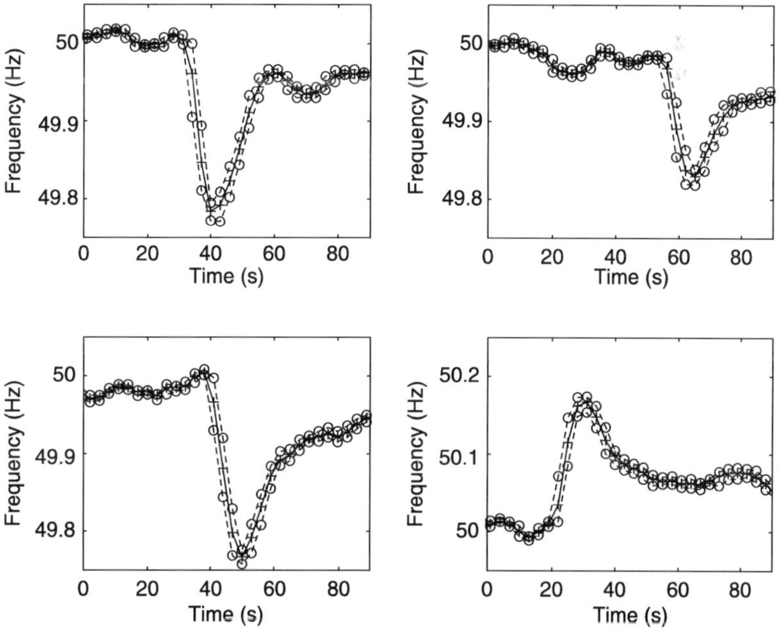


**Figure 5.36** Four examples of frequency events.

a 3-s window). The circles indicate 3-s maximum and minimum values. In three cases the frequency suddenly dropped, whereas in one case the frequency showed a sudden rise. The power–frequency control was in all four cases capable of bringing the frequency back to a value close to 50 Hz. In this case these four events were selected by visual inspection of the frequency-versus-time plots. A more precise definition of frequency events would require a triggering level, for example, based on frequency deviation from 50 Hz, frequency deviation from a sliding average, or rate of change in frequency.

#### 5.3.2.4 *Integrated Frequency Deviation*

The main consequence of frequency deviations is an error in the time indicated by clocks using the power system frequency as a time signal. If the frequency is too high, clocks run too fast; if the frequency is too low, clocks run too slow. The error between clock time and real time is equal to the integrated frequency deviation according to (2.19)

$$\Delta t = \int_0^T \frac{f(t) - f_0}{f_0} dt \tag{5.22}$$

with $f(t)$ the actual frequency, $f_0$ the nominal frequency, and $T$ the measurement window. The integrated frequency deviation has been calculated from the 3-s frequency values. The results are shown in Figure 5.37. The error in clocks varies between $-15$ and $+15$s.

### 5.3.3 Voltage Magnitude

The rms voltage, the frequency, and the spectrum were obtained over every 3-s interval during most of a one-week period at the end of July 2003. The measurements

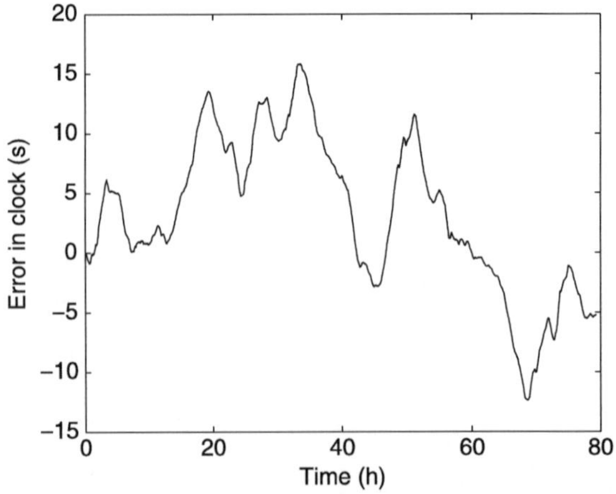


**Figure 5.37** Integrated frequency deviation during 80-h period.

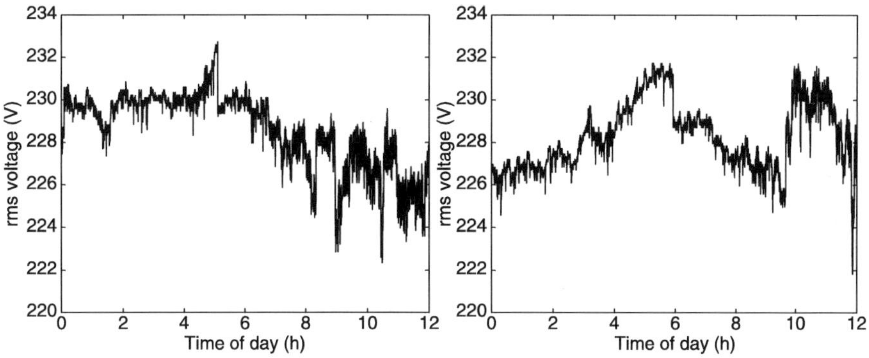


**Figure 5.38** The 3-s fundamental voltage during night and morning on two different days.

were single phase and were obtained at the wall outlet in an apartment in an area characterized by high-rise apartment blocks. There were a few gaps in the measurement. Whereas such gaps may not be acceptable for contractual applications, the measurements still serve very well as an example. We will present some of the results obtained from the 3-s rms voltage in this section. The harmonic results will be presented in the Section 5.3.5.

***5.3.3.1 Basic Measurements: 3-s Values*** During the whole measurement period, over 220,000 basic measurements (3-s values in this case) were obtained. The measurements were synchronized to clock time, not to power system frequency. The 3-s values of the rms voltage during a 12-h period (midnight to noon) are plotted in Figure 5.38. A number of tap-changer operations are visible: The left-hand plot shows a tap down around 5 in the morning and a tap up around 8 in the morning. The low-voltage system is supplied from the 10-kV system via a transformer with fixed turns ratio. The tap-changer operations point to a decrease or increase in voltage at the 10-kV side of an HV/MV transformer. The rise in voltage between 4 and 5 in the morning is thus due to a change in load over a wide area. However, the drop and subsequent rise in voltage between 1 and 2 at night may be a local phenomenon. A few days later (right-hand plot) the pattern is completely different. Although there is again a rise in voltage starting at about 4 in the morning, the tap down only takes place around 6. There are two taps up between 9 and 10. The tap-changer operations cause the rms voltage versus time to look different every day even though the load pattern is most likely very similar. We will come back to daily load patterns when discussing harmonic distortion in Section 5.3.5.

***5.3.3.2 Time Aggregation: 1- and 10-min Values*** The 3-s values have been aggregated into 1- and 10-min values. The 1-min values are not defined in any international standard, but the techniques in IEC 61000-4-30 can be applied to obtain

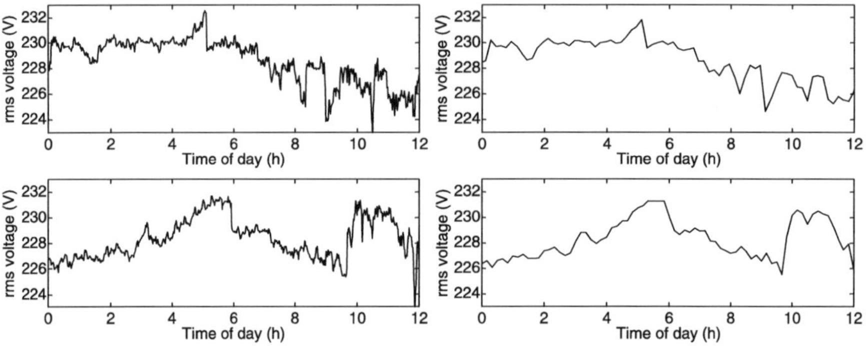


**Figure 5.39** Aggregated rms voltages: 1-min values (left) and 10-min values (right) for measurement periods shown in Figure 5.6.

them. The use of an additional 1-min window will result in a more equal spread of the time scales. The 1- and 10-min values for the two half-day periods shown before are presented in Figure 5.39. The 1-min values show the same general pattern as the 3-s values but they do not show the fast "noise" that characterizes the 3-s values. Using the 10-min values, on the other hand, seems to lead to the loss of a significant amount of detail. The use of 1-min values is somewhat of a reasonable compromise. Where needed, the 10-min values can later be obtained by aggregation of the 1-min values.

When plotting the rms voltages as a function of time over a whole week, as in Figure 5.40, the 1-min value will result in about 10,000 points. The result is still a rather noisy curve, as can be seen in the upper plot in Figure 5.40. When plotting

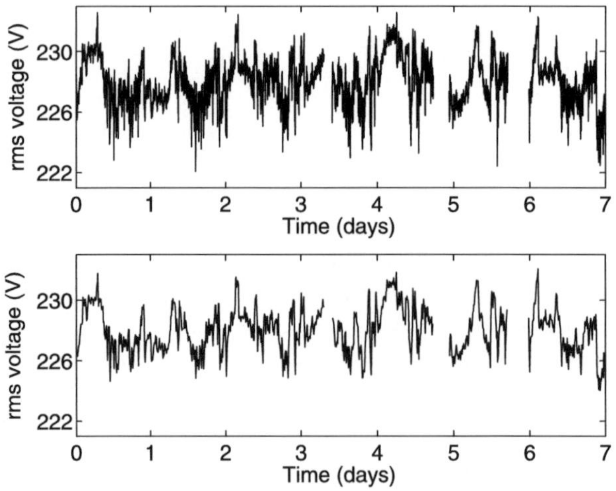


**Figure 5.40** The 1-min (top) and 10-min (bottom) rms voltage versus time.

over a one-week period, the 10-min values appear a better choice than the 1-min values. The additional information present in the 1-min or even the 3-s values may be presented in the form of VSVs or by counting the number of voltage steps.

***5.3.3.3 Daily Patterns in rms Voltage*** Variations in voltage magnitude (rms voltage) are due to variations in load. As the load variations show a clear daily pattern at most locations (large industrial installations are the main exception), such daily patterns are also expected in the voltage magnitude. To check for any daily patterns, the rms voltage has been plotted versus hour of the day in the left-hand plot of Figure 5.41. There is no obvious daily pattern visible in the voltage with the possible exception of a peak around 8 p.m. The tap-changer operations, random with respect to the time of day, blur the variations in voltage drop due to load variations. However, the rms voltages are somewhat lower during daytime. To further enhance any daily pattern, the 24-h day has been divided into 48 half-hour intervals. For each of these intervals the average has been taken over all 10-min values measured during that interval of the day. The result is shown in the right-hand plot of Figure 5.41. After this additional aggregation a daily pattern appears: The rms voltage is on average about 2 V lower at daytime than at night. The lowest voltage occurs in the evening, which corresponds to the peak load for a residential area. Note that the transformer tap changer controls the voltage at the MV level, not at LV. No clear explanation is available for the higher voltages between 7 and 8 in the evening. This may be due to a difference in load patterns for the local low-voltage grid as compared to the medium-voltage grid.

***5.3.3.4 rms Voltage and Statistics*** The probability distribution function of the rms voltage at the three time scales is shown in Figure 5.42. Only the upper and lower 10% are presented to better visualize the differences between the three functions. The steps in the 3-s values are due to the resolution of the monitor. After averaging these steps disappear and are no longer present in the 1- and 10-min values.

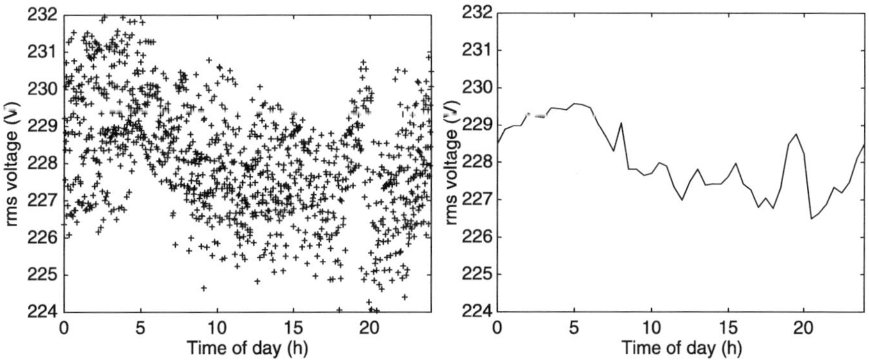


**Figure 5.41** The 10-min fundamental voltage versus time of day (left) and averaged over every half-hour interval (right).

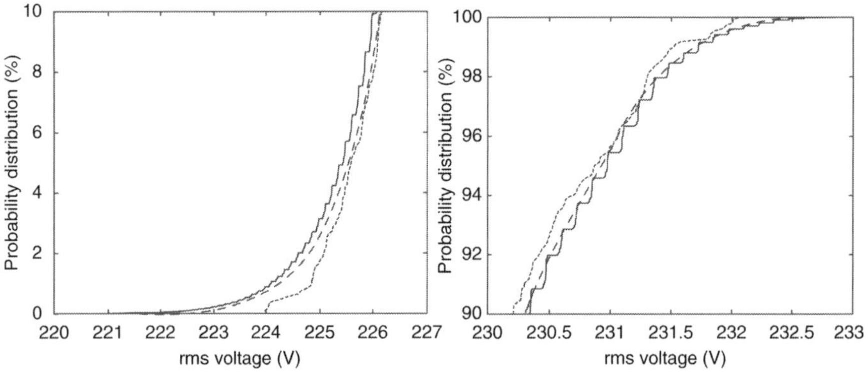


**Figure 5.42** Probability distribution function of rms voltage: 3-s values (solid line); 1-min values (dashed line); 10-min values (dotted line).

**TABLE 5.3 Statistics for Fundamental Voltage**

| Statistic | From 3-s Values (V) | From 1-min Value (V) | From 10-min Value (V) |
|---|---|---|---|
| 50% value | 228.1 | 228.1 | 228.1 |
| 90% range | 225.4–231.0 | 225.5–230.6 | 226.0–230.9 |
| 95% range | 224.8–231.3 | 224.9–231.3 | 225.6–231.1 |
| 99% range | 223.6–232.0 | 223.8–231.9 | 224.5–231.6 |
| 100% range | 220.2–233.0 | 221.1–232.6 | 224.0–232.1 |

The probability distribution function of the 3-s and 1-min rms voltages are very similar. After time aggregation to 10-min values a number of the most extreme values have disappeared. This is mentioned by many as an argument against the use of the 10-min values to characterize the supply performance. Using the 10-min values would remove the most extreme values, which are exactly the ones that lead to equipment damage. By using the 10-min values the supply would "look better than it is." We will come back to this discussion when presenting site indices in Section 5.4.

Some statistics obtained from the 3-s, 1-min, and 10-min values for this site are presented in Table 5.3. The median values are the same, which could be expected from the aggregation method used. Also the 90% range is very similar. However, for the higher percentiles the range of 3-s and 1-min values is larger than for the 10-min values. This is related to the removal of the extreme values by the aggregation process, as mentioned above.

### 5.3.4 Very Short Variations

The VSVs were introduced in Section 5.2.4 as a method to include variations in voltage at the time scale between 3 s and 10 min. A number of plots of the

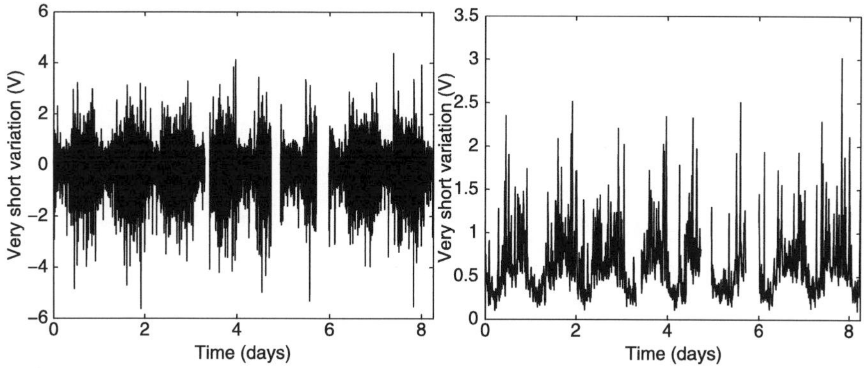


**Figure 5.43** Very short time variations obtained over an eight-day period: 3-s values (left) and 10-min values (right).

characteristic versus time were already shown in that section. In this section we will apply the method to the eight-day measurement in the residential apartment. The 3-s and 10-min VSVs as a function of time for the whole measurement period are shown in Figure 5.43.

The plots in Figure 5.43 show a continuous level with a number of spikes superimposed. As we mentioned before, the spikes are due to steps in the voltage associated with capacitor switching and transformer tap-changer operation. The continuous level shows a clear daily pattern for both the 3-s and the 10-min values. To enhance this pattern, the 10-min values have been plotted as a function of the time of day in Figure 5.44. The daily pattern in the continuous level is very clearly visible now. The spikes due to voltage steps are more common during daytime and evening than during nighttime. This is understandable as the load variations are more during daytime. In the right-hand plot in Figure 5.44 the data are

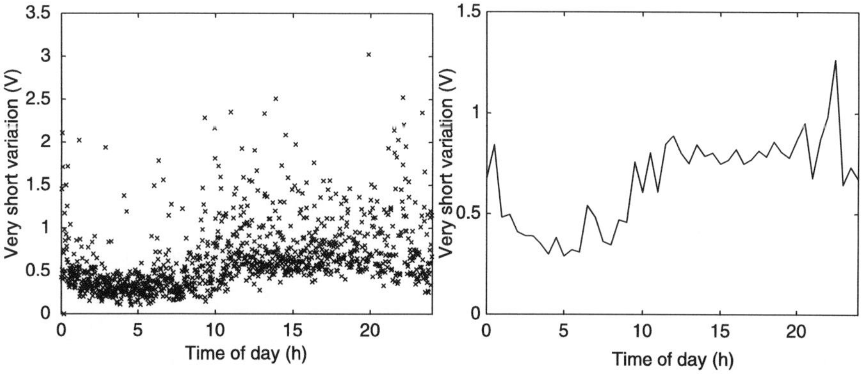


**Figure 5.44** Very short time variations as a function of time of day (left) and half-hour averages of VSVs (right).

even more compressed leading to a further enhancement of the daily pattern. The right-hand plot is obtained by taking the average of all values within each half-hour interval of the day (i.e., of the left-hand plot). The peak around 10:30 P.M. is due to transformer tap-changer operations when the load starts to drop in the late evening.

### 5.3.5 Harmonic Distortion

***5.3.5.1 Basic Measurement Window: 3 s*** The same measurement discussed before also resulted in the spectrum of the voltage over each 3-s interval. The 3-s values for the third and fifth harmonic during a 24-h (midnight to midnight) interval are shown in Figure 5.45. Both the third and the fifth harmonics reach their maximum in the evening.

It is immediately noticeable from this figure that the harmonic distortion shows a very large variation at short time scales. These fast variations are due in part to the limited resolution of the monitor, about 0.1% in harmonic voltage, but also in part to actual changes in the harmonic voltage. The harmonic distortion is due to the combined effect of large and small, local and remote, distorting loads. A detailed study of these short-term variations in harmonic distortion is needed to get an understanding of the underlying phenomena. An approach similar to the VSVs for rms voltage may be applied.

***5.3.5.2 Time Aggregation: 1-min and 10-min Values*** The time aggregation algorithm used for the rms voltages has been applied to the 3-s harmonic voltages resulting in 1- and 10-min values. The aggregated values are shown in Figure 5.46 for the data in Figure 5.45. The 1-min values still contain some fast fluctuations, but the 10-min values show a rather smooth behavior.

The 10-min values of the third- and fifth-harmonic voltages are plotted in Figure 5.47 for the whole one-week measurement period. Despite the gap in the

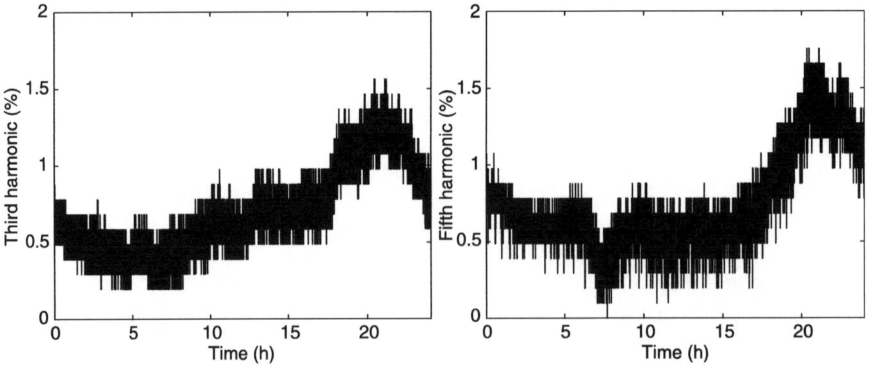


**Figure 5.45** Harmonic distortion: 3-s third- (left) and fifth- (right) harmonic voltages versus time.

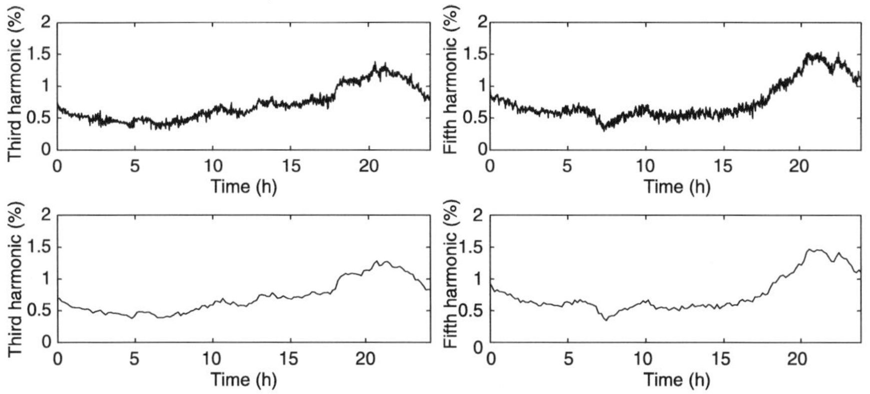


**Figure 5.46** The 1-min (top) and 10-min (bottom) third- (left) and fifth- (right) harmonic voltages versus time.

measurements, it is clear that both harmonic components show a daily pattern with their peak in the evening. We will come back to daily pattern in harmonic distortion in this chapter.

***5.3.5.3 Spectra*** The 3-s and 10-min values of each harmonic have been used to obtain percentile spectra in the same way as for the arc-furnace installation before. For each time scale three percentile spectra have been obtained: 90, 95, and 99%. The results are shown in Figure 5.48. The 3-s spectra show only slightly higher values than the 10-min spectra. The high spikes in the 3-s values occur less than

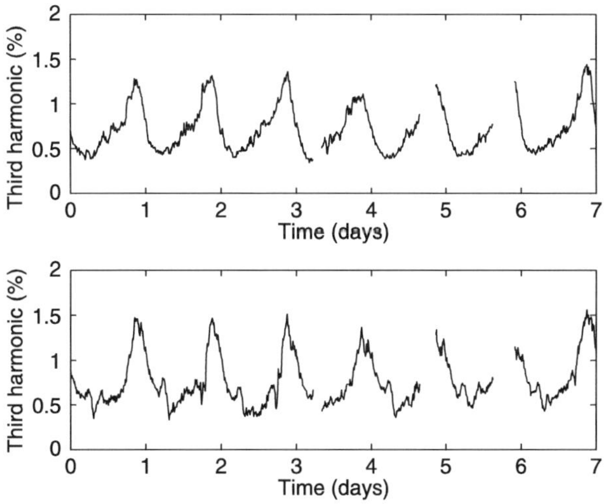


**Figure 5.47** Third- (top) and fifth- (bottom) harmonic voltages during one-week period.

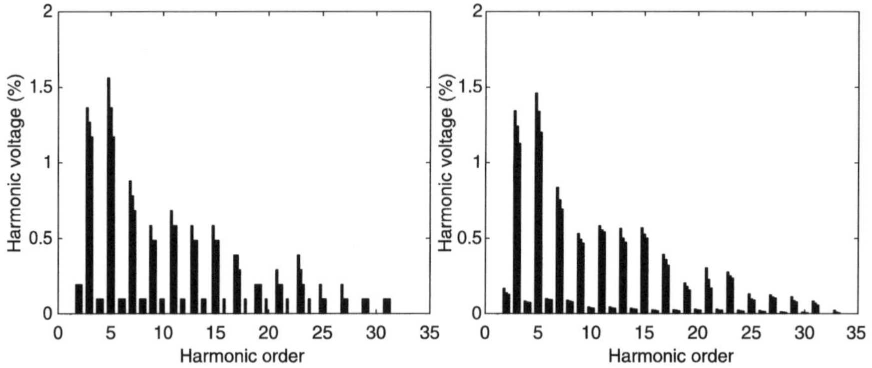


**Figure 5.48** The 3-s (left) and 10-min (right) percentile voltage spectra: (left to right) 99, 95, and 90%.

1% of time. The third and fifth harmonics dominate followed by the seventh harmonic. Also interesting is the relatively high values of the 9th through 15th harmonics. After the 15th harmonic the distortion decays quickly. Even-harmonic distortion is small.

***5.3.5.4 Daily Patterns*** The harmonic distortion has been measured during a three-month period (May 30 through September 4, 2004). The average harmonic level was recorded for every 1-min interval. The 10-min values were obtained by taking the rms over ten 1-min values. The 10-min third-harmonic voltages are shown in Figure 5.49. The periodicity of the distortion is visible in the left-hand figure, the daily pattern in the right-hand figure.

To further enhance the daily pattern, the average of the 10-min values over each half-hour time of the day has been calculated. The results are shown in Figure 5.50

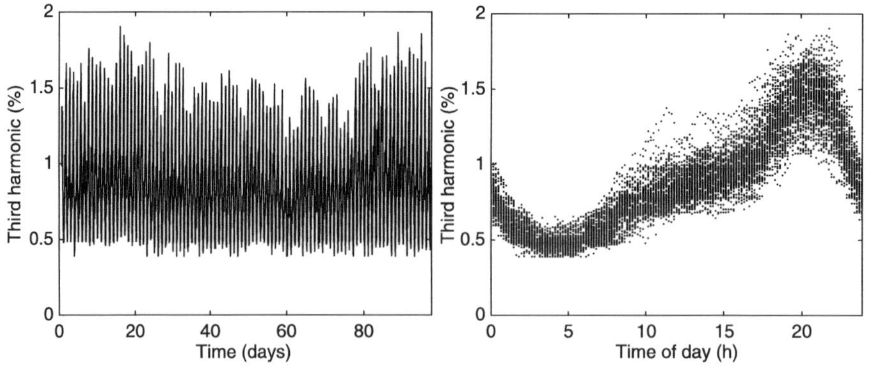


**Figure 5.49** Third-harmonic voltage for three-month period as function of time (left) and time of day (right).

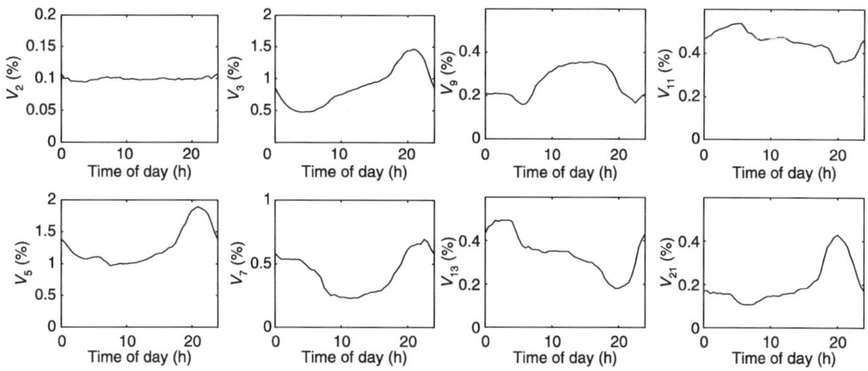


**Figure 5.50** Average daily pattern for harmonics 2, 3, 5, 7 (left) and 9, 11, 13, and 21 (right); note the differences in vertical scale.

for harmonics 2, 3, 5, 7, 9, 11, 13, and 21. The daily pattern is different for different harmonics. Comparing these with the daily load patterns could help in identifying the source or the main contribution to the different harmonic components. Both third and fifth harmonics have a peak in the evening, the so-called television peak that can be seen in harmonic distortion pattern at most locations around the world. The third harmonic is a "local harmonic": It does not propagate through the distribution transformer. The presence of a high third-harmonic voltage is in almost all cases due to load supplied from the local distribution transformer. The fifth harmonic is however a "global harmonic": It spreads through transformers and is thus related to the amount of harmonic load over a much wider area. The measured television peak for the third and fifth harmonics occurred for most days at about the same time. However, for some days the third-harmonic distortion started to drop up to 1 h before the fifth harmonic, pointing to a difference in local viewing patterns compared to the average over a wider area.

The second harmonic is low and shows almost no daily pattern. The cause of the second-harmonic component is unknown, but most likely it is not due to equipment that is operated according to a daily pattern. Drawing conclusions from the measurement of the second harmonic is further complicated as its level was for many measurement windows (1 min duration) below the resolution limit of the monitor.

The seventh harmonic is high in the evening and at night but low during daytime. The latter may be due to three-phase equipment compensating the distortion by single-phase equipment during daytime. The daily pattern of the ninth harmonic is the opposite of that of the seventh harmonic. Interestingly, the ninth harmonic does not show a television peak but instead is high during working hours and low in the evening and night.

Harmonics 11 and 13 follow a similar pattern: They have their maximum in the early morning and their minimum in the evening. Note that the minimum in the 11th and 13th harmonics corresponds to the maximum in the 3rd and 5th harmonics. This is possibly due to the background distortion from a permanent (industrial) source

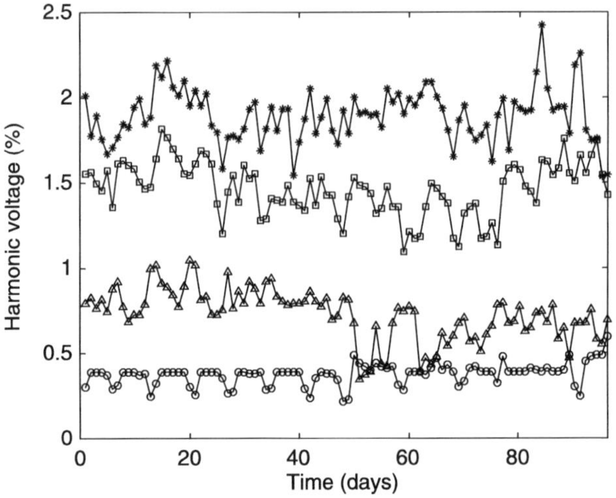


**Figure 5.51** Daily 95 percentile of harmonics 3 (square), 5 (star), 7 (triangle), and 9 (circle) during three-month period; June 1 = day 2; July 1 = day 32; August 1 = day 63; September 1 = day 94.

being compensated by domestic load. The 21st harmonic shows a very clear peak in the late evening, probably again due to local televisions. Note that the 21st harmonic is a local harmonic, like harmonics 3 and 9. The pronounced peak of harmonic 21 is probably related to a parallel resonance in the local low-voltage network.

***5.3.5.5 Variations over a Longer Period*** When the harmonic distortion (or any other characteristic) is to be plotted over a period longer than several weeks, a further level of time aggregation may be needed. For the measurement being discussed here, we calculated for every day (midnight to midnight) the 95 percentile of the 10-min harmonic voltages. The results for harmonics 3, 5, 7, and 9 are plotted in Figure 5.51. The distortion is rather constant and no clear general pattern is visible. A weekly pattern is visible in the ninth harmonic during the first two months after which it disappears. The measurement period has been too short to see if this weekly pattern reappears. The third harmonic appears to become somewhat higher in the middle of August (end of holiday for many in this part of the world), but the pattern is not clear.

## 5.4 SITE INDICES

The previous sections defined methods to obtain basic and aggregated characteristics as a function of time over a certain measurement interval. In this section a number of methods are discussed for quantifying the voltage or current quality at a given site by a small number of indices. Those indices are appropriately referred to as site indices. We will start with a general description of the methods used followed by a more detailed discussion for the most common voltage variations. The discussion will mainly be based on existing international and national standards and recommendations. Where further research or development work is needed, this will be pointed out in the text and some suggestions for additional or different methods will be given.

### 5.4.1 General Overview

Site indices are obtained as a statistic from the basic or aggregated characteristics recorded at one site over a certain measurement interval. Typical periods over which site indices are calculated are one day, one week, and one year. Some examples of site indices were already introduced in the examples in Section 5.3, even though the term *site indices* was not used there.

The aim of site indices is to obtain a representative value to quantify the performance of a specific site concerning one or more power quality disturbances. A whole range of methods are being used or proposed and even more have been or are under discussion. The derivation of site indices falls typically into one of the three following methods:

- The site index is obtained as the average or median value of the aggregated characteristic values.
- The site index is obtained as the maximum value of the aggregated characteristic values.
- The site index is obtained as a high percentile (e.g., 95%) of the aggregated characteristic values.

The latter value is preferred by many and in most international standards, although the average value remains commonly used as a site index. The advantage of using the average is that it is an easily understood index. Its use as a quantifier of power quality remains limited, however. Most power quality problems only occur when the disturbance level exceeds certain values. Thus, even if the disturbance is low most of the time, one or two periods of high distortion may cause severe problems. The maximum value is not deemed a good index as it may be due to one exceptional case or even to a measurement or interpretation error.

Where there is a growing consensus about the use of high percentiles of aggregated characteristics as a site index, there is no agreement at all about which percentile to use and which aggregation window to use. The 99 percentile of the 3-s characteristic may give a much higher value of an index than the 90 percentile of the 10-min characteristic. The choice of site index may be the difference between having to make a costly improvement of the network (or of the equipment for emission indices) or not. The choice of index has to be made in agreement between the partners in case of bilateral contract or set by a regulator or standard-setting organization. By the latter some strong recommendations have already been

**TABLE 5.4 Overview of Possible Site Indices**

| | Mean | 50% | 95% | 99% | 100% |
|---|---|---|---|---|---|
| 200 ms | | | | | |
| 3 s | | | $U_{vs95}$ | $U_{vs99}$ | |
| 1 min | | | | | |
| 10 min | | | $U_{sh95}$ | $U_{sh99}$ | |
| 2 h | | | $U_{lt95}$ | $U_{lt99}$ | |

made, with the emphasis being on the 95 percentile of the 10-min values during one week.

Some of the options for calculating site indices are summarized in Table 5.4. The notation in some of the cells is according to IEC 61000-3-6 and IEC 61000-3-7, where vs (very short) refers to the 3-s values, sh (short) to the 10-min values, and lt (long time) to the 2-h values. There are obviously more possibilities (99.9%, 90%, 10 s, 3 min), but the ones in the table are the ones most commonly used. For completeness the table should contain a third dimension: the period over which the statistic is calculated, which may be one day, one week, or one year. For longer or permanent measurement campaigns a multistage approach may be used. A *hypothetical* example to explain the multistage concept is as follows:

- The 99 percentile of the 3-s values is calculated for each day.
- The highest daily index is used as a weekly index.
- The annual index is the average of the weekly indices.

Site indices can be used to observe trends in power quality, for example, an annual variation in weekly indices, or a slow increase in annual indices pointing to a deterioration of the power quality. In most cases the percentile values are calculated over a one-day or a one-week period and updated every day or week. The site index over a longer period is typically calculated as the highest of the one-day or one-week values. Site indices are thus another level of aggregation. The daily 95 percentile of the 10-min harmonic voltages, as visualized in Figure 5.51, is an example of a site index as a function of time.

Site indices may also be used to compare different sites and to calculate system indices. System indices will be discussed in Section 5.5. First we will discuss site indices for some power quality variations. For each variation we will first discuss the calculation of site indices according to international standards followed by a discussion of some of the issues from other standard documents and other publications.

### 5.4.2 Frequency Variations

The basic characteristic for frequency variations is, according to IEC 61000-4-30, the average frequency over a 10-s window. The 10-s values form the basis for the calculation of site indices. No further time aggregation is used. The average frequency will quickly become very close to the nominal frequency when time aggregation is used.

An informative annex with IEC 61000-4-30 proposes a number of site indices for contractual applications. A one-week minimum measurement period is recommended as well as lower and upper limits to be agreed upon by the contract partners. The following indices are proposed:

- The number or percent of values exceeding the limits
- The worst case values
- A high percentile (e.g., 95%)
- The number of consecutive values that exceed the limit
- The integration of the frequency deviation

In the European voltage quality standard EN 50160 [106] objective values are given for the voltage quality. The objectives given in that document holds for all sites. An objective value requires the definition of a site and/or system index. The following site indices are used in EN 50160:

- The 99.5% range of the 10-s values during one year
- The maximum frequency
- The minimum frequency

No specific time interval is given for the measurement of maximum and minimum frequency, as these objectives hold for any period. To assess compliance with EN 50160 concerning frequency, a one-year minimum measurement period is needed and a total of $365 \times 24 \times 360 = 3{,}153{,}600$ values are needed. Fortunately the frequency is very similar everywhere in an interconnected system so that one monitor would cover a very wide area. In fact four monitors would cover the frequency variations for most of Europe and the same holds for North America. For the other power quality variations lots of measurement locations are needed, but EN 50160 requires only a one-week measurement for those.

As the frequency varies around a nominal value, the 99.5% interval is not uniquely defined. From an objective viewpoint this is not a concern as the requirement that 99.5% of values should be within a certain range is a unique requirement. However, the corresponding site index is not uniquely defined: The interval excluding the 0.5% highest values fits the requirement equally well as the interval excluding the 0.5% lowest values. A "balanced interval" would appear most appropriate, with the lower limit being the 0.25 percentile and the upper limit the 99.75 percentile. Alternatively, an interval centered around the nominal frequency could be used.

To separately study underfrequency and overfrequency, the (frequency) underdeviation and overdeviation have been introduced. As site indices a high percentile of underdeviation and overdeviation may be used. Note that the 99.75 percentiles of under- and overdeviation correspond to the limits of the 99.5% interval of the frequency as defined in the previous paragraph. Underdeviation and overdeviation may be of use when aggregating frequency variation into longer intervals. The average values or nonzero average values of under- and overdeviation give an impression of the range of frequency values within a certain time interval. This would allow tracking longer term variations in the amount of frequency variations, for example, the difference between day and night or between summer and winter. Another example where longer term variations in frequency need to be quantified is when assessing the impact of increasing penetration of distributed generation on the system. The first application of this will be needed for small insular systems with large amount of wind power. Wind power does not normally contribute to frequency control, and increasing penetration of wind power is likely to lead to an increased level of frequency variations.

In the South African standard NRS048.2 [225] different indices and limits are introduced for *grid networks* and *island networks*. For grid networks the 99.5 percentile of the frequency deviation should be used, for island networks the 95 percentile. The corresponding limits are 2% and 2.5% of the nominal frequency, respectively. For all network types, the maximum frequency is used as a second index, with limits equal to 2.5% for grid networks and 5% for island networks.

The whole exercise of calculating frequency indices remains rather theoretical in most countries. Due to the nature of large interconnected systems the frequency variations are normally very small and they do not cause any problems. Even for small systems (as on remote islands) or in large systems with a consistent shortage of power, the consequences of frequency variations appear to be limited to clocks indicating the wrong time. This again calls for the use of the integrated frequency deviation as an index.

As mentioned before, the level of frequency variations may be used as a measure of the security of the system. A larger range in frequency variations will typically point to a lower system security. Additional indices may have to be developed to quantify the relation between frequency variations and system security. An increasing penetration of distributed generation without frequency control capabilities may call for such indices.

Site indices have been calculated for the one-week measurement presented in Section 5.3.2. A 3-s basic measurement window was used as well as a 3-min aggregated window. The 3-min values were obtained as the average over sixty 3-s values. Site indices obtained for this one week are summarized in Table 5.5. As the frequency is the same through the interconnected system, these values can also be used as a system index for the whole interconnected system.

Note that the average 3-s nonzero overdeviation is higher than the 3-min aggregated value. For the nonzero underdeviation the 3-s value is lower. For the other averages the 3-min and 3-s values are the same. The fact that the 3-min and 3-s averages are the same is easy to understand when one realizes that aggregation is done by taking the (arithmetic) average. However, that does not immediately explain the difference for the nonzero overdeviation and underdeviation. To understand this, we have to consider the way in which the averages are taken. The average over the 3-s values considers each nonzero value as equal. The 3-min average is

**TABLE 5.5 Site Indices Obtained from Frequency Measurement over One-Week Period**

| Statistic | 3-s Value | 3-min Value |
|---|---|---|
| Average nonzero overdeviation | +32.3 mHz | +23.5 mHz |
| Average overdeviation | +15.5 mHz | +15.5 mHz |
| Average frequency | 50.0005 Hz | 50.0005 Hz |
| Median frequency | 49.999 Hz | 49.998 Hz |
| Average underdeviation | −15.0 mHz | −15.0 mHz |
| Average nonzero underdeviation | −29.4 mHz | −22.5 mHz |
| 100% range | 49.767–50.186 Hz | 49.873–50.151 Hz |
| 99% range | 49.906–50.104 Hz | 49.918–50.095 Hz |
| 95% range | 49.928–50.079 Hz | 49.935–50.071 Hz |

taken over the nonzero values within a 3-min interval. If there are only a few nonzero values, the resulting 3-min value has the same weight as when all values are nonzero. But in the latter case the values are typically higher. Thus the lower 3-s values get a higher weight than the higher values, resulting in a reduction of the overall average.

Another observation from the statistics listed above is that the range of 3-min values is less than the range in 3-s values. This holds especially for the 99 and 100% intervals. One of the effects of aggregation is that the extreme values are removed. Extreme values of variations are normally not considered as relevant. As we will see later, the 95% range is often used to quantify the voltage quality at a certain location.

### 5.4.3 Voltage Variations

According to IEC 61000-4-30 [158, page 25], the basic characteristic for voltage variations should be "the rms value of the voltage magnitude over a 10-cycle time interval for 50 Hz power system or 12-cycle time interval for 60 Hz power system" (referred to as the 200-ms value even though the interval is not exactly 200 ms). Aggregation should be done in 3-s values (more precisely 150 or 180 cycles), 10-min values, and 2-h values according to the aggregation algorithm discussed before. The 10-min value is the one most commonly used to quantify voltage variations. As we saw in Figure 5.40, this leads to a loss of information on short-time variations. The daily variations remain, however. Fast fluctuations are covered by the flickermeter standard, but no standard document covers the range between 1 s and 10 min. The VSVs introduced before can be used as a basis for site indices covering this range.

There is a growing feeling that a 10-min window to quantify the supply performance is too long. An important argument is that equipment may be damaged by shorter duration over- or undervoltages than 10 min. Instead of using the 3-s values, an additional

1-min aggregation interval could be introduced. The measurements in Section 5.3.3 show that a lot of information is lost when aggregating from 1 to 10 min. These kinds of studies will of course have to be repeated for many more sites before a general conclusion can be drawn.

There is another reason for introducing an additional aggregation window of 1 min. This is related to the means of voltage control used in transmission and distribution systems. A distinction of the different methods based on a time scale would result in the following [86]:

- The primary control operates at time scales up to a few seconds. The primary control includes the voltage control (excitation control) of generator units and synchronous condensers. Also power-electronic devices such as SVCs could be classified here.
- The secondary control operates at times scales between a few seconds and about 1 min. This includes transformer tap-changer operation, automatic switching of capacitor banks, and modification of voltage-control set points.
- The tertiary control operates at times scales of 1 min and longer. This includes all other measures, such as manual capacitor bank switching and optimization to reduce transmission system losses and increase security against voltage collapse.

Using an additional time scale (window length) of 1 min would give a better mapping to the control means. The performance of primary, secondary, and tertiary control would be covered by the 3-s, 1-min, and 10-min indices, respectively.

The 1-min time scale also comes back in a number of power quality documents and publications to distinguish between dips and undervoltages as well as between short and long interruptions. The original choice of the 1-min value was made to distinguish between automatic means of voltage recovery (protection and automatic reclosing) versus manual means. This corresponds to the above distinction between secondary and tertiary voltage control.

There is a third and more fundamental argument for an additional time scale. Looking at the ratios between the time intervals in IEC 61000-4-30, we find the following values:

- 200 ms to 3 s: 15 times
- 3 s to 10 min: 200 times
- 10 min to 2 h: 12 times

The second step is much bigger than the other ones and is thus more likely to cover different phenomena: This may refer to different control means as in the above discussion but also to different mechanisms for equipment damage due to voltage variations. Equally spreading the steps would require and additional interval at $\sqrt{3\,\mathrm{s} \times 10\,\mathrm{min}} = 42.5\,\mathrm{s}$, but 1 min is close enough. Choosing 1 min results in the following ratios between the intervals:

- 200 ms to 3 s: 15 times
- 3 s to 1 min: 20 times
- 1 min to 10 min: 10 times
- 10 min to 2 h: 12 times

An informative annex with IEC 61000-4-30 proposes the following indices for contractual applications, to be calculated from the 10-min characteristics:

- The number or percentage of values exceeding the limits
- The maximum and minimum values
- The 95 percentile values or any other high percentile
- The number of consecutive values exceeding the limits

In EN 50160 the following site indices are used:

- The 95% range of the 10-min rms voltages during one week
- The maximum 10-min rms voltage
- The minimum 10-min rms voltage

As the rms voltage varies around its nominal value, there may be undervoltages as well as overvoltages. The consequences of overvoltages are different from those of undervoltages so it may be appropriate to consider them separately. The resulting ambiguity in the choice of the 95% window was already mentioned when discussing frequency variations in the previous section. For this purpose the (voltage) underdeviation and overdeviation have been defined. The basic characteristic is the 200-ms value; aggregated characteristics are the 3-s, 10-min, and 2-h values. No information is available on the use of under- and overdeviation for the calculation of site indices. However, in processing rms values, one may decide to use 95% or 97.5% and extreme 10-min values. As discussed before, there remains some uncertainty over the aggregation of over- and underdeviation: should the average taken be over all values or only over the nonzero values. As shown for frequency variations, the differences are significant.

The South African standard NRS048-2 [225] considers all three phases in the calculation of site indices. For solidly grounded networks the three phase-to-neutral voltages are used, for other networks the phase-to-phase voltages. The 10-min rms voltages are obtained for each phase during one week. Next the 95 percentile values of the voltage deviations are obtained for each phase. The highest of the three deviations is the site index.

Next to the 95 percentile, the maximum voltage is also used as an index. The corresponding limits for the maximum deviation are wider than for the 95 percentile. For higher voltage levels two upper limits for the maximum voltage are in

**TABLE 5.6 Limits for Voltage Variations in South Africa**

| | | 100% Limit | |
|---|---|---|---|
| | 95% Limit | Declared Voltage | Nominal Voltage |
| Below 500 V | 10% | 15% | — |
| Other LV | 5% | 10% | — |
| 11–275 kV | 5% | 10% | +10% |
| 400 kV | 5% | 10% | +5% |

use: one related to the declared voltage and one related to the nominal voltage. The lowest upper limit should not be exceeded. For the lower limit only the one relative to the declared voltage is used. The various limits are (somewhat simplified) summarized in Table 5.6.

For measurements of duration longer than one week, a sliding weekly index is calculated: index over days 1 through 7, index over days 2 through 8, index over days 3 through 9, and so on.

In [206] a multistep procedure as used by the French transmission operator is explained. The method results in one value to describe voltage variations which is less than 100% in all cases when no limit is exceeded. The indices are obtained by using the voltage deviation calculated from the 10-min rms voltage $U_{rms}$ in each of the three phases through the expression

$$\Delta U_{rms} = \frac{U_{rms} - U_0}{U_0} \tag{5.23}$$

with $U_0$ the declared or nominal voltage according to the contract. Next the highest and lowest values of the voltage deviation over the three phases are obtained over a one-week period: $\Delta_{max}$ and $\Delta_{min}$. If the maximum value is negative (thus all 10-min rms values are below the contractual level), $\Delta_{max}$ is set to zero. In the same way $\Delta_{min}$ is set to zero when all rms values are above the contractual value. The extreme values are compared with the upper and lower limits $\lambda_{max}$ and $\lambda_{min}$ and the highest relative value is used as a site index:

$$I_T = \max\left\{\frac{\Delta_{max}}{\lambda_{max}}, \frac{\Delta_{min}}{\lambda_{min}}\right\} \tag{5.24}$$

There is some merit in calculating such a site index: The value immediately shows how far from the limit the voltage quality is. It also makes it easier to compare, even aggregate, different types of variations. Similar site indices have been proposed for other variations.

A proposal for voltage quality requirements in the Norwegian system [273] uses the EN 50160 measurement method (10-min rms voltages during one week) and limits, but the limits hold for 100% of time. The proposal is to no longer allow the rms voltage to exceed the limits during 5% of the time. The combination of a

10-min time window, only normal operation, and excluding the worst 5% of data was considered too much by the authors of [273]. An additional proposal is that the one-week average voltage should be between 94 and 106%.

### 5.4.4 Very Short Variations

Very short variations were introduced in Section 5.2.4; some additional measurements were presented in Section 5.3.3. As these are newly introduced characteristics, there is no experience yet in their use to quantify site performance.

The 10-min VSVs have been calculated for 10 sites in 6 different cities in 5 countries: domestic sites and hotels in 4 cities and offices in 2 cities. The probability distribution functions have been calculated for the 10-min VSVs of all 10 measurements. The results are plotted in Figure 5.52. A number of site indices are shown in Table 5.7.

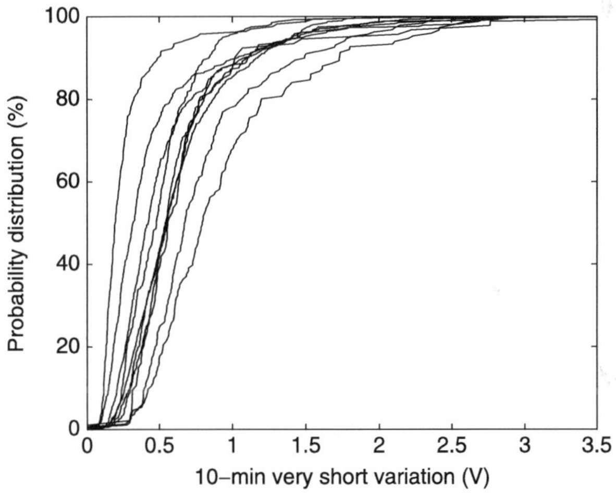


**Figure 5.52** Probability distribution function of 10-min VSV, for 10 measurement locations.

**TABLE 5.7 Site Indices for 10-min VSVs**

| Location | 50% (V) | 90% (V) | 95% (V) | 99% (V) |
|---|---|---|---|---|
| 1. Hotel | 0.67 | 1.49 | 1.82 | 2.35 |
| 2. Domestic | 0.80 | 1.73 | 2.20 | 2.59 |
| 3. Office | 0.54 | 1.11 | 1.41 | 2.50 |
| 4. Office | 0.19 | 0.49 | 0.76 | 1.50 |
| 5. Domestic | 0.54 | 1.17 | 1.45 | 2.09 |
| 6. Domestic | 0.42 | 1.08 | 1.45 | 2.63 |
| 7. Hotel | 0.56 | 0.92 | 1.01 | 1.92 |
| 8. Hotel | 0.55 | 1.10 | 1.48 | 2.70 |
| 9. Domestic | 0.32 | 1.02 | 1.41 | 1.75 |
| 10. Hotel | 0.48 | 0.84 | 0.93 | 1.32 |

**TABLE 5.8 Average and Standard Deviation of VSV Site Indices**

| Percentile | Average (V) | Standard Deviation (V) | 95% Value (V) |
|---|---|---|---|
| 50% | 0.51 | 0.17 | 0.97 |
| 90% | 1.10 | 0.34 | 1.66 |
| 95% | 1.39 | 0.42 | 2.09 |
| 99% | 2.13 | 0.50 | 2.95 |

There is a certain risk in attempting to draw general conclusions from a limited number of measurements, even though they are obtained at 10 significantly different locations. We do however take that risk at this moment while stressing the need to collect further measurements in the future.

The mean and standard deviation from the percentile values are given in Table 5.8. These values have next been used to calculate a 90% value over all sites under the assumption that the site indices form a normal distribution (the 95% value is for such a distribution equal to the average plus 1.65 times the standard deviation). In the absence of more measurements, these values may serve as a reference for measurements at other locations.

From the measurements presented in Sections 5.2.4 and 5.3.3 we concluded already that the 10-min VSVs consist of two "components":

- A continuous level up to about 1 V due to the combined effect of local load switching. The measurements at two of the domestic locations clearly showed a daily pattern.
- A number of spikes up to 3 or 4 V due to switching actions in the power system.

Two different performance indices are proposed here: the 50 percentile and the 95 percentile of the 10-min VSV voltage. The 50 percentile limits the continuous level, whereas the 95 percentile limits the number of severe spikes. Note that the setting of a 95 percentile level allows for at most seven 10-min values per day above the threshold.

### 5.4.5 Voltage Unbalance

The basic characteristic for voltage unbalance, as prescribed in IEC 61000-4-30, is the ratio between negative- and positive-sequence voltages calculated over a 200-ms interval. Aggregated values are obtained over the standard intervals of 3 s, 10 min, and 2 h.

The before-mentioned informative annex with IEC 61000-4-30 proposes the following site indices for contractual applications (over a one-week measurement period):

- The number of 10-min values that exceed the limit
- The highest 10-min value

- The number of 2-h values that exceed the limit
- The highest 2-h value
- The 95 percentile of the 10-min values
- The 95 percentile of the 2-h values

Only one site index is used in EN 50160: the 95 percentile of the 10-min values over one week. The issue of unbalance indices was taken up in two CIGRE working groups: working group 36.05 and working group 36.07 (later renamed in C4.07). The former working group recommended the following site indices in its 1992 report [254]:

- The maximum daily 95 percentile of the 3-s values
- The maximum 10-min value

The observation period should be at least a few days, including one weekend.

The issue was further addressed in CIGRE working group C4.07 [64]. The aim was to develop a procedure for processing voltage unbalance measurements along the same lines as in IEC 61000-3-6 and IEC 61000-3-7 for distortion and flicker. The final recommendation of the working group is to use the following site indices:

- The 95 percentile of the 3-s values over one day
- The 95 percentile of the 10-min values over one week
- The 99 percentile of the 3-s values over one day (in exceptional cases only)

Compared with IEC 61000-4-30 we can observe a shift to shorter time scales: from 10 min and 2 h to 3 s and 10 min. The work on defining indices and objectives for unbalance has been taken over by another CIGRE working group, CIGRE C4.1.03.

The South African standard NRS 048-2 [225] uses the 95 percentile of the 10-min values during one week as a site index. The minimum measurement period is one week. The final site index is the highest weekly value over the measurement period. As a second index the maximum 10-min value is used.

Flagged data (due to interruptions, sags, and swells) are to be removed before site statistics are calculated. The issue of fuse failure in one single phase is specifically mentioned in the standard document as it may cause rather high unbalance values. But these unbalance values are flagged and do not affect the site index. For a discussion on voltages during single phasing see [33, Section 3.6]. Note that single phasing is not removed from the reporting of supply performance, but it is reported as an interruption (i.e., the opening of an interrupting device), not as an unbalance.

The EN 50160 measurement windows (10 min) and limits are used as regulatory requirements in The Netherlands, but instead of the 95% values, the unbalance shall not exceed the limit for 99.5% of one week. The limit may thus be exceeded for five data points, that is, for less than 1 h.

### 5.4.6 Voltage Fluctuations and Flicker

The calculation of aggregated characteristics for flicker proceeds along completely different lines as for other variations such as voltage variations or distortion. However, once the aggregated characteristics have been calculated, the processing is very similar.

From the flickermeter standard IEC 61000-4-15 two aggregated characteristics result:

- The *short-term flicker severity* $P_{st}$ calculated over a 10-min window
- The *long-term flicker severity* $P_{lt}$ calculated from the short-term values over a 2-h window

The further processing of flicker leading to site indices is discussed in technical report IEC 61000-3-7, where the following site indices are recommended:

- The 99 percentile of the 10-min ($P_{st}$) values during one week
- The 99 percentile of the 2-h ($P_{lt}$) values during one week

(Note the difference between a "standard" and a "technical report" within IEC. A standard defines one method to be followed by all, whereas a technical report defines one or more methods which users are free to follow. Within IEEE standards the latter are referred to as "recommended practices" or "guide" where a recommended practice is stronger than a guide.)

As one week contains only eighty-four 2-h values, the 99 percentile of the 2-h values corresponds to the highest or the highest but one (depending on the rounding of $0.99 \times 84$). As there could be a large difference between the highest and the highest-but-one value (especially when no flagging is used), the 99 percentile is not a good choice. A better alternative would be to use either the maximum 2-h value if flagging is used or the 95 percentile if no flagging is used.

Even if the flagging concept is applied there may be cases in which short-duration events not commonly considered as voltage fluctuations lead to a high flicker severity. Examples are voltage transients for which a flagging concept has not been defined. This brings us again to the basic aim of the flicker severity indices. If they are aimed at quantifying the perceived light flicker, all data (also the flagged ones) should be considered. Voltage dips cause real light flicker after all. But if the flicker severity is aimed only at quantifying the voltage fluctuations, flagged data should not be included in the calculation of site indices because voltage dips are not considered voltage fluctuations. More research is needed on how events such as dips and transients affect the flicker indices as well as a discussion on what should be included in the flicker indices.

There is a concern among network operators of "double counting" where a high voltage-dip frequency will also lead to a high flicker index. With the possibility of heavy fines and other economic incentives being introduced in the future for exceeding objectives, the concern is understandable.

An informative annex with IEC 61000-4-30 suggests the following site indices for flicker:

- The number of 10-min values that exceed the limit during the measurement period
- The number of 2-h values that exceed the limit during the measurement period
- The 99 percentile of the 10-min values during one week
- The 95 percentile of the 2-h values during one week

The measurement period should again be a multiple of one week.

The European voltage quality standard EN 50160 uses the following site index:

- The 95 percentile of the 2-h value during one week

An overview of site indices in international standards is given in Table 5.9. Most national standards also use 95 and 99 percentiles of $P_{st}$ and $P_{lt}$, although the methods are not always clearly defined in the documents. Fortunately more and more national standards include the methods defined in international standards or simply refer to the relevant international standard. This not only allows for interchange of equipment and comparison of results but also leads to better defined national standards.

The lowest values are obtained for $P_{lt95}$ (the 95 percentile of the 2-h values). The voltage characteristics according to EN 50160 are based on this index because they have to hold for all sites (see the discussion on voltage characteristics, compatibility levels, etc., in Section 1.9). The highest values are obtained for $P_{st99}$. For the three values used in international standards, the following ranking holds:

$$P_{lt95} \leq P_{lt99} \leq P_{st99} \tag{5.25}$$

The use of all three indices could give a reasonable impression of the voltage fluctuations at a given site. If $P_{lt95}$ is much lower than $P_{lt99}$, this points to voltage fluctuations that are only present during a limited period, up to 5% of the week (not more than four 2-h intervals). If $P_{st99}$ is much higher than $P_{lt99}$, the voltage fluctuations come in short-duration bursts (several minutes duration) that are spread throughout the week.

Flicker concerns, however, are linked to a limited number of sources, with the ac arc furnace being the dominant one. As the operation of different arc furnaces is rather similar, one may expect a relation between the indices in Table 5.9. Such a

**TABLE 5.9 Overview of Flicker Site Indices in International Standards**

| | 95 Percentile | 99 Percentile |
|---|---|---|
| $P_{st}$, 10 min | — | IEC 61000-3-7, IEC 61000-4-30 |
| $P_{lt}$, 2 h | IEC 61000-4-30, EN 50160 | IEC 61000-3-7 |

relation would justify the use of just one site index. A comparison of indices at 37 sites from LV to extra high voltage (EHV), all with arc furnaces as the main flicker source, was made in [64]. Those data have been used to obtain the relations between 12 different site indices, as presented in Table 5.10, for example, $P_{st99} = 1.43 \times P_{lt90}$.

The CIGRE working group C4-07 recommends the use of the following site indices [64]:

- The 95 percentile of the 10-min values; to be compared with the planning level
- The 99 percentile of the 10-min values; to be compared with a value equal to 1 to 1.5 times the planning level
- The 95 percentile of the 2-h values; to be compared with the voltage characteristics

The South African standard NRS 048-2 [225] uses the 95 percentile and maximum value of $P_{st}$ and $P_{lt}$ obtained over one week as site indices. The highest value of the three phases should be used. Flagged data should not be considered when calculating the indices.

### 5.4.7 Voltage Distortion

Voltage distortion is quantified by a much larger number of indices than the other variations. One may of course stick to the THD only, but in most surveys and standard documents indices are used for each harmonic, sometimes even for interharmonics. The THD alone would give only limited information about the distortion.

The basic characteristics according to IEC 61000-4-30 are the harmonic and interharmonic subgroups obtained over a 10/12-cycle window. The subgroups are defined in such a way that there is no redundancy in energy between harmonics and interharmonics. The basic characteristics are to be aggregated into 3-s, 10-min, and 2-h values (aggregated characteristics). The aggregated characteristics are next to be used to calculate site indices.

In an informative annex with IEC 61000-4-30, the following site indices are suggested for contractual applications:

- The number of 3-s values that exceed the limit during the measurement period
- The number of 10-min values that exceed the limit during the measurement period
- The maximum 3-s value during the measurement period
- The maximum 10-min value during the measurement period
- The 95 percentile of the 3-s values during one week
- The 95 percentile of the 10-min values during one week

**TABLE 5.10 Relation Between Different Flicker Indices**

| | | | | | | | | | | | | |
|---|---|---|---|---|---|---|---|---|---|---|---|---|
| $P_{st50}$ | 1 | 0.62 | 0.57 | 0.52 | 0.50 | 0.39 | 0.68 | 0.84 | 0.71 | 0.65 | 0.62 | 0.58 |
| $P_{st90}$ | 1.61 | 1 | 0.92 | 0.84 | 0.80 | 0.62 | 1.10 | 1.35 | 1.14 | 1.04 | 1.00 | 0.94 |
| $P_{st95}$ | 1.75 | 1.09 | 1 | 0.92 | 0.87 | 0.68 | 1.19 | 1.47 | 1.24 | 1.13 | 1.08 | 1.03 |
| $P_{st98}$ | 1.91 | 1.18 | 1.09 | 1 | 0.95 | 0.74 | 1.30 | 1.66 | 1.35 | 1.24 | 1.18 | 1.12 |
| $P_{st99}$ | 2.02 | 1.25 | 1.15 | 1.06 | 1 | 0.78 | 1.37 | 1.69 | 1.43 | 1.30 | 1.24 | 1.18 |
| $P_{st100}$ | 2.60 | 1.61 | 1.48 | 1.36 | 1.29 | 1 | 1.76 | 2.17 | 1.84 | 1.68 | 1.60 | 1.52 |
| $P_{lt95}$ | 1.47 | 0.91 | 0.84 | 0.77 | 0.73 | 0.57 | 1 | 1.23 | 1.04 | 0.95 | 0.91 | 0.86 |
| $P_{lt50}$ | 1.19 | 0.74 | 0.68 | 0.62 | 0.59 | 0.46 | 0.81 | 1 | 0.84 | 0.77 | 0.74 | 0.70 |
| $P_{lt90}$ | 1.41 | 0.88 | 0.81 | 0.74 | 0.70 | 0.55 | 0.96 | 1.19 | 1 | 0.91 | 0.87 | 0.83 |
| $P_{lt98}$ | 1.55 | 0.96 | 0.88 | 0.81 | 0.77 | 0.60 | 1.05 | 1.30 | 1.09 | 1 | 0.95 | 0.91 |
| $P_{lt99}$ | 1.62 | 1.00 | 0.92 | 0.85 | 0.80 | 0.62 | 1.10 | 1.36 | 1.15 | 1.05 | 1 | 0.95 |
| $P_{lt100}$ | 1.71 | 1.06 | 0.98 | 0.89 | 0.85 | 0.66 | 1.16 | 1.43 | 1.21 | 1.10 | 1.05 | 1 |
| × | $P_{st50}$ | $P_{st90}$ | $P_{st95}$ | $P_{st98}$ | $P_{st99}$ | $P_{st100}$ | $P_{lt95}$ | $P_{lt50}$ | $P_{lt90}$ | $P_{lt98}$ | $P_{lt99}$ | $P_{lt100}$ |

The contract should define limits for each harmonic or for groups of harmonics. As we discussed in Section 3.3.1, one may define THD values for different combinations of harmonics. A hypothetical example is as follows.

- Harmonic 3
- Harmonics 5 and 7
- Even harmonics 2 through 10
- Triplen harmonics 9, 15, 21, 27, 33, and 39
- Odd harmonics 11, 13, 17, 19, 23, 25, 29, 31, 35, and 37
- Even harmonics 12 through 40
- Interharmonic 0.5 (subharmonics)
- Interharmonics 1.5 and 2.5
- Interharmonics 3.5 through 39.5

Technical report IEC 61000-3-6 recommends the use of the following site indices for comparison with planning levels:

- The 95 percentile of the 3-s values over one day
- The maximum 3-s value over one week
- The maximum 10-min value over one week

European standard EN 50160 uses the following site index as a voltage characteristic:

- The 95 percentile of the 10-min values over one week

An overview of indices used or recommended in international standards is given in Table 5.11.

The CIGRE working group C4-07 recommends the use of the following indices for harmonics [64].

- The 95 percentile of the 3-s values over one day
- The 99 percentile of the 3-s values over one week
- The 95 percentile of the 10-min values over one week

**TABLE 5.11 Overview of Distortion Site Indices in International Standards**

| | 95 Percentile | 99 Percentile | Maximum Value |
|---|---|---|---|
| 3 s | IEC 61000-4-30, IEC 61000-3-6, CIGRE C4-07 | CIGRE C4-07 | IEC 61000-4-30, IEC 61000-3-6 |
| 10 min | IEC 61000-4-30, EN 50160, CIGRE C4-07 | — | IEC 61000-4-30, IEC 61000-3-6 |

The first two indices are used to verify compliance with planning levels, whereas the third one is to be used for voltage characteristics.

The South African standard NRS048-2 [225] uses the following site indices for voltage distortion (the highest of the values for the three phases should be used):

- The 95 percentile of the 10-min values during one week
- The maximum of the 10-min values during one week
- The 99 percentile of the 3-s values during one day
- The maximum of the 3-s values during one day

The French transmission operator uses the maximum 10-min values over one week to calculate site indices for use in their power quality information system [206]. For harmonics 2 through 25 and for the THD the maximum weekly value (highest of the values in the three phases) is divided by the limit. The maximum of the 25 resulting values is used as the harmonic site index (the term *global indicator relative to harmonic distortion* is used in [206]). The use of relative values makes it easier to interpret the indices. It also makes it easier to aggregate the indices for individual harmonics into one value and to merge them with indices for other disturbances. The disadvantage is that comparisons between voltage levels and between customers with different limits are difficult. Additional difficulties in interpretation may arise when the limits are changed in the future.

The Spanish approach is discussed in [5]. The 95 percentile of the 10-min THD values is obtained over each one-week period. An annual index is calculated as the average of the weekly indices. The aim is to obtain these indices for 10% of the MV substations and for all customers of the transmission network (i.e., including the distribution networks).

A proposal for voltage quality requirements in the Norwegian system [273] uses the EN 50160 measurement method (10-min rms voltages during one week) and limits, but the limits hold 100% of the time. An additional proposal is that the one-week average THD should be less than 5% (vs. 8% as the limit for the maximum value).

The EN 50160 limits and measurement methods are used for voltage quality regulation in The Netherlands, but the limit holds for 99.5% of the 10-min values [77].

### 5.4.8 Combined Indices

In all the standards and other examples discussed before, the different variations were treated as independent from each other. Each variation resulted in one or more site indices which are next compared with a limit value. However, there are some calls for further simplification of the process, resulting in an even smaller number of indices. In [206] a proposal is presented in which the different indices are merged into one "global index." For each of the disturbances voltage dips, voltage variations, distortion, and unbalance, an index is defined by comparing the actual values with a limit. The maximum of these four values is used as a global index covering "all" disturbances.

In [123, 124] the whole merging process is taken a few steps further: A scheme is proposed resulting in one *unified power quality index* quantifying the power quality for a whole network. Both variations and events are included in the overall single index. Two steps after the calculations of site indices are *normalization* and *consolidation*. Normalized site indices are obtained by dividing the value by its limit. Thus if the fifth-harmonic index (e.g., the 95 percentile of the 10-min values over one week) is 4.5% and the limit (e.g., as in EN 50160) is 6%, then the normalized index equals 0.75. Normalization does not reduce the number of indices, but it allows the next step: consolidation. If a disturbance is described by more than one index (waveform distortion being the best example), a consolidated normalized index is calculated as the maximum of all the individual normalized indices. From the consolidated normalized index one can immediately conclude if a limit is exceeded or not. The consolidated normalized indices for all variations and events are combined into one *site unified power quality index* or *site UPQI*.

## 5.5 SYSTEM INDICES

Site indices for a number of sites can be aggregated into system indices. A system index quantifies the quality of supply over a part of the power system: part of a town but also a whole country. Indices typically apply to only one voltage level or to a group of similar voltage levels. It rarely makes sense to merge site indices from completely different voltage levels. Below first we discuss the general issues in calculating system indices and then some specific issues for the most common power quality variations. Also some interesting examples from the literature are presented. The difference in approach between the different variations becomes small when discussing system indices.

### 5.5.1 General

The basic principle for calculating system indices was shown in Figure 5.1: some kind of statistical processing is applied to the site indices to obtain the system index. The difference between the different indices is in the statistical processing. We can distinguish between four methods of calculating system indices:

- The system index is obtained as the average or median value of the site indices.
- The system index is obtained as the 95 percentile (or any other high percentile such as 90 or 99%) of the site indices. Note that this requires a minimum number of sites. To calculate the 95 percentile at least 20 sites are needed, for the 99 percentile at least 100 sites. If measurements are only available at a small number of sites, it is best to choose either the average or the maximum of the site indices. However, generally speaking the 95 percentile is seen as the most appropriate index to quantify the power quality of a system.
- The system index is the maximum of the site indices.

- The system index is the number or fraction of sites where the site index exceeds a given limit value. Such a system index is suitable when comparing with a threshold, for example, as defined in a contract. The method is however not of much use when the limit value is so high that it is rarely exceeded. Such is the case for the limits in EN 50160 when applied to most European sites, with the possible exception of the fifth-harmonic distortion or the flicker indices.

With reference to electromagnetic compatibility standards two system indices are obvious choices: the 95 percentile of the 95% site indices corresponds to the compatibility level; the maximum of all 95% site indices corresponds to the voltage characteristic. These two system indices are therefore commonly used.

Two problems occur when calculating system indices, neither of which has been satisfactorily solved yet. The first problem is that not all sites are monitored. In most surveys the vast majority of sites are not monitored at all. Still the objective is to calculate an index that is representative for the whole system, not just for the monitored sites. The second problem is that not all sites are equal. Sites differ in number of customers, size of customers, and type of customers. Despite all these differences between sites, a representative statistical value is required for the whole system. A solution often mentioned is to apply weighting factors to the sites. The discussion is about which weighting factors to use. Three options appear available here:

- Weighting is based on the number of customers supplied from each site. The only place where weighting is already widely used for calculation site indices is in the calculation of reliability indices (see chapter 10). In the vast majority of cases the weighting is based on the number of customers. In this method, a small domestic customer and a large industrial customer would get the same weighting.
- An alternative which appears somewhat more popular for power quality indices is to base the weighting on the average load or the rated load of the site. This takes better into account the size of the customer. If the sites are all within one company and at similar voltage levels, the load is likely to be rather similar, so that one may decide to give all sites equal weighting. In practice this is often the method used.
- The most appropriate weighting would be one based on importance of the loads. This would require a detailed study of the load connected to each site, the kind of equipment used, and the (economic) consequences of equipment failure or early replacement. This is however not considered feasible and to our knowledge not used.

An additional issue related to weighting is to determine the amount of load (or the number of customers) supplied from a given site. If monitoring takes place on a substation bus, there is not much doubt about this. But if the monitoring takes place somewhere along a feeder, it may be more difficult. Should only the load downstream of the monitor location be considered or all load supplied from the same substation? This issue is similar to the first issue mentioned: not all sites are monitored.

It is commonly known that the disturbance level varies throughout a system. Some examples of system surveys will be given below which clearly show this. However, monitoring typically only covers a small fraction of all sites. Ideally one would want measurements for all sites, but the investment and operational costs become very high. There are two solutions to this problem, apart from installing a very large number of monitors. One may simply assume that the monitor locations represent a statistically relevant sample of the whole system. The system indices are calculated from the monitored sites as if these were all sites. Under the assumption that the sites represent a statistically relevant sample from a homogeneous population, the calculated system index is a true estimate of the actual system index (i.e., the hypothetical one obtained from all sites). The good news is that the accuracy of the estimation depends on the number of monitored sites, not on the total number of sites. The bad news is that the assumptions are rather dubious and very difficult to prove without installing more monitors. Even if the assumption is correct, most customers are at more remote locations (thus probably with higher disturbance levels) than the monitor locations.

The alternative is to use simulation techniques to estimate the disturbance levels at those sites for which no measurements are available. This will probably require the measurements of voltages as well as currents, but the extra investment required for this is limited. This will also allow an estimation of the disturbance levels closer to the customer than the location of the monitor.

More data and studies are needed on the distribution of site indices through the system. This should include studies on the impact of the measurement location on the resulting indices. Generally speaking the distortion level increases toward the load; exceptions are rotating machines in general and distributed-generation units in particular. The variation of site indices (*propagation* and *penetration* are terms used as well) throughout the system is an important field for further research. It is important thereby to distinguish between statistical variations among sites (e.g., all LV sides of distribution transformers) and propagation (e.g., MV vs. LV side of distribution transformers).

In the future, developments in telecommunication and data processing may make it possible to measure at a much larger number of locations. This could completely change the way in which site and system indices are being calculated.

### 5.5.2 Frequency Variations

The frequency is very similar for all locations within an interconnected system. Therefore there is no need to calculate system indices. An exception are regions consisting of a number of (geographical and electrical) islands. Examples that come to mind are the southern parts of Greece and the northern parts of Scotland.

### 5.5.3 Voltage Variations

Voltage variations are very much location dependent. The voltage may even drop a few more volts between the metering point and the location of sensitive equipment. This means that the monitor location should be defined very precisely before relevant system indices can be calculated. Using the definition of power quality as the interaction between the customer and the network, the metering point would be the most appropriate monitoring point. However, everybody agrees that this is not practical. Instead a typical monitoring location at LV is the secondary side of an MV/LV transformer. Some kind of correction must be made for the voltage drop between the monitoring point and the metering point. Note that this drop is highest during peak load, and thus when the voltage typically is low already. The choice of monitoring point is thus likely to affect the lowest rms voltage to a higher degree than the highest rms voltage.

As the rms voltage varies around a nominal value, both overvoltages and undervoltages will occur. This will require some additional choices to be made. One may decide to use separate indices for undervoltages and overvoltages or to only consider the absolute value of the deviation. Using separate indices will provide more information as equipment is affected in a different way by undervoltage than by overvoltage.

The European voltage quality standard EN 50160 gives values that should not be exceeded for any site. The site index for the worst site would be a corresponding system index. The fact that the rms voltage varies around a nominal value will again require some additional rules to be set when defining system indices. The site index is a range of voltage, so is the system index. The straightforward way is to consider upper and lower limits separately, and thus to take the 97.5% upper limit and the 2.5% lower limit of the site indices. The result may be referred to as the range of voltage not exceeded for 95% of the sites. This however assumes that no site exceeds both the upper and lower limits of the range for the system index.

Consider as an example the following situation: One site has a (95%) voltage range of 0.97 to 1.12 pu; for another site the range is 0.89 to 1.07 pu. There is no straightforward way to decide which site is worse. Using separate undervoltage and overvoltage indices will solve this problem but would require rewriting of several standard documents.

The Spanish distribution association UNESA developed a number of system indices (*zonal quality indices*) for use in distribution networks [5]. The site index for voltage variations is the number of 10-min periods that the rms voltage is outside the limits. System indices are obtained by weighting based on the power supplied by each monitored bus. Let $T_{\text{out}(i)}$ be the amount of time the rms voltage is outside of the limits at bus $i$ and $P_{a(i)}$ the annual load of bus $i$; then the *equivalent time outside of thresholds* is defined as

$$\text{TEFU} = \frac{\sum_i P_{a(i)} T_{\text{out}_{(i)}}}{\sum_i P_{a(i)}} \tag{5.26}$$

The associated limits are 7 or 5.6% of declared voltage depending on the monitor location.

### 5.5.4 Voltage Fluctuations

Voltage fluctuations show a much higher range of site indices than most other variations. The flicker may exceed 5 for locations close to an arc furnace and be less than 0.2 for most of the rest of the system. Apart from the average value, all other indices will result in a low value. An additional, more practical problem is that flicker measurements are normally only done at locations where a high flicker value can be expected. Measurements of flicker levels at randomly chosen locations are rare. Therefore CIGRE working group C4-07 recommends to not use system indices for voltage fluctuations [64].

The CIGRE report also gives measurement results of randomly located power quality monitors. From the measurements at 112 MV sites a probability distribution function has been obtained for the 95 and 99% value of $P_{st}$ during one week. The results are shown in Figure 5.53. The median values from the measurement campaign are 0.2 for the 95% value and 0.3 for the 99% value. The 90 percentiles are 0.6 and 0.95, respectively; the 95 percentiles are 0.8 and 1.2, respectively.

The spread of modern monitoring equipment being able to calculate flicker levels at no extra cost may provide more information on the distribution of flicker severity indices throughout the power system. This in turn may help in defining appropriate system indices for voltage fluctuations. The number or fraction of sites exceeding a limit could be an appropriate choice if a sufficient number of monitors are present and if the location of the monitors is not affected by the expected flicker level.

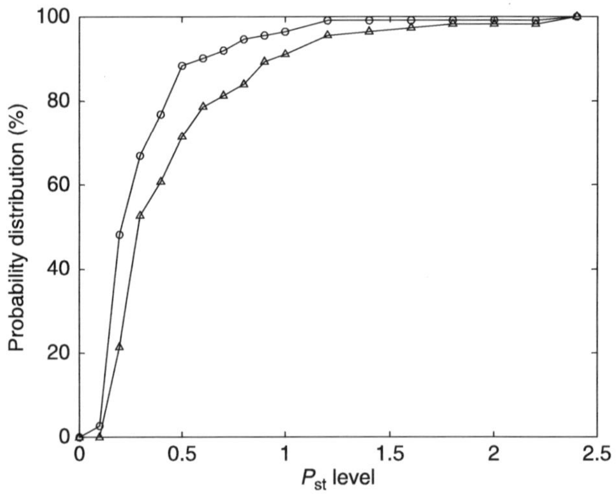


**Figure 5.53** Probability distribution function of 95% $P_{st}$ (circles) and 99% $P_{st}$ (triangles) for 122 MV sites.

Flicker measurements are treated as single-phase measurements in the literature and standard documents. The argument is that flicker concerns light bulbs and as such is a single-phase phenomenon. This is correct when the measurements are taken at the terminals of the light bulb, which is rarely the case. Most flicker measurements are done in the public supply, in many cases even at another voltage level. The measured flicker level thus does not directly represent the flicker as experienced from a light bulb. The concept of transfer coefficient has been introduced to correct for the difference in flicker level between the monitor location and the terminals of the light bulb. However, this concept considers a single-phase system. This assumption is not true as there may be Dy-connected transformers between the monitor location and the light bulb. Further, the transfer is likely to be different for positive- and negative- and zero-sequence components. A possible method is to apply the flickermeter algorithms separately to the symmetrical components and to apply different transfer coefficients to the different components. This requires some further development of the flickermeter algorithm. Note that this method does not solve the problem of how to merge the three phases. After applying the transfer coefficients the symmetrical components may be merged again, resulting in three single-phase values. Any of the phase aggregation methods discussed in Section 5.2.6 may be applied next.

### 5.5.5 Unbalance

The calculation of system indices for voltage unbalance is more straightforward than for voltage magnitude or flicker. The value of unbalance is ideally zero, and its value does not vary as much as the flicker levels between different sites.

System indices for unbalance may be defined as the 95 percentile or the maximum of the 95 or 99% site values. CIGRE working group C4-07 further recommends using the following index for comparison with voltage characteristics limits [64]:

- The percentage of sites exceeding the limit

The before-mentioned indices developed by the Spanish distribution association UNESA [5] also include an index for unbalance. The 95 percentile of the 10-min averages is calculated every week. The annual index $U_{a95(i)}$ for bus $i$ is obtained as the average of the weekly indices. The annual system index is the weighted average of the annual site indices:

$$\mathrm{GDE} = \frac{\sum_i P_{a(i)} U_{a95(i)}}{\sum_i P_{a(i)}} \qquad (5.27)$$

with $P_{a(i)}$ the power supplied by the busbar at which monitor $i$ is located.

### 5.5.6 Distortion

System indices for waveform distortion are very similar to those for unbalance. The difference is in the number of them. Where there is only one unbalance index, there are up to 80 distortion indices (e.g., harmonic and interharmonic subgroups). For a given site it makes sense to give the whole spectrum as different frequencies affect equipment in different ways. For a whole system it is not clear if this level of detail is needed. Using only the THD may be too much compression of information. A small number of indices need to be defined to quantify the distortion for different types of equipment: for example, weighted for capacitors, weighted for induction motors, or even harmonics only.

Most work currently concentrates on indices for voltage distortion. As waveform distortion is mainly a problem coming from the load (a "current quality" issue) current distortion indices need to be defined as well. Again there is no need to consider all 80 individual indices. One obvious choice is an index based on the $K$-factor to quantify the effect of the load on the transformers. Most of the information available on existing levels concerns (integer) harmonics. Information on interharmonics and noise is rare and would require a widespread measurement campaign to obtain.

Reference [258] discusses a number of possible site and system indices for harmonic distortion and applies them to the U.S. DPQ survey results. The basic characteristics are obtained from one-cycle snapshots of the waveform taken every 30 min. However, the procedure is general enough to allow 3-s and 10-min aggregated values according to IEC standards. For three-phase measurements the average over the three phases is used for further processing. The following site and system indices are defined:

- The 95 percentile of the individual samples over the whole monitoring period is used as a site index. The *system total harmonic distortion CP95* (STHD95) is the 95 percentile of the site indices. To calculate the 95 percentile, a weighting is applied based on the apparent power supplied by each monitored bus or feeder. In this way an estimation is made of the distortion experienced by 95% of the load.
- Alternatively the site index is the average of the individual samples over the whole monitoring period. The *system average total harmonic distortion* (SATHD) is the weighted average of the site indices. Weighting is done based on the total apparent power supplied from each site.
- A third index is an estimation of the amount of time the distortion exceeds a certain limit. The fraction of samples exceeding the limit is used as a site index. The *system average excessive total harmonic distortion ratio index for a limit x* ($\text{SAETHDRT}_x$) is the weighted average of the ratio over all monitored sites. Obviously the whole exercise only makes sense when a limit is used well above the average value but still sufficiently below the maximum value.

The indices have been applied to two years of monitoring data for 277 sites at distribution levels (mainly 10 to 30 kV). The values for THD are summarized in Table 5.12. These values may be used as reference values to compare individual site measurements. A site with a level close to or exceeding the 95% index could be classified as a high-distortion site. The wide range in values for the six different

**TABLE 5.12 Harmonic System Indices for U.S. Distribution Systems**

| | System Index | | |
|---|---|---|---|
| Site Index | Average | 95% | Maximum |
| Average | 1.6% | 2.6% | 4.6% |
| 95% | 2.2% | 4.0% | 6.4% |

system indices shows that it is very important to define system indices in a clear manner. Also when presenting measurement results it is very important to clearly describe the methods used to come to the final result.

In [304] the results are shown of a measurement campaign in The Netherlands at 269 sites over six years. For each site the yearly 95 percentile was calculated. Two system indices were calculated from the site indices for each year: the average over all sites and the maximum over all sites. Some of the results are summarized in Figure 5.54. None of the curves show an increasing or decreasing trend. Despite the increase in the amount of electronic equipment used, the harmonic distortion remains constant. The maximum value of the 7th harmonic even shows a decrease. The same holds for the 11th and 13th harmonics. This is probably due to improvements made at one or a small number of sites. Interpretation of the different system indices remains an area that requires further study. A wide-scale measurement campaign or detailed simulation studies are needed.

From the results presented in [304] two system spectra have been calculated. Both spectra are based on 95% averages over time for each site. The average over all sites and the maximum value over all sites were used to give an impression

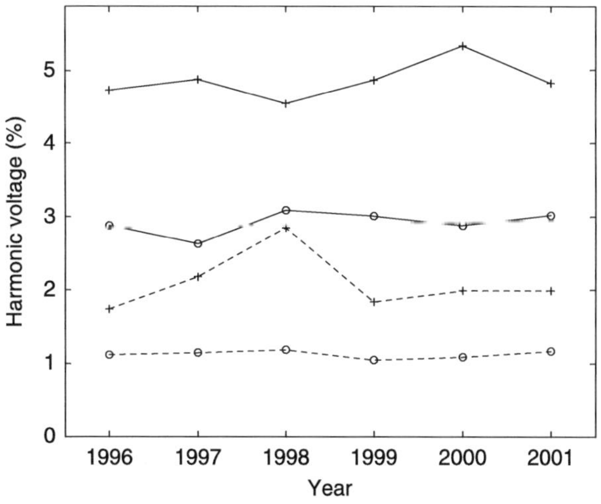


**Figure 5.54** Six-year trend in harmonic distortion: solid lines: 5th harmonic; dotted lines: 7th harmonic; plus sign: maximum site; circle: average site.

of the spread of values. The results are presented in Figure 5.55. For 5th and 7th harmonics, the difference between the average and maximum is less than a factor of 2, but especially for 11th and 13th harmonics the spread is very large. The average site is thus not a good indicator of the power quality of individual sites. The 95 percentile or maximum covers a much wider range of sites (and thus customers).

Reference [5] discusses the issue of weighting of individual sites to obtain system indices as used in Spain. The weighting factors are proportional to the amount of load supplied by the substation bus. For each site the 95 percentile of the THD is calculated for each week. The annual site index $\mathrm{THD}_{a95(i)}$ is the average of the weekly indices. The system index (the *equivalent total harmonic distortion*) is calculated from

$$\mathrm{THDE} = \frac{\sum_i P_{a(i)} \mathrm{THD}_{a95(i)}}{\sum_i P_{a(i)}} \tag{5.28}$$

with $P_{a(i)}$ the power supplied by the busbar at which monitor $i$ is located.

A survey of the harmonic distortion in 20 LV substations in France is presented in [24]. The basic characteristics are obtained over a 200-ms window. Aggregated values for THD, 3rd, 5th, 7th, 9th, 11th, and 13th harmonics are obtained by taking the arithmetic average over each 10-min window. From the 10-min values the following annual site indices are calculated:

- The median (50 percentile) of the 10-min values
- The 95 percentile of the 10-min values
- The maximum of the 10-min values

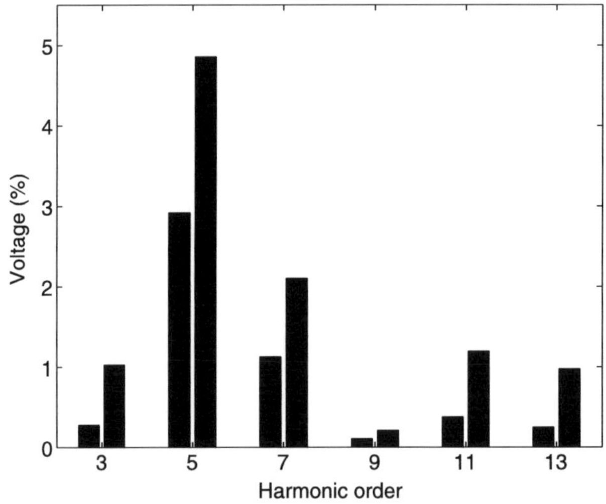


**Figure 5.55** System spectra based on 95% of time: average site (left) and worst site (right).

**TABLE 5.13 System Indices for LV Substations in France Based on 95% Site Indices**

| Harmonic | Median | 90% | Maximum |
|---|---|---|---|
| Third | 0.8% | 1.2% | 1.6% |
| Fifth | 3.5% | 4.5% | 5.5% |
| Seventh | 1.5% | 2.0% | 2.5% |
| Ninth | 0.3% | 0.5% | 0.9% |

**TABLE 5.14 System Indices for LV Substations in France Based on 50% Site Indices**

| Harmonic | Median | 90% | Maximum |
|---|---|---|---|
| Third | 0.4% | 0.6% | 0.6% |
| Fifth | 2.5% | 3.0% | 3.5% |
| Seventh | 1.0% | 1.0% | 2.0% |
| Ninth | 0.2% | 0.3% | 0.4% |

**TABLE 5.15 System Indices for LV Substations in France Based on 100% Site Indices**

| Harmonic | Median | 90% | Maximum |
|---|---|---|---|
| Third | 1.2% | 1.6% | 1.8% |
| Fifth | 6.0% | >6% | >6% |
| Seventh | 2.0% | 2.0% | 4.5% |
| Ninth | 0.5% | 0.7% | 1.1% |

Tables 5.13 through 5.15 give the median, 90 percentile, and maximum of the 95% site indices. The 90 percentile was chosen instead of the more common 95 percentile because results were only available for 16 sites. The choice of percentile makes a big difference in the results.

The paper also shows an interesting seasonal variation. The maximum over all sites of the weekly 95 percentile values varied between 4 and 5% during winter

**TABLE 5.16 Voltage Distortion in European Residences**

| | 50% | 95% |
|---|---|---|
| THD | 1.9% | 3.8% |
| Third | 0.4% | 1.5% |
| Fifth | 1.3% | 3.5% |
| Seventh | 0.7% | 1.4% |
| Eleventh | 0.2% | 0.7% |

and about 6% during summer. Seasonal variations in harmonic distortion are also presented in [304, 258]. This would indicate that a one-week monitoring period is not sufficient. In this case the seasonal variation is especially important because the summer values are around the EN 50160 limit (6% for the 5th harmonic). Note that the values in the tables above were obtained as full-year site statistics, whereas the maximum of weekly site statistics is more in line with EN 50160. The latter would result in higher values. Using maximum weekly 95% values for each site would result in the following system indices (compare with 5th-harmonic row in Table 5.13):

- 50% site: 4.5%
- 90% site: 7.0%
- 100% site: 7.5%

This again shows that great care should be taken in defining site and system indices.

In [214] the results are presented of harmonic measurements conducted during one week at 65 residences spread over six European countries. Some of the results are shown in Table 5.16.

We can compare the values in Table 5.16 with the results from the three-month measurement presented in Figure 5.51. That measurement campaign resulted in the following range for the daily 95% 10-min values:

- Third harmonic: 1.5 to 1.8%
- Fifth harmonic: 2.0 to 2.5%
- Seventh harmonic: 0.7 to 1.0%
- Ninth harmonic: 0.4 to 0.5%
- Eleventh harmonic: 0.5 to 0.7%

The conclusion from the comparison is that the 3rd-harmonic value is on the high side, the 11th-harmonic is sometimes on the high sides, but the 5th and 7th harmonics are within their typical range.

## 5.6 POWER QUALITY OBJECTIVES

In this section we will discuss a number of power quality standards and other documents that place limits on disturbance levels. The term *objective* is used to denote limits for power quality characteristics, site indices, or system indices. Objectives may apply to voltage or to current, to customers or to individual equipment, to generators or to loads. The main discussion in this section will concern harmonic standards. Objectives for some other power quality variations are presented at the end of this section.

### 5.6.1 Point of Common Coupling

Several power quality standards apply to the so-called *point of common coupling* (PCC) with other customers. It is the bus closest to a polluting source from which other customers are supplied. In a radial system it is rather straightforward to find the PCC. An example is shown in Figure 5.56, where the PCC corresponds to the metering point. This is not always the case, as is shown in the left part of Figure 5.57. A dedicated substation for the polluting load is owned and operated by the network operator. The PCC remains at the same location as before as there are no other customers supplied from the same bus. However, the connection of even a small customer to the same bus as the polluting load would move the PCC to this bus, as shown on the right in Figure 5.57. This is not just a theoretical issue but is very important in determining the amount of current distortion that is allowed. A connection to a higher voltage level will typically reduce the demands on the load. The connection of even a small customer to the same bus as the polluting load could lead to the requirement to install additional filters or otherwise reduce the distortion.

In meshed systems it is not always possible to uniquely define a PCC. An example is shown in Figure 5.58. As there is no load supplied directly from bus 1, this cannot be considered the PCC. The voltage distortion at buses 2 and 3 due to the polluting load will be less than at bus 1 so such a distinction is important in deciding the emission requirements for the polluting load.

### 5.6.2 Voltage Characteristics, Compatibility Levels, and Planning Levels

We saw in Section 1.4 that the basic way of achieving electromagnetic compatibility (EMC) is by coordinating emission limit, compatibility level, and immunity level. We also saw that the EMC standards cannot set emission limits for the power system and that the relation between equipment emission and disturbance level is not always straightforward. This is the main reason the compatibility level for most conducted disturbances coming from the power system is based on the existing disturbance levels. The disturbance level will be different for different locations and

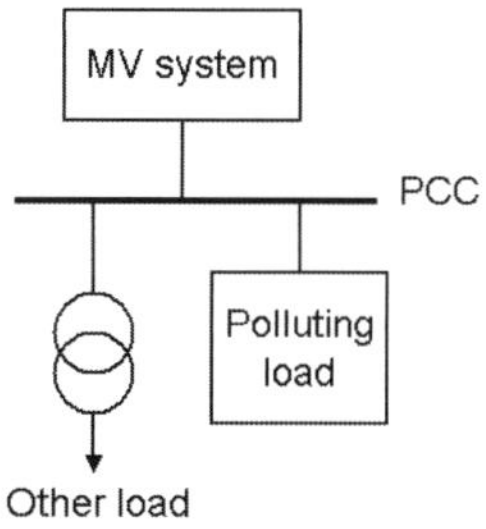


**Figure 5.56** Definition of PCC in radial system.

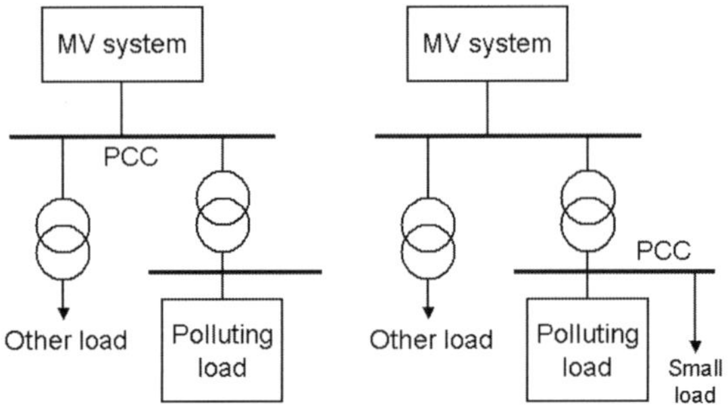


**Figure 5.57** Change in location of PCC due to addition of additional customer.

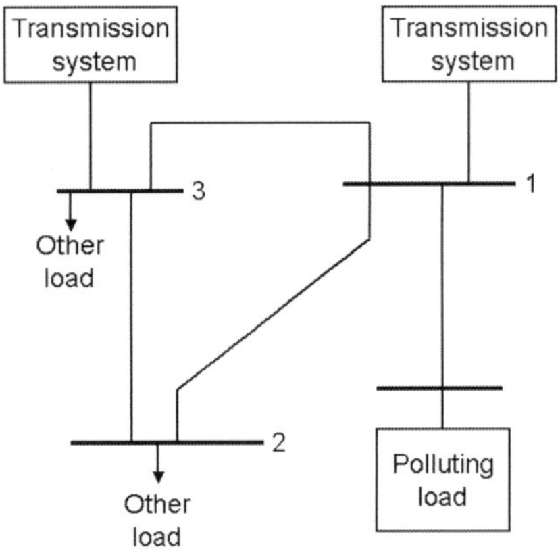


**Figure 5.58** Point of common coupling in meshed system.

will vary with time for each location. The disturbance level is thus a stochastic variable with a probability distribution function of location and time. At this stage it is important to distinguish between three different objectives for the disturbance levels:

- *Compatibility Levels* Levels not exceeded for 95% of locations for 95% of time. Compatibility levels for the LV public supply are given in IEC 61000-2-2, whereas IEC 61000-2-4 gives values for industrial networks.
- *Voltage Characteristics* Levels not exceeded for 100% of locations for 95% of time. Voltage characteristics for Europe are given in EN 50160.
- *Planning Levels* Levels used by network operators as an internal design and operation value. Planning levels are an internal matter for a network operator, although a regulator may prescribe planning levels. The IEC standards only give recommended values for planning levels (e.g., in IEC 61000-3-6 for harmonics).

More recently a fourth category has bean added: the regulatory requirements and the voltage quality guarantees given by some network operators to their customers. We came across a number of these when discussing site and system indices in Sections 5.4 and 5.5. These requirements and guarantees are often based on EN 50160, but a more fundamental discussion may be needed to either extend the definition of planning level or introduce a new category altogether.

From the definition it is clear that the compatibility level is less than or equal to the voltage characteristics. The planning level should be lower than the compatibility level as there should be some space for unexpected future growth, modeling errors, and so on. To obtain compatibility levels and voltage characteristics exactly according to the above definitions would require a long measurement period for a large number of locations. In practice the levels have been determined based on a limited number of measurements over a shorter period of time. The estimation of 95 and 100% values is a combination of engineering judgment and speculation. This rather inaccurate estimation is less of a concern than one may think. The aim of the whole exercise is to set immunity levels which should be above the compatibility level in any case. Also in the immunity level there is an uncertainty, as we discussed before. Therefore a margin is needed between immunity and compatibility levels. Some additional uncertainty in the actual 95% level does not affect the setting of any of the limits.

There has been a lot of discussion about the 95% values, especially where it concerned the European voltage characteristics standard EN 50160. That the value during the remaining 5% of time "could be anything" was an often-heard argument. The voltage characteristics standard was even initially rejected by some countries as it did not address safety issues. In the second edition of the document some values have been defined for 100% of time and 100% of location. The 95% issue is discussed in detail in [33, Section 1.4.3]. Another common criticism of EN 50160 is that it is extremely conservative: It excludes a whole range of abnormal operating conditions; it takes averages over 10-min intervals, thus excluding all short-duration phenomena; and it further excludes the 5% worst data [see, e.g., 273]. As long as the document is exclusively used as a voltage characteristic, this is expected. However, the document is regularly used as a basis for regulatory requirements or for quality guarantees to the customers. The document may not be the most appropriate choice for that application. As mentioned before, a discussion should be started on defining a fourth category of objectives, one more suited to these kinds of applications.

### 5.6.3 Voltage Characteristics EN 50160

European standard EN 50160 [106] describes electricity as a product, including its shortcomings. It gives the main characteristics of the voltage at the customer's supply terminals in public LV and MV networks under normal operating conditions. Some disturbances are just mentioned, for others a wide range of typical values are given, and for some disturbances actual voltage characteristics are given. Next to voltage characteristics, the document also defined the measurement method in much more detail than any of the then existing standard documents (the document

**TABLE 5.17 Harmonic Voltage Limits According to EN 50160**

| Order | Relative Voltage | Order | Relative Voltage | Order | Relative Voltage |
|---|---|---|---|---|---|
| 5 | 6% | 3 | 5% | 2 | 2% |
| 7 | 5% | 9 | 1.5% | 4 | 1% |
| 11 | 3.5% | 15 | 0.5% | 6–24 | 0.5% |
| 13 | 3% | 21 | 0.5% | | |
| 17 | 2% | | | | |
| 19 | 1.5% | | | | |
| 23 | 1.5% | | | | |
| 25 | 1.5% | | | | |

was first published in the early 1990s). Many utilities started to measure "in accordance with EN 50160." The measurement methodology later became the basis for the IEC standard on power quality measurements, IEC 61000-4-30.

European standard EN 50160 prescribes that the measurement of harmonic distortion shall be based on the 10-min values of the harmonic components up to order 25 obtained over a one-week period. For each harmonic up to order 25 a value is defined which shall not be exceeded for 95% of the 10-min values obtained in one week. The THD shall not exceed 8% during 95% of the week. The limits for harmonics have been reproduced in Table 5.17.

The third-harmonic voltage component is due to single-phase equipment; the fifth and seventh harmonics are also due to three-phase equipment. The fifth-harmonic voltage distortion is generally higher than the seventh-harmonic distortion. The distortion due to single-phase rectifiers (two-pulse) decreases rather quickly with increasing harmonic order. Relatively high distortion is also present for the doublets due to six-pulse rectifiers: Next to 5/7 are present 11/13, 17/19, and 23/25. Even-harmonic distortion is generally low, apparently with the exception of some second- and fourth-harmonic components. (This may be more of a "safety factor" than actual high values being observed.) The voltage characteristics in accordance with EN 50160 are shown as a spectrum in Figure 5.59. Note that this spectrum is a combination of the worst-case values for each of the harmonic orders. The spectrum in Figure 5.59 would never be measured in reality as the THD (11.3%) would exceed the voltage characteristic (8%).

The voltage characteristic levels in Table 5.17 originate from a study of harmonic distortion performed by a CIGRE working group [63] resulting in two values for the harmonic voltage distortion:

- *Low Value* The value likely to be found in the vicinity of large disturbing loads and associated with a low probability of causing disturbing effects.
- *High Value* The value rarely found in the network and with a higher probability of causing disturbing effects.

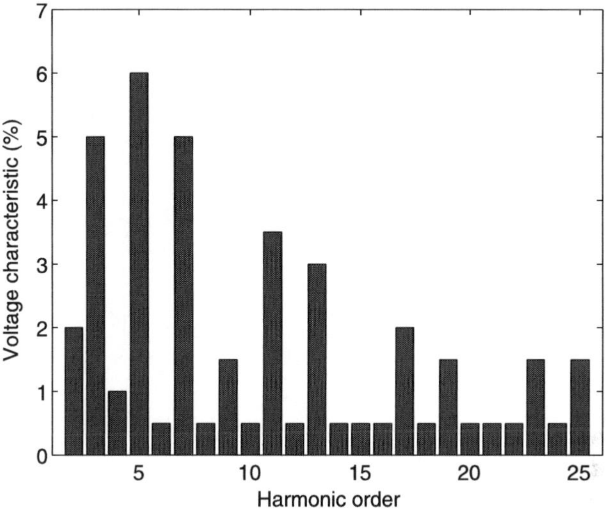


**Figure 5.59** Voltage characteristics according to EN 50160.

The values found by the CIGRE working group have been summarized in Table 5.18. The values used in EN 50160 are obviously the values rarely exceeded anywhere in Europe. This is exactly what is intended by the term *voltage characteristics.*

### 5.6.4 Compatibility Levels: IEC 61000-2-2

Compatibility levels for low-frequency conducted disturbances originating in the power system are given in IEC 61000-2-2 [149]. The compatibility levels for harmonics in systems up to 1 kV are reproduced in Table 5.19.

All percentage values are relative to the nominal supply voltage. The compatibility level for interharmonics is set to 0.2% of the nominal supply voltage. With some exceptions, the limits are the same as in EN 50160. No measurement method is

**TABLE 5.18 Harmonic Voltage Levels in Europe [63]**

| Order | Low | High | Order | Low | High |
|---|---|---|---|---|---|
| 3 | 1.5% | 2.5% | 15 | ≤0.3% | |
| 5 | 4% | 6% | 17 | 1% | 2% |
| 7 | 4% | 5% | 19 | 0.8% | 1.5% |
| 9 | 0.8% | 1.5% | 21 | ≤0.3 | |
| 11 | 2.5% | 3.5% | 23 | 0.8% | 1.5% |
| 13 | 2% | 3% | 25 | 0.8% | 1.5% |

**TABLE 5.19 Compatibility Levels for Public LV Systems According to IEC 61000-2-2**

| Order | Level | Order | Level | Order | Level |
|---|---|---|---|---|---|
| 5 | 6% | 3 | 5% | 2 | 2% |
| 7 | 5% | 9 | 1.5% | 4 | 1% |
| 11 | 3.5% | 15 | 0.3% | 6 | 0.5% |
| 13 | 3% | 21 | 0.2% | 8 | 0.5% |
| 17 | 2% | >21 | 0.2% | 10 | 0.5% |
| 19 | 1.5% | | | 12 | 0.2% |
| 23 | 1.5% | | | >12 | 0.2% |
| 25 | 1.5% | | | | |
| >25 | $0.2 + 1.3 \times 25/n$ | | | | |

associated with the compatibility levels. When measurements are performed, this either involves voltage characteristics or planning levels.

### 5.6.5 Planning Levels: IEC 61000-3-6

Indicative planning levels for harmonic distortion for voltage levels of 1 kV and higher are given in IEC 61000-3-6 [153]. Planning levels are internal quality criteria that are set by a network operator. As such they cannot be defined by an international standard. Therefore only indicative planning levels are given in the document. Different indicative levels are given for MV systems (between 1 and 35 kV) and for HV and EHV systems (higher than 35 kV). These levels are reproduced in Tables 5.20 and 5.21.

Technical report IEC 61000-3-6 also recommends site indices to assess if the planning level is met. The following rules hold for any of the harmonics in Tables 5.20 and 5.21:

- The highest daily 95 percentile of the 3-s values should not exceed the planning level.

**TABLE 5.20 Indicative Planning Levels for 1 to 35 kV**

| Order | Level | Order | Level | Order | Level |
|---|---|---|---|---|---|
| 5 | 5% | 3 | 4% | 2 | 1.6% |
| 7 | 4% | 9 | 1.2% | 4 | 1% |
| 11 | 3% | 15 | 0.3% | 6 | 0.5% |
| 13 | 2.5% | 21 | 0.2% | 8 | 0.4% |
| 17 | 1.6% | >21 | 0.2% | 10 | 0.4% |
| 19 | 1.2% | | | 12 | 0.2% |
| 23 | 1.2% | | | >12 | 0.2% |
| 25 | 1.2% | | | | |
| $h > 25$ | $0.2 + 1.0 \times 25/h$ | | | | |

**TABLE 5.21 Indicative Planning Levels Above 35 kV**

| Order | Level | Order | Level | Order | Level |
|---|---|---|---|---|---|
| 5 | 2% | 3 | 2% | 2 | 1.5% |
| 7 | 2% | 9 | 1% | 4 | 1% |
| 11 | 1.5% | 15 | 0.3% | 6 | 0.5% |
| 13 | 1.5% | 21 | 0.2% | 8 | 0.4% |
| 17 | 1% | >21 | 0.2% | 10 | 0.4% |
| 19 | 1% | | | 12 | 0.2% |
| 23 | 0.7% | | | >12 | 0.2% |
| 25 | 0.7% | | | | |
| >25 | $0.2 + 0.5 \times 25/n$ | | | | |

- The maximum weekly 10-min value should not exceed the planning level.
- The maximum weekly 3-s value should not exceed 1.5 to 2 times the planning level.

### 5.6.6 Current Distortion by Customers: IEC 61000-3-6; IEEE Standard 519

The planning levels for voltage distortion as recommended in IEC 61000-3-6 are intended as a basis for determining emission limits for individual customers. For LV customers, the responsibility lies not with the customer but with the equipment manufacturer. The emission of LV equipment is limited in product standards based on IEC 61000-3-2 and IEC 61000-3-4, both of which will be discussed later. For customers connected to MV levels and higher, emission limits should be set by the network operator to prevent the voltage distortion from exceeding the limits. A rather complicated set of procedures is set in the technical report to determine limits for the current distortion emitted into the network. The procedure is based on the actual source impedances, the contribution from different voltage levels, and the share of emission limits over the different voltage levels. To compare the actual emission with the emission limit, the following site indices and objectives are recommended:

- The highest daily 95 percentile of the 3-s values should not exceed the limit.
- The maximum weekly 10-min value should not exceed the limit.
- The maximum weekly 3-s value should not exceed 1.5 to 2 times the limit.

These objectives hold for every harmonic component. Interharmonics should be limited to 0.2% in all cases. The document also states that bursts of harmonics should be limited but gives neither any limits nor a method for calculating indices.

The IEEE harmonic standard IEEE 519 [162] gives limits for the harmonic current distortion by individual customers and for the harmonic distortion of the voltage supplied by the network operator. The standard document can be interpreted as a good example of relating compatibility level and emission limits in a system with multiple emitters. However, the document does not refer to the term *compatibility level*; instead it only refers to *voltage limits* and *current limits*.

Note the fundamental difference with the IEC emission standard, to be discussed below, where limits are given for each individual device but not for the total emission by a customer. The IEEE emission limits apply to the customer as a whole but do not limit the emission by individual equipment. Under the IEEE standard a customer can buy any type of equipment as long as the total emission does not exceed certain limits. To maintain the limits the customer may obtain devices which cancel each other, add resistive equipment that provides damping to the distortion, or install harmonic mitigation equipment such as filters. Under the IEC rules this is not allowed as each individual device has to fulfill the standard.

The basic philosophy of IEEE 519 is that the voltage distortion in systems up to 69 kV shall not exceed 3% for individual harmonics and 5% for THD. The limits for other voltage levels are given in Table 5.22. In EMC terms these limits can be interpreted as *compatibility levels* or *planning levels*.

The next step is to allow each customer a certain contribution to the voltage distortion. The larger a customer, the more contribution to the voltage distortion (the 3% value for individual harmonics for voltage levels up to 69 kV) is allowed. If only one customer is connected to a bus, this customer will be solely responsible for the harmonic voltage distortion. The other way around: The customer has an allowance of 3% harmonic voltage distortion. If there are many customers connected to the same bus, the individual distortions will add but not linearly with the number of customers. There will be some cancellation effects between the customers. A small customer will be allowed a smaller contribution to the voltage distortion, but the relative allowance (in relation to the load size) is larger. The allowance per customer is not based on the number of customers connected to a bus as this would make the standard unworkable. Instead the ratio between the short-circuit capacity of the source and the load size is used. The resulting allowance per customer is given in Table 5.23, where SCR stands for short-circuit ratio: the ratio between the short-circuit capacity of the source and the load size.

Table 5.23 gives emission limits for individual customers, but these limits are in terms of voltage, not in terms of current as is normally the case. It is not impossible to define emission limits in terms of voltage. The flicker standards work according to this principle after all, but it is more practical to give limits in terms of current. The

**TABLE 5.22 Voltage Limits According to IEEE 519**

| Bus Voltage at PCC | Individual Voltage Distortion (%) | Voltage THD (%) |
|---|---|---|
| 69 kV and below | 3.0 | 5.0 |
| Above 69 kV up to 161 kV | 1.5 | 2.5 |
| Above 161 kV | 1.0 | 1.5 |

**TABLE 5.23 Basic Harmonic Current Limits in IEEE 519**

| SCR at PCC | Individual Voltage Distortion (%) | Related Assumption |
|---|---|---|
| 10 | 2.5–3.0 | Dedicated system |
| 20 | 2.0–2.5 | One to two large customers |
| 50 | 1.0–1.5 | A few relatively large customers |
| 100 | 0.5–1.0 | Five to 20 medium-size customers |
| 1000 | 0.05–0.10 | Many small customers |

relation between the current distortion and the voltage distortion is simply Ohm's law. The actual source impedance as a function of frequency is rather difficult to obtain and varies from location to location and in many cases even with time. Therefore a gross simplification is made: The source impedance is purely inductive. All capacitances in the system are neglected. The (inductive) source impedance can easily be obtained from the source impedance as used to determine the relative source size. Combining all this information gives current limits for different voltage levels, different load sizes, and different harmonic orders. The result is given in Table 5.24 for voltage levels up to 69 kV. The stronger the supply (or the smaller the load), the more harmonic current is allowed. The ratio between the fault current and the load current (the SCR) is the determining factor. Even harmonics are limited to 25% of the odd-harmonic limits.

The emission limits for small customers (SCR above 1000) are shown in Figure 5.60.

It is important to realize that the emission limits in Table 5.24 are expressed not as a percentage of the fundamental current but as a percentage of the *maximum load current*. The TDD (total demand distortion) plays the role of the THD in other standards. The reason for this is that it is the distortion in amperes that affects the voltage waveform. High relative distortion during low load is normally not a concern, but moderate distortion during high load may well be. According to the standard, the maximum demand (maximum load current) should be obtained over a certain observation period. Alternatively, and easier to put in a contract, the subscribed demand (or the rating of the main fuse) can be used as a reference to determine the relative harmonic distortion.

**TABLE 5.24 Current Distortion Limits According to IEEE 519**

| | Harmonic Order (Odd Harmonics) | | | | | |
|---|---|---|---|---|---|---|
| Current Ratio | <11 | 11–16 | 17–22 | 23–34 | >34 | TDD |
| <20 | 4.0% | 2.0% | 1.5% | 0.6% | 0.3% | 5.0% |
| 20–49.9 | 7.0% | 3.5% | 2.5% | 1.0% | 0.5% | 8.0% |
| 50–99.9 | 10.0% | 4.5% | 4.0% | 1.5% | 0.7% | 12.0% |
| 100–999 | 12.0% | 5.5% | 5.0% | 2.0% | 1.0% | 15.0% |
| >1000 | 15.0% | 7.0% | 6.0% | 2.5% | 1.4% | 20.0% |

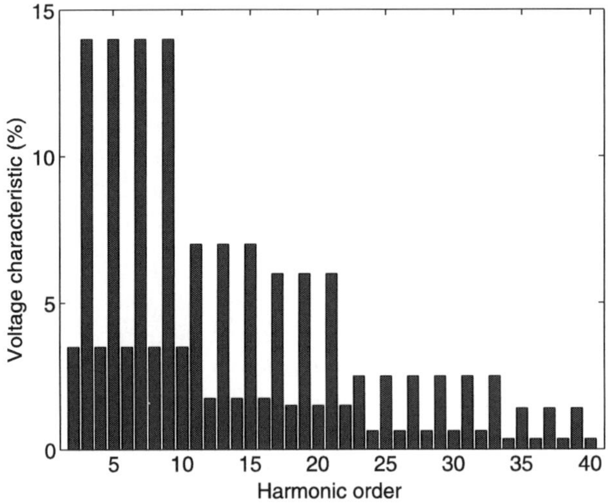


**Figure 5.60** Emission limits according to IEEE 519 for small customers.

Harmonic standard IEEE 519-1992 does not define any measurement method, which makes it difficult to verify compliance with the standard. The working group responsible for this document aims at including a measurement protocol based on IEC 61000-4-7 and IEC 61000-4-30 in the next revision of the document [135].

### 5.6.7 Current Distortion by Equipment: IEC 61000-3-2

Emission limits for small equipment are defined in IEC 61000-3-2 [150]. The document distinguishes between four classes of equipment:

- Class A includes three-phase equipment, motor-driven equipment, and all equipment that does not fall in any of the other classes. Roughly speaking this class contains the larger equipment.
- Class B contains all portable loads.
- Class C contains lighting equipment and dimming devices.
- Class D includes equipment with a special wave shape. This class contains equipment which takes current pulses from the supply instead of continuous current.

Equipment belongs to class D when its current has a “special waveshape”: a waveform that stays below the rectangular curve in Figure 5.61 for more than 95% of the cycle duration. Note that only one half-cycle is shown. The second (negative) half-cycle has the opposite form. The requirement can be interpreted as the current has a special “waveform” when the main conducting period lasts less than 60° per half-cycle. An example is shown with two stylized waveforms. The solid curve has a

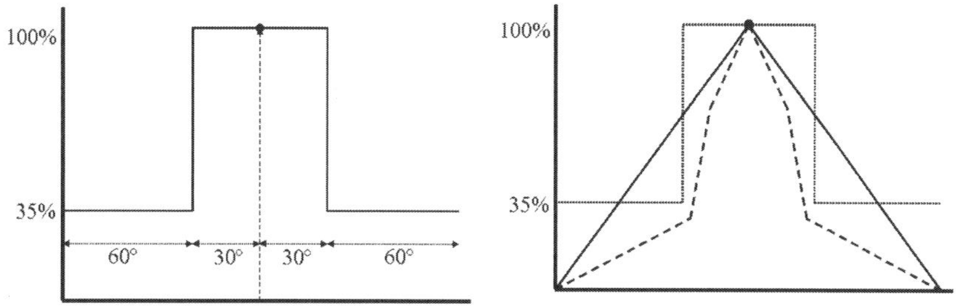


**Figure 5.61** Left: Special wave shape criterion according to IEC 61000-3-2. The black dot indicates the position of the current peak. Right: the dashed curve is a special wave shape, the solid curve not.

triangular shape and it stays above the rectangular criterion for a substantial part of the half-cycle (i.e., more than 5%). The dashed curve stays below the criterion for the whole half-cycle.

The standard document gives emission limits as absolute values (in amperes) and relative to the power consumption of the device (in milliamperes per watt). The main emission limits are summarized in Table 5.25. For class A equipment only the absolute limits apply. For class D equipment both relative and absolute limits apply. The relative limits apply, roughly, for class D equipment with a power consumption up to 600 W. This power limit varies slightly with harmonic order. The emission limits for class B equipment are 1.5 times the absolute limits.

The limits for class C (lighting) equipment are different. They are given in Table 5.26. Limits for even harmonics only apply to class A equipment. The emission limits for even harmonics with other equipment classes are rather simple: They should not produce any even-harmonic currents. The requirements for harmonic distortion of the supply source used in testing the equipment allow for 0.2%

**TABLE 5.25 Current Limits According to IEC 61000-3-2**

| Order | Relative Limit (mA/W) | Absolute Limit (A) |
|---|---|---|
| 2 | | 1.08 |
| 3 | 3.4 | 2.30 |
| 4 | | 0.43 |
| 5 | 1.9 | 1.14 |
| 6 | | 0.30 |
| 7 | 1.0 | 0.77 |
| 8–40 | | 0.23 $(8/n)$ |
| 9 | 0.5 | 0.4 |
| 11 | 0.35 | 0.33 |
| 13 | 0.30 | 0.21 |
| 15 and above | $3.85/n$ | $2.25/n$ |

**TABLE 5.26 Current Limits for Lighting Equipment According to IEC 61000-3-2**

| Harmonic | Current Limit |
|---|---|
| 2 | 2% |
| 3 | Power factor times 30% |
| 5 | 10% |
| 7 | 7% |
| 9 | 5% |
| 11, 13, 15, . . . , 39 | 3% |

even-harmonic distortion up to orders 10 and 0.1% for orders 12 through 40. One could conclude from this that 0.2 and 0.1%, respectively, are allowed as even-harmonic current emission.

The power factor in Table 5.26 is defined as the ratio of the active power and the product of rms voltage and rms current:

$$\lambda = \frac{P}{V_{\mathrm{rms}} \times I_{\mathrm{rms}}} \tag{5.29}$$

The emission limits for class A equipment are given in Figure 5.62. The harmonics with the highest emission limits are the third, fifth, and seventh. Note that the emission limit is highest for the third harmonic, contrary to the voltage distortion (Fig. 5.59), where the fifth harmonic dominates. For higher harmonic order the limits for even-and odd-harmonic distortion become about equal.

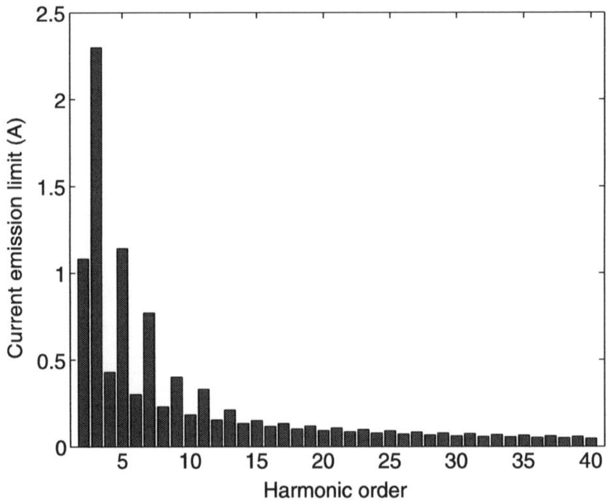


**Figure 5.62** Emission limits according to IEC 61000-3-2, class A equipment.

The standard document not only contains the limits in Tables 5.25 and 5.26 but also a detailed description of the test equipment and the circumstances of the tests. For example, with reference to testing washing machines, the document states, "The washing machine is tested in a normal 60° laundry program. It is filled with a normal load of cotton cloths, size 70 cm × 70 cm, dry weight from 140 $g/m^2$ to 175 $g/m^2$." [150, page 45]

A common misunderstanding is that compliance with IEC 61000-3-2 will automatically imply compliance with IEEE 519; in other words that the IEC standard is stricter than the IEEE standard. It can however be easily shown that this is not the case. Consider class A equipment with a fundamental current of 16 A (the highest current for which IEC 61000-3-2 applies). According to Table 5.25, a third-harmonic current of 2.3 A is allowed. In relative terms this is 14%. If we compare this to Table 5.23, we see that medium-size users and larger would not meet the IEEE limits if they would only have this type of equipment.

The smaller the equipment, the larger the relative distortion that is allowed. It is the absolute harmonic current that is limited after all. For equipment taking 5 A fundamental current, the limit is 46% third harmonic. Even a small customer would not meet IEEE 519. Obviously, this assumes that all equipment has an emission level just below the emission limit and there is no cancellation between equipment. This is obviously not a very likely situation so the use of equipment that complies with the IEC emission limits will likely also mean that the customer will pass the IEEE emission limits. Note again the difference between IEC and IEEE emission limits. The IEC gives emission limits per device; the IEEE gives emission limits per customer.

For equipment with a rated current exceeding 16 A per phase standard IEC 61000-3-4 [151] applies. The emission limits depend on the ratio between the rated power of the load and the short-circuit capacity of the source (as in IEEE 519 this is referred to as the short-circuit ratio). An overview of the emission limits is given in Table 5.27 and Figure 5.63. For larger SCR, that is, smaller equipment, a higher percentage of harmonic distortion is allowed. These emission limits are significantly larger than the limits in IEEE 519, but then again the IEC gives equipment emission limits whereas the IEEE gives emission limits per customer.

**TABLE 5.27 Emission Limits for Large Equipment According to IEC 61000-3-4**

| Short-Circuit Ratio | Third | Fifth | Seventh | Ninth | Eleventh | Thirteenth |
|---|---|---|---|---|---|---|
| >33 | 21.6% | 10.7% | 7.2% | 3.8% | 3.1% | 2% |
| >66 | 23% | 11% | 8% | 6% | 5% | 4% |
| >120 | 25% | 11% | 10% | 7% | 6% | 5% |
| >175 | 29% | 14% | 11% | 8% | 7% | 6% |
| >250 | 34% | 18% | 12% | 10% | 8% | 7% |
| >350 | 40% | 24% | 15% | 12% | 9% | 8% |
| >450 | 40% | 30% | 20% | 14% | 12% | 10% |
| >600 | 40% | 30% | 20% | 14% | 12% | 10% |

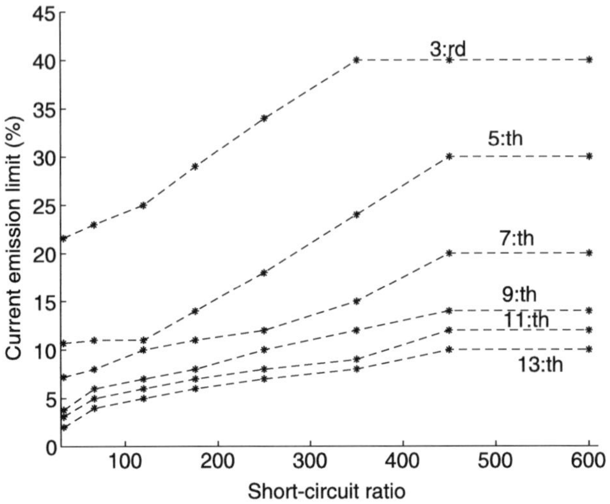


**Figure 5.63** Emission limits according to IEC 61000-3-4 as function of SCR for different harmonic orders.

As one customer rarely has only one piece of equipment, the rest of the load will likely compensate at least for some of the distortion from the large polluting equipment.

The fact that both IEC 61000-3-2 and IEC 61000-3-4 give emission limits that are significantly larger than the IEEE limits has another consequence as well. The IEEE limits were based on a compatibility level for the voltage distortion of 3%. As the IEC gives higher emission limits, the resulting voltage distortion would be significantly higher when all equipment would have an emission level close to the limit and there would be no cancellation. Fortunately not all equipment emits harmonic distortion (although the fraction that does is growing) and there is always cancellation between equipment, even if it concerns equipment of the same type.

### 5.6.8 Other Power Quality Objectives

***5.6.8.1 Frequency Variations*** The European voltage characteristic standard EN 50160 [106, page 11] defines the "frequency of the supply voltage as the repetition rate of the fundamental wave of the supply voltage measured over a given interval of time." The nominal value of the frequency should be 50 Hz.

The standard also gives a method of measuring the frequency variation. As shown in the measurement examples, the frequency may vary significantly even within 1 min. Therefore the measurement window has to be defined. The standard document EN 50160 prescribes a measurement window of 10 s. The average frequency is used as a characteristic to quantify the frequency variation. This 10-s value is limited to a certain range of values. In standard terms: "Under normal operating conditions the mean value of the fundamental frequency measured over 10 s shall be within a range of. . . ." [106, page 11]

The range of values is given for 99.5% of the year and for 100% of the time. Different values are given for large systems (*systems with synchronous connection to an interconnected system*) and for small systems (*systems with no synchronous connection to an interconnected system*).

For large systems the 10-s average is between 49.5 and 50.5 Hz ($\pm 1\%$) during 99.5% of the year and between 47 and 52 Hz ($\pm 4\%$) all of the time. For small systems the variation is larger: $\pm 2\%$ during 95% of a week and $\pm 15\%$ during 100% of the time.

In Great Britain the frequency limits at transmission level are 49.5 to 50.5 Hz (99 to 101%) of the nominal voltage. The transmission operator responsible for frequency control National Grid Company, NGC reports annually the number of times per year that the frequency is outside of these limits for longer than 60 s [93].

Within the Nordel system (Finland, Sweden, Norway, and part of Denmark) the frequency under normal operation should be between 49.9 and 50.1 Hz, with a standard deviation of 0.03 Hz. The error between the integrated frequency deviation and the actual time should not exceed 10 s [280].

In the South African standard NRS 048.2 [225] different indices and limits are introduced for *grid networks* and *island networks*. For grid networks the 99.5 percentile of the frequency deviation should be used, for island networks the 95 percentile. The corresponding limits are 2 and 2.5% of the nominal frequency, respectively. For all network types, the maximum frequency is used as a second index, with limits equal to 2.5% for grid networks and 5% for island networks.

***5.6.8.2 Voltage Variations*** The voltage characteristic standard EN 50160 states that 95% of the 10-min rms voltages during one week should be between 90 and 110% of the nominal voltage. This holds for LV and MV systems. For LV systems an additional requirement is that all 10-min values should be between 85 and 110% of nominal.

Most network operators have internal design rules concerning the voltage magnitude that is acceptable for different voltage levels. These rules may be interpreted as planning levels and were in the past even presented as semiguaranteed levels. However, recently any guarantees given are no longer based on internal design rules but are based on (the much less strict) EN 50160.

A recommendation of voltage quality requirements in the Norwegian system gives a permitted range of 94 to 106% for the average voltage over a one-week period. This should be used as an additional requirement above the ones in EN 50160 [273].

A rather strict objective is used for underground distribution networks in Argentina: The voltage should be in the range 95 to 105% of nominal for 97% of each seven-day period. This objective is linked to a complicated compensation scheme [88].

In Great Britain the voltage limits at transmission level are 90 to 110% of the nominal voltage. Each transmission operator reports annually the number of times per year that the voltage is outside of these limits for longer than 15 min [93]. The statutory limits at distribution level are 94 to 110% of nominal. Distribution

**TABLE 5.28 Recommended Limits for Repetitive Voltage Changes in MV and HV Networks**

| Repetition Rate | MV Limit | HV Limit |
|---|---|---|
| Less than 1 per hour | 4% | 3% |
| 1–10 times per hour | 3% | 2.5% |
| 10–100 times per hour | 2% | 1.5% |
| 100–1000 times per hour | 1.25% | 1% |

companies report the number of verified voltage complaints due to voltages outside of the limits.

For capacitor banks that are switched several times a day the permissible step is 1 or 2%. For banks that are switched once a day or less, the permissible step is higher but should certainly not exceed 5%.

The technical report on emission limits for fluctuating loads, IEC 61000-3-7, also gives limits for voltage steps based on how often they occur. The recommended limits are as in Table 5.28.

#### 5.6.8.3 *Voltage Fluctuations*

The permissible range of voltage fluctuations is defined in a number of international standards: IEC 61000-3-3, IEC 61000-3-5, IEC 61000-3-7, and IEC 61000-4-15. Older standards limit voltage fluctuations by using a flicker curve. This method may be applicable in a design stage but has not much value when assessing the impact of a measured voltage or current. The short-term flicker severity and long-term flicker-severity are more suitable to characterize a measured supply. The definition of the short-term flicker severity $P_{st}$ is such that a value exceeding unity corresponds to annoying light flicker with standard incandescent lamps. Thus $P_{st} = 1$ forms a natural limit.

The voltage characteristic standard EN 50160 states that the long-term flicker severity should not exceed unity during 95% of time. This is clearly a less severe requirement than the "natural limit"; after all it allows for disturbing flicker levels during 5% of time. Compatibility limits for voltage fluctuations up to 35 kV are given in IEC 61000-3-7:

- The short-term flicker severity $P_{st}$ should be less than 1.0.
- The long-term flicker severity $P_{lt}$ should be less then 0.8.

The indicated planning levels (for 99% of time) according to IEC 61000-3-7 are reproduced in Table 5.29.

An interesting ongoing discussion concerns the suitability of flicker levels measured at higher voltage levels. What is needed is a relation between the flicker level at a higher voltage level and the flicker level at the terminals of the lamp. This ratio is referred to as the *transfer coefficient*. The indicative planning levels in IEC 61000-3-7 are based on a transfer coefficient equal to unity. In many cases the flicker severity at low voltage due to a source at a higher voltage

**TABLE 5.29 Indicated Planning Levels in IEC 61000-3-7**

| | 1–35 kV | Above 35 kV |
|---|---|---|
| $P_{st}$ | 0.9 | 0.8 |
| $P_{lt}$ | 0.7 | 0.6 |

level is less than at the higher voltage level. In that case a higher flicker severity is allowed at the higher voltage level. But the contribution of the higher voltage levels to the flicker severity at low voltage should not exceed the values given before. In that way some margin remains for low-voltage sources of voltage fluctuations without the flicker level exceeding unity.

Limits for equipment emission are given in IEC 61000-3-5 [152], where it should be noted that the emission limits are also expressed in terms of $P_{st}$ and $P_{lt}$ even though these are limits on the fluctuation in the current. To obtain voltage fluctuation from the measured current fluctuations, an impedance value has to be chosen. For small equipment (rated current up to 16 A) a standard reference impedance is used. The test impedance is $0.24 + j0.15\ \Omega$ in each phase and $0.16 + j0.10\ \Omega$ in the neutral. For single-phase equipment a reference impedance of $0.40 + j0.25\ \Omega$ should be used [154]. For medium-size equipment a test impedance less than the standard reference impedance is used; the larger the equipment, the smaller the impedance. The standard document states that the impedance should be such that the voltage drop due to the equipment should be 3 to 5%, with an $R/X$ ratio of 0.5 to 0.75. For large equipment (rated current above 75 A) the test impedance is equal to the actual source impedance.

For equipment up to 75 A rated current, the resulting $P_{st}$ should be less than 1.0 and the resulting $P_{lt}$ less than 0.65. For equipment with rated current exceeding 75 A the limits depend on the size of the load $S_L$ in relation to the size of the MV/LV transformer $S_{TR}$:

$$P_{st} \leq \sqrt[3]{\frac{S_L}{S_{TR}}} \tag{5.30}$$

A load with a size equal to the transformer size is given the whole range of flicker, $P_{st} = 1$, as this is a dedicated transformer. A load half the transformer size is allowed to produce a short-term flicker severity of 0.79. The cubic law for addition of independent flicker sources will cause a flicker level equal to unity in two of these loads. The limit is never less than 0.6. The long-term flicker severity limit is 65% of the short-term flicker severity limit.

#### 5.6.8.4 Voltage Unbalance

The European voltage characteristic standard EN 51060 [106, page 15] gives the following limit for the negative-sequence unbalance in public LV and MV systems: "Under normal operating conditions, during each period of one week, 95% of the 10 minute mean rms values of the negative phase sequence component of the supply voltage shall be within the range 0 to 2% of

**TABLE 5.30 Limits for Voltage Unbalance in Different Countries**

| Country | Low Voltage | Medium Voltage | High Voltage | Extra High Voltage |
|---|---|---|---|---|
| Belgium | 2% | 2% | 2% | |
| Denmark | — | — | 2% | |
| France | — | 2% | 1% | 1% |
| Germany | 2% | | | |
| Italy | 2% | 2% | 1% | |
| Spain | 2% | | | |
| United Kingdom | 2% | 2% | 2% | 2% |

the positive phase sequence component.” In other words: the 10-min average negative-sequence unbalance should be less than 2% during 95% of a week.

This holds for three-phase systems. For systems with significant amounts of unbalanced load, or single-phase feeders, the unbalance can be larger: “In some areas with partly single phase or two phase connected customers’ installations, unbalances up to about 3% at three-phase supply terminals occur.” [106, page 15]

Different countries use somewhat different limits, partly due to the different ways of statistical assessment of the values (see Chapter 5). An inventory of limits in different European countries [302] shows that the limits are in all cases 1 or 2%. The results are presented in a highly simplified way in Table 5.30. The table cannot be used to compare the various countries as no consistent definition has been used to determine the levels.

The 95% 10-min unbalance index should not exceed 2% according to EN 50160. The latter is becoming a point of discussion for two reasons. The standard EN 50160 refers to LV and MV networks only. However, a number of countries apply the values to higher voltage levels as well. The result is that generator companies become concerned about the potential consequences of voltage unbalance on their synchronous machines (which according to IEC 60034 should be able to tolerate 1% unbalance). Similar concerns appear with the installation of generation plants at distribution levels. The whole discussion centers around the responsibility question: Should the network operator provide a voltage quality that is acceptable to (the majority of) the customers or is it the customer’s responsibility that their equipment is compatible with the voltage? This again calls for the start of a discussion on which indices and objectives to use for regulatory requirements and voltage quality guarantees.

## 5.7 SUMMARY AND CONCLUSIONS

Time aggregation is introduced as a method for merging measurements obtained over a short measurement window into values that hold for a longer measurement window. The standard method, as defined in IEC 61000-4-30, uses a basic measurement window of 200 ms duration which is aggregated into 3-s, 10-min, and 2-h values. Next to the standard method two new methods for time aggregation are introduced: the number of steps per hour and the VSVs. Both methods are introduced for the rms voltage but may also be applied to other characteristics such as unbalance or harmonic distortion. The VSV is a method of quantifying the level of variations on a time scale between 3 s and 10 min without the need to store 3-s values. The method results in one value every 10 min so that it will not significantly increase the data storage needs. Further work is needed on the development of these new methods, including the collection of sufficient data to allow the definition of suitable indices and objectives. Further work is also needed to study the definition of a 1-min aggregation step, for which a number of arguments are given in this chapter.

Phase aggregation is a somewhat forgotten subject in the calculation of power quality indices. It is best to consider the three-phase character of the system in the calculation of the basic characteristics. This is however in many cases not done so that three different values for the three phases result. Somewhere along the path to system indices the three values have to be merged into one three-phase value. Further development of standard methods is needed as well as a study after the impact of the different aggregation methods on the indices. Further research is especially needed in appropriate three-phase indices for flicker and waveform distortion. Such methods should be based on symmetrical components, an existing and well-proven tool.

Flagging of basic and aggregated time windows is used to prevent events from affecting the indices for variations. The triggering of an event will flag the variation characteristics over the basic measurement window and all aggregated windows containing the basic window. Flagging helps to prevent double counting and to prevent erroneous values for variation characteristics during events. Further study is needed to better understand the potential impact of events on variation characteristics and indices. Such a study should also include the role of the triggering method and level. Until now the flagging method has been defined for dips, swells, and interruptions. Standard methods need to be developed on how to further process flagged data. It is important that no more data than needed are removed for it would make the supply look better than it is. Flagging methods also need to be defined for transients.

After time aggregation and removing flagged data where appropriate, the characteristics over the measurement interval result. These are in most cases presented as a function of time. From the time plot it is possible to detect daily and weekly patterns by visual inspection and by applying appropriate pattern recognition methods. Only some very simple pattern recognition methods have been presented in this chapter. Further research and more advanced and better methods are needed. Such methods should aim not only at detecting patterns but also at correlating them with load and production patterns. Instead of as a function of time a characteristic may also be presented in the form of a probability distribution function. Such is more appropriate when no clear patterns are present in the time function and/or when the characteristic shows a large variation.

Site indices are used to obtain a representative value of the disturbance level at an individual site over a certain period of time (e.g., one day, one week, or one year).

Typical methods use the average, 95%, or maximum value of the aggregated characteristics over the measurement period. The choice of aggregation window (3 s, 10 min, etc.) and of the statistical value (mean, 95%, etc.) will strongly influence the result. When setting limits for regulatory purposes or as quality guarantees, it is important that the associated indices be defined in an appropriate way. Further standard development is needed here. Further research is needed in the relations between site indices using different aggregation window and different statistical value. Such relations would allow a better comparison between different sites. Further research is also needed in the relation between site indices and the impact of power quality disturbances on equipment. This research could possibly result in characteristics and indices that give a better relation with the impact of a disturbance on the performance and aging of equipment.

The calculation of site indices for voltage magnitudes and frequency suffers some difficulties. The fact that voltage magnitude and frequency may deviate in two directions makes it difficult to uniquely define site indices. Splitting up the disturbance in underdeviation and overdeviation will solve this problem, but it will require new indices and objectives.

Further research is needed in the relation between monitor location and site index. An important question to be addressed is how relevant the monitor location is for the voltage quality as experienced by the end costumer or by sensitive equipment. Such studies are needed for almost all power quality disturbances. Flicker propagation could be a special concern because of its origin at transmission levels and its impact at the remote end of the LV network (i.e., at the terminals of incandescent lamps). Such studies are also needed of harmonics, unbalance, and voltage variations to assist in the choice of appropriate monitor locations. Knowledge of the relation between monitor location and site index is also important when calculating system indices.

System indices are obtained as the mean, 95%, or maximum value of the site indices obtained for all monitor locations. The calculation method is straightforward, but two important practical issues remain. The first is that the system indices are strictly speaking not exactly calculated but estimated from a limited sample of all possible measurement sites. Further research is needed to better understand the errors made here. This includes the accuracy of the estimation and the presence of any bias in the estimation. The first is a statistical problem, the latter a power system problem that is very much related to the before-mentioned relevance of the monitor location for the end customer or sensitive equipment. The second issue concerns the choice of weighting factors. This is partly related to the first issue but mainly a matter of development of standard methods. System indices further require more experience in the interpretation of 95% values from a limited set of data points. This interpretation is strongly related to the statistics of the sample and the accuracy of the estimation.

Power quality objectives come in the form of limits on voltage and current quality. The voltage requirements in standard documents are voltage characteristics, planning levels, and compatibility levels. A fourth category may be needed: regulatory requirements and quality guarantees. A discussion needs to be started to define appropriate indices and objectives.

The approach developed for power quality indices can be applied to quantify power system performance in general. It could, for example, be used to study the impact of increasing penetration of wind power or of a continued load growth without investment in new lines and generators. New performance indices will have to be defined to relate voltage and frequency variations with system security.

Further information on these subjects can be found in various IEC standard documents and in documents published by network operators and national regulators. The issues are also a recurring theme at CIGRE and CIRED conferences. No general overview work is available yet on the subjects discussed in this chapter.

## CHAPTER 6

# ORIGIN OF POWER QUALITY EVENTS

This chapter is the first of five chapters on power quality events: significant deviations from the normal steady-state voltage or current. In this chapter we will discuss the origin of three important types of events: interruptions, voltage dips, and transients. The subdivision in interruptions, dips, and transients is mainly based on the general appearance of the voltages at the equipment terminals. For each event type it is however possible to point out a whole range of causes. These various causes of the events will be discussed as well as their impact on the voltage waveforms. This information will form the basis for the forthcoming chapters in which the processing of these waveforms will be discussed.

This chapter consists of three main sections, one for each type of event. The contents of each section will be a combination of measurement results and theoretical considerations. When presenting the measurement results we will occasionally be using characterization methods that will only be explained in further detail in later chapters.

This chapter concludes with a section containing summary and conclusions, in which special emphasis is placed on the need for further research and development. The final section also contains some recommendations for additional reading on the subject of this chapter.

*Signal Processing of Power Quality Disturbances.* By Math H. J. Bollen and Irene Yu-Hua Gu

## 6.1 INTERRUPTIONS

### 6.1.1 Terminology

Within the power quality literature an interruption is defined as a situation in which the voltage magnitude is zero or close to zero. Typical thresholds to detect an interruption are 1 and 10% of the nominal voltage. From a network point of view and in reliability studies, an interruption is more often defined as a situation in which there is no longer any galvanic connection between a customer and the rest of the power system. The interruption corresponds to the "opening of an interruption device." This is the definition used in IEEE Standard 1366 [169] and recommended by CIGRE working group C4-07 [64]. Obtaining interruption statistics involves in almost all cases keeping records of the opening and especially the closing of interrupting devices. We will come back to this in Section 10.1.

For most customers the two definitions are equivalent as a disconnection of the supply very quickly leads to zero voltage (we will see some examples in the remainder of this section). However, for customers having their own generation the situation becomes different. The opening of the interruption device will disconnect the generator unit plus a certain amount of load from the rest of the system. A new steady state with an acceptable voltage magnitude and frequency will be reached when the local generation has a suitable rating and is equipped with voltage and frequency control (reactive and active power control, respectively). The resulting operational state is referred to as *insular operation* or *microgrid operation*. Such a way of operation is used for large industrial systems to guarantee the supply to the industrial process when the public supply fails. With the increased penetration of distributed generation, microgrid operation is being discussed as a feasible option for smaller customers as well. This so-called *intentional islanding* should not be confused with *unintentional (or uncontrolled) islanding*, which may occur by accident upon the tripping of a feeder with distributed generation.

**TABLE 6.1 Interruption Definitions in Standard Documents**

| Standard | Term | Definition |
|---|---|---|
| IEEE 1159 | Interruption | Voltage below 10% of nominal |
| | Sustained interruption | Longer than 3 s |
| | Momentary interruption | 0.5 cycles to 3 min |
| | Temporary interruption | 3 s to 1 min |
| IEEE 1250 | Instantaneous interruption | Shorter than 30 cycles |
| | Momentary interruption | 0.5 to 2 s |
| | Temporary interruption | 2 s to 2 min |
| | Sustained interruption | Longer than 2 min |
| EN 50160 | Short interruption | Shorter than 3 min |
| | Long interruption | Longer than 3 min |
| IEEE 1366 | Momentary interruption | Shorter than 5 min |
| | Sustained interruption | Longer than 5 min |

Unintentional islanding may lead to very dangerous situations for equipment and for persons.

A further subdivision of interruptions is often made based on their duration; an overview of the terminology as used in several standard documents is given in Table 6.1. There is a general agreement that events longer than a few minutes classify as *long* or *sustained* interruptions. This distinction is based on the difference between automatic reclosing and manual reclosing. For durations less than this, there are however confusing definitions. The distinction between *instantaneous*, *momentary*, and *temporary* is also based on different reclosing methods, but this terminology is seldom used. The most recent version of the reliability index standard IEEE 1366 therefore only uses two terms: momentary and sustained interruptions, which are (almost) synonym with the IEC terms *short* and *long interruptions*, respectively.

### 6.1.2 Causes of Interruptions

Interruptions are due to the disconnection of one or more customers from the power network. The causes for this disconnection are as follows:

- Opening of a circuit breaker or a fuse due to a short circuit or earth fault. This is the most common cause for interruptions in radial parts of the network, thus for most customers.
- Inadvertent operation of a circuit breaker or another switching device. This may be due to an incorrect tripping signal being generated by a protection relay, a mechanical failure in the switching device, or an incorrect manual operation.
- Intentional disconnection of part of the system, typically with the aim of performing maintenance or repair on the disconnected part of the system.

Interruptions due to the first and second causes are referred to as *forced interruptions*; those due to the third cause as *planned interruptions* or *scheduled interruptions*.

The discussion on the *cause of an interruption* generally goes much further to include what led to the fault or to the inadvertent opening of the interrupting device. Obviously such information is of high importance for the network operator. In [51] a whole chapter is dedicated to these root causes of interruptions. That discussion is however beyond the scope of this book. The resulting voltage waveform and thus the impact on the customer and its equipment are the same for an interruption due to a lightning stroke as for one caused by a squirrel.

An example of an interruption due to a single-phase fault is shown in Figure 6.1. This measurement was done in a 120-V, 60-Hz system. The single-phase fault caused a large voltage drop in one phase and a slight drop in the nonfaulted phases. After about five cycles the fault is cleared. Unfortunately this also leads to a complete loss of voltage for the customer. The voltage decays to zero within two cycles after the opening of the interrupting device. Any interruption

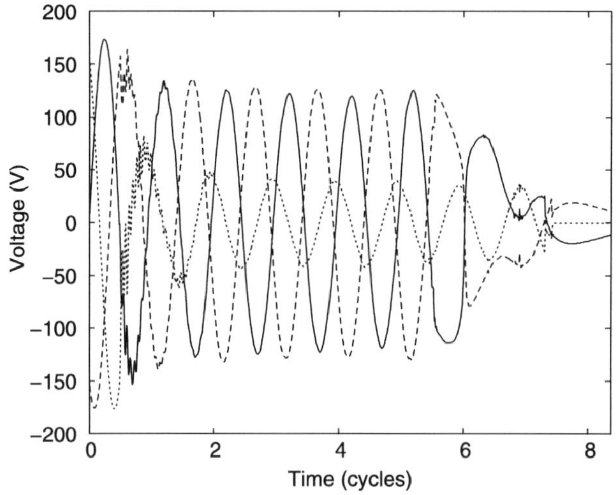


**Figure 6.1** Interruption due to clearing of single-phase fault.

motors connected to the supply will have lost most of their air gap field already during the fault so they cannot contribute much more to the voltage after the fault is cleared.

The start of a planned interruption is shown in Figure 6.2. The measurement was done in a university laboratory when the power supply was shut down for several hours due to building activities nearby. The switching took place on the primary side of the 10-kV/400-V transformer supplying the laboratories. The transformer is delta connected on the primary side and star connected and grounded on the secondary side—a typical configuration for distribution transformers. When the

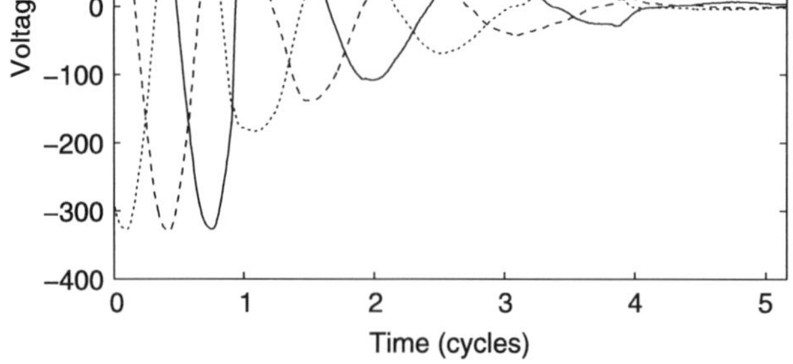


**Figure 6.2** Interruption due to intentional opening of switching device.

first phase opens on the primary side, the voltage in two phases on the secondary side drops to a value opposite in phase but half the magnitude of the voltage in the third phase. This third phase is not affected. Upon opening the second and third phases, (which happens at the same time because there is no ground return), the secondary side is completely on its own [33, Section 3.6]. The three phases cannot open at the same time because a switching device can only interrupt at current zero. The current zero crossings come 60° separated in a three-phase system.

After the load is disconnected, it takes a few cycles before the voltage becomes zero. This is due to the rotating load that is connected to the system. Induction and synchronous motors will start to operate in a generation mode when the supply disappears. The amount of energy present is rather small, however. Induction motors can only supply for a few cycles as the air gap field decays. Without an air gap field there can be no energy transfer between the rotor and the stator. A synchronous machine with independent field voltage will maintain its air gap field and can maintain the voltage longer, up to several seconds, depending on the amount of other load connected to the system.

Another example of the start of an interruption is shown in Figure 6.3. Again the interruption is due to the opening of a switching device without a fault. The decay of the voltage takes much longer than in Figure 6.2, which indicates the presence of a significant number of rotating machines in the system. The machines slow down during the interruption, which in turn results in a reduction in voltage frequency. The frequency of the voltage has been estimated for the waveform shown in Figure 6.2 by measuring the time between voltage zero crossings in phase $a$. The

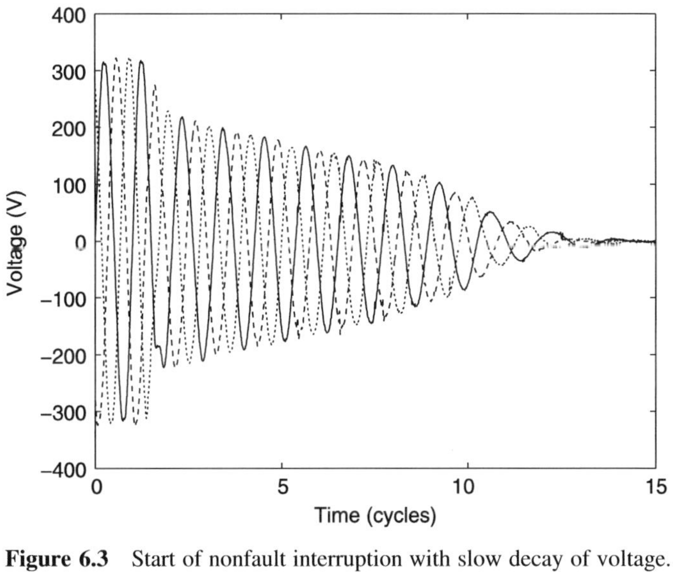


**Figure 6.3** Start of nonfault interruption with slow decay of voltage.

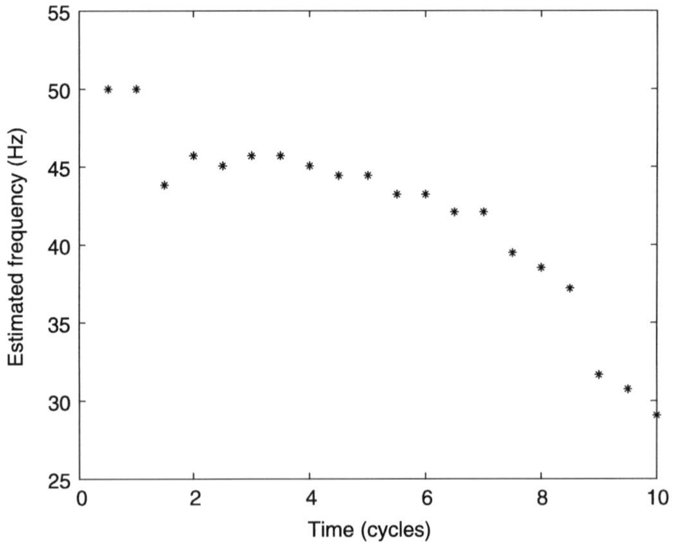


**Figure 6.4** Decrease in frequency during beginning of interruption.

results are shown in Figure 6.4. The frequency drops to 45 Hz almost immediately and drops below 40 Hz about five cycles later.

In [282] a distinction is made between nonfault interruptions (those merely due to the opening of a switching devices) and fault-related interruptions (those due to the operation of a protection device after a fault). Two causes are given for overvoltages associated with the start of a nonfault interruption: transformer saturation and current chopping. The former phenomenon occurs when transformers are connected to the feeder being tripped. The feeder capacitance discharges via the transformer magnetizing reactance. The phenomenon is very similar to ferroresonance but without the chaotic behavior. Current chopping typically occurs when a small current is interrupted. The breaker arc becomes unstable, leading to interruption before the natural current zero. This in turn leads to overvoltages when the inductor energy transfers to the system capacitance around the breaker location.

Two measurements with overvoltages after opening of the interrupted device are shown in Figure 6.5. Especially in the right-hand plot the overvoltage is rather severe. Such overvoltages are related to the presence of stored energy on the load side of the interrupting device. This can be in the form of electrical energy in a capacitor, magnetic energy in an inductor or transformer, or rotational energy in electrical machines. The presence of distributed generation will add a source of energy and will thus increase the risk of overvoltages during the start of an interruption. This is one of the reasons that distributed generation is typically equipped with very sensitive *islanding detecting*.

When interruptions are defined as a voltage magnitude below 10% of nominal, a fourth cause occurs. A short circuit or earth fault close to a substation will force the

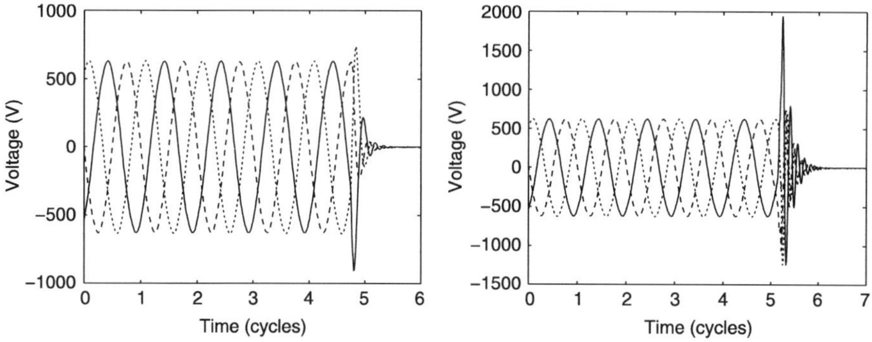


**Figure 6.5** Two examples of interruption due to opening of switching device, with overvoltage at beginning of event.

voltage to a very low value in one or more phases. Even though a monitor would classify such an event as an "interruption" from an origin view, it is similar to a voltage dip. The likelihood of such an event is however very small. The fault would have to be just outside the substation or inside the substation without leading to an interruption. For the purpose of this chapter we will refer to those events as voltage dips.

A number of terms are in use that are related to the term interruption. A *blackout* is an interruption that affects a wide geographical area or a large number of customers for a long time. There is no clear distinction between a blackout and an ordinary interruption and the term blackout is sometime incorrectly used for any interruption. In some cases it is needed to temporarily disconnect part of the load to prevent overloading of the system. The term *rotating interruption* is used as different customers are disconnected part of the time to spread the burden. This is also sometimes referred to confusingly as a *brownout* or a *rotating blackout*. Finally, it should be noted that the term *outage* is still often used as a synonym for *interruption*. An outage is the forced or planned removal of a component in the power system; this does not have to lead to an interruption, although at the distribution level, component outages and supply interruptions are closely related.

### 6.1.3 Restoration and Voltage Recovery

The cause of the interruption determines the way in which the voltage drops to zero, as shown in the previous section. The restoration process determines when the voltage recovers. One may distinguish between three types of restoration:

- Automatic switching, leading to interruption durations up to a few minutes
- Manual switching, leading to interruption durations up to a few hours
- Repair or replacement of the faulted component, which may last several days

The earlier mentioned distinction between short (momentary) and long (sustained) interruptions is related to the difference in restoration time between automatic and manual restoration. Short interruptions are typically due to faults on overhead lines that are restored by autoreclosure actions.

The voltage and current waveforms upon restoration depend to some extent on the duration of the interruption. Restoration after a short interruption will show more dynamic effects from motor loads and transformers. A problem after long interruptions is the heavy steady-state current. Equipment with a thermostat, such as air conditioners, electric heaters, and refrigerators, have a natural diversity under normal operation. But after a long interruption all equipment will be in the on state: All rooms will be too cold on a winter's day or too hot on a summer's day. This *postinterruption inrush* is a serious problem during the restoration after an interruption affecting a large area [4]. It will also have to be considered when planning rotating interruptions.

Three dynamic phenomena take place on a shorter time scale immediately after an interruption:

- Many electronic devices are equipped with a small capacitor on the ac side of the diode rectifier. This capacitor is often part of the *EMC filter*. Upon voltage recovery after an interruption, this capacitor takes a large current. Currents up to 100 A have been measured for ordinary computers and televisions. The total current to a room full of computers may be such that the circuit breaker or fuse will immediately trip again. The transient seen elsewhere (i.e., for those customers not experiencing an interruption) is similar to a capacitor-switching transient [297].
- Starting of an induction motor takes a current of five to six times nominal. In systems with a large amount of motor load the total current after an interruption may be such that the voltage does not recover. This is one of the reasons why such installations are equipped with undervoltage protection tripping the motors when the voltage stays too low for too long: Typical settings for undervoltage relays are around 70% for 1 to 5 s. Contactors typically trip when the voltage drops below 50% for a few cycles. (The other reason for undervoltage protection is safety: When the voltage comes back unexpectedly, the sudden start of the motors could cause personal injuries.) For very short interruptions the motor loads may still be connected, which may lead to severe dynamic transients in current as well as in voltage. If the current does not lead to tripping of a breaker or fuse, the voltage waveform will show a reduced voltage for a certain period after the interruption.
- If the interruption occurs at a medium-voltage level, the restoration may lead to the saturation of transformers. The saturation lasts longer when the transformers are lightly loaded. For short interruptions there may be some remanent magnetism present in the core, leading to very severe inrush currents. As with motor inrush this may result in a new interruption or a period of lower voltage after the restoration of the supply.

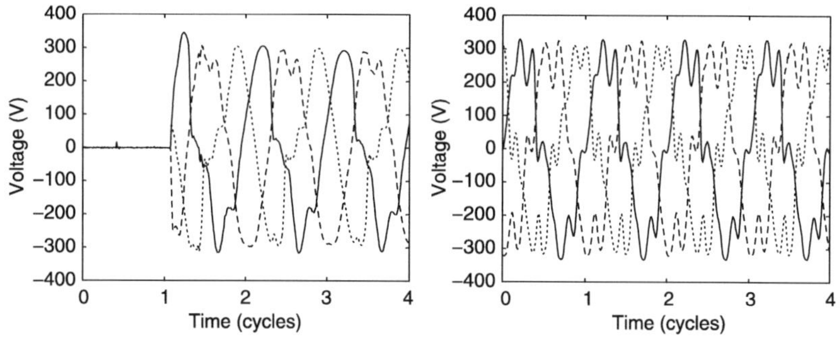


**Figure 6.6** End of interruption with heavy transformer saturation: immediately after reclosing (left); 2 s later (right).

An example of transformer saturation associated with the voltage recovery after a short interruption is shown in Figure 6.6. The voltage waveform at the start of this interruption is shown in Figure 6.3. The duration of this interruption was about 15 s. In this case the transformer saturation lasted for several seconds. The second waveform is obtained about 2 s after the first one. Although the waveform has changed, there remains heavy distortion. This sustained phenomenon could be due to a resonance between the capacitance connected to the bus and the magnetizing reactance of the transformer. We will come back to this phenomenon when discussing transformer saturation after a voltage dip in Section 6.2.2.8.

Another recording of the voltage recovery after a short interruption is shown in Figure 6.7. This recording was obtained about 8 s after the one in Figure 6.1. The

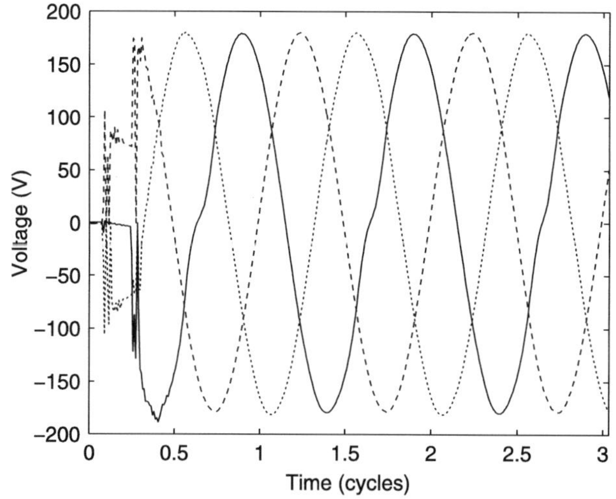


**Figure 6.7** End of interruption with difference in reclosing instants and mild transformer saturation.

voltage waveform shows some mild transformer saturation and a delay in the reclosing between the different phases. The reclosing took place at the primary side of a delta–star-connected transformer.

### 6.1.4 Multiple Interruptions

Due to automatic switching actions an interruption can be ended within a short time. The practice of fast autoreclosure is rather common for both distribution and transmission lines. For transmission lines it may save the system from instability. At the distribution level it will limit the duration of an interruption. A large fraction of the faults on overhead lines are of a transient nature. They do not require any repair, just a short time for the arc to disappear. After that the line can be energized again. For distribution systems a minor dip may occur even for a successful autoreclosure due to the inrush current of transformers and motors connected to the distribution feeder.

But the fault has not in all cases disappeared upon autoreclosure. In that case the customers will experience a second interruption, typically a longer one. An example of such an event is shown in Figure 6.8. This event is due to a fault on a 130-kV line that was in the radial operation at that moment due to the outage of a nearby line a few minutes earlier. A heavy lightning storm was moving through the area when these recordings were made. The measurement is performed at 10 kV. The figure plots the half-cycle rms voltages updated every half-cycle.

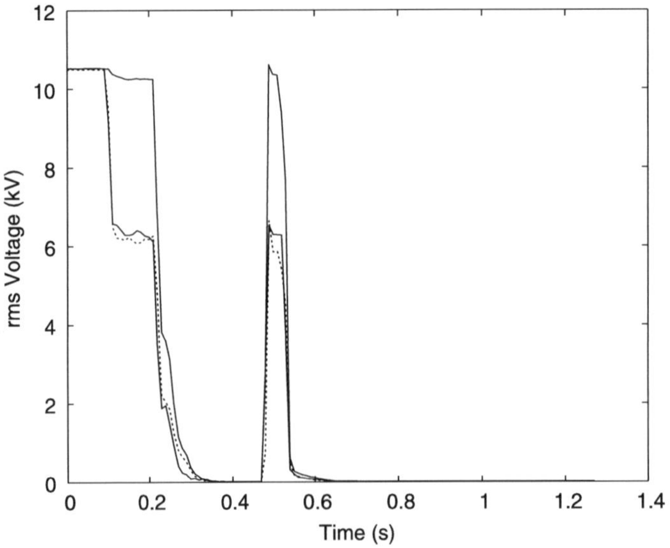


**Figure 6.8** Short interruption followed by long interruption: unsuccessful autoreclosure of 130-kV line.

Upon fault initiation two of the voltages drop in rms value. A few cycles after that all three voltages drop to zero due to the opening of an upstream circuit breaker. After about 300 ms a reclosing is attempted. But as the fault remains present, two of the rms voltages are depressed. Note that the voltages during the reclosing attempt are the same as during the initial voltage dip. The recurring overcurrents lead to permanent tripping of the line. The result for the end users is a long interruption, which in this case lasted about 20 min.

## 6.2 VOLTAGE DIPS

### 6.2.1 Causes of Voltage Dips

Voltage dips are short-duration reductions in voltage magnitude. Their duration is typically less than 1 s. The residual voltage (the voltage magnitude during the event) may be anywhere between close to zero and close to the nominal voltage. Typically a monitor is triggered for a voltage-dip recording when the rms voltage drops below 90% of nominal.

Voltage dips are currently one of the main power quality issues. Especially industrial customers suffer from regular production stoppages due to voltage dips. We will not go into further detail of the consequences of voltage dips, nor will we discuss mitigation methods. An extended overview of consequences and mitigation methods for voltage dips is given in [33, Chapters 5 and 7].

Voltage dips are due to short-duration increases in current magnitude, in most cases somewhere else in the system. Consider the simplified system shown in Figure 6.9.

The short-duration increase in current gives an additional voltage drop at the indicated bus. This bus is referred to as the *point of common coupling* (*between the customer and the fault*), or PCC. Note that the PCC may be different for different customers. All customers supplied from the indicated bus will experience a reduced

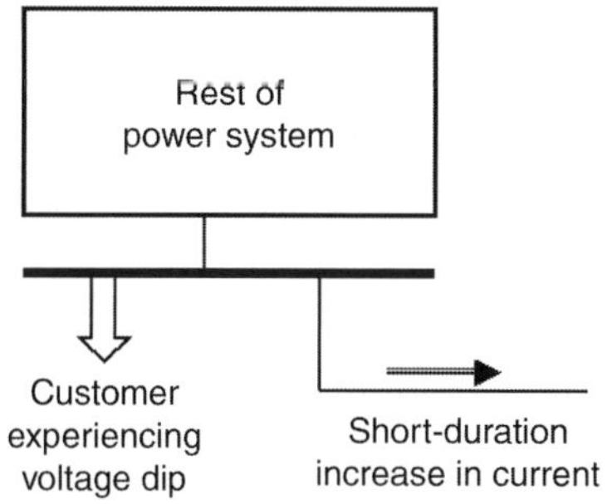


**Figure 6.9** Origin of voltage dips.

voltage during the increased current—thus a *voltage dip*. There are many causes for temporary increase of current, but the three main causes are as follows:

- Short circuits and earth faults
- Starting of induction motors
- Energizing of transformers

A fourth cause, capacitor energizing, would have to be added here for completeness. However, the duration of the overcurrent due to capacitor energizing is only a few milliseconds and the resulting voltage event is typically classified as a transient. We will discuss capacitor energizing in detail in Section 6.3.

Faults are the main cause of voltage dips discussed in the power quality literature. Many publications, including several by the authors of this book, do not even discuss other causes of dips. The reason is that the majority of equipment malfunctions are due to dips caused by faults. However, a power quality monitor will capture dips due to all causes, so that it is important to know the difference between dips due to different causes.

Voltage-dip duration is mostly in the range between a few cycles and several seconds. It is not possible to give an exact range of what is a voltage dip, but a widely accepted range in duration is from one half-cycle up to 1 min. Based on the definition of a voltage dip given above, it is easy to realize that the main characteristics describing a voltage dip are the residual voltage and the duration. These two terms are defined for measurements in the standard document IEC 61000-4-30. Next to that some other characteristics have been introduced in the literature, such as phase-angle jump, point on wave, and three-phase characteristics. The main reason for introducing these additional characteristics is that the behavior of load during voltage dips cannot be fully explained by magnitude and duration only. The additional characteristics also provide additional information on the power system event that led to the voltage dip.

### 6.2.2 Voltage-Dip Examples

***6.2.2.1 Three-Phase Faults in Transmission Systems*** An example of a voltage dip due to a three-phase fault is shown in Figure 6.10. The curve on the left shows the voltages in the three phases as measured by a power quality monitor. The voltages show an equal drop in the three phases (a balanced voltage dip) followed by a recovery about five cycles after the start of the dip. The initial voltage drop corresponds to fault initiation; the recovery starts when the fault is cleared by the protection. The actual drop in voltage due to the fault is better visible on the right where the rms value of the voltage is shown as a function of time. The rms is obtained over a 10-ms (one-half-cycle) rectangular window which shifts through the waveform. Most power quality monitors will calculate not the fundamental component but the rms value over a one-cycle or half-cycle window. The result will be close to the magnitude of the fundamental component.

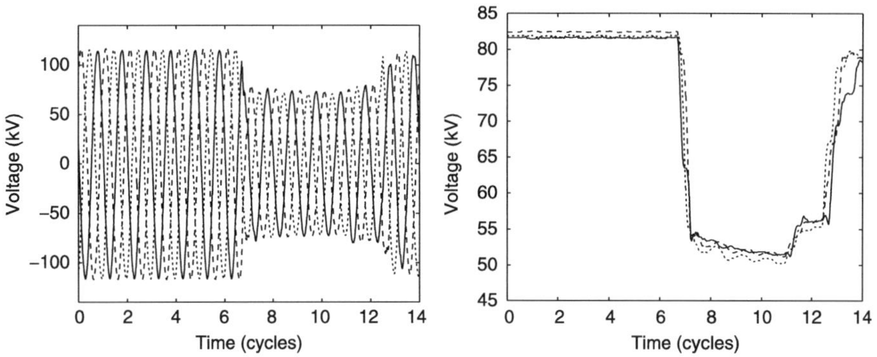


**Figure 6.10** Voltage dip due to three-phase fault in transmission system: voltage waveforms (left); rms value (right).

The method for calculating the rms voltage and other methods for obtaining voltage-dip magnitude are discussed in more detail in Chapter 8.

Due to the combined load of the system, the fundamental voltage changes slightly during the fault. After the fault the voltage does not recover immediately, again due to load effects. Additionally the voltage recovery may lead to widespread transformer saturation causing even-harmonic distortion in the voltages. Some examples will be shown later.

Also visible in Figure 6.10 is the two-stage recovery in voltage. This is due to the opening of the circuit breakers at the different sides of the transmission line at slightly different instants. The presence of two-stage recovery is an indication that the dip is due to a fault at the transmission or subtransmission level.

The recovery of the voltage is not instantaneous due to the dynamic behavior of the load. Especially induction motor load takes a larger current after a fault than during normal operation. Also transformers have been shown to lead to a similar phenomenon, with the difference that the postfault dip due to motor reacceleration is balanced whereas the postfault dip due to transformer saturation is unbalanced with a high level of (odd and even) harmonic distortion. An example of a dip due to a three-phase fault followed by a slow recovery is shown in Figure 6.11. As the recovery is balanced it is mainly due to the postfault inrush currents taken by rotating machines. Also a synchronous generator, as used in large power stations, will show a slow recovery after a fault: The slow recovery is due not to a high inrush current but to the slow recovery of the terminal voltage of synchronous machines. With increasing penetration of distributed generation this may also appear at distribution levels.

***6.2.2.2 Three-Phase Faults in Distribution Systems*** An example of a voltage dip due to a three-phase fault in a distribution system is shown in Figure 6.12. Next to the rms voltage, the phase angle is shown as a function of time. The rms voltage is obtained over a half-cycle rectangular window. The

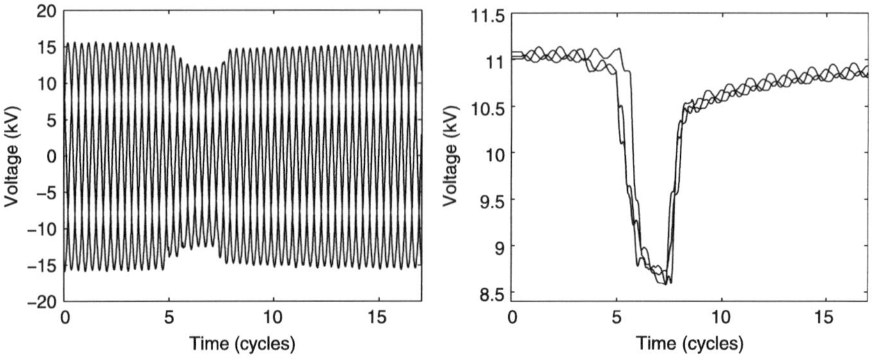


**Figure 6.11** Voltage dip due to three-phase fault at transmission level: waveform (left); half-cycle rms voltage (right).

phase angle is obtained from the fundamental (50-Hz) component taken over a one-cycle rectangular window.

The phase-angle jump in Figure 6.12 has a rather large value. Such a large phase-angle jump points to a fault at the distribution level. In distribution systems, the source impedance is typically formed by a transformer with a rather

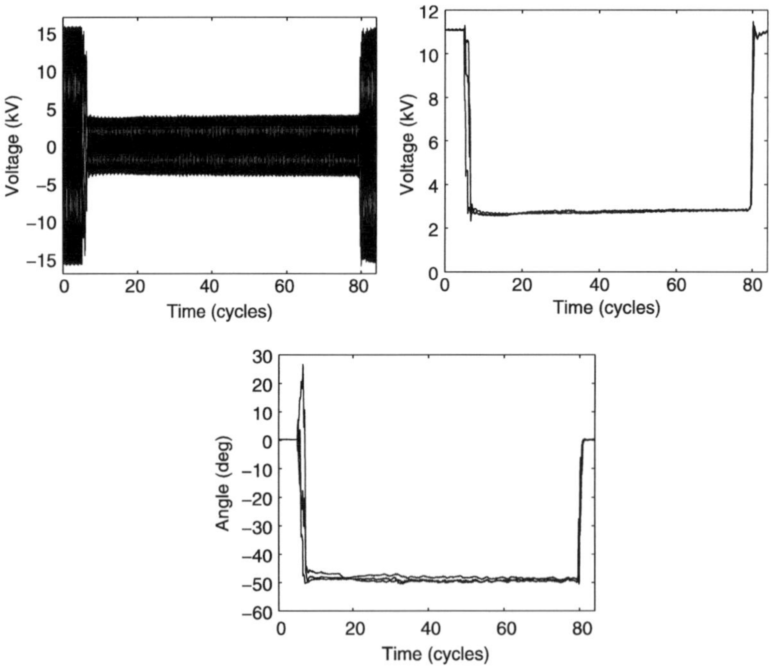


**Figure 6.12** Voltage dip due to three-phase fault in distribution system: waveforms (left); rms voltages (right); phase angles (bottom).

large $X/R$ ratio, whereas the impedance to the fault is formed by lines or cables with a much smaller $X/R$ ratio. This will lead to a significant change in phase angle, especially for cable faults. Phase-angle jumps are discussed in detail in [33, Section 4.5], where it is shown that there is a relation between the drop in voltage and the phase-angle jump for any given feeder. The measured residual voltage and phase-angle jump may be used to identify the faulted feeder.

#### 6.2.2.3 Nonsymmetrical Faults in Transmission Systems

A voltage dip measured at 11 kV due to a nonsymmetrical fault at the transmission level is shown in Figure 6.13. Both the waveforms and the half-cycle rms values are shown for the three line voltages. As before, the character of the event is much better visible from the rms voltages than from the waveforms. One voltage shows a significant drop in magnitude whereas the other two show a minor drop. The drop in the two latter phases is very similar. For almost all dips due to nonsymmetrical faults at the transmission level two of the three voltages have about the same magnitude. (We will see later that this is not the case for dips due to distribution system faults.) The two-stage recovery is another indication that the dip is due to a fault at a higher voltage level.

A fault at the transmission level causes a dip at the transmission substation, which next propagates down to the equipment terminals at a much lower voltage level. The magnitudes of the three voltages do not stay the same during this downward propagation, as is shown in Figure 6.14, which shows the voltage dips due to the same single-phase fault as recorded at three voltage levels. Only the rms voltages as a function of time are shown. The voltage dip at 132 kV shows a large drop in voltage magnitude in one phase, together with a small drop in the two other phases. The voltage drops in the latter two phases are the same. We will learn later in this chapter that this is a clean type D dip.

The voltage dip as measured at 10 kV also shows a large drop in one phase and a minor drop in the two other phases. There are however two differences compared to

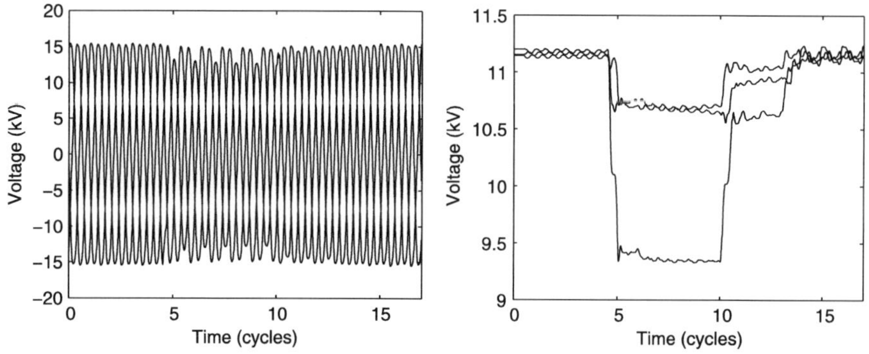


**Figure 6.13** Voltage dip due to nonsymmetrical fault: waveforms (left); rms voltages (right).

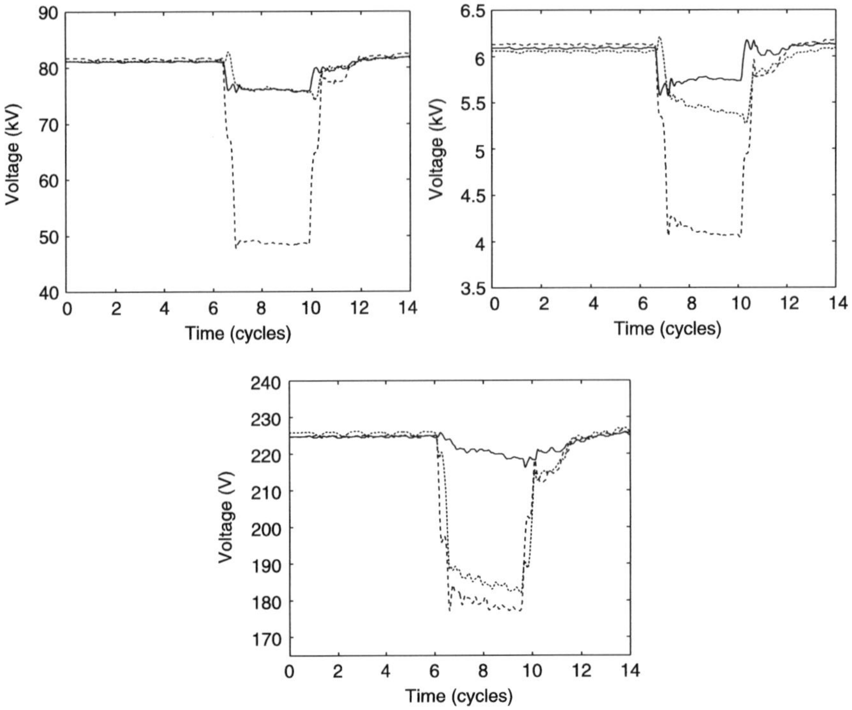


**Figure 6.14** Voltage dip of single-phase fault at 400 kV: measured at 132 kV (left), at 10 kV (right), and at 400 V (bottom).

the measurement at 132 kV. The first difference is that the rms voltages are no longer constant during the fault. The second difference is that the rms voltage in the two least affected phases are no longer the same. Both effects are due to the fact that the load currents vary during the fault.

The voltage dip at 400 V looks completely different: It shows a large drop in two phases and almost no drop in the third phase. We will later refer to this as a type C dip. The voltages show a slow further drop during the fault in all three phases. After the fault the voltage does not fully recover. The difference in type between 10 kV and 400 V is due to the Dy transformer between these voltage levels. The slow drop in voltage during the fault and the slow recovery are due to the dynamic behavior of the load. A general model for the propagation of voltage dips from transmission to the equipment has been discussed in [33, Section 4.4] and will also be taken up in Section 6.2.3. The impact of load on the voltage-dip propagation has be discussed in [41, 39].

Another example of a voltage dip due to a nonsymmetrical fault at the transmission level is shown in Figure 6.15. In the curves on the left we see that the

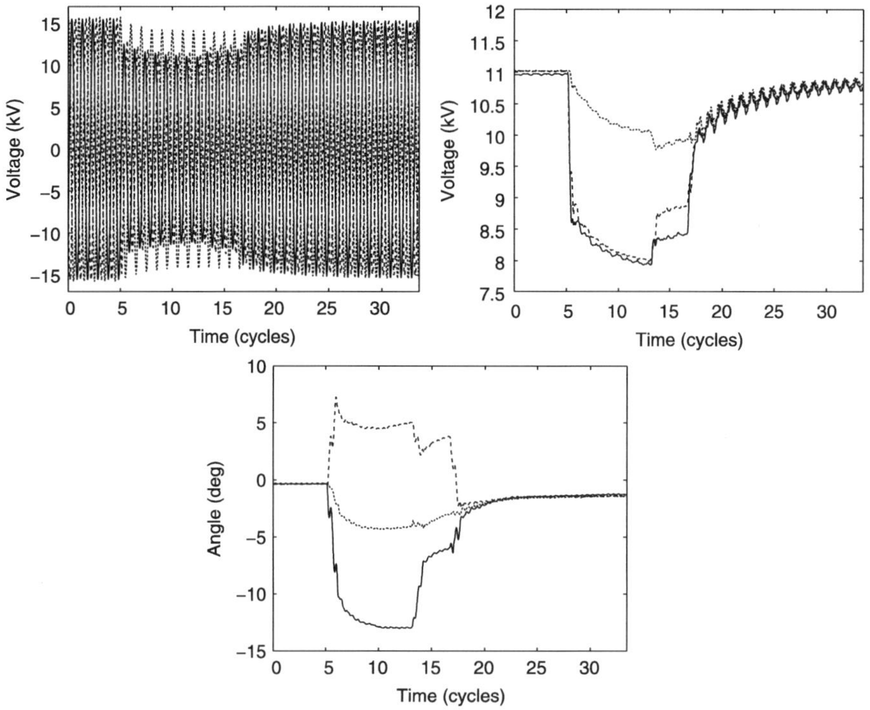


**Figure 6.15** Unbalanced voltage dip with heavy load influence: waveforms (left); rms voltages (right); phase-angle jump (bottom).

rms voltages are equal in two phases and that the third phase shows only a small drop in voltage. During the course of the fault all three voltages slowly drop further. This is due to the load, mainly induction motors, gradually taking more current from the supply. After about eight cycles the faulted transmission line is opened on one side. This leads to an immediate rise of the voltage in two of the phases and a small drop in the third phase. Another three cycles later the second breaker also clears the fault and the system recovers. It is interesting to note that the load already starts to recover after the opening of the first breaker (observed as the slow rise in rms voltage in all three phases). After clearing of the fault it still takes rather long for the voltage to completely recover. This is most likely due to the presence of significant amounts of motor load near the fault or near the monitor location. The oscillations in the rms voltage after fault clearing are due to the presence of even-harmonic distortion in the voltage waveforms caused by the saturation of transformers upon voltage recovery.

The voltage dip in Figure 6.16 is due to a two-phase-to-ground fault at the transmission level. Compared with Figure 6.14, this shows a significantly larger initial drop in rms voltage in the phase with the smallest drop. This type of dip will later

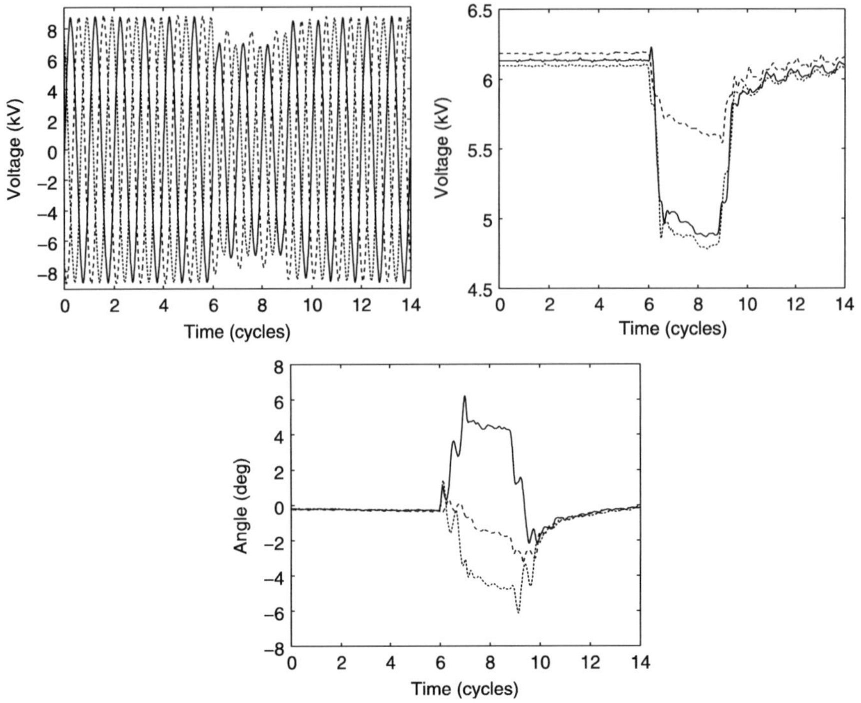


**Figure 6.16** Voltage dip at 10 kV due to two-phase-to-ground fault at the transmission level: waveforms (left); rms voltages (right); phase angles (bottom).

be referred to as type G. The effect of the load on the voltages is clearly visible. After fault clearing we see a slow recovery and some minor transformer saturation.

The voltage dip in Figure 6.17 is also due to a two-phase-to-ground fault at the transmission level. But in this case the voltage shows a large drop in two phases and a minor drop in the third phase. This dip will later be referred to as a type F dip. Note also the two-stage recovery clearly visible in the rms voltage but not in the phase angle. The rms voltage is obtained over a half-cycle window and has thus a higher time resolution that the phase angle which is obtained over a one-cycle window.

#### *6.2.2.4 Nonsymmetrical Faults in Distribution Systems*

A nonsymmetrical fault in a distribution system results in a nonbalanced dip with an additional phase-angle jump. We will discuss the theory behind this in more detail in Section 6.2.4 (see also [33, Section 4.6]), but for now it is sufficient to know that this leads to an additional unbalance between the phases. For a fault at the transmission level, that is, one without a phase-angle jump, the resulting rms voltage is about the same in two of the phases. The phase-angle jump (the term *characteristic phase-angle jump* will be used later) leads to an additional unbalance between

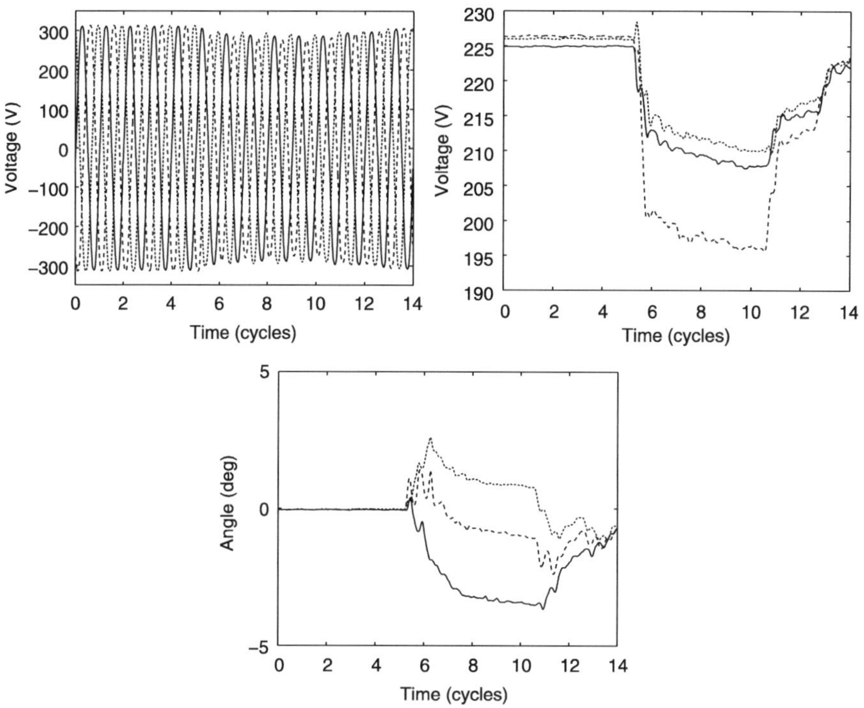


**Figure 6.17** Voltage dip at 400 V due to two-phase-to-ground fault at transmission level: waveforms (left); rms voltages (right); phase angles (bottom).

the phases. The result is that the rms voltages are different in the three phases. Figure 6.18 shows a recorded voltage dip due to a nonsymmetrical fault at the distribution level. This is still a voltage dip of type D, that is, with a large voltage drop in one phase and a minor drop in the two other phases. The phase with the lowest voltage also shows the largest phase-angle jump ($-40^\circ$), which is in this case equal to the characteristic phase-angle jump. A three-phase fault at the same location would have resulted in a phase-angle jump of $-40^\circ$ in all three phases.

Another example is shown in Figure 6.19; in this case the dip shows a large drop in two phases and almost no drop in the third phase. The rms voltages in the two faulted phases are no longer the same, neither is the phase-angle jump. Note that the phase with the largest drop in voltage shows a smaller phase-angle jump.

A special category of voltage dips are those due to single-phase-to-ground faults. Any single-phase fault leads to a zero-sequence voltage at the fault location and at the same voltage level as the fault. However, the zero-sequence component does not propagate through a Dy transformer and is also not measured by delta-connected equipment or monitors. In a solidly grounded system, a single-phase fault leads to a zero-sequence voltage that is about equal to the negative-sequence voltage and to the drop in positive-sequence voltage. The result in terms of phase voltages is

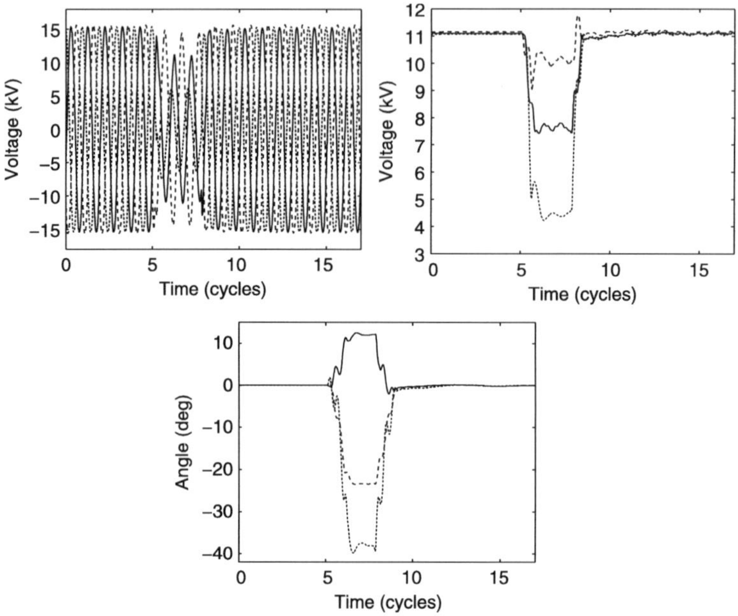


**Figure 6.18** Voltage dip at 10 kV due to nonsymmetrical fault at distribution level: waveforms (left); rms voltages (right); phase angles (bottom).

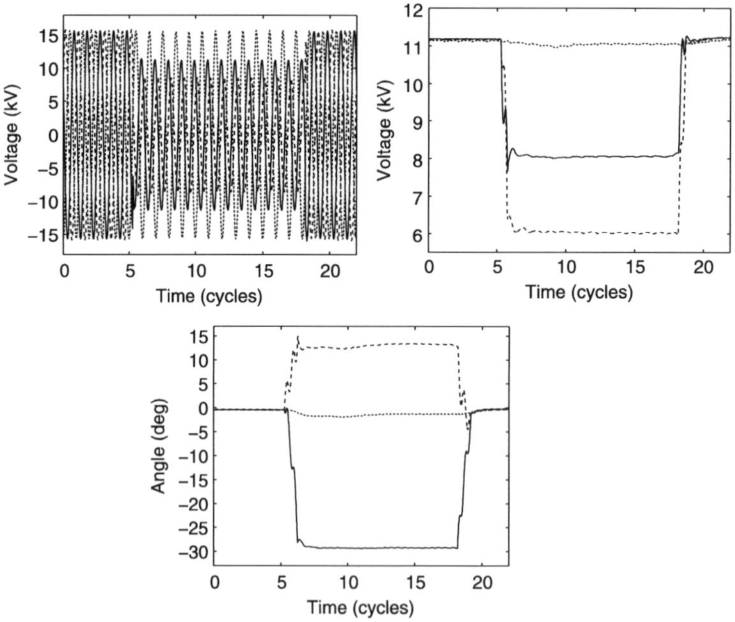


**Figure 6.19** Voltage dip at 10 kV due to nonsymmetrical fault at distribution level: waveforms (left); rms voltages (right); phase angles (bottom).

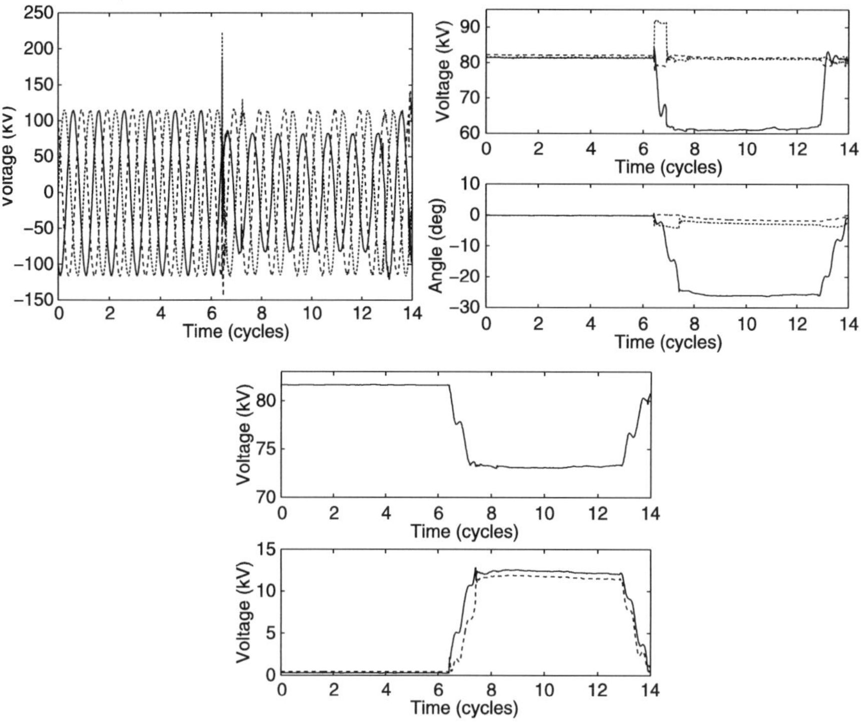


**Figure 6.20** Voltage dip due to single-phase fault recorded at same voltage level as fault: voltage waveforms (left); rms voltage (right top); phase angles (right bottom); positive-sequence voltage (bottom top); negative-sequence voltage (lower bottom, solid line); zero-sequence voltage (lower bottom, dashed line).

either a small drop or a small rise of the voltages in the nonfaulted phases. An example of such a dip is the one preceding the interruption in Figure 6.1.

An example of a voltage dip due to a single-phase fault in a solidly earthed system is shown in Figure 6.20. Both the fault and the monitor are located in the same 130-kV system. The waveform shows an overvoltage at the start of the dip in all three phases. This overvoltage is almost certainly related to the cause of the fault. During the dip the voltage in one phase (the *faulted phase*) is suppressed with a large phase-angle jump, whereas the voltage in the other two phases (the *nonfaulted phases*) is almost not affected. In terms of symmetrical components, we see a drop in (the magnitude of the) positive-sequence voltage and a rise in negative- and zero-sequence voltages. The magnitudes of negative- and zero-sequence voltages are similar, which is typical for single-phase faults in solidly earthed systems. It is easy to prove that no impact of the fault on the nonfaulted phases is equivalent to zero- and negative-sequence voltages being equal.

Another example of a voltage dip due to a single-phase fault is shown in Figure 6.21. The recording was made at the same location as the dip in

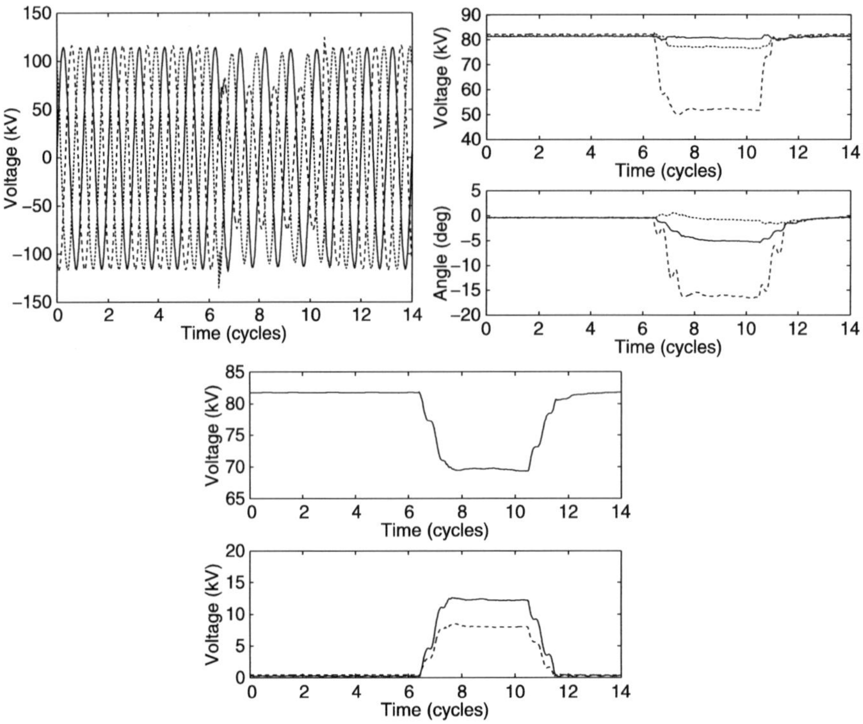


**Figure 6.21** Voltage dip due to single-phase fault at higher voltage level: voltage waveforms (left); rms voltage (right top); phase angles (right bottom); positive-sequence voltage (bottom, top); negative-sequence voltage (lower bottom, solid line); zero-sequence voltage (lower bottom, dashed line).

Figure 6.20, but the fault occurred one voltage level up, at 400 kV. A number of interesting differences can be observed between the two events, partly due to the different origin of the event, partly due to its propagation to a lower voltage level. The fault-clearing time for the 400-kV fault is less than four cycles, whereas it is about seven cycles for the 130-kV fault. The 400-kV grid is used for the bulk transport of power where fast fault clearing is essential to maintain stability after a fault. The phase-angle jump in the faulted phase is smaller for the dip originating at 400 kV even though the residual voltage is much lower. This is related to the difference in $X/R$ ratio between 400- and 130-kV lines, as will be explained in more detail in Section 6.2.4. The transformer between the 400- and 130-kV systems is an autotransformer with a delta winding. The effect of the delta winding is a reduction of the zero-sequence voltage to about two-thirds of its original value. The negative-sequence voltage is not affected by the transformer; it is still equal to the drop in positive-sequence voltage. For this location, the ratio between negative- and zero-sequence voltages can be used to find the voltage level at which the fault took

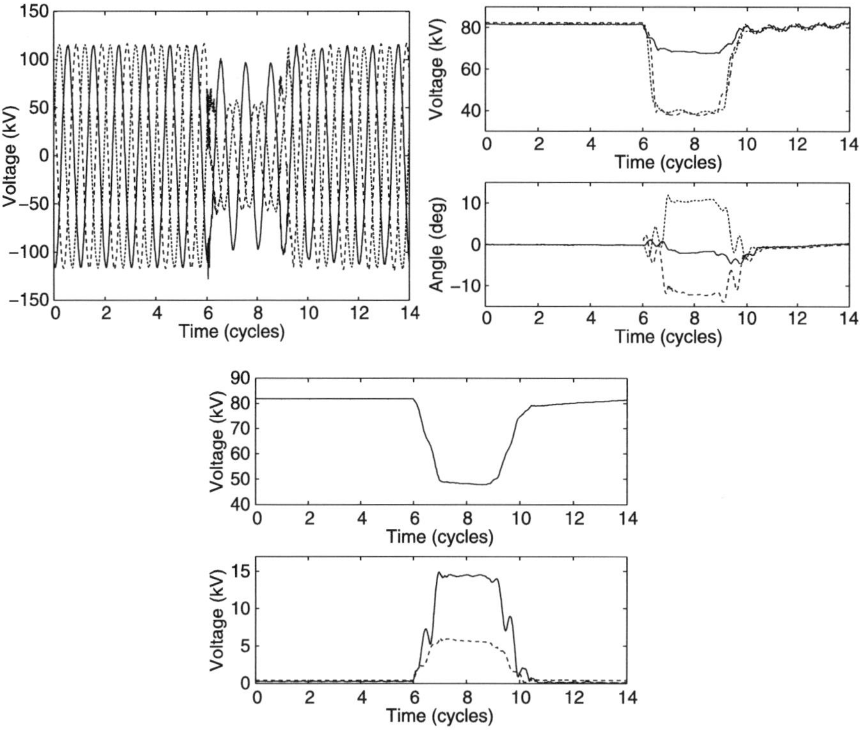


**Figure 6.22** Voltage dip due to two-phase-to-ground fault at higher voltage level: voltage waveforms (left); rms voltage (right top); phase angles (right bottom); positive-sequence voltage (bottom top); negative-sequence voltage (lower bottom, solid line); zero-sequence voltage (lower bottom, dashed line).

place. The result of the reduction in zero-sequence voltage is that the nonfaulted phases show a different magnitude and phase angle during the fault. In terms of the dip classification to be introduced in Section 6.2.3, the dip is between type B and type D.

A voltage dip due to a two-phase-to-ground fault is shown in Figure 6.22. The dip was recorded at the same location as the one in Figure 6.21 and took place at 400 kV, about 20 min earlier in time. The negative-sequence voltage is about half the drop in positive-sequence voltage and the zero-sequence voltage is less than the negative-sequence voltage. In terms of dip classification this one is between type E and type G.

The situation becomes completely different for single-phase faults in high-impedance grounded systems. As the zero-sequence source impedance is much higher than the positive- and negative-sequence impedances, the zero-sequence voltage dominates in the dip. The result is a significant change in the voltages in the nonfaulted phases. For a Peterson coil earthed system [193, Section 3.7.3] the voltages in the nonfaulted phases reach a value equal to the square root of three

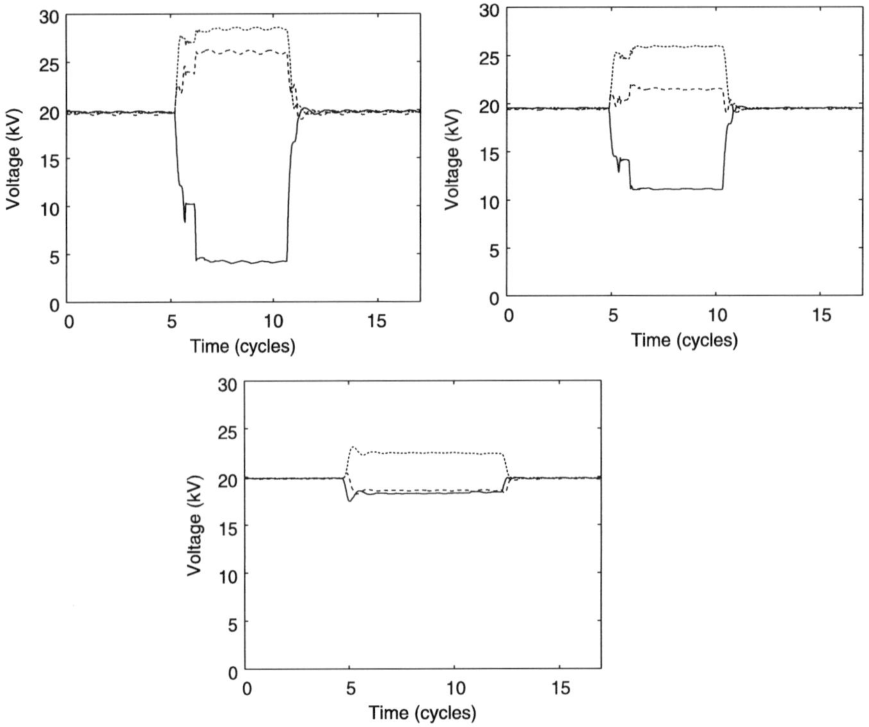


**Figure 6.23** Voltage dips at three different locations due to same single-phase fault; half-cycle rms voltages versus time.

times their normal value. The same holds for a resistance-earthed system with a very small fault current (so-called *high-impedance earthed systems*). The most interesting disturbances occur, however, for single-phase-to-ground faults in *low-resistance earthed systems*. The voltage drop over the earthing resistance results in a zero-sequence voltage that is at a large angle to the positive- and negative-sequence voltages. The result may be a voltage rise in one or two phases, depending on this angle. The same fault may cause a voltage rise in one phase at one location but a rise in two phases at another location. An example of this is shown in Figure 6.23: The three dips were all recorded at the same instant but the relation between the voltage magnitudes is different.

Another example of a fault in a resistance-earthed system is shown in Figure 6.24. The fault developed from a single-phase-to-ground fault into a two-phase-to-ground fault after about five cycles. The single-phase fault results in a voltage drop in one phase and a rise in two other phases. The two-phase-to-ground fault (second stage of the event) leads to a drop in two phases and a voltage rise in one phase. When plotting the symmetrical components the picture becomes completely different and much easier to interpret. During the single-phase fault, positive- and

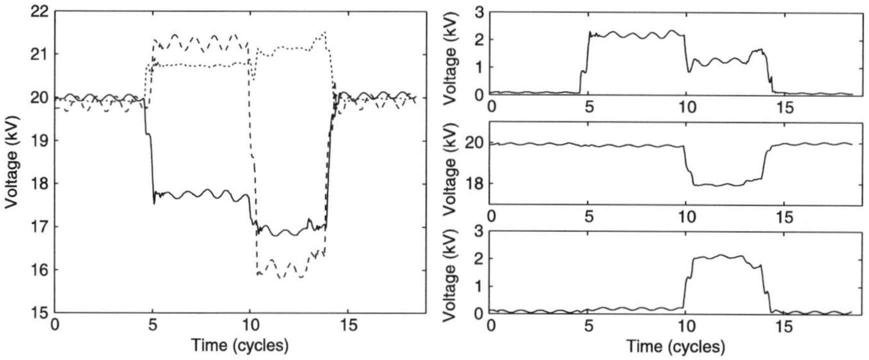


**Figure 6.24** Voltage dip due to developing fault: rms voltage (left); zero-, positive-, and negative-sequence voltages (right, top to bottom).

negative-sequence voltages are not affected. Only in the second stage do we see a drop in positive-sequence voltage and a nonzero negative-sequence voltage. Any end-user equipment supplied from the substation at which the recording is made would be located behind a Dy transformer or be delta connected. The equipment would thus not experience the zero-sequence voltage at all. Only the second stage of the dip may affect end-user equipment. The first stage of the event would not be classified as a voltage dip by the classification presented in Section 6.2.3.

***6.2.2.5 Multistage Voltage Dips*** Multistage voltage dips are dips during which the rms voltage shows sudden changes. Those changes may be due to changes in the fault (developing faults) or changes in the system (fault clearing). A fault on a transmission line is cleared by opening the breakers on both sides of the faulted line. In most cases this will not happen at exactly the same instant. The result is that the voltage recovery takes place in two stages. This two-stage recovery is very common for transmission system faults. It can be used to recognize voltage dips due to transmission system faults. Some examples were already presented in Figures 6.10, 6.13, and 6.15.

In some cases the difference in opening time for the breakers on both sides of the line is very big. Such a situation occurs, for example, when a fault takes place near one of the line terminals. An example is shown in Figure 6.25. The fault-clearing time is about 18 cycles for one line terminal and about 24 for the other one. The voltage dip is a type C in the line voltages.

In the previous sections a number of voltage dips due to symmetrical and nonsymmetrical faults were shown. A comparison of the different recordings shows that different fault types cause different dips. We will come back to this relation in more detail in Section 6.2.3. In some cases the fault type changes. This may be due to an arcing fault in which the arc spreads to other phases or to a cable fault where the insulation breaks down in different phases at different instants. These changes in fault type in turn lead to changes in dip type. An example of a dip due to a

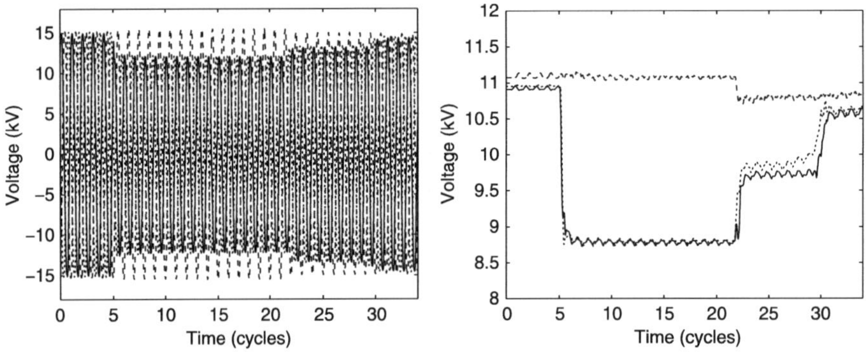


**Figure 6.25** Voltage dip with two-stage recovery: voltage waveforms (left); half-cycle rms voltages (right).

developing fault was shown in Figure 6.24. Another example is shown in Figure 6.26, again concerning a fault developing from a single phase to two phases to ground. The rms voltage has been calculated over a half-cycle window. The rms calculation has been updated at every sample, which in this case corresponds to 120 times per 60-Hz cycle.

Figure 6.27 shows a voltage dip due to a developing fault recorded at the medium-voltage level. The prefault rms voltage is about 27 kV. Another interesting feature is the heavy oscillation every time there is a change in the system: at fault initiation, when the fault develops into a three-phase fault, and at fault clearing. Note also the typical recovery sequence for a three-phase-to-ground fault, with oscillations for the clearing of every individual phase. The different recovery instants are discussed further in Section 6.2.5.

Figure 6.27 also shows the currents measured at the same location. Both at fault initiation and when the fault develops into a three-phase fault does the load current

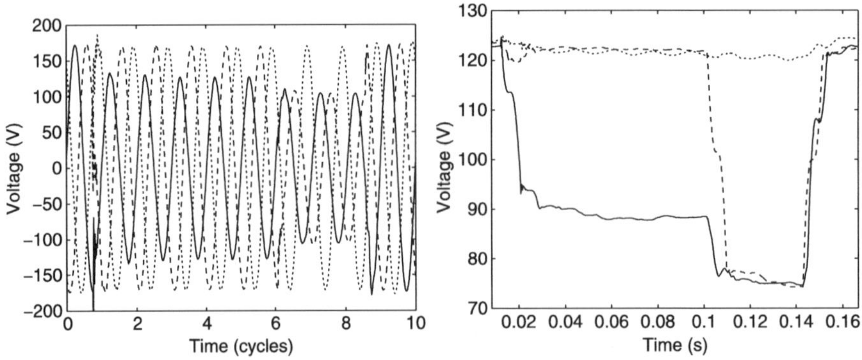


**Figure 6.26** Voltage dip due to developing fault: voltage waveforms (left); half-cycle rms voltages (right).

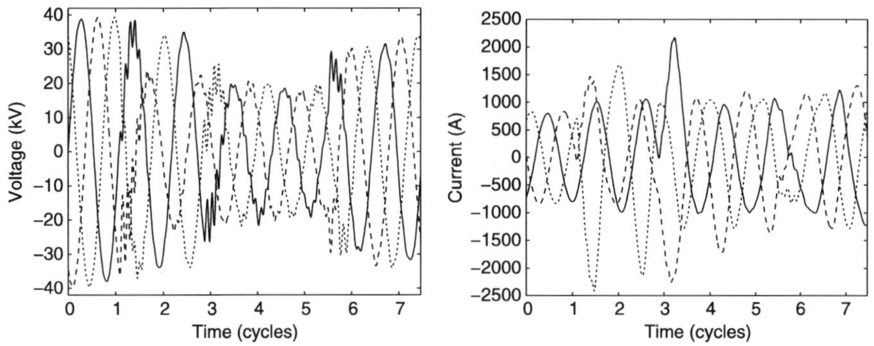


**Figure 6.27** Voltage dip due to developing fault: voltages (left); load currents (right).

show a phase reversal in one of the phases. This is due to induction motor load feeding into the fault. The heavy oscillations in voltage do not show up in the current. This indicates that there is not much capacitance connected to the system downstream of this location.

Figure 6.28 shows the voltage dip due to a fault that develops from a single phase through two phases to ground to three phases. Next to the half-cycle rms voltages, the half-cycle positive- and negative-sequence voltages are shown. Note that the negative-sequence voltage disappears when the fault develops to three phase. The appearance of a negative-sequence voltage with voltage recovery is related to the difference in recovery instant in the different phases.

A final example is shown in Figure 6.29. As in the previous example, the fault develops from a single phase to three phases. In this case the negative-sequence voltage remains the same when the fault develops from single phase to two phase. This can be explained from the expressions in [33, Sections 4.4.1 and 4.4.3] by assuming that positive- and zero-sequence impedances are equal. Note also the large motor influence at the beginning of the dip, after every development

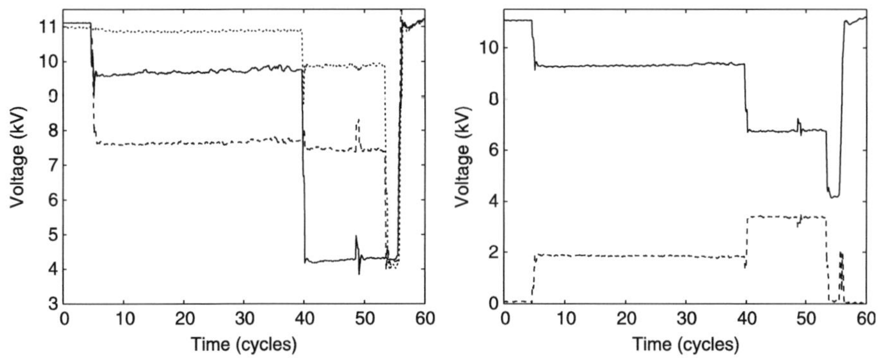


**Figure 6.28** Voltage dip due to developing fault: rms voltages (left); positive-sequence (right, solid line) and negative-sequence (right, dashed line) voltages.

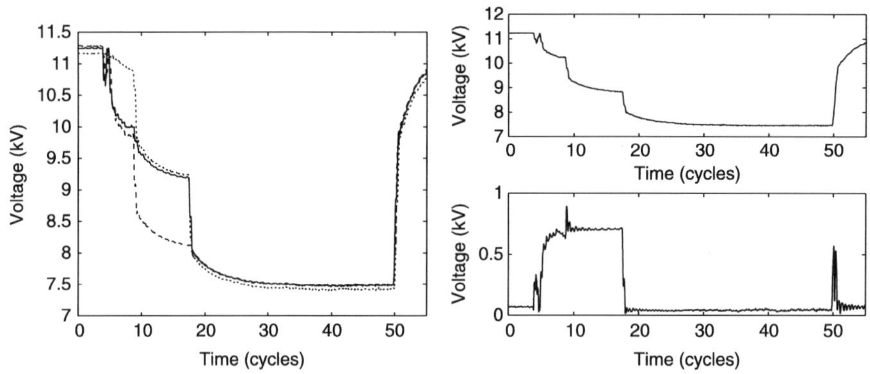


**Figure 6.29** Voltage dip due to developing fault: rms voltages (left); positive-sequence (right top) and negative-sequence (right bottom) voltages.

in the fault, and upon fault clearing. The motor influence is only visible in the positive-sequence voltage.

***6.2.2.6 Self-Clearing Faults*** All the dips shown earlier were terminated by the intervention of the power system protection. The vast majority of faults is cleared by this protection, but in some cases faults disappear before the protection gets the chance to intervene. Such faults are typically of very short duration and of a rather unstable character. An example of the voltage dip due to a self-clearing fault is shown in Figure 6.30. The voltage waveforms are shown for two of the phases; the third phase was less affected by the disturbance. The dotted lines indicate the preevent waveforms. This makes it easier to see where the voltages deviate from the normal voltage waveforms. The voltage waveforms show a repetition of very short duration faults: first in one phase, next in two phases, and finally in one

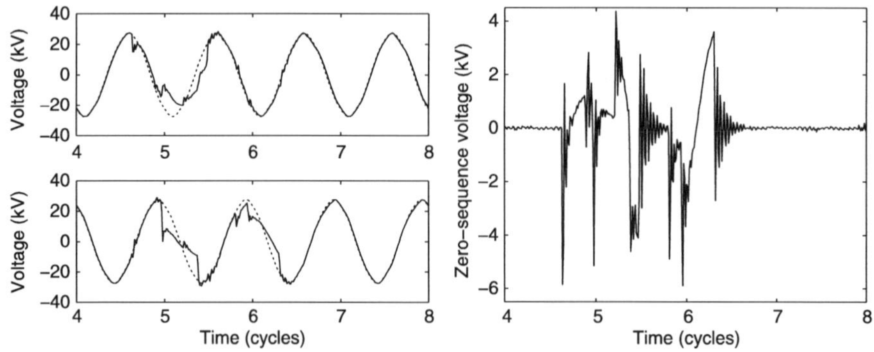


**Figure 6.30** Voltage dip due to self-clearing fault. Voltage waveforms (left) and instantaneous zero-sequence voltage (right).

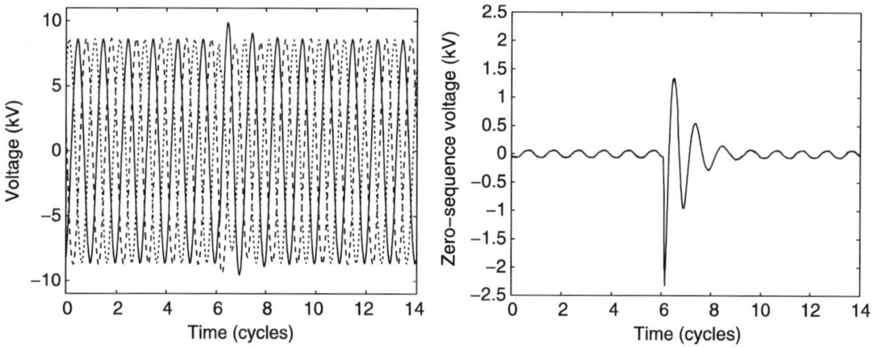


**Figure 6.31** Voltage dip due to self-clearing fault in reactance-earthed system. Voltage waveforms (left) and instantaneous zero-sequence voltage (right).

phase again. The instantaneous zero-sequence voltages (the average of the instantaneous voltages in the three phases) shows an oscillation after every fault clearing. This oscillation takes place between the zero-sequence source impedance and the total phase-to-ground capacitance in the system. This oscillation is very weakly damped because there is no load connected phase to ground that can provide any significant damping. The fast oscillations are only visible in the phase-to-ground voltages, not in the phase-to-phase voltages.

Another example of a voltage dip due to a self-clearing fault is shown in Figure 6.31. In this case the fault occurs in a reactance-earthed system. In such a system the grounding reactance is tuned to the total capacitance of the system so that the zero-sequence source impedance becomes very large. A single-phase fault will self-clear almost immediately, resulting only in a transient in zero-sequence voltage. This transient leads to an increase or decrease in the voltage magnitudes for the phase-to-ground voltages. The end customers are however not affected by this event. Note that the zero-sequence transient has a frequency close to 50 Hz. A similar event was recorded at the same location about 25 min earlier. About 5 min after the event in Figure 6.31 a dip due to a three-phase fault was measured, probably at the same location as the transient faults.

***6.2.2.7 Motor Starting*** Motor starting leads to a sudden simultaneous voltage drop in the three phases followed by a slow recovery. As an induction motor takes the same current in the three phases, the voltage drop is the same in the three phases; in other words, motor starting results in a balanced dip. Other large three-phase equipment will likely show similar voltage dips during starting.

In low-voltage systems, large single-phase loads may lead to short-duration voltage reductions during starting or even during normal operation. Examples that the authors came across during measurements at the point of utilization include refrigerators, vacuum cleaners, and water cookers.

**6.2.2.8 Transformer Energizing** Transformer energizing also results in a sudden voltage drop followed by a slow recovery, the same as motor starting. However, the drop is different in the three phases and the event is associated with heavy even-harmonic distortion.

An example of a voltage dip due to the energizing of a nonloaded transformer is shown in Figure 6.32. The figure on the left shows the voltage waveform during the first few cycles. The figure on the right shows the one-cycle rms voltages during about 2 s. The character of the event is best visible in the rms voltage. The energizing leads to a large inrush current that slowly decays. The inrush current causes the observed voltage drop, which therefore shows the same slow decay. The unbalance between the phases is clearly visible. This is an important part of the signature of voltage dip due to transformer energizing. Another part of the signature is the exponential recovery, determined by the time constant with which the flux offset in the transformer core decays.

A third characteristic of voltage dips due to transformer energizing is harmonic distortion, especially the high level of even-harmonic components in the voltage. Figure 6.33 shows harmonics 1 through 8 for the voltage in phase $c$. The harmonic spectrum was obtained by applying a DFT algorithm to a one-cycle window and updating the result every cycle. Whereas the second-harmonic voltage has its highest value at the start of the transient, the other harmonics have their maxima later. The result is that the harmonic spectrum changes during the transient. Initially the second harmonic dominates but gradually the third and fifth harmonics take over.

The measurement of the voltages due to transformer energizing also captured the actual energizing currents. These are shown in Figure 6.34 for the first seven cycles of the transient. The currents due to transformer energizing are however notoriously difficult to measure due to the dc component they include. After a few cycles the currents show very sudden changes in derivative. This is the moment at which the core of the instrument transformer goes into saturation. Although methods

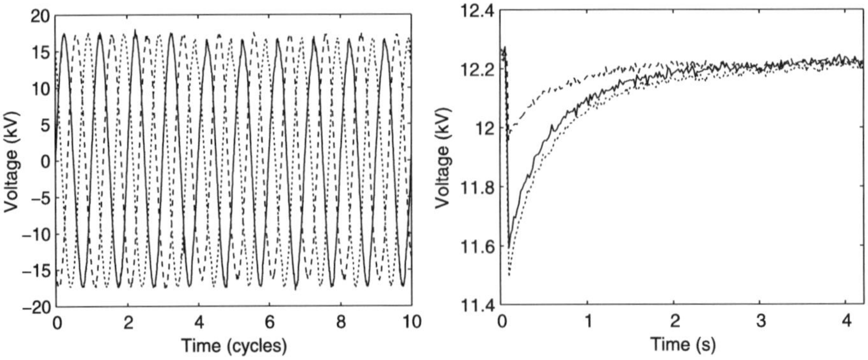


**Figure 6.32** Voltage dip due to transformer energizing: voltage waveforms (left) and rms voltages (right); note the difference in horizontal scale.

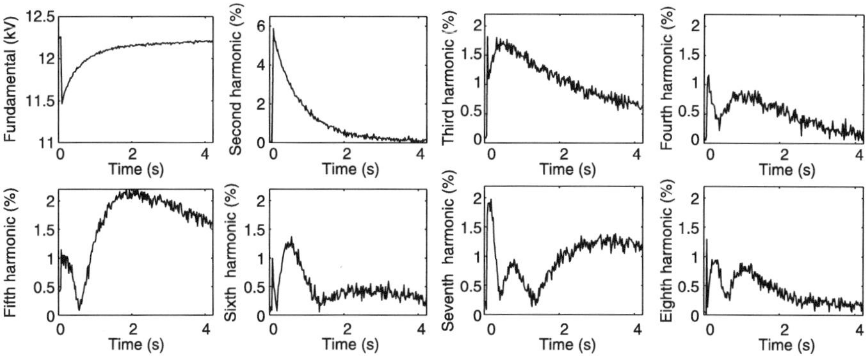


**Figure 6.33** Harmonics 1 through 8 due to transformer energizing.

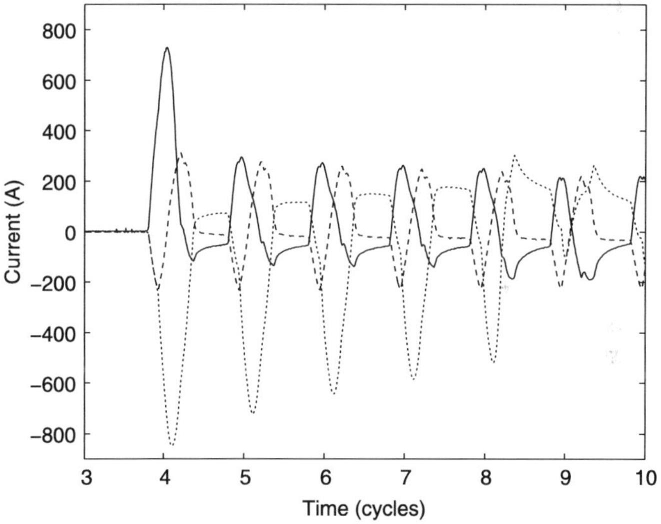


**Figure 6.34** Transformer inrush currents leading to dip in Figure 6.32.

have been proposed to extract characteristics from the output of a saturated instrument transformer, we will not further analyze this current waveform.

Transformer-energizing voltage dips also appear when loaded transformers are energized. The impact of the load is some reduction in the peak of the inrush current but mainly a reduction in the time constant with which the current decays. Energizing of loaded transformers takes place with the energizing of a distribution feeder after an interruption. An example of such an event, measured downstream of the breaker, was shown before in Figure 6.6. An upstream measurement during autoreclosing of an 11-kV feeder is shown in Figure 6.35. The autoreclosing leads to saturation of all MV/LV transformers along the feeder, the result is a significant inrush current as shown in the right-hand plot. The current measured is

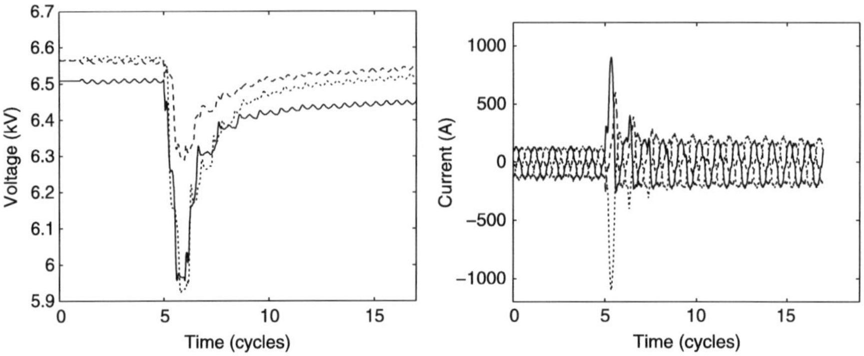


**Figure 6.35** Voltage dip due to autoreclosure of distribution feeder (left) and current during autoreclosure (right).

the total current through a number of feeders. The rise in current corresponds to the load of the feeder being energized. The current transient should be compared to this load.

The harmonic voltage distortion associated with transformer-energizing voltage dips is due to the harmonic distortion of the transformer inrush current. Certain harmonic components become amplified by a phenomenon called *harmonic resonance*; see Section 2.5.5. Mainly due to the presence of capacitor banks or underground cables, the source impedance becomes very large for certain frequencies. A current containing these frequencies will lead to severe distortion of the voltage waveform. An example of such a case is shown in Figure 6.36. Transformer energizing leads in this case to overvoltages in the waveform even though the fundamental voltage shows a decrease in magnitude.

Transformer energizing results in unbalanced dips, making interpretation of the characterization results more complicated than for motor-starting dips. Understanding transformer-energizing dips requires a closer look at transformer energizing. We will first consider a single-phase transformer and next extend the results to three phases. The voltage to which the transformer is exposed during energizing can be

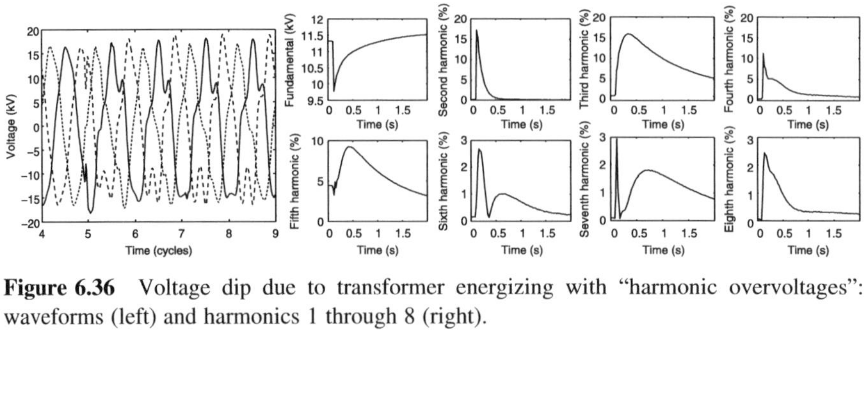


**Figure 6.36** Voltage dip due to transformer energizing with “harmonic overvoltages”: waveforms (left) and harmonics 1 through 8 (right).

written as follows:

$$v(t) = \begin{cases} 0 & t < 0 \\ V\sqrt{2}\cos(\omega_0 t + \psi) & t < 0 \end{cases} \tag{6.1}$$

The step in voltage causes a flux through the transformer core, determined by

$$v(t) = \frac{d\phi}{dt} \tag{6.2}$$

which results in the following expression, assuming the flux to be zero before energizing:

$$\phi(t) = \int_0^t v(\tau)\,d\tau = \frac{V\sqrt{2}}{\omega_0}\sin(\omega_0 t + \psi) - \frac{V\sqrt{2}}{\omega_0}\sin(\psi) \tag{6.3}$$

The first term in (6.3) represents the steady-state flux. The design of a transformer is such that the maximum value of the steady-state flux corresponds to the knee point in the flux–field (*BH*) characteristics. In other words, at the normal-operation maximum flux, the transformer core starts to go into saturation. Any flux above this maximum leads to saturation and thus to large currents and drops in voltage. The second term in (6.3) causes a shift in the flux wave, thus leading to saturation either with the maximum or the minimum values of the flux wave. Assume that the magnetizing current is proportional to the amount with which the flux exceeds the saturation point. This corresponds to a core model in which the permeability is infinite below the saturation point and equal to the permeability in vacuum ($\mu_0$) above the saturation point. The relation between flux and current is as follows:

$$i(t) = \begin{cases} k(\phi(t) - \phi_{\text{sat}}) & \phi(t) > \phi_{\text{sat}} \\ 0 & -\phi_{\text{sat}} < \phi(t) < \phi_{\text{sat}} \\ k(-\phi(t) + \phi_{\text{sat}}) & \phi(t) < -\phi_{\text{sat}} \end{cases} \tag{6.4}$$

The flux through the transformer core and the resulting current from the model in (6.3) are shown in Figure 6.37 for two different switching angles $\psi$. The left-hand figure has been calculated for $\psi = 190^\circ$, resulting in mild saturation. The right-hand figure, $\psi = 270^\circ$, shows severe saturation. The value of $k$ in (6.4) has been taken equal to 10 to obtain the figures.

If we assume that $\sin(\psi)$ is less than zero, there will be a positive inrush current peak. The starting and ending points of the saturation are obtained from

$$\phi(t) = \phi_{\text{sat}} = \frac{V\sqrt{2}}{\omega_0} \tag{6.5}$$

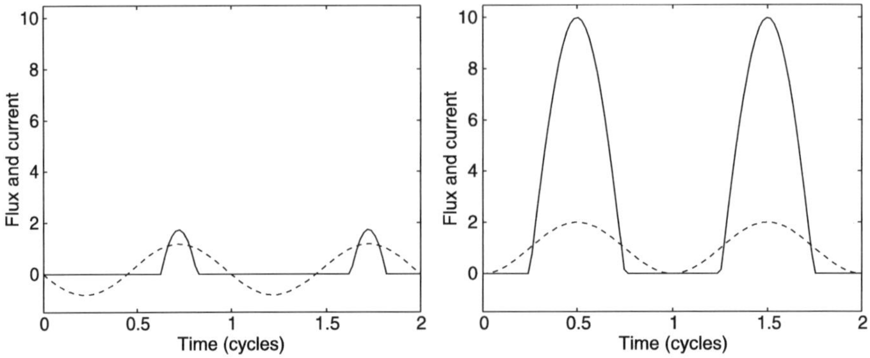


**Figure 6.37** Transformer energizing flux (dashed line) and resulting inrush current (solid line) for two different switching angles; mild saturation (left) and severe saturation (right).

Combining (6.5) with (6.3) enables the calculation of the instant $t_1$ when the transformer goes into saturation and the instant $t_2$ when the transformer comes out of saturation. The result is

$$\omega_0 t_1 = 2\pi - \psi + \arcsin(1 + \sin\psi) \tag{6.6}$$

$$\omega_0 t_2 = 3\pi - \psi - \arcsin(1 + \sin\psi) \tag{6.7}$$

The duration of the saturation period is obtained by subtracting (6.6) and (6.7), resulting in

$$\omega_0(t_2 - t_1) = \pi - 2\arcsin(1 + \sin\psi) \tag{6.8}$$

The duration of the saturation period is shown in Figure 6.38 as a function of the switching angle.

The inrush current $i(t)$ leads to a drop in voltage over the source impedance. The resulting voltage waveform for a source impedance $R + j\omega L$ is obtained from

$$v(t) = e(t) - L\frac{di}{dt} - Ri \tag{6.9}$$

with $e(t) = V\sqrt{2}\cos(\omega_0 t + \psi)$ the source voltage (the voltage waveform for zero load). It will be obvious to the reader that this is a rather simplified model, as the inrush current affects the terminal voltage of the transformer and thus the flux in accordance with (6.2). A more accurate calculation is beyond the scope of this book. The interested reader is advised to use one of the several power system analysis packages, which do a good job in calculating voltage and current wave shapes.

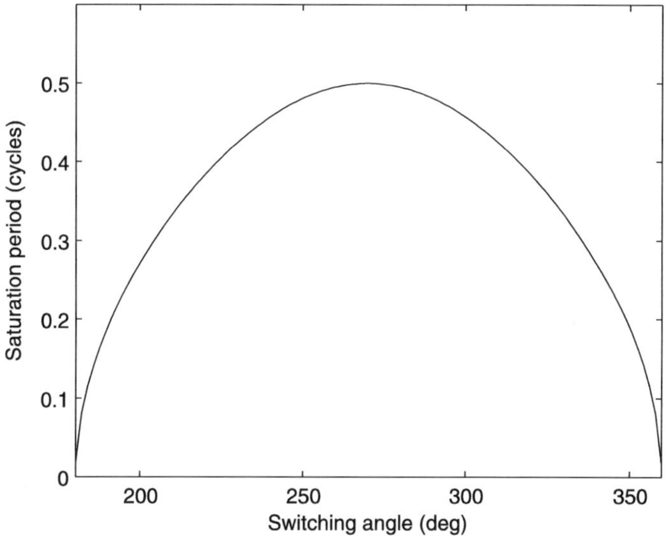


**Figure 6.38** Duration of saturation period for transformer energizing as function of switching angle.

From the earlier discussion we find that the current during saturation is

$$i(t) = \frac{kV\sqrt{2}}{\omega_0}[\sin(\omega_0 t + \psi) - \sin\psi - 1] \tag{6.10}$$

Combining (6.10) with (6.9) gives the following expression for the voltage waveform during saturation:

$$v(t) = V\sqrt{2}(1 - kL)\cos(\omega_0 t + \psi) - \frac{kRV\sqrt{2}}{\omega_0}[\sin(\omega_0 t + \psi) - \sin\psi - 1] \tag{6.11}$$

If we neglect the voltage drop due to the resistance, we obtain the following simple expression:

$$v(t) = V\sqrt{2}(1 - kL)\cos(\omega_0 t + \psi) \tag{6.12}$$

During the saturation period, the voltage drops simply to a lower value but does not change in waveform or phase angle. At the start and end of the saturation the voltage waveform will show sudden steps due to the sudden change in the rate of change of the current. The resulting voltage waveforms are shown in Figure 6.39 for two different switching angles. To obtain the plots, we have chosen a value of $kL = 0.4$. We see that saturation takes place in an interval symmetrical around the

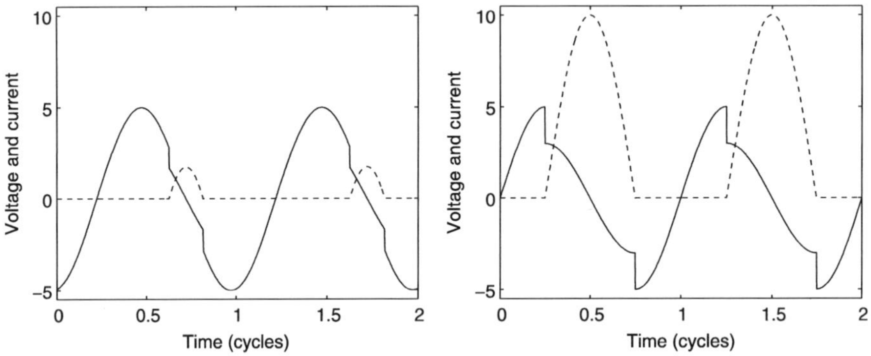


Figure 6.39 Voltages (solid) and currents (dashed) during transformer energizing for two switching angles: mild saturation (190°, left) and heavy saturation (270°, right).

voltage zero crossing. Measured voltage dips due to transformer saturation show very similar waveforms.

When characterizing voltage dips, the rms voltage over one cycle is typically used. This value has been calculated as a function of the switching angle for the above model. The results are shown in Figure 6.40. The lowest rms voltage is less than the drop in voltage according to (6.12) as the voltage drop only takes place during part of the cycle. Note that it has been assumed here that there is no change in saturation level. In reality the saturation level slowly drops, thus reducing the current and bringing the voltage gradually back to normal.

During energizing of a three-phase transformer each phase will experience a different switching angle. This results in different levels of saturation in the three phases and thus in different rms voltage. This unbalance is a well-known

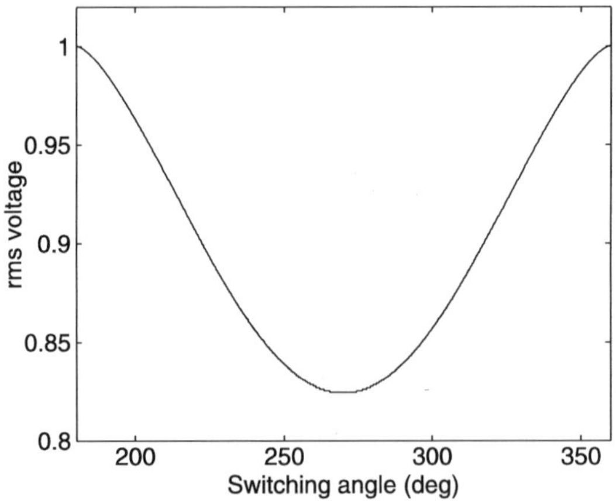


Figure 6.40 Residual voltage for dips due to transformer energizing as function of switching angle.

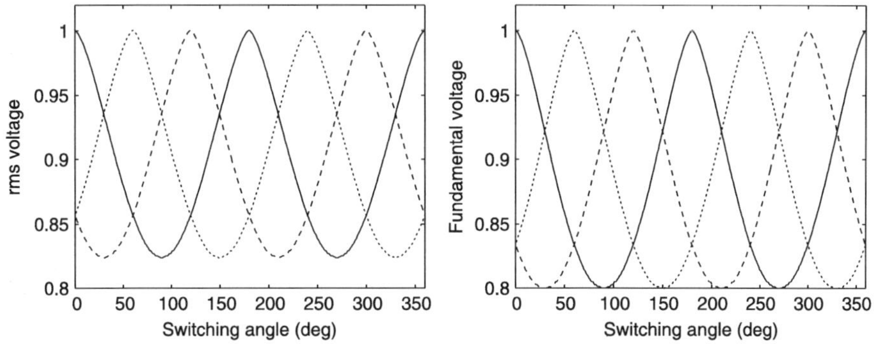


**Figure 6.41** Residual voltages (left) and fundamental voltages (right) for three phases during transformer energizing.

characteristic of voltage dips due to transformer energizing. The calculations resulting in Figure 6.40 have been repeated for energizing of a three-phase transformer. The rms voltage in the three phases are shown in Figure 6.41 as a function of the switching angle in phase $a$. We see that the lowest rms voltage is rather independent of the switching angle. We also see that there is never a situation in which the three rms voltages are the same; voltage dips due to transformer energizing are always unbalanced.

Figure 6.41 also shows the absolute value of the fundamental voltage. From the fact that the fundamental voltage is significantly lower than the rms voltage, it can be concluded that the waveform is heavily distorted. For the switching angle with the lowest rms voltage the total harmonic distortion is about 20%.

***6.2.2.9 Other Voltage Dips*** Some voltage events do not fit in any of the categories mentioned before. They may be due to some rare load events or a combination of events. Some examples are given below to satisfy the curiosity of the reader. But one should keep in mind that these are rare events and that the majority of events are covered very well by the examples given earlier.

An example of a very remarkable event is shown in Figure 6.42. During about 25 cycles (0.5 s) the voltage shows an almost square wave with heavy transients. Even though the rms voltage did not drop significantly, the event could still seriously affect equipment. This event was due to a failure on the line side of a small UPS. As all load in the neighborhood of this faulted UPS was also supplied through a UPS, no problems occurred.

Figure 6.43 shows a voltage dip due to a fault cleared by a fuse. The terminology "dip due to fuse clearing" is strictly speaking not correct, as it is the fault that leads to the dip. The fuse ends the fault and thus the dip. This terminology is however widely used, more compact than the more accurate description, and not confusing as long as the underlying phenomenon is kept in mind. Because the fuse clears the fault within a few milliseconds, the voltage dip is very short, less than one cycle. According to most definitions, this event would not even classify as a

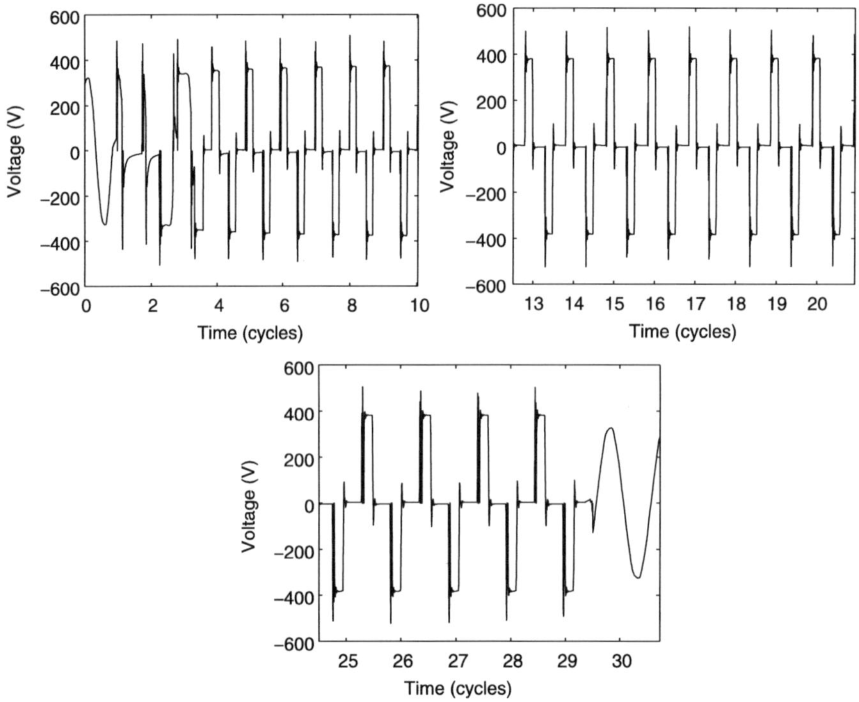


**Figure 6.42** Heavy-distorted voltage at terminals of small UPS.

voltage dip. The half-cycle rms voltage is also shown. The fault takes place close to $t = 0.5$ cycle, with only a very small drop in the half-cycle rms voltage. The drop in full-cycle rms voltage will be even less. But a few milliseconds after fault clearing the rms voltage starts to drop again. This is due to the saturation of transformers

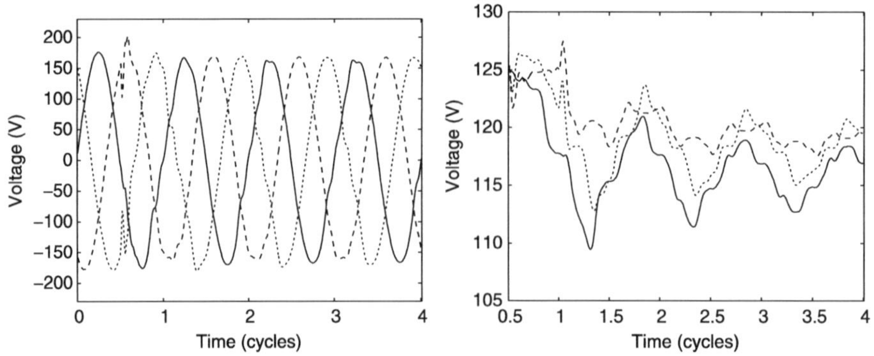


**Figure 6.43** Voltage dip due to fuse clearing with postfault transformer saturation: voltage waveforms (left) and half-cycle rms voltages (right).

close to the fault location. The oscillation in the half-cycle rms that is clearly visible here is typical for transformer saturation.

### 6.2.3 Voltage Dips in Three Phases

Most methods for characterizing voltage dips use two parameters to quantify the severity of a voltage dip: magnitude (or *residual voltage*) and duration (see Chapter 8 for more details). For a single-phase measurement this is a reasonable approximation. Some assumptions are made, most notably that the rms voltage can be described through a single value. This is not always the case. Also this method neglects the phase-angle jump and some other characteristics. However, more severe is the fact that measurements are rarely single phase. Most power systems are three phases, and a large fraction of equipment tripping due to voltage dips concerns three-phase loads.

The commonly used method is to characterize these *multi-channel measurements* through the lowest residual voltage and the longest duration of all the channels. This has a number of remarkable consequences:

- A voltage drop in one phase is characterized as equally severe as a voltage drop in three phases, whereas the latter event is typically much more severe for equipment.
- The dip due to an earth fault in a high-impedance grounded network will be seen as equally severe (or even more severe) as the dip due to a short-circuit fault, whereas the former has hardly any effect on equipment.
- There is no clear relation between the dip characteristics at both sides of a transformer or between a star-connected and a delta-connected monitor.
- An additional drawback of this commonly used method is that information is lost, making it harder to draw conclusions, for example, on fault type and location.

These four points mean that additional characterization effort is needed for these so-called three-phase unbalanced dips (i.e., voltage dips due to nonsymmetrical faults). A number of proposals for their characterization were compared by [85]. Further contributions to the problem were made by [75, 192, 216, 278]. Most of this earlier work was directed toward obtaining dip characteristics from measurements, without considering the basic circuit theory behind the phenomenon. An alternative approach was followed in [32]. From analyzing basic fault types in idealized systems, a classification in four dip types was proposed. Although the classification can be used for stochastic prediction [36] and equipment testing [37, 46, 266], it could not directly be used to classify measured voltage-dip events. The work presented in [334] generalized the classification and proposed an algorithm for extracting dip type and characteristics from sampled voltage waveforms. Simplified algorithms for extracting dip type and characteristics are presented and tested in

[35, 38]. We will discuss methods for extracting the dip characteristics in Chapter 8. The underlying theory will be discussed in this chapter.

**6.2.3.1 The ABC Classification** The so-called ABC classification of three-phase unbalanced voltage dips is to a large extent intuitive and therefore easy to understand. Measurements as well as a fundamental model confirm the validity of the ABC classification. The fundamental model is discussed in detail in the next Section.

The ABC classification distinguishes between seven types of three-phase unbalanced voltage dips. Expressions for the complex voltages for these seven types are given in Table 6.2. The complex prefault voltage in phase $a$ is indicated by $E_1$. The voltage in the faulted phase or between the faulted phases is indicated by $V^*$. As will be shown later, this is equal to the characteristic voltage in the symmetrical-component classification for all types except for type B. The relation for type B will be discussed further below. The waveform, rms voltage, and phase angle for the seven types of dip are shown in Figures 6.44 through 6.50. A short description of the dip types is

**TABLE 6.2 Seven Types of Three-Phase Unbalanced Voltage Dips According to ABC Classification**

| Type | Voltage Phasors | Type | Voltage Phasors |
|---|---|---|---|
| A | $U_a = V^*$ | E | $U_a = E_1$ |
| | $U_b = -\frac{1}{2}V^* - \frac{1}{2}jV^*\sqrt{3}$ | | $U_b = -\frac{1}{2}V^* - \frac{1}{2}jV^*\sqrt{3}$ |
| | $U_c = -\frac{1}{2}V^* + \frac{1}{2}jV^*\sqrt{3}$ | | $U_c = -\frac{1}{2}V^* + \frac{1}{2}jV^*\sqrt{3}$ |
| B | $U_a = V^*$ | F | $U_a = V^*$ |
| | $U_b = -\frac{1}{2}E_1 - \frac{1}{2}jE_1\sqrt{3}$ | | $U_b = -\frac{1}{2}V^* - (\frac{1}{3}E_1 + \frac{1}{6}V^*)j\sqrt{3}$ |
| | $U_c = -\frac{1}{2}E_1 + \frac{1}{2}jE_1\sqrt{3}$ | | $U_c = -\frac{1}{2}V^* + (\frac{1}{3}E_1 + \frac{1}{6}V^*)j\sqrt{3}$ |
| C | $U_a = E_1$ | G | $U_a = \frac{2}{3}E_1 + \frac{1}{3}V^*$ |
| | $U_b = -\frac{1}{2}E_1 - \frac{1}{2}jV^*\sqrt{3}$ | | $U_b = -(\frac{1}{3}E_1 + \frac{1}{6}V^*) - \frac{1}{2}jV^*\sqrt{3}$ |
| | $U_c = -\frac{1}{2}E_1 + \frac{1}{2}jV^*\sqrt{3}$ | | $U_c = -(\frac{1}{3}E_1 + \frac{1}{6}V^*) + \frac{1}{2}jV^*\sqrt{3}$ |
| D | $U_a = V^*$ | | |
| | $U_b = -\frac{1}{2}V^* - \frac{1}{2}jE_1\sqrt{3}$ | | |
| | $U_c = -\frac{1}{2}V^* + \frac{1}{2}jE_1\sqrt{3}$ | | |

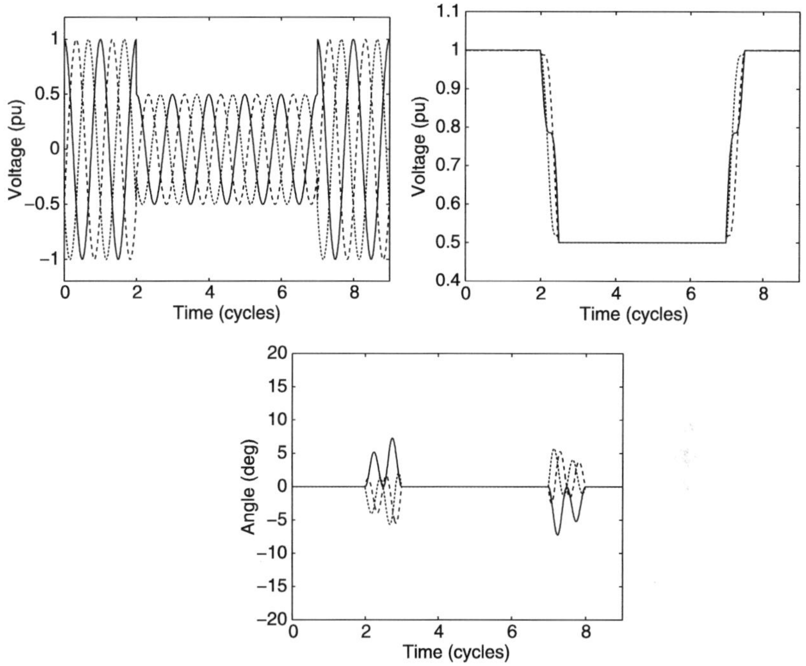


**Figure 6.44** Voltage dip of type A: voltage waveforms (left); half-cycle rms voltage (right); phase angle (bottom).

given below; for more details the reader is referred to [33] or a book on power system analysis.

Voltage dips due to three-phase faults (or symmetrical faults) lead to the same voltage dip in the three phases (or balanced dips). In our classification of dips we refer to these as type A; a synthetic example is shown in Figure 6.44. The voltage in the faulted phase equals 0.5 pu and there is no phase-angle jump in the faulted-phase voltage. The phase-angle jump is zero in all three phases. The oscillation in the phase-angle plot (right) is an artifact due to the half-cycle window moving through the start of the event. These artifacts will be discussed in more detail in Section 8.2.

A single-phase fault results in a voltage drop in the faulted phase. The nonfaulted phases may show a rise in voltage, a drop in voltage, or no change at all, depending on the ratio between positive-sequence and negative-sequence source impedances. For the synthetic dip shown in Figure 6.45 it has been assumed that the voltage in the nonfaulted phases is not affected. The dip becomes unbalanced with a drop in one phase only but no impact in the other two phases.

A phase-to-phase fault leads to a voltage drop between the faulted phases, as shown in Figure 6.46. In terms of phase voltages this translated into a drop in rms voltage for two phases and a change in phase angle for these two phases as well. Although the voltage between the faulted phases does not show any phase-angle

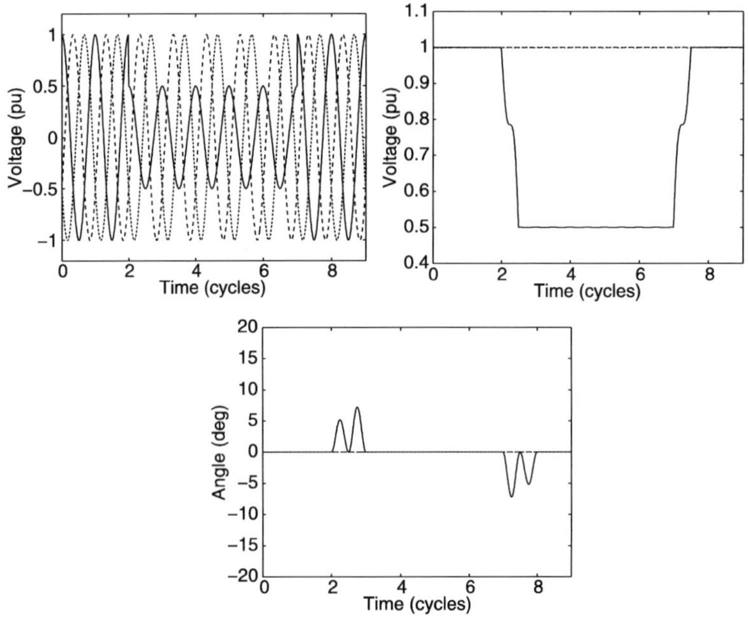


**Figure 6.45** Voltage dip of type B: voltage waveforms (left); half-cycle rms voltage (right); phase-angle (bottom).

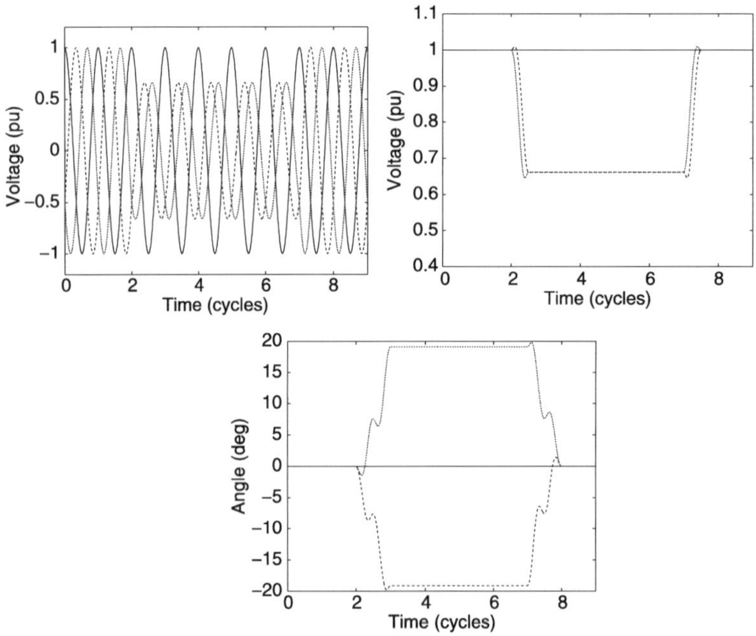


**Figure 6.46** Voltage dip of type C: voltage waveforms (left); half-cycle rms voltage (right); phase angle (bottom).

**TABLE 6.3 Dips at Different Voltage Levels Due to Different Fault Types**

| Fault Type | Location I | Location II | Location III |
|---|---|---|---|
| Three phases | A | A | A |
| Two phases to ground | E | F | G |
| Two phases | C | D | C |
| Single phase to ground | B | C | D |

jump, two of the phase voltages show a phase-angle jump. In this example the phase-angle jump is about 20°. For a zero *characteristic phase-angle jump* (the phase-angle jump in the voltage between the faulted phases, $V^*$ in Table 6.2), the phase-angle jump in the two faulted phases is of equal size but opposite direction. The third phase is not affected. A type C voltage dip may also be due to a single-phase fault at a higher voltage level, according to Table 6.3.

Removing the zero-sequence component from a type B dip or transferring a type C dip through a Dy-connected transformer results in a type D dip, as shown in Figure 6.47. A type D dip is characterized by a major voltage drop in

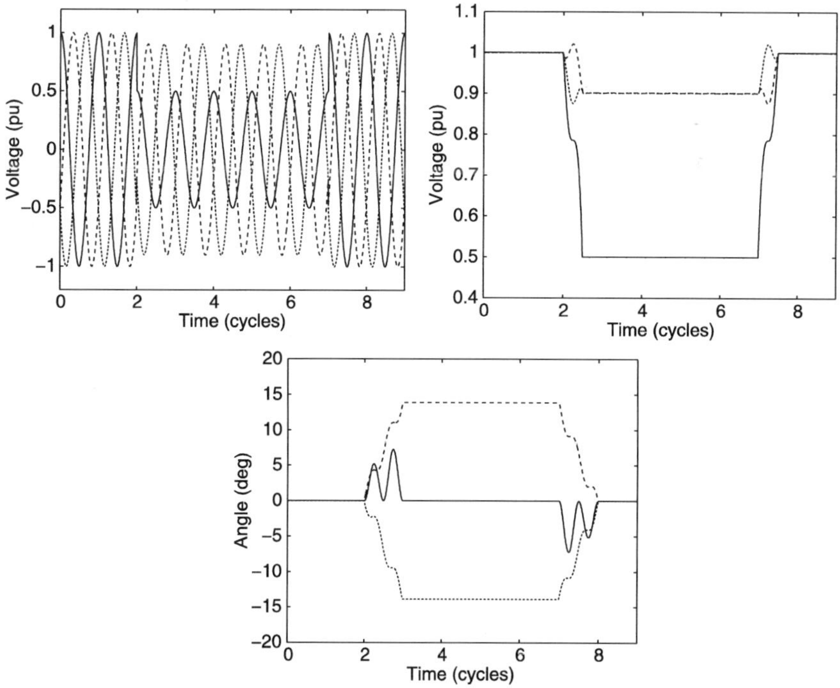


**Figure 6.47** Voltage dip of type D: voltage waveforms (left); half-cycle rms voltage (right); phase angle (bottom).

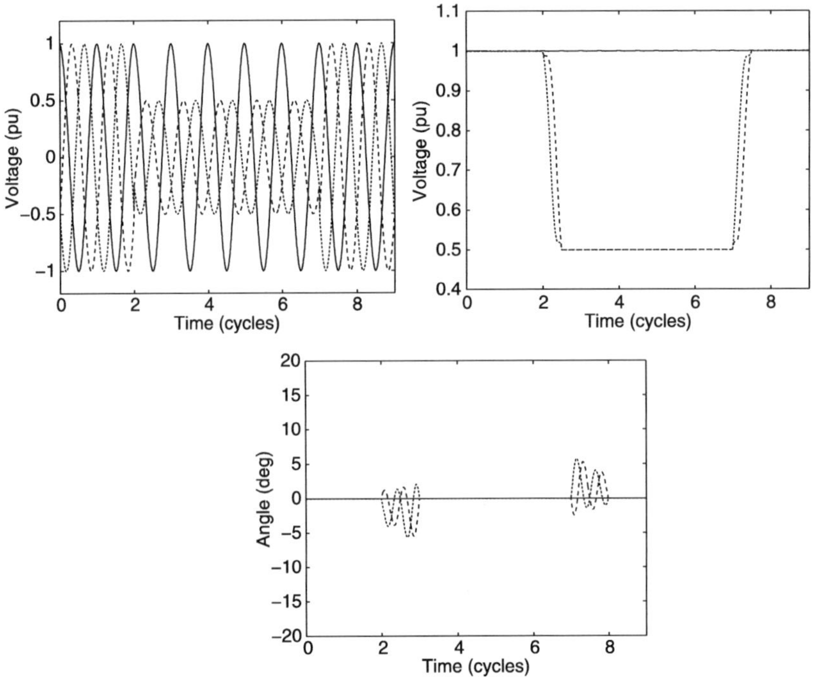


**Figure 6.48** Voltage dip of type E: voltage waveforms (left); half-cycle rms voltage (right); phase angle (bottom).

one phase and a minor drop in the other two phases. For a zero characteristic phase-angle jump, the voltages in the "*nonfaulted phases*" have the same rms voltage and phase-angle jumps that are identical in size but opposite in sign.

A voltage dip of type E, as shown in Figure 6.48, is the first of three types due to two-phase-to-ground faults. A type E dip occurs for a two-phase-to-ground fault at the same voltage level as the dip measurement. It has in common with a type B dip that it contains a zero-sequence component. As with type B, this zero-sequence component depends on the ratio between the zero-sequence and positive-sequence source impedances. For unknown fault position, this ratio appears as the voltage in the nonfaulted phase. For the basic dip types in Table 6.2 it has been assumed that the voltage in this phase is not affected by the fault. For a type E dip, the rms voltage shows a drop in two phases without phase-angle jump.

After a Dy-connected transformer, a type E dip changes into a type F dip. As with a type D dip, it shows a large voltage drop in one phase and a minor drop in the other two phases. But the latter voltage drop is more for a type F than for a type D dip. The phase-angle jump in these "not-faulted phases" is smaller; compare Figure 6.47 with Figure 6.49.

A second Dy-connected transformer results in a type G dip for a two-phase-to-ground fault, as shown in Figure 6.50. The same dip results by removing the

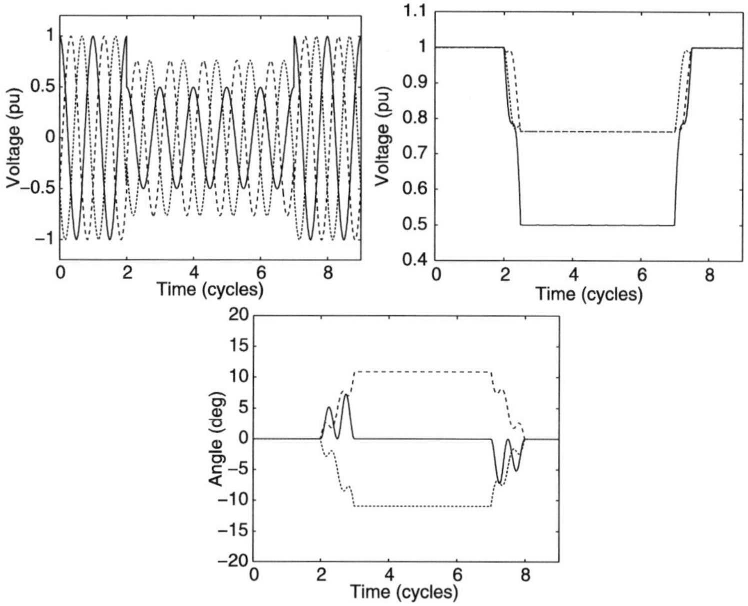


**Figure 6.49** Voltage dip of type F: voltage waveforms (left); half-cycle rms voltage (right); phase-angle (bottom).

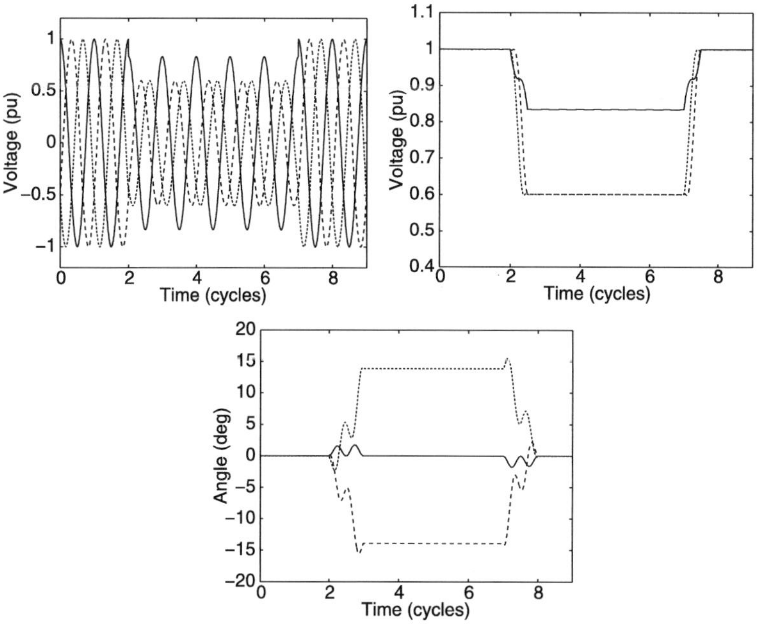


**Figure 6.50** Voltage dip of type G: voltage waveforms (left); half-cycle rms voltage (right); phase-angle (bottom).

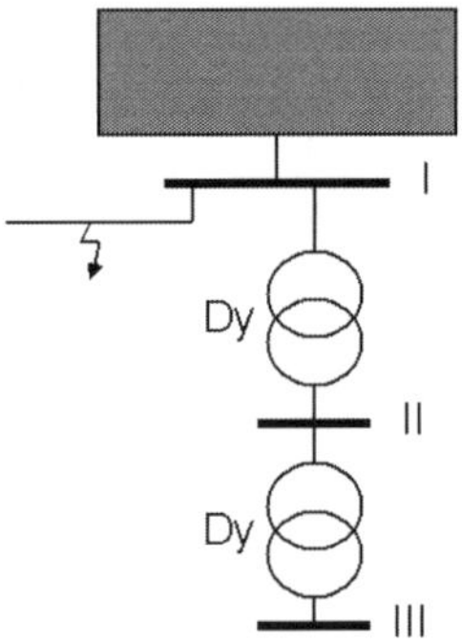


**Figure 6.51** Dips at different voltage levels due to different fault types.

zero-sequence component from a type E dip. A type G dip shows a large drop in rms in two phases and a minor drop in the third phase. The latter phase shows no phase-angle jump, whereas the other two phases show opposite phase-angle jumps of identical size. The ratio between zero-sequence and positive-sequence source impedances not only impacts type E but also type F and type G. The higher is this ratio, the more types F and G will resemble types D and C, respectively.

One of the reasons for introducing this classification was to describe the propagation of dips through transformers. The origin and transformation of the seven types are given in Table 6.3 and Figure 6.51. A balanced dip (type A) does not change when propagating through the system, but all other types are affected by the transformer winding connections. A Dy-connected transformer changes type C into type D, type F into type G, and the other way around. The removal of the zero-sequence component means that type B becomes type D and type E becomes type G.

This method was originally developed to be part of the stochastic prediction of voltage dips. From known statistics on faults it is possible to calculate the frequency of occurrence of the different types of voltage dips. This classification can also be used for testing of equipment against voltage dips. By using this classification it is possible to generate the dips that can be expected at the terminals of three-phase equipment. A testing protocol based on IEC 61000-4-11 is proposed in [266]. The method has also been used to extract information on fault type from voltage-dip recordings [283].

The weak point of this method is that it is only simulation-based. Extraction of the dip type from measured voltage waveforms is not immediately possible. Recent work has however shown that the dip type may be estimated from the three rms voltages only. But this method only works for phase-angle jumps that are not too big and thus mainly for faults in transmission systems [38, 90].

***6.2.3.2 Symmetrical-Component Classification*** The symmetrical-component classification does not suffer from the same limitation as the ABC classification. The symmetrical-component classification distinguishes between dips with the main voltage drop in one phase and dips with the main voltage drop between two phases. Dips due to three-phase faults, with an equal drop in the three phases, are a limiting case for both *single-phase drops* and *two-phase drops*. The zero-sequence voltage is treated as a separate characteristic from the beginning and in many studies not even considered. The two other characteristics are the *characteristic voltage* $V$ and the *PN factor* $F$. The general expression for the *non-zero-sequence part* of a voltage dip with the main drop in phase $a$ (referred to as dip type Da) is

$$\begin{aligned} U_a &= V \\ U_b &= -\tfrac{1}{2}V - \tfrac{1}{2}jF\sqrt{3} \\ U_c &= -\tfrac{1}{2}V + \tfrac{1}{2}jF\sqrt{3} \end{aligned} \tag{6.13}$$

The general expression for a voltage dip with the main drop between phases $b$ and $c$ (dip type Ca) is

$$\begin{aligned} U_a &= F \\ U_b &= -\tfrac{1}{2}F - \tfrac{1}{2}jV\sqrt{3} \\ U_c &= -\tfrac{1}{2}F + \tfrac{1}{2}jV\sqrt{3} \end{aligned} \tag{6.14}$$

where $|F| \geq |V|$. Note that a balanced dip, due to a three-phase fault, is obtained for $F = V$. Similar expressions hold for the other four dip types: Db (drop in phase $b$), Dc (main drop in phase $c$), Cb (main drop between phases a and $c$), and Cc (main drop between phases $a$ and $b$). The somewhat illogical notation was used to make the classification consistent with the older ABC classification. For $F = E_1$ and $V = V^*$, (6.13) and (6.14) are identical to types D and C, respectively, as defined for the ABC classification in Table 6.2. The definition of characteristic voltage and PN factor and the algorithm for obtaining the characteristics from measured voltage wave shapes are both based on symmetrical components. The underlying mathematics will be described in detail below.

Consider a three-phase fault somewhere in a three-phase power network. If the fault impedance is zero, the residual voltage is zero at the fault location and nonzero but less than the prefault voltage for all other locations in the network. To determine the residual voltage at a given location in the network, the PCC between the fault and the dip location is determined [33]. The residual voltage at the PCC is obtained from

$$U_1 = \frac{Z_{F1}}{Z_{S1} + Z_{F1}} E_1 \tag{6.15}$$

where the subscript 1 indicates positive-sequence quantities: $U_1$ is the (positive-sequence) residual voltage at the PCC; $E_1$ is the prefault voltage at the PCC (or at the monitor location); $Z_{F1}$ is the (positive-sequence) impedance between the PCC and the fault position; and $Z_{S1}$ is the source impedance at the PCC.

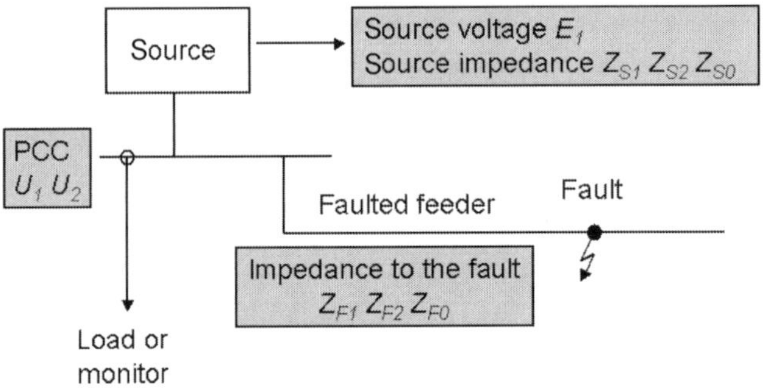


**Figure 6.52** Voltage-divider model for three-phase unbalanced voltage dips.

Expression (6.15) gives a direct relation between the residual voltage and the fault location. This expression is the basis for the characterization of voltage dips due to nonsymmetrical faults. The fault impedance is normally assumed zero in voltage-dip calculations. Including it requires adjusting the value of the impedance to the fault to include the fault impedance.

In nonsymmetrical faults, what need to be known are not only positive-sequence quantities such as in (6.15) but also negative-sequence quantities, index 2, and zero-sequence quantities, index 0. See Figure 6.52 for a definition of the various quantities used below.

### 6.2.3.3 Phase-to-Phase Faults

For a phase-to-phase fault between phases $b$ and $c$, the following expressions are obtained for the positive- and negative-sequence components of the residual voltage:

$$U_1 = \left(1 - \frac{Z_{S1}}{Z_{S1} + Z_{S2} + Z_{F1} + Z_{F2}}\right) E_1 \tag{6.16}$$

$$U_2 = \frac{Z_{S2}}{Z_{S1} + Z_{S2} + Z_{F1} + Z_{F2}} E_1 \tag{6.17}$$

To characterize the event, the sum and difference of positive- and negative-sequence voltages are used:

$$U_1 + U_2 = \left(1 - \frac{Z_{S1} - Z_{S2}}{Z_{S1} + Z_{S2} + Z_{F1} + Z_{F2}}\right) E_1 \tag{6.18}$$

$$U_1 - U_2 = \frac{Z_{F1} + Z_{F2}}{Z_{S1} + Z_{S2} + Z_{F1} + Z_{F2}} E_1 \tag{6.19}$$

Note that (6.19) is similar to the expression for the residual voltage during a three-phase fault, (6.15). The difference is that the sum of positive- and negative-sequence impedances determines the residual voltage. Due to this similarity, (6.19) is used as a generalization of the residual voltage. Expression (6.19) is referred to as the *characteristic voltage*. Expression (6.18) in turn is referred to as the PN factor as its value is determined by the difference between positive- and negative-sequence source impedances [333, 334].

In most cases positive- and negative-sequence impedances are equal. This holds especially for the impedance between the PCC and the fault and to a lesser extent for the source impedance at the PCC. Under these assumptions, the characteristic voltage is the same for a three-phase fault as for a phase-to-phase fault, and the PN factor becomes equal to the prefault voltage $E_1$.

This event is referred to as a type Ca dip, that is, the main drop in voltage is in phases $b$ and $c$, with characteristic voltage according to (6.19) and PN factor according to (6.18).

For a phase-to-phase fault between phases $a$ and $b$ the resulting component voltages at the PCC are

$$U_1 = \left(1 - \frac{Z_{S1}}{Z_{S1} + Z_{S2} + Z_{F1} + Z_{F2}}\right) E_1 \tag{6.20}$$

$$U_2 = \frac{a^2 Z_{S2}}{Z_{S1} + Z_{S2} + Z_{F1} + Z_{F2}} E_1 \tag{6.21}$$

with $a = -\frac{1}{2} + \frac{1}{2}j\sqrt{3}$ a rotation over 120° in the complex plane. The resulting voltage dip is of type Cc (main drop between phases $a$ and $b$) with characteristic voltage

$$U_1 - aU_2 = \frac{Z_{F1} + Z_{F2}}{Z_{S1} + Z_{S2} + Z_{F1} + Z_{F2}} E_1 \tag{6.22}$$

and PN factor

$$U_1 + aU_2 = \left(1 - \frac{Z_{S1} - Z_{S2}}{Z_{S1} + Z_{S2} + Z_{F1} + Z_{F2}}\right) E_1 \tag{6.23}$$

The difference is not in the value of the characteristics but in the way they are calculated from the positive- and negative-sequence voltages. For a fault between phases $a$ and $c$ the result is a dip of type Cb with characteristic voltage

$$U_1 - a^2 U_2 = \frac{Z_{F1} + Z_{F2}}{Z_{S1} + Z_{S2} + Z_{F1} + Z_{F2}} E_1 \tag{6.24}$$

and PN factor

$$U_1 + a^2U_2 = \left(1 - \frac{Z_{S1} - Z_{S2}}{Z_{S1} + Z_{S2} + Z_{F1} + Z_{F2}}\right)E_1 \tag{6.25}$$

**6.2.3.4 *Transfer Through Transformers*** Transformers between the PCC and the monitoring location change the character of three-phase unbalanced voltage dips, as shown in Table 6.3. Different winding connections affect the voltages in different ways. However, by considering symmetrical components the transfer of three-phase unbalanced voltage dips through transformers can be studied in a systematic way. The transfer of the zero-sequence voltage depends on the neutral grounding of the transformer. When both sides are solidly grounded, the zero-sequence impedance is not affected, unless there is a tertiary delta winding present. In the latter case the zero-sequence voltage is damped. The zero-sequence voltage is not considered in the classification of three-phase unbalanced voltage dips. It is a completely separate characteristic. Positive- and negative-sequence voltages remain the same absolute value (in per-unit) but are rotated over a multiple of 30°, where the rotation of the negative-sequence voltage is opposite to the rotation of the positive-sequence voltage.

Consider as an example, a Dy11 transformer between the PCC and the monitoring location. This transformer rotates the positive-sequence voltage over +30° and the negative-sequence voltage over −30°. Using the preevent voltage in phase *a* as a reference on both sides of the transformer, the result is a rotation over −60° of the negative-sequence voltage whereas the positive-sequence voltage is not affected. This leads to three more types, all with the same positive-sequence voltage as before but with different argument (direction) of the negative-sequence voltage. For a *bc* fault behind a Dy11 transformer, the negative-sequence voltage reads as follows (note that $-a$ corresponds to a rotation over −60°).

$$U_2 = \frac{-aZ_{S2}}{Z_{S1} + Z_{S2} + Z_{F1} + Z_{F2}}E_1 \tag{6.26}$$

The result is a type Db dip (main drop in phase *b*) with characteristic voltage

$$U_1 + a^2U_2 = \frac{Z_{F1} + Z_{F2}}{Z_{S1} + Z_{S2} + Z_{F1} + Z_{F2}}E_1 \tag{6.27}$$

and PN factor

$$U_1 - a^2U_2 = \left(1 - \frac{Z_{S1} - Z_{S2}}{Z_{S1} + Z_{S2} + Z_{F1} + Z_{F2}}\right)E_1 \tag{6.28}$$

In a similar way the characteristic voltage and PN factor for a type Dc dip are obtained from $U_1 + aU_2$ and $U_1 - aU_2$, respectively, and for a type Da from $U_1 + U_2$ and $U_1 - U_2$, respectively.

This whole exercise results in six dip types that differ only in the phase angle of the negative-sequence voltage. Any transformer will rotate the positive-sequence voltage over an integer multiple of 30°. Using the local voltage as a reference on both sides, the negative-sequence voltage will be rotated over an integer multiple of 60°. As the directions of the negative-sequence voltage for the six types are 60° apart, transformers will not lead to new types.

***6.2.3.5 Dip Characteristics*** The results of the analysis in the previous section are summarized in Table 6.4: There are six dip types characterized by a characteristic voltage $V$ and a PN factor $F$. In the first column the dip types are given. The second column gives the phases that show a severe drop in voltage; thus type Cb is a severe drop in phases $a$ and $c$, and so on. The resulting characteristic voltage and PN factor are, for dips due to phase-to-phase faults, in all cases given by the following expressions:

$$V = \frac{Z_{F1} + Z_{F2}}{Z_{S1} + Z_{S2} + Z_{F1} + Z_{F2}} E_1 \tag{6.29}$$

$$F = \left(1 - \frac{Z_{S1} - Z_{S2}}{Z_{S1} + Z_{S2} + Z_{F1} + Z_{F2}}\right) E_1 \tag{6.30}$$

However, the relation between these characteristics and positive- and negative-sequence voltages is different for the different dip types. The result of the definitions is that characteristic voltage and PN factor are independent of the type of dip and thus independent of the phases involved in the fault and of the transformer winding connections between the fault and the measurement location. The characteristics indicate the severity of the event, whereas the type of dip indicates which phases are involved in the event.

For equal positive- and negative-sequence impedances (6.29) and (6.30) simplify to

$$V \approx \frac{Z_{F1}}{Z_{S1} + Z_{F1}} E_1 \tag{6.31}$$

$$F \approx E_1 \tag{6.32}$$

**TABLE 6.4 Definition of Characteristics for Three-Phase Unbalanced Voltage Dips**

| Type | Drop in Phases | Characteristic Voltage | PN Factor |
|---|---|---|---|
| Ca | $bc$ | $U_1 - U_2$ | $U_1 + U_2$ |
| Cb | $ac$ | $U_1 - a^2 U_2$ | $U_1 + a^2 U_2$ |
| Cc | $ab$ | $U_1 - aU_2$ | $U_1 + aU_2$ |
| Da | $a$ | $U_1 + U_2$ | $U_1 - U_2$ |
| Db | $b$ | $U_1 + a^2 U_2$ | $U_1 - a^2 U_2$ |
| Dc | $c$ | $U_1 + aU_2$ | $U_1 - aU_2$ |

Thus the characteristic voltage for a dip due to a phase-to-phase fault is the same as for a three-phase fault at the same fault location. The PN factor for a dip due to a phase-to-phase fault is about equal to the prefault positive-sequence voltage.

**6.2.3.6 Monitoring** From the definitions in Table 6.4, the characteristic voltage and PN factor can be obtained once the dip type is known. The dip type can be obtained from the positive- and negative-sequence voltages. Consider, for example, dip type Ca (main drop between phases $b$ and $c$), for which the following relation is obtained from (6.17) and (6.18):

$$U_2 = \frac{Z_{S2}}{Z_{S1}}(E_1 - U_1) \approx E_1 - U_1 \qquad (6.33)$$

For a Cc-type dip, (6.20) and (6.23) give

$$U_2 = a^2 \frac{Z_{S2}}{Z_{S1}}(E_1 - U_1) \approx a^2(E_1 - U_1) \qquad (6.34)$$

The difference between a Ca dip and a Cc dip is brought back to its essence: the angle between the negative-sequence voltage $U_2$ and the drop in positive-sequence voltage $E_1 - U_1$. Each type of dip has its own specific angle. Knowing the complex positive- and negative-sequence voltages (which can be obtained from the measured voltage waveforms), it is thus straightforward to determine the dip type. Knowing the dip type its characteristics can be obtained from Table 6.4. The limitation of this method is in the factor $Z_{S2}/Z_{S1}$ in (6.33) and (6.34). Immediately after fault initiation this ratio is close to unity so that the approximation holds very well. But due to dynamic loads, this ratio changes during the dip: not only in magnitude but also in phase angle. This change in phase angle makes it more difficult to distinguish between the different dip types. This requires some modification of the algorithm as described in [35]. Note that the effect of dynamic load can also be described as a change in source voltage $E_1$.

An alternative algorithm for obtaining the dip type without the need for a symmetrical-component transformation is given in [33, 35]. This algorithm is less affected by load effects, but it will give incorrect results for larger phase-angle jumps with shallow dips [35]. Methods for extracting the dip type and characteristics from measured voltage waveforms will be discussed in further detail in Section 8.3.

**6.2.3.7 Single-Phase Faults** Consider a single-phase fault in phase $a$ at the fault location in Figure 6.52. The resulting positive- and negative-sequence voltages at the PCC are obtained from

$$U_1 = \left(1 - \frac{Z_{S1}}{Z_{S1} + Z_{S2} + Z_{S0} + Z_{F1} + Z_{F2} + Z_{F0}}\right) E_1 \qquad (6.35)$$

$$U_2 = -\frac{Z_{S2}}{Z_{S1} + Z_{S2} + Z_{S0} + Z_{F1} + Z_{F2} + Z_{F0}} E_1 \qquad (6.36)$$

Using the definitions from Table 6.4 the characteristic voltage and PN factor are obtained from the following expressions (note that this is a Da event):

$$V = U_1 + U_2 = \left(1 - \frac{Z_{S1} + Z_{S2}}{Z_{S1} + Z_{S2} + Z_{S0} + Z_{F1} + Z_{F2} + Z_{F0}}\right)E_1 \tag{6.37}$$

$$F = U_1 - U_2 = \left(1 - \frac{Z_{S1} - Z_{S2}}{Z_{S1} + Z_{S2} + Z_{S0} + Z_{F1} + Z_{F2} + Z_{F0}}\right)E_1 \tag{6.38}$$

These expressions are obviously different from those for a Da dip due to a phase-to-phase fault at the same location, as shown in (6.29) and (6.30). To relate the characteristic voltages for dips due to single-phase and phase-to-phase faults, we introduce the voltage in the faulted phase at the PCC in Figure 6.52:

$$V^* = \left(1 - \frac{Z_{S1} + Z_{S2} + Z_{S0}}{Z_{S1} + Z_{S2} + Z_{S0} + Z_{F1} + Z_{F2} + Z_{F0}}\right)E_1 \tag{6.39}$$

(Note that this is the same voltage as indicated by $V^*$ in the expression for dip type B in Table 6.2.) The voltage in the faulted phase at the PCC can be obtained from the same single-phase voltage-divider model as for three-phase faults, but with the sum of positive-, negative-, and zero-sequence impedances. Together with (6.37) this results in

$$V = \frac{Z_{S0}}{Z_{S0} + Z_{S1} + Z_{S2}} E_1 + \frac{Z_{S1} + Z_{S2}}{Z_{S0} + Z_{S1} + Z_{S2}} V^* \tag{6.40}$$

The characteristic voltage contains a nonzero component that is independent of the fault position. Even for a fault at the PCC ($V^* = 0$) a nonzero characteristic voltage results. Also note that (6.40) is only influenced by the source impedance at the PCC. The voltage dip due to a single-phase fault is less severe than due to a phase-to-phase fault at the same location. However, when the fault location is not known, a dip due to a single phase fault cannot be distinguished from a dip due to a phase-to-phase fault. This justifies the use of the same classification for dips due to single-phase and phase-to-phase faults.

For a solidly grounded distribution system, one may approximate $Z_{S0} = Z_{S1} = Z_{S2}$, so that (6.40) becomes

$$V = \tfrac{1}{3}E_1 + \tfrac{2}{3}V^* \tag{6.41}$$

The same expression was derived in [32] in a different way. For a solidly grounded transmission system, the zero-sequence source impedance is typically two to three times the positive-sequence value. Using $Z_{S0} = 2.5Z_{S1}$ and $Z_{S1} = Z_{S2}$ gives

$$V = \tfrac{5}{9}E_1 + \tfrac{4}{9}V^* \tag{6.42}$$

For such a system the characteristic voltage of a dip due to a single-phase fault is always higher than 0.55 pu. For a high-impedance grounded system, $Z_{S0} \gg Z_{S1}$, so that (6.40) reads as

$$V \approx E_1 \tag{6.43}$$

In other words, single-phase faults in high-impedance grounded systems do not lead to voltage dips. This should not come as a surprise because it is one of the reasons for using high-impedance grounding. Note, however, that the commonly used method of characterizing the dip through the lowest phase voltage would give a very low residual voltage for dips due to single-phase faults in impedance-grounded systems.

#### *6.2.3.8 Two-Phase-to-Ground Faults*

For a fault between phases *b*, *c*, and ground, positive- and negative-sequence voltages at the PCC are as follows:

$$U_1 = \left(1 - \frac{Z_{S1}(Z_{S0} + Z_{F0} + Z_{S2} + Z_{F2})}{D}\right)E_1 \tag{6.44}$$

$$U_2 = \frac{Z_{S2}(Z_{S0} + Z_{F0})}{D}E_1 \tag{6.45}$$

with

$$D = (Z_{S0} + Z_{F0})(Z_{S1} + Z_{F1} + Z_{S2} + Z_{F2}) + (Z_{S1} + Z_{F1})(Z_{S2} + Z_{F2}) \tag{6.46}$$

Comparing this with the results for the definitions of the basic dip types in Table 6.4 shows that this is a dip of type Ca with the following characteristics:

$$V = \left(1 - \frac{(Z_{S1} + Z_{S2})(Z_{S0} + Z_{F0}) + Z_{S1}(Z_{S2} + Z_{F2})}{D}\right)E_1 \tag{6.47}$$

$$F = \left(1 - \frac{(Z_{S1} - Z_{S2})(Z_{S0} + Z_{F0}) + Z_{S1}(Z_{S2} + Z_{F2})}{D}\right)E_1 \tag{6.48}$$

The characteristic voltage is again a measure of the fault location and thus of the severity of the dip. This can best be seen by considering equal positive- and negative-sequence impedance, $Z_{S1} = Z_{S2}$ and $Z_{F1} = Z_{F2}$, after which (6.47) and

(6.48) are simplified to

$$V = \frac{Z_{F1}}{Z_{F1} + Z_{F2}} \tag{6.49}$$

$$F = \left(1 - \frac{Z_{S1}}{Z_{S1} + Z_{F1} + 2(Z_{S0} + Z_{F0})}\right)E_1 \tag{6.50}$$

The expression for the characteristic voltage is again the same as for three-phase and phase-to-phase faults. The difference is only in the value of the PN factor, which is less than the prefault voltage $E_1$ but larger than the characteristic voltage $V$. The other dip types from Table 6.4 are again obtained for faults between other phases and after transfer of the dip through a transformer.

***6.2.3.9 Examples*** Consider as an example an 11-kV resistance-grounded system with the following source and feeder parameters (all values are on a 100-MVA base):

- $Z_{S0} = 787 + j220\%$
- $Z_{S1} = Z_{S2} = 4.94 + j65.9\%$
- $Z_{F0} = 18.4 + j112\%\ \text{km}^{-1}$
- $Z_{F1} = Z_{F2} = 9.7 + j26\%\ \text{km}^{-1}$

The characteristic voltage and PN factor have been calculated as a function of the distance between the PCC and the fault for single-phase, phase-to-phase, and two-phase-to-ground faults. The results are shown in the left-hand plot of Figure 6.53: The solid line gives the characteristic voltage for dips due to phase-to-phase and

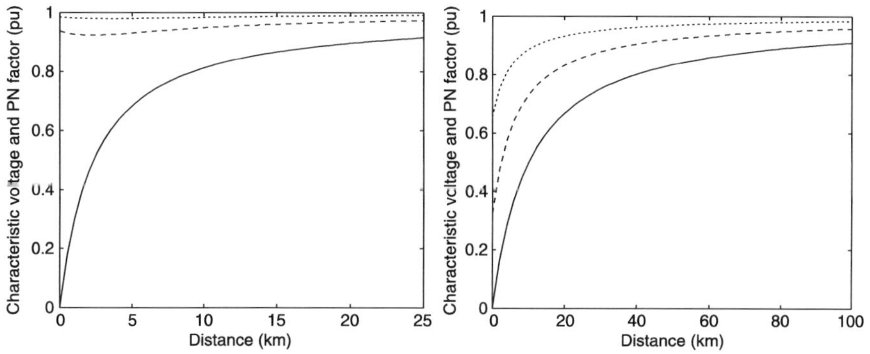


**Figure 6.53** Characteristics of voltage dips due to nonsymmetrical faults in resistance-grounded system (left) and solidly grounded system (right). Solid line: characteristic voltage for phase-to-phase and two-phase-to-ground faults; dashed line: characteristic voltage for single-phase faults; dotted line: PN factor for two-phase-to-ground faults.

two-phase-to-ground faults according to (6.29) and (6.47), respectively; the dotted line gives the PN factor for dips due to two-phase-to-ground faults according to (6.48); and the dashed line gives the characteristic magnitude for dips due to single-phase faults according to (6.37). Due to the large zero-sequence source impedance (the result of the resistance grounding), the PN factor for two-phase-to-ground faults and the characteristic voltage for single-phase faults are close to 1 pu.

The calculations are repeated for a solidly grounded 132-kV system with the following parameters:

- $Z_{S0} = 0.047 + j2.75\%$
- $Z_{S1} = Z_{S2} = 0.09 + j2.86\%$
- $Z_{F0} = 0.23 + j0.65\%\ \text{km}^{-1}$
- $Z_{F1} = Z_{F2} = 0.101 + j0.257\%\ \text{km}^{-1}$

The results are shown in the right-hand plot of Figure 6.53, with the different curves having the same meaning as in the plot on the left. Compared with the resistance-grounded system, the characteristic voltage for a dip due to a single-phase fault and the PN factor for a dip due to a two-phase-to-ground fault are significantly lower, their lowest values being about $\frac{1}{3}$ and $\frac{2}{3}$, respectively.

***6.2.3.10 Comparison Between ABC Classification and Symmetrical-Component Classification*** The symmetrical-component classification as described in the previous section is a systematic approach to the analysis of three-phase unbalanced voltage dips. Being a systematic approach it covers all cases and is therefore preferable above the ABC classification, which is a more intuitive approach that requires certain approximations which are not directly clear from the model used. The symmetrical-component classification also leads to a well-defined algorithm for extracting dip type and characteristics from measured voltage waveforms; it further allows a better quantification of the load effects on voltage-dip characteristics [39, 40].

The ABC classification does have its merits as well. Being a more intuitive classification, it is easier to grasp without the need to study symmetrical-component theory. It especially gives a very easy graphical interpretation of the transfer of three-phase unbalanced dips through transformers. Another clear advantage of the ABC classification is that it limits the number of possible cases. This makes it more appropriate as a basis for testing protocols, for example, during the development of control algorithms for grid-connected power-electronic converters.

The two classification methods should certainly not be seen as two independent different methods. The ABC classification is merely a special case of the more general symmetrical-component classification. The seven dip types from the ABC classification can be obtained from the different fault types (three-phase, two-phase-to-ground, phase-to-phase, and single-phase-to-ground) under the assumption that positive-, negative-, and zero-sequence impedances are equal. The relation between the dip types for the two classification methods is summarized in

**TABLE 6.5 Relation Between ABC Classification and Symmetrical-Component Classification for Three-Phase Unbalanced Voltage Dips**

| | Symmetrical-Component Classification | | | |
|---|---|---|---|---|
| | Type | Characteristic Voltage | PN Factor | Zero-Sequence Voltage |
| A: | any | $V = V^*$ | $F = V^*$ | $U_0 = 0$ |
| B: | Da | $V = \frac{1}{3}E_1 + \frac{2}{3}V^*$ | $F = E_1$ | $U_0 = \frac{1}{3}V^* - \frac{1}{3}E_1$ |
| C: | Ca | $V = V^*$ | $F = E_1$ | $U_0 = 0$ |
| D: | Da | $V = V^*$ | $F = E_1$ | $U_0 = 0$ |
| E: | Ca | $V = V^*$ | $F = \frac{2}{3}E_1 + \frac{1}{3}V^*$ | $U_0 = \frac{1}{3}E_1 - \frac{1}{3}V^*$ |
| F: | Da | $V = V^*$ | $F = \frac{2}{3}E_1 + \frac{1}{3}V^*$ | $U_0 = 0$ |
| G: | Ca | $V = V^*$ | $F = \frac{2}{3}E_1 + \frac{1}{3}V^*$ | $U_0 = 0$ |

Table 6.5. For each of the seven types defined in Table 6.2, the corresponding types and characteristics according to (6.13) and (6.14) are given. Note that both expressions lead to a type A dip for $F = V$. For all dip types it is assumed that phase $a$ is the symmetrical phase. Extending the ABC classification to include other symmetrical phases is straightforward and would lead to a total of 19 types.

As mentioned before, an advantage of the symmetrical-component classification is that its characteristics can be easily extracted from measured voltage waveforms. Knowing the type (i.e., C or D) and the characteristics the measured dip can be classified into one of the seven types of the ABC classification according to Table 6.6. Such a classification can be used, for example, to obtain statistics on voltage-dip types or to extract information on the type of fault that caused the measured voltage dip.

**TABLE 6.6 ABC Classification Based on Measurements Obtained Through Symmetrical-Component Classification**

| ABC | Symmetrical Components |
|---|---|
| Type A | Any type with $F \approx V$ |
| Type B | Type D with $F \approx 1$ pu and $U_0 > 0$ |
| Type C | Type C with $F \approx 1$ pu and $U_0 \approx 0$ |
| Type D | Type D with $F \approx 1$ pu and $U_0 \approx 0$ |
| Type E | Type C with $F < 1$ pu and $U_0 > 0$ |
| Type F | Type D with $F < 1$ pu and $U_0 \approx 0$ |
| Type G | Type C with $F < 1$ pu and $U_0 \approx 0$ |

### 6.2.4 Phase-Angle Jumps Associated with Voltage Dips

Most voltage dips are associated with a phase-angle jump: The voltage waveform not only drops in magnitude, but also shows a shift in phase angle. In the voltage waveform this is visible as a shift in the voltage zero crossings. There are two different phenomena that lead to a phase-angle jump in the voltage. For a three-phase fault, the difference between the $X/R$ ratio of the source and of the faulted feeder leads to an identical phase-angle jump in the three phases. An example of such a balanced dip is shown in Figure 6.54.

The second cause of phase-angle jumps occurs only with voltage dips due to nonsymmetrical faults. The expressions for dip types C, D, F, and G show that phase-angle jumps may occur even if the voltage in the faulted phase or between the faulted phases does not show any phase-angle jump. Examples of this are discussed in detail in [33]. The situation becomes complicated, however, for unbalanced dips in case of nonequal $X/R$ ratios. In such a case the phase-angle jump will occur in the characteristic voltage $V^*$ in the equations in Table 6.2. For the voltage waveforms and rms voltages this results in an additional unbalance: All three voltage magnitudes will be different for dip types C, D, F, and G

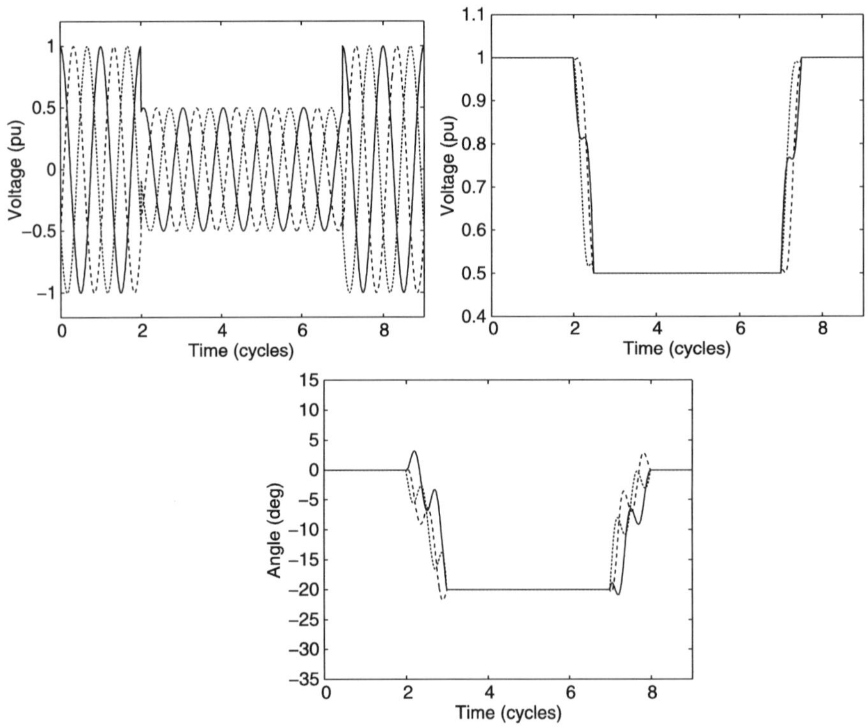


**Figure 6.54** Voltage dip of type A: voltage waveforms (left); half-cycle rms voltage (right); phase angle (bottom).

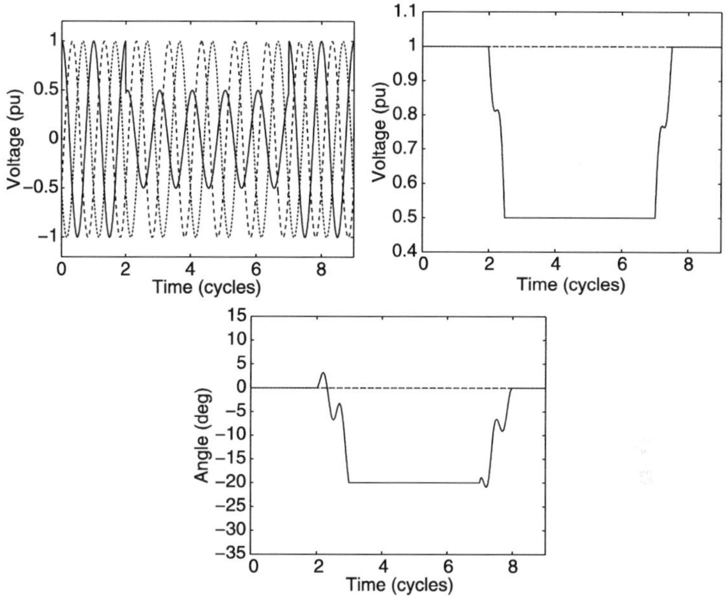


**Figure 6.55** Voltage dip of type B: voltage waveforms (left); half-cycle rms voltage (right); phase angle (bottom).

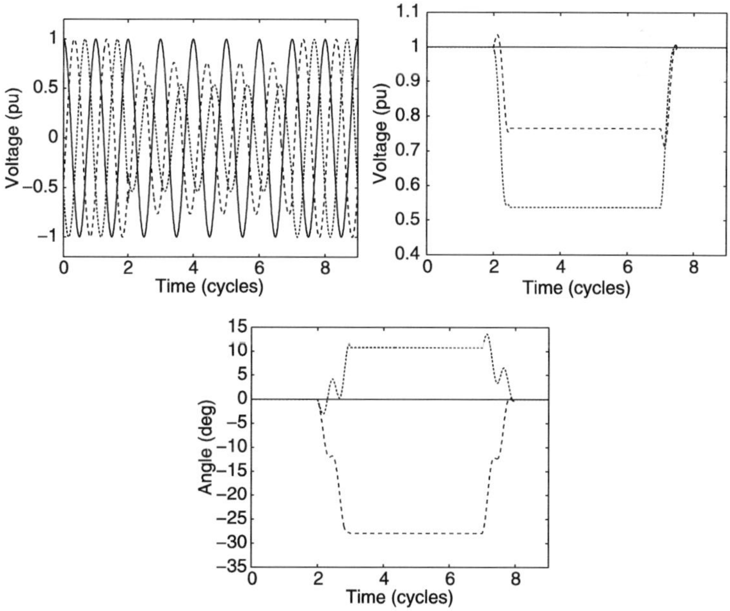


**Figure 6.56** Voltage dip of type C: voltage waveforms (left); half-cycle rms voltage (right); phase angle (bottom).

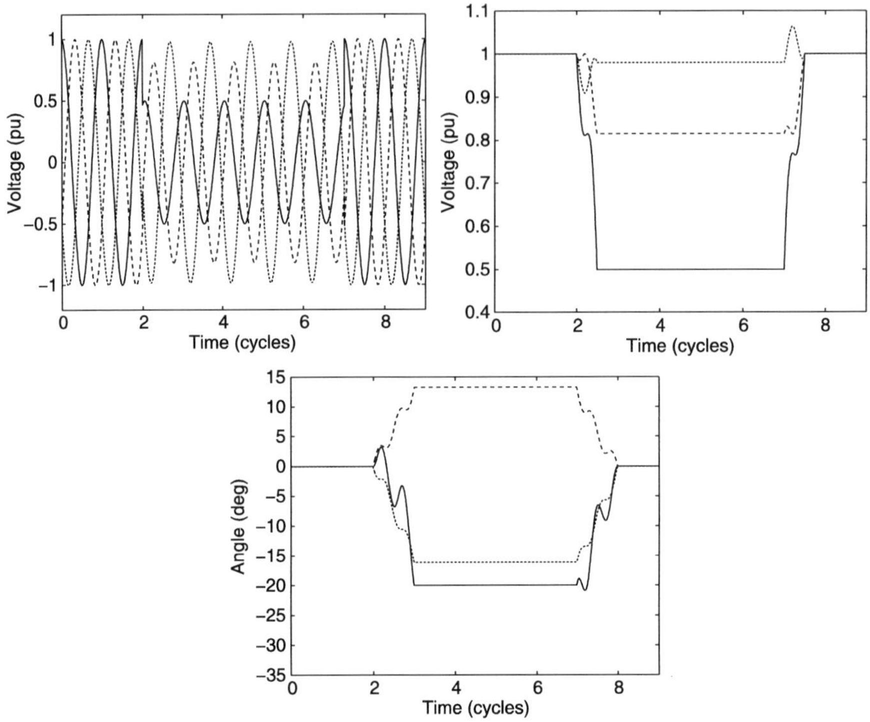


**Figure 6.57** Voltage dip of type D: voltage waveforms (left); half-cycle rms voltage (right); phase angle (bottom).

(see Figs. 6.56, 6.57, 6.60, and 6.61, respectively). All the figures contain synthetic dips with a characteristic voltage of 50% and a phase-angle jump of $-20^{\circ}$. Note that such a phase-angle jump only occurs for faults in cables or low-capacity overhead lines, typically at distribution level. Comparing the difference with Figures 6.44 through 6.50

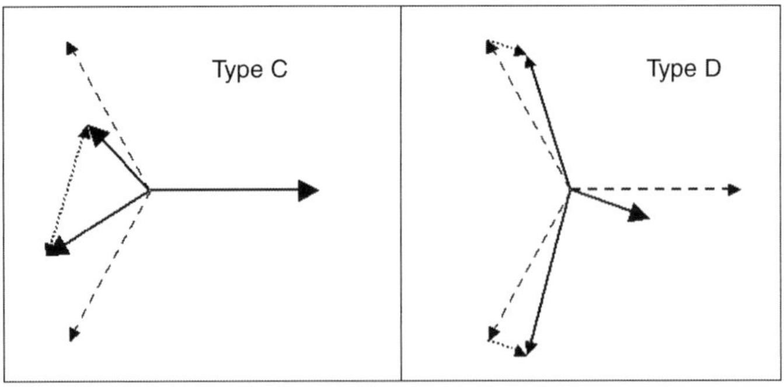


**Figure 6.58** Type C and D unbalanced dips with large characteristic phase-angle jumps.

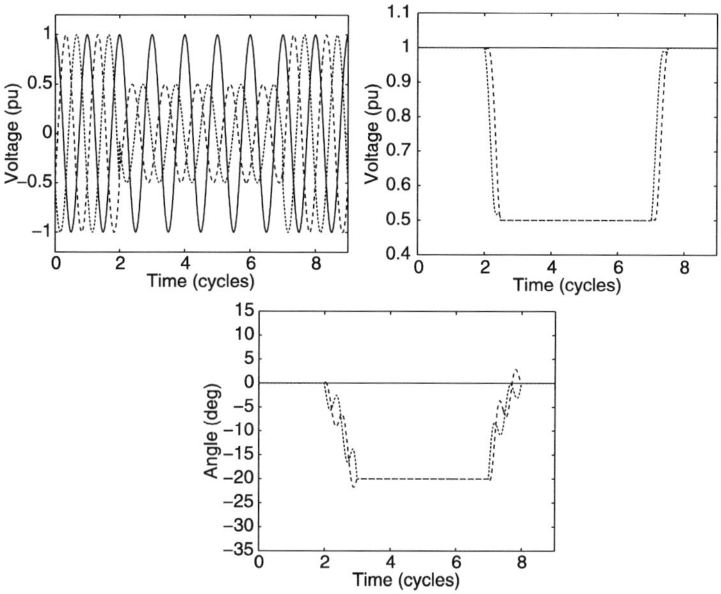


**Figure 6.59** Voltage dip of type E: voltage waveforms (left); half-cycle rms voltage (right); phase angle (bottom).

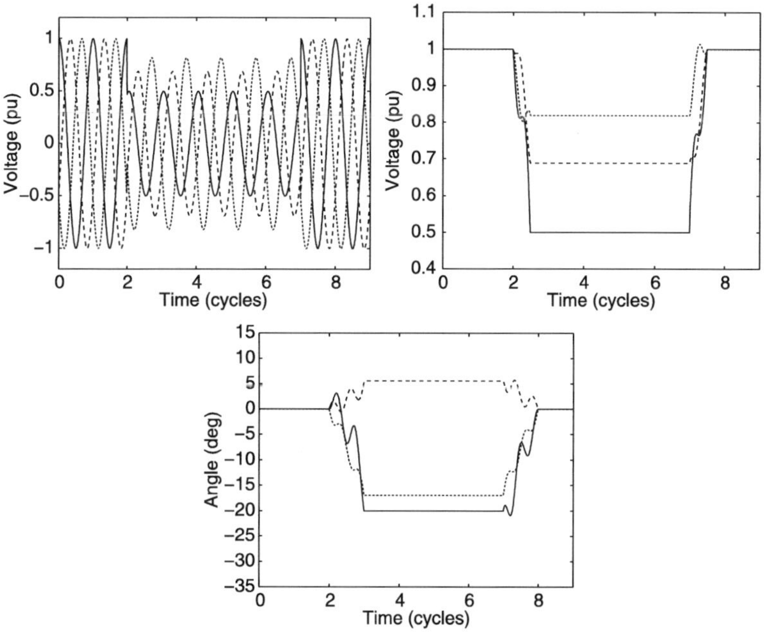


**Figure 6.60** Voltage dip of type F: voltage waveforms (left); half-cycle rms voltage (right); phase angle (bottom).

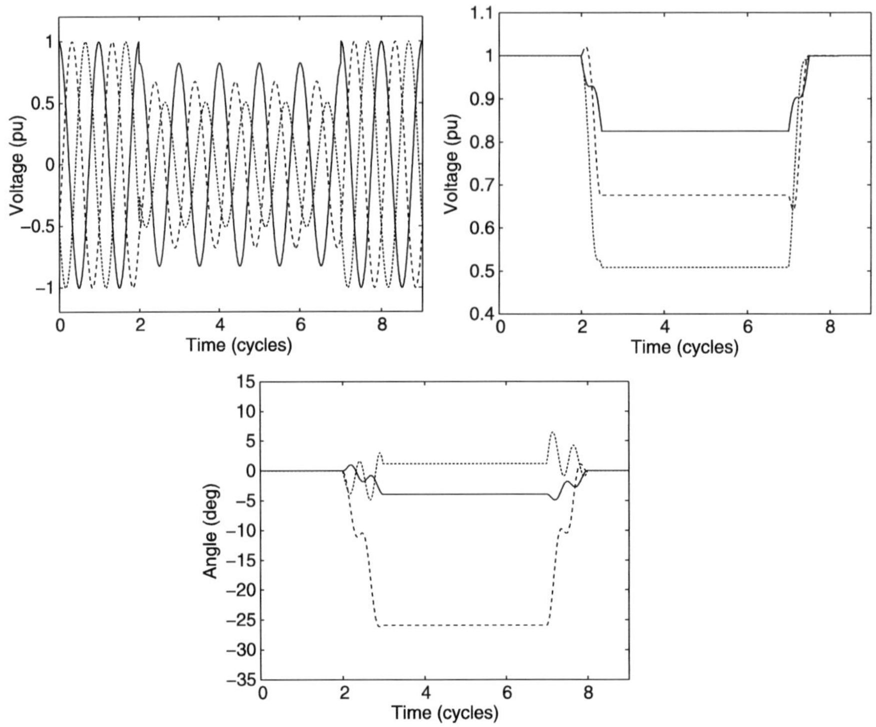


**Figure 6.61** Voltage dip of type G: voltage waveforms (left); half-cycle rms voltage (right); phase angle (bottom).

clearly shows the effect of the characteristic phase-angle jump on the voltage waveform and rms. Note that it becomes very difficult now to distinguish between dips of type C and type D from the rms voltages only [38]. As mentioned before, the symmetrical-component method is still able to give the correct answer.

For dips of type A associated with nonzero characteristic phase-angle jumps, as shown in Figure 6.54, the rms voltages remain the same in the three phase. Also the phase-angle jumps remain the same in the three phases, although they are no longer zero.

Also for a type B dip (Fig. 6.55), the presence of a nonzero characteristic phase-angle jump does not impact the rms voltage. The only impact is the nonzero phase-angle jump in the faulted phase.

For voltage dips of types C and D, the nonzero characteristic phase-angle jump also impacts the rms voltages, as shown in Figure 6.56 and 6.57. For a type C dip, the rms voltages in the nonfaulted phases become different, as does the size of the phase-angle jump. The phase with the largest drop in voltage shows the smallest jump in phase angle. The same occurs for the two nonfaulted phases during a type D dip.

Explaining the impact of the nonzero characteristic phase-angle jump on type C and D dips requires a phasor diagram, as in Figure 6.58. The principal impact of the characteristic phase-angle jump on a type C dip is the rotation of the voltage between the faulted phases. The result is that one of the phase voltages becomes smaller whereas the other increases in magnitude.

The impact on a type D dip is the rotation of the voltage in the faulted phase as well as the rotation of the complex change in voltage in the nonfaulted phases. This latter rotation is linked to the rotation in the faulted phase because the sum of the phasors remains zero (or in other words a type D dip does not contain a zero-sequence component).

The impact of a nonzero characteristic phase-angle jump on dip types E, F, and G (due to two-phase-to-ground faults) follows along the same line of reasoning. For a type E dip (Fig. 6.59), the voltage in the faulted phases shows a phase-angle jump equal to the characteristic phase-angle jump. The two rms voltages are not affected and remain the same. Type F and G dips will show three different rms voltages and three different phase-angle jumps (Figs. 6.60 and 6.61).

### 6.2.5 Voltage Recovery After a Fault

The instant of the main voltage recovery after a dip corresponds to the moment when the circuit breaker clears the fault. Circuit breakers clear a fault when the fault current has a zero crossing. Let $\cos\psi$ be the power factor of the fault current. Then the current zero crossing takes place at an angle $\psi$ for the prefault voltage or 180° later. The point on wave of voltage recovery after a dip due to a fault is thus $\psi$ or $\psi + 180°$ with reference to the upward zero crossing of the preevent voltage. In the remainder only the first value will be considered, knowing that there are always two possible points of recovery per cycle, half a cycle apart.

The voltage after recovery can be written as

$$v(t) = \sin(\omega t + \psi) \tag{6.51}$$

with $t = 0$ the recovery instant. Fault clearing does not always take place at the same instant for the three phases. The angle of the fault current will be different for different types of faults, but generally speaking they are found in the same range of values: 45° through 60° for faults in distribution systems; 75° through 85° for faults in transmission systems.

We will only consider solidly grounded systems in this Section. For high-impedance grounded systems, the angle is different for single-phase-to-ground faults. Also single-phase-to-ground faults do not cause any significant dip, and two-phase-to-ground faults cause the same dip as phase-to-phase faults.

#### *6.2.5.1 Voltage Dips of Type B*

Voltage dips of type B are due to single-phase faults. Let $\cos\psi_1$ be the power factor of the fault current for a single-phase fault. Then the voltage after recovery is, in the three phases,

$$\begin{aligned} v_a(t) &= \sin(\omega t + \psi_1) \\ v_b(t) &= \sin(\omega t + \psi_1 - 120^\circ) \\ v_c(t) &= \sin(\omega t + \psi_1 + 120^\circ) \end{aligned} \tag{6.52}$$

The voltage recovers in the three phases at the same instant, with the point on wave in the faulted phase equal to the angle between source voltage and fault current ($\psi_1$). When the nonfaulted phases show a swell, this also recovers together at the same instant as the dip in the faulted phase.

***6.2.5.2 Voltage Dips of Type C*** Voltage dips of type C are due to single-phase-to-ground or phase-to-phase faults. Consider a phase-to-phase fault between phases $b$ and $c$. The fault current is driven by the voltage difference between phases $b$ and $c$ and has a zero crossing at an angle $\psi_2$ compared to the prefault voltage, with $\cos\psi_2$ the power factor of the fault current for a phase-to-phase fault. An angle $\psi_2$ for the $bc$ phase voltage corresponds to an angle $\psi_2 + 90^\circ$ for the voltage in phase $a$. The voltages after recovery can be written as

$$\begin{aligned} v_a(t) &= \sin(\omega t + \psi_2 + 90^\circ) \\ v_b(t) &= \sin(\omega t + \psi_2 - 30^\circ) \\ v_c(t) &= \sin(\omega t + \psi_2 - 150^\circ) \end{aligned} \tag{6.53}$$

As with type B the voltage recovery takes place in all phases at the same time. The point on wave of recovery is equal to $\psi_2$ for the voltage difference between the faulted phases.

For a voltage dip of type C due to a single-phase fault, the expressions in (6.52) have to be transformed according to the Dy transformer. This results in the following expressions:

$$\begin{aligned} v_a(t) &= \sin(\omega t + \psi_1 + 90^\circ) \\ v_b(t) &= \sin(\omega t + \psi_1 - 30^\circ) \\ v_c(t) &= \sin(\omega t + \psi_1 - 150^\circ) \end{aligned} \tag{6.54}$$

The only difference with (6.53) is in the angle $\psi$. But as this angle is similar for single-phase-to-ground and phase-to-phase faults, the recovery after a type C dip will be the same for a single-phase-to-ground and for a phase-to-phase fault.

***6.2.5.3 Voltage Dips of Type D*** Voltage dips of type D are due to a single-phase-to-ground fault after two Dy transformers or due to a phase-to-phase fault after one Dy transformer. The voltage after recovery is the same as (6.52):

$$\begin{aligned} v_a(t) &= \sin(\omega t + \psi_k) \\ v_b(t) &= \sin(\omega t + \psi_k - 120^\circ) \\ v_c(t) &= \sin(\omega t + \psi_k + 120^\circ) \end{aligned} \tag{6.55}$$

**TABLE 6.7 Voltage Recovery for Voltage Dips Due to Single-Phase and Phase-to-Phase Faults**

| | Point-on-Wave of Voltage Recovery | |
|---|---|---|
| Dip Type | Phase voltages | Line voltages |
| Ba | $\phi_a = \psi_1$ | |
| | $\phi_b = \psi_1 - 120^\circ$ | |
| | $\phi_c = \psi_1 + 120^\circ$ | |
| Ca | $\phi_a = \psi_k + 90^\circ$ | $\psi_{bc} = \psi_k$ |
| | $\phi_b = \psi_k - 30^\circ$ | $\psi_{ca} = \psi_k - 120^\circ$ |
| | $\phi_c = \psi_c - 150^\circ$ | $\psi_{bc} = \psi_k + 120^\circ$ |
| Da | $\phi_a = \psi_k$ | |
| | $\phi_b = \psi_k - 120^\circ$ | |
| | $\phi_c = \psi_k + 120^\circ$ | |

with $k = 1$ for dips due to single-phase-to-ground faults and $k = 2$ for dips due to phase-to-phase faults. Recovery again takes place in the three phases at the same time, with the point on wave of recovery equal to $\psi_k$ in the phase with the lowest voltage (the "faulted phase").

The results for single-phase and phase-to-phase faults are summarized in Table 6.7. For these two fault types it is possible to define a point on wave of voltage recovery, but the definition is different for the different fault types. For types B and D it is the angle of the lowest phase voltage; for type C it is the angle of the lowest line voltage.

***6.2.5.4 Two-Phase-to-Ground Faults: Fault Clearing*** The clearing of a two-phase-to-ground fault takes place in two steps. After the clearing of the first phase of the fault current, the two-phase-to-ground fault develops into a single-phase-to-ground fault and the amplitude and phase angle of the fault current will typically be different.

Consider a fault between phases $b$, $c$, and ground. Let $0^\circ$ be an upward zero crossing in the phase $b$ fault current. Then there are four occasions per cycle in which a circuit breaker may clear the fault current in one phase: $0^\circ$ and $180^\circ$ in phase $b$; $120^\circ$ and $300^\circ$ in phase $c$. After the clearing of the first phase, the second phase will clear when the current in that phase gets through zero. If we neglect the difference in phase angle for single-phase-to-ground and two-phase-to-ground fault current, this results in four possible clearing sequences:

- Phase $b$ at $0^\circ$, phase $c$ at $120^\circ$
- Phase $c$ at $120^\circ$, phase $b$ at $180^\circ$
- Phase $b$ at $180^\circ$, phase $c$ at $300^\circ$
- Phase $c$ at $300^\circ$, phase $b$ at $360^\circ$

The second clearing takes place 120° after the first clearing when the first clearing is in phase $b$, and 60° later when the first clearing is in phase $c$. This implies that there will be two different recovery sequences after voltage dips due to two-phase-to-ground faults.

Taking into account the difference in phase angle of the fault current for different fault types, the duration of the intermediate stage is $120^\circ + \psi_1 - \psi_3$ when the first clearing takes place in phase $b$ and $60^\circ + \psi_1 - \psi_3$ when the first clearing takes place in phase $c$, with $\cos \psi_3$ the power factor of the current for a two-phase-to-ground fault and $\cos \psi_1$ for a single-phase-to-ground fault.

Note that it is assumed here that the second clearing takes place at the first occasion after the first clearing. This is not necessarily the case. Statistical analysis of voltage dip and/or fault current recordings is needed to decide how often the first occasion is missed by the circuit breaker.

***6.2.5.5 Two-Phase-to-Ground-Faults: Voltage Recovery*** The resulting voltage dip and recovery sequence depend on the location of the voltage dip-measurement compared to the fault location. With reference to Figure 6.51 and Table 6.3, the dip types before the first clearing are E, F, and G at locations I, II, and III, respectively.

The two fault-clearing sequences for a two-phase-to-ground fault are as follows:

- *bcn* becomes *cn* becomes normal in 120°
- *bcn* becomes *bn* becomes normal in 60°

This translates, with help of Table 6.3 in the following voltage recovery sequences at location I in Figure 6.51:

- Ea becomes Bc becomes normal in 120°, first recovery at angle $\psi_3$ in phase $b$
- Ea becomes Bb becomes normal in 60°, first recovery at angle $\psi_3$ in phase $c$

After one Dy transformer, that is, at location II in Figure 6.51, the voltage recovery sequences are as follows:

- Fa becomes Cc becomes normal in 120°, first recovery at angle $\psi_3$ for phase difference *ac*
- Fa becomes Cb becomes normal in 60°, first recovery at angle $\psi_3$ for phase difference *ab*

And after two Dy transformers (location III in Figure 6.51):

- Ga becomes Dc becomes normal in 120°, first recovery at angle $\psi_3$ in phase $b$
- Ga becomes Db becomes normal in 60°, first recovery at angle $\psi_3$ in phase $c$

The results for two-phase-to-ground faults are summarized in Table 6.8.

**TABLE 6.8 Voltage Recovery for Voltage Dips Due to Two-Phase-to-Ground Faults**

| Dip Type | Point on Wave of First Recovery | Intermediate Stage and Duration | Point on Wave of Second Recovery |
|---|---|---|---|
| Ea | $\phi_a = \psi_3 + 120^\circ$<br>$\phi_b = \psi_3$<br>$\phi_c = \psi_3 - 120^\circ$ | Bc: $120^\circ + \psi_1 - \psi_3$ | $\phi_a = \psi_1 - 120^\circ$<br>$\phi_b = \psi_1 + 120^\circ$<br>$\phi_c = \psi_1$ |
| Ea | $\phi_a = \psi_3 - 120^\circ$<br>$\phi_b = \psi_3 + 120^\circ$<br>$\phi_c = \psi_3$ | Bb: $60^\circ + \psi_1 - \psi_3$ | $\phi_a = \psi_1 - 60^\circ$<br>$\phi_b = \psi_1 + 180^\circ$<br>$\phi_c = \psi_1 + 60^\circ$ |
| Fa | $\phi_a = \psi_3 - 150^\circ$<br>$\phi_b = \psi_3 + 90^\circ$<br>$\phi_c = \psi_3 - 300^\circ$ | Cc: $120^\circ + \psi_1 - \psi_3$ | $\phi_a = \psi_1 - 30^\circ$<br>$\phi_b = \psi_1 - 150^\circ$<br>$\phi_c = \psi_1 + 90^\circ$ |
| Fa | $\phi_a = \psi_3 + 150^\circ$<br>$\phi_b = \psi_3 + 30^\circ$<br>$\phi_c = \psi_3 - 90^\circ$ | Cb: $60^\circ + \psi_1 - \psi_3$ | $\phi_a = \psi_1 - 150^\circ$<br>$\phi_b = \psi_1 + 90^\circ$<br>$\phi_c = \psi_1 - 30^\circ$ |
| Ga | $\phi_a = \psi_3 + 120^\circ$<br>$\phi_b = \psi_3$<br>$\phi_c = \psi_3 - 120^\circ$ | Dc: $120^\circ + \psi_1 - \psi_3$ | $\phi_a = \psi_1 - 120^\circ$<br>$\phi_b = \psi_1 + 120^\circ$<br>$\phi_c = \psi_1$ |
| Ga | $\phi_a = \psi_3 - 120^\circ$<br>$\phi_b = \psi_3 + 120^\circ$<br>$\phi_c = \psi_3$ | Db: $60^\circ + \psi_1 - \psi_3$ | $\phi_a = \psi_1 - 60^\circ$<br>$\phi_b = \psi_1 + 180^\circ$<br>$\phi_c = \psi_1 + 60^\circ$ |

***6.2.5.6 Three-Phase Faults*** Balanced dips (type A) are due to three-phase and three-phase-to-ground faults. As shown in Table 6.3 the dip before fault clearing is the same for all locations. The voltage recovery however depends on the presence of an earth connection with the fault and on the presence of transformers between the fault location and the place where the dip is measured.

The clearing of a three-phase fault (without earth connection) takes place in two stages. After the clearing in the first phase, a phase-to-phase fault results. This one is cleared in the two residual phases at the same time, about $90^\circ$ after the first fault clearing. The recovery of the voltage dip depends on the presence of transformers between the fault and the dip. The fault-clearing sequence is

- *abc* becomes *bc* becomes normal in $90^\circ$

For location I in Figure 6.51 this results in the following voltage recovery sequence:

- A becomes Ca becomes normal in $90^\circ$, first recovery at angle $\psi_4$ in phase *a*

At location II the recovery sequence is

- A becomes Da becomes normal in $90^\circ$, first recovery at angle $\psi_4$ for the phase difference *bc*

At location III the recovery sequence is the same as at location I. As there is no zero-sequence component present, the voltages behind two Dy transformers are the same as at the fault location.

**6.2.5.7 Three-Phase-to-Ground Faults** For a three-phase-to-ground fault the fault clearing takes place in three stages: from three-phase-to-ground to two-phase-to-ground to single-phase-to-ground to full recovery. The duration of each of the two intermediate stages is 60°. Assume that phase $a$ is the first to clear, then phase $c$ will recover 60° later, and phase $b$ another 60° later. The fault-clearing sequence is

- *abcn* becomes *bcn* becomes *bn* becomes normal in two times 60°

This results in the following voltage recovery sequences, with reference to Figure 6.51:

- A becomes Ea becomes Bb becomes normal at location I
- A becomes Fa becomes Cb becomes normal at location II
- A becomes Ga becomes Db becomes normal at location III

The results for three-phase and three-phase-to-ground faults are summarized in Table 6.9. All dips are of type A, but the recovery can take place in five different ways depending on the location of the fault and the presence of an earth connection with the fault.

**6.2.5.8 Simulations and Measurements** The first case to be studied is a three-phase-to-ground fault in the transmission system. The fault-clearing angle $\psi$ is 85° for all fault types and there is no phase shift during the dip. Figure 6.62 shows the voltage recovery measured at location I in Figure 6.51 (no transformer). The dip develops from type A to type Ea to type Bb as the voltage recovers in phases $a$, $c$, and $b$. The expressions in Table 6.2 have been used to calculate the complex voltages, with $V = 0.50$. The voltage recovery after a Dy transformer is shown in Figure 6.63. The dip develops in this case from type A to type Fa to type Cb.

As shown in Figure 6.63, phase $a$ (solid line) recovers in two stages 60° apart, phase $b$ (dashed line) recovers in two stages 120° apart, and phase $c$ (dotted line) recovers in two stages 60° apart. The last recovery of phase $c$ corresponds to the first one in phase $a$.

The calculations have been repeated for a three-phase-to-ground fault at the distribution level. The fault-clearing angle is taken as 45° and there is a phase shift of −20° during the dip. The results are shown in Figures 6.64 and 6.65. Note that in all cases the first fault clearing takes place at time zero. This partly explains the change in phase angle for the prerecovery voltages. Also the difference in phase-angle jump contributes to this.

**TABLE 6.9 Voltage Recovery for Dips Due to Three-Phase and Three-Phase-to-Ground Faults**

| Type | First Recovery | Stage 1 and Duration | Second Recovery | Stage 2 and Duration | Third Recovery |
|---|---|---|---|---|---|
| A | $\phi_a = \psi_4$ | Ca: $60° + \psi_2 - \psi_4$ | $\phi_a = \psi_2 + 90°$ | NA | NA |
| | $\phi_b = \psi_4 - 120°$ | | $\phi_b = \psi_2 - 30°$ | | |
| | $\phi_c = \psi_4 + 120°$ | | $\phi_c = \psi_2 - 150°$ | | |
| A | $\phi_a = \psi_4 + 90°$ | Da: $60° + \psi_2 - \psi_4$ | $\phi_a = \psi_2 + 180°$ | NA | NA |
| | $\phi_b = \psi_4 - 30°$ | | $\phi_b = \psi_2 + 60°$ | | |
| | $\phi_c = \psi_4 - 150°$ | | $\phi_c = \psi_2 - 60°$ | | |
| A | $\phi_a = \psi_4$ | Ea: $60° + \psi_3 - \psi_4$ | $\phi_a = \psi_3 + 60°$ | Bb: $60° + \psi_1 - \psi_3$ | $\phi_a = \psi_1 + 120°$ |
| | $\phi_b = \psi_4 - 120°$ | | $\phi_b = \psi_3 - 60°$ | | $\phi_b = \psi_1$ |
| | $\phi_c = \psi_4 + 120°$ | | $\phi_c = \psi_3 + 180°$ | | $\phi_c = \psi_1 - 120°$ |
| A | $\phi_a = \psi_4 + 90°$ | Fa: $60° + \psi_3 - \psi_4$ | $\phi_a = \psi_3 + 150°$ | Cb: $60° + \psi_1 - \psi_3$ | $\phi_a = \psi_1 - 150°$ |
| | $\phi_b = \psi_4 - 30°$ | | $\phi_b = \psi_3 + 30°$ | | $\phi_b = \psi_1 + 90°$ |
| | $\phi_c = \psi_4 + 150°$ | | $\phi_c = \psi_3 - 90°$ | | $\phi_c = \psi_1 - 30°$ |
| A | $\phi_a = \psi_4$ | Ga: $60° + \psi_3 - \psi_4$ | $\phi_a = \psi_3 + 60°$ | Db: $60° + \psi_1 - \psi_3$ | $\phi_a = \psi_1 + 120°$ |
| | $\phi_b = \psi_4 - 120°$ | | $\phi_b = \psi_3 - 60°$ | | $\phi_b = \psi_1$ |
| | $\phi_c = \psi_4 + 120°$ | | $\phi_c = \psi_3 + 180°$ | | $\phi_c = \psi_1 - 120°$ |

*Note*: NA = not applicable.

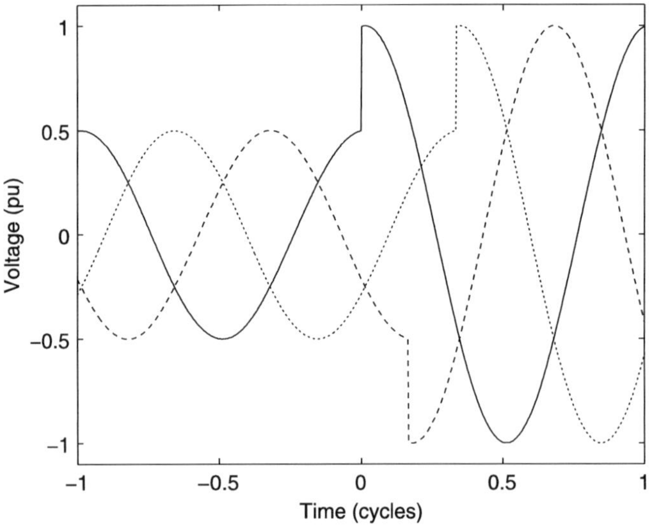


**Figure 6.62** Voltage recovery for dip due to three-phase-to-ground fault at transmission level: solid line: phase *a*; dotted line: phase *b*; dashed line: phase *c*.

Even though the voltage dip before fault clearing is the same for these four cases, the wave shapes during fault clearing are significantly different. Equipment that is sensitive to the voltage recovery after a dip may show completely different behavior for the different recovery sequences.

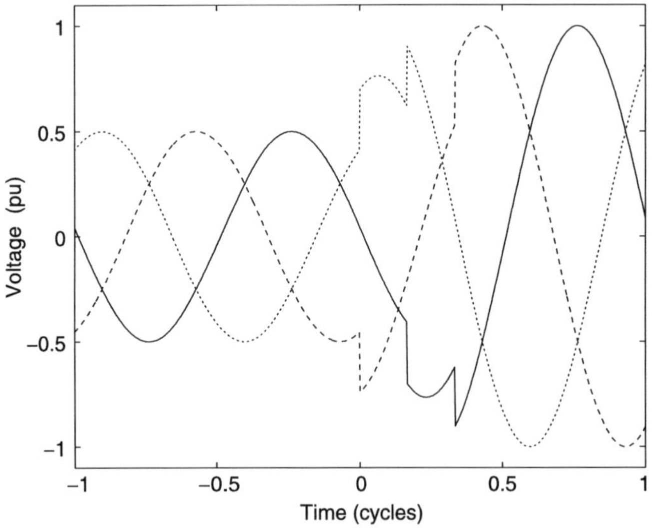


**Figure 6.63** Voltage recovery for dip due to three-phase-to-ground fault at transmission level measured behind Dy transformer.

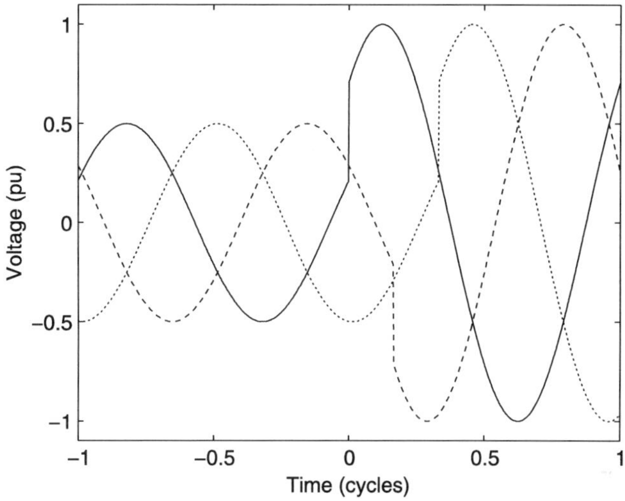


**Figure 6.64** Voltage recovery for dip due to three-phase-to-ground fault at distribution level.

The voltage recovery after a dip due to a two-phase-to-ground fault is shown in detail in Figure 6.66; The voltage recovery (and hence the fault clearing) appears to take place in two stages about 40° separated in time.

Two examples of voltage recovery after a three-phase fault are shown in detail in Figure 6.67: For the event on the left the recovery takes place in two stages, for the

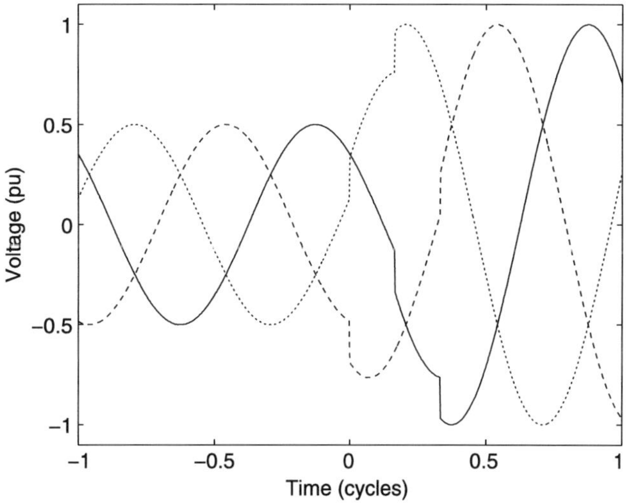


**Figure 6.65** Voltage recovery for dip due to three-phase-to-ground fault at distribution level measured behind Dy transformer.

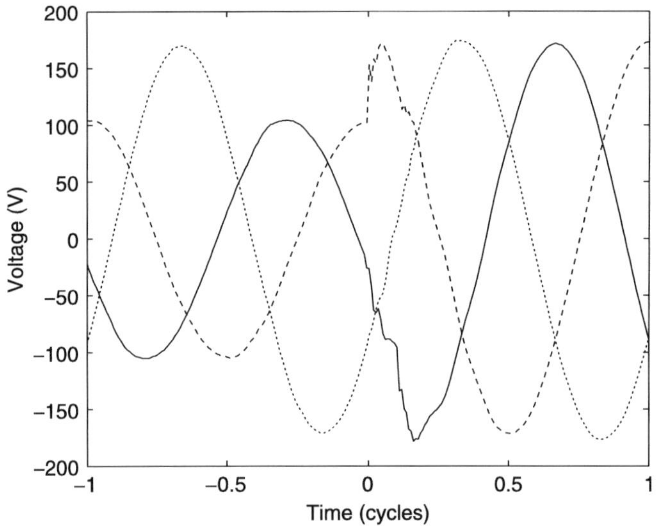


**Figure 6.66** Voltage recovery after two-phase-to-ground fault.

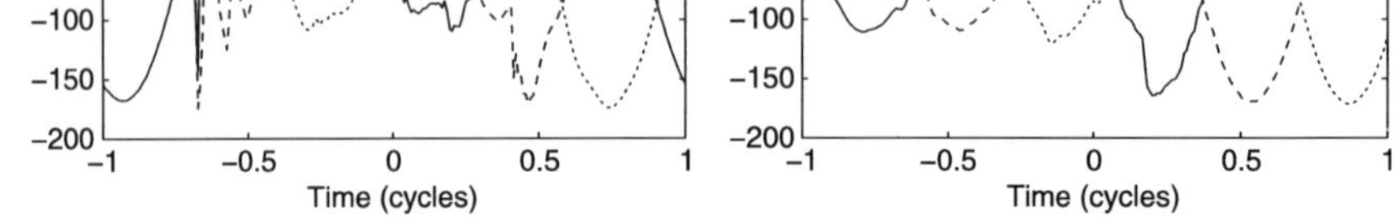


**Figure 6.67** Voltage recovery instants for dip due to three-phase fault: two examples.

one on the right in three stages. The oscillations in the left-hand plot occur when the phase-to-phase fault develops in a three-phase fault. Fault clearing takes place soon after that.

## 6.3 TRANSIENTS

### 6.3.1 What Are Transients?

There are a number of definitions of *transient* all of which have in common that the phenomenon is of short duration. The term transient originates from electric circuit theory where it denotes the component of voltage and current that is due to the transition between two steady states and which is of limited duration (the so-called transient component). In mathematical terms the transient component is the solution to the homogeneous form of the linear differential equation describing the system (more correctly the solution that fits the initial conditions); the new steady state is a particular solution to the nonhomogeneous differential equation.

In power systems analysis the term transient is used more generally to denote any voltage or current event that has a short duration. A duration less than one half-cycle or one cycle is generally considered as short in this context. The IEEE standard 1100 uses the term "subcycle disturbance" [164]. The definition of transient as commonly in use in the power engineering community is thus not exactly the same as the one used in electric circuit theory. The main difference is that the latter one refers only to one of the components of the (voltage or current) signal, whereas the former one refers to the whole waveform. Also the circuit theory definition allows for any duration of the phenomenon. A transformer inrush current, which may last several seconds, would be classified as a transient according to the circuit theory definition, but in power engineering it is generally viewed as a temporary overcurrent. In this book we will use the power engineering definition as a starting point, allowing ourselves some freedom to move beyond the subcycle nature of transients.

Some related definitions can be found in the literature. The term '*surge*' is sometimes used as a synonym for transient or more specifically transient overvoltage or overcurrent (e.g., in *surge arrestor*). The term *notch* is used to denote a voltage transient in which the instantaneous voltage moves toward zero for up to one half-cycle [164, 165].

A commonly made distinction is between *impulsive transients* and *oscillatory transients*. The distinction is based on the appearance of the event, with impulsive transients being unidirectional and oscillatory transients oscillating around a new steady-state value [165]. This distinction is however somewhat artificial and should not be used as an absolute classification. Many transients show impulsive as well as oscillatory behavior and may even change character when propagating through the system.

The classification we will use in this book is based on the origin of the event. We will distinguish between *lightning transients*, *normal switching transients*, and *abnormal switching transients*. We will further only consider those transients where the preevent steady state is similar to the postevent steady state. The start and end of an interruption or voltage dip also contain a transient component, but from a power quality viewpoint those transients are completely overshadowed by the interruption or dip.

The distinction between normal and abnormal switching transients is based on Greenwood [125]. Greenwood defines normal switching transients as those transients where the instantaneous value cannot exceed a value equal to twice the steady-state maximum. Abnormal transients are associated with trapped energy somewhere in the system. They are uncommon but of importance because of the very high overvoltages or overcurrents they can cause. Even though the distinction is according to Greenwood [125] somewhat artificial, we may state that normal switching transients are due to the switching of capacitors or inductors in linear circuits by ideal switches. Strictly speaking such a definition would make any transient abnormal as neither linear circuits nor ideal switches exist in the real world. But in most cases these can be used as an acceptable approximation.

### 6.3.2 Lightning Transients

A lightning stroke to an overhead line or in the neighborhood of the line will lead to a high overvoltage on the line. Such overvoltage often leads to an earth fault or to a short-circuit fault. The resulting event is a voltage dip or interruption, as discussed before. When the fault occurs (electrically) close to the measurement location a short-duration overvoltage may be observed before the dip, but despite having studied many voltage-dip recordings, only one recording was found that might be due to this. A possible explanation for this lack of lightning overvoltages associated with faults is that the fault occurs already in the initial stages of the voltage rise so that the duration of the overvoltage is very short, probably not more than 1 μs. An overvoltage of such short duration is damped very quickly when it propagates through the system. Next to that it will be filtered by the anti-aliasing filter in the measurement equipment, having typical sampling frequencies around 10 kHz.

Of more interest for this section are lightning transients that do not cause a fault. In [305] a distinction is made between the following lightning-induced transients.

- Lightning strokes in the vicinity of the overhead lines, which do not hit the conductors themselves
- Direct lightning strokes on the line conductors injecting an electromagnetic wave into the line
- Lightning strokes to the towers or to the shielding wires

A direct stroke to a line conductor causes a very high overvoltage. A current $I_l$ will cause a voltage equal to $\frac{1}{2}Z_w I_l$ with $Z_w$ the wave impedance of the line. Even a small lightning current of 5 kA will lead to an overvoltage of over 500 kV that can only be withstood by a line at the highest voltage level. According to data presented in [125], only 0.5% of lightning strokes have a current less than 5 kA. The conclusion is that a direct stroke to a line conductor will in almost all cases lead to a fault.

A lightning stroke to a tower or to a shielding wire will cause part of the lightning current to flow to ground, thus reducing the overvoltage. The lower the *footer impedance* of the tower, the less the resulting overvoltage. A low footer impedance is an efficient way of limiting the risk of lightning-induced faults in transmission lines.

A lightning stroke in the vicinity of a line will induce a transient overvoltage in the line. Also a stroke between two clouds may cause such an event. According to [99], the majority of lightning-related damage to equipment is due not to direct strokes to the equipment but to strokes in the vicinity. The transient overvoltage may reach the equipment via the power system or as a transient rise in the earth potential.

The shape of the lightning-induced transient is more or less the same as the shape of the lightning current. The lightning current typically shows a fast sharp rise followed by a slower decay. Often a second stroke occurs at the same location after 20 to 50 ms. The actual shape varies significantly; however, a standard lightning shape has been defined as a curve with a rise time of 1.2 μs and a decay time of 50 μs (the so-called 1.2/50 wave). This curve is used for testing of equipment against lightning-induced voltages and currents. The curve can be described mathematically through the following expression:

$$v(t) = V_0\{e^{-t/\tau_b} - e^{-t/\tau_a}\} \tag{6.56}$$

with $\tau_a = 71$ μs and $\tau_b = 0.2$ μs.

### 6.3.3 Normal Switching Transients

Normal switching transients can be modeled as opening or closing of ideal switches in linear *RLC* circuits. The four switching actions of importance are as follows:

- Capacitor energizing
- Capacitor deenergizing
- Inductor energizing
- Inductor deenergizing

The most severe case from a power quality viewpoint is the energizing of a capacitor. It will lead to an initial change in the voltage waveform toward zero followed by an oscillation with a frequency of a few hundred hertz. This case will be discussed in more detail below. The overvoltage can become even more severe when more capacitors are present in the system, a phenomenon referred to as *voltage magnification*. The overvoltage (with our without magnification) may lead to the incorrect tripping of adjustable-speed drives. This direct relation with equipment tripping means that this phenomenon is well described in the power quality literature [3, 89, 119, 213, 215, 311].

***6.3.3.1 Capacitor-Energizing Transients*** The connection of a capacitor to the supply leads to a damped oscillation superimposed on the new steady-state voltage. An expression for this oscillation is obtained by using the circuit shown in Figure 6.68:

$$e(t) = u_L(t) + u_R(t) + u_C(t) = L\frac{di}{dt} + Ri + u(t) \tag{6.57}$$

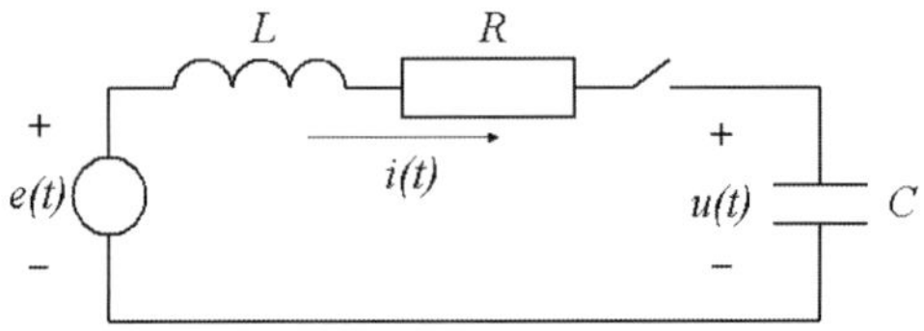


**Figure 6.68** Capacitor-energizing circuit.

Filling in the expression for the source voltage,

$$e(t) = \sqrt{2}E\cos(\omega_0 t + \phi) \tag{6.58}$$

and for the current through the capacitor,

$$i(t) = C\frac{du}{dt} \tag{6.59}$$

results in the following second-order differential equation for the voltage:

$$\sqrt{2}E\cos(\omega_0 t + \phi) = LC\frac{d^2u}{dt^2} + RC\frac{du}{dt} + u(t) \tag{6.60}$$

The initial conditions are that the current through the inductor and the voltage over the capacitor are zero at time zero:

$$\lim_{t\downarrow 0} i(t) = 0 \tag{6.61}$$

$$\lim_{t\downarrow 0} u(t) = 0 \tag{6.62}$$

The two expressions are more commonly written as $i(0+) = 0$ and $u(0+) = 0$. Using (6.59) changes (6.61) into the following initial condition for the voltage over the capacitor:

$$\left.\frac{du}{dt}\right|_{t=0+} = 0 \tag{6.63}$$

The particular solution to this differential equation gives the new steady-state voltage:

$$U_{ss}(t) = U\sqrt{2}\cos(\omega_0 t + \psi) \tag{6.64}$$

The solution to the homogenous equation

$$LC\frac{d^2u}{dt^2} + RC\frac{du}{dt} + u(t) = 0 \tag{6.65}$$

is the voltage transient. The solution to a second-order differential equation is a damped oscillation. To obtain oscillation frequency and damping, we need to solve the characteristic equation of the differential equations (6.60) and (6.65):

$$LC\lambda^2 + RC\lambda + 1 = 0 \tag{6.66}$$

with solution for $R^2 < 4L/C$:

$$\lambda = -\frac{R}{2L} \pm j\sqrt{\frac{1}{LC} - \frac{R^2}{4L^2}} \tag{6.67}$$

Note that in theory there are also nonoscillatory solutions to this expressions but they do not occur in practical systems. The resistance is always low in a power system in order to reduce losses. Therefore the condition for undercritical damping, $R^2 < 4L/C$, holds in the vast majority of cases. To limit the magnitude of the transient, some capacitor switches are equipped with damping resistors. In that case the oscillation may be overcritically damped. The resistors are only present in the circuit for a short time. As such switching will not lead to any serious transient, it will not be considered here.

The general expression for the transient solution is as follows:

$$u_{tr} = A_{tr}e^{-t/\tau}\cos(\omega_{tr}t + \phi_{tr}) \tag{6.68}$$

with $\tau = 2L/R$ and $\omega_{tr} = \sqrt{1/LC - R^2/4L^2} \approx \sqrt{1/LC}$. In short, capacitor switching causes a transient with a frequency

$$f_{tr} = \frac{1}{2\pi\sqrt{LC}} \tag{6.69}$$

and a damping time constant equal to

$$\tau = \frac{2L}{R} \tag{6.70}$$

Rewriting these expressions in terms of fault level, capacitor bank size, and $X/R$ ratio proceeds as follows. The capacitor bank size for star-connected capacitors is

$$Q = \omega_0 CU^2 \tag{6.71}$$

The fault level is

$$S_k = \frac{U^2}{\omega_0 L} \tag{6.72}$$

Substituting (6.71) and (6.72) in (6.69) gives the following expression for the frequency of the transient:

$$f_{tr} = f_0 \sqrt{\frac{S_k}{Q}} \tag{6.73}$$

The damping time constant is obtained directly from the $X/R$ ratio:

$$\tau = \frac{X/R}{\omega_0} \tag{6.74}$$

This expression only gives an upper limit for the damping time constant. Any parallel resistance, for example, due to a resistive load, has been neglected. Also the increase in resistance with frequency should be considered. Accurately determining the damping remains one of the most challenging problems in power system analysis. The complete solution of the differential equation is obtained by adding the transient to the new steady state:

$$u(t) = u_{ss}(t) + u_{tr}(t) = U\sqrt{2}\cos(\omega_0 t + \psi) + A_{tr}e^{-t/\tau}\cos(\omega_{tr}t + \phi_{tr}) \tag{6.75}$$

$$\frac{du}{dt} = -\omega_0 U\sqrt{2}\sin(\omega_0 t + \psi) - \frac{1}{\tau}A_{tr}e^{-t/\tau}\cos(\omega_{tr}t + \phi_{tr}) - \omega_{tr}A_{tr}e^{-t/\tau}\sin(\omega_{tr}t + \phi_{tr}) \tag{6.76}$$

The amplitude $A_{tr}$ and phase angle $\phi_{tr}$ of the transient solution are obtained from the initial conditions:

$$u_{ss}(t) + u_{tr}(t) = 0 \qquad \text{for } t = 0+ \tag{6.77}$$

$$\frac{du_{ss}}{dt} + \frac{du_{tr}}{dt} = 0 \qquad \text{for } t = 0+ \tag{6.78}$$

For $t = 0+$ we obtain the following expressions using (6.75), (6.76), and the initial conditions (6.77) and (6.78):

$$u(t)|_{t=0+} = U\sqrt{2}\cos\psi + A_{tr}\cos\phi_{tr} = 0 \tag{6.79}$$

$$\left.\frac{du}{dt}\right|_{t=0+} = -\omega_0 U\sqrt{2}\sin\psi - \frac{1}{\tau}A_{tr}\cos\phi_{tr} - \omega_{tr}A_{tr}\sin\phi_{tr} = 0 \tag{6.80}$$

Rewriting (6.79) and (6.80) gives

$$A_{\mathrm{tr}} \cos \phi_{\mathrm{tr}} = -U\sqrt{2} \cos \psi \tag{6.81}$$

$$A_{\mathrm{tr}} \sin \phi_{\mathrm{tr}} = -\frac{\omega_0}{\omega_{\mathrm{tr}}} U\sqrt{2} \sin \psi + \frac{U\sqrt{2}}{\omega_{\mathrm{tr}} T} \cos \psi \tag{6.82}$$

Using $\sin^2 + \cos^2 = 1$ gives the following expression for the amplitude of the transient:

$$A_{\mathrm{tr}} = \sqrt{2}\sqrt{(U \cos \psi)^2 \left(\frac{\omega_0}{\omega_{\mathrm{tr}}} U \sin \psi\right)^2 + \left(\frac{1}{\omega_{\mathrm{tr}} T} U \cos \psi\right)^2 - 2\frac{\omega_0}{\omega_{\mathrm{tr}}^2 T} U^2 \sin \psi \cos \psi} \tag{6.83}$$

For $\omega_{\mathrm{tr}} \gg \omega_0$ and $\omega_{\mathrm{tr}} T \gg 1$ we get

$$A_{\mathrm{tr}} \approx U\sqrt{2} |\cos \psi| \tag{6.84}$$

The magnitude of the transient is large for $\cos \psi = 1$, that is, when switching takes place near the maximum of the system voltage. A small magnitude is obtained when the switching takes place around voltage zero. The latter is called *synchronized switching*.

In Figure 6.69 the exact expression (6.83) is compared with the approximated expression (6.84) for different switching angles $\psi$. The calculation was done for

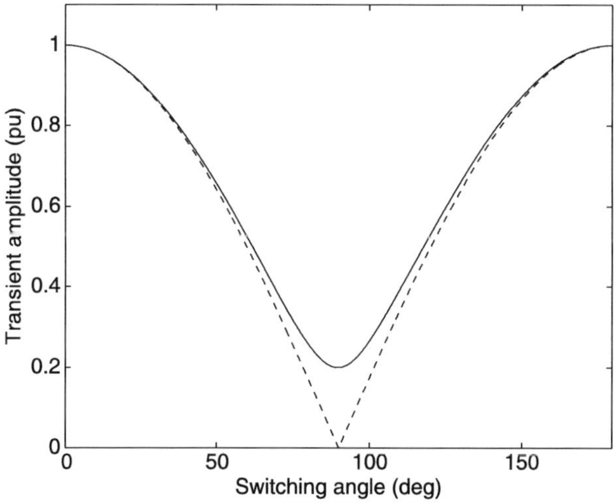


**Figure 6.69** Comparison between exact expression (solid line) and approximation (dashed) for transient amplitude due to capacitor energizing.

$f_{tr} = 250\,\text{Hz}$ and $X/R = 10$. Note that 1 pu corresponds to $U\sqrt{2}$ in (6.83). We see that the approximated expression incorrectly predicts a zero transient for switching at the voltage zero crossing. But the minimum transient is still obtained with voltage zero crossing.

#### 6.3.3.2 Capacitor Energizing: Measurement Examples

Figure 6.70 shows the voltages measured during synchronized capacitor energizing. The transient is indeed very small, but not completely zero. In this case the three-phase capacitor bank was star connected but not grounded. Two phases are energized near a zero crossing of the phase-to-phase voltage; the third phase is energized 5 ms later. The same figure also shows the currents measured at the same location. The measurement was performed at the 10-kV side of a 130/10-kV transformer. The current thus contains the load current as well as the capacitor current. Although the transient in voltage is small, the transient in current is still relatively big but much smaller than it would have been for nonsynchronized energizing. Note also the rise in voltage magnitude and the increased distortion in the current after the event.

Four examples of a voltage transient due to nonsynchronized capacitor energizing measured at the same location are shown in Figure 6.71. The first transient (top left) is due to the energizing of the capacitor in phase *a*. The main transient is in phase *a* (solid line); minor transients are visible in phases *b* (dashed line) and *c* (dotted line). The transient in phase *a* is due to the energizing of the phase-to-ground connected capacitor in that phase. The transients in the other two phases are due to the mutual coupling between the phases.

The second transient (top right) occurred a few cycles after the first one. The main transient occurs in phase *c* (dotted line) with minor transients in the other two phases. The transient is due to the energizing of the capacitor in phase *c*. It is interesting to notice that phase *a* is affected in a different way than phase *b*. This is due to the fact that phase *a* contains a capacitor whereas phase *b* does not.

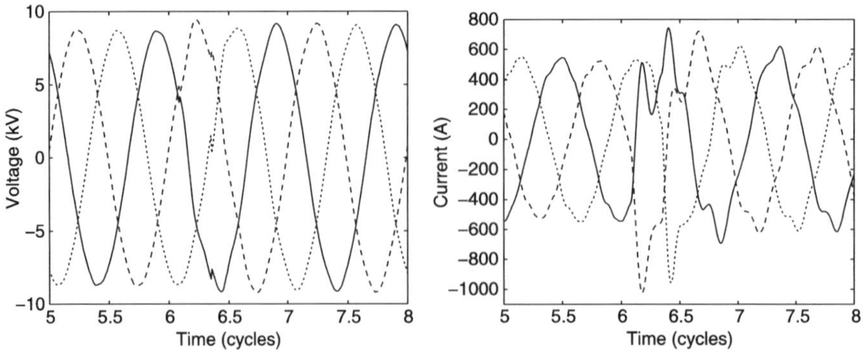


**Figure 6.70** Voltages (left) and currents (right) during synchronized energizing of three-phase capacitor bank.

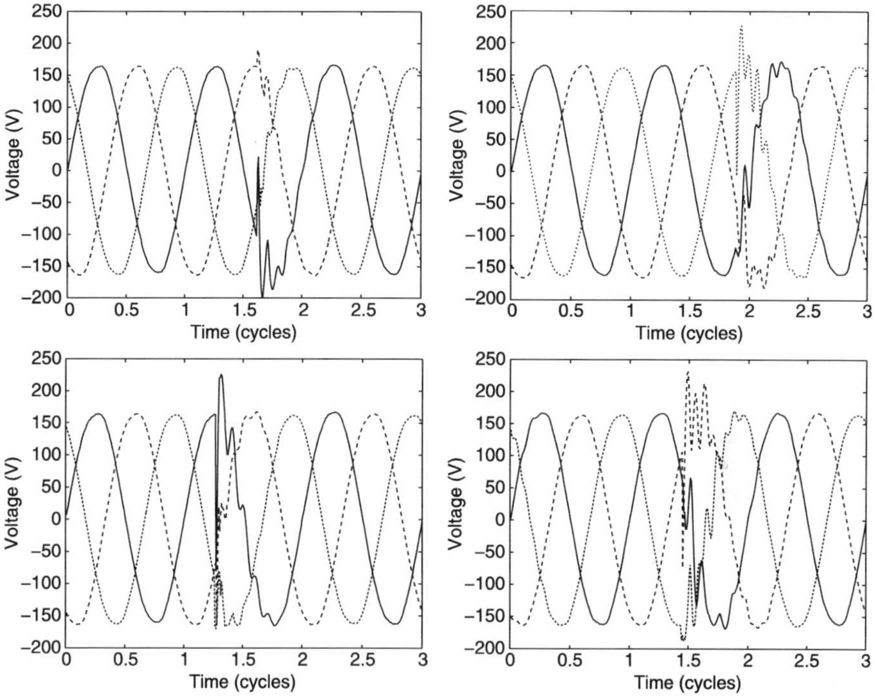


**Figure 6.71** Four capacitor-energizing transients measured at the same location.

The third example (bottom left) is very severe in phase $a$, with the voltage even showing a zero crossing. As shown above this cannot be explained with the single-capacitor model. The severe transient is probably due to voltage magnification, as will be described below.

The fourth transient (bottom right) shows a very severe oscillation in phase $b$ (dashed line), with the double-frequency character of the waveform clearly visible.

Previously in this chapter we saw how the rms voltage is a more common way of presenting voltage dips and interruptions than the waveforms as a function of time. Also for transients the waveform is not always the best way of presenting the event. The fundamental-frequency component in many cases dominates so that it is difficult to see the "actual transient". Several methods will be discussed in detail in Section 8.10.1 to extract the transient. In this chapter we will use a very basic method: subtracting the signal value one cycle back in time. The extracted transients for the four capacitor-energizing events in Figure 6.71 are presented in Figure 6.72. The first (top left) and the third (bottom left) event are similar despite the total waveform looking completely different. Both transients are due to the energizing of phase $a$ only. The transient in phase $a$ is double the magnitude and of opposite sign as the transients in the other two phases. The fact that the first and the third transients are

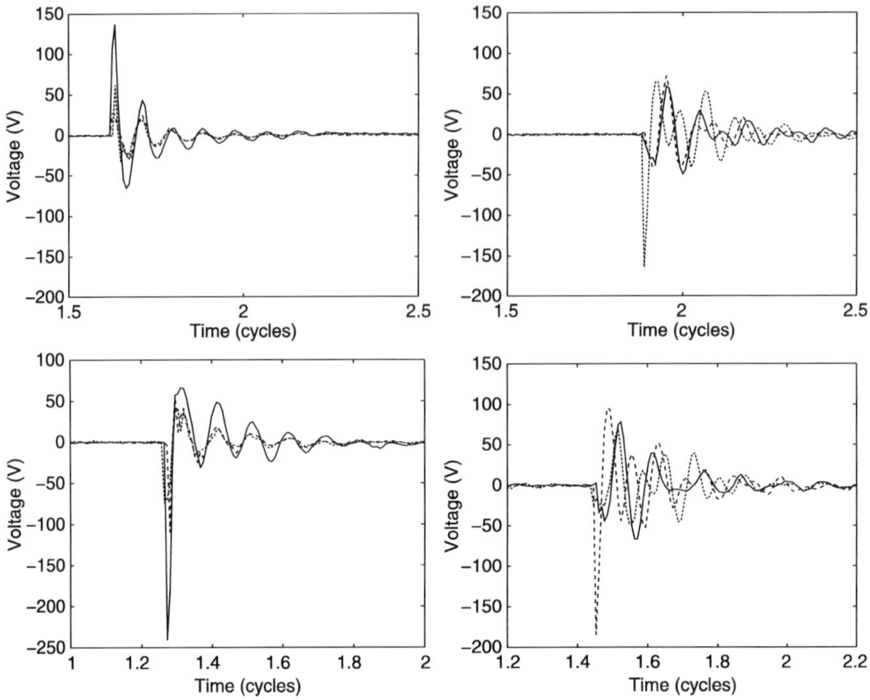


**Figure 6.72** Extracted transients for four capacitor-energizing events in Figure 6.71.

mirrored is because they occur at different positions of the fundamental sine wave. The second and fourth extracted transients are also very similar: Both are due to a second phase being energized. There is now no longer an easy relation between the transients in the three phases.

A next step in the analysis of the transient event is estimating the spectrum of the extracted event. Any of the spectrum-estimating methods from Chapter 3 can be applied here. The spectrum estimation method is applied to a window in time containing the transient. In this case we applied a DFT to a one-cycle window starting slightly before the start of the transient. The results are shown in Figure 6.73. The signals that appeared similar in time the domain also appear similar in the frequency domain. All four spectra contain a small amount of low-frequency components. These are due to the change in fundamental-frequency voltage associated with the capacitor energizing. The method used here for extracting the transient only removes the preevent value of the fundamental component. If we neglect the low-frequency contents of the spectrum, we see that events 1 and 3 (due to the first phase being energized) contain only one frequency component (around 600 Hz) whereas events 2 and 4 (due to the second phase being energized) contain two frequency components (around 500 and 800 Hz).

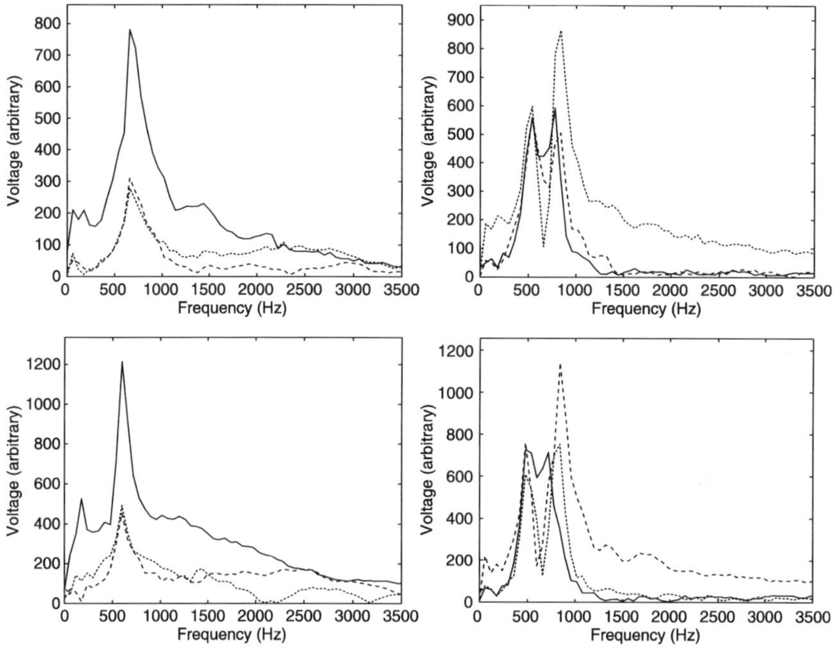


**Figure 6.73** Spectrum of extracted transient for four capacitor-energizing events in Figure 6.71.

#### *6.3.3.3 Voltage Magnification with Capacitor Energizing*

The presence of multiple capacitor banks may lead to more severe transients elsewhere in the system than at the terminals of the capacitor bank. This phenomenon is called voltage magnification and can lead to serious problems with the operation of equipment and even to physical damage to the equipment.

The phenomenon can be explained by means of Figure 6.74, where $C_1$ is the capacitor being energized and $C_2$ a capacitor elsewhere in the system, typically at a lower voltage level. The inductance $L_1$ represents the source impedance at the switching location; $L_2$ represent the impedance between the switching location and the other capacitor, typically a transformer. This model is obviously not complete: Losses in the system and resistive load will determine the damping and thus to a large extent the severity of the overvoltage.

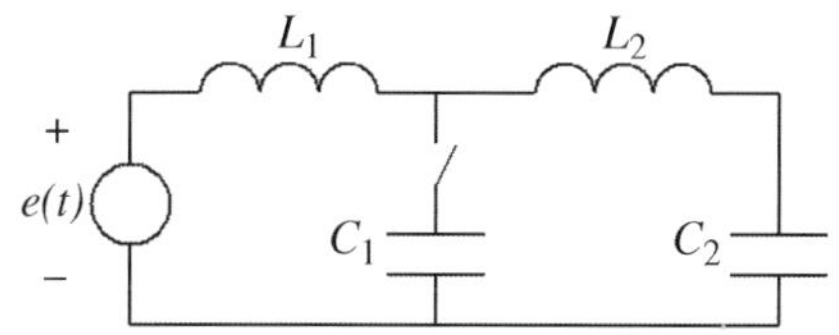


**Figure 6.74** Circuit for describing voltage magnification with capacitor energizing.

To get a qualitative description of the phenomenon, assume that $L_2$ and $C_2$ do not affect the switching transient at the switching location. This is in fact the same assumption made in the previous section, where the rest of the system was simply not considered. The result of the capacitor energizing will be a (damped) oscillation with an angular velocity equal to

$$\omega_1 = \frac{1}{\sqrt{L_1 C_1}} \tag{6.85}$$

and a magnitude $V_1$ depending on the point on wave at which the switch is closed. The voltage over the capacitor $C_2$, that is, the voltage at the other location, is found from the voltage-divider equation:

$$V_2 = \frac{1/(j\omega_1 C_2)}{j\omega_1 L_2 + 1/(j\omega_1 C_2)} V_1 \tag{6.86}$$

The amplification factor is equal to

$$\frac{V_2}{V_1} = \frac{1}{1 - \omega_1^2 L_2 C_2} \tag{6.87}$$

Using (6.85) we obtain the following expression:

$$\frac{V_2}{V_1} = \frac{1}{1 - L_2 C_2/(L_1 C_1)} \tag{6.88}$$

The voltage amplification is shown in Figure 6.75 as a function of the component ratio $L_2 C_2/(L_1 C_1)$. For a component ratio close to unity strong amplification occurs. The actual amplification is much less than would be suggested from this

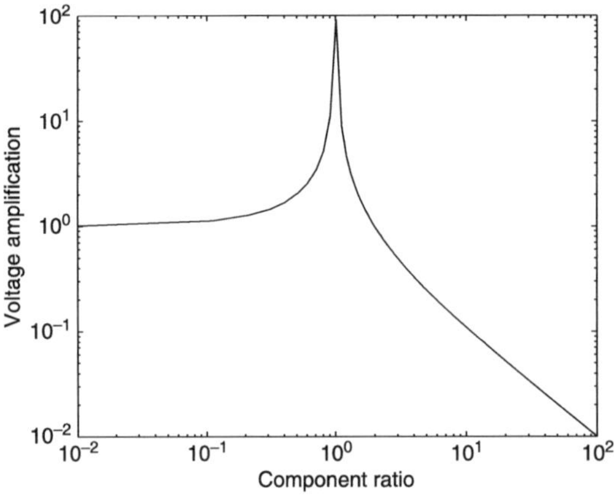


**Figure 6.75** Voltage amplification due to capacitor energizing in lossless circuit.

figure due to losses both in the system and with the load connected to the system; overvoltages up to 4 pu have been measured.

From Figure 6.75 it can also be concluded that the amplification factor is larger than unity for component ratio up to 2, thus for

$$L_2C_2 < 2L_1C_1 \tag{6.89}$$

The amplification factor can also be written in terms of the resonance frequencies at the two locations:

$$\frac{V_2}{V_1} = \frac{1}{1 - \omega_1^2/\omega_2^2} \tag{6.90}$$

with $\omega_2 = 1/\sqrt{L_2C_2}$ the oscillation frequency when energizing capacitor $C_2$ when the effect of $L_1$ on the oscillation frequency is neglected. The criterion (6.89) can be rewritten as

$$\omega_1 < 4\omega_2 \tag{6.91}$$

There is another reason, apart from the damping, why the transient cannot reach very high values. The impedance seen at the switching location is formed by the series connection of $L_2$ and $C_2$. When the component ratio is close to unity, this impedance is close to zero, which will severely damp any oscillation at the switching location.

A case study of voltage magnification due to capacitor energizing is shown in [282]. This example is reproduced here. The system studied is shown in Figure 6.76.

The system of Figure 6.76 is simulated in EMTP for energizing the 12.5-kV capacitor bank when a capacitor bank is connected to the 480-V busbar ($CB_{CL}$ is closed). The line between the two busbars is modeled using lumped elements; therefore the influence of traveling waves is neglected. This simplification is acceptable for lines that have short length compared to the wavelength of the transients [305]. The transformer is also modeled using lumped elements. The short circuit level of

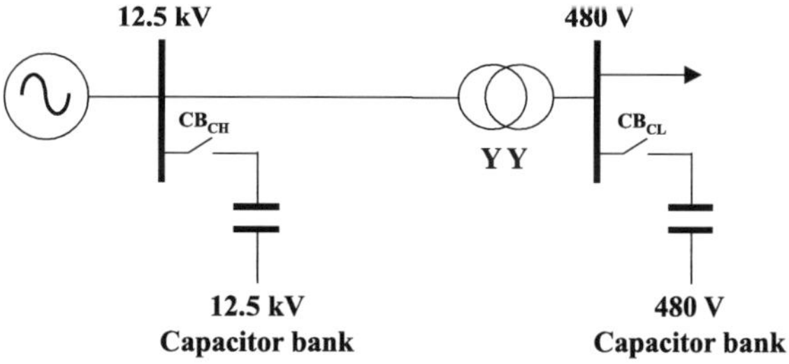


**Figure 6.76** Distribution system for simulation of voltage magnification due to capacitor energizing.

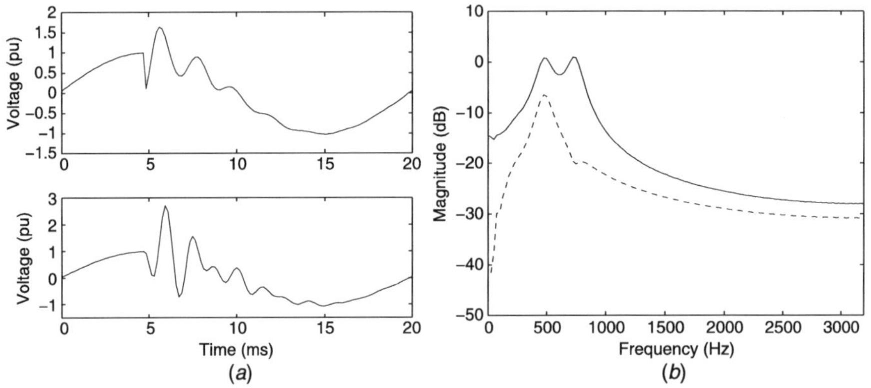


**Figure 6.77** Voltage waveforms (left) and spectrum (right) during capacitor energizing: at 12.5-kV bus (top) and at 480-V bus (bottom). EMTP simulation from [282].

the source is 250 MVA. The capacitor being switched is 2.5 Mvar and the capacitor bank at 480 V is 170 kvar delta connected. The resulting transients at the high- and low-voltage busbars are shown in Figure 6.77. The peak voltage at the low-voltage busbar is 2.5 pu, significantly higher than the peak voltage at the high voltage busbar.

Figure 6.77 also shows the spectrum of the transients. The spectrum of the transient at the 12.5-kV bus shows a peak close to 450 Hz. The spectrum of the transient at 480 V shows two peaks: one at the same frequency as the transient at the 12.5-kV bus and another around 700 Hz. Their amplitudes are almost equal. As described in [100], the presence of a second capacitor in the vicinity of the capacitor that is being switched produces a transient which is a combination of two frequency components. These two frequencies cannot be calculated using simple formulas such as (6.69). Formulas for the calculation of these frequencies are given in [100]. Furthermore, more resonance frequencies appear if more capacitors are close.

***6.3.3.4 Back-to-Back Capacitor Energizing*** The influence of other capacitor banks on the energizing transient is also important in the case of back-to-back capacitor energizing. Back-to-back capacitor energizing is the energizing of a capacitor bank from a bus on which other capacitor banks are already connected. This type of switching produces transients with two frequency components: A very rapid transient brings about an exchange of charge between the capacitors so they are brought to a common voltage; during a slower transient the two capacitor banks reach the source potential [125]. The first transient is of high frequency, as $C$ in (6.69) is the series combination of the capacitances of the capacitor banks and $L$ is the inductance between them. The latter is very small for two capacitors connected to the same bus. The second transient has significantly lower frequency: $C$ in (6.69) is in this case the parallel combination of the capacitances and $L$ is the source inductance [125].

For example, the connection of a first 2-Mvar capacitor bank at 12.5 kV, 60 Hz with a short-circuit capacity of 200 MVA gives a transient with a frequency equal to 600 Hz. Connecting a second identical capacitor bank with an inductance of 20 μH separating the two (corresponding to about 20 m of busbar) leads to a double-frequency transient. The high-frequency oscillation is due to the oscillation between the series connection of the two banks (of 68 μF each) and the inductance connecting them. The low-frequency oscillation is due to the combined oscillation of two 2-Mvar banks and the 200-MVA source. The two oscillation frequencies are 6.1 kHz and 425 Hz, respectively.

***6.3.3.5 Capacitor-Deenergizing Transients*** Capacitor deenergizing causes only a minor transient. The magnitude of the transient is determined by the difference in steady-state voltage before and after the switching action, thus by the size of the capacitor bank. The frequency of oscillation is determined by the inductive part of the source impedance and the capacitance on the source side of the interrupting device [305]. This capacitance could be another capacitor bank or the combined capacitance of all lines and cables connected to the substation.

***6.3.3.6 Inductor-Energizing Transients*** Switching of linear inductors does not create any severe transients. The circuit representation is shown in Figure 6.78, where $L_1$ is the source inductance at the switching location, $C$ is the capacitance at the switching location, and $L_2$ is the inductance being switched. Energizing of inductances (reactors) takes place to compensate for the reactive power generated by long transmission lines during low load. The capacitance $C$ in Figure 6.78 therefore typically is the capacitance of the transmission line that is compensated. But it may include the capacitance of other transmission lines connected to the switching location.

Let $u(t)$ be the voltage at the source side of the switch, $i_1(t)$ the current through $L_1$, $i_2(t)$ the current through $L_2$, and $i_C$ the current through $C$. Then the differential equation can be written as

$$e(t) = L_1 \frac{di_1}{dt} + L_2 \frac{di_2}{dt} \tag{6.92}$$

$$i_1 = i_2 + i_C \tag{6.93}$$

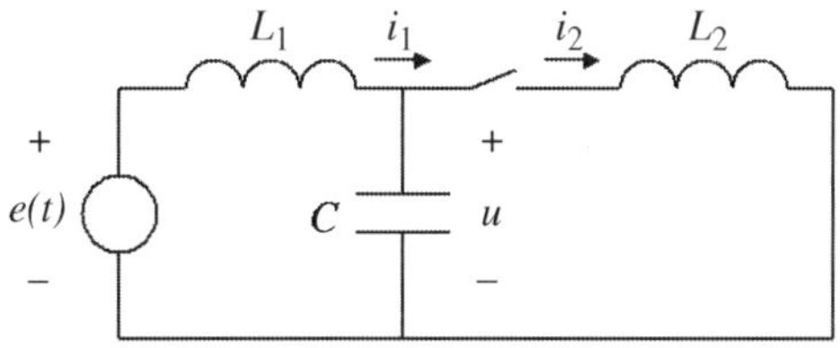


**Figure 6.78** Circuit for inductor energizing.

$$i_C = C\frac{du}{dt} \quad (6.94)$$

$$u(t) = L_2\frac{di_2}{dt} \quad (6.95)$$

Substituting (6.93) and (6.94) in (6.92) and using (6.95) give the following second-order linear differential equation for the voltage on the source side of the switch:

$$e(t) = \frac{L_1 + L_2}{L_2}u(t) + L_1 C\frac{d^2u}{dt^2} \quad (6.96)$$

The frequency of the resulting transient can be found easily from the characteristic equation:

$$f_{\mathrm{tr}} = \frac{1}{2\pi}\sqrt{\frac{L_1 + L_2}{L_1 L_2 C}} \quad (6.97)$$

The oscillation frequency of the transient is determined by the capacitance $C$ and the parallel connection of the switched inductor and the source inductance. In most cases the source inductance will dominate (i.e., be much smaller), so that (6.97) can be approximated as

$$f_{\mathrm{tr}} \approx \frac{1}{2\pi\sqrt{L_1 C}} \quad (6.98)$$

which is the same frequency as when the capacitance $C$ would be energized. The size of the reactance does not affect the frequency of the switching transient. The amplitude of the transient is determined by the difference between the preswitching and the postswitching steady-state voltage, which is typically only a few percent.

***6.3.3.7 Inductor-Deenergizing Transients*** When the inductance $L_2$ in Figure 6.78 is deenergized, the switch opens at current zero. (We will discuss current chopping in the next section.) This means that no severe transients are to be expected. The oscillation frequency is again determined by the source inductance $L_1$ and the total capacitance $C$ connected to the substation.

### 6.3.4 Abnormal Switching Transients

***6.3.4.1 Current Chopping*** A circuit breaker or any other switching device normally interrupts a current only at a natural-current zero crossing. There are two exceptions to this rule: Current-limiting fuses create a forced-current zero crossing; with some circuit breakers in some cases the current suddenly drops to zero shortly before or shortly after the zero crossing. The difference between a normal-current interruption and a current chopping is shown in Figure 6.79. With current chopping the current shows a sharp drop toward zero.

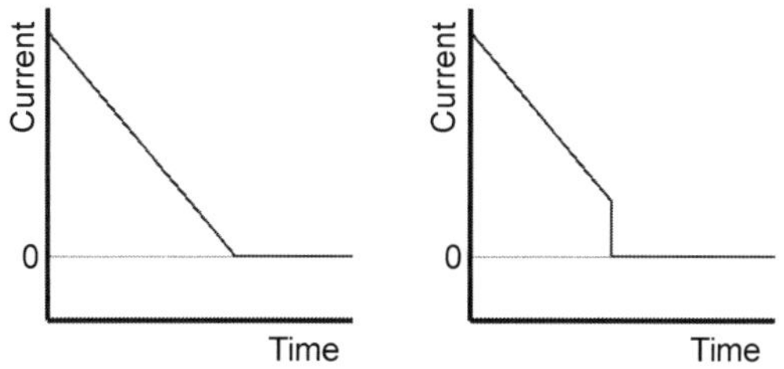


**Figure 6.79** Normal-current interruption (left) and current chopping (right).

Current chopping can only take place in a circuit with a certain amount of capacitance close to the circuit breaker. The current through a pure inductance cannot suddenly become zero after all. The circuit diagram needed to explain current chopping is shown in Figure 6.80.

Current chopping especially takes place for so-called small inductive currents switched by a circuit breaker. A circuit breaker is designed to clear large currents. A small current could be forced to zero before the current zero crossing. The current through the inductances in Figure 6.80 cannot become zero and will continue to flow. But as the circuit breaker no longer forms part of the circuit, the inductor current is forced to close through the capacitance instead. On the source side the current through the source inductance $L_S$ closes through the source-side capacitance $C_S$. On the load side the current through the load inductance $L$ closes through the load-side capacitance $C$. These currents will charge the capacitance to a value sometimes much higher than the nominal voltage of the capacitor. Such current-chopping events with their associated high overvoltages have led to damage to equipment such as transformer and induction motors. Flashover outside of the circuit breaker is another possible result of overvoltages due to current chopping.

Consider the following source voltage:

$$e(t) = E\sqrt{2}\cos(\omega_0 t + \phi) \tag{6.99}$$

With the switch opening (i.e., the breaker interrupting) at time zero, the chopped current is equal to

$$I_{\text{chop}} = \frac{E\sqrt{2}}{\omega_0 L}\sin\phi \tag{6.100}$$

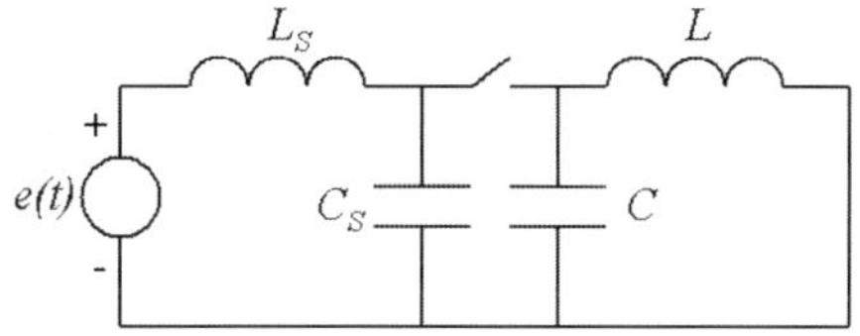


**Figure 6.80** Electric circuit explaining current chopping and resulting overvoltages.

where we have neglected the voltage drop over the source impedance and the capacitive part of the current. This seems a reasonable assumption as this phenomenon occurs for small inductive currents and thus for large values of $L$. The voltage $v(t)$ over the load-side capacitor is obtained from the differential equation

$$LC\frac{d^2v}{dt^2} + v(t) = 0 \tag{6.101}$$

with initial conditions

$$v(0) = E\sqrt{2}\cos\phi \tag{6.102}$$

and

$$\left.\frac{dv}{dt}\right|_{t=0} = \frac{E\sqrt{2}}{\omega_0 LC}\sin\phi \tag{6.103}$$

The solution to this differential equation is

$$v(t) = E\sqrt{2}\,\cos\phi\cos(\omega_r t) + \sqrt{\frac{L}{C}}\,I_{\text{chop}}\,\sin(\omega_r t) \tag{6.104}$$

with $\omega_r = 1/\sqrt{LC}$ the resonance frequency. The first term is the transient due to normal circuit breaker operation (the so-called transient recovery voltage). As current chopping takes place typically close to the natural zero crossing, $\cos\phi \approx 1$ and the amplitude of the first term is about equal to the amplitude of the source voltage. This term will thus never lead to overvoltages for the switching of small inductive currents.

On the other hand, the second term has no relation to the source voltage but is instead proportional to the chopped current and the *wave impedance* $\sqrt{L/C}$. For large inductance with small capacitance (e.g., transformers in no load or reactances) the overvoltage can become significant even for small values of the chopped current.

A faster way of obtaining the maximum overvoltage is by using conservation of energy [125]. At the instant of current chopping the magnetic energy in the inductor is equal to $\frac{1}{2}LI_{\text{chop}}^2$. This energy is transformed into electrical energy in the capacitor. If we neglect the losses, the maximum voltage $V_{\max}$ over the capacitor is obtained from

$$\tfrac{1}{2}LI_{\text{chop}}^2 = \tfrac{1}{2}CV_{\max}^2 \tag{6.105}$$

resulting in

$$V_{\max} = \sqrt{\frac{L}{C}} I_{\text{chop}} \tag{6.106}$$

which corresponds to the amplitude of the second term in (6.104). The first term in (6.104) means that the total overvoltage will be higher. However, the losses have been neglected in this whole discussion, which could result in a considerable difference at higher frequencies. Therefore (6.106) is an acceptable estimation of the overvoltage due to current chopping.

The above reasoning only considered the overvoltage on the load side of the circuit breaker. This overvoltage is obviously of serious concern for the load as well as for the circuit breaker, but it does not propagate to other customers. The voltage on the system side of the circuit breaker is obtained from the differential equation

$$e(t) = v(t) - L_S C_S \frac{d^2 v}{dt^2} \tag{6.107}$$

with the initial conditions (the same as on the load side of the breaker)

$$v(0) = E\sqrt{2} \cos \phi \tag{6.108}$$

and

$$\left.\frac{dv}{dt}\right|_{t=0} = \frac{E\sqrt{2}}{\omega_0 L C_S} \sin \phi \tag{6.109}$$

where the prechopping voltage drop over the inductor and the prechopping current through the capacitor have been neglected. Note that the second initial condition is determined by the load-side inductance $L$ (through the current value at the instant of chopping) and by the system-side capacitance $C_S$ (as this determines the rate of rise of the system-side voltage when the load current is commutated into the capacitance).

Neglecting also the posttransient voltage drop over the inductor, so that the steady-state voltage over the capacitor equals the source voltage, gives the following solution to the above differential equation with initial conditions:

$$v(t) = E\sqrt{2} \cos(\omega_0 t) + I_{\text{chop}} \sqrt{\frac{L_S}{C_S}} \sin(\omega_r t) \tag{6.110}$$

with $\omega_r = 1/\sqrt{L_S C_S}$ the oscillation frequency. It has further been assumed that $\phi \approx 0$ and $\omega_r \gg \omega_0$. From (6.110) it follows that the oscillatory transient adds to the maximum of the source voltage, leading to maximum overvoltages. However, the peak voltage is much less severe than at the load side of the breaker. At the source side the inductance $L_S$ is smaller and the capacitance $C_S$ is in most cases significantly larger. Therefore no severe overvoltages are to be expected at the system side of the breaker.

#### *6.3.4.2 Restrike During Capacitor Deenergizing*

Deenergizing a capacitor bank (i.e., capacitor deenergizing) does not usually result in any transient oscillation, as discussed before. The bus voltage will drop up to a few percent due to the loss of voltage support provided by the capacitor. During the deenergizing process, the contactor of a capacitor opens and discontinues the current flow. If the contactor does not open successfully, an arc will be established between the contacts and the capacitor reignites or restrikes. The event is called reignition if current conduction is reestablished within half a cycle of current interruption. If current conduction occurs later, the event is called a restrike [125]. The contactor might open and restrike later. The same event might be repeated several times (multiple restrikes).

Figure 6.81 shows a damped oscillatory transient caused by restrike during capacitor deenergizing. The voltages at both the line side and the capacitor side are shown as obtained by an EMPT simulation. The capacitor is disconnected at current zero when voltage is at its maximum (current leads voltage by approximately 90°). Therefore, the capacitor is charged to maximum voltage. Half a cycle later the voltage difference between the line side and the capacitor side is 2.0 pu and this could lead to restrike, as shown in Figure 6.81. The case of multiple restrikes is shown in Figure 6.82.

For the cases shown in Figures 6.81 and 6.82, the voltage initially changes sign due to the fact that the capacitor is charged with a voltage of opposite sign than the

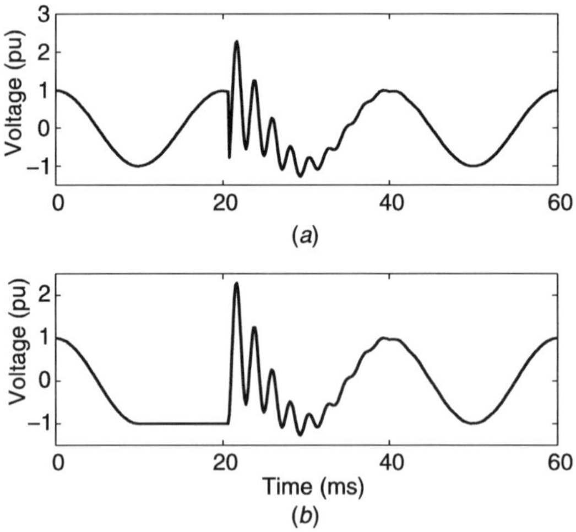


**Figure 6.81** Voltage waveform of restrike during capacitor deenergizing: line side (top) and capacitor side (bottom). From [282].

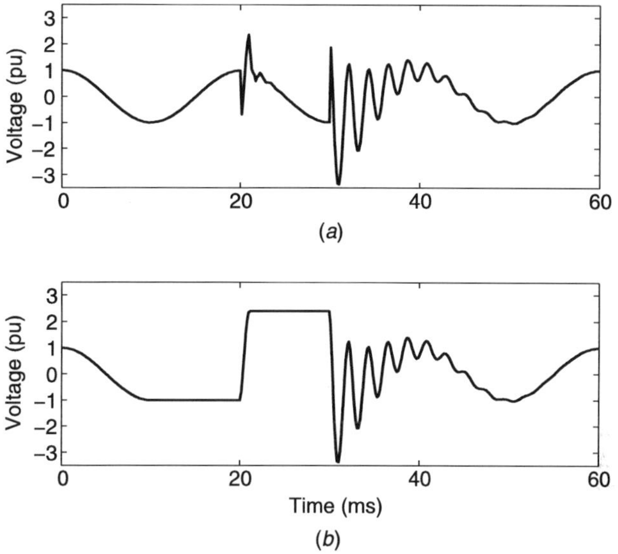


**Figure 6.82** Voltage waveform for multiple restrikes during capacitor deenergizing: line side (top) and capacitor side (bottom), from [282].

line voltage at the moment of restrike. The amplitude of the overvoltage due to this phenomenon depends on the trapped charge in the capacitor. The voltage at the instant of restrike will oscillate at the natural frequency characterized by the capacitance of the capacitor and the system inductance, as in the case of capacitor energizing.

In both examples shown here the current does not interrupt after the last restrike. Often the contactor is able to interrupt the current in the end, so that all that is seen at the line side of the breaker is a series of oscillations with increasing amplitude.

***6.3.4.3 Restrike During Inductor Deenergizing*** Also during the deenergizing of a large inductance restrike may occur. As the current amplitude is small, it can be interrupted at the first current zero after contact separation. If the breaker does not build dielectric strength fast enough, a dangerous sequence of events may occur [125, 305].

The sequence of events can be described approximately as follows:

- For an inductive current, the current interruption takes place at voltage maximum. After the separation of system and load the load-side capacitor will discharge via the inductance of the load. This will occur with a frequency much higher than the system frequency, typically several kilohertz. As the contact separation is still small, the dielectric strength may not be sufficient to withstand the fast-growing voltage difference over the contact gap.

- When the voltage difference over the contact gap exceeds the dielectric strength, reignition occurs. This can happen at any instant in the oscillation, depending only on the dielectric strength. The load-side capacitor is now part of the system again and it will start a high-frequency oscillation with the source inductance. As the source inductance is much smaller than the load inductance, the frequency of this recharging oscillation is much higher than the frequency of the discharging oscillation.
- During the recharging the current through the breaker contains two components: a high-frequency component due to the above-mentioned recharging oscillation and a ramp current due to the source voltage driving the current through the inductor. The rate of rise of the current is such that the source voltage is equal to $L(di/dt)$. Current zero-crossing takes place when these two components are opposite to each other.
- The circuit breaker again interrupts the current on one of these zero crossings. But at that moment the current through the inductor is not zero and is instead commutated into the capacitor. This leads to high overvoltage similar to those due to current chopping. The term *virtual current chopping* is sometimes used for this phenomenon.
- The new oscillation on the load side of the breaker will again at a certain moment reach such a level that the contact gap experiences a dielectric breakthrough. At this moment the whole sequence will start again.
- The reignitions will continue to occur until the dielectric strength becomes higher than the highest voltage difference over the breaker contacts.

From the system side of the breaker the phenomenon is only visible when the breaker is conducting, that is, during the very high frequency oscillations after each restrike. Other customers thus experience a series of very high frequency oscillations, typically with increasing amplitude. An example of the measurement of this phenomenon is shown in [20].

#### 6.3.4.4 Ferroresonance

Ferroresonance is the series resonance between a capacitor and a nonlinear inductor. The latter is in almost all cases the magnetizing inductance of a transformer. The phenomenon was discovered in 1908 and first described analytically in 1920 [177]. There are some examples of such series circuits under normal operation, but most cases of ferroresonance are associated with a fault in the system.

The following causes of ferroresonance are listed in [177]:

- One or two phases in the supply to a transformer show an open-circuit fault. This may be due to the inadvertent clearing of a fuse or a circuit breaker or due to a transient fault.
- A transformer is connected to a series-compensated transmission line. Ferroresonance may occur here under certain (low-load) operating conditions.

- The voltage transformer connected to a system with an isolated neutral may experience ferroresonance with the total (zero-sequence) capacitance of the system.
- The phenomenon may also occur with a capacitive voltage transformer.
- When a transformer is connected to a deenergized line running in parallel with an energized line, the coupling capacitance between the lines forms a series circuit with the transformer.
- Ferroresonance can also occur when a transformer is connected to a long transmission line or cable with low short-circuit power.

With a ferroresonance circuit the currents and voltages attempt to find such a value that the resonance frequency of the inductor and the capacitor is equal to the excitation frequency (50 or 60 Hz). As the inductance is a function of the current through the inductor, so is the resonance frequency of the series circuit. This reasoning is used as the basis for a quantitative analysis in [125, 256]. However, in both publications the complex calculation method is used, which does not hold for nonlinear circuits. It is possible to obtain a differential equation for the voltage in the time domain, given the inductance as a function of the current, but that equation contains several nonlinear terms which are neglected in the analysis in [125, 256]. It is unlikely that it will be possible to find analytical expressions for the voltage waveforms during ferroresonance. The waveforms obtained from measurements and from simulations are not only heavily distorted but also heavily nonstationary, whereas classical analysis predicts stationary solutions such are rarely observed in practice.

### 6.3.5 Examples of Voltage and Current Transients

Figure 6.83 shows two voltage transients recorded at the wall outlet in an office building. For the event shown on the left the most severe transient is in phase *c* (dotted line) with minor transients in the two other phases. Note, however, that the overvoltage is highest in phase *b* (dashed line). These kinds of transients are very common in an office or domestic environment. They are mainly due to switching

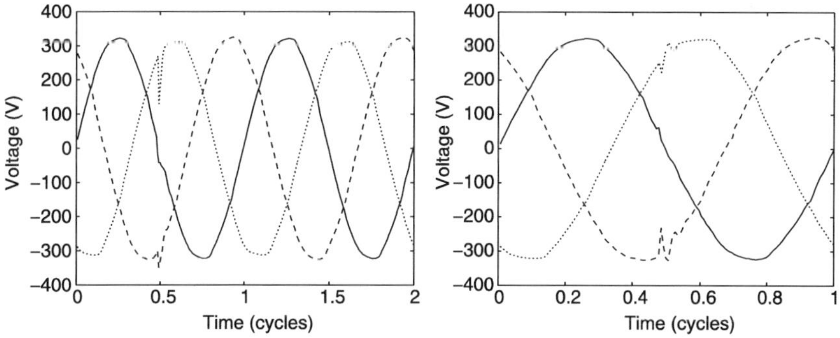


**Figure 6.83** Voltage transient in one phase (left) and between two phases (right).

of electronic equipment, in this case probably the switching of a computer. Many such devices have a capacitor over the terminals on the grid side, often part of an EMC filter. The connection transient is mainly due to the energizing of the capacitor. These kinds of transients are discussed in detail in [297].

The second event, measured at the same location about 20 min later, is of a different character. It is no longer dominant in just one phase but takes place between phases *b* and *c*. In this case there was no high-frequency component present in the signal. We will discuss transients in three-phase systems in more detail in Section 8.10.3.

The measurement examples shown in Figure 6.84 were obtained in a laboratory environment. The supply was dedicated to a number of laboratory experiments. At times the load was very small (when no experiments were running). The voltage waveform was therefore very "clean," as can be seen from the preevent waveform in the figures below. Compare this with Figure 6.83, where the preevent waveforms are already heavily distorted due to the presence of large numbers of electronic equipment (mainly computers and fluorescent lighting). The recordings all show multiple transients. They are due to one of the experiments in the laboratories, but even though the experiment was the same, the transients are all different.

An example of an oscillatory transient is shown in Figure 6.85. The measurement was performed in a 10-kV substation in an industrial suburb of a medium-sized city. The extracted transient is obtained by subtracting the waveform during five cycles

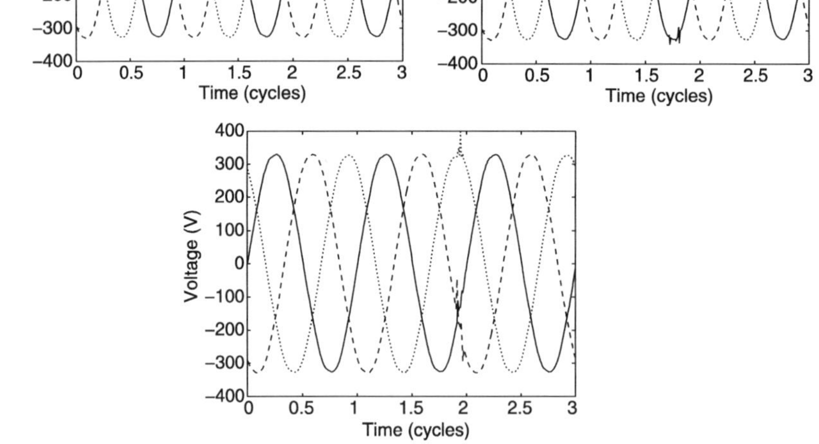


**Figure 6.84** Three examples of complex-voltage transients.

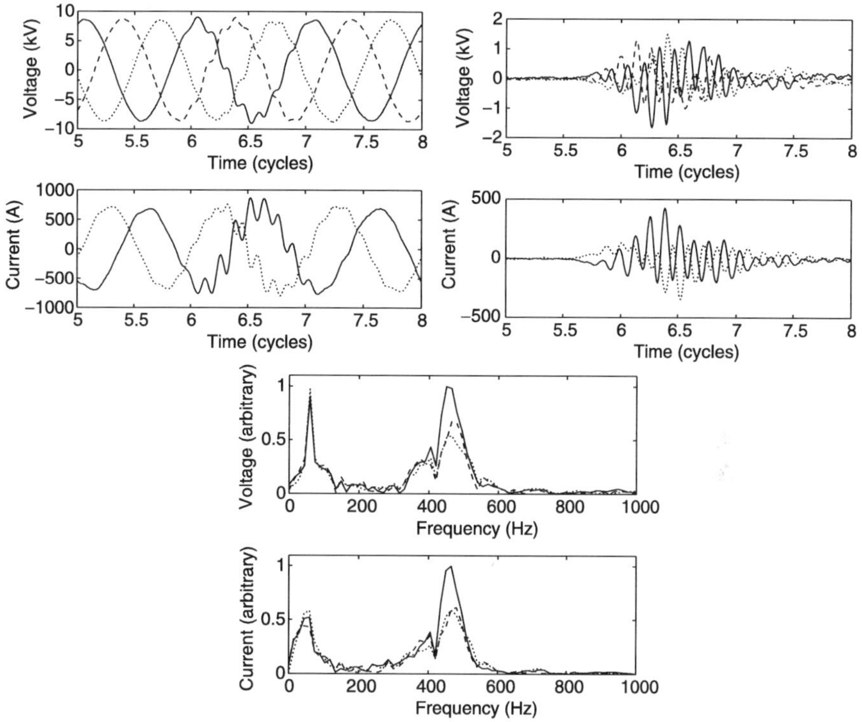


**Figure 6.85** Oscillations in voltage (top) and current (bottom) measured at 10 kV: waveforms (left); extracted transients (right); spectrum of extracted transients (bottom).

before the start of the transient. The spectrum is obtained by applying a DFT to a four-cycle window. Both voltage and current show an oscillation with a frequency around 400 Hz. The spectrum reveals, however, that the transient contains two frequencies: around 400 and 450 Hz. The result is a modulation (with about 50 Hz) in the amplitude of the oscillation. The low-frequency component in the spectrum of the extracted transient is again due to the change in fundamental and harmonic components. Comparing voltages and current in the same phase shows that they are in opposite phase throughout the event. This indicates that the source of the disturbance should be looked for downstream of the monitor location.

Another example of an oscillatory transient is shown in Figure 6.86. This transient is measured at 132 kV in the same substation as the previous transient. This transient, however, is less regular than the previous one but shows a much clearer start of the event. The frequency spectrum shows two distinct components in the current but an almost continuous spectrum for the voltage. The spectrum was again obtained over a four-cycle period.

The character of the event becomes clearer from the extracted transient: A switching action takes places in phase *b* (dashed line) first, followed by phase

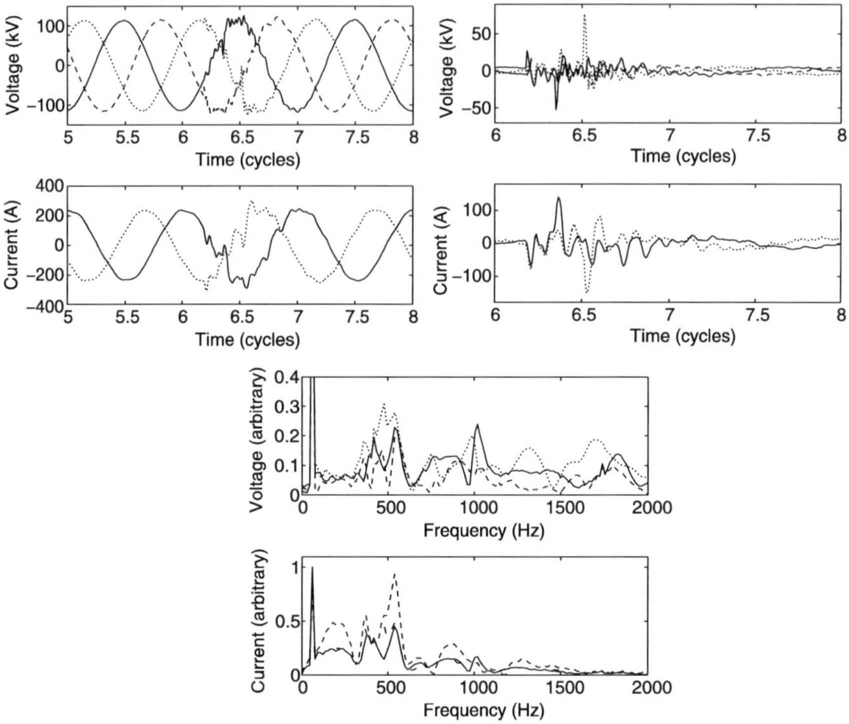


**Figure 6.86** Oscillations in voltage (top) and current (bottom) measured at 132 kV: waveforms (left); extracted transients (right); spectrum of extracted transients (bottom).

*a* (solid line) about 4 ms later, followed by phase *c* (dotted line) another 4 ms later. The initial change in current is opposite to the initial change in voltage. This again indicates that this is a *downstream event*: The change in current causes the change in voltage.

Figure 6.87 shows another transient event due to capacitor energizing. In this case the switching action takes place at the customer premises. From the network operator viewpoint this event is classified as a *current event.* This event was associated with a decrease in reactive power consumption of 2 kvar in phase *c* and of 1 kvar in phases *a* and *c*. The active power consumption was not affected. The voltage and currents are in opposite direction for the extracted transient. This indicates that this is a downstream event: where the current transient causes the voltage transient.

A transient due to an upstream switching action is shown in Figure 6.88. The transient shows similarities with the one in Figure 6.87. However, there was no measurable change in active or reactive power consumption associated with the transient. The extracted transients show that the change in current is in phase with the change in voltage. This points to a source of disturbance which is located upstream. In other words, in this case the current transient is due to the voltage

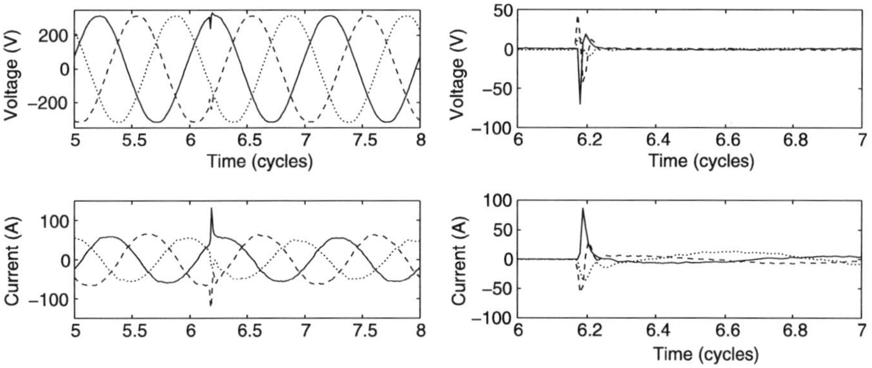


**Figure 6.87** Currents (top) and voltages (bottom) due to downstream capacitor energizing: waveforms (left) and extracted transients (right).

transient. The fact that the current transient is much more severe than the voltage transient is normal for many electronic loads. Such a load contains a small capacitor at the grid side of the power supply, for example, as part of an EMC filter. A small amplitude but fast variation in voltage will lead to a large variation in current. The heavily distorted current clearly indicates the presence of electronic load.

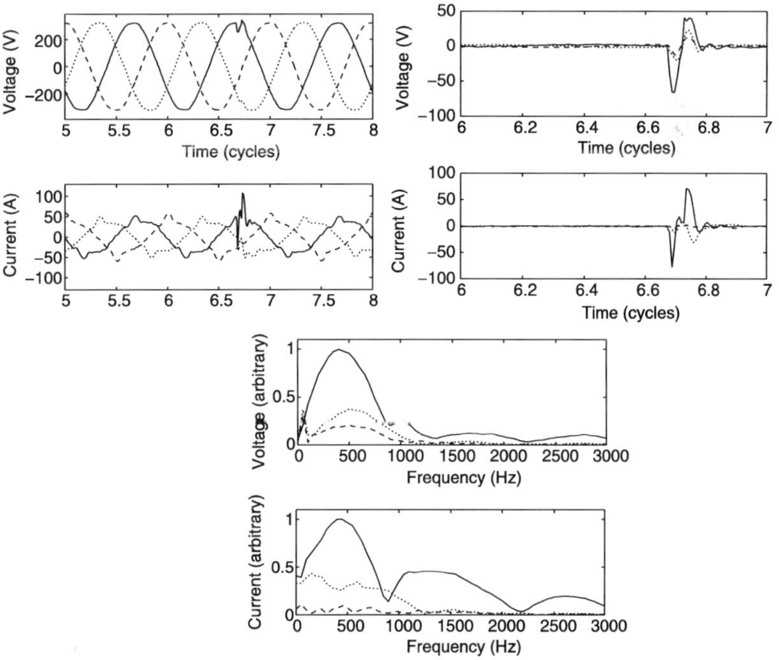


**Figure 6.88** Currents (top) and voltages (bottom) due to upstream switching event: waveforms (left); extracted transients (right) and spectrum (bottom).

## 6.4 SUMMARY AND CONCLUSIONS

### 6.4.1 Interruptions

Interruptions can be defined in two ways: as the opening of an interrupting device or from the voltage magnitude becoming close to zero. This distinction is very important when determining interruption statistics, as the two methods may give different results. This is especially the case when distributed generation is present on the load side of the interrupting device.

Interruptions are mainly due to a correct intervention by the protection to remove a fault. Such interruptions are generally viewed as *necessary interruptions*: some of the customers are disconnected from the supply to save the remaining customers. Interruptions can also be due to the incorrect opening of an interrupting device or due to the intentional opening of an interrupting device to perform maintenance or repair. In the latter case the term *planned interruption* or *scheduled interruption* is used.

A further subdivision of interruptions is made based on their duration. Various terminology is in use, but more recently the main terms used are *short* or *momentary* interruptions versus *long* or *sustained* interruptions. The border between short/momentary and long/sustained is between 1 and 5 min. The distinction is based on the restoration process being used: automatic versus manual restoration.

Interruptions due to the (intentional or unintentional) opening of an interrupting device are sometimes associated with transient overvoltages on the load side of the interrupting device. It is not clear how widespread this phenomenon is and whether there is any adverse effect on equipment. A further study of the voltages on the load side of the interrupting device is needed to better understand the requirements placed on islanding detecting with distributed generation. With an increasing penetration of distributed generation it becomes important that the generation units only disconnect from the system during a genuine interruption, not during any other disturbance.

Interruptions, especially long interruptions, are not generally seen as a power quality issue. Instead they are typically treated as part of *power system reliability.* Information on the causes of interruptions and on their stochastic prediction can be found in the handful of books on power system reliability [26, 27, 28, 51, 107]. Also [25, Chapter 6; Chapters 2 and 3; 99, Chapter 3] discuss interruptions and reliability issues.

### 6.4.2 Voltage Dips

Voltage dips are a less severe version of an interruption. The voltage magnitude does not drop all the way down to zero and the voltage recovers after a few cycles to several seconds (typical durations of interruptions range from a few seconds up to several hours). Although voltage dips are less severe than interruptions, they are of more concern, especially to industrial customers, because of their much higher frequency of occurrence.

Voltage dips are due to a short-duration increase in current, typically somewhere else in the system. Most severe dips are due to faults, at either the transmission level or the distribution level. Other causes of dips are motor starting and transformer energizing. Different causes of dip lead to different properties of the voltage waveform. Several examples are given in the chapter. A more detailed study is needed on the properties of dips due to different causes. The results of such a study are an important basis for an automatic classification method for power quality events. Such a method does provide statistics that may be of help with deciding about mitigation methods and also may be important when using power quality measurements as a system diagnostics tool. Severe deviations from the normal steady-state voltage or current are not only a concern for the customer but also point to events in the system that may endanger the reliability or security of the system. Voltage dips due to faults are an important class of events in this as short-circuit faults are one of the most severe threats to security and reliability.

A large part of the section on voltage dips discusses a comprehensive model for voltage dips in three-phase systems. Two versions of the model are presented: an intuitive model (the ABC classification) and a more theoretical model (the symmetrical-component classification). The former has applications in equipment testing and in stochastic prediction whereas the latter can be used to extract dip characteristics from measurements. Further research is needed to understand the limitations of both models, especially where it concerns the impact of load on the voltages during and after the dip. It is shown that the symmetrical-component model is a good basis for such a study.

Voltage dips have recently also attracted attention in the study of large wind parks. Their tripping is not just a problem for the operator of the park. The sudden loss of a large amount of generation may cause stability problems for the transmission operator, especially as the fault causing the dip already led to the tripping of a line or busbar. The same problem will occur with distributed generation when its penetration increases. Theoretical models like the ones presented in this chapter and elsewhere will have to be applied to study the impact of dips on distributed generation and wind turbine interfaces.

The impact of voltage dips on equipment and mitigation of voltage dips is not part of this book. For this the reader is referred to the other literature on voltage dips, where these subjects are discussed in sufficient detail. A large part of the further development and research on voltage dips will be found in improving equipment immunity and other mitigation methods. Note that some of the mitigation methods may in turn lead to other power quality problems, such as high-frequency components in the voltage due to the use of active converters.

Voltage dips are an important part of most books on power quality. Several chapters in [33] are dedicated to voltage dips, including impact on equipment and mitigation methods. An excellent, more practical, text on voltage dips is found in [99, Chapter 3]. Voltage dips are also discussed in [161, Chapter 9; 263, Chapter 2] and play an important role in [95]. Series converters to mitigate voltage dips are discussed in [118, Chapter 9].

### 6.4.3 Transients

Transients are short-duration deviations from the steady-state voltage or current waveform. There is no clear borderline between transients and dips or swells, but generally speaking any event with a duration less than half a cycle is viewed as a transient.

Transients are due to lightning strokes, self-clearing faults, and switching actions. Some examples of transients due to self-clearing faults were presented in Section 6.2. An important property of transients is their high-frequency contents. Many transients contain a damped oscillation with frequencies ranging from a few hundred hertz up to the megahertz range. The examples shown in this chapter are however all limited to the lower frequency ranges. The reason for this is the limited frequency range of the measurement instruments used. Several papers and the books mentioned below also present high-frequency transients. These may be more dangerous to equipment and may be associated with much higher overvoltages than the ones shown here.

A distinction in switching transients can be further made between normal and abnormal switching transients. The main normal switching transient in the grid is the energizing of capacitor banks. This event is rather well described in the power quality literature due to its impact on the operation of adjustable-speed drives. Abnormal transients are, for example, due to restrike with capacitor or inductor energizing and due to current chopping. The propagation of transients through the system remains an ill-understood subject. The main studies are performed after the amplification of a capacitor-energizing transient due to capacitor banks elsewhere. During a phenomenon known as voltage magnification overvoltages up to 4 pu may appear over the terminals of low-voltage capacitors.

A large fraction of the transients at equipment terminals are due to the switching of local load. Capacitors over the terminals of equipment and power-electronic devices are common causes of transients. A direct consequence of this is that there is a large variation in the statistics of transients at different sites.

Several examples of simulated and measured transients are shown in this chapter. In most cases, the waveforms, the extracted transients, and the frequency spectra were shown. More experience is needed in the interpretation of the various features seen in these three ways of presenting the transient. The features should be related to the origin and propagation of the transient through the system.

Voltage and current transients show a wide range of properties but can roughly be classified into impulsive and oscillatory transients. Oscillatory transients typically show a damped oscillation but in some cases also a lower frequency oscillation modulated on the amplitude of the higher frequency oscillation. Transients in three-phase systems are often very complicated, especially when switching actions in the three phases take place close together in time. Further research is needed toward a better understanding of transients in three-phase systems. Further research is also needed to distinguish between upstream and downstream transients.

Transients do not cause the amount of widespread inconvenience posed by interruptions or voltage dips. Also they possess a much wider range of origins than dips and interruptions. Because of the shorter duration, transients require a wider bandwidth measurement circuit and a higher sampling rate. Therefore large-scale data collection of transients is much less common than of dips and interruptions. All this has led to an underexposure of transients in the power-quality literature. There are a number of books on power system transients of which [125] and more recently [305] should be mentioned, but neither of them addresses transients as a power quality issue. Transients are also an important part of the power quality mitigation issues discussed in [185]. An interesting overview of standard test waveforms and their origin is presented in [296, Chapter 10]. Transients are further treated in [99, Chapter 4; 263, Chapter 3]. Also [95] shows several examples of measured transients.

### 6.4.4 Other Events

In this chapter we discussed three types of events: voltage dips, interruptions, and transients. As mentioned in Chapter 1 a power quality event is in fact defined by its triggering mechanism. For dips and interruptions suitable triggering mechanisms are easy to find. For transients the situation becomes more complex, as we will see in Chapter 7. By using additional triggering mechanism, other power quality events may be defined.

Voltage swells are not explicitly mentioned in this chapter. Swells with durations up to a few seconds are caused by single-phase faults. We saw some examples of those when discussing voltage dips due to faults in Section 6.2.2. Longer duration swells may be due to sudden loss of load or to capacitor and reactor switching. The processing of voltage swells is very similar to the processing of voltage dips.

Frequency events were briefly mentioned in Section 5.3.2. They are rarely mentioned in the power quality literature but are an important aspect of transmission system studies. The deregulation of the electricity industry and the intended introduction of large amounts of distributed generation and large wind parks in some countries have resulted in renewed interest in frequency transients [91, 194].

Ferroresonance and other short-duration increases of waveform distortion may also be treated as power quality events. Further study would be needed of triggering algorithms, or other ways of defining these events. The triggering methods and the subsequent characterization would allow more detailed studies resulting in better understanding of the phenomena.

# CHAPTER 7

# TRIGGERING AND SEGMENTATION

This chapter addresses two closely related issues: finding triggering points for events and segmenting disturbance data sequences that may be due to different underlying power system events. A range of methods/schemes that may provide solutions to these issues will be described. These methods/schemes are in many cases a direct application of the fundamental signal-processing methods introduced in Chapters 3 and 4.

A triggering point is the time instant at which a power quality event starts or ends. The triggering method detects the presence of an event and in most cases also the starting and ending instants. The chapter starts with a description of existing methods for triggering: simple methods that detect changes directly from the waveform or from an rms sequence. More sophisticated methods are needed for triggering points that are not associated with obvious waveform changes or require high time resolution. Methods are introduced which exploit the prominent residuals from time-varying models or find singular points from multiscale signal components.

Partitioning a data sequence into disjoint segments (segmentation) is a necessary preprocessing step before effective data analysis methods can be applied. Although nonstationary signals can be analyzed by using fixed-size block-based methods, as described in Chapter 4, it is desirable that the size of data blocks be adaptive according to some given criteria. Segmentation methods aim at optimally choosing variable-size data blocks, each associated with a duration of time within which the system does not change its state or with a transition process between two system states. Segmentation methods introduced in this chapter can be subdivided into two categories: methods based on using the residuals of different signal models and methods directly using rms sequences or bandpass signal components.

*Signal Processing of Power Quality Disturbances.* By Math H. J. Bollen and Irene Yu-Hua Gu

## 7.1 OVERVIEW OF EXISTING METHODS

### 7.1.1 Dips, Swells, and Interruptions

Voltage dips, voltage swells, and voltage interruptions are detected as a significant deviation of the rms voltage from its nominal value. Typical threshold values are 90% of the nominal voltage for voltage dips, 110% for voltage swells, and 10% for voltage interruptions. There are two different approaches for the definition of the three types of events. The first approach defines dips, swells, and interruptions as events for which the rms voltage is below 90%, above 110%, and below 10%, respectively. This approach is prescribed in detail by the power quality monitoring standard IEC 61000-4-30. The second approach first determines magnitude and duration of the event and next determines the type of event based on its location in the magnitude–duration plane. An example of the relation between the location in the magnitude–duration plane and the type of event is presented in Figure 1.6 in Section 1.2.4. This is the approach used in the IEEE power quality monitoring standard 1159. Under this second approach the triggering (i.e., an rms voltage outside the normal operating range) leads to the detection of a so-called rms variation. After the magnitude (residual voltage) and duration of the event are known, the event is classified as a dip, a swell, an interruption, and so on. Every event thus only belongs to one type. An event is either a dip or a swell or an interruption and so on. The following classification rules are recommended in IEEE 1159:

- An event with a residual voltage between 0.1 and 0.9 pu and a duration between 0.5 cycle and 1 min is a (voltage) sag.
- An event with a residual voltage less than 0.1 pu is an interruption.
- An event with a residual voltage above 1.1 pu is a swell.

The document further recommends the use of the following classifiers to indicate the duration of the event:

- The classifier "instantaneous" is used for events with a duration between 0.5 and 30 cycles.
- The classifier "momentary" is used for events with a duration between 30 cycles and 3 s.
- The classifier "temporary" is used for events with a duration between 3 s and 1 min.
- The classifier "sustained" is used for events with a duration longer than 1 min.

The use of the classifiers is not exactly the same for interruptions as for dips and swells. The complete subdivision of the magnitude–duration plane is shown in Figure 7.1.

The IEEE power quality monitoring standard does not define how the magnitude (residual voltage) and duration of an event should be obtained. The general understanding is, however, that the magnitude is the lowest rms voltage in any of the phases in case of a voltage dip or interruption and the highest rms voltage in case of a swell. Note that this results in a circular definition: To define the event magnitude, the type of event has to be known, but the type of event is determined from the magnitude and duration. For single-phase measurements this is rarely a problem; however, earth faults may lead to a swell in one and a dip in another phase. Neither IEEE 1159 nor any of the later IEEE documents resolves this problem.

The duration of a dip, swell, or interruption is defined as the time below or above the threshold (below the threshold for dips and interruptions, above the threshold for swells). The IEC standard allows for the use of different dip-ending and dip-starting thresholds. The difference between the two thresholds (the *hysteresis*) is only a few percent (IEC 61000-2-8 recommends a value of 1 to 2% or nominal).

As mentioned before, each event belongs to only one type according to the *IEEE approach*. However, the *IEC approach* for definition of events allows for events that belong to more than one type. This is especially the case for three-phase measurements. Consider as an example the event shown in Figure 7.2: a synthetic event due to a single-phase fault that develops into a three-phase fault that leads to the load being disconnected from the supply. The resulting power quality event is, according to IEC 61000-4-30, a dip with a duration $T_1 + T_2 + T_3$ but at the same

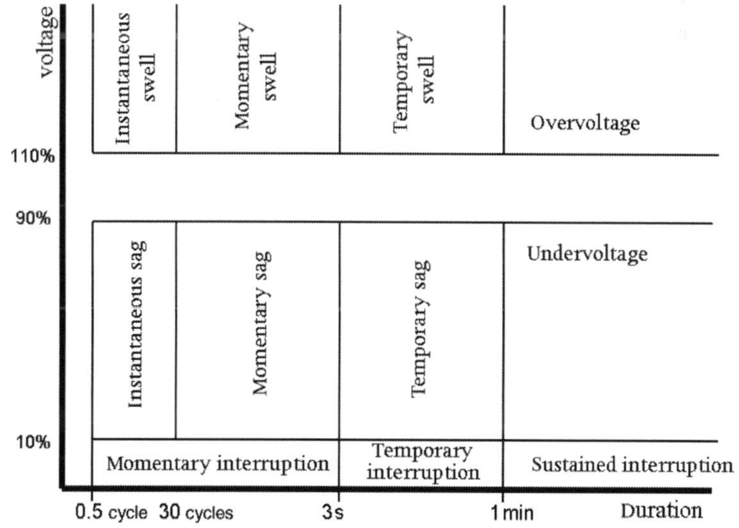


**Figure 7.1** Subdivision of magnitude–duration plane for classification of events as recommended by IEEE 1159.

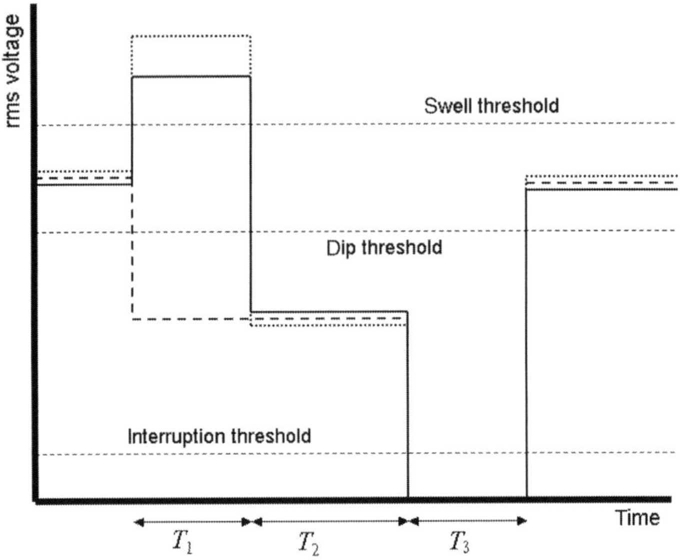


**Figure 7.2** Example of an event that can be classified in a number of ways.

time a swell with a duration $T_1$ and an interruption with a duration $T_3$. Note that this event would also pose a problem for IEEE 1159.

The approach in IEC 61000-4-30 not only may classify the same event into different types but also may classify a different number of events based on the type of event. Consider the recording (rms voltage as a function of time) shown in Figure 7.3. This is a synthetic event representing the voltages recorded downstream of a nonsuccessful reclosing action. The first drop in voltage is due to a fault. Next the breaker is opened and the voltage drops to zero. Upon reclosing the fault remains present and the voltage only recovers partly for a short time. In the end the voltage is restored by switching to another supply. The voltage is below the dip threshold between the fault initiation and the restoration of the supply. This event is thus classified as one dip, with a duration $T_3$. The voltage recovers, however, above the interruption threshold during the reclosing attempt. Hence it is classified as two interruptions, with durations $T_1$ and $T_2$.

When presenting and interpreting power quality statistics in the form of site and system indices (see Chapter 10), the presence of these "multiple events" has to be considered to prevent double counting. For example, every voltage interruption is also at the same time a voltage dip. Most, but not all, voltage swells are also voltage dips. The treatment of these kinds of details becomes very important when power quality statistics are used as performance indicators in contractual and regulatory documents. We will come back to some of these issues in Chapter 10.

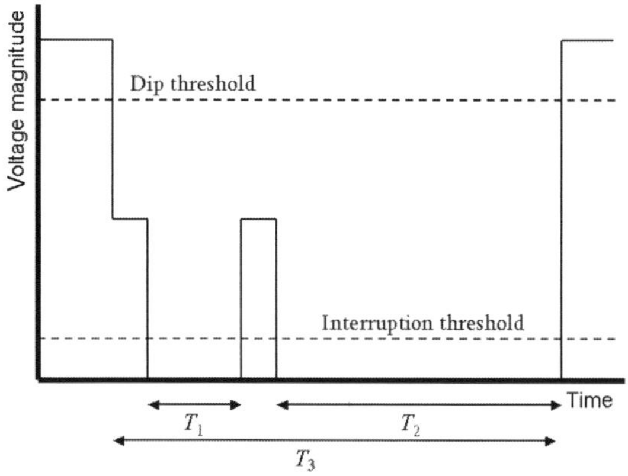


**Figure 7.3** Event that is classified as two interruptions but one dip.

### 7.1.2 Transients

Triggering methods for voltage and current transients are less standardized than for dips, swells, and interruptions. A number of methods are being used, none of which is dominating. As most monitor manufacturers do not publish their methods, it is difficult to get a good overview of all the methods being used. The methods mentioned here should thus only be seen as examples; the list is almost certainly not complete and some of the methods mentioned may not be used in practice.

Triggering for transients can be based either on the actual voltage waveform or on the extracted transient. When the actual voltage waveform is used, the major method compares the absolute value of the voltage with a threshold. This threshold can be set in most power quality monitors. Typical settings are in the range between 1.1 and 2 pu. Note that a 1.1-pu setting will also capture voltage swells. In fact, many of the triggering methods for transients also detect voltage dips, swells, and interruptions. It would require some further processing to remove these events from the statistics on transients.

When the extracted transient is used, triggering typically takes place when the absolute value of the extracted transient exceeds a certain threshold. The difference between the different methods is in the way in which the transient is extracted. A number of methods are available for this:

- The transient is extracted as the output of a high-pass filter. The high-pass filter can be implemented in hardware so that looking for transients does not require any additional computation time. The disadvantage of this method is that high levels of harmonics could lead to false triggering. The threshold has to be set higher than the highest harmonic level. Alternatively a notch filter can be used; this allows for the detection of transients close to the fundamental frequency as well.

- The transient is extracted as the difference with the previous cycle of the power system frequency. This will require a certain amount of computational time. A circular buffer can be used to store one cycle of samples. This method allows for a low threshold setting and thus for a very sensitive detection of transients. However, large frequency deviations may lead to erroneous trips as the buffer length is typically based on the nominal frequency. In the Electric Power Research Institute (EPRI) distribution power quality survey a transient (a so-called wave shape fault) is detected when 10% of the cycle deviates more than 8% from the previous cycle [259].
- Triggering takes place when an instantaneous value deviates more than the threshold from the average sine wave. The average sine wave is obtained by averaging over a number of cycles. This method is sensitive to high levels of harmonics and to deviations in the power system frequency.

We will come back to these methods in Section 8.10 when discussing event indices for transients. A disadvantage of most of the above methods is that they continuously require computational effort. Therefore simple methods have been developed to detect transients. These methods typically are based on the measured voltage waveform, thus skipping the need to extract the transient. The peak voltage was already mentioned, but this method will miss a large number of transients. Even a rather significant transient may not cause any significant instantaneous overvoltage. The simplest method uses the rate of change of the voltage or current as the triggering criterion: Triggering takes place in that case when

$$|v_n - v_{n-1}| > \delta \tag{7.1}$$

An alternative method uses three measurement points and compares the middle point with the average of the two neighboring points [57]. The resulting triggering criterion is

$$|v_n - \tfrac{1}{2}(v_{n-1} + v_{n+1})| \tag{7.2}$$

Note, however, that the difference between two consecutive values is very much determined by the anti-aliasing filter used. An anti-aliasing filter with a very sharp cutoff will make it more difficult to detect transients. In such a case it is recommended to use values two or three samples apart rather than consecutive samples.

### 7.1.3 Other Proposed Methods

The power quality literature contains several papers that propose methods to detect power quality events. Most of these methods use a set of wavelet filters, where typically the output of the highest frequency band is used for detecting the event.

The use of wavelets for power quality analysis was originally proposed by Ribeiro [250]. This paper also proposes the use of neural networks and expert systems. The proposed methodology has been the basis for a significant number of publications in the following years [e.g., 59, 116, 145, 241, 242, 267, 268, 269, 270, 299, 323]. A good overview of the various wavelet applications to power quality is given by [116].

The advantage of using a wavelet-based method over rms methods is that it allows for a much more precise time localization of the event. The wavelet transform can be interpreted as a set of filter banks with increasing bandwidth. In most publications the highest frequency band is used to detect sudden changes in the waveform. This filter has the shortest filter length and thus gives the best time localization. When the change is faster than the time resolution of the signal (i.e., it happens within one time step), the highest frequency band gives the best result. However, in many signals the transition takes several samples. One of the intermediate frequency bands gives a better result in that case. By using a set of wavelet filter banks, the optimum time localization can be chosen for any signal.

Different authors use different mother wavelets. The two most commonly used ones are the Morlet wavelet and the Daubechies wavelet. The Morlet mother wavelet is a sine wave multiplied by a Gaussian window. In [242] a complex version of the Morlet wavelet is used. The advantage of using a sine wave is that the results can be interpreted in a similar way as the results of the short-term Fourier transform. The Daubechies mother wavelets have a somewhat triangular shape, which makes its interpretation more difficult. These wavelets do however have some useful mathematical properties.

A wavelet transform can be used to decompose a signal into components as a function of frequency and time. It can be used to generate a spectrogram (or, more precisely, the so-called scalogram, where the scale is associated with the frequency in logarithmic scale) which is a magnitude spectrum in time. A spectrogram can also be obtained from a short-time Fourier transform, but with a wavelet transform the time resolution increases with the increase of frequency. In [8] a Morlet waveform is used to obtain the time–frequency representation of signal. Local maxima of magnitudes are connected into chains using a method proposed by [205]. After removing short chains related to noise, the power quality event is located in time. The method is applied to three synthetic events with added noise. The method is ingenious and uses the wavelet transform in a clever way, but more work is needed to find out if it will give useful results for measured signals.

A problem in the practical implementation of the wavelet-based triggering method is the noise in the filter outputs when using actual measurements. This "noise" is a combination of harmonic waveform distortion, cycle-to-cycle variations in the fundamental and harmonic contents, and high-frequency components in the signal. For large or quick changes in the original signal, the singular points from the event/variation will be significantly larger than those from the noise. Being able to reduce the noise would make the method also applicable for the detection of very small changes and for the detection of transients in heavily distorted signals. Methods for reducing the noise level of the filter output are proposed in [61, 274, 331]. These three proposed methods can be applied to any signal with a zero expected value before the event, not just to the output of a wavelet filter. In [61] the proposed method is applied to the output of a prediction error filter. All three methods have been applied to measured signals, but none of them have been applied to a large set of measurements to test the performance of the method.

## 7.2 BASIC CONCEPTS OF TRIGGERING AND SEGMENTATION

Triggering is used to detect the occurrence of an event. In Section 1.2.3 we introduced the difference between variations and events. A very important property of events is that they require a method of triggering and one or more threshold settings. The original distinction between variations and events [33, Section 1.3] was that variations are small, slow deviations from an ideal or nominal value whereas events are large, fast deviations. Triggering is the method used to distinguish between *small and slow* and *large and fast*. It will be immediately clear to the readers that any choice of a triggering method and level will be somewhat arbitrary. This dilemma can be solved by simply defining an event by its triggering method and threshold setting. Such has been the basic principle for a number of power quality standards. The power quality monitoring standard IEC 61000-4-30, for example, defines a voltage dip as a situation where the rms voltage is below the voltage-dip threshold.

Triggering may be used to detect both the beginning and the end of the event. This information can next be used to estimate the duration of the event, as will be discussed in more detail in Chapter 8. The information is also used to decide the amount of data to be stored: the so-called event recording. In many cases a power quality monitor not only records the fact that an event has occurred and its basic characteristics but also stores the waveform or rms voltage versus time for the event. This allows for further analysis of the events, either automatically to obtain site and system statistics or visually to understand the origin of the event. The window for which data are stored consists of the period between the two triggering instants (start and end of the event) plus a certain number of samples before the start of the event (the *pretrigger*) plus a certain number of samples after the end of the event (the *posttrigger*). Most of the processing presented in this book was done in the form of postprocessing of such event recordings.

To explain the concept of triggering, consider the voltage dip due to a fault shown in Figure 7.4. The plot on the left shows the three voltage waveforms: The voltage goes down in one of the phases and up in the two nonfaulted phases. This is a typical waveform for a voltage dip due to a single-phase fault at the same voltage level as where the voltage is measured. The plot on the right shows the three rms voltages as a function of time. In this case the rms voltage has been calculated every half-cycle and updated every sample. Note that this is not exactly the method defined in the IEC standards; the principle remains the same however. The reader is referred to Section 8.1.2.

Triggering takes place by comparing the rms voltage with a *voltage-dip threshold*. In this case the dip threshold has been set to 90% of the nominal

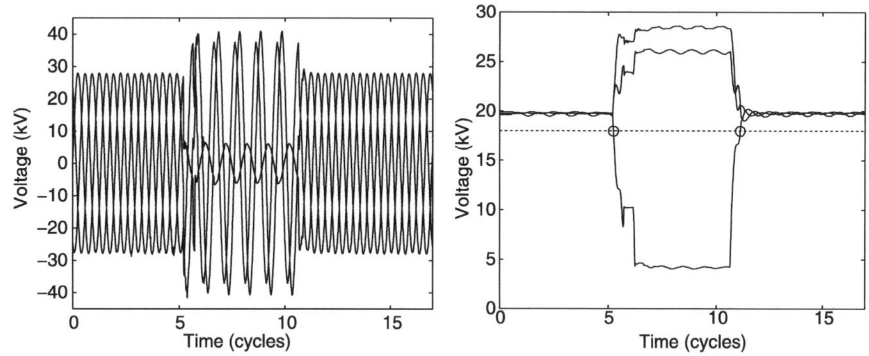


**Figure 7.4** Voltage dip due to single-phase faults: voltage waveforms (left) and rms voltage (right) as function of time.

voltage (18 kV). The threshold has been indicated by the horizontal dotted line in Figure 7.4, the triggering instants by the circles at the locations where the rms voltage crosses the threshold. In a similar way a threshold at 110% of the nominal voltage (22 kV) can be used for detecting voltage swells.

A more fundamental approach, the *power system approach*, is to consider what actually happens in the power system causing the signal to deviate significantly from its normal magnitude. Every (significant) change in voltage or current is due to a change in the power system. We will refer to this change as the *underlying system event* (or, sometimes also referred to as the *underlying cause of the disturbance*), as contrary to the power quality (voltage, current) event. If we consider again the voltage dip shown in Figure 7.4, the underlying events are *fault initiation* and *fault clearing*. We can describe the triggering process as an estimation of the underlying system events. The same dip as in the previous figures is shown again in Figure 7.5. The two vertical bands indicate the estimated regions in which fault

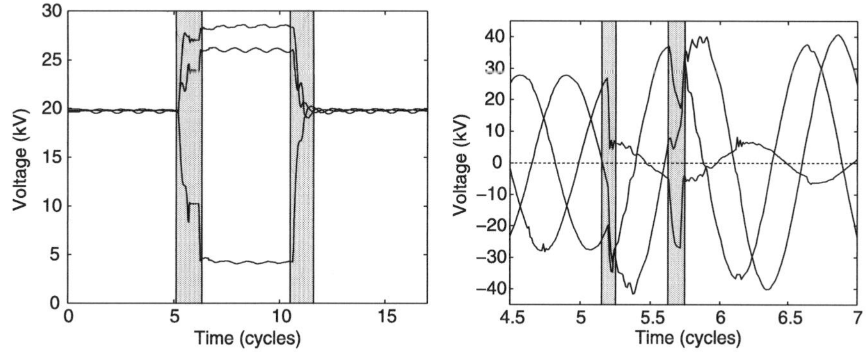


**Figure 7.5** Voltage dip due to a fault: event and transition segments (left); detail of first transition segment (right).

initiation and fault clearing take place. We refer to these intervals as *transition segments*. The interval between the transition segments is referred to as an *event segment*. The intervals before the first and after the last transition segment can also be referred to as event segments, or alternatively as the *preevent segment* and the *postevent segment*, respectively. Instead of one exact triggering point, the segmentation methods to be discussed in this section result in a time interval during which the underlying system event takes place.

Other signal-processing methods or visual inspection can be used to analyze transition segments in more detail. The right-hand plot of Figure 7.5 shows the result of a visual inspection of the first transition segment. The first transition segment can be subdivided into two transition segments and one event segment. The new transition segments can be labeled as *initial fault initiation* and *self-clearing of fault followed by reignition*. Note that the small oscillation about one cycle after the reignition could be due to another instability in the fault arc, which in this case did not lead to self-clearing of the fault. The location of the fault clearing gives the $X/R$ ratio of the source impedance at the fault location. This information can in turn be used to estimate the fault location.

A more complicated example is shown in Figure 7.6. This recording shows four transition segments, corresponding to (from left to right) initiation of a phase-to-phase fault, fault develops into a three-phase fault, first fault clearing, and second fault clearing. It is not immediately clear that the initial fault is phase to phase. This requires a study of the three-phase event characteristics during the first and second event segment. The three-phase classification method is discussed in Section 6.2.3.

From Figure 7.6 it follows that the first fault-clearing segment (the third transition segment) is much longer than the second one. A detailed inspection of these two segments, as shown in Figure 7.7, shows why. The three phases of the first breaker do open with a delay of more than one cycle, whereas for the second breaker to open this takes only one-third of a cycle, as is more common. See Section 6.2.5 for a discussion on the spread in fault-clearing instants for three-phase faults.

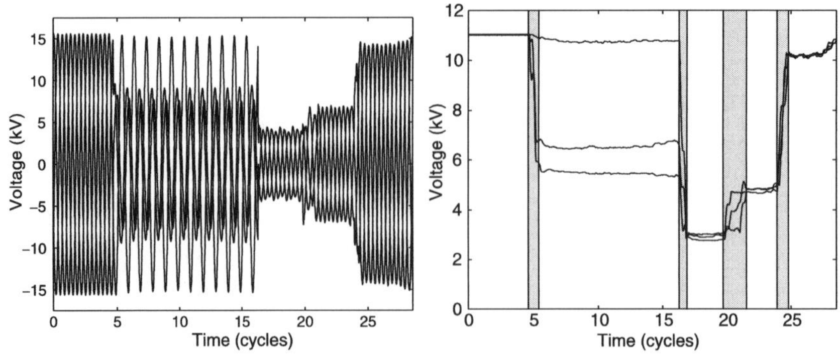


**Figure 7.6** Voltage dip due to developing fault with delayed fault clearing: waveform (left) and rms voltages (right).

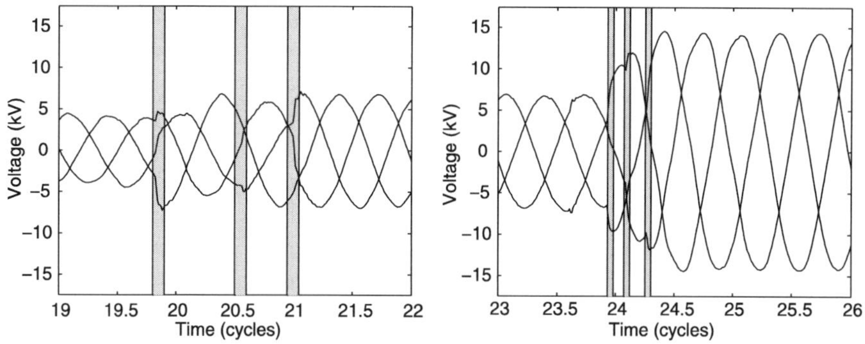


**Figure 7.7** Further detail of first and second fault-clearing segments in Figure 7.6.

In this section we mainly discussed manual methods ("visual inspection") of finding the transition segments. In the following sections we will discuss a number of automatic signal-processing methods.

## 7.3 TRIGGERING METHODS

### 7.3.1 Changes in rms or Waveforms

An important class of events are those associated with sudden changes in voltage or current waveforms. Examples are voltage dips or swells, but also overcurrents and sudden loss of load. A simple way to detect a triggering point for these kinds of events is by using a time-dependent rms sequence. A time-dependent voltage/current rms sequence is an approximation of the time-dependent voltage/current magnitudes at the power system frequency as introduced in Section 3.2.2. From the shape of the rms sequence, one can find some useful information, including the approximate time instant where the underlying system event is likely to have happened.

Figure 7.8 shows an example where four triggering points are detected from the rms sequences of a three-phase measurement. The original three-phase measurement is shown in Figures 7.8*a–c*, where the signal is sampled at $f_s = 4800$ Hz and the power system frequency is 50 Hz. This measurement shows a voltage dip due to a three-stage developing fault: at the $t_1 \approx 320$ sample position an earth fault occurs in phase *a*. After a while, at the $t_2 \approx 650$ sample position, the single-phase fault develops into a two-phase-to-ground fault (in phases *a* and *b*). This leads to a further development to a three-phase-to-ground fault at the $t_3 \approx 1500$ sample position. These three underlying system events are clearly visible in the rms sequences in Figure 7.8*d*. One can also observe from Figure 7.8*d* that there is another triggering point at a somewhat later instant of time $t_4 \approx 4700$ sample position where the voltages at all three phases quickly recover to their normal values. This indicates the instant at which the protection

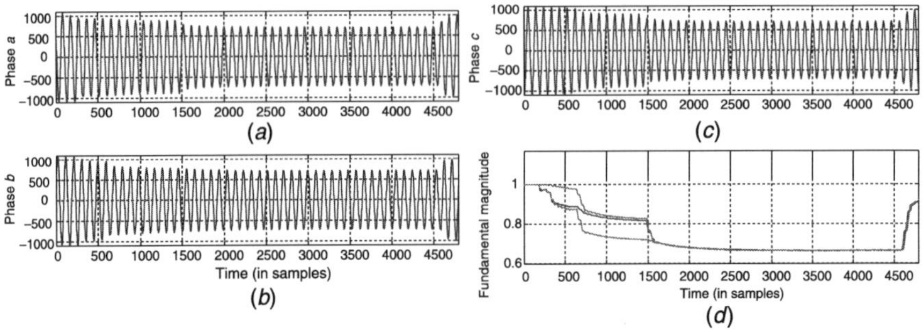


**Figure 7.8** Detecting triggering points by using estimated fundamental magnitude (or estimated rms) sequences: (*a*–*c*) waveforms of original three-phase measurement data; (*d*) estimated fundamental voltage magnitude sequences.

relay clears the fault. All three phases are cleared at almost the same time. There is always a small difference in fault-clearing time between the phases (see Section 6.2.5), but in most cases that cannot be detected from rms sequences due to their low time resolution.

It should be mentioned that although rms sequences are found useful in detecting triggering points, there is a serious limitation of the method; that is, the detected triggering points have rather low accuracy in terms of their time positions. This is due to the nature of the rms where a half-cycle or one-cycle window of data is typically used for computing the rms, hence the time resolution of the rms sequence is associated with this window size. One should bear in mind that the detected triggering points typically have an accuracy in the scale of one half-cycle or one cycle depending on the averaging window used in computing the rms.

### 7.3.2 High-Pass Filters

Another simple triggering method is to use a high-pass filter followed by a threshold. This is based on the idea that a high-pass filter can detect the quick changes caused by most underlying system events. Figure 7.9 shows an example where a high-pass filter is applied to a measured signal. The sample frequency of data is $f_s = 4800$ Hz and the power system frequency is 50 Hz. The measured signal contains a voltage dip due to a fault. The high-pass filter applied is an equiripple linear-phase FIR filter (or the Parks-McClellan optimal FIR filter generated by the MATLAB function *Remez*) with the following specification: stop-band edge frequency is 70 Hz, stop-band minimum attenuation is 50 dB, passband edge frequency is 300 Hz, and passband maximum ripple is 0.1 dB. The high-pass filter order $N = 52$ (or 53 in length).

From the peak locations in Figure 7.9*b*, one can easily find two triggering points: One is located at the $t_1 \approx 487$ sample position corresponding to the voltage dip

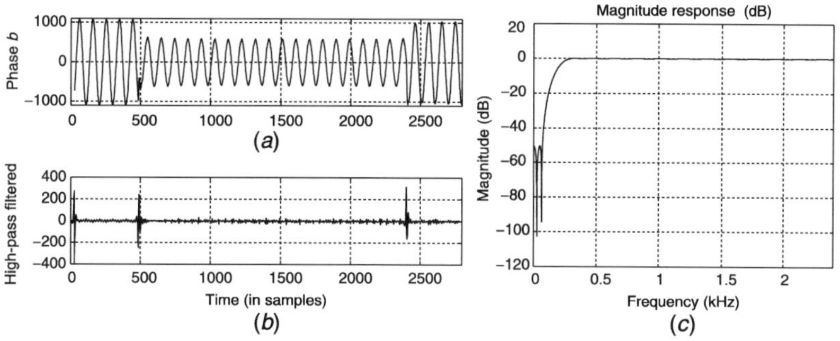


**Figure 7.9** Applying high-pass filter for detecting triggering points: (*a*) waveforms of original signal (phase *b*) containing voltage dip due to a fault; (*b*) high-pass filtered signal; (*c*) magnitude response of high-pass FIR filter.

initiation and another at the $t_2 \approx 2402$ sample position corresponding to the voltage recovery. Note that the time delay of the FIR filter, which is equal to half of its order $N$, has already been taken into account.

In the above example, the triggering points are obvious since the voltage dip is associated with a *step change*; hence a simple high-pass filter serves well for the detection purpose. However, the method does not always work well, especially when the changes are not obvious (e.g., a slope change instead of the step change used in Fig. 7.9). Further the method is sensitive to noise. Therefore some more sophisticated methods will be studied in the next several sections.

### 7.3.3 Detecting Singular Points from Wavelet Transforms

Singular points are those points where signal discontinuities are present. For power disturbance data, a significant singular point is often associated with a sudden change in the system; hence it can be considered as a candidate triggering point.

In Section 4.3, dyadic DWTs were described for decomposing signal into subband components in the time–scale domain where scales are associated with frequencies in the logarithmic scale. Due to the multiresolution nature, DWTs are found to be particularly attractive in the automatic detection of singular points.

Contrary to this, STFTs described in Section 4.2 can fail to detect singular points if the time resolution does not correspond to that of the singular point. The time resolution of a singular point can roughly be described as a signal discontinuity being presented in a short or a long time duration, for example, a step change or a ramp change. Since the required resolution is usually not known in advance, one needs to try a set of different resolution analyses by redoing a STFT with a sliding window of different size, making STFTs less attractive in such an application.

In Section 4.3 we have shown examples for both wavelet transforms and STFTs. In one case, a signal containing a voltage dip due to a fault is analyzed by DWTs where the points of dip initiation and recovery can clearly be found from the decomposed signal components (see Example 4.3). The same case is also analyzed by STFTs (see Example 4.1), where a STFT with a sliding window of size $L = 64$ is found appropriate for detecting the initiation and recovery points of the voltage dip. However, a STFT with a window size $L = 256$ has shown a too low time resolution in analyzing the triggering points from the given signal.

### 7.3.4 Prominent Residuals from Models

In Chapter 3 we introduced a number of model-based methods to estimate the characteristics (e.g., the spectrum) of a signal. For each sample instant the residual is calculated as the difference between the measured value and the value as predicted by the model. Small residual values indicate an accurate model.

The start of an event is associated with a sudden change in the system. It is natural to expect that large residuals occur at around the time instant of the underlying system event, even though the selected model is appropriate in representing the system status associated with the previous data samples. One can hence explore the model residuals for the triggering purposes. The same holds for the end of the event when this is associated with another change in the system. However, for example, a voltage dip due to motor starting will only give one triggering point when using this method.

In Chapters 3 and 4 several model-based methods have been described in which analyzing and characterizing power system disturbances become the issue of choosing a most relevant model and then estimating the model parameters. In the next two sections we will further discuss the use of model residuals for detecting triggering points of power quality events.

**7.3.4.1 *Models Associated with Broadband Spectrum*** In Section 4.3 we studied block-based AR models, where the model is associated with a broadband spectrum of the signal. Apart from the estimation of model parameters, the residuals from the model can be explored for detecting power quality events. Some previously given examples of block-based AR models also show such a potential: In one case, the detected singular points are associated with the beginning and end point of a voltage dip (see Example 4.4). In another case, the detected singular points are the switching time instants of the capacitor bank in a power system (see Example 4.5). The use of residuals from AR models for triggering and segmentation will be discussed in more detail in Section 7.4.3.

**7.3.4.2 *Models Associated with Line Spectrum*** In Section 3.5 we studied sinusoidal models, where the models are used to estimate the line spectrum of the signal with high frequency resolution. A particular interest of applying such models in power quality data analysis is that the frequencies and magnitudes of spectral lines can be estimated (as in the MUSIC and ESPRIT methods; see Chapter 3) or can be set at the power system frequency and its harmonics (as in the example of Kalman filters; see Section 3.5.4). We have applied Kalman filters (Sections 3.5.4 and 4.5) and the sliding-window MUSIC and ESPRIT methods (Section 4.4.4) for estimating the unknown model parameters associated with the power disturbance data. Apart from estimating the parameters of a sinusoidal model, an equally important application is to explore the residuals from the model for the detection of power quality events.

Figure 7.10 shows an example in which the residuals of a Kalman filter are used for triggering. The measured signal (phase $a$) shown in Figure 7.10$a$ contains a voltage dip due to the transformer energizing followed by a maltrip of the protection; that is, the transformer protection trips on the inrush current. The data sampling rate is $f_s = 4800$ Hz and the power system frequency is 50 Hz. The fundamental-frequency component estimated from the Kalman filter is used to approximate the rms sequence in Figure 7.10$b$, and the residuals from the Kalman filter are shown in Figure 7.10$c$. From Figure 7.10$c$ one can observe that there is a large residual at the time instant $t_1 \approx 313$ sample position, indicating the triggering where a voltage dip starts; another large residual is at the time $t_2 \approx 1285$ sample position, indicating the triggering associated with the maltrip by the protection relay.

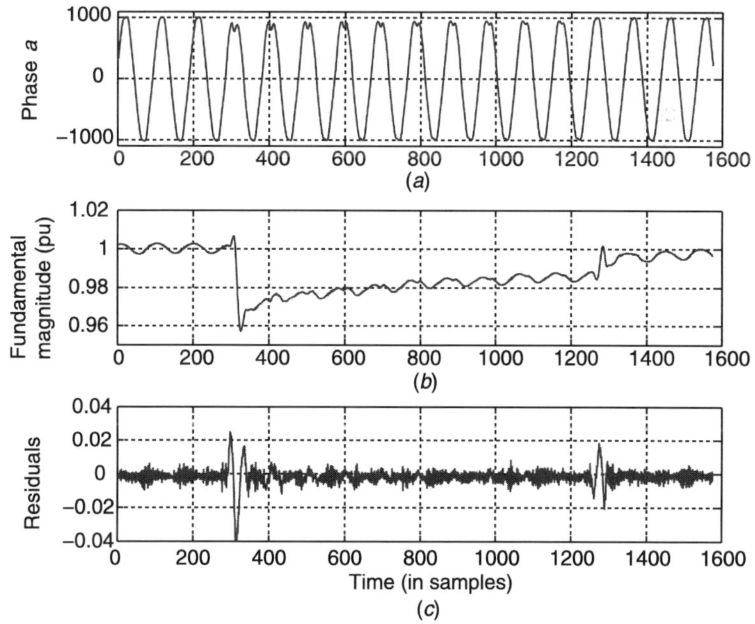


**Figure 7.10** Detecting triggering points using residuals from Kalman filter: ($a$) waveforms of measured signal (phase $a$) containing voltage dip due to transformer energizing followed by maltrip of protection relay; ($b$) estimated fundamental-voltage magnitudes from Kalman filter; ($c$) residual sequence from Kalman filter.

A more detailed discussion on using Kalman filter residuals can be found in Section 7.4.2. Although in those examples the residuals are exploited for the purpose of segmentation, one may also examine the locations of prominent peaks in the residual sequences and find the corresponding triggering points. In Section 7.4.2 we will also introduce a method to suppress the noise in the residual sequence (bottom plot in Fig. 7.10) before, during, and after the event. The segmentation method may be used to obtain an estimation of the time location of the underlying system events (the so-called transition segments). Next, the residuals shown in Figure 7.10*c* can be used for a relatively accurate time location.

In order to gain more insight into the residuals of Kalman filters, Figure 7.11 shows three synthetic signals to which a Kalman filter is applied. Each can be considered as the signal obtained by multiplying a sinusoidal signal with a rectangular window (or a repetitive rectangular window) but at different phase-angle location.

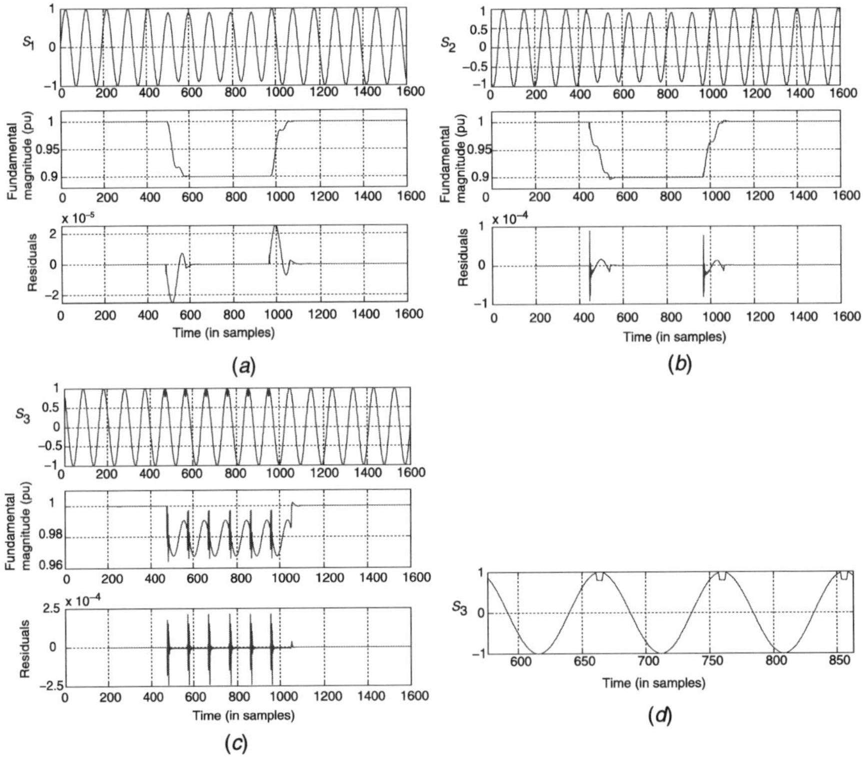


**Figure 7.11** Using residuals of Kalman filters for detection of triggering points: From top to bottom in (*a*–*c*): synthetic waveforms, estimated fundamental-voltage magnitudes, and residuals of Kalman filter from synthetic signals $S_1(n)$, $S_2(n)$, and $S_3(n)$; (*d*) waveforms zoomed in from $S_3(n)$.

Note that these are only hypothetical cases aimed at explaining the performance of a signal-processing method. These signals should not be interpreted as typical signals. For more typical signals, several measurement examples will be discussed in this chapter.

The first synthetic signal top (Fig. 7.11*a*, top) is generated according to

$$S_1(n) = \begin{cases} \sin(2\pi f_0 n) & n \in \{\mathcal{N}_0 - \mathcal{N}_1\} = \{[1, 479], [961, 1800]\} \\ 0.9 \sin(2\pi f_0 n) & n \in \mathcal{N}_1 = [480, 960] \end{cases} \tag{7.3}$$

where

$$\mathcal{N}_0 = [1, 1800] \qquad \mathcal{N}_1 = [480, 960]$$

This signal represents a voltage dip starting and ending at zero voltage positions (this could be a fault being initiated and cleared at the zero crossing of a voltage). The voltage waveform does not show any sudden step, only a change in slope. When the Kalman filter moves near the start of the dip, the deviation between the new data sample and the predicted data from the model using the previous data samples is small. Hence, the residual does not show any large values. Gradually the deviation becomes larger and the residuals start to increase. As the Kalman filter adjusts to the change, the value of the residual decreases again. The whole transition process takes about one cycle time. This is related to the periodic nature of the signal: the dominating 50-Hz component takes 20 ms (96 samples in this case) to be detected.

The second synthetic signal (Fig. 7.11*b*) is generated according to

$$S_2(n) = \begin{cases} \sin(2\pi f_0 n - 2\pi/3) & n \in \{\mathcal{N}_0 - \mathcal{N}_1\} \\ 0.9 \sin(2\pi f_0 n - 2\pi/3) & n \in \mathcal{N}_1 \end{cases} \tag{7.4}$$

where

$$\mathcal{N}_0 = [1, 1800] \qquad \mathcal{N}_1 = [440, 960]$$

This signal represents a dip where the change in magnitude takes place at the voltage maximum, resulting in a large step in instantaneous value. The result is a large mismatch to the model and hence a large residual value. Next to this short-duration spike, the residual sequence shows a longer oscillation with a total duration of about one cycle. The explanation for this oscillation is the same as for the first example.

The third synthetic signal (Fig. 7.11*c*), is generated according to

$$S_3(n) = \begin{cases} \sin(2\pi f_0 n + 2\pi/3) & n \in \{\mathcal{N}_0 - \mathcal{N}_1\} \\ 0.8 \sin(2\pi f_0 n + 2\pi/3) & n \in \mathcal{N}_1 \end{cases} \tag{7.5}$$

where

$$\mathcal{N}_0 = [1, 1800],$$
$$\mathcal{N}_1 = \{[470, 474], [566, 570], [662, 666], [758, 762], [854, 858], [950, 954]\}$$

Comparing the first two synthetic signals, we conclude that the Kalman filter residual allows us to accurately allocate the time instants of events with a significant step in voltage. However, a change in slope is detected but with less accuracy in time.

The third example is more difficult to describe. The waveform shows a small notch around the maximum (see the zoomed-in waveform in Fig. 7.11*d*). This example represents the distortion that often appears in the waveform due to the transformer energizing. An essential property of such a waveform is the difference between its positive and negative half-cycle (see Fig. 2.33 in Section 2.5.3). This type of waveform distortion is normally described as *even-harmonic distortion.* For example, using a one-cycle DFT will show the presence of the second-, fourth-, and other even-harmonic components. We will give some examples in Section 8.1.2.

However, the Kalman filter works differently. The periodic occurrence of the notches causes a periodic pattern in the estimated signal magnitudes and in the residual sequence. The oscillation in the residual sequence lasts only a few samples. A longer period (one cycle) is present in the estimated signal magnitude. This one is also present in the first two examples but hidden in the step in magnitude. The Kalman filter residuals can be used to detect individual notches. Alternatively a segmentation method can be used to determine the approximate location of an underlying system event.

In all three cases the power system frequency $f_0 = 50$ Hz and the number of samples is 96 per cycle. The duration of the event is approximately six cycles in each case. The drop in voltage magnitude is 10% for the first two signals and 20% during the notches for the third signal.

## 7.4 SEGMENTATION

### 7.4.1 Basic Idea for Segmentation of Disturbance Data

In Section 4.4 analysis of nonstationary data was performed by first dividing the data into fixed-size blocks before the parameters of the model are estimated. Each block of data is considered as nearly stationary and is characterized by one set of time-invariant model parameters. Obviously, the main reason for choosing a fixed block size is for simplicity. This is perfectly acceptable if the changes in characteristics are small and slow. For power quality events, however, it is desirable that the block size changes accordingly, so that each block of data (or data segment) can be associated with one underlying cause in the power system. See Section 1.2.3 for the difference between power quality variations and events.

Segmentation is a common approach used in speech signal processing where a speech signal is segmented into disjoint blocks, each representing a basic sound unit such as vowels, consonants, or diphones (e.g, sound between the central stationary points of two vowels) [246]. A similar concept has been introduced in [282, 283] to analyze the power system disturbance where a data sequence is divided into disjoint events. As explained in Section 7.2 transition segments correspond to the sudden changes in voltage or current waveform or magnitude and are correlated with the underlying system events. An event segment is located between the two adjacent transition segments. A power quality event is characterized by the event segments. The first step of segmentation consists of locating the transition segments. The event segments are bordered by the two transition segments:

- A transition segment corresponds to a segment of data between the two (quasi-stationary) event segments (or between a preevent/postevent segment and an event segment), where fast and abrupt disturbances are often present in the signal. These abrupt changes are caused by system events such as fault initiation, fault clearing by protection operation, induction motor starting, transformer energizing, reclosing following fault clearing, load switching, and capacitor switching, and among others. Due to the fast changes between the power system states and the short duration usually involved in a transition segment, data in a transition segment present different characteristics as compared with those in an event segment. It is often difficult to analyze or model the data samples in a transition segment. Obviously, they should be treated differently from the event segments. Methods with a higher time resolution may be applied to transition segments, for instance to locate the time instant of an event with high precision. Methods with a high time resolution are often very sensitive, leading to a high risk of false alarm. However, by applying those methods only to the transition segment, the risk of false alarm is significantly reduced. After all it is now known with a high probability that a system event takes place within the segment.
- An event segment contains disturbance data that can be roughly explained by a single underlying cause. It is therefore often possible to model and characterize the data samples within an event segment. The model parameters are either stationary or quasi-stationary, corresponding to a steady state in the power system. As the signal is quasi-stationary during the event segments, the characterization methods from Chapters 3 and 4 can be applied to the event segments. The main difference is that a much shorter window length is needed.

Apart from these two types of segments, a data sequence often contains a *preevent segment* (before the disturbance) and a *postevent segment* (after the system recovers to the normal state).

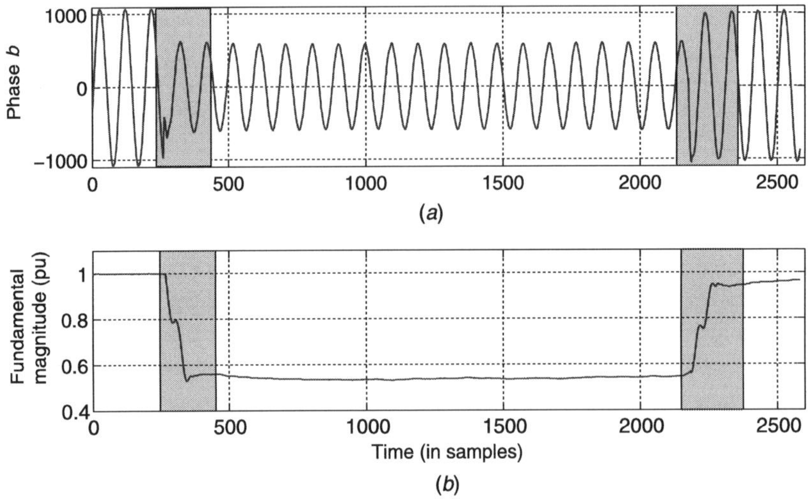


**Figure 7.12** Dividing data sequence into segments: a preevent segment, two transition segments, an event segment, and a postevent segment. (*a*) Waveforms of a measured signal. (*b*) Segments that are marked on the estimated fundamental-voltage magnitude sequence corresponding to the waveforms in (*a*). The shaded areas indicate the transition segments.

An example of segmentation is shown in Figure 7.12*a*, where the original measurement consists of a voltage dip caused by a fault. The data sequence is subdivided into five regions: a preevent segment (before the first shaded area), the first transition segment or fault initiation (the first shaded region), the first event segment or during a fault (between the two shaded regions), the second transition segment or fault clearing (the second shaded region), and the postevent segment (after the second shaded area).

In the remaining part of this section, we shall describe some methods that can be used for automatic segmentation. These methods include the use of residuals from different models, such as from Kalman filters under sinusoidal models (Section 7.4.2) and from block-based AR models (Section 7.4.3) and the direct use of signal magnitude component(s) without signal modeling, such as from fundamental-voltage magnitude (or estimated rms) sequences (Section 7.4.4) and from subband signal components using wavelet decomposition (Section 7.4.5). In fact, many more methods can be possibly found. One of our main emphases here is on the concept and the principle of segmentation itself, rather than the detailed implementation issues.

### 7.4.2 Using Residuals of Sinusoidal Models

In this section, we will describe an automatic segmentation method based on using the residuals of Kalman filters under sinusoidal signal modeling.

Although segmentation according to the principles described in the previous sections can be done manually, it is desirable that such segmentation be performed by an automatic method. This is especially true when the number of data recordings to be analyzed is large. This section will contribute to methods and algorithms that perform such automatic segmentation.

The main idea here is to exploit the residuals of a time-varying model or the singular points from the recordings. Since a model usually requires some time to adapt to the sudden changes in signal characteristics associated with an underlying system event, the model residuals become large during the time of those changes. Large residual values can thus be used to associate with transition segments. We will describe the procedure in more detail below.

Let data samples $v(n)$ be modeled as sinusoids in additive noise. We repeat (3.158) below for the sake of convenience,

$$v(n) = \sum_{k=1}^{K} a_k \cos(n\omega_k + \phi_k) + w(n)$$

We assume that the model order $K$ is chosen sufficiently high so that $w(n)$ is nearly white. Applying a Kalman filter to a signal under a sinusoidal model described in Section (3.5.4) results in the residual sequence $\varepsilon(n)$ (or the estimation errors) in (3.218):

$$\varepsilon(n) = z(n) - \hat{z}(n|n)$$

**7.4.2.1 Hypothesis Test** In order to detect the boundaries of segments, we first define the so-called *measure of changes* as the squared short-time mean values of Kalman filter residuals,

$$\mathcal{M}(t_k) = \left( \frac{1}{L} \sum_{t_i = t_k}^{t_{k+L-1}} \varepsilon(t_i) \right)^2 \qquad t_k = t_1, \ldots, t_N \tag{7.6}$$

where $L$ is the length of the short-time *sliding window* within which the mean of the residuals is computed and $N$ is the total length of the data recording. Here, $\mathcal{M}(t_k)$ provides a measure of the difference between the original and the predicted signal waveforms within a short-time duration. In other words, it roughly indicates how large the signal deviates from the model within an average window of size $L$.

We will see later, from the examples, that the measure of changes is a much better indicator for an underlying system event than the nonprocessed residual sequence. Previous examples of residual sequences have shown that a high noise level may present even when there is no system events. Therefore, directly using residuals may prove difficult for detecting most system events under the constraint of small false-alarm rates. However, the square of averages used in computing the measure of changes can suppress the random variations caused by the estimation noise but amplify the values associated with an underlying system change.

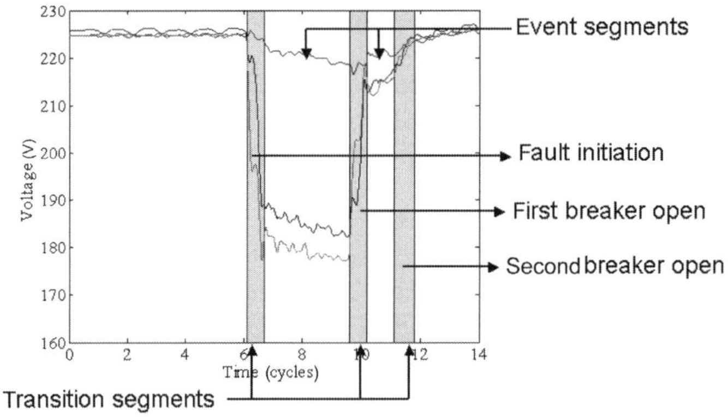


**Figure 7.13** Segmentation of data according to underlying causes (events) or transition between two adjacent events. The vertical lines, marked on the rms voltage of data, indicate the segment boundaries.

To detect the sudden changes in signal waveforms, two hypotheses are applied,

$$\begin{aligned} &H_0 \text{ (no sudden change):} && f(\mathcal{M}(t_k)) < \delta \\ &H_1 \text{ (with sudden change):} && f(\mathcal{M}(t_k)) \geq \delta \end{aligned} \tag{7.7}$$

where $f(\cdot)$ is a function of $\mathcal{M}(t_k)$ and $\delta$ is a threshold determined empirically. The concept of segmentation is presented in Figure 7.13, where the segments are marked on the rms voltage.

When using Kalman filters for segmentation, one should set the order of the sinusoidal model sufficiently high so as to take into account the phenomena and spectral characteristics of all possible events under consideration.

If candidate causes of a disturbance might include transformer saturation and arcing which can result in a high number of harmonics, then the model order should be set sufficiently high so that the segmentation result will still be reliable. The same holds when the segmentation algorithm is applied to current signals with high harmonic distortion, for example, the supply to electrical drives or systems with high computer or lighting load. Also see Section 2.5.4 for more details on harmonic distortion.

A Kalman filter can result in the estimated magnitudes for the fundamental and a number of harmonics depending on the chosen harmonic order in the model. These can be used for signal characterization in the event segments. The estimated values during the transition segments are not reliable and should not be used. However, as mentioned before, other methods can be explored to extract some useful information from the transition segments.

***7.4.2.2 Detecting Boundary Positions*** From the definition of transition segments and event segments, the residuals from Kalman filters are partitioned into two types of segments, each having the following characteristics:

- *Event Segments* When the measure of changes $\mathcal{M}(t_k)$ in (7.6) is below a given threshold.

- *Transition Segments* When the measure of changes $\mathcal{M}(t_k)$ exceeds the given threshold.

More specifically, this is performed by detecting the boundaries of transition segments.

An extra requirement needs to be added to the above: that the measure of changes remain below the threshold for a minimum time. Depending on the shape of the residuals, the corresponding measure of changes may show an oscillating behavior. Alternatively, the situation may occur where two underlying system changes happen in a short time interval. This may result in an event segment of very short duration. The characterization and interpretation of such a segment of data can be difficult. Therefore, a minimum duration for an event segment is set. This results in the following method to define transition segments.

Boundary points are defined at the border between the transition segments and the event segments. To detect the boundary point, the following hypothesis test is applied to $f(\mathcal{M}(t_k))$ in (7.7):

- The starting point of a transition segment (or the end boundary point of an event segment): $t_k$ is the starting boundary point of a transition segment (or equivalently the end boundary point of an event segment) if it satisfies

$$t_k: \quad \begin{cases} \mathcal{M}(t) < \delta & \text{for } t < t_k \\ \mathcal{M}(t) \geq \delta & \text{for } t = t_k \end{cases} \tag{7.8}$$

- The end point of a transition segment (or equivalently the beginning boundary point of an event segment) $t_k$ is detected if it satisfies

$$t_k: \quad \begin{cases} \mathcal{M}(t) \geq \delta & \text{for } t < t_k \\ \mathcal{M}(t) < \delta & \text{for } t = t_k, \ldots, t_{k+s} \end{cases} \tag{7.9}$$

where a short-time duration $\Delta t = t_{k+s} - t_k$ is applied to recognize the end of the transition.

#### 7.4.2.3 Discussion

*Segmentation of Three-Phase Signals* For three-phase voltage or current data, the segmentation is obtained by modifying the above method as follows:

1. Apply the sinusoidal model and a Kalman filter to the signal in each individual phase.
2. Compute the measure of changes $\mathcal{M}^j(t_k)$ in each phase, $j = a, b, c$, where $\varepsilon(t_i)$ used in (7.6) is replaced by $\varepsilon_j(t_i)$.
3. Compute the measure of overall changes in three phases by

$$\mathcal{M}(t_k) = \max\{\mathcal{M}^a(t_k), \mathcal{M}^b(t_k), \mathcal{M}^c(t_k)\} \tag{7.10}$$

4. Perform the remaining steps similar to those in the single-phase case, that is, applying (7.8) and (7.9) to find the boundary points of transition segments.

### Principles for Selecting Parameters

- *Selection of Model Order K.* The model order should be set sufficiently high to be able to accommodate the principal line spectral components contained in the possible types of events (e.g., transformer saturation, arcing, capacitor switching, and other events).
- *Selection of Window Size L in $\mathcal{M}(t_k)$.* To select $L$ in (7.6) that is used to compute short-time mean residual values, one must consider two issues: the accuracy (or the time resolution) of segment boundary positions and the segmentation error (e.g., false-alarm and missed segments). Choosing a large $L$ yields a relatively smooth $\mathcal{M}(t_k)$, but the time resolution is low. Conversely, a small $L$ implies a high time resolution but a high variance (or more fluctuations) in $\mathcal{M}(t_k)$. From the signal-processing viewpoint a low time resolution implies less reliable segment boundary positions. On the other hand, $\mathcal{M}(t_k)$ tends to fluctuate more if $L$ is small, hence a higher probability that $\mathcal{M}(t_k)$ would accidently exceed the threshold $\delta$ just because of the fluctuations rather than due to an underlying system event. Consequently, more segment errors (e.g., *false segment*) may appear. Obviously, choosing $L$ must be a trade-off between the time resolution of segment boundaries and the false alarm of segments.

  One way is to determine $L$ empirically. A reasonable choice is to set $L$ equal to a half-cycle or one cycle of power system frequency (i.e., 50 Hz in Europe and 60 Hz in North America). For a half-cycle $L$, fluctuations in $\mathcal{M}(t_k)$ are shown to be acceptable for performing a relatively reliable segmentation; meanwhile the time resolution of a half-cycle is rather acceptable. However, for disturbance data containing fast repetitive changes, such as the signal in (7.5) (see the top in Fig. 7.11*c*), a one-cycle window is required to smooth out the periodic repetition of spikelike residuals in order to obtain a meaningful segmentation result.
- *Selection of Time Duration $\Delta_t$ in (7.9).* Selection of $\Delta_t$ may influence the detection of end boundaries in transition segments. It is selected such that the fluctuations in $\mathcal{M}(t_k)$ due to evolving events do not (or are less likely to) lead to a false detection of the end of transition segments. Obviously, introducing $\Delta_t$ reduces the fluctuations. Time duration $\Delta_t$ is determined empirically, for example, equal to a half-cycle time duration. As mentioned before, $\Delta_t$ also sets the minimum duration of an event segment. As it is difficult to obtain characteristics from very short segments, a half-cycle minimum duration appears to be a reasonable value.
- *Selection of Threshold $\delta$.* The threshold $\delta$ in (7.7), (7.8), and (7.9) is determined empirically. One may observe that a low threshold makes the detection

scheme more sensitive to changes but can also lead to incorrect detection of segment boundaries.

Based on the theoretical analysis of simple hypothesis-testing problems [184], $\delta$ should be determined according to the detection probability and false-alarm rate of the segments if the probability density function (pdf) of $\mathcal{M}$ for each hypothesis is known. For the addressed segmentation issue, once the pdf's for the two hypotheses $H_0$ and $H_1$ are given, $\delta$ can be determined according to the Neyman–Pearson method, which maximizes the detection probability under the constraint that the false alarm is below a prespecified value (see Section 10.3.2).

However, one often encounters some practical problems that prevent us from following this approach. This can be due to the lack of data or the lack of power system configuration/setting information, making estimation of pdf's from data difficult. The issue of learning data statistics and applying statistical detection theory to classify the causes of power system disturbances can be found in Chapter 3 and will be further addressed in Section 8.3.

*Accuracy and Time Resolution of Segmentation Using Kalman Filter–Based Method* It is obvious that the time resolution of segmentation is directly influenced by the size of the averaging window $L$ in (7.6) and $\Delta_t$ in (7.9).

It is also worth noticing that the accuracy of segmentation is indirectly limited by the order $K$ of the harmonic model and the convergence speed of the Kalman filter. Note that the basic Kalman filter algorithm (see Section 3.5.4 and Section 4.5) computes the error covariance matrix $\mathbf{P}(n|n)$ [or $\text{MSE} = \text{tr}\{\mathbf{P}(n|n)\}$] by using *all* the available data up to $n$ with an equal weight. This may slow down the convergence and result in poor estimates if the signal is highly nonstationary. This can be improved by applying a forgetting factor (or a sliding window) that scales down (or removes) the old data samples in the estimation process, which is known to yield an improved convergence and a better estimation for a dynamic system/nonstationary data (see Section 3.5.4).

*Limitations of Segmentation Methods* In a residual-based segmentation method, it is assumed that the model associated with each event is either stationary (i.e., the model parameters are time invariant within the event) or nearly stationary. The criterion exploiting the sudden changes in model residuals essentially looks for detecting temporally unfitted models due to the transition or change between two events.

Further, there exist a number of intrinsic limitations to the segmentation concept. The main limitation concerns the evaluation of the performance of the method:

- *The "ground truth" (i.e., the true states in the power systems) of correct segmentation is difficult to obtain.* In fact, for automatic diagnostics using measured data, often little is known in advance about the nature of the data (e.g., the number and types of variations/events that take place in the system and their impact on the voltage or current waveform) since the configuration or the settings of power systems and some other user-related factors are difficult to obtain. Power quality monitoring becomes especially of interest when unexpected events are captured. As a result, the issue of whether or not the segmentation boundaries are correctly detected can only be justified at the end of the diagnostic system based on whether the results are self-explainable according to the phenomena occurring in the power system at the time of measurement. The results of an automatic segmentation algorithm can be further compared with the visual inspection results of a power system expert. It is also important to extract as many characteristics as possible and look for contradictions that would indicate errors in the segmentation results. If no such contradiction can be found, then the method can be assumed correct until proven otherwise. Note that this requires not only the segmentation itself but also the interpretation of the underlying system changes and events.

- *Whether to choose a global or a local criterion for the segmentation.* In principle, the criterion selection in any automatic segmentation method should be based on the end result of the application, for example, in a diagnostic system a good criterion results in a correct finding of the underlying changes and events in the system. Obviously, this may be influenced by a series of processing steps; for example, in a diagnostic system this could be dependent on the performance of multiple steps such as segmentation and classification, the rules used for an expert system, the selection of an artificial neural network (e.g., the structure), and coverage of statistics in the training data. The best criterion should be based on the global, or the "end-to-end," performance. Practically, one often sets a criterion based on local optimization to limit the scope of consideration, for example, the boundary-finding criterion in the above segmentation method. Therefore, one should be aware of the possible pitfall that such a local optimization in each processing step may not naturally guarantee the global optimum performance.
- *Performance evaluation: a "chicken-and-egg" problem?* In addition to the above, it should be noted that with the unknown (or partially given) power system configuration or settings and the uncertainties that could happen during the time of measurement (e.g., lighting, arc furnace, trigger of multiple power stations), even a power system expert could find it difficult to manually set the boundaries of segments or explain the underlying causes. Evaluating whether a segmentation method yields a global optimization can become a chicken-and-egg problem. Therefore, data learning or a trial-and-error philosophy in engineering would be more realistic.

***7.4.2.4 Examples*** In this section five segmentation examples based on using Kalman filter residuals will be given. All data used in these examples were obtained from measurements with the sampling rate $f_s = 4800$ Hz and 50 Hz power system frequency (or 20 ms per cycle).

In all examples, the sinusoidal model order in the Kalman filter was set to $K = 20$ and the frequencies in the sinusoidal model were chosen to be the power system frequency $f_0 = 50$ Hz and its harmonics $f_m = mf_0$, $m = 2, \ldots, K$. Further, the model

**TABLE 7.1 Settings Used in Segmentation Based on Kalman Filter Residuals**

| Kalman Filter Settings | Detector Settings |
|---|---|
| Model order $K = 20$ | Window size $L$ = half-cycle |
| Variance of model noise $\sigma_q^2 = 0.4$ | Window size $(\Delta_t)$ = half-cycle |
| Variance of measurement noise $\sigma_v^2 = 10^{-5}$ | Threshold $\delta = 0.2 \times 10^{-4}$ |

noise in Kalman filters was set to be high, $\sigma_q^2 = 0.4$, to ensure a fast response if a change in the signal occurs, and the measurement noise was set to $\sigma_v^2 = 10^{-5}$. For simplicity, the initial values for the error covariance matrix was set to be $\mathbf{P}(0 \mid 0) = \sigma_q^2 \mathbf{I}$. The time index of $\mathcal{M}(t)$ corresponded to the beginning sample location in the sliding window. It is shown in [282] that setting the window size $L$ and $\Delta t$ equal to a half-cycle and the threshold as $\delta = 0.00002$ enables the detection of voltage magnitude drops $\geq 0.05$ pu. The parameters used are summarized in Table 7.1.

**Example 7.1** Voltage Dip Caused by Transformer Energizing. The measured signal contains a voltage dip caused by transformer energizing. The underlying system event is closing the switching device connecting the transformer to the system (the actual energizing instant). Figure 7.14 shows the three measured waveforms, the residuals from the Kalman filters, the measure of changes $\mathcal{M}^j(t)$ for each individual phase, and the transition segment marked on both the three-phase estimated magnitude sequences and on $\mathcal{M}(t)$. The magnitude of the signal has been estimated as the absolute value of the estimated fundamental component from the Kalman filter. The same holds for all other examples presented below.

From Figure 7.14*b* one can observe large fluctuations in the residuals. This may be due to an insufficient model order of Kalman filters during the disturbance and also due to continuous small changes in the power system. However, averaging the residual values over a short-time window cancels most of these fluctuations and results in $\mathcal{M}(t)$ in Figure 7.14*c* mainly containing the prominent changes around the transition region. This leads to correct identification of the transition segment in Figure 7.14*d*.

Also, from the shape of the magnitude-versus-time curve in Figure 7.14*d* one can observe that there is a sharp drop in voltage magnitude for all three phases simultaneously. However, the size of the drop is different as there is an unbalance among the three phases. Following the initial drop, there is a slow voltage recovery. From this pattern one can conclude that the voltage dip is caused by the transformer energizing. As a result, the segmentation shows one transition segment in Figure 7.14*d* (the shaded area) followed by one event segment. The region before the transition segment can be regarded as a preevent segment. Note that the event segment is not limited by another transition segment. The transformer inrush current gradually decays to a small value (the magnetizing current); there is thus no end to the event. However, the end of the recording is typically reached when the voltage magnitude recovers above a certain value.

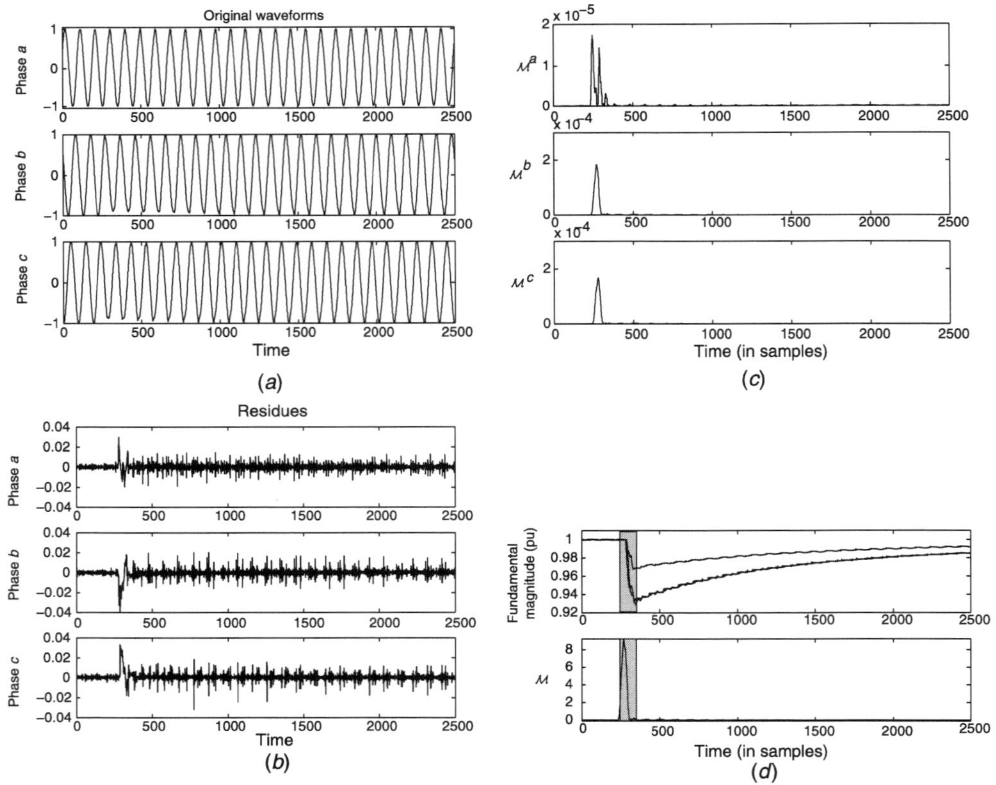


**Figure 7.14** Using Kalman filter residuals for segmentation of voltage dip caused by transformer energizing. All horizontal axes show the time in samples (where each cycle contains 96 samples or 20 ms). (*a*) Measured voltage waveforms for the three phases. (*b*) Residuals from Kalman filters. (*c*) Measure of changes $\mathcal{M}^j(t)$ in each individual phase. (*d*) Transition segment (shaded area) marked on estimated magnitude sequences from Kalman filters and on measure of total changes $\mathcal{M}(t)$ in the three phases.

**Example 7.2.** End of Interruption (Reenergizing System). This measured data sequence contains the process of "end of interruption" where transformer saturation occurs while the system is reenergized. As can be seen from the voltage waveform in Figure 7.15*a*, the voltage at the measurement location recovers from zero to its normal magnitude (1 pu). This sudden onset leads to large residuals in the Kalman filters shown in Figure 7.15*b* and hence the large $\mathcal{M}(t)$ in Figure 7.15*c*. This leads in the segmentation result shown in Figure 7.15*d* containing one transition segment (the shaded area).

Note that the measure of changes in phase *a* shows two peaks. This could be incorrectly interpreted as a small event segment bordered by two transition segments. To prevent this, a minimum length for an event segment is introduced, as discussed before.

**Example 7.3** Voltage Dip Due to Developing Fault. The measured voltage sequence contains the information of a developing fault consisting of three stages. From the estimated voltage magnitudes shown in Figure 7.16*d*, one can see that at around $t_1 \in [143, 404]$ an earth fault occurs in phase *a*. After about four cycles (around $t_2 \in [602, 766]$) the single-phase fault develops into a two-phase-to-ground fault (in phases *a* and *b*). This is followed by a further development into a three-phase-to-ground fault at around $t_3 \in [1445, 1587]$. Afterward the protection clears the fault, causing the voltages of all three phases to quickly recover to their normal values (at around $t_4 \in [4555, 4791]$). The segmentation method has successfully detected all four underlying system changes, as shown in the four transition segments (shaded areas) in Figure 7.16*d*. Between these transition segments, there are three event segments. The region prior to the first transition segment is the preevent segment, and the region after the last transition segment is a postevent segment.

The first transition segment (corresponding to fault initiation) is longer than the second and third transition segments (2.7 cycles versus 1.7 and 1.5 cycles). Many dips show a short transient or notch short before the fault fully develops. Close inspection of the measure of changes shows a small spike followed by a larger spike during the first transition segment. The fourth transition segment (corresponding to fault clearing) has a length of 2.5 cycles. The last transition segment is typically somewhat longer due to the difference in fault-clearing instant between the phases to three-phase and two-phase-to-ground faults.

Note that the first transition is only visible in phases *a* and *b*, the second mainly in phases *b* and *c*, and the third only in phases *a* and *c*. This is completely in agreement with an underlying cause in which a single-phase fault develops first into a two-phase-to-ground fault and next into a three-phase fault. Note that the measurements were obtained from phase-to-phase connected instrument transformers.

**Example 7.4** Voltage Dip Due to Transformer Energizing Followed by Protection Maltrip. This measurement data sequence contains a voltage dip. From the shape of

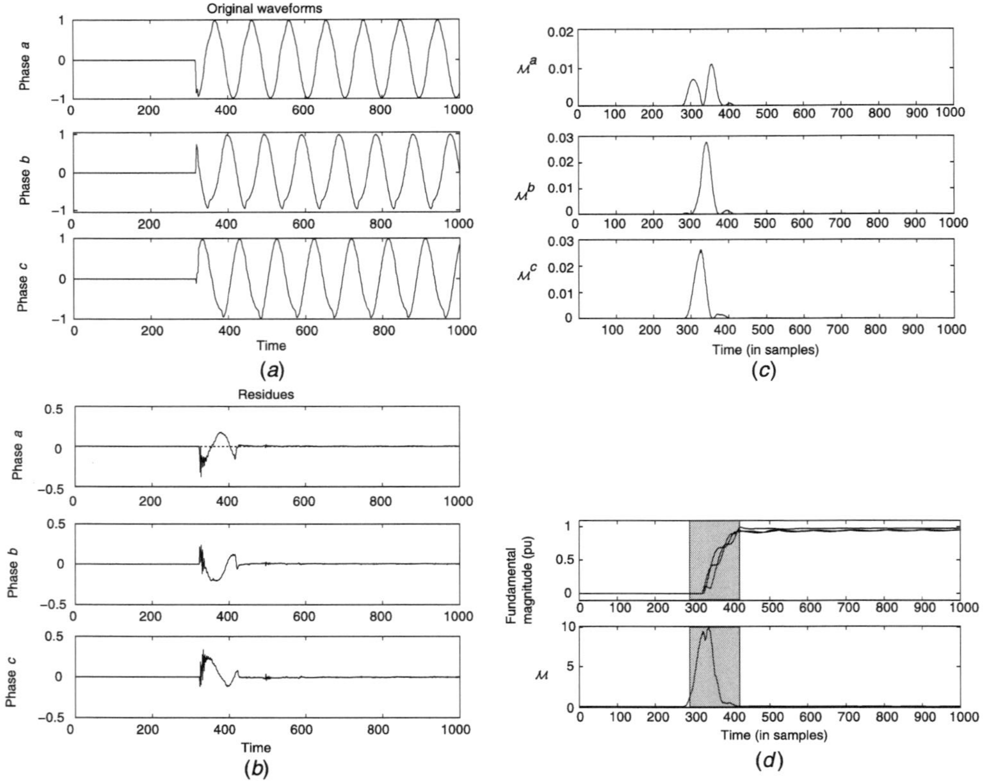


**Figure 7.15** Using Kalman filter residuals for segmentation of measurement data containing *end of an interruption*. All horizontal axes show the time in samples (where each cycle contains 96 samples or 20 ms). (*a*) Waveforms of three phase measurements. (*b*) Residuals from Kalman filters. (*c*) Measure of changes $\mathcal{M}^j(t)$ in each individual phase. (*d*) Transition segment (shaded area) marked on fundamental-voltage magnitude sequences obtained from Kalman filters as well as on measure of total changes $\mathcal{M}(t)$ in three phases.

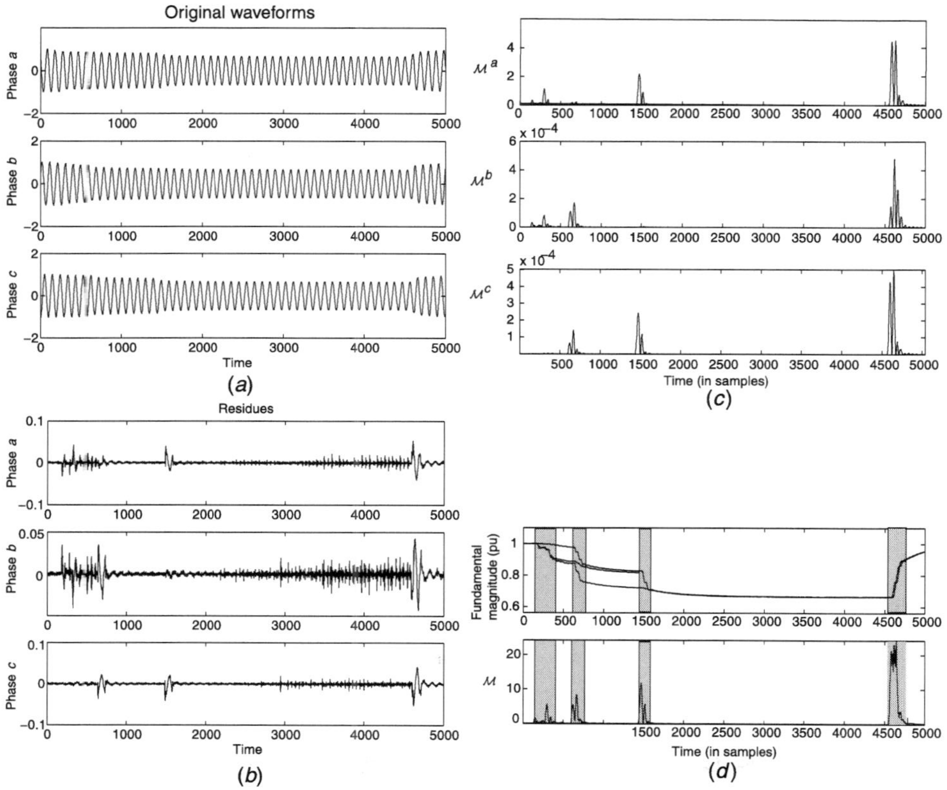


**Figure 7.16** Using Kalman filter residuals to segment voltage dip due to developing fault. All horizontal axes show the time in samples (where each cycle contains 96 samples or 20 ms). (*a*) Waveforms of three phase measurements. (*b*) Residuals from Kalman filters. (*c*) Measure of changes $\mathcal{M}^j(t)$ in each individual phase. (*d*) Transition segment marked on estimated fundamental-voltage magnitude sequences from Kalman filters and on measure of total changes $\mathcal{M}(t)$ in three phases.

the voltage magnitude versus time in Figure 7.17*d*, that is, a sudden drop in voltage different in all three phases followed by a slow recovery, one can conclude that the voltage dip is caused by transformer energizing. This explains the first transition segment at around time $t = t_1$ in Figure 7.17*d*. Comparing this with Example 7.1, one can see that initially the two examples are exactly the same. Contrary to the in Example 7.1, however, this transformer-energizing dip contains a second transition segment. This is explained as an incorrect operation of the protection relay during the energizing of the transformer. More precisely this event was due to the fast autoreclosure of an MV feeder after a fault (a voltage dip due to a fault was recorded at the same location shortly before). All distribution (MV/LV) transformers connected to the feeder experience saturation during the reclosing. The combined inrush current can lead to maltrip of the relay protecting the feeder.

In this example segmentation results in a total of two transition segments and one event segment (between the transition segments). Similarly, the region before the first transition segment is the preevent segment, and the region after the second transition segment is the postevent segment.

**Example 7.5** Voltage Dip Due to Fault. The measured data sequence contains a voltage dip caused by a fault. From the rectangular shape of the estimated fundamental voltage-magnitudes in Figure 7.18*d* one can conclude that the dip is caused by a fault from Figure 7.18*a* it can be seen that the dip only occurs in phases *b* and *c*. It further shows that large residuals only appear at the corresponding locations in phases *b* and *c* in Figure 7.18*b*. Note the much smaller scale for phase *a* in both Figures 7.18*b* and *c*: The noiselike changes in phase *a* do not influence the segmentation result. As shown in Figure 7.18*d*, the segmentation results in two transition segments (one corresponding to fault initiation and another to fault clearing) and one event segment between the transition segments. There is also one preevent segment and one postevent segment.

### 7.4.3 Using Residuals of AR Models

Applying the block-based AR models (see Section 4.3) yields a residual sequence of the errors between the data and the models. This residual sequence can be exploited for the purpose of segmentation, although it often does not work as well as using the residuals of sinusoidal models by a Kalman filter. This does not come as a surprise as the sinusoidal models fit better with the voltage and current waveforms in an electric power system where the main disturbances are often in the form of power system harmonics. A disadvantage of AR models is that they do not provide sufficiently high frequency resolution as compared with those in the sinusoidal models. Therefore, the spectrum from an AR model cannot be used to examine closely spaced harmonic (or inter harmonic) components. Despite that, a time-varying AR model is able to characterize the entire range of a spectrum while a sinusoidal model only estimates the line spectrum in a number of (selected) frequencies.

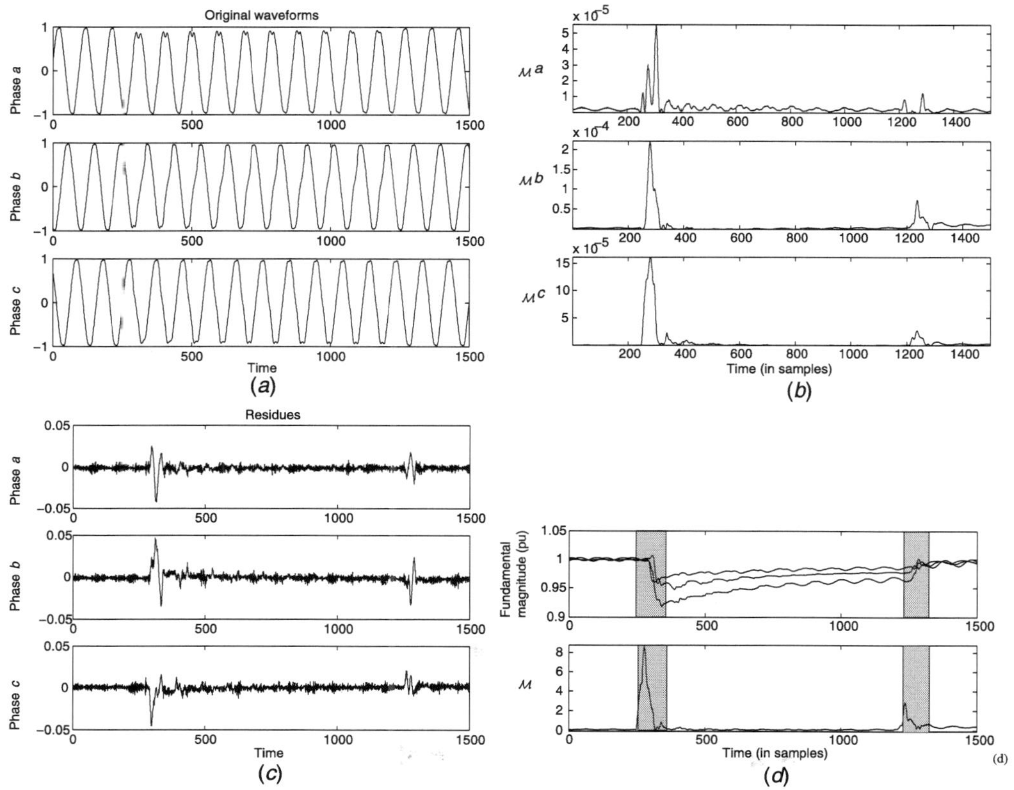


**Figure 7.17** Using Kalman filter residuals for segmentation of voltage dip due to transformer energizing followed by protection maltrip. All horizontal axes show the time in samples (where each cycle contains 96 samples of 20 ms). (*a*) Waveforms of three phase measurements. (*b*) Residuals from Kalman filters. (*c*) Measure of changes $\mathcal{M}^j(t)$ in each individual phase. (*d*) Transition segment marked on estimated fundamental-voltage magnitude sequences and on measure of total changes $\mathcal{M}(t)$ in three phases.

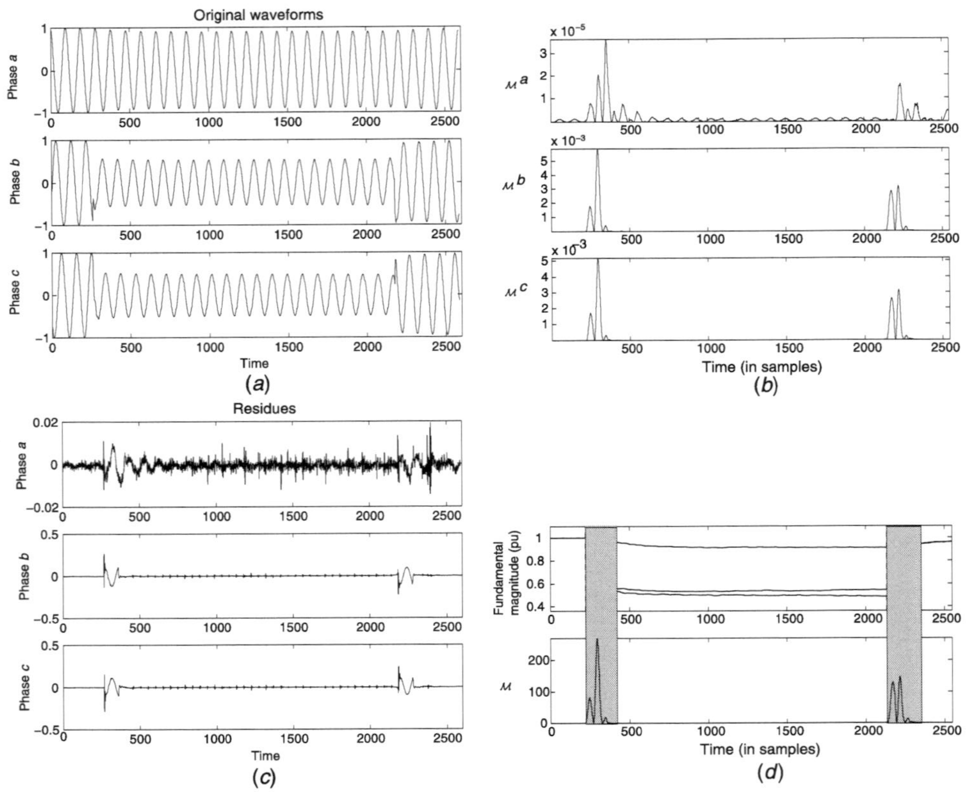


**Figure 7.18** Using Kalman filter residuals for segmentation of voltage dip due to fault. All horizontal axes show the time in samples (where each cycle contains 96 samples or 20 ms). (*a*) Waveforms of three phase measurements. (*b*) Residuals from Kalman filters. (*c*) Measure of changes $\mathcal{M}^j(t)$ in each individual phase. (*d*) Transition segment marked on estimated fundamental-voltage magnitude sequences from Kalman filters and on measure of total changes $\mathcal{M}(t)$ in three phases.

The estimation errors (or residuals) from an AR model contain rich information and can be further exploited. For the same reason as in the Kalman filter, the model usually cannot adapt fast enough to sudden changes due to the disturbance. As a result, the model will yield some relatively large estimation errors (or residuals) around the time of such changes.

In the following, we shall give two examples. These examples use same original data as in Examples 7.3 and 7.4. We shall examine the possibility of segmentation by exploiting the residual sequences from the block-based AR models.

**Example 7.6** A Developing Fault. This example uses the same measured signal (see Fig. 7.16*a*) as in the Example 7.3. Figures 7.19*a* and *b* show the residual sequences from the block-based AR models and the corresponding measure of changes $\mathcal{M}^j(t)$ in each individual phase. The parameters used are as follows: AR model order in (4.44) is $N = 10$, the size of the data blocks for AR models in (4.42) is $L = 2$ cycles (96 samples per cycle), the step size between two consecutive data blocks is 10 samples [i.e., the overlap in (4.42) is $K = 182$ samples], and the sliding window for obtaining $\mathcal{M}^j(t)$ in (7.6) is two cycles. Figure 7.19*c* includes the estimated voltage amplitudes extracted from the time-dependent AR magnitude spectra at 50 Hz frequency.

One can observe from the residuals in phases *a* and *b* that there is a transition segment at around $t_1 \in [180, 830]$; however, the residual in phase *c* indicates the existence of another segment at around $t_2 \in [630, 830]$. These segments can be confirmed by examining Figure 7.19*c*. In addition, at around $t_3 \in [1480, 1680]$ the residuals in both phases *a* and *c* present large peaks indicating a transition segment (however, it is not obvious from the residuals of phase *b*). These seem to be consistent to the voltage magnitudes in the three different phases. Finally, at the time interval $t_4 \in [4600, 4800]$ the large residuals in all three phases indicate another transition segment. Interestingly, the observations are somewhat close to the transition segments found in the Example 7.3, except that due to using a large block size (two cycles) there is an insufficient time resolution for detecting the event segment between the two transition segments around [180, 830]. The spikelike fluctuations in $\mathcal{M}^j(t)$ in Figure 7.19*b* are smoothed out by applying a one-cycle averaging window to $\mathcal{M}^a(t) + \mathcal{M}^b(t) + \mathcal{M}^c(t)$, resulting in $\mathcal{M}(t)$ shown in Figure 7.19*d* as the improved measure of overall changes.

Comparing this with the Kalman filter–based method one can observe that the residual sequences from AR models need to be examined carefully in order to obtain the required segment information. The automatic detection method used in the Kalman filter case cannot be directly applied to the AR model case. A more sophisticated detection method will be required in this case.

**Example 7.7** Voltage Dip Due to Transformer Energizing Followed by Protection Maltrip. This example uses the same data as in Example 7.4 (see Fig. 7.17*a*). The parameters used are as follows: the AR model order in (4.44) is $N = 10$; the size of data blocks for AR models in (4.42) is $L = 2$ cycles (96 samples per cycle); the step

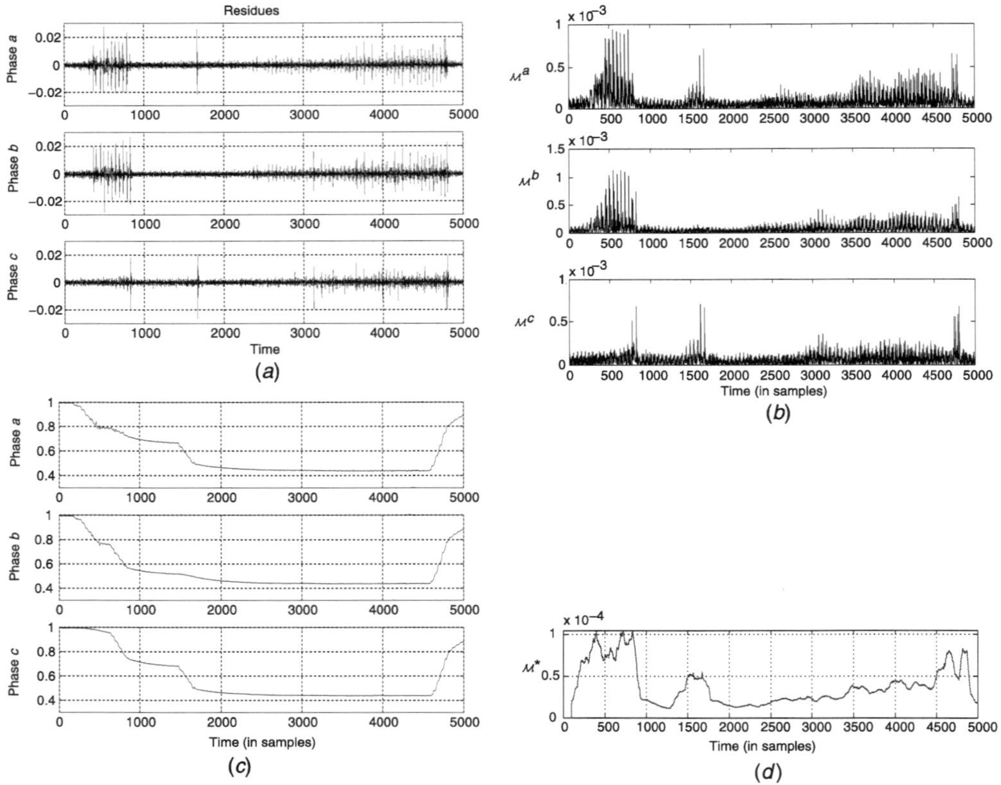


**Figure 7.19** Using residuals from block-based AR models for detecting transition segments from measurement containing developing fault. All horizontal axes show time in samples (where each cycle contains 96 samples or 20 ms). (*a*) Residuals from AR models. (*b*) Measure of changes $\mathcal{M}^j(t)$ in each individual phase. (*c*) Estimated fundamental-voltage magnitudes extracted from block-based AR magnitude spectra at power system frequency. (*d*) Improved $\mathcal{M}^*(t)$ obtained by applying one-cycle averaging window to $\mathcal{M}^a(t) + \mathcal{M}^b(t) + \mathcal{M}^c(t)$ shown in (*b*).

size between two consecutive data blocks is 10 samples [i.e., the overlap in (4.42) is $K = 182$ samples]; $\mathcal{M}(t) = \sum_{j=a,b,c} \mathcal{M}^j(t)$ is used to replace (7.10); and $L$ in (7.6) and $\Delta_t$ in (7.9) both equal one cycle.

In Figure 7.20*a* $\mathcal{M}^j(t)$ clearly indicates the existence of two transition segments. Applying the same segmentation method as in the Kalman filter case leads to the correct segmentation in Figure 7.20*b*. One can also observe from Figure 7.20*c* that the voltage magnitudes obtained from the time-dependent AR magnitude spectrum at power system frequency 50 Hz shows a significantly different behavior that for any of the earlier discussed estimation methods (e.g., Fig. 7.17*d*). The drop in voltage magnitude takes about two cycles and neither the drop nor the recovery is smooth. The slow drop is related to the two-cycle data block size. The irregular behavior indicates that the selected AR models do not fit well to the data.

*Discussion* From the above two examples using AR model residuals, we have shown that explaining the residuals from AR models requires careful observations: The explanation is sometimes not straightforward. Having some prior knowledge in our case about Kalman filter segmentation can help in the interpretation. Results are less consistent in different cases as compared with those obtained from Kalman filters using sinusoidal models. Our conclusion is that the AR model–based segmentation method is not always suitable for automatic segmentation.

### 7.4.4 Using Fundamental-Voltage Magnitude or rms Sequences

Instead of using signal modeling and extracting the information from the model residuals as described above, it is also possible to segment a signal by directly extracting the information from the waveforms. In this section, we shall describe the use of fundamental-voltage magnitude sequences (or the approximated rms sequences) for the segmentation. Five examples are included in this section in which the segmentation is performed using magnitude sequences.

The sole usage of time-dependent magnitude sequences for segmentation is itself attractive since many power quality monitors only choose to store rms sequences of the disturbances rather that the entire waveforms in order to save the memory space. It is therefore also important to see whether such a choice is acceptable for data segmentation and subsequent analysis, especially in terms of inexpensive alternatives for future power quality monitoring equipment. A time-dependent rms sequence can be viewed as being approximately equal to the voltage or current magnitudes at the fundamental frequency (50 Hz in Europe and 60 Hz in North America) and is defined as

$$\mathcal{V}_{\text{rms}}(t_k) = \sqrt{\frac{1}{N} \sum_{t=t_{k-N+1}}^{t_k} v^2(t)} \qquad t_k = t_0, t_1, \ldots \tag{7.11}$$

where $v(t_k)$ is the voltage/current sample and $N$ is the size of the sliding window used for computing the rms (see Section 3.2.2). An rms sequence can be considered

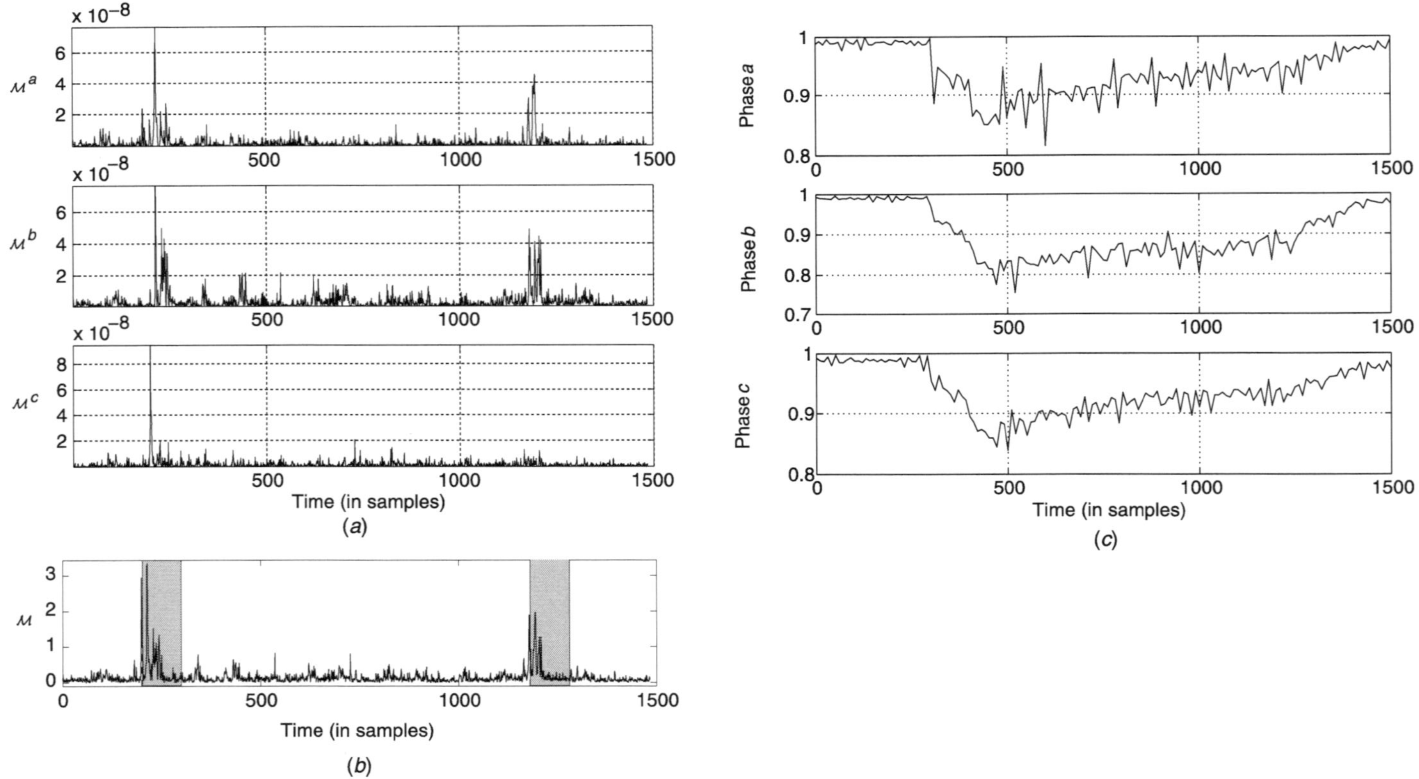


**Figure 7.20** Using residuals from block-based AR models for detecting transition segments from disturbance data containing voltage dip due to transformer energizing followed by protection maltrip. All horizontal axes show time in samples (where each cycle contains 96 samples or 20 ms). (*a*) Measure of changes $\mathcal{M}^j(t)$ in each individual phase. (*b*) Transition segments (shaded areas) marked on measure of overall changes $\mathcal{M}(t)$. (*c*) Estimated fundamental-voltage magnitudes extracted from block-based AR magnitude spectra at power system frequency.

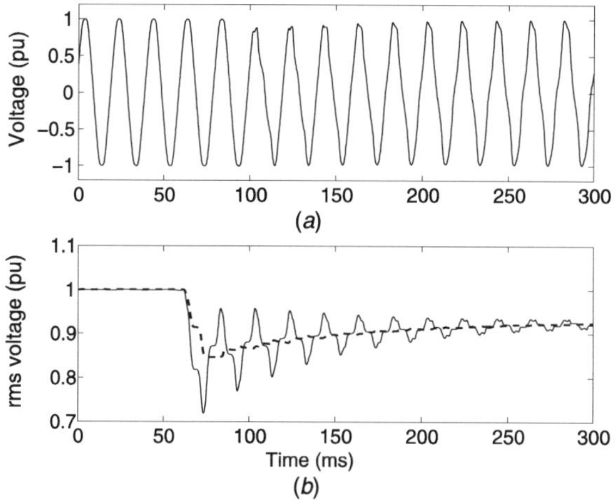


**Figure 7.21** Two rms sequences computed using a one-cycle and a half-cycle respectively. (*a*) Original signal waveform containing a voltage dip due to transformer energizing. (*b*) Dashed line: rms sequence obtained using one-cycle window; solid line: rms sequence obtained using half-cycle window.

as the output of a nonlinear low-pass filter whose input is the original data sequence. It is worth mentioning that the window or block size $N$ influences the shape of the resulting rms sequence. For example, the rms obtained from using a half-cycle window is associated with a higher time resolution but with more fluctuation as compared with that from a one-cycle window. Figure 7.21 compares the rms sequences obtained from using a half-cycle and a one-cycle window, where the original signal contains a voltage dip due to transformer energizing. Figure 7.21*b* shows that using a half-cycle window results in a high level of oscillations in the rms sequence as compared with a much smoother rms sequence from a one-cycle window. As mentioned in Section 7.4.2 such fluctuations can trigger a false segmentation. Hence, the choice of window size $N$ should be a trade-off between the time resolution of segmentation boundaries and the false alarm of segmentation. Note that the large difference can be traced back to the presence of a significant harmonic distortion in the signal (see Figure 2.33 in Section 2.5.3 and Section 8.1.2).

Instead of using an rms sequence, the time-dependent magnitude sequence at the power system frequency can be used as well. That is, instead of using (7.11) one may estimate the voltage magnitudes by first estimating the time-dependent magnitude spectra (e.g., by STFT, subband and wavelet filters, AR models, and sinusoidal model) followed by extracting the time-dependent magnitude component centered at the power system frequency. For example, a STFT (short-time Fourier transform or sliding-window FFT) can be regarded as a bank of bandpass filters (see Section 4.2). The advantage is that such a voltage magnitude estimation can achieve higher time resolution than that of the rms sequences from (7.11). Table 7.2 contains a short MATLAB program where the voltage magnitude is estimated by extracting the

**TABLE 7.2 MATLAB program Computing Voltage-Fundamental Magnitude Sequence Using STFT (or Sliding-Window FFT)**

```
%data: the original data sequence
%magn: the estimated voltage magnitude
N=No_of_samples_per_cycle;
for i=1: length(data)-N
  Pyy=fft(data(i:i+N-1), N);
  magn(i)=abs(Pyy(2,:))/(N/2);
end
```

*Note*: Size of sliding window, $N$, is equal to one cycle.

magnitude component from the corresponding band in the STFT centered at the power system frequency. *For simplicity, we shall refer to the "voltage magnitude sequence" as the "rms sequence" or "approximated rms sequence" in the remaining of this section.*

The rms sequence-based segmentation uses a similar strategy as that in Section 7.4.1; however, the *measure of changes* is computed from the derivatives of rms values instead of from model residuals. The segmentation can be described by the following steps:

1. *Down Sampling an rms Sequence* Since the time resolution of an rms sequence is low and the difference between two consecutive rms samples is relatively small, the rms sequence is first down sampled before computing the derivatives. This will reduce both the sensitivity of the segmentation to the fluctuations in rms derivatives and the computational cost. In general, an rms sequence with a larger down-sample rate will result in fewer false segments (or split segments) but with a lower time resolution of segment boundaries. Conversely, for an rms sequence with a smaller down-sample rate the opposite holds. Practically the down-sample rate $m$ is often chosen empirically; for the examples in this section, $m \in [N/16, N]$ is chosen ($N$ is the number of rms samples in one cycle). In many cases only a limited number of rms values per cycle are stored (two according to the IEC standard method) so that further down sampling is not needed. For notational convenience we denote the down-sampled rms sequence as

$$\mathcal{V}_{\text{rms}}(\tilde{t}_k) \qquad \tilde{t}_k = \frac{t_k}{m}$$

2. *Computing First-Order Derivatives* A straightforward way to detect the segmentation boundaries is from the changes of rms values, for example, using the first-order derivative,

$$\mathcal{M}^j_{\text{rms}}(\tilde{t}_k) = \left|\mathcal{V}^j_{\text{rms}}(\tilde{t}_k) - \mathcal{V}^j_{\text{rms}}(\tilde{t}_{k-1})\right| \tag{7.12}$$

where $j = a, b, c$ indicates the different phases. Consider either a single-phase or a three-phase measurement; the measure of changes $\mathcal{M}_{\text{rms}}$ in rms values is defined by

$$\mathcal{M}_{\text{rms}}(\tilde{t}_k) = \begin{cases} \mathcal{M}^{a}_{\text{rms}}(\tilde{t}_k) & \text{for single phase} \\ \max\left(\mathcal{M}^{a}_{\text{rms}}(\tilde{t}_k),\, \mathcal{M}^{b}_{\text{rms}}(\tilde{t}_k),\, \mathcal{M}^{c}_{\text{rms}}(\tilde{t}_k)\right) & \text{for three phases} \end{cases} \tag{7.13}$$

3. *Detecting Boundaries of Segments* A simple step is used to detect the boundaries of segments under two hypotheses,

$$\begin{array}{ll} H_0 \text{ (event segment):} & \mathcal{M}_{\text{rms}}(\tilde{t}_k) < \delta \\ H_1 \text{ (transition segment):} & \mathcal{M}_{\text{rms}}(\tilde{t}_k) \geq \delta \end{array} \tag{7.14}$$

where $\delta$ is a threshold. A transition segment starts at the first $\tilde{t}_k$ for which $H_1$ is satisfied and ends at the first $\tilde{t}_k$ for which $\mathcal{M}_{\text{rms}}(\tilde{t}_k) < \delta$ occurs after a transition segment is detected.

The five examples below use the measurement sequences as in Section 7.4.2. However, segmentation is performed by using the estimated fundamental-voltage magnitudes extracted from the corresponding band in the STFT (see Table 7.2). It is worth mentioning that whether to use a rms sequence from (7.11) or an approximated rms sequence from another subband method is not essential for the segmentation result.

**Example 7.8** Voltage Dip Due to Transformer Energizing. This example uses the same original data sequence as in Example 7.1 (see Fig. 7.14*a*). The measured signal is sampled at $f_s = 4800\,\text{Hz}$ and the power system frequency is 50 Hz (i.e., $N = 96$ samples per cycle). The voltage magnitude sequences for three phases are estimated by using the STFT with a one-cycle window. The down sample of voltage magnitude sequence is set to be $m = \frac{1}{4}$ cycle (i.e., 24 samples). The threshold is $\delta = 0.02$. This will enable the detection of fast changes larger than 0.04 pu in the voltage/current magnitudes. Noting that for a fast change of $k$ pu in the magnitude, this method will capture a change at least $k/2$ pu (i.e., half of the actual change) assuming that the window center is located at the point of change. For changes at any other location in the window, the change in the estimated voltage magnitude will be greater than $k/2$ pu. Figure 7.22 shows the resulting $\mathcal{M}_{\text{rms}}(t_k)$ using (7.13) and the segments (shaded areas) marked on the estimated voltage magnitude sequences. The measure of overall changes $\mathcal{M}_{\text{rms}}(t_k)$ does show a large peak around the region of transition, leading to a transition-segment. This result can be further confirmed by comparing with Figure 7.14*d* in the Example 7.1.

**Example 7.9** End of Interruption (Re-energizing System). This example uses the same original measured signal as in Example 7.2 (see Fig. 7.15*a*) where the

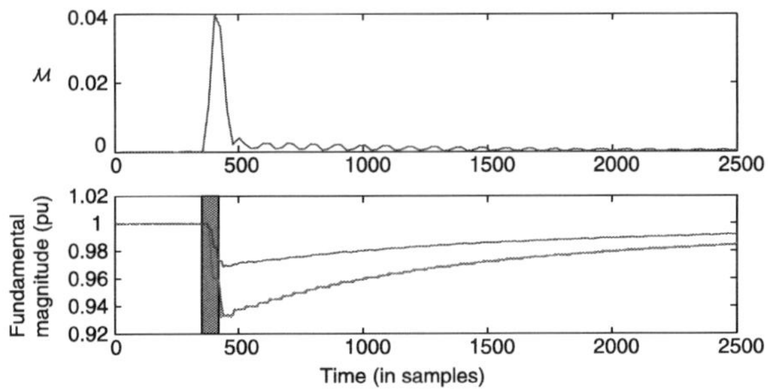


**Figure 7.22** Using estimated three-phase voltage magnitudes for segmentation of voltage dip due to transformer energizing. All horizontal axes show time in samples (where each cycle contains 96 samples or 20 ms). Top: measure of overall changes. Bottom: transition segment (shaded area) marked on estimated rms sequences.

sample rate is $f_s = 4800\,\text{Hz}$ and the power system frequency is 50 Hz. The three-phase voltage magnitude sequences are estimated by using the STFT with a one-cycle sliding window. The down-sample rate of rms sequence is $m = \frac{1}{8}$ cycle (resulting in 12 samples per cycle) and the threshold is set to be $\delta = 0.02$. Figure 7.23 shows the resulting measure of changes $\mathcal{M}_{\text{rms}}(t_k)$ from using (7.13) and the resulting segments marked on the estimated voltage magnitude sequences. As can be seen from the bottom figure in Figure 7.23, the segmentation result (two event segments and one transition segment) and the boundaries of the transition segment are similar to those in Example 7.2.

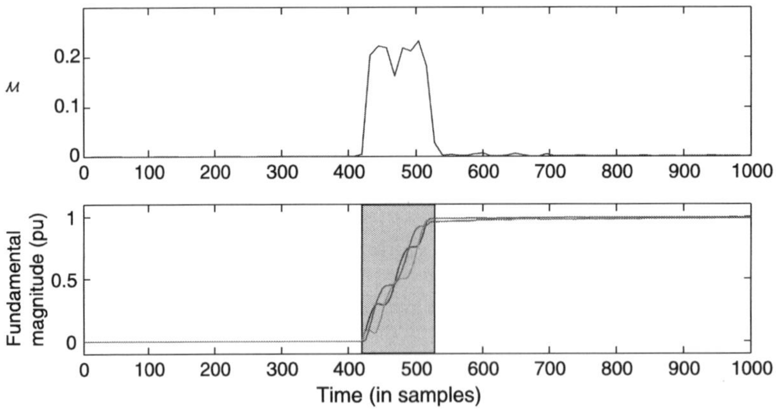


**Figure 7.23** Using estimated three-phase voltage magnitude sequences for segmentation of data sequence containing end of interruption. All horizontal axes show time in samples (where each cycle contains 96 samples, or 20 ms). *Top*: measure of overall changes. Bottom: transition segment (shaded area) marked on rms sequences.

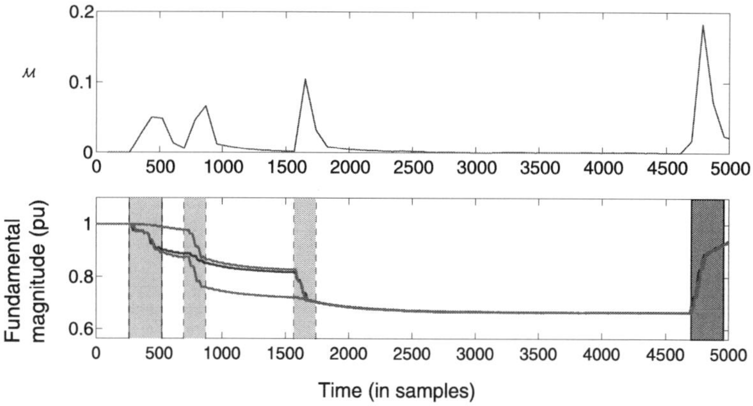


**Figure 7.24** Using estimated three-phase voltage magnitude sequences for segmentation of developing fault. All horizontal axes show time in samples (where each cycle contains 96 samples or 20 ms). Top: measure of overall changes. Bottom: transition segments (shaded areas) marked on rms sequences.

**Example 7.10** A Developing Fault. This example uses the same original measured signal as in Example 7.3 (see Fig. 7.16*a*), with the sample rate $f_s = 4800\,\text{Hz}$, 50 Hz power system frequency, and a one-cycle sliding window in the STFT. For the segmentation, the threshold is set to $\delta = 0.02$, and a relatively small down-sample rate $m = 96/1.1$ is applied to the voltage magnitude sequences in order to maintain a high time resolution in the segment boundaries.

Figure 7.24 shows the resulting measure of changes $\mathcal{M}_{\text{rms}}(t_k)$ using (7.13) and the resulting transition segments marked on the estimated rms sequences. As can be seen from Figure 7.24, good segmentation results (three event segments and four transition segments) are obtained even for this rather complicated event. Further, the boundaries of transition segments are comparable to those obtained from the Kalman filter–based method. This example shows the strength of the magnitude-based segmentation method.

**Example 7.11** Voltage Dip Due to Transformer Energizing Followed by Protection Maltrip. This example uses the same measured signal as in Example 7.4 (see Fig. 7.17*a*), with the sample rate $f_s = 4800\,\text{Hz}$, 50 Hz power system frequency, and a one-cycle sliding window in the STFT. For the segmentation, the down-sample rate used is $m = \frac{1}{2}$ cycle and the threshold $\delta = 0.02$.

As shown in Figure 7.25, two transition segments are correctly detected with good accuracy. This result can be compared with that in Figure 7.17*d*, where the first transition segment corresponds to the transformer energizing and the second transition segment corresponds to the maltrip of the protection relay.

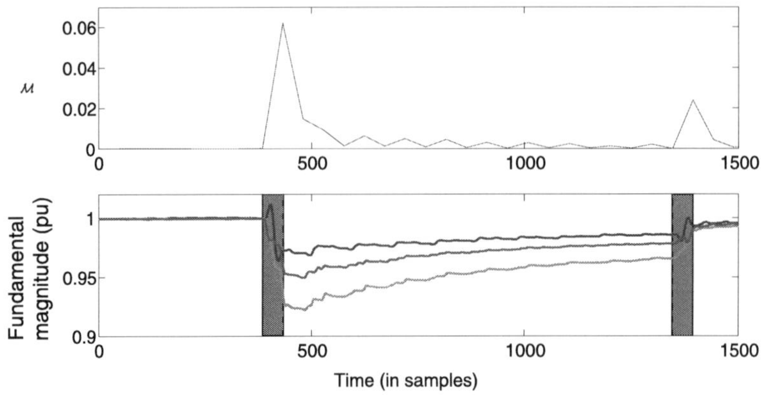


**Figure 7.25** Using estimated three-phase voltage magnitude sequences for segmentation of disturbance data containing voltage dip due to transformer energizing followed by protection maltrip. All horizontal axes show time in samples (where each cycle contains 96 samples or 20 ms). Top: measure of overall changes. Bottom: transition segments (shaded areas) marked on rms sequences.

**Example 7.12** Voltage Dip Due to a Fault. This example uses the same original measured signal as in Example 7.5 (see Fig. 7.18*a*), with the sample rate $f_s = 4800\,\text{Hz}$, 50 Hz power system frequency, and a one-cycle sliding window in the STFT. For the segmentation, the threshold is set to $\delta = 0.02$ and the down-sample rate is $m = \frac{1}{4}$ cycle. Good segmentation results are again obtained as shown in Figure 7.26.

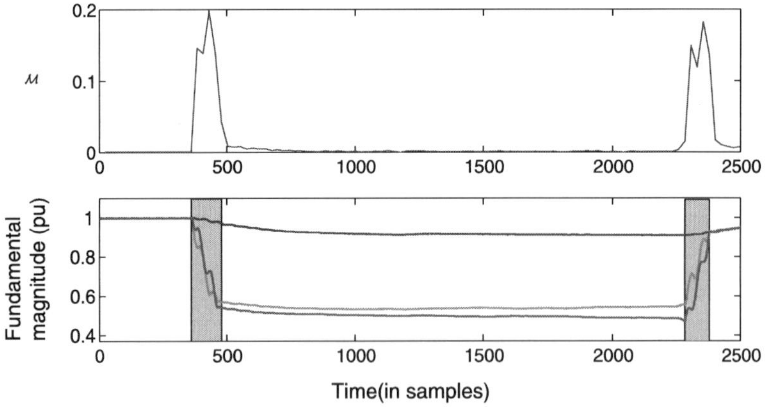


**Figure 7.26** Using estimated three-phase voltage magnitude sequences for segmentation of voltage dip caused by fault. All horizontal axes show time in samples (where each cycle contains 96 samples or 20 ms). Top: measure of overall changes. Bottom: transition segments (shaded areas) marked on rms sequences.

Summarizing the fundamental-voltage magnitude-based segmentation, the obvious advantage is that no complicated computation is needed. The fundamental-voltage magnitude is easy to compute and the segmentation is based on the difference between the magnitude values. The method can thus easily be implemented in a power quality monitor.

The method does have its limitations, however. Only steps in voltage magnitudes lead to transition segments: phase-angle jumps, changes in harmonic level, and transients are not detected by this method. The time resolution of the estimated voltage magnitude sequence may limit the option of closely looking at the transition segments. Obviously, a more advanced method may be applied to transition segments when waveform data are available.

### 7.4.5 Using Time-Dependent Subband Components from Wavelets

Instead of using the model residuals or rms sequences for the segmentation, one can also exploit information from several signal components, for example, subband signal components from the DWTs or the STFT. In Section 4.3, we described the use of wavelets and STFTs for signal decomposition. To further exploit the application of using multiresolution subband signal components, we shall study their potential in segmentation with a number of examples. Four examples (7.13 to 7.16) are included in this section to demonstrate the potential of segmentation where the weaknesses and the strengths of using subband components are discussed.

In order to define the measure of changes $\mathcal{M}(t_k)$ from the signal components, let us first denote the subband signal components from a $M$-scaled dyadic-structured DWT as

$$S_{b_j}(t_k) \qquad b_j = 1, 2, \ldots, M+1 \tag{7.15}$$

where $b_j$ indicates the subband from the lowest ($b_j = 1$ low pass) to the highest ($b_j = M + 1$ high pass). Daubechies or biorthogonal wavelets can be used to generate the mother wavelet and the corresponding two-band filter kernels (see Section 4.3).

To obtain the measure of changes in each subband, one may compute the first-order derivative of the bandpass component followed by computing the energy of changes,

$$\mathcal{M}_{b_j}(t_k) = |S_{b_j}(t_k) - S_{b_j}(t_{k-1})|^2 \tag{7.16}$$

To measure the changes from a single-phase measured signal, the changes are accumulated over a selected number of subbands where the changes are prominent,

$$\mathcal{M}(t_k) = \sum_{b_j=b_{j1}}^{b_{j2}} \mathcal{M}_{b_j}(t_k) \tag{7.17}$$

If the measurement consists of three phases, then (7.10) can be applied to obtain the measure of overall changes $\mathcal{M}(t_k)$.

In the following four examples one may observe that detection of segment boundaries requires a more sophisticated method than using a simple threshold-based method in (7.8) and (7.9) in the Kalman filter case. This is clearly a disadvantage of the wavelet-based method. However, one may observe a rather high time resolution in such a multiscale signal component-based detection as compared with the previous segmentation examples. As a result, this method may provide a good potential in yielding the more accurate boundary locations of segments.

**Example 7.13** Voltage Dip Due to a Fault. This example contains one of the test signals from the IEEE project group 1159.2. The sample rate of the signal is $f_s =$ 15,360 Hz and the power system frequency is 60 Hz (i.e., 256 samples per cycle). We have studied this signal in Example 4.3 of Section 4.3 for wavelet-based signal decomposition. In this example, we shall further study this signal for the purpose of segmentation.

After applying the same wavelet signal decomposition as in Figure 4.6*b*, that is, Daubechies wavelet db4 (from the wavelet toolbox in MATLAB) with a total number of scales $M = 7$ applied to the data sequence, the bandpass filter outputs are examined to see whether they can be exploited for detecting the underlying system changes: fault initiation and fault clearing. Figures 7.27*a* and *b* show the original signal and the decomposed components from the subbands $b_0$ to $b_8$. Figure 7.27*c* shows $\mathcal{M}_{b_j}(t_k)$ using (7.16), $b_j = 3, \ldots, 7$ and the measure of overall changes $\mathcal{M}(t_k) = \sum_{b_j=3}^{7} \mathcal{M}_{b_j}(t_k)$. From Figure 7.27*c* one can clearly see the existence of two clusters of peaks in $\mathcal{M}(t_k)$ corresponding to two transition segments (dip initiation and voltage recovery) and an event segment between and with high time resolution. Obviously, an automatic algorithm can be applied to $\mathcal{M}(t_k)$ to detect these two transition segments.

Note that the fault initiation is mainly indicated in bands 4 through 7, whereas fault clearing is better indicated in band 3. Fault initiation takes place instantaneously. The steepness of the voltage step in the waveform is mainly determined by the cutoff frequency of the anti-aliasing filter. Such a step is best indicated in the higher frequency bands. The voltage recovery after fault clearing is however slowed down by the presence of capacitance in the system. Especially capacitor banks lead to a slower voltage recovery. This means that the second transition is not visible in the highest frequency bands and limits the time resolution that can be obtained.

**Example 7.14** A Developing Fault. This example uses the same original measurement as in Example 7.3 (see Fig. 7.16*a*) except there is a slight shift in data sample indexes. From Figures 7.28*a* and *b*, one observes a sudden change at the time instant $t_1 \approx 520$ sample position corresponding to the initial single-phase fault causing a dip in phases *a* and *b* (phase-to-phase voltages were measured). This single-phase fault appears to be associated with a high level of higher frequency

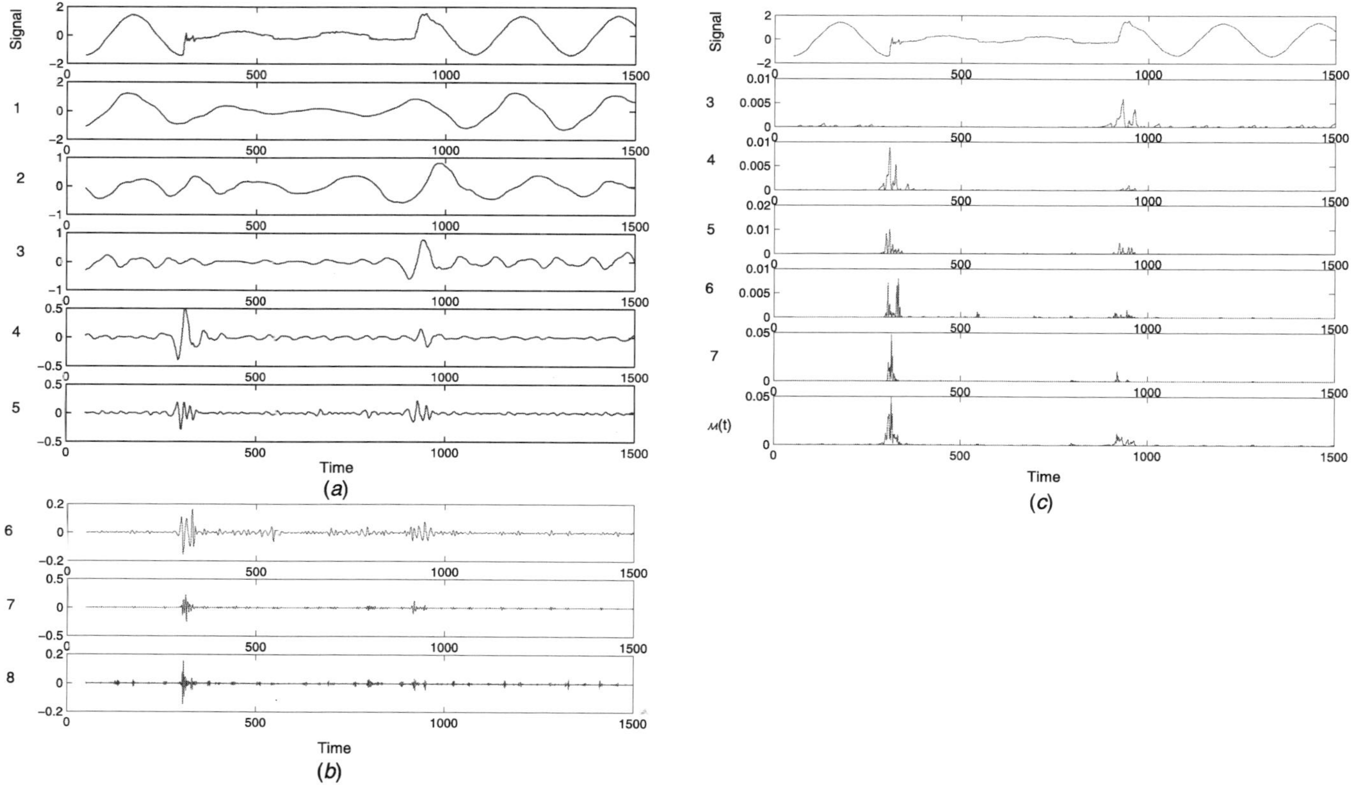


**Figure 7.27** Using signal components from wavelet decomposition to detect transition segments from voltage dip caused by a fault. All horizontal axes show time in samples [where each cycle contains 256 samples or 16.67 ms (60 Hz power system frequency). (*a*) Top to bottom: waveforms from original measurement; signal components from subbands 1 to 5. (*b*) Signal components from subbands 6 to 8. (*c*) Top to bottom: waveforms from original measurement; measure of changes $\mathcal{M}_{b_j}$ corresponding to subbands $b_j = 3, \ldots, 7$; measure of changes $\mathcal{M}$ by summing up $\mathcal{M}_{b_j}$ in these five bands.

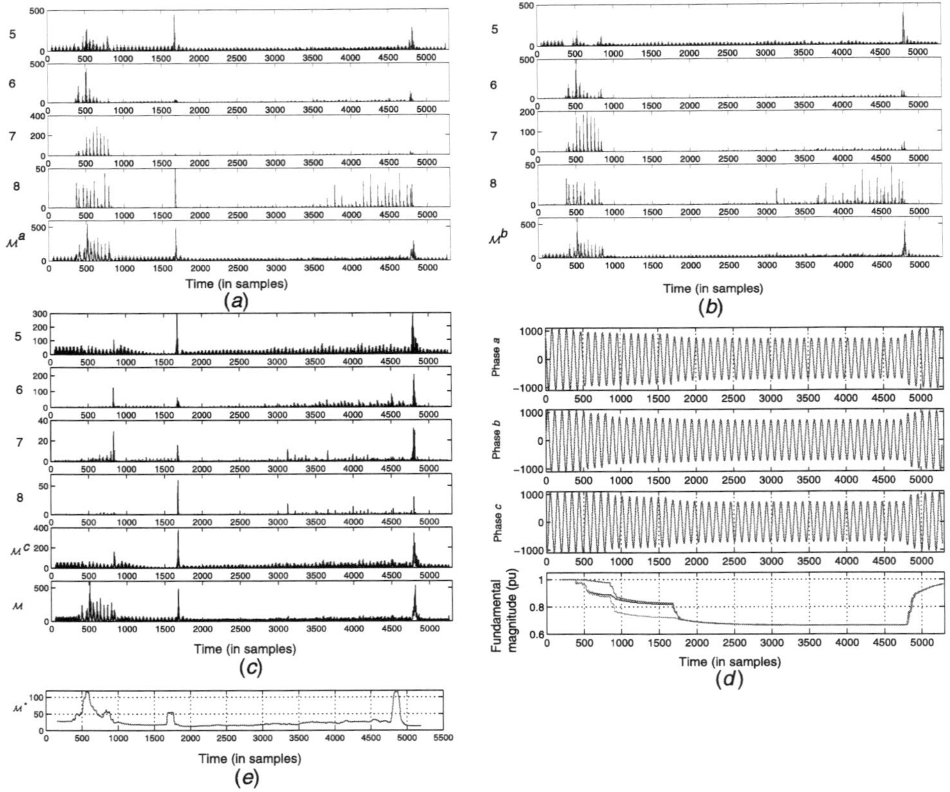


**Figure 7.28** Using signal components from wavelet decomposition for detecting transition segments from developing fault. (*a–c*) Resulting measures of changes $\mathcal{M}^i_{b_j}$ correspond to subbands $b_j = 5, \ldots, 8$ and $\mathcal{M}^i$ for individual phases, $i = a,b,c$, respectively. The bottom figure in (*c*) is measure of overall changes $\mathcal{M}$ for three-phase data. (*d*) Original three-phase measurement data and estimated fundamental-voltage magnitude sequences. (*e*) $\mathcal{M}^*(t)$ obtained by applying one-cycle averaging to $\mathcal{M}(t)$. Each cycle contains 96 samples or 20 ms.

distortion leading to a high value for the measure of changes over a longer period of time. These high-frequency components may be due to transients associated with fault initiation or due to an unstable arc. The observation that these high-frequency components reappear toward the end of the dip points to the latter explanation. Noting the relatively low sampling frequency of 96 samples per cycle, the highest frequency band is located at 1200 to 2400 Hz. The overall measure of change does however have a sharp maximum at the time location closely corresponding to the fault initiation.

Note the almost continuous background level in the fifth band. The frequency range in the fifth band is from 150 to 300 Hz, thus including the fifth harmonic. The continuous background level is therefore most likely due to this fifth harmonic. As the fifth harmonic is often the dominating harmonic in the voltage, the use of any frequency bands containing this frequency should be avoided.

Figure 7.28*c* clearly indicates another transition at the $t_2 \approx 845$ sample position (although changes in Fig. 7.28*b* around $t_2$ are less obvious) when the single-phase fault develops into a two-phase-to-ground fault. This transition is very sharp, the high-frequency contents of the voltage has disappeared by then, and time allocation of the underlying system change can take place with high accuracy.

Further, Figures 7.28*a* and *c* clearly indicate a third transition at the $t_3 \approx 1675$ sample position. Finally, all three phases in Figures 7.28*a* to *c* show a transition at the $t_4 \approx 4815$ sample position indicating a voltage recovery. This is consistent with the voltage magnitude sequences shown in Figure 7.28*d*. However, due to the spikelike $\mathcal{M}$ in Figure 7.28*c*, a simple threshold-based method as used in the Kalman filter case cannot be directly applied. Such spikes can be smoothed out by applying a one-cycle averaging window to $\mathcal{M}$, resulting in an improved measure of changes $\mathcal{M}^*$ shown in Figure 7.28*e*. All four transition segments are then associated with clear and distinguishable peaks. Obviously, a simple threshold-based method can then be successfully applied for the segmentation. However, this is at the expense of the high time resolution that would have been obtained otherwise.

**Example 7.15** Voltage Dip Due to Transformer Energizing Followed by Protection Maltrip. This example uses the same original measured signals as in Example 7.4 (see Fig. 7.17*a*). One can clearly observe from subbands 5, 6, and 7 in Figures 7.29*a–b* the increase in amplitudes of these signal components, indicating an increase in disturbances. This is more obvious in the measure of changes indicated in Figure 7.29*c*. The high amplitudes, especially in bands 5 and 6, are due to the harmonic components associated with transformer energizing. Note again the rather low sampling frequency: bands 5 and 6 correspond to frequency ranges 150–300 and 300–600 Hz, respectively, thus containing harmonic orders 3 through 12. Note the decrease in amplitude in band 6 and the increase in amplitude in band 5. This corresponds to the well-described phenomenon that the frequency contents of the transformer energizing shifts from high-frequency harmonics to lower frequency harmonics [12] (also see Section 6.2.2.8).

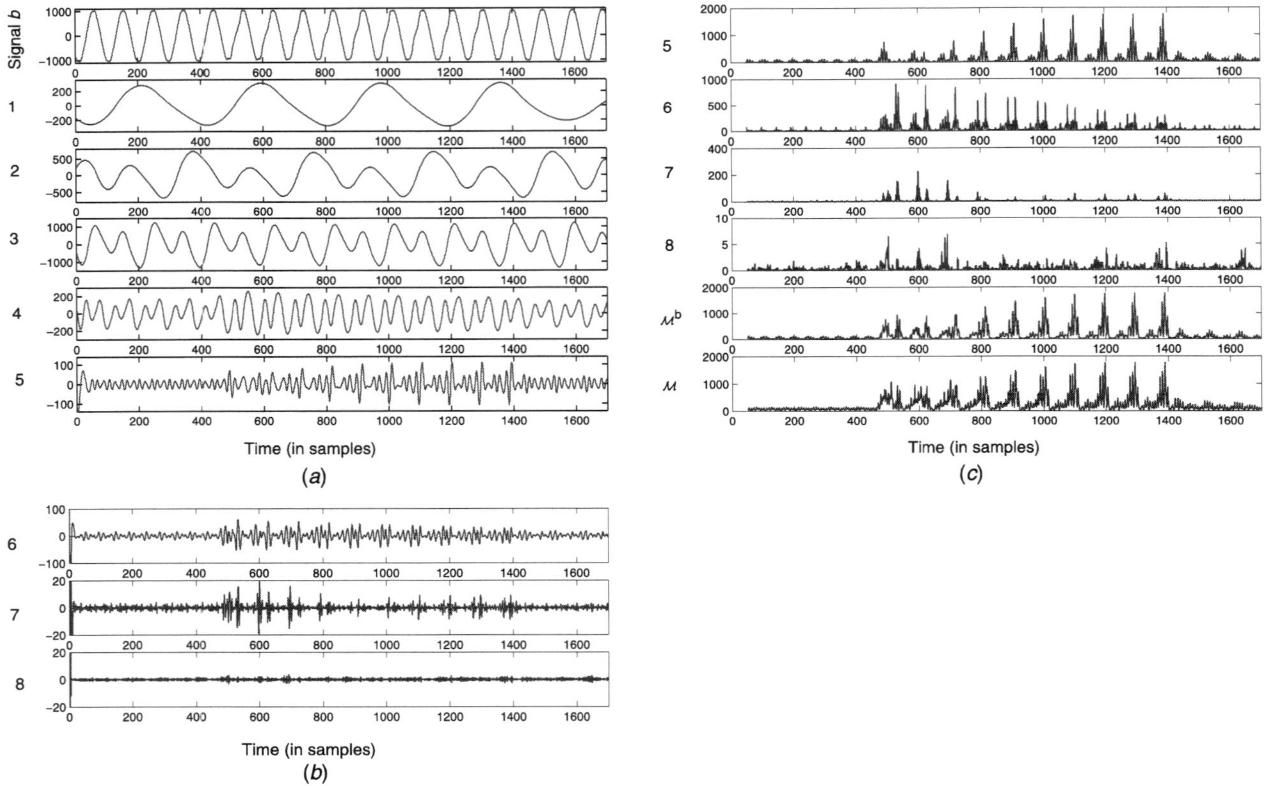


**Figure 7.29** Using signal components from wavelet decomposition for detecting transition segments from voltage dip due to transformer energizing followed by protection maltrip. All horizontal axes show time in samples (where each cycle contains 96 samples, or 20 ms). (*a*) Top to bottom: waveforms from original measurement phase $b$, signal components in phase $b$ from subbands 1 to 5. (*b*) Signal components in phase $b$ from subbands 6 to 8. (*c*) Top to bottom: measure of changes $\mathcal{M}^b b_j$ corresponding to subbands $b_j = 5, \ldots, 8$ in phase $b$, measure of changes $\mathcal{M}^b$ in phase $b$ by summing up these four bands, and measure of overall changes for three phases $\mathcal{M}$.

Observing $\mathcal{M}(t)$ from the bottom figure in Figure 7.29*c* could lead to the impression that there is only one transition segment: After all, the measure of changes remains high for a long time. This would lead to an incorrect segmentation. The high harmonic content of the signal together with the rather low sampling rate means that the signal in most of the frequency bands is dominated by harmonics. The harmonic distortion only changes slowly from cycle to cycle, so that model-based methods are more successful at suppressing it. The harmonic content in voltage signals is normally relatively low, so that subband-based segmentation methods will work well in most cases. However, for segmentation of current signals more problems are to be expected. A subband-based method may however still be used for a detailed study of the transition segments detected by another method.

Note that in this example of a voltage dip caused by transformer energizing a more sophisticated *measure of overall changes* needs to be defined to first remove the spikelike repetition before a simple threshold-based method can be successfully applied for the segmentation.

**Example 7.16** Voltage Dip Due to a Fault. This example uses the same original measurement as in the Example 7.5 (see Fig. 7.18*a*, phase *b*). One can clearly observe from subbands 4 to 8 in Figures 7.30*a*,*b* the sudden increase in amplitudes at around the $t_1 \approx 460$ sample location corresponding to the fault initiation and at the $t_2 \approx 2380$ sample location corresponding to the fault clearing. These two peaks are further enhanced by the measure of changes in the bottom figure of Figure 7.30*c*. It is obvious that a simple threshold-based method can be served for segmentation purpose in this case.

## 7.5 SUMMARY AND CONCLUSIONS

Existing triggering methods for voltage dips, swells, and interruptions are based on the one-cycle rms voltage. The event is detected when the rms voltage exceeds a threshold: typically 90% of nominal voltage for dips, 10% for interruptions, and 110% for swells. These methods are well defined in an IEC standard. For the detection of transients different methods are in use, including cycle to cycle variations and high-pass filters, but no standard method exists.

Both triggering and transition segments correspond to a transition between system states. Methods for detecting triggering points and for detecting transition segments are therefore strongly linked with one another. Segmentation should be seen as a more general form of triggering. Whereas triggering results in individual points (the triggering points), segmentation results in intervals (the transition segments). The triggering point is often embedded in such a transition segment. One should bear in mind that the accuracy of a triggering point is constrained by the time resolution of the selected detection method. The accuracy of segment boundaries is often of secondary importance.

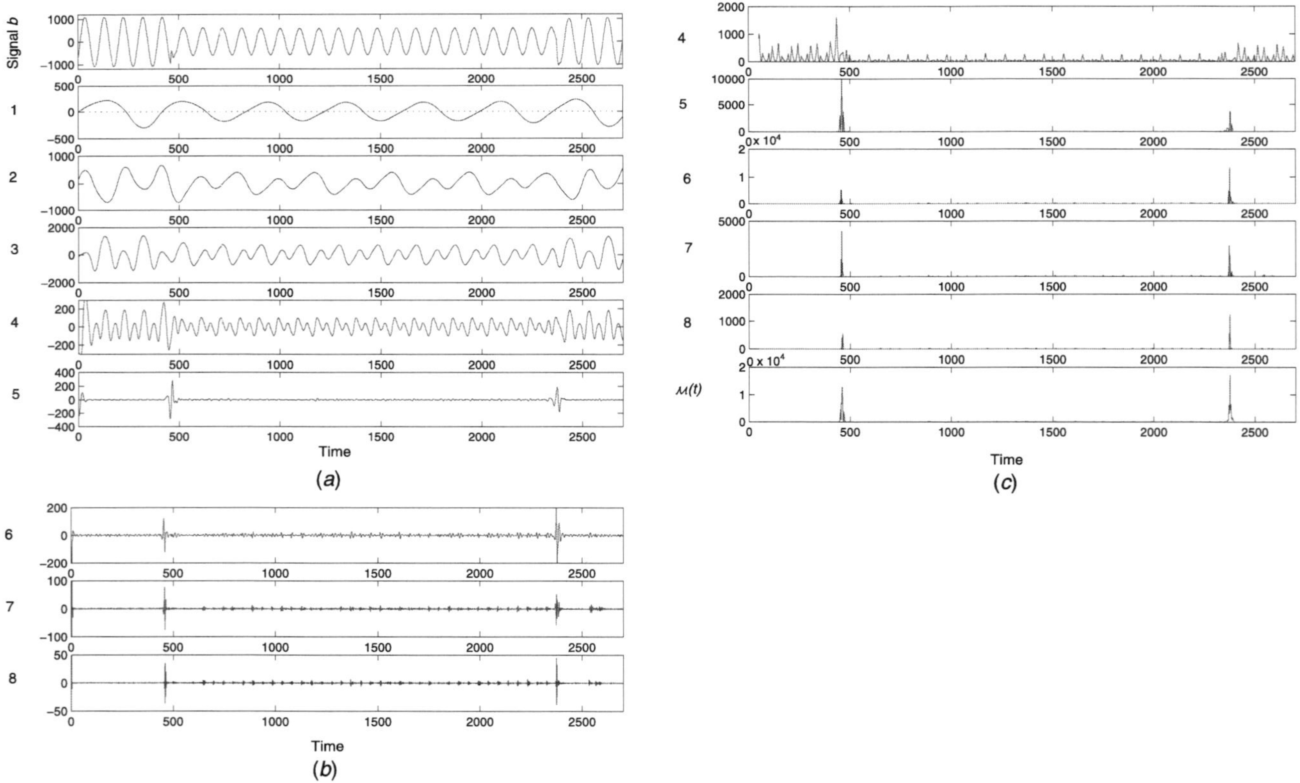


**Figure 7.30** Using signal components from wavelet decomposition for detecting transition segments from voltage dip due to a fault. All horizontal axes show time in samples (where each cycle contains 96 samples, or 20 ms). (*a*) Top to bottom: waveforms from original measurement phase $b$, signal components in phase $b$ from subbands 1 to 5. (*b*) Signal components in phase $b$ from subbands 6 to 8. (*c*) Top to bottom: measure of changes $\mathcal{M}_{b_j}^b$ corresponding to subbands $b_j = 4, \ldots, 8$ in phase $b$ measure of changes $\mathcal{M}$ in phase $b$ by summing up these five bands.

A practical difference between triggering and segmentation is that triggering is always done in real time. The triggering method is used by the power quality monitor to start the storage and/or further process the event. Practical triggering methods will thus be rather simple. Segmentation is usually performed as off-line processing. The requirements of simplicity and computational speed are therefore much less.

Triggering and segmentation methods require in almost all cases the setting of one or more thresholds. The setting of these thresholds is ruled by the classical trade-off between false-alarm rate and detection rate. With segmentation the length of the transition segments becomes a third parameter in the trade-off. Methods exist for optimally choosing a threshold (some of these will be discussed in Chapter 9), but it requires a fundamental discussion on the requirements placed on triggering and segmentation. Currently, power quality events are mainly defined through their triggering points. In such a case, the accuracy of a triggering method is not essential. A discussion of accuracy becomes useful when the triggering and segmentation are viewed as methods for detecting the underlying events in a power system or for predicting the behavior of equipment connected to the system. This discussion has not yet started in the power quality community. The current availability of many advanced signal-processing tools for triggering and segmentation may stimulate such a discussion. Development of data-processing hardware makes it more likely that even these advanced methods can be implemented for practical use.

The triggering and segmentation methods described in this chapter can roughly be classified into two categories. The *first category* is based on the prominent residuals of a selected model. This is done by first selecting a suitable signal model (e.g., sinusoidal models, AR models) followed by examining the prominent residuals from the model. Examples are given for detecting the triggering points as well as for the segmentation. Among these methods Kalman filters using sinusoidal models are found to be rather successful for segmentation. Residuals from AR models are found useful; however, sometimes a more sophisticated measure may be required for detecting the significant changes, especially when repetitive spikes or notches are present in the signal.

The *second category* is based either on using signal magnitude component sequences at the power system fundamental frequency (or rms sequences) that can be regarded as a linear or nonlinear low-pass filtered signal or on using signal component sequences from several subbands. This category does not require an initial step of signal modeling. The simple method based on using component magnitude sequences is found rather robust to a range of disturbance types, although it suffers from the limitation of low time resolution. Replacing an rms sequence with a fundament-component magnitude sequence can improve the poor time resolution in an rms sequence. Methods using multiscale subband components from wavelets offer higher time resolution. Successful segmentation is shown to be achievable; however, one often requires to define a more sophisticated measure for significant changes, for example, to apply a preprocessing that smooths out the repetitive spikes and notches.

Among the range of methods described in this chapter, the Kalman filter–based method (using sinusoidal models) is found to be most robust in the first category. The rms-based and fundamental-magnitude-based methods are the simplest and most effective in the second category. Also, the other methods described in this chapter show some advantages and hence offer complementary solutions to the segmentation problem. Several of the methods discussed have shown a good potential for practical implementation, but more systematic studies are required. This holds for the segmentation of dip, swell, and interruption recordings and for the triggering and segmentation of voltage and current transients.

# CHAPTER 8

# CHARACTERIZATION OF POWER QUALITY EVENTS

This chapter presents methods for extracting information from individual power quality events, captured by any of the triggering or segmentation methods discussed in Chapter 7. Extracting information from individual events can be a first step in the quantification of the supply by means of site or system indices. The general procedure for calculating characteristics and indices for power quality events is summarized in Figure 8.1. The first step after obtaining the sampled waveforms is to calculate characteristics as a function of time. From the characteristics versus time, single-event indices (or *single-event characteristics*) are obtained. The aim of the event characteristics is to quantify the severity of the event. From the single-event indices obtained at one site over a certain period (typically one year), site indices are calculated. Finally, the system indices are calculated from the site indices of all sites within a system. This chapter will discuss the calculation of characteristics versus time and of single-event indices. The calculation of site and system indices is discussed in Chapter 10.

The first part of this chapter will discuss characteristics versus time for voltage dips, swells, and interruptions. Standard methods as well as new proposals will be discussed, including an appropriate method for voltage dips in three-phase systems. Next, single-event indices will be introduced for voltage dips, swells, and interruptions. The last part of the chapter discusses characteristics versus time and single-event indices for voltage and current transients. Also for transients a method is proposed that includes the three-phase character of the system. For dips and interruptions as well as for transients some methods will be discussed that enable the extraction of additional information from the recordings.

*Signal Processing of Power Quality Disturbances.* By Math H. J. Bollen and Irene Yu-Hua Gu

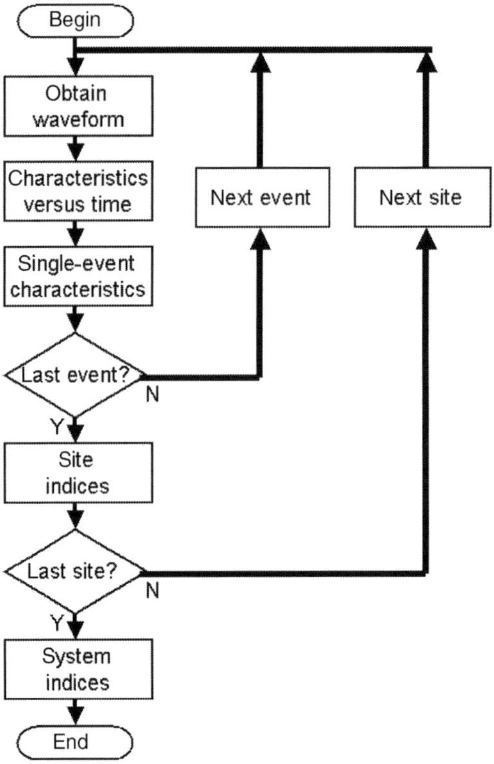


**Figure 8.1** General procedure for calculating event characteristics and indices.

## 8.1 VOLTAGE MAGNITUDE VERSUS TIME

Interruptions, voltage dips, and swells are all characterized from the voltage magnitude as a function of time. Because one typically uses the rms voltage to estimate the voltage magnitude, these events are also referred to as *rms variations*. We will not use this term in this book to prevent confusion with the term *variations* as defined in Section 1.2.3. In [33] these events are referred to as *voltage magnitude events*. In this section we will mainly discuss the standard method for characterizing voltage magnitude: the rms voltage. A number of alternatives (including the peak voltage) will be more briefly discussed as well. A method for characterizing voltage magnitude based on the three-phase character of the power system will be discussed in Section 8.3.

### 8.1.1 rms Voltage

The IEC power quality measurement standard 61000-4-30 [158] prescribes a very precise method for obtaining the voltage magnitude as a function of time. This method is also recommended in the latest draft of IEEE standard 1564 [172]. The first step in this procedure is to obtain the rms voltage over a window with a length exactly equal to one cycle of the power system frequency.

The calculation of this *one-cycle rms voltage* is repeated every half cycle; in other words: the window is shifted one half cycle in time. What results is a discrete function with a time step equal to one half-cycle of the power system frequency. This function is referred to as the *basic measurement* in IEC 61000-4-30 and as the *characteristic versus time* in draft IEEE 1564 [172].

The one-cycle sliding-window rms value has been calculated for a number of synthetic voltage dips with residual voltage 50% and phase-angle jumps from $-60^\circ$ through $+60^\circ$ in $30^\circ$ steps. The results for the start of the event are shown in Figure 8.2. The rms voltage takes one cycle for the transition from the preevent value to the during-event value. It is also shown that some oscillations occur during this transition. For dips around voltage zero crossing these oscillations may result in rms values outside the actual range.

The calculations have been repeated in Figure 8.3 for a residual voltage of 90% and phase-angle jumps between $-20^\circ$ and $+20^\circ$ in $10^\circ$ steps. The relative error is bigger for a smaller drop in voltage. This is partly due to the difference in vertical scale, but also due to the relatively larger phase-angle jump.

As we will see later, the severity of a voltage dip is characterized among others from the lowest of the rms voltages calculated during the event. In the same way, the highest rms voltage is used as a characteristic for a voltage swell. The rms voltage may obtain values outside the actual amplitude range during the transition periods at the beginning or end of the dip. This may lead to an error in estimation of the event characteristics. To quantify this error, the highest and lowest values of the rms voltage have been calculated for a number of synthetic voltage-dip events.

The results for a voltage dip with residual voltage 90% and a one-cycle window to calculate the rms voltage are shown in Figure 8.4. The range of phase-angle jumps is somewhat larger than can be expected in practical systems, but still the error can

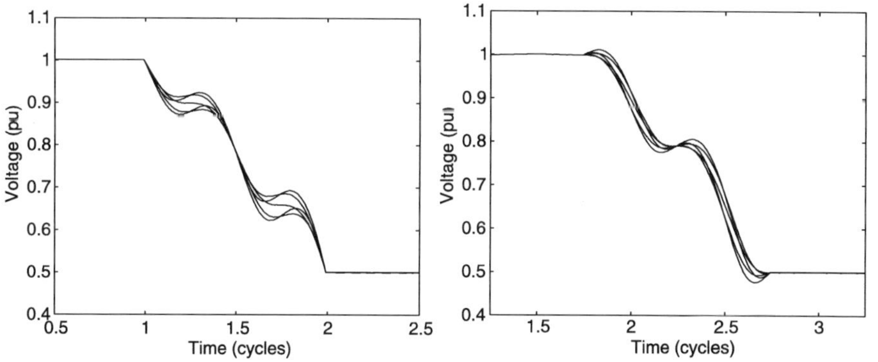


**Figure 8.2** One-cycle rms voltage for dips with residual voltage 50% and different phase-angle jumps for dips at maximum voltage (left) and at voltage zero crossing (right).

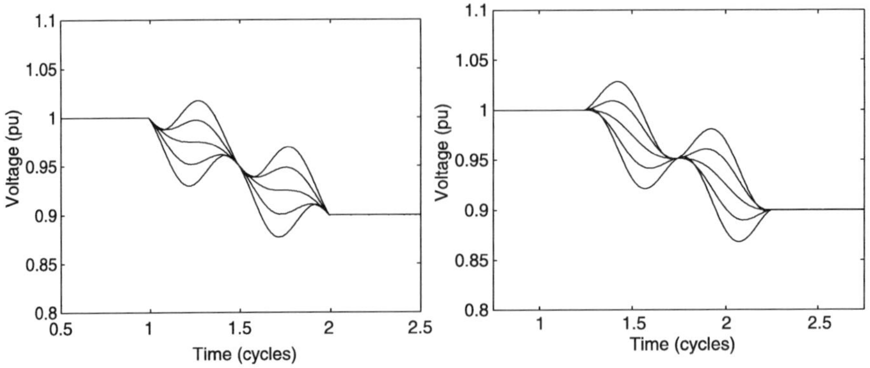


**Figure 8.3** One-cycle rms voltage for dips with residual voltage 90% and different phase-angle jumps for dips at maximum voltage (left) and at voltage zero crossing (right).

be relatively high. It is especially important to realize that with most measurement campaigns events with a residual voltage above 90% are not considered in the statistics. The events studied for Figure 8.4 would be at the borderline of detection when their magnitude would be correctly estimated. We see however that these events can incorrectly end up in event statistics. This is not a random error as there are no cases in which the residual voltage will be estimated as too high. The estimation error will always result in an increase in the apparent number of dips. The estimated dip duration is only one half-cycle so that this deviation is especially of concern when the events are aggregated over a wide range of dip durations.

One can also conclude from Figure 8.4 that some events lead to an apparent voltage swell. The apparent residual voltage will be rather close to 1 pu so that this will probably not affect the statistics on voltage swells: Events with a residual voltage less than 110% are typically not counted.

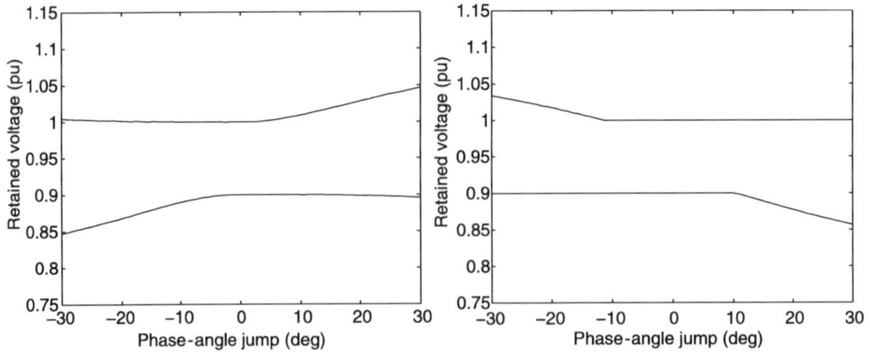


**Figure 8.4** Influence of phase-angle jump on apparent residual voltage for synthetic event with actual residual voltage of 90% when using one-cycle window to calculate rms voltage: at zero crossing (left) and at voltage maximum (right).

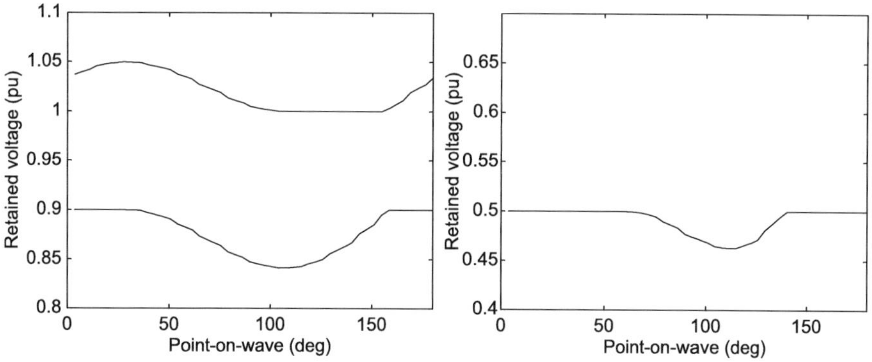


**Figure 8.5** Apparent residual voltage as function of point on wave for event with actual residual voltage of 90% and a phase angle jump of $-30^\circ$ (left); 50% and $-60^\circ$ (right).

The lowest and highest rms voltages at the start of the event are plotted as a function of the point on wave in Figure 8.5. The rms voltage has been calculated by using a one-cycle window. The synthetic event had a residual voltage of 90% and a phase-angle jump of $-30^\circ$. For comparison the lowest voltage has also been calculated for a synthetic event with a residual voltage of 50% and a phase-angle jump of $-60^\circ$. The results are plotted with the same vertical scale, showing that the effect becomes less severe for lower during-dip voltage, despite the phase-angle jump being twice as big.

The calculations resulting in Figure 8.5 have been repeated for a range of phase-angle jumps. The lowest and highest rms voltages have been plotted as a function of the phase-angle jump in Figure 8.6 for an actual residual voltage of 90 and 50%. For the latter case, no overshoot occurs. Note that the range in phase-angle jump is larger for the latter case.

The algorithm for calculating the rms voltage has been applied to the voltage dip shown in Figure 8.7. The one-cycle rms voltage has been calculated every sample as

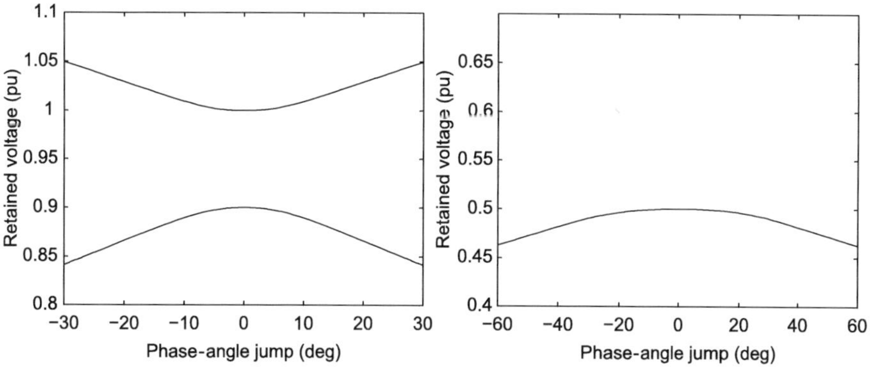


**Figure 8.6** Apparent residual voltage for worst-case point on wave as function of phase-angle jump for synthetic events with residual voltage equal to 90% (left) and 50% (right).

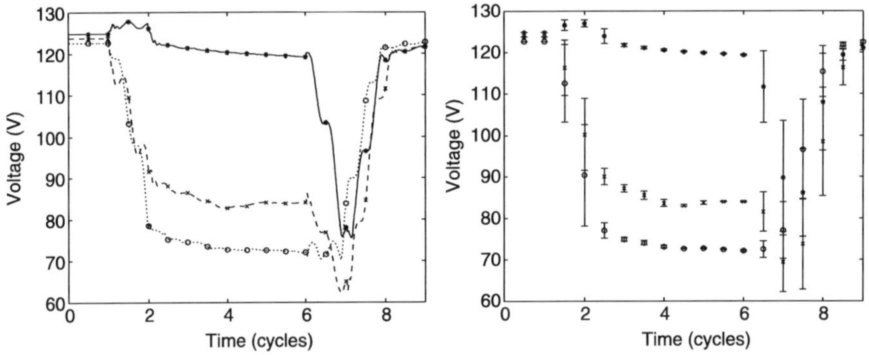


**Figure 8.7** One-cycle rms voltage for voltage dip due to developing fault: complete function with half-cycle updates (left); error margin in half-cycle updates (right).

well as every half-cycle. The results are shown in the left-hand plot in Figure 8.7 for the three phases. The smooth curves show the values updated every sample (in this case 120 times per cycle), whereas the markers indicate the resulting function when an update is only made once every half-cycle (as prescribed in IEC 61000-4-30). We see in the figure that the one-cycle rms voltage may vary significantly during one half-cycle. The result is that the value depends on where in the half-cycle interval the update is done. To show the resulting uncertainty the highest and lowest rms values over each half-cycle interval have been calculated and plotted as error bars in the right-hand part of Figure 8.7. We see that especially in the second stage of the event, where the fault develops into a three-phase fault and is cleared shortly after, the uncertainty in the results is very big. Different monitors may result in significantly different values for the estimated characteristics of such an event.

Another example of the application of the one-cycle rms voltage is shown in Figure 8.8. The waveforms for this dip are shown in Figure 6.24. Again we see large uncertainties during the transition periods, but because the second stage of

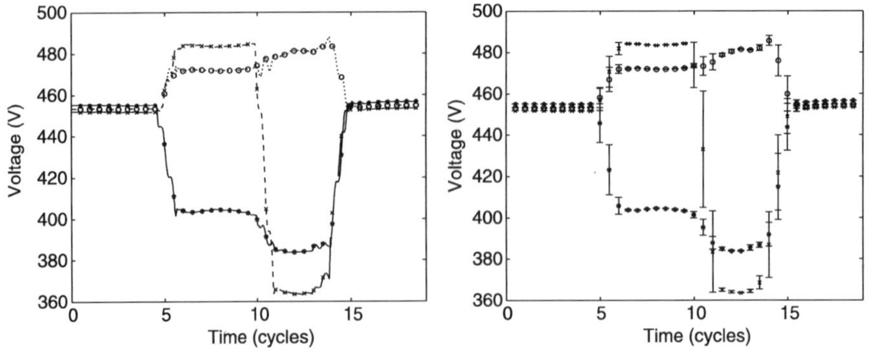


**Figure 8.8** One-cycle rms voltage for dip shown in Figure 6.24: complete function with half-cycle updates (left); error margin in half-cycle updates (right).

the dip lasts longer, the uncertainty does not affect the lowest and the highest values of the rms voltage. As we will see below, these are the values used to characterize the dip.

### 8.1.2 Half-Cycle rms

In some cases it may be more appropriate to use a half-cycle window to calculate the rms voltage. The main advantage of using a half-cycle window is a faster transition from the prefault voltage to the during-fault voltage and from the during-fault voltage to the postfault voltage. If the actual magnitude changes instantaneously, as is typically the case with fault initiation, it will take one cycle for the one-cycle rms to follow. By using the half-cycle rms this transition period can be brought back to one half-cycle. This is especially of interest when studying short dips or dips with different stages shortly after each other. The two different window lengths have been applied to a measured voltage dip at 10 kV caused by a three-phase fault at a higher voltage level. The results are shown in Figure 8.9, where the dashed line indicates the one-cycle rms voltage. The half-cycle rms voltage shows a faster transition, thus showing that the voltage recovery actually takes place in two stages. But the half-cycle rms voltage also shows a more unstable character during and after the event. This is due to its sensitivity to even-harmonic distortion.

The sensitivity to even-harmonic distortion is especially noticeable for voltage dips due to transformer energizing and for dips associated with postfault transformer saturation. An example of such a voltage dip is shown in Figure 8.10. Whereas the one-cycle rms voltage shows a rather smooth recovery (dashed line in the figure), the half-cycle rms voltage oscillates heavily around an average value due to the

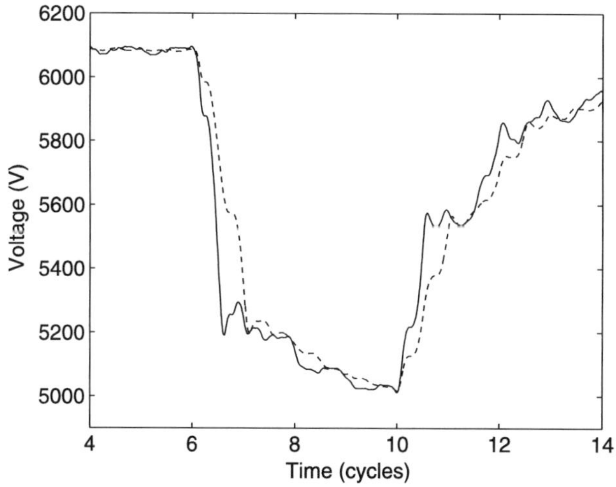


**Figure 8.9** Comparison of one-cycle rms voltage (dashed line) and half-cycle rms voltage (solid line).

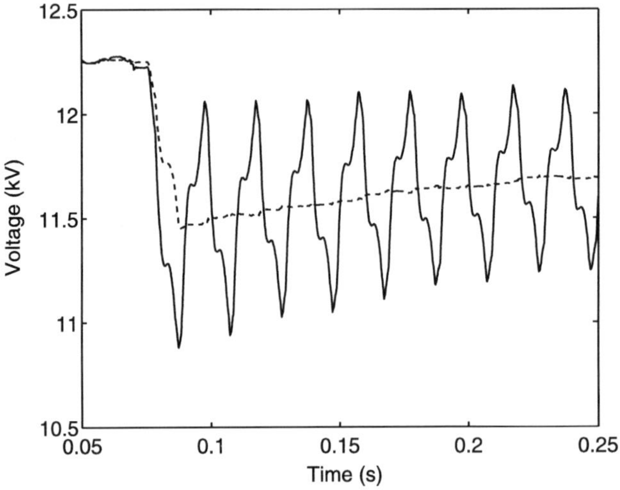


**Figure 8.10** Comparison of one-cycle rms voltage (dashed line) and half-cycle rms voltage (solid line).

even-harmonic distortion. This is obviously a disadvantage for characterizing purposes, but it may be seen as an advantage when trying to extract additional information from the voltage recording. Using the half-cycle rms provides an easy method to detect transformer energizing and postfault transformer saturation.

Instead of recording the one-cycle rms voltage updated every half-cycle, a monitor may record the half-cycle rms voltage updated every half-cycle. The voltage-dip duration and the residual voltage can be calculated later by averaging two consecutive half-cycle rms values into a one-cycle rms value (where the rms of the two one-cycle rms values should be taken). However, this does not guarantee the detection of transformer saturation. It depends on the instants at which the half-cycle values are updated. If this occurs when the half-cycle rms is equal to the one-cycle rms (which happens about once every half-cycle, as shown in Fig. 8.10), there will be no oscillation in the half-cycle rms values. Updating the half-cycle rms value every quarter-cycle would be more reliable in detecting transformer saturation.

Another disadvantage of using the half-cycle rms voltage is that the oscillations at the start and end of the event are larger. To show this, the calculations resulting in Figure 8.6 have been repeated in Figure 8.11 for the rms voltage being calculated over a half-cycle window.

### 8.1.3 Alternative Magnitude Definitions

We saw in Section 3.2.2 that there are different ways of quantifying the magnitude of the voltage, the principal ones being rms voltage, fundamental component of the voltage, and peak voltage. We saw that the rms voltage and fundamental component

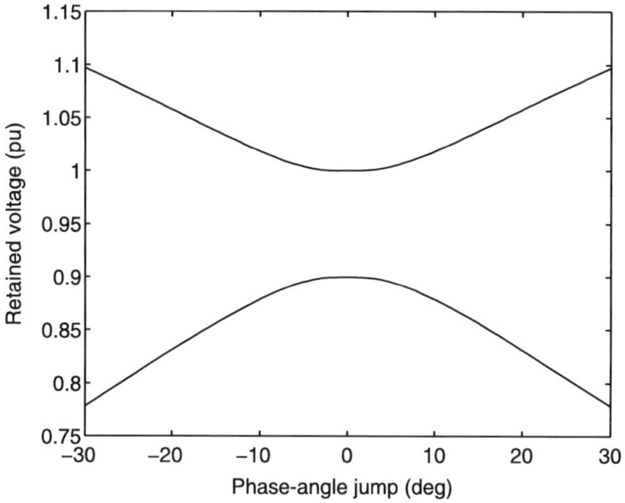


**Figure 8.11** Apparent residual voltage calculated using a half-cycle rms voltage for worst-case point on wave as a function of phase-angle jump for synthetic events with residual voltage equal to 90%.

of the voltage give very similar results. For voltage dips the harmonic voltage distortion is sometimes significantly higher than for normal operation. This would make the difference between rms and fundamental bigger, but one does, even during voltage dips, not expect situations where the harmonic distortion is of such a level that rms and fundamental differ significantly. The rms voltage $U_{\mathrm{rms}}$ and the fundamental-voltage component $U_{\mathrm{fund}}$ are related by the THD according to the following expression:

$$U_{\mathrm{rms}} = U_{\mathrm{fund}}\sqrt{1 + \mathrm{THD}^2} \tag{8.1}$$

The class A accuracy requirement of IEC 61000-4-30 implies that the fundamental-voltage component can be used instead of the rms voltage as long as the THD is less than 6.3%. For class B instruments the limit is a THD level of 20%. The 6% limit may be exceeded for some dips, but it is very unlikely that the 20% limit will be exceeded. We will discuss the harmonic distortion during dips in more detail in Section 8.4.

The difference between fundamental and rms voltage is merely of academic interest. After all, the main aim of voltage-dip characterization is to obtain a measure for the impact of the dip on sensitive equipment. The difference between the rms and the fundamental voltage is too small to be able to study the difference in impact on equipment. A large fraction of sensitive equipment consists of a diode rectifier and a smoothing capacitor on the dc side of the rectifier. The voltage-dip tolerance of such equipment is determined mainly by the voltage over this capacitor,

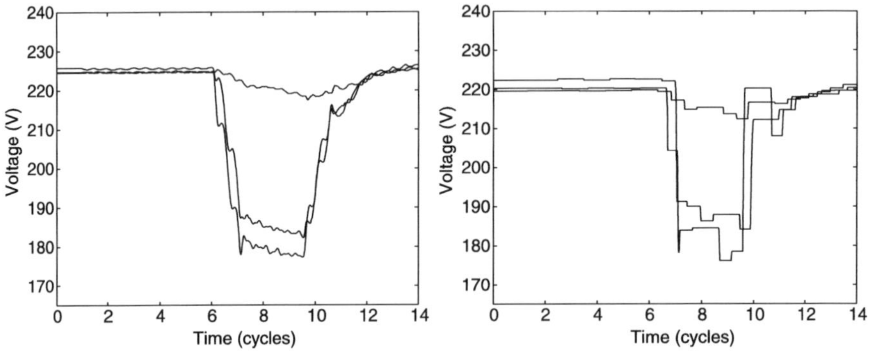


**Figure 8.12** Voltage magnitude versus time for typical voltage dip: one-cycle rms voltage (left) and one-cycle peak voltage (right).

which in turn is determined mainly by the peak value of the voltage on the ac side of the rectifier. Using the peak voltage instead of the rms voltage would thus be justifiable from an equipment performance viewpoint. This becomes an even more important issue because the peak voltage may differ from the rms voltage by more than a few percent.

A comparison between rms voltage and peak voltage is shown in Figure 8.12 for a typical voltage dip measured at low voltage. Note that this is the same event as in Figure 6.14. The peak voltage is calculated as the maximum absolute value of the voltage over the preceding cycle. The result is divided by the square root of 2 to obtain a value that can be directly compared to the rms voltage. The sudden steps in voltage magnitude are typical for the peak voltage calculated in this way. Their origin becomes obvious when one considers the way in which the window shifts through the waveform. This makes the peak voltage very sensitive to short-duration transients. A kind of low-pass filter would have to be applied to the rectified waveform to obtain a suitable characteristic.

The lowest voltage magnitudes have been calculated for the dip in Figure 8.12. The results are summarized in Table 8.1.

The difference between the two methods of calculating the residual voltage is of the order of 1%. This implies an uncertainty of about 1% in predicting the behavior of sensitive equipment. One may conclude from this that an accuracy in determining the residual voltage of less than 1% is not needed when monitoring is performed

**TABLE 8.1 Residual Voltage Based on rms and Peak Voltage for Typical Voltage Dip**

| Phase | rms Voltage | Peak Voltage |
|---|---|---|
| A | 94.6% | 92.4% |
| B | 77.1% | 76.6% |
| C | 79.3% | 80.1% |

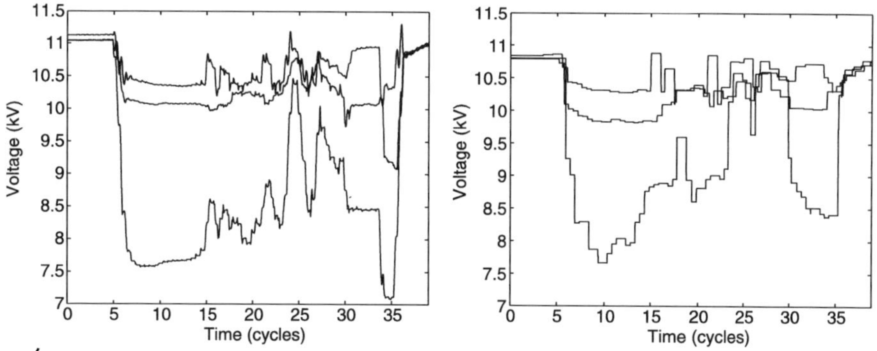


**Figure 8.13** Voltage magnitude versus time for dip with high distortion: one-cycle rms voltage (left) and one-cycle peak voltage (right).

with reference to equipment performance. More accurate requirements may be needed for contractual applications. It is interesting to note that the difference is largest in the phase that is least affected by the fault (fault A). No immediate explanation for this is available.

In some rare cases the difference between rms voltage and peak voltage can become significant. Such an example is shown in Figure 8.13. This event was associated with high waveform distortion (see Fig. 8.37 later and the surrounding text for more information on this event). During the first part of the event, when the waveform distortion was low, rms and peak voltages were about the same. However, at a certain moment the voltage distortion increased significantly, leading to an increase in peak voltage. The result is that the rms-voltage-based magnitude becomes significantly lower than the peak-voltage-based magnitude. The difference is especially striking during the last few cycles of the event, when the rms-based magnitude drops whereas the peak-based voltage remains the same.

The authors are not aware of any publication in which the peak voltage is used to characterize a voltage dip. Some further development work is needed before this becomes possible. One of the issues to be solved is the influence of short-duration transients on the resulting single-event indices.

## 8.2 PHASE ANGLE VERSUS TIME

The origin of the phase-angle jump was discussed in Section 6.2.4. As we saw the phase angle of the voltage contains additional information on the cause of the event. A further reason for using the phase-angle jump as a characteristic is that it affects certain types of equipment. The literature remains somewhat divided on this issue, but there are sufficient indications of such a relation to justify using the phase-angle jump as an additional single-event index.

As with any single-event index the first step is to define the characteristic versus time from which the index is determined. For phase-angle jumps we start from the argument of the complex voltage versus time. Consider the situation that both the amplitude and the phase angle of the sine wave are a function of time:

$$v(t) = \sqrt{2}V(t)\cos[\omega t + \phi(t)] \tag{8.2}$$

Extracting the fundamental component (e.g., by means of a DFT algorithm) over the period $(t - T, t)$ with $T = 2\pi/\omega$ one cycle of the fundamental frequency gives the following complex value:

$$V_{\text{fund}} = V(t)e^{j(\omega t + \phi(t))} \tag{8.3}$$

where it has been assumed that both magnitude and phase angle change slowly, so that we can assume the signal to be quasi-stationary within the one-cycle window used to determine the fundamental voltage. When the window is shifting through the signal, the initial phase of the signal, at the start of the window, changes continuously. Plotting the argument (phase angle) of the complex fundamental voltage would result in a linear increase with time:

$$\arg[V_{\text{fund}}(t)] = \omega t + \phi(t) \tag{8.4}$$

The second term $\phi(t)$ after subtraction of its preevent value is referred to as the phase-angle jump as a function of time:

$$\Delta\phi = \phi(t) - \phi(0) \tag{8.5}$$

The above reasoning gives a method for extracting the phase-angle jump from a measured voltage waveform. Combining (8.4) and (8.5) and using that $\phi(0) = \arg[V_{\text{fund}}(0)]$ gives the following expression:

$$\Delta\phi = \arg[V_{\text{fund}}(t)] - \arg[V_{\text{fund}}(0)] - \omega t \tag{8.6}$$

This expression, although mathematically correct, is very difficult to implement. The problem is that the argument, being an angle, has a limited range. The last term in (8.6) increases linearly with time. It is possible to correct for this by shifting the result to a value between $-180^\circ$ and $+180^\circ$, but rewriting the expression results in a more practical expression. For this we use that $\omega t = \arg(e^{j\omega t})$ and that $\arg(A) - \arg(B) = \arg(A/B)$. Inserting this gives the following expression

$$\Delta\phi = \arg\left(\frac{V_{\text{fund}}(t)}{V_{\text{fund}}(0)} \times e^{-j\omega t}\right) \tag{8.7}$$

The result of this expression is always a value between $-180^\circ$ and $+180^\circ$. Expression (8.7) also illuminates the main problem in extracting phase-angle information: Accurate knowledge of the power-system frequency is needed. An error in frequency will result in an apparent shift in phase angle. An absolute error $\Delta f$ in frequency results in the following phase angle as a function of time.

$$\Delta\phi = \phi(t) - \phi(0) + 2\pi\Delta ft \tag{8.8}$$

The error in phase angle increases linearly with time. Especially for longer events this could result in significant deviations. Consider an error of 0.1 Hz in frequency. This results in an error in phase angle that increases with 0.628 rad/s, or 36°/s (360° for one cycle of the 0.1-Hz frequency error). For calculating phase-angle jumps it is essential to use an accurate value of the frequency. A phase-locked loop is essential in such a case. This obviously poses a problem with postprocessing of recorded waveforms. If one or more cycles of preevent waveform are available, the frequency has to be estimated from these. An appropriate frequency estimation can be obtained from the average time derivative of the argument versus time of the preevent complex voltage.

The method is explained graphically in Figure 8.14 for a synthetic event with a residual voltage of 50% and a phase-angle jump of −90°. Note that the extremely large value of the phase-angle jump was only chosen to explain the method; such a large phase-angle jump does not occur in practice. The figure on the left shows the argument (or phase angle) of the complex voltage $V_{\text{fund}}(t)$ as obtained from a sliding-window DFT. The dotted line is the argument of the complex voltage in case there would not be any voltage dip: $V_{\text{fund}}(0) \times e^{j\omega t}$. The voltage dip causes a deviation of the actual phase angle from the synchronous frame (the dotted line). The difference between the solid line and the dotted line gives the phase angle versus time. This function, calculated according to (8.7), is plotted on the right in Figure 8.14. The terms *stationary frame* and *rotating frame* are part of the concept that a sine wave is the projection of a circular movement. The observer in the stationary frame will only see a small (hardly noticeable) change in the

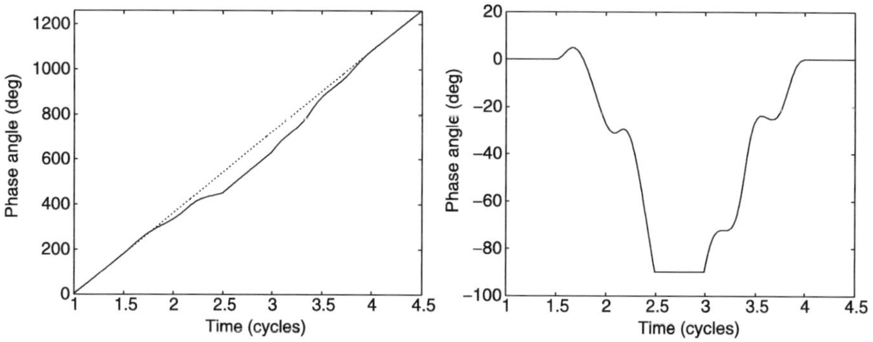


**Figure 8.14** Phase angle of complex voltage in stationary frame (left) and in rotating frame (right) for synthetic voltage dip. The dotted line on the left indicates the phase angle of the synchronous frame.

rotation, whereas an observer in the rotating frame will see what we refer to as the *phase-angle jump*.

The phase angle as a function of time has been calculated for a range of synthetic dips. The results for a dip with residual voltage 50% are shown in Figure 8.15. The range of phase-angle jump has again been chosen in accordance with the expected range of residual voltage and phase-angle jump as presented in [33, Fig. 4.113]. As with the calculation of the rms voltage versus time, we see an oscillation at the start of the event. A similar oscillation can be observed at the end of the event. The values during the transition have no physical meaning and are just an artifact of the method used to estimate the phase angle. When using the maximum or minimum value of the phase angle as a single-event index, care must be taken to not consider the values during the transition. When using the method for event segmentation introduced in Section 7.4 the phase-angle values during the transition segments should be removed from further analysis.

The artifacts in the phase angle versus time at the beginning and end of the dip can be understood by considering the path of the complex voltage in the complex plane. The complex voltage in the rotating frame, $V_{\text{fund}}(t) \times e^{-j\omega t}$, has been plotted in Figure 8.16 for a voltage dip with residual voltage 50% and three different phase-angle jumps ($-30°$, $0°$, and $+30°$). The fact that the path from the preevent voltage (circle) to the during-event voltage (cross) is not a straight line accounts for the oscillations in the phase angle as a function of time.

The method for estimating the phase angle as a function of time has been applied to a typical voltage dip due to a fault. The same measurement was used earlier to generate Figures 6.14 and 8.12. The event shows a large voltage drop in two phases and a minor drop in the third phase; it can be classified as a type C dip according to the classification introduced in Section 6.2.3. The phase angle as a function of time is shown in Figure 8.17. When calculating the phase angle in the rotating frame, a frequency equal to 50 Hz has been assumed; thus $\omega = 2\pi 50$ in (8.7). The phase angle shows a steady drift which already starts before the event and continues

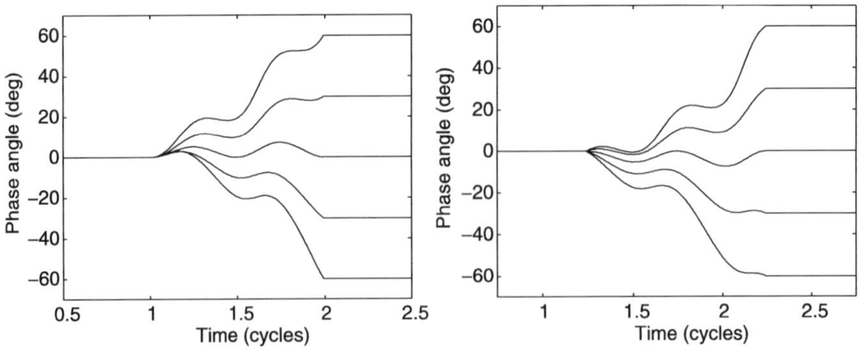


**Figure 8.15** Phase angle as function of time for synthetic events: residual voltage 50%; phase-angle jump $-60°$, $-30°$, $0°$, $+30°$, and $+60°$ (bottom to top); dip starting at voltage maximum (left) and at voltage zero crossing (right).

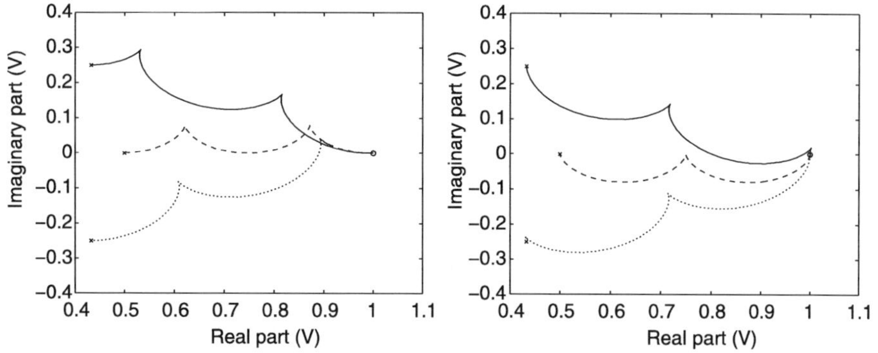


**Figure 8.16** Path of complex voltage in rotating frame at start of voltage dip for phase-angle jump equal to $-30^\circ$ (solid curve), $0^\circ$ (dashed curve), and $+30^\circ$ (dotted curve) for dips starting at voltage maximum (left) and at voltage zero crossing (right).

after the event. This is due to the difference between the actual frequency and the frequency value used in the estimation (50 Hz in this case).

When the phase angle is calculated in real time in a power quality monitor, a phase-locked loop can be used to determine the frequency. Some monitors use a sampling frequency that is in all cases an integer multiple of the actual power system frequency. This is again typically implemented by using a phase-locked loop. In such a case the actual frequency, when expressed in samples per cycle, is always exactly known. This allows for extraction of the exact phase angle during postprocessing of the data.

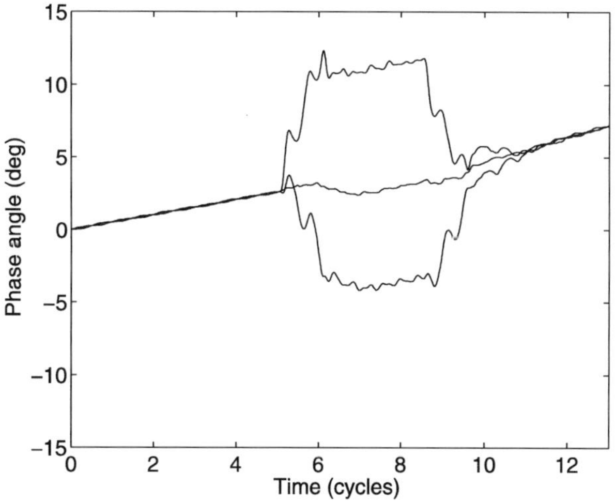


**Figure 8.17** Phase angle as function of time for typical unbalanced voltage dip assuming frequency equal to 50 Hz.

For most of the measurements used in this book, the sampling frequency was constant and not locked to the actual power system frequency. The result is the drift in apparent phase angle visible in Figure 8.17. This can be corrected by estimating the actual frequency from the drift in phase angle before the event occurs. For the measurement shown in Figure 8.17 the phase-angle drift is 2.6° in 100 ms. Using (8.8) with $\phi(t) = \phi(0)$, we find for the frequency error, $\Delta f = \Delta\phi/(2\pi t) = 0.072$ Hz. The actual frequency is thus 50.072 Hz. The resulting phase angle as a function of time, when using the correct frequency, is shown in Figure 8.18. The use of the actual frequency has been implemented by adding a correction term to the estimated phase angle as in (8.8).

In accordance with the theory for three-phase unbalanced dips (see Section 6.2.3), the change in phase angle is of the same size but opposite for the two phases with the highest drop in voltage. The phase with the minor drop in voltage also shows a minor change in phase angle. After the major dip it takes a few cycles for the phase angle to recover.

The largest phase-angle jumps occur for voltage dips due to distribution system faults. An example of such a dip is shown in Figure 8.19. The rms voltage versus time for the same event is shown in Figure 6.12. Next to the large change in phase angle directly attributable to the fault, we can observe a small overshoot at the beginning of the dip, a slow increase in phase angle during the dip, and a slow recovery afterward. The gray areas at the beginning and end of the dip indicate the *transition segments*, as introduced in section 7.4. In this case the duration of the transition segments has been estimated. The values during the transition segments have no physical meaning, similar to the spike at the beginning of the dip. Indicating the transition segments makes it easier to see the effect of the load on the phase

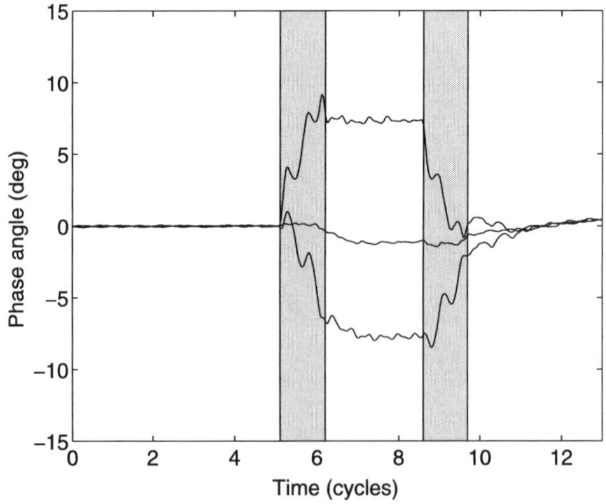


**Figure 8.18** Phase angle as function of time for typical unbalanced voltage dip using accurate estimation of frequency value. The gray rectangles indicate transition segments.

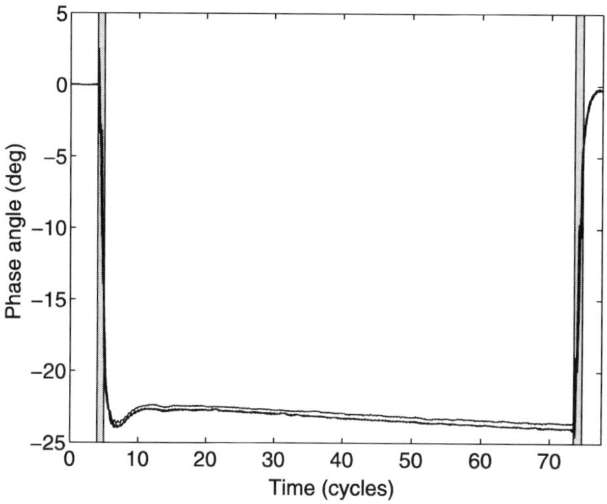


**Figure 8.19** Phase-angle versus time for voltage dip due to three-phase fault at distribution level.

angle. The decay in phase angle continues even after the transition segment. This is probably due to induction motor load feeding into the fault. The slow recovery after the fault is due to those same induction motors taking energy from the supply to rebuild their air gap magnetic field.

The induction machine influence is even more severe in Figure 8.20. As the dip is shorter than the one in the previous figure, the transition segments (gray areas) are relatively large. The transition segment at the end of the dip is additionally long as the voltage recovers in two stages about one cycle separated in time. The figure shows both rms voltage and phase angle as a function of time. Disregarding the transition segments shows clear similarities between this event and the one shown in Figure 8.19.

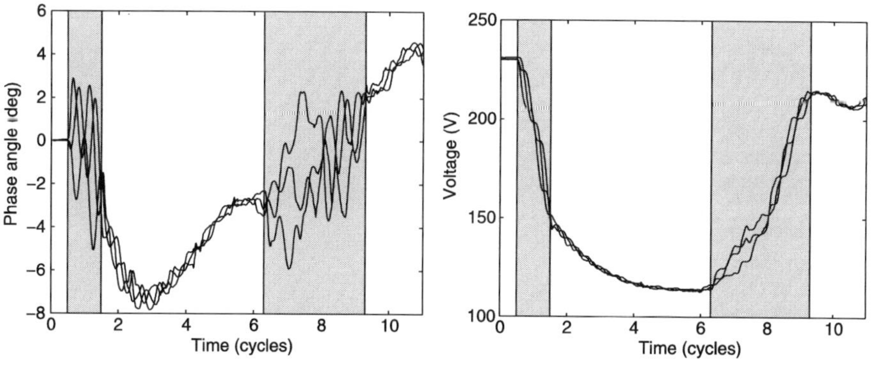


**Figure 8.20** Phase angle (left) and rms voltage (right) versus time for balanced dip with very large induction machine influence.

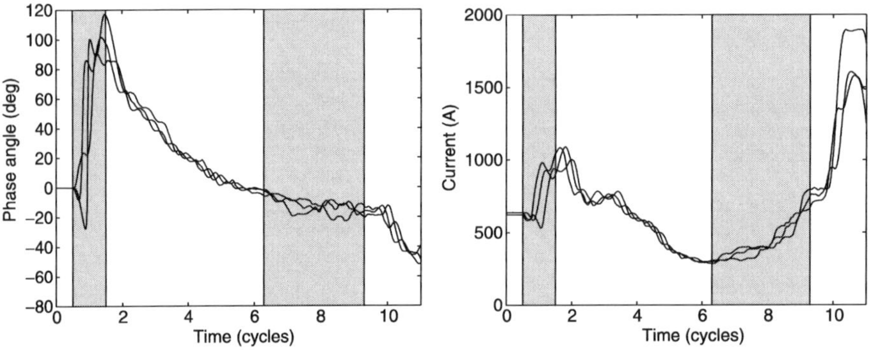


**Figure 8.21** Phase angle (left) and rms (right) of current versus time for balanced dip with very large induction machine influence.

The calculation of phase angle and rms is not limited to the voltages; the methods can equally be applied to the measured currents. The resulting curves for the event in Figure 8.20 are shown in Figure 8.21. The gray areas again indicate the transition segments. We see a large increase in current magnitude at the beginning of the dip and at the end of the dip. At the beginning of the dip the current is leading the voltage and the induction machines support the voltage in the system. At the end of the dip the current is lagging and the machines lead to a drop in the system voltage. Note that the transition segments were based on the voltages. They are not necessarily the same for the currents.

The same signal processing has been applied to the current measured during synchronized capacitor energizing at a 10-kV substation. The results are shown in Figure 8.22. The current was measured at the 10-kV side of the substation transformer, thus including both the load current and the current to the capacitor bank. We see that the current magnitude stays about the same. The reactive current

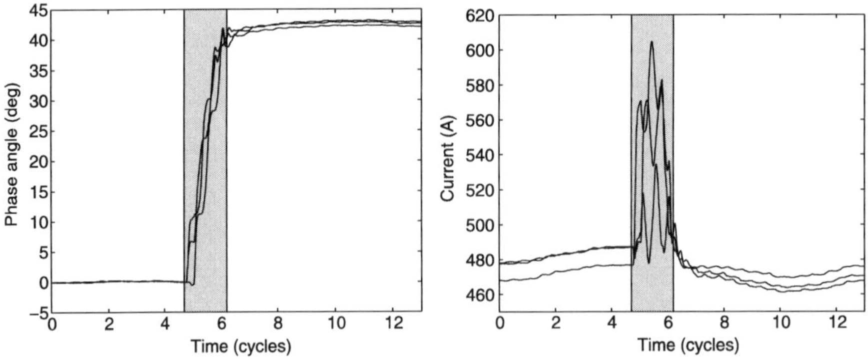


**Figure 8.22** Phase angle (left) and rms (right) of current versus time during capacitor energizing.

through the transformer changes from 3.6 Mvar inductive to 3 Mvar capacitive, so that the current magnitude (or the apparent power) stays about the same. In this case the switching of the capacitor bank took place before the early morning rise in power consumption. The transition segment is again indicated as a gray rectangular area. The transient itself cannot be analysed using these methods, only the transition between preevent and postevent steady state can.

## 8.3 THREE-PHASE CHARACTERISTICS VERSUS TIME

In section 6.2.3 a method was introduced to classify and characterize balanced and unbalanced voltage dips in a consistent manner. Each dip in a three-phase system is characterized by a dip type, a (complex) residual voltage, and a duration. To extract these single-event indices from measurement, the first step is again to obtain a voltage magnitude as a function of time. This the so-called characteristic voltage is defined in [334]. The method is based on the method of symmetrical components and follows in a rather straightforward way from the symmetrical-component classification. An alternative method is proposed in [35]. The latter method is less accurate but significantly easier to implement and understand. Both methods will be described in detail below.

### 8.3.1 Symmetrical-Component Method

The algorithm proposed in [334] determines the dip type from the positive- and negative-sequence voltages. From Section 6.2.3 the expressions for the complex phase voltages of a type Ca dip are as follows:

$$
\begin{aligned}
V_a &= F \\
V_b &= -\tfrac{1}{2}F - \tfrac{1}{2}jV\sqrt{3} \\
V_c &= -\tfrac{1}{2}F + \tfrac{1}{2}jV\sqrt{3}
\end{aligned}
\tag{8.9}
$$

with $F$ the PN factor and $V$ the characteristic voltage. The expressions for a type Da dip are

$$
\begin{aligned}
V_a &= V \\
V_b &= -\tfrac{1}{2}V - \tfrac{1}{2}jF\sqrt{3} \\
V_c &= -\tfrac{1}{2}V + \tfrac{1}{2}jF\sqrt{3}
\end{aligned}
\tag{8.10}
$$

Similar expressions have been derived for dip types Cb, Cc, Db, and Dc. The aim of the characterization algorithm is to determine the dip type and to estimate the value of the characteristic voltage and PN factor. A first look at (8.9) and (8.10) shows that additional information is required: (8.9) can be obtained from (8.10) by exchanging the position of $V$ and $F$. The additional information is given in the form of the requirement that the characteristic voltage in absolute value is less than the PN factor.

The characterization algorithm is based on the expressions for the positive- and negative-sequence voltages for the six dip types. The positive-sequence voltage (with reference to phase $a$ prefault voltage) is the same for all dip types and is only a function of the characteristic voltage and the PN factor:

$$V_1 = \tfrac{1}{2}(F + V) \tag{8.11}$$

The negative-sequence voltage is the same in magnitude but different in argument:

$$\begin{aligned}
V_2 &= \tfrac{1}{2}(F - V) && \text{Type Ca} \\
V_2 &= \tfrac{1}{2}a(F - V) && \text{Type Cb} \\
V_2 &= \tfrac{1}{2}a^2(F - V) && \text{Type Cc} \\
V_2 &= -\tfrac{1}{2}(F - V) && \text{Type Da} \\
V_2 &= -\tfrac{1}{2}a(F - V) && \text{Type Db} \\
V_2 &= -\tfrac{1}{2}a^2(F - V) && \text{Type Dc}
\end{aligned} \tag{8.12}$$

where $a$ constitutes a rotation over 120°. For dips due to single-phase and phase-to-phase faults, the PN factor is close to unity (in per-unit with the preevent voltage as a base). If we assume that $F = 1$, the angle between the drop in positive- and negative-sequence voltages is an integer multiple of 60:

$$\begin{aligned}
&\text{angle}(1 - V_1, V_2) = 0 && \text{Type Ca} \\
&\text{angle}(1 - V_1, V_2) = 120^\circ && \text{Type Cb} \\
&\text{angle}(1 - V_1, V_2) = -120^\circ && \text{Type Cc} \\
&\text{angle}(1 - V_1, V_2) = 180^\circ && \text{Type Da} \\
&\text{angle}(1 - V_1, V_2) = -60^\circ && \text{Type Db} \\
&\text{angle}(1 - V_1, V_2) = 60^\circ && \text{Type Dc}
\end{aligned} \tag{8.13}$$

It has been assumed here that the preevent voltage is 1 pu and along the positive real axis. The angles are multiples of 60° and different for the different dip types. The dip type can be estimated by comparing the measured angle with the values in (8.13). The angle is obtained by calculating the positive and negative-sequence voltages by any of the methods introduced in Section 3.2.4 and next determining the argument of the ratio between the complex positive- and negative-sequence voltages.

This results in the following algorithm for obtaining the dip type $T$:

$$T = \frac{1}{60^\circ} \times \arg\left(\frac{V_2}{1 - V_1}\right) \tag{8.14}$$

where $T$ is rounded to an integer. We will come back later to the method of rounding. Initially we will assume that the rounding takes place to the nearest integer. The value of $T$ is related to the earlier classification in accordance with the following

$$\begin{array}{ll} T = 0 & \text{Type Ca} \\ T = 1 & \text{Type Dc} \\ T = 2 & \text{Type Cb} \\ T = 3 & \text{Type Da} \\ T = 4 & \text{Type Cc} \\ T = 5 & \text{Type Db} \end{array} \tag{8.15}$$

Knowing the dip type, the other characteristics can be obtained, for example, from the sum and difference of positive- and negative-sequence voltages according to (8.11) and (8.12). For example, for type Ca we can obtain the characteristic voltage and PN factor from the expressions

$$V = V_1 - V_2 \qquad F = V_1 + V_2 \tag{8.16}$$

and for type Dc from

$$V = V_1 + aV_2 \qquad F = V_1 - aV_2 \tag{8.17}$$

A general expression can be obtained by using the dip type $T$ as an additional parameter:

$$V = V_1 - b^{6-T} V_2 \qquad F = V_1 + b^{6-T} V_2 \tag{8.18}$$

with $b = -a^2 = \frac{1}{2} + \frac{1}{2} j\sqrt{3}$ a rotation over 60°.

### 8.3.2 Implementation of Symmetrical-Component Method

***8.3.2.1 Frequency Estimation*** The first step in calculating the characteristics versus time is to create a synchronous signal. For this the accurate frequency is needed. When postprocessing is used of signals for which the sample frequency was not synchronized to the actual frequency, the frequency needs to be estimated. The method used here is only possible when at least two cycles of preevent waveform data are available. It is possible to estimate the frequency from a shorter preevent window, but the results could become very inaccurate. The first step is to estimate the length of one cycle of the power system frequency. An integer number of samples with a duration as close as possible to 20 ms (for a 50-Hz system) is used for this. Let the length of this estimation be $T$. The initial angle $\alpha_1$ of the fundamental component is determined over the window [0, $T$] by using a DFT algorithm. The initial angle over the next "period", [$T$, $2T$], is $\alpha_2$.

The second angle is obtained by increasing the first angle by $2\pi fT$ with $f$ the actual frequency and $T$ the estimated period. The actual frequency can be obtained from

$$2\pi + \alpha_2 = \alpha_1 + 2\pi fT \tag{8.19}$$

resulting in the explicit expression

$$f = \frac{1}{T}\left(1 + \frac{\alpha_2 - \alpha_1}{2\pi}\right) \tag{8.20}$$

***8.3.2.2 Complex Voltages Versus Time*** The second step is to determine the complex phase voltages in the rotating frame. We could again use a DFT algorithm, but using synchronous sine and cosine waves results in an algorithm that is easier to implement and faster when only one frequency is of interest. Consider the following signal:

$$v(t) = \sqrt{2}X\cos(2\pi ft) - \sqrt{2}Y\sin(2\pi ft) = \mathrm{Re}[\sqrt{2}(X + jY)e^{j2\pi ft}] \tag{8.21}$$

with $f$ the power system frequency. Two new signals are obtained from this signal as follows:

$$\begin{aligned} v_d(t) &= \sqrt{2}v(t) \times \cos(2\pi ft) \\ v_q(t) &= \sqrt{2}v(t) \times \sin(2\pi ft) \end{aligned} \tag{8.22}$$

Combining (8.21) and (8.22) results in

$$\begin{aligned} v_d(t) &= X + X\cos(2\pi ft) + Y\sin(2\pi ft) \\ v_q(t) &= -Y + Y\cos(2\pi ft) + X\sin(2\pi ft) \end{aligned} \tag{8.23}$$

Averaging these two signals over a multiple of one half-cycle of the power system frequency results in the following expression for the complex voltage as a function of time:

$$X(t) + jY(t) = \overline{v_d(t)} - j\overline{v_q(t)} \tag{8.24}$$

In practical implementations it is recommended to take the averaging over one cycle of the power system frequency. Using a half-cycle average will make the result sensitive to even-harmonic distortion. The amplitude and phase angle as a function of time can be obtained from the absolute value and argument of the complex voltage. In most cases the preevent voltage in phase $a$ is used as reference: Its complex value is along the real axis. The complex phase voltages are in that case obtained by dividing the value of $X + jY$ by the value for phase $a$ at time zero:

$$\begin{aligned} \underline{V}_a(t) &= \frac{X_a(t) + jY_a(t)}{X_a(0) + jY_a(0)} \\ \underline{V}_b(t) &= \frac{X_b(t) + jY_b(t)}{X_a(0) + jY_a(0)} \\ \underline{V}_c(t) &= \frac{X_c(t) + jY_c(t)}{X_a(0) + jY_a(0)} \end{aligned} \tag{8.25}$$

Note that the phase $a$ voltage is used as a reference for all three phases. The result is a phase difference of approximately 120° between the three complex voltages. When the three phase voltages are treated as three single-phase measurements the $b$ and $c$ phase complex voltages should be obtained through dividing by the initial complex voltage in phases $b$ and $c$, respectively.

#### 8.3.2.3 Symmetrical-Component Voltages

The symmetrical-component voltages are typically calculated from the complex phase voltages by using the standard expressions for complex voltages as a function of time:

$$\begin{aligned} \underline{V}_1(t) &= \tfrac{1}{3}[\underline{V}_a(t) + a\underline{V}_b(t) + a^2\underline{V}_c(t)] \\ \underline{V}_2(t) &= \tfrac{1}{3}[\underline{V}_a(t) + a^2\underline{V}_b(t) + a\underline{V}_c(t)] \\ \underline{V}_0(t) &= \tfrac{1}{3}[\underline{V}_a(t) + \underline{V}_b(t) + \underline{V}_c(t)] \end{aligned} \tag{8.26}$$

Alternatively, the positive- and negative-sequence voltages may be obtained directly from the voltage waveforms by using the $dq$-transform, as defined in (3.67) and (3.69) in Section 3.2.4.4. This still requires an accurate frequency reference, in this case for the transformation from $\alpha\beta$-voltages to $dq$-voltages. The first step is calculating the $\alpha\beta$-voltages:

$$\underline{v}_{\alpha\beta} = \tfrac{1}{3}[v_a(t) + av_b(t) + a^2v_c(t)] \tag{8.27}$$

with $a = -\frac{1}{2} + \frac{1}{2}\sqrt{3}$ a rotation over 120°. Next, positive and negative $dq$-voltages are calculated as follows:

$$\underline{v}_{dq}^{+} = \underline{v}_{\alpha\beta}e^{-j\omega t} \tag{8.28}$$

$$\underline{v}_{dq}^{-} = \underline{v}_{\alpha\beta}e^{+j\omega t} \tag{8.29}$$

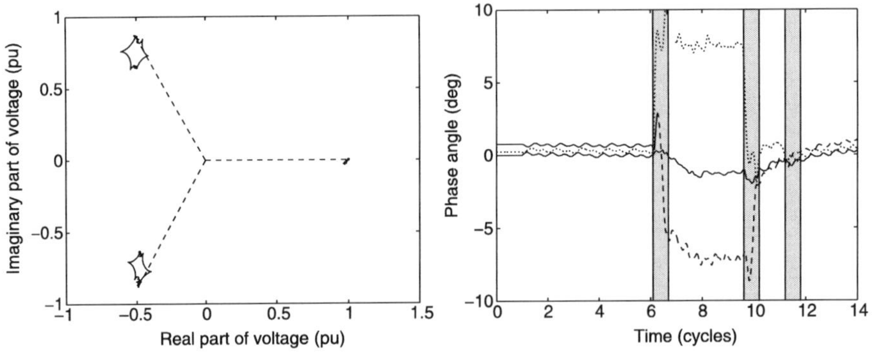


**Figure 8.23** Complex voltages (left) and phase angle versus time (right) for three-phase unbalanced voltage dip.

The positive- and negative-sequence voltages are obtained from the following expressions:

$$\underline{U}^+ = \overline{\underline{v}_{dq}^+} \tag{8.30}$$

$$\underline{U}^- = \overline{\underline{v}_{dq}^-}^* \tag{8.31}$$

The complex positive- and negative-sequence voltages are obtained by averaging the $dq$-voltages over at least one-quarter of a cycle. To prevent oscillations due to even-harmonic distortion a longer averaging window is recommended.

The algorithm for calculating complex voltages and symmetrical-component voltages has been applied to the voltage dip shown in Figure 6.14. The resulting complex voltages are shown in Figure 8.23, where each dot represent one value. The dashed lines indicate the voltage phasors at time zero, that is, the preevent phasors. The complex voltages in the two faulted phases ($b$ and $c$) move toward each other during the dip. The third phase ($a$) is only slightly affected by the fault. The phase angles of the complex voltages as a function of time are shown on the right in the same figure. We see opposite phase-angle jumps in the two faulted phases and a minor change in phase angle in the third phase.

#### 8.3.2.4 Angle Between Positive- and Negative-Sequence Voltages

The symmetrical-component voltages are shown in Figure 8.24. The positive-sequence voltage (solid line) shows a drop during the event, whereas the negative-sequence voltage (dashed line) increases. The zero-sequence voltage (dotted line) is very close to zero. The figure on the right shows the so-called dip-type angle: the angle between the negative-sequence voltage and the drop in positive-sequence voltage as defined in (8.14). In that expression it was assumed that the preevent positive-sequence voltage is equal to 1 pu and is along the positive real axis. To create a similar situation in measurements the preevent positive-sequence voltage

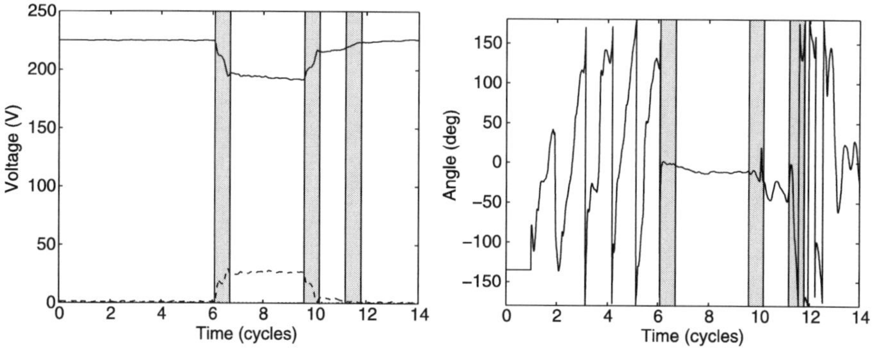


**Figure 8.24** Symmetrical-component voltage (left) and dip-type angle (right) for three-phase unbalanced voltage dip.

is used as a reference to calculate the complex per-unit values of positive- and negative-sequence voltages:

$$\underline{U}^+_{\text{pu}} = \frac{\underline{U}^+}{\underline{U}^+_{\text{preevent}}} \tag{8.32}$$

$$\underline{U}^-_{\text{pu}} = \frac{\underline{U}^-}{\underline{U}^+_{\text{preevent}}} \tag{8.33}$$

after which the dip-type angle is obtained as follows:

$$\alpha_T = \arg\left(\frac{\underline{U}^-_{\text{pu}}}{1 - \underline{U}^+_{\text{pu}}}\right) \tag{8.34}$$

Note that (8.34) can be rewritten as

$$\alpha_T = \arg\left(\frac{\underline{U}^-}{\underline{U}^+_{\text{preevent}} - \underline{U}^+}\right) \tag{8.35}$$

When the complex phase voltages are calculated with reference to the preevent phase *a* voltage, as in (8.25), the resulting positive- and negative-sequence voltages do not need to be corrected.

The dip-type angle as plotted on the right-hand side of Figure 8.24 shows heavy oscillations before and after the actual voltage dip. These oscillations have no physical meaning. The main oscillations are due to small variations in the positive-sequence voltage. This means that $1 - \underline{U}^+_{\text{pu}}$ oscillates around zero with its angle being ill-defined. When including this algorithm in an automatic analysis of power quality events, it is important to block out the part of the recording outside the actual dip. The segmentation method discussed in Section 7.4 can be used for

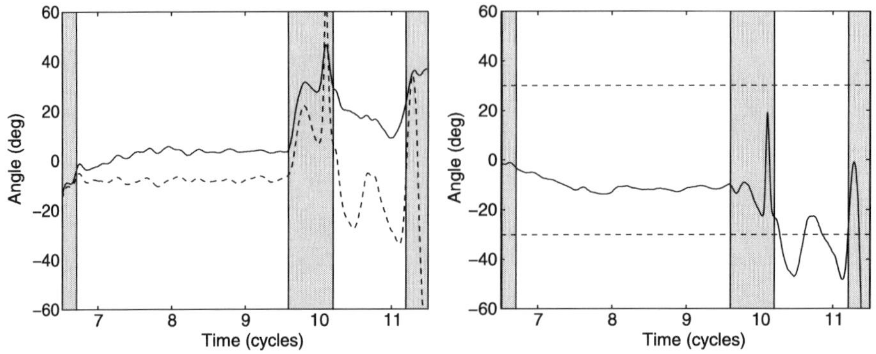


**Figure 8.25** Left: angle of drop in positive-sequence voltage (solid line) and in negative-sequence voltage (dashed line). Right: dip-type angle.

this. The dip-type angle, the dip type, and the characteristics are only calculated for event segments. Alternatively, the values can be disregarded when the positive- or negative-sequence voltages are too close to their normal values (being 1 pu and zero, respectively).

It is very important that an accurate frequency value is used. An error in frequency will cause the same rotation in positive- and negative-sequence voltages, but in opposite directions. Especially for longer dips this can easily lead to the estimation of an incorrect dip type.

The dip-type angle during the dip shown in Figure 8.24 has been enlarged in Figure 8.25. The same figure also shows the individual angles that make up the dip-type angle: the negative-sequence voltage and the drop in positive-sequence voltage. The transition segments are again indicated through gray areas. We see that the dip-type angle is close to zero at the beginning of the dip and close to the theoretical value for a type Ca dip. But during the first few cycles of the fault, the angle gradually decreases to about $-14^\circ$, after which it stays constant until the first recovery. During the second stage of the dip, the dip-type angle varies between $-22^\circ$ and $-45^\circ$. The right-hand plot also shows the limits for dip-type Ca used in the algorithm: $-30$ and $+30^\circ$. During part of the second stage of the dip it would be incorrectly classified as type Db. The changes in dip-type angle can be explained by inspecting the individual angles. During the initial cycles of the dip, the angle of the drop in positive-sequence voltage increases slowly, whereas the angle of the negative-sequence voltage remains constant. During the second stage the negative-sequence voltage retards whereas the drop in positive-sequence voltage advances. Further study is needed to find an explanation for this behavior.

***8.3.2.5 Dip Type*** Knowing the phase angle between negative-sequence voltage and drop in positive-sequence voltage (the dip-type angle), the dip type is calculated using expression (8.14), where 1 pu corresponds to the preevent positive-sequence voltage. Here it should be noted that the expression assumes that the angle is in the right interval. The resulting value has to be taken modulus 6 when it is outside of the interval 0 through 5.

**TABLE 8.2 Relation Between ABC Classification and Symmetrical-Component Classification**

| ABC Type | Type | PN Factor | Zero-Sequence Voltage |
|---|---|---|---|
| A | C or D | $F \approx V$ | $U_0$ Small |
| B | D | $F \approx E$ | Significant $U_0$ |
| C | C | $F \approx E$ | No $U_0$ |
| D | D | $F \approx E$ | No $U_0$ |
| E | C | $V < F < E$ | Significant $U_0$ |
| F | D | $V < F < E$ | No $U_0$ |
| G | C | $V < F < E$ | No $U_0$ |

Note that there are only 6 possible dip types according to this method (Ca, Cb, Cc, Da, Db, and Dc) versus 19 types under the ABC classification. It is possible to define the additional dip types in accordance with the ABC classification. This requires some further processing, for example, as given in Table 8.2. The two rules used are based on the value of the zero-sequence voltage $U_0$ and the value of the PN factor $F$ compared to the characteristic voltage $V$ and the preevent voltage $E$. A complete implementation of this algorithm requires the choice of a number of threshold settings, for example, for the PN factor. Appropriate values can be obtained by studying the theoretical values as discussed in Section 6.2.3.

***8.3.2.6 Dip Characteristics*** Knowing the dip type, the characteristic voltage and the PN factor can be obtained from expressions (8.18) above. Again the result has only physical meaning when the drops in positive- and negative-sequence voltages are large enough to obtain an accurate value for the angle between them. The results are shown in Figure 8.26. The result of the calculation is that this is a dip of type Ca (dip type 0), that is, a drop between phases $b$ and $c$.

During the dip we see a slow shift in dip-type angle. This is due to the effect of, especially the induction motor, load on the during-fault voltages. If this shift

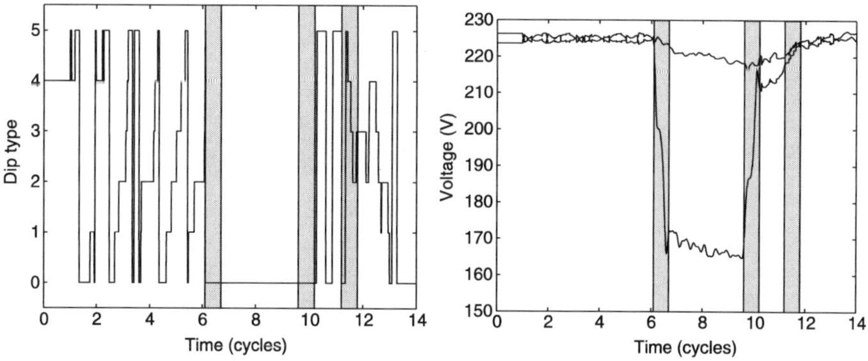


**Figure 8.26** Dip type (left) and characteristics versus time (right) for three-phase unbalanced dip.

becomes too large, it may lead to an incorrect estimation of dip type. This can be corrected by changing the algorithm such that changes in dip type not associated with sudden changes in complex voltage are not allowed. In terms of the earlier discussion on segmentation, the dip type remains the same within each event segment.

***8.3.2.7 Another Example*** Another example of the application of the symmetrical-component algorithm to a three-phase unbalanced dip is shown in Figures 8.27 through 8.29. The amplitude of the complex voltages shows that this is a two-stage event, probably a developing fault. The symmetrical components show the characteristic drop in positive-sequence voltage and the rise in negative-sequence voltage. Dip-type angle and dip type are shown in Figure 8.28. The dip changes from type 0 (Ca) to type 5 (Db). The characteristics versus time are shown in Figure 8.29. Both characteristic voltage and PN factor become smaller at the transition from the first to the second stage of the event.

The before-introduced theory on three-phase unbalanced voltage dips can be used to extract information on the type of fault. The second stage shows a low PN factor,

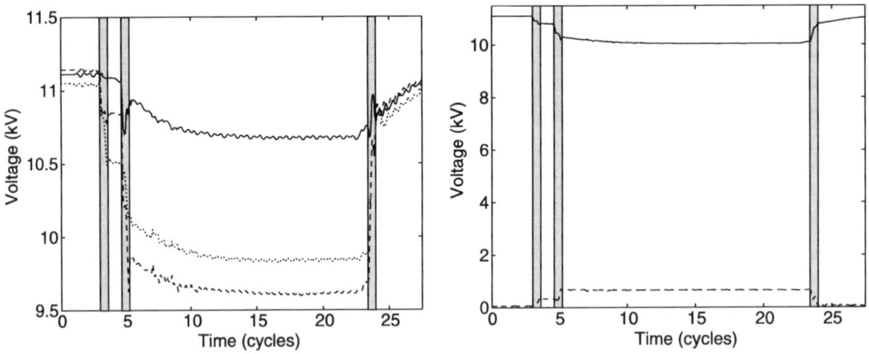


**Figure 8.27** Amplitude of complex voltages (left) and symmetrical component voltages (right) as function of time for three-phase unbalanced voltage dip.

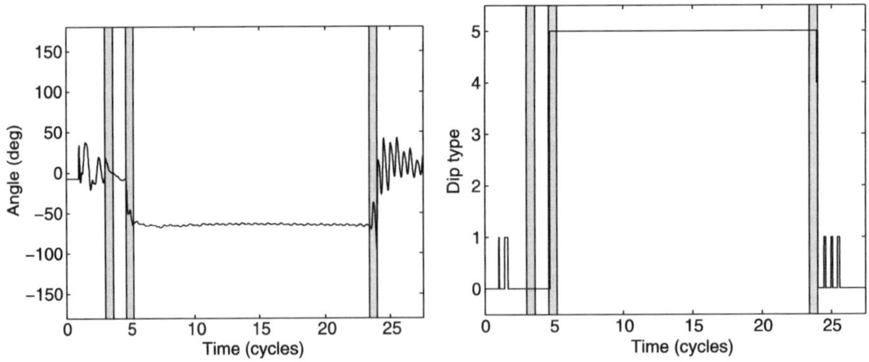


**Figure 8.28** Dip-type angle (left) and dip type (right) as function of time for three-phase unbalanced voltage dip.

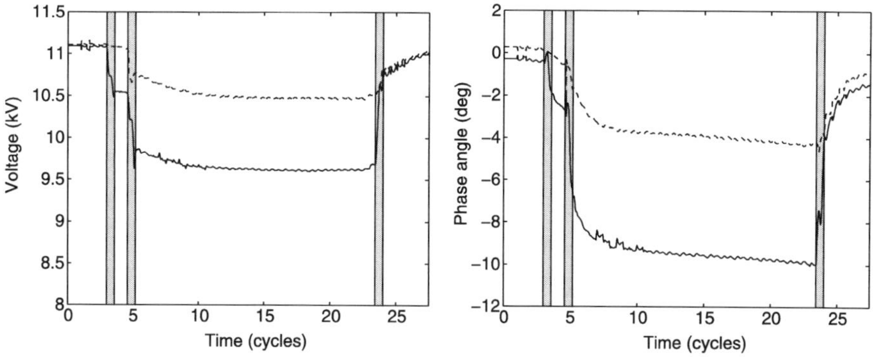


**Figure 8.29** Absolute value (left) and phase angle (right) as function of time of characteristic voltage (solid lines) and PN factor (dashed lines) for three-phase unbalanced voltage dip.

which indicates that the dip is due to a two-phase-to-ground fault. The initial event is either single phase or phase to phase, based on the PN factor. However, the characteristic voltage shows a drop at the transition, indicating that the first stage is a single-phase-to-ground fault. The conclusion is thus that this dip is due to a single-phase-to-ground fault that develops into a two-phase-to-ground fault.

The change in dip type (from Ca to Db) can be understood better when it is realized that the line voltages have been recorded. In terms of phase voltages, the change in dip type would be from Da to Cb or from *a* drop in phase *a* to a drop in phases *a* and *c*. This is consistent with the development from a single phase to a two phase to ground fault.

### 8.3.3 Six-Phase Algorithm

A simplified algorithm for extracting three-phase characteristics is described in [43]. The first stage in this so-called six-phase algorithm is the removal of the zero-sequence voltage. The zero-sequence voltage does not affect the dip type and is therefore treated separately. After subtraction of the zero-sequence voltage, the rms voltage is obtained for the three phase voltages and the three phase-to-phase voltages:

$$\begin{aligned} V_A &= \text{rms}\{v_a - \tfrac{1}{3}(v_a + v_b + v_c)\} \\ V_B &= \text{rms}\{v_b - \tfrac{1}{3}(v_a + v_b + v_c)\} \\ V_C &= \text{rms}\{v_c - \tfrac{1}{3}(v_a + v_b + v_c)\} \end{aligned} \tag{8.36}$$

$$\begin{aligned} V_{AB} &= \text{rms}\left\{\frac{v_a - v_b}{\sqrt{3}}\right\} \\ V_{BC} &= \text{rms}\left\{\frac{v_b - v_c}{\sqrt{3}}\right\} \\ V_{CA} &= \text{rms}\left\{\frac{v_c - v_a}{\sqrt{3}}\right\} \end{aligned} \tag{8.37}$$

**TABLE 8.3 Dip Type from Six-Phase Algorithm**

| Lowest rms Voltage | Dip Type |
|---|---|
| $V_A$ | Da |
| $V_B$ | Db |
| $V_C$ | Dc |
| $V_{BC}$ | Ca |
| $V_{CA}$ | Cb |
| $V_{AB}$ | Cc |

The dip characteristics are obtained directly from these six rms voltages: The characteristic voltage is the lowest of the six rms voltages, the PN factor the highest of the six. Note that this algorithm results in the dip characteristics as a function of time. We will come to the calculation of single-event indices in Section 8.8.

The dip type is determined from the voltage with the lowest rms value, as in Table 8.3. The six-phase algorithm can also be used to obtain the arguments of the complex voltages $V$ and $F$. The argument of the characteristic voltage is the phase angle of the voltage that gives the lowest rms value: $v_a - v_0$ for type Da, $v_b - v_c$ for type Cc, and so on. The argument of the PN factor is the phase angle of the voltage that gives the highest rms value.

Two examples of the application of the six-phase algorithm are shown here. In both cases the same data files have been used as for an earlier example in which the symmetrical-component method has been applied. The results for the first example are shown in Figures 8.30 and 8.31, corresponding to Figure 8.23 through 8.26. The first figure shows the six rms voltages as a function of time, as defined in (8.36) and (8.37). A half-cycle window has been used to obtain the rms voltage. A one-cycle window will give a longer transition time but less uncertainties

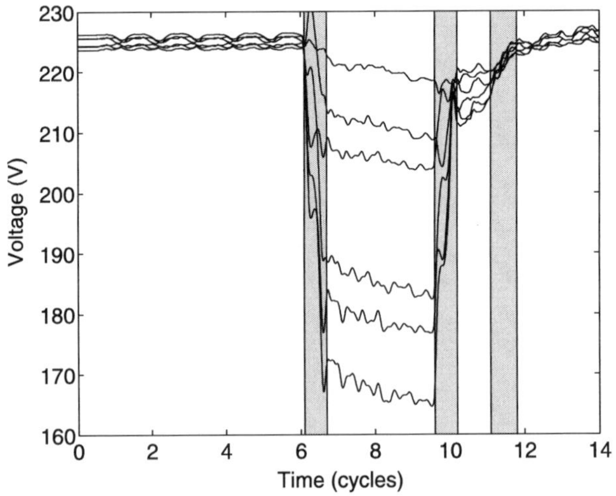


**Figure 8.30** The six rms voltages as function of time.

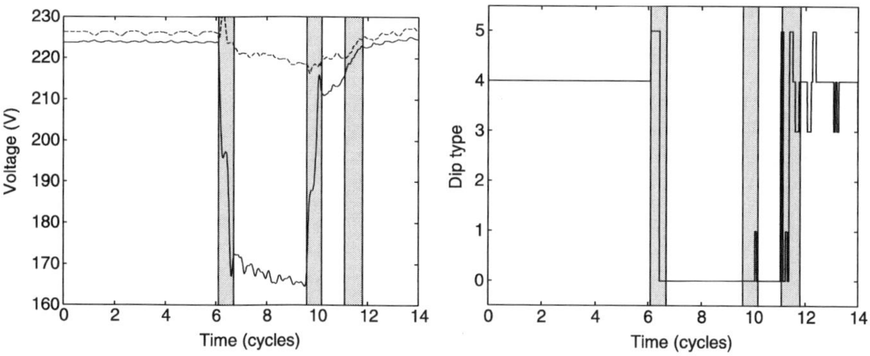


**Figure 8.31** Dip type and characteristics: characteristic voltage (left, solid line), PN factor (left, dashed line), and dip type (right).

in dip characteristics. We see in Figure 8.30 that during the main part of the dip the difference between the lowest rms voltage and the lowest but one is rather large. There is thus no risk for an incorrect estimation of the dip type.

The dip type and the characteristics as a function of time are shown in Figure 8.31. The consequence of this algorithm is always that the characteristic voltage is less than the PN factor, even before and after the actual dip. The transition segments are again indicated as gray areas; in these areas the results are not reliable. This is best visible for the dip type: Before and after the fault and during the first-stage fault clearing, its values deviate from the actual value (zero in this case). Another example is shown in Figure 8.32 for the same dip used to obtain Figure 8.27. For these examples, the two algorithms result in the same characteristics versus time. The differences between the two algorithms will be studied in more detail in the next section.

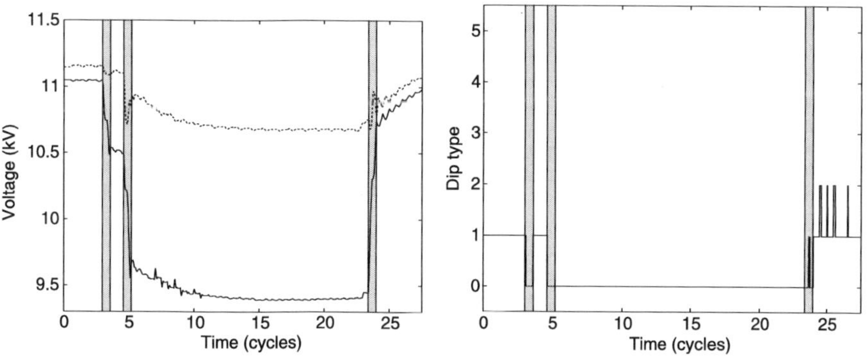


**Figure 8.32** Dip type and characteristics: characteristic voltage (left, solid line), PN factor (left, dashed line), and dip type (right).

### 8.3.4 Performance of Two Algorithms

To show the overall performance of the two algorithms and their limitations, the complex phase voltages are, for a number of cases, calculated from given dip characteristics. The two algorithms are next applied to the complex phase voltages and the resulting characteristics compared with the known input values.

***8.3.4.1 Single-Phase Fault*** Consider a drop of voltage in phase $a$ down to 50% of its preevent value. It is assumed that the voltage in phases $b$ and $c$ remains as before the fault. According to the classification introduced before, this is a dip of type Da with $F = 1$, $V = 0.67$, $V_0 = 0.33$. The results for the symmetrical-component algorithm are shown in the first column of Table 8.4. The remaining columns give the results for a voltage drop in one phase, including a phase-angle jump of $-20^\circ$, $-30^\circ$, $-40^\circ$. As the PN factor equals exactly 1, the angle between drop in positive- and negative-sequence voltages is exactly 180°. The dip type is obtained correctly and so are the other characteristics.

The same synthetic dips as in Table 8.4 have also been applied to the six-phase algorithm. The results are shown in Table 8.5. For moderate values of the phase-angle jump, also the six-phase algorithm gives the correct type and characteristics. For a $-40^\circ$ phase-angle jump the six-phase algorithm results in a type Cc instead of Da. The rotation in the phase $a$ voltage is so large that the phase difference $ab$ becomes less in absolute value (after removal of the zero-sequence component) than the phase $a$ voltage.

**TABLE 8.4 Performance of Symmetrical Component Algorithm During Single-Phase Faults**

| | 50% 0° | 50% 20° | 50% 30° | 50% 40° |
|---|---|---|---|---|
| $V_a$ | 0.50 | $0.47 - 0.17j$ | $0.43 - 0.25j$ | $0.38 - 0.32j$ |
| $V_b$ | $-0.50 - 0.87j$ | $-0.50 - 0.87j$ | $-0.50 - 0.87j$ | $-0.50 - 0.87j$ |
| $V_c$ | $-0.50 + 0.87j$ | $-0.50 + 0.87j$ | $-0.50 + 0.87j$ | $-0.50 + 0.87j$ |
| $V_0$ | $-0.17$ | $-0.18 - 0.06j$ | $-0.19 - 0.08j$ | $-0.21 - 0.11j$ |
| $V_1$ | 0.83 | $0.82 - 0.06j$ | $0.81 - 0.08j$ | $0.79 - 0.11j$ |
| $V_2$ | $-0.17$ | $-0.18 - 0.06j$ | $-0.19 - 0.08j$ | $-0.21 - 0.11j$ |
| Angle | 180° | 180° | 180° | 180° |
| $V$ | 0.67 | $0.65 - 0.11j$ | $0.62 - 0.17j$ | $0.59 - 0.21j$ |
| $\|V\|$ | 0.67 | 0.66 | 0.64 | 0.63 |
| $F$ | 1.00 | 1.00 | 1.00 | 1.00 |

**TABLE 8.5 Performance of Six-Phase Algorithm During Single-Phase Faults**

| | 50% 0° | 50% 20° | 50% 30° | 50% 40° |
|---|---|---|---|---|
| Type | 3 (Da) | 3 (Da) | 3 (Da) | 4 (Cc) |
| $\|V\|$ | 0.67 | 0.66 | 0.64 | 0.60 |
| $\|F\|$ | 1.00 | 1.00 | 1.00 | 1.02 |

**TABLE 8.6 Performance of Symmetrical-Component Algorithm During Phase-to-Phase Faults**

| | 50% 0° | 50% 20° | 50% 30° | 50% 40° |
|---|---|---|---|---|
| $V_a$ | 1.00 | 1.00 | 1.00 | 1.00 |
| $V_b$ | $-0.50 - 0.43j$ | $-0.65 - 0.41j$ | $-0.72 - 0.38j$ | $-0.78 - 0.33j$ |
| $V_c$ | $-0.50 + 0.43j$ | $-0.35 + 0.41j$ | $-0.28 + 0.38j$ | $-0.22 + 0.33j$ |
| $V_1$ | 0.75 | $0.73 - 0.09j$ | $0.72 - 0.13j$ | $0.69 - 0.16j$ |
| $V_2$ | 0.25 | $0.27 + 0.09j$ | $0.28 + 0.13j$ | $0.31 + 0.16j$ |
| Angle | 0° | 0° | 0° | 0° |
| $V$ | 0.50 | $0.47 - 0.17j$ | $0.43 - 0.25j$ | $0.38 - 0.32j$ |
| $F$ | 1.00 | 1.00 | 1.00 | 1.00 |

**TABLE 8.7 Performance of Six-Phase Algorithm During Phase-to-Phase Faults**

| | 50% 0° | 50% 20° | 50% 30° | 50% 40° |
|---|---|---|---|---|
| Type | 0 (Ca) | 0 (Ca) | 1 (Dc) | 1 (Dc) |
| $\|V\|$ | 0.50 | 0.50 | 0.47 | 0.40 |
| $\|F\|$ | 1.00 | 1.00 | 1.01 | 1.04 |

***8.3.4.2 Phase-to-Phase Fault*** The same testing of the algorithms has been performed for drops in the voltage between phases $b$ and $c$ to 50%, thus representing a phase-to-phase fault at the same voltage level or a single-phase fault at another voltage level. For this the algorithm should result in a dip of type Ca with $F = 1$ and $V = 0.5$. The results for the symmetrical-component algorithm are shown in Table 8.6, the results for the six-phase algorithm in Table 8.7. Again the symmetrical-component algorithm gives the correct result in all cases, whereas the six-phase algorithm gives a wrong dip type and characteristics for large phase-angle jumps.

***8.3.4.3 Impact of Load: Drop in Voltage*** Measurements as well as simulations [33, 43, 283, 330, 333] have shown that the voltages in the three phases drop by an equal factor due to the load. In terms of our three-phase classification, the PN factor $F$ drops to a value less than unity. See also the discussion on load effects on voltage dips in [39, 41].

The load impact has been modeled for the phase-to-phase faults (type Ca dips) studied before. It is assumed that the voltage drops by 15% in all three phases due to the load. The input values of the dip characteristics are type Da, $V = 0.43$, and $F = 0.85$. The effect of the phase-angle jump is a rotation in $V$ but not in $F$. The results are shown in Table 8.8. As the PN Factor is no longer equal to 1, the angular difference between the drop in positive- and negative-sequence voltages is no longer exactly an integer multiple of 60°. But after rounding to the nearest integer multiple of 60° the algorithm still gives the correct results. As all three phases are affected in the same way, the six-phase algorithm shows the same behavior as for unity PN factor.

**TABLE 8.8 Performance of Symmetrical-Component Algorithm with Additional Drop in Voltage Due to Load Effects**

| | 50% 0° | 50% 20° | 50% 30° | 50% 40° |
|---|---|---|---|---|
| $V_a$ | 0.85 | 0.85 | 0.85 | 0.85 |
| $V_b$ | $-0.43-0.37j$ | $-0.55-0.35j$ | $-0.61-0.32j$ | $-0.66-0.28j$ |
| $V_c$ | $-0.43+0.37j$ | $-0.30+0.35j$ | $-0.24+0.32j$ | $-0.19+0.28j$ |
| $V_1$ | 0.64 | $0.62-0.07j$ | $0.61-0.11j$ | $0.59-0.14j$ |
| $V_2$ | 0.21 | $0.23+0.07j$ | $0.24+0.11j$ | $0.26+0.14j$ |
| Angle | 0° | 6.9° | 8.6° | 9.18° |
| $V$ | 0.43 | $0.40-0.15j$ | $0.37-0.21j$ | $0.33-0.27j$ |
| $F$ | 0.85 | 0.85 | 0.85 | 0.85 |

***8.3.4.4 Impact of Load: Phase Shift*** The change in load currents not only leads to a drop in the three voltages, it typically also causes a phase shift (rotation) of the three voltages. It is again assumed that the load effect is the same for the three phases: a drop of 15% and a rotation of 20°; thus $F = 0.85\exp(-j20^\circ)$. The results for this case are shown in Table 8.9. It turns out that the symmetrical-component algorithm gives an erroneous result for small phase-angle jumps in the voltage. The rotation in PN factor severely affects the angles of $1 - V_1$ and $V_2$, leading to a wrong dip type. For larger phase-angle jumps the error in angle is less than 30° so the correct dip type results after rounding to the nearest multiple of 60°. The problem with the symmetrical-component algorithm can be relatively easily solved. The dip type is determined by rounding off the angle. Modifying the rounding somewhat will solve the problem: The range $-50^\circ$, $+10$ becomes dip type 0, and so on. The resulting equation, replacing (8.14), reads as

$$T = \frac{1}{60^\circ} \times \left[\arg\left(\frac{V_2}{1 - V_1}\right) + 20^\circ\right] \tag{8.38}$$

**TABLE 8.9 Performance of Symmetrical-Component Algorithms with Additional Voltage Drop and Rotation Due to Load Effects**

| | 50% 0° | 50% 20° | 50% 30° | 50% 40° |
|---|---|---|---|---|
| $V_a$ | $0.80-0.29j$ | $0.80-0.29j$ | $0.80-0.29j$ | $0.80-0.29j$ |
| $V_b$ | $-0.53-0.20j$ | $-0.64-0.14j$ | $-0.68-0.09j$ | $-0.78-0.04j$ |
| $V_c$ | $-0.27+0.49j$ | $-0.16+0.43j$ | $-0.12+0.38j$ | $-0.08+0.33j$ |
| $V_1$ | $0.60-0.22j$ | $0.56-0.28j$ | $0.54-0.31j$ | $0.51-0.33j$ |
| $V_2$ | $0.20-0.07j$ | $0.24-0.01j$ | $0.26+0.02j$ | $0.29+0.04j$ |
| Angle | −48.5° | −34.9° | −29.8° | −26.2° |
| $V$ | $0.56-0.01j$ | $0.45-0.07j$ | $0.27-0.33j$ | $0.21-0.37j$ |
| $F$ | $0.64-0.53j$ | $0.67-0.49j$ | $0.80-0.29j$ | $0.80-0.30j$ |
| Type | 0 (Ca) | 0 (Ca) | 1 (Dc) | 1 (Dc) |
| $\|V\|$ | 0.43 | 0.43 | 0.40 | 0.34 |
| $\|F\|$ | 0.85 | 0.85 | 0.86 | 0.89 |

**TABLE 8.10 Performance of Adjusted Symmetrical-Component Algorithms with Additional Voltage Drop and Rotation Due to Load Effects**

| | 50% 0° | 50% 20° | 50% 30° | 50% 40° |
|---|---|---|---|---|
| Angle | −28.5° | −14.9° | −9.8° | −6.2° |
| $V$ | $0.40 - 0.15j$ | $0.33 - 0.27j$ | $0.27 - 0.33j$ | $0.21 - 0.37j$ |
| $F$ | $0.80 - 0.29j$ | $0.80 - 0.29j$ | $0.80 - 0.29j$ | $0.80 - 0.30j$ |

where $T$ is again rounded to the nearest integer. The results for this revised symmetrical-component algorithm are shown in Table 8.10.

#### 8.3.4.5 Six-Phase Algorithm

In the previous paragraphs, the six-phase algorithm has been applied to a small number of synthetic dips with known characteristics. The same process has been repeated for a large number of synthetic dips covering a wide range of magnitude and phase angle of the characteristic voltage. (As mentioned before, the PN factor does not affect the performance of the six-phase algorithm.) The results are presented graphically in Figure 8.33. The black dots indicate the combinations of magnitude and phase angle (of the characteristic voltage) for which the algorithm gives incorrect results. Also indicated in the figure, by the solid curves, is the range of magnitude and phase-angle jump that can be expected. The upper and lower curves are for impedance angles $+10°$ and $-60°$, respectively. The black dots are generated by varying the impedance angle between $-90°$ and $+90°$. It follows from Figure 8.33 that the six-phase algorithm gives incorrect results for events with a large (negative) phase-angle jump and moderate drops in voltage.

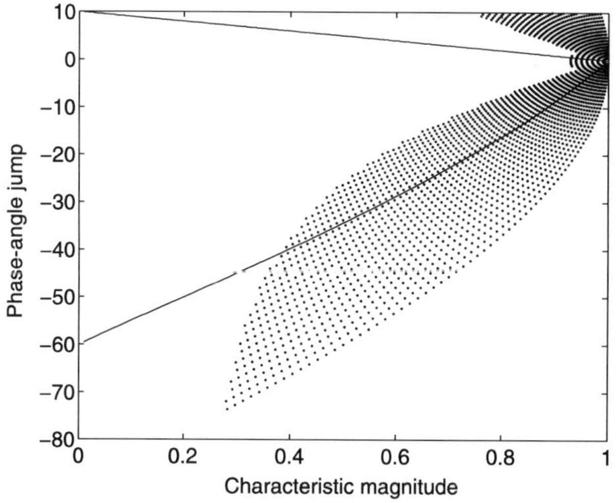


**Figure 8.33** Whole-range testing of six-phase algorithm: Black dots indicate where algorithm gives incorrect result. The area between the solid lines indicates the range of complex characteristic voltage that occurs in practice.

**8.3.4.6 *Perfect Algorithm*** To understand why there is no perfect algorithm to determine the dip characteristics, consider the following set of synthetic events:

$$
\begin{aligned}
&\text{Dip type} = \text{Ca} \\
&V = \frac{xe^{-j60^\circ}}{1 + xe^{-j60^\circ}} \\
&F = (0.7 + 0.3|V|)e^{-j40^\circ(1-|V|)}
\end{aligned}
\tag{8.39}
$$

with $0 < x < \infty$. The characteristic voltage is given for an impedance angle of $-60^\circ$. For the PN factor, both magnitude and phase angle are assumed to depend linearly on the magnitude of the characteristic voltage. This represents dips with a large (characteristic) phase-angle jump, a large drop in PN factor, and a large phase shift in PN factor. The symmetrical-component algorithm according to (8.14) has been applied to these events. The resulting values are shown in Figures 8.34 and 8.35. For $x < 0.6$ the algorithm results in an incorrect dip type (Db in this case). The incorrect dip type in turn results in incorrect values for characteristic voltage and PN factor.

The estimated magnitudes of the characteristic voltage and PN factor are shown in Figure 8.34, the phase angles in Figure 8.35. In both cases the magnitude of the actual characteristic voltage is given along the horizontal axis. For low values of the characteristic magnitude the algorithm results in significant errors in both magnitude and phase angle, where the magnitude of the PN factor and the phase angle of the characteristic voltage reach unrealistic values. The presence of these unrealistic values can be used as an indication that the algorithm has given an incorrect result.

The "incorrect values" are mathematically speaking not incorrect. There are simply two combinations of dip type, $F$ and $V$, that result in the same phase voltages.

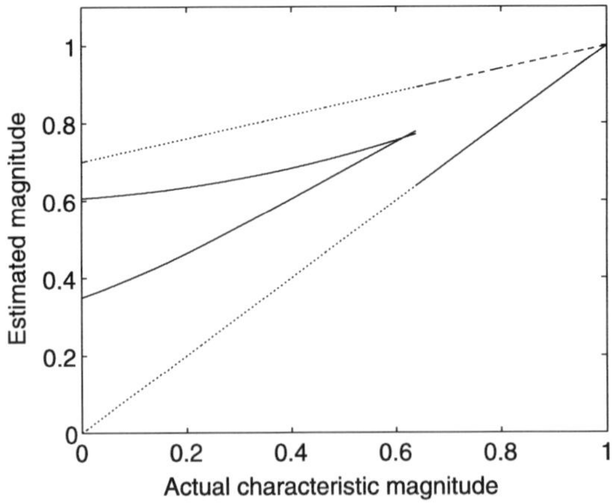


**Figure 8.34** Estimated magnitude of characteristic voltage (solid line) and PN factor (dashed line). The actual values are indicated by dotted lines.

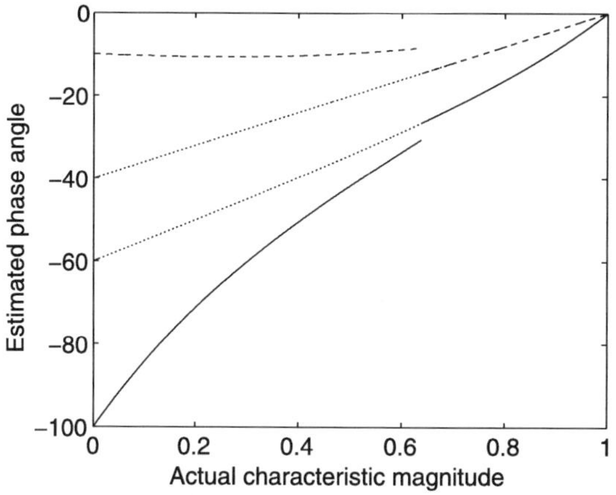


**Figure 8.35** Estimated phase angle of characteristic voltage (solid line) and PN factor (dashed line). The actual values are indicated by dotted lines.

For example, the voltage dip, according to (8.39), with characteristic magnitude 50%, has the following phase voltages:

$$
\begin{aligned}
V_a &= 0.85\angle -20^\circ \\
V_b &= 0.68\angle -162^\circ \\
V_c &= 0.53\angle +107^\circ
\end{aligned} \tag{8.40}
$$

These phase voltages were generated from a type Ca dip with the following characteristics:

$$V = 0.50\angle -34^\circ \qquad F = 0.85\angle -20^\circ \tag{8.41}$$

But exactly the same phase voltages are obtained for a type Db dip with characteristics

$$V = 0.68\angle -42^\circ \qquad F = 0.72\angle -10^\circ \tag{8.42}$$

The only way of distinguishing between them is by realizing that one of the solutions gives nonrealistic values of the characteristics. Being able to make such a decision requires knowledge of dip characteristics as they may occur in practice. This calls for a further analysis of available monitoring data and for detailed simulations of unbalanced faults in realistic systems. The effect of the load on the complex voltages plays an important role in the dip characteristics so that special emphasis has to be placed on the load model.

#### *8.3.4.7 Dips Due to Motor Starting or Transformer Energizing*

The method for characterization of three-phase unbalanced voltage dips is based on the assumption that the dip is due to a short circuit of the earth fault. But also motor starting and transformer energizing lead to voltage dips. It is possible to make a first selection of dips based on their origin. The method described above would then only be applied to dips due to faults, and dips due to motor starting and dips due to transformer energizing would be characterized in a different way.

The two methods for extracting the dip type and characteristics do not require any assumptions on the cause of the dip. The methods result in a dip type and characteristics for any three waveforms. The methods can therefore be applied to any voltage-dip recording, independent of its cause. This does of course not imply that it will be possible to interpret the results of these two methods.

Motor starting results in balanced dips that will be classified as type A. As with dips due to three-phase faults, the characteristic voltage and PN factor will be equal to the voltage in any of the affected phases.

Transformer energizing results in unbalanced dips, making interpretation of the characterization results more complicated than for motor-starting dips. Understanding the way in which transformer-energizing dips are characterized requires knowledge of voltage dips due to transformer energizing as presented in Section 6.2.2.8.

From the (complex) fundamental voltages presented in Figure 6.41 in Section 6.2.2.8 it is possible to determine positive- and negative-sequence voltages and their angles. The positive- and negative-sequence voltages remain about the same in absolute value for varying switching angle. But the angle between negative-sequence voltage and the drop in positive-sequence voltage can get any value between zero and 360°. For certain switching angles, the dip-type angle is close to a transition between two dip types; for example, 30° is the transition between type Ca and type Dc. For those switching angles the output of the three-phase classification algorithms may oscillate between two values. The symmetrical-component algorithm introduced before has been applied to the transformer-energizing model presented in Section 6.2.2.8. The resulting dip type and characteristics are shown in Figure 8.36. Both characteristic voltage and PN factor are rather independent

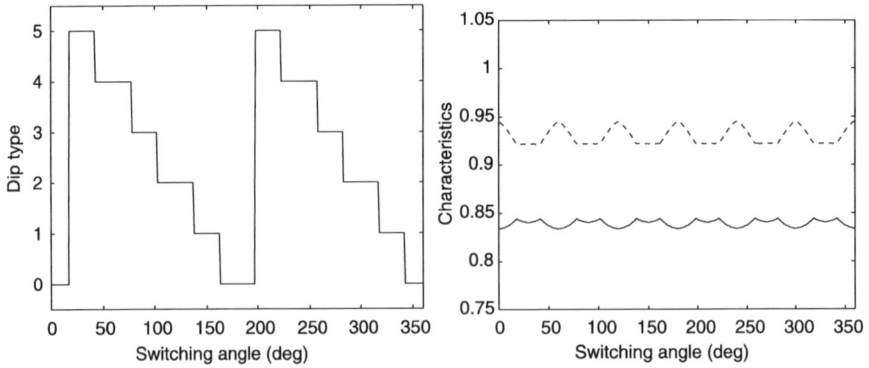


**Figure 8.36** Type of three-phase unbalanced dip (left), characteristic voltage (right, solid line) and PN factor (right, dashed line) for dips due to transformer energizing as a function of the switching angle.

of the switching angle. The dip type in turn gives an indication of the phases that are most affected by the event.

The model assumptions that form the basis for the classification of three-phase unbalanced dips do not hold for dips due to transformer energizing. Despite this, the results of applying a classification algorithm do give a good indication of the phases that are involved in the saturation. An alternative classification for transformer-energizing dips could use the absolute value of the positive-sequence voltage and the angle between the negative-sequence voltage and the drop in positive-sequence voltage. This angle can be used to obtain an approximation of the switching angle.

## 8.4 DISTORTION DURING EVENT

The waveform distortion during a voltage dip or swell may provide some additional information on the cause of the event as well as some diagnostics on the system. The obvious examples are voltage dips due to transformer energizing, where especially the even-harmonic distortion is high. During dips due to faults, the waveform distortion remains at moderate values, with the exception of arcing faults. The presence of a high distortion level during the dip could be used as additional information on the cause of the event.

But even in those cases there are no indications that the distortion has a significant effect on equipment. The actual drop in voltage is still the main impact. Even-harmonic distortion could have some adverse influence on the operation of power-electronics converters with very fast controllers. The event could be classified as a *burst of even-harmonic distortion* with as characteristics the highest value of $THD_{even}$ and the time during which its value exceeds a certain threshold. Note, however, that the start and end of a dip as well as any transient will cause an apparent burst in even-harmonic distortion. The distortion during the transition segments will need to be removed from the data before the single-event indices are determined.

An example of calculating distortion-related characteristics is shown in Figures 8.37 through 8.40. The voltage waveforms and the half-cycle rms voltages for the three phases are shown in Figure 8.37. The circuit breaker that was supposed to clear the fault did not operate properly. The faulty operation of the circuit breaker manifests itself through severe waveform distortion in the voltages.

The second- and third-harmonic components as a function of time are shown in Figure 8.38 for the three phases. The figure indicats the transition segments estimated from the waveform and the rms voltage versus time. The spectrum has been determined by applying a standard DFT algorithm to a one-cycle window. The harmonic components are expressed as a percentage of the nominal voltage (11 kV). No conclusions can be drawn from the high distortion levels during the transition segments; only the values between the transition segments can be used for further processing. The waveform distortion is low initially, even lower than before the event. The harmonic distortion increases significantly about 10 cycles after the start of the dip. The distortion appears to start a bit earlier for the second

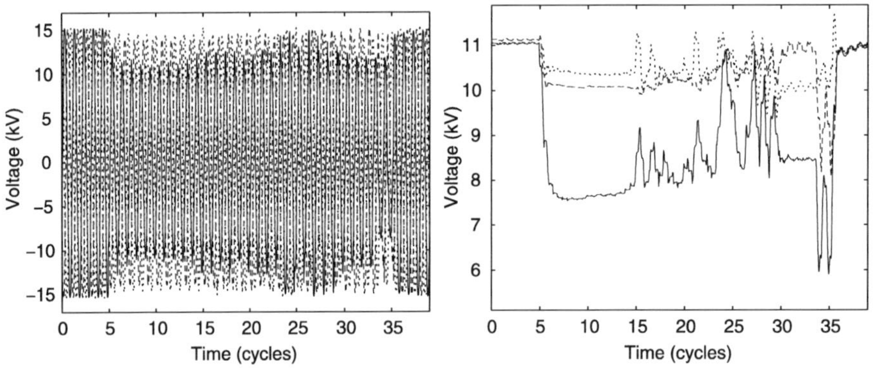


**Figure 8.37** Voltage waveforms (left) and half-cycle rms voltage (right) for voltage dip with high level of waveform distortion.

harmonic than for the third harmonic. The harmonic distortion fluctuates heavily during the remainder of the event with distortion levels exceeding 10%. After the event the distortion level returns to normal, with the second-harmonic component remaining high for a few cycles due to transformer saturation common after voltage recovery.

As a next step the harmonic voltages in the three phases are aggregated into one value. The highest absolute value in the three phases is used as the aggregated value for each sample point. The resulting harmonic components versus time are shown in Figure 8.39 for harmonics 2 through 7. The distortion is higher in the lower harmonic components than in the higher harmonic components, but their overall patterns are the same. Before and after the event, the fifth-harmonic component dominates at 2.4%, which is a normal situation for medium-voltage systems. After the event, the fifth-harmonic distortion is slightly higher (at 2.8%) than before the event. This may

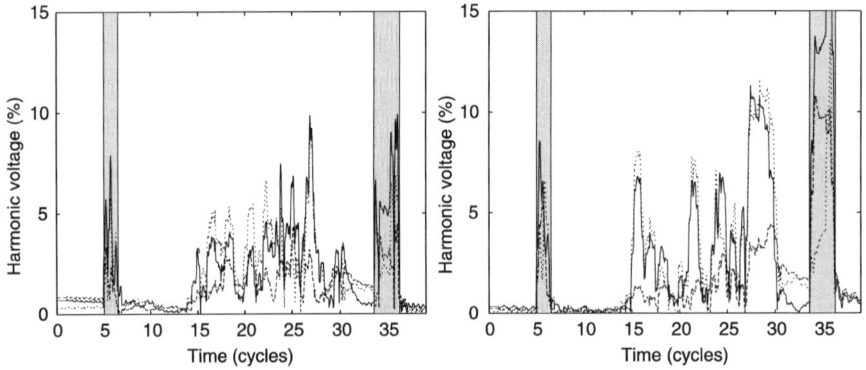


**Figure 8.38** Third-harmonic component (left) and second-harmonic component (right) as function of time for three phases.

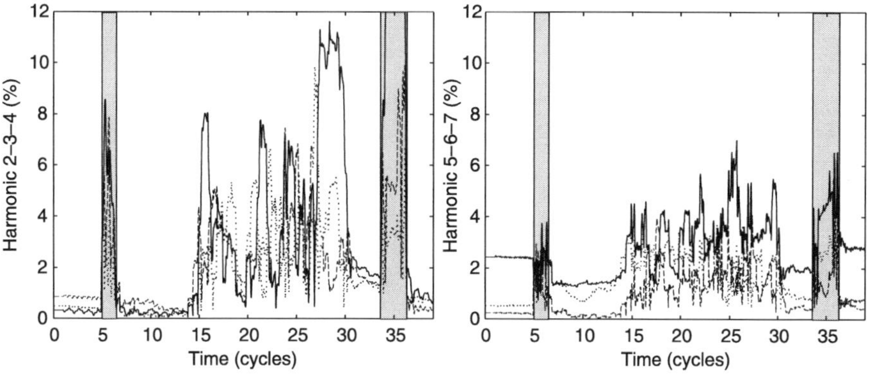


**Figure 8.39** Three-phase aggregated harmonic components as function of time. Left: second (solid), third (dashed), and fourth (dotted) harmonics. Right: fifth (solid), sixth (dashed), and seventh (dotted) harmonics.

be related to electronic equipment recharging after the dip and to postfault saturation of transformers.

To obtain single-event indices, the highest and average value for each harmonic has been calculated. Only values between the transition segments indicated before have been considered in calculating the indices. The results are shown in Figure 8.40. The average values are significantly lower than the highest values due to the heavily fluctuating character of the distortion.

The calculations leading to Figure 8.39 have been repeated for the voltage dip shown in Figure 6.15. The results are shown in Figure 8.41. The even harmonics are small before the fault, slightly higher during the fault, but significantly higher after the fault, with a maximum around 2% of nominal immediately after fault clearing. The odd harmonics are already present before the fault, with the fifth harmonic

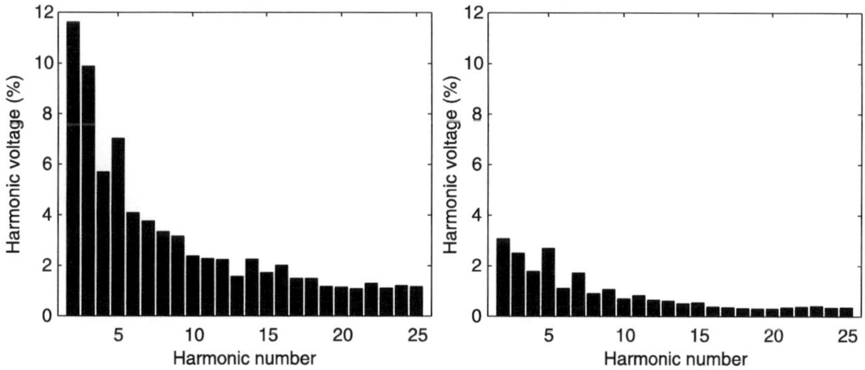


**Figure 8.40** Worst-case harmonic spectrum (left) and average harmonic spectrum (right) during the dip.

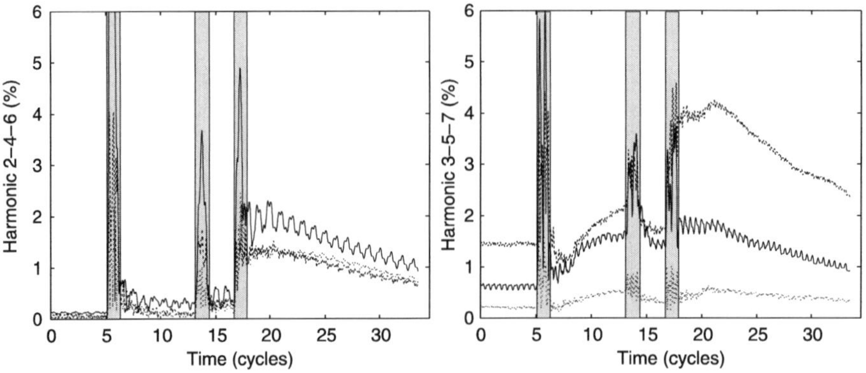


**Figure 8.41** Three-phase aggregated harmonic components as function of time. Left: second (solid), fourth (dashed), and sixth (dotted) harmonics. Right: third (solid), fifth (dashed), and seventh (dotted) harmonics.

dominating as normal. During the event the odd harmonics show a slow increase. After the event especially the fifth harmonic increases to about 4% due to the saturation of transformers upon voltage recovery.

The method of calculating harmonic distortion is not limited to voltage dips; it can also be applied to other types of events. The difference between the harmonic spectrum before and after a transient is an important characteristic for determining the cause of transients. As an example the three-phase aggregated third-, fifth-, and seventh-harmonic components have been calculated for a capacitor-energizing transient. The results are shown in Figure 8.42. Note again the transition segment

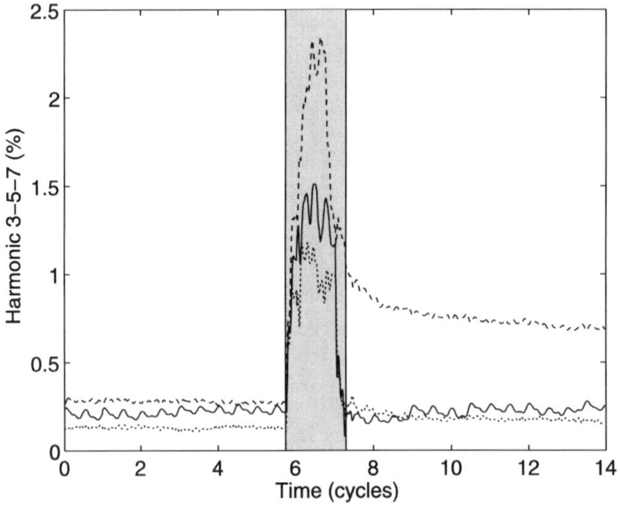


**Figure 8.42** Three-phase aggregated harmonic components as function of time for capacitor: energizing transient, third (solid), fifth (dashed), and seventh (dotted) harmonics.

indicated by a shaded rectangle. The harmonic distortion values during the transition segment should be discarded.

## 8.5 SINGLE-EVENT INDICES: INTERRUPTIONS

The main index of an interruption is its duration. The voltage magnitude is compared with a so-called interruption threshold, as shown in Figure 8.43. The duration of the interruption is the amount of time during which the magnitude of the voltage is less than the interruption threshold. Note that Figure 8.43 is not to scale: The interruption threshold is typically 1 to 10% of the nominal voltage magnitude, and the duration is several orders of magnitude longer than the time it takes for the voltage to drop from its preevent value to zero.

The voltage magnitude as a function of time may be calculated by using any of the methods discussed in Chapters 3 and 4. By far the most commonly used characteristic is the rms voltage obtained typically over a one-cycle window. The power quality measurement standard IEC 61000-4-30 defines this process in a very clear way:

- The rms voltage is calculated over a one-cycle window and updated every half-cycle.
- The start of the interruption is when the rms voltage drops below the interruption threshold. The end of the interruption is when the rms voltage recovers above the interruption threshold.
- The duration of the interruption is the difference in time between the start and the end of the interruption.

This way of definition can lead to some unexpected and most likely unintended results. Consider, for example, the event that was shown in Figure 7.3. Such an event should be described, according to IEC 61000-4-30, as two interruptions,

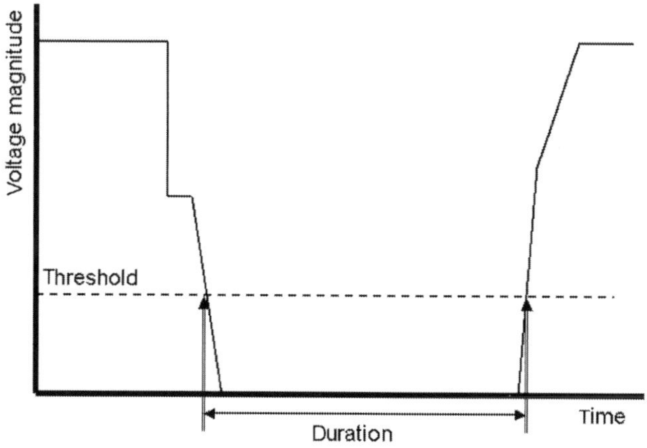


**Figure 8.43** Definition of interruption duration.

one with a duration $T_1$, the second with a duration $T_2$. This is not so much something wrong with the measurement standard but more a problem in interpreting the results of the measurements. One way of solving this is by using so-called time aggregation, to be discussed in Section 10.3.

Not much can be said about the voltage during an interruption because it is zero. Therefore the duration is in most cases the only event index that is of interest. The start and end of the interruption may however be studied in more detail, which could lead to a number of additional event indices. The initial stage of an interruption is often a voltage dip. This dip can be characterized as any other voltage dip due to a fault (see Section 6.3). The voltage recovery sometimes shows significant even-harmonic distortion due to transformer saturation. This may also be used as an additional characteristic.

For three-phase measurements, all three phases need to be considered when determining the start and end of an interruption. Many different definitions are possible here. According to the IEC standard the interruption starts when all three rms voltages drop below the threshold and ends when at least one of them rises above the threshold. In other words, the supply is only interrupted when all three voltages are (close to) zero. For single-phase measurements (e.g., for low-voltage domestic customers) only one rms voltage needs to be considered.

## 8.6 SINGLE-EVENT INDICES: VOLTAGE DIPS

### 8.6.1 Residual Voltage and Duration

Voltage dips are described by two main indices: duration and residual voltage. Both are obtained from the voltage magnitude as a function of time, as shown in Figure 8.44. The duration of a voltage dip is the amount of time during which the voltage magnitude is below the dip threshold. The dip threshold is typically chosen as 90% of the nominal voltage magnitude, although IEC 61000-4-30 suggests the use of a 70% threshold for contractual purposes. In the vast majority of cases the rms voltage is used as voltage magnitude. The power quality standard IEC 61000-4-30 prescribes the use of a one-cycle rms value that is updated every half-cycle.

The second event index, the so-called residual voltage, is defined by IEC 61000-4-30 as the lowest value of the voltage magnitude during the event. Other standard documents use the term *dip magnitude* or *sag magnitude* instead of *residual voltage*.

For measurements in three-phase systems the three rms voltages have to be considered to determine residual voltage and duration of the dip. The voltage dip starts when at least one of the rms voltages drops below the dip-starting threshold. The dip ends when all three voltages have recovered above the dip-ending threshold. The residual voltage of the voltage dip is the lowest one-cycle rms voltage in any of the three phases. Note the difference with the definition for interruptions in three-phase systems. The duration of an interruption is the time during which all three rms voltages are below the threshold. For a dip the duration is the time during

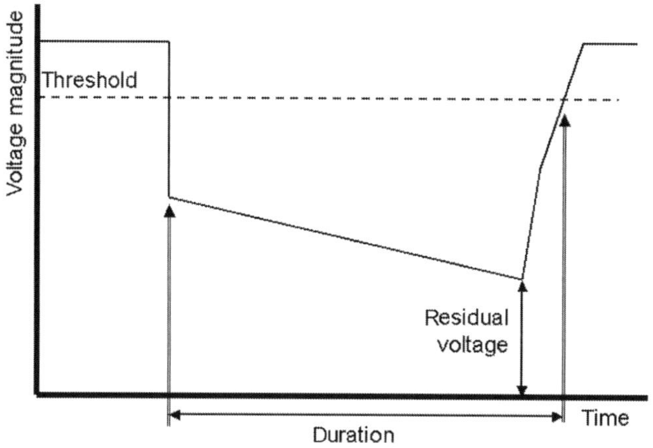


**Figure 8.44** Definition of duration and residual voltage of voltage dip.

which at least one of the rms voltages is below the threshold. It is thus possible that an event is classified as a dip with zero residual voltage but not as an interruption. This is the case when the voltage is zero in only one or two phases. If the voltage is zero in all three phases the event is classified as a voltage dip and as an interruption.

### 8.6.2 Depth of a Voltage Dip

Many of the earlier studies on voltage dips use the actual drop in voltage instead of the residual voltage to characterize the event. The advantage is that a small value of the drop corresponds to a mild dip. When using residual voltage the somewhat strange situation occurs that a small value corresponds to a severe event. The split has long been between European publications which referred to voltage dips and used the drop in voltage and American publications that used the term *voltage sag* and the residual voltage (referred to as *sag magnitude*) as a characteristic. However, the last few years the general tendency has been to use the residual voltage instead of the drop in voltage.

The depth of a voltage dip is the difference between the residual voltage and a reference voltage. To be able to use the depth, it is thus essential to define whether the nominal voltage or the preevent voltage is used as a reference. But as both the depth and the residual voltage are typically expressed in percent, one needs a reference voltage in both cases anyway.

### 8.6.3 Definition of Reference Voltage

The thresholds needed to detect the beginning and end of a voltage dip are typically given as a percentage of a reference voltage. This reference voltage may be the nominal voltage or the preevent voltage. The choice of reference voltage is far from trivial and has already resulted in many discussions in standard-setting working groups. For measurements close to the terminals of sensitive equipment the situation is still rather clear. The equipment is designed to operate at a certain voltage. Setting the threshold as a percentage of this voltage would be most appropriate. For low-voltage equipment in Europe the nominal voltage of 230 V should be used. In the United States a reference value of 120 V is more appropriate. The first problems, however, already occur here. Some equipment is capable of operating for a wide range of supply voltages, for example, from 90 to 250 V. For such equipment a 70% residual voltage in a 120-V system is more severe than a 70% residual voltage in a 250-V system. The former will likely lead to tripping of the equipment, the latter not. For low-voltage systems there is however wide agreement that the voltage-dip threshold should be expressed as a percentage of the nominal voltage.

For MV and HV systems the situation becomes more complicated. The measured voltage is no longer the same as the voltage at the equipment terminals. The effect of transformer-winding connections and the load influence on the voltages during the fault were already discussed in Section 6.2.3. That leaves us with the effect of transformer on-load tap changers on the dip. The tap changers do not directly affect the voltage dip as they react far too slowly to be of any influence. But they do affect the reference voltage. The control principle of on-load tap changers is such that the voltage in the distribution system is always around the same range independent of the voltage in the transmission system. Even off-load tap changers are set according to this principle. If this principle would be valid in all cases, the preevent voltage at the equipment terminals would always be close to the nominal voltage. The residual voltage at the equipment terminals, in percentage of the nominal voltage, would thus be equal to the residual voltage at the measurement location, in percentage of the preevent voltage. This reasoning is used sometimes to set the voltage-dip threshold as a percentage of the preevent voltage. An additional reasoning for doing this is the practice of some network operators to have typical operating voltages that differ from the nominal voltage, for example, 140 kV typical operating voltage in a 132-kV network. Discussion on the choice of the preevent voltage will continue for a while. Currently there is simply not enough objective material available to give general guidelines. Therefore the IEC power quality measurement standard allows for both methods but indicates that for LV measurements the nominal voltage should in all cases be used as a reference.

### 8.6.4 Sliding-Reference Voltage

Deciding to use the preevent voltage as a reference for the voltage-dip threshold does not immediately solve all problems as the preevent voltage is not constant. Using the last cycle before the dip would result in another value than using the last 10-min average before the dip. An important aim of the IEC power quality measurement standard is to ensure that different monitors give the same result. The term *sliding-reference voltage* has been introduced to define the way in which the preevent voltage should be measured. The basis for the calculation of the preevent voltage is formed by the 10/12-cycle rms voltages, as discussed in Chapter 3. The sliding-reference voltage $U_{sr}$ is the output of a first-order filter with a time constant equal to 1 min and the 10/12-cycle rms voltages $U_{10/12}$ as input:

$$U_{sr}(n) = \left(1 - \frac{1}{N}\right) \times U_{sr}(n-1) + \frac{1}{N} \times U_{10/12}(n) \tag{8.43}$$

with $N = 300$ the average number of 10/12-cycle values in 1 min. If a number of dips occur shortly behind each other, it is important to not let a dip affect the preevent voltage for the next dip. Using (8.43) as it is would result in a reduced value for the preevent voltage as estimated for the next dip. Therefore any flagged 10/12-cycle value is not considered in (8.43). If $U_{10/12}(n)$ is flagged, $U_{sr}(n) = U_{sr}(n-1)$. As mentioned in Section 5.2.5 a 10/12-cycle interval is flagged when any of the one-cycle rms values in the interval exceed the dip, swell, or interruption threshold.

When using a low value of the voltage-dip threshold (e.g., 70%), shallow dips will not lead to flagging of the 10/12-cycle interval. During periods with high voltage-dip activity (e.g., a lightning storm) the sliding-reference voltage may be reduced due to shallow dips. Not enough statistical data are available to assess how serious this problem is. If one is only interested in dips with a residual voltage below 70%, one may decide to still use a 90% threshold and make a selection later based on residual voltage.

### 8.6.5 Multiple-Threshold Setting

In IEC 61000-4-30 the possibility is given to use two threshold values: a dip-starting threshold and a dip-ending threshold. The choice of the dip-ending threshold can affect the estimated duration of the dip. It may even affect the number of dips that are detected. The need for multiple threshold setting will be discussed when treating multiple events in Section 10.3.

The *interruption threshold* (typically 10%) is used to determine the duration of an interruption. The *voltage-dip threshold* (typically 90%) is used to determine the duration of a voltage dip. According to the most recent definitions, any event with a residual voltage below a dip threshold is called a voltage dip, even an event with zero residual voltage. This could lead to events being classified as a voltage dip and as an interruption. Consider as an example the hypothetical event that was shown in Figure 7.3. From a voltage dip viewpoint this is one event with duration $T_3$ and zero residual voltage, whereas from an interruption viewpoint these are two events with durations $T_1$ and $T_2$. A power quality monitor that complies with IEC 61000-4-30 should keep separate records for voltage dips and for interruptions. When an event is both a dip and an interruption, the duration of the dip is not necessarily the same as the duration of the interruption.

### 8.6.6 Uncertainty in Residual Voltage

The IEC power quality standard sets requirements for the uncertainty in the measurement of residual voltage and maximum swell voltage. The maximum permitted uncertainty is 0.2% for class A instruments and 1.0% for class B instruments. The standard document is not clear on what is included in the uncertainty. It makes sense to not include any uncertainty in the actual measurement process (the instrument transformers). We saw in Section 8.1.1 that the location of the rms calculation window with reference to the start or end of the dip can result in an error in rms voltage and thus in residual voltage. We will assume that this phenomenon is not considered in the uncertainty requirements. The uncertainty requirements in residual voltage are thus the same as the uncertainty requirements in rms voltage.

The two contributions to the uncertainty are the resolution due to the number of bits used in representing the rms voltage and the error in estimated rms due to the error in frequency. The latter term can be estimated from (3.17):

$$\Delta Y_{\text{rms}} \approx \tfrac{1}{2}\delta_T f \cos(4\pi f t) \tag{8.44}$$

with $\Delta Y_{\text{rms}}$ the relative error in estimated rms value, $\delta_T$ the absolute error in the window length (i.e., the difference between the window length used to calculate the rms value and the actual length of one cycle of the power system frequency), and $f$ the power system frequency. Expression (8.44) can be rewritten as the relative error in cycle length equals the relative error in frequency. This results in the following surprisingly simple relation between the relative error in rms voltage and the relative error in frequency:

$$\frac{\Delta U}{U} = \frac{1}{2}\frac{\Delta f}{f} \tag{8.45}$$

The above-mentioned uncertainty requirements thus translate into a 0.4% uncertainty in frequency for class A instruments and a 2% uncertainty for class B instruments. In large interconnected systems the frequency stays most of the time within 0.2% of its nominal value but during severe disturbances it may deviate up to a few percent. See the discussion on frequency variations in Section 2.1. The standard document does, however, clearly mention the conditions under which an instrument should fulfill the accuracy requirements. The range of frequencies given in the document is 42.5 to 57.5 Hz for 50-Hz systems and 51 to 69 Hz for 60-Hz systems, which is 15% in both cases. To fulfill the standard requirements it is necessary to include a phase-locked loop or another accurate method to estimate the actual power system frequency, even for a class B instrument. However, the accuracy requirements appear rather severe on this point considering all the other sources of error.

### 8.6.7 Point on Wave

The duration of a voltage-dip event is defined as the time below threshold for the rms voltage. However, there remains some interest in more accurate measures of the duration of the event as well as information on where in the voltage sine wave a dips starts and ends. We will discuss both in more detail below.

The standard method for calculating dip duration has two obvious and well-known shortcomings. The first is the uncertainty in the starting and ending instants due to the use of a one-cycle rms voltage updated every half-cycle. The resulting uncertainty in duration is not easy to determine, partly because there is no agreement about what the "real duration" of a voltage-dip event is. However, it is not hard to see that the uncertainty in duration will be around one cycle for most dips. The second shortcoming of the standard method concerns events with a slowly recovering voltage. The ending instant becomes an arbitrary point depending strongly on the threshold setting. One can no longer talk about an uncertainty in duration because there is nothing with which to compare the results.

There is some interest in more accurate measures of the time between fault initiation and fault clearing and any changes in fault character in between. Note that this information is only of concern for dips due to faults. For motor-starting and transformer-energizing dips there is no single recovery point. Note also that this information is not a characteristic for the purpose of obtaining performance indicators, but merely some additional information related to individual events. This information may be used for network operators to improve their supply. For such applications the need for a precision less than one cycle is however limited.

Another reason for the interest in point-on-wave information is related to the immunity of equipment against voltage dips. The performance of some equipment depends on the point on wave at which the voltage drops or recovers. The commonly quoted example concerns motor contactors. A dip at voltage maximum corresponds to minimum magnetic energy in the contactor, whereas a voltage zero crossing corresponds to maximum magnetic energy [70, 71, 130, 239, 300]. Studies of the static transfer switch for voltage-dip mitigation have also shown a dependence of the transfer time on the point on wave of fault initiation [264, 265]. Much sensitive equipment takes a rather large inrush current upon voltage recovery. This current peak could lead to damage to the equipment, which is one of the reasons for the equipment's sensitivity to dips. The inrush current upon voltage recovery depends on the speed with which the voltage recovers and thus on the point on wave of voltage recovery. Statistics on point on wave are therefore important to predict equipment performance.

From the above discussion it becomes clear that the concept of dip duration needs some more precise definition before it can be measured more accurately. This discussion has been somewhat missing from the discussion on dip characteristics. The only points that can be accurately defined, because they are physically existing instants, are fault initiation, fault clearing, and any changes in fault character in between. To uniquely determine a point-on-wave index it has to be assumed that there is only one such point. There are however a number of cases in which there are multiple points:

- Some faults develop from single-phase to two- or three-phase faults. If the fault development takes place over a time scale of several cycles, the point on wave can be determined for each change in fault character. However sometimes the fault develops very quickly, that is, within a few milliseconds. A method for obtaining the point on wave should be able to detect all the steps. For the interest of obtaining performance statistics a unique definition is needed so that only one of these steps is considered.

- For two- and three-phase faults fault clearing takes place in a number of steps (see Section 6.2.5). The method either has to find all steps or a clear definition is needed.
- Typical sampling frequencies for voltage-dip recordings are between 100 and 250 samples per cycle. The anti-aliasing filter used means that the time resolution will be limited to about two samples. The accuracy of the point-on-wave characteristic will thus not exceed about $5^\circ$.
- Fault initiation takes place very fast, but voltage recovery is slowed down by the presence of capacitance in the system. This will make it even more difficult to obtain accurate values for the point-on-wave indices. The frequency limitation in the voltage transformers used for the measurements may further limit the time resolution.
- Fault clearing at the transmission level takes place at two sides of the line at different instants. The number of recovery instants doubles in that case.
- Voltage dips due to motor starting and transformer energizing do not have a recovery point.
- Even during a fault, the voltage is not really constant.

Several methods have been proposed to extract more accurate information on starting and ending points. Some methods use a cycle-by-cycle comparison of the voltage waveform to find where the change in waveform occurs. These methods require some rather tedious administration, but their results have a reasonable repeatability. An example of such a method is discussed in detail in [117]. The method uses the difference between preevent and during-event waveforms to determine the initiation instant.

Other methods are based on high-frequency components (e.g., wavelets) or on the residue of model-based methods (e.g., the Kalman filter). Such methods are more straightforward but lead more easily to spurious detections. These methods have been discussed in detail in various sections in Chapter 7.

None of the methods discussed in the literature are able to provide satisfactory results. A possible approach would be to determine the transition segments and event segments as discussed in Section 7.4 and to look closer within the transition segments for the actual instant at which the changes take place. An expert system can be used to give information on the number and character of transitions that are expected. For example, if the transition segment is the fault clearing of a two-phase-to-ground fault, two or four recovery points are expected.

Next to the development of such an algorithm, more work is needed to determine the relation between the voltage recovery details and equipment performance.

Characteristics need to be defined that have a direct relation to the performance of equipment.

### 8.6.8 Phase-Angle Jump

From the phase angle as a function of time, a single-event index (or single-event characteristic) can be calculated. This can be done in a similar way as for the residual voltage, which is defined as the lowest value of the rms voltage versus time. Currently, no standard document exists that defines how the phase angle versus time can be used to obtain single-event indices. For single-phase measurements the extreme values of the phase angle versus time can be used, where it is important to exclude any values during the transition segments. Especially for events with a small phase-angle jump the error may be large otherwise.

Summarizing we may state that defining the phase-angle jump as a single-event index for a voltage dip in a consistent way is even more difficult than for the residual voltage. Coming up with a unique definition is the task of standard-setting organizations. However, any definition should be structured along the following lines:

- The phase angle of the voltage is calculated as a function of time. The exact frequency is used, for example, by means of a phase-locked loop or the pre-event phase-angle drift is corrected afterward.
- The values during the transition segments are removed from further assessment.
- The phase-angle jump is the maximum or minimum value of the phase angle versus time.
- An appropriate method of phase aggregation is used.

For the event in Figure 8.19 this will result in phase-angle jumps equal to $-24.5^\circ$, $-23.9^\circ$, and $-24.2^\circ$ for phases $a$, $b$, and $c$, respectively. But for some events, such as the one in Figure 8.20, both positive and negative phase angles occur outside the transition segments. In such a case both positive and negative phase-angle jumps will have to be defined. In this case the negative phase-angle jumps are $-7.5^\circ$, $-7.8^\circ$, and $-7.8^\circ$ for phases $a$, $b$, and $c$, respectively. The positive phase-angle jumps are $+4.6^\circ$, $+4.4^\circ$, and $+4.6^\circ$ for phases $a$, $b$, and $c$ respectively.

The situation becomes even more complicated for three-phase events. Both examples in the previous section were for balanced dips where the difference between the phases is small. An unbalanced dips would typically result in three different values for the phase-angle jump. Consider, for example, the event in Figure 8.18. Disregarding the values during the transition segment as well as before and after the event leads to the following phase-angle jumps: $-1.3^\circ$ for phase $a$, $-8.0^\circ$ for phase $b$, and $+7.7^\circ$ for phase $c$. Combining these three values into one value for the three-phase event is not straightforward. In line with the method used for residual voltage, the most extreme value should be used: $-8.0^\circ$ in this case. When both positive and a negative phase-angle jumps are defined,

**TABLE 8.11 Phase-Angle Jump for Selected Dips**

| Event | Positive | | | Negative | | | Three-Phase Event | |
|---|---|---|---|---|---|---|---|---|
| | *a* | *b* | *c* | *a* | *b* | *c* | Positive | Negative |
| Figure 8.19 | 0 | 0 | 0 | −24.5° | −23.9° | −24.2° | 0 | −24.5° |
| Figure 8.20 | +4.6° | +4.4° | +4.6° | −7.5° | −7.8° | −7.8° | +4.6° | −7.8 ° |
| Figure 8.18 | 0 | 0 | +7.7° | −1.3° | −8.0° | 0 | +7.7° | −8.0° |

the values for the three-phase event would become −8.0° and +7.7°. The various indices for the three examples are summarized in Table 8.11.

Instead of calculating the phase angle versus time for each phase, the phase angle versus time may be calculated for the characteristic voltage. This results in a more consistent method for defining the phase-angle jump of a three-phase event. We will discuss single-event indices based on three-phase characteristics in Section 8.8.

The rms voltage and the phase angle as a function of time during a voltage dip due to transformer energizing are shown in Figure 8.45. The voltages were measured in a 10-kV substation and caused by the reenergizing of a distribution feeder. The voltage magnitude drops in all three phases and slowly recovers. The exponential recovery and the unbalance between the phases are a typical signature of transformer saturation. The phase angles are small, not more than a few degrees. This is understandable when realizing that the current taken by a saturated transformer is mainly reactive. This current does not cause any significant phase-angle jump over the (reactive) source impedance.

Note that the source impedance is mainly formed by the transformer from a higher voltage level. To energize MV/LV transformers (e.g., 10 kV/400 V), for example during fast auto reclosure, the source impedance is mainly formed by the HV/MV transformer (e.g., 130/10 kV).

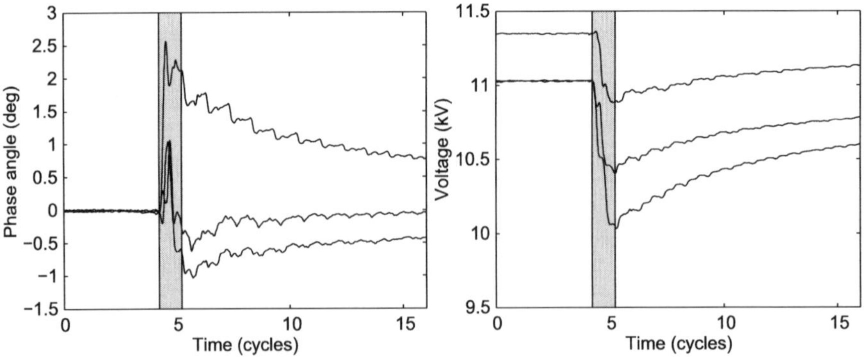


**Figure 8.45** Phase angle (left) and rms voltage (right) as function of time for transformer-energizing dip.

### 8.6.9 Single-Index Methods

Some characterization schemes have been proposed in which only one single-event index (instead of residual voltage and duration) is used. This has advantages in comparing event severity and in calculating site and system indices. We will discuss below two single-index methods that are currently being considered for inclusion in the IEEE standard on voltage-dip indices, IEEE 1564 [172].

***8.6.9.1 Voltage-Dip Energy Index*** The *voltage-dip energy* $\boldsymbol{E}_{\text{dip}}$ is defined as the integral of the drop in signal energy over the duration of the event:

$$\boldsymbol{E}_{\text{dip}} = \int_0^T \left[ 1 - \left( \frac{U(t)}{U_{\text{nom}}} \right)^2 \right] dt \tag{8.46}$$

where $U(t)$ is the rms voltage versus time as obtained from any of the methods discussed before and $U_{\text{nom}}$ is the rated voltage. Alternatively the "declared voltage" as in IEC 61000-4-30 can be used where this deviates from the nominal voltage. An important issue in any index that is based on an integral is to very clearly define the integration window. An appropriate rule is to perform the integration over the duration of the event, that is, for all values of the rms voltage below the voltage-dip threshold. A direct consequence of this is that events for which the rms voltage does not drop below the dip threshold have zero voltage-dip energy.

The rms voltage in (8.46) can be expressed in any unit (volts, kilovolts, per-unit, percent) as long as the nominal voltage is expressed in the same unit. The voltage-dip energy has the unit of time and may be expressed in cycles, milliseconds, or seconds.

When the rms voltage is calculated over a one-cycle window updated every half-cycle (as in IEC 61000-4-30), the integration in (8.46) becomes a summation:

$$\boldsymbol{E}_{\text{dip}} = \frac{1}{2f_0} \sum_{k=1}^{N} \left[ 1 - \left( \frac{U_{\text{rms}(1/2)}(k)}{U_{\text{nom}}} \right)^2 \right] \tag{8.47}$$

with $f_0$ the power system frequency. The summation is again taken over all values of the rms voltage below the threshold. The lowest possible nonzero value is obtained for an event with one rms sample just below the threshold and all other samples above the threshold. If we assume the threshold to be at 90% of the nominal voltage, we get 0.095 cycle (1.58 ms at 60 Hz; 1.9 ms at 50 Hz) for the lowest nonzero value.

If only residual voltage and duration of an event are available (as may be the case for many databases and for older generation monitoring equipment), the voltage-dip energy can be estimated by assuming that the rms voltage is constant over the duration of the event. This results in the following expression for the voltage-dip energy:

$$\boldsymbol{E}_{\text{dip}} = \left[1 - \left(\frac{V}{U_{\text{nom}}}\right)^2\right] \times T \tag{8.48}$$

with $T$ the duration and $V$ the residual voltage of the event.

The voltage-dip energy can be interpreted as the duration of an equivalent interruption leading to the same loss of energy for an impedance load as the voltage dip. This points immediately to a potential disadvantage in the use of this index. As the unit is the same as for the duration of interruptions, somebody may decide to simply add the results with the duration of interruptions to obtain a site index over a certain period. The contribution of all voltage dips will then easily be dominated by just one single interruption. In Table 8.12 a comparison has been made for a number of typical events, using (8.48), where it should be noted that the voltage-dip energy for interruptions is equal to their duration. One interruption is equivalent to 2000 severe dips or 50,000 minor dips. When using the voltage-dip energy to calculate site and system indices, it is extremely important that dips, short interruptions, and long interruptions be treated separately. Alternatively a maximum duration or a maximum value for the voltage-dip energy may be defined for any individual events. From Table 8.12 it follows that 1 s would be an appropriate value for such a limit.

To interpret the voltage-dip energy consider a constant-impedance load with active power consumption $P_0$ at nominal voltage $V_{\text{nom}}$. When the voltage drops to a value $V$, the active power consumption by the load drops to

$$P = \left(\frac{V}{U_{\text{nom}}}\right)^2 P_0 \tag{8.49}$$

**TABLE 8.12 Voltage-Dip Energy for "Typical Events"**

| Event | Residual Voltage (%) | Duration | Voltage-Dip Energy (s) |
|---|---|---|---|
| Shallow, short dip | 80 | 100 ms | 0.036 |
| Deep, short dip | 30 | 100 ms | 0.091 |
| Long, shallow dip | 80 | 1 s | 0.36 |
| Long, deep dip | 30 | 1 s | 0.91 |
| Short interruption | 0 | 3 s | 3 |
| Long interruption | 0 | 30 min | 1800 |

The reduction in power delivered to the load is equal to

$$\Delta P = P_0 \left[ 1 - \left( \frac{V}{U_{\text{nom}}} \right)^2 \right] \tag{8.50}$$

Integration over the duration $T$ of the voltage dip gives an expression for the nondelivered energy:

$$\Delta E = \int_0^T \Delta P \; dt = P_0 \int_0^T \left[ 1 - \left( \frac{V}{U_{\text{nom}}} \right)^2 \right] dt \tag{8.51}$$

Using the definition for voltage-dip energy in (8.46), the following expression is obtained:

$$\Delta E = P_0 E_{\text{dip}} \tag{8.52}$$

Thus the voltage-dip energy is equal to the nondelivered energy to a constant-impedance load, expressed in per-unit with the rated power of the load as a base.

A *voltage-swell energy* can be defined in the same way as the voltage-dip energy:

$$\boldsymbol{E}_{\text{swell}} = \int_0^T \left[ \left( \frac{U(t)}{U_{\text{nom}}} - 1 \right)^2 \right] dt \tag{8.53}$$

where the variables have the same meaning as in (8.46). The same expressions as in (8.47) and (8.48) can be obtained for the voltage-swell energy.

#### 8.6.9.2 Voltage-Dip Severity

The voltage-dip severity is calculated from the residual voltage and the duration of a voltage-dip event in combination with a reference curve. In the current draft of IEEE 1564 [172] the SEMI (Semiconductor Equipment and Materials International) curve is recommended as a reference, but the method works equally well with other methods.

From the residual voltage $V$ and the event duration $T$ the voltage-dip severity of the event is calculated from the following expression:

$$S_e = \frac{1 - V}{1 - V_{\text{curve}}(T)} \tag{8.54}$$

where $V_{\text{curve}}(T)$ is the residual-voltage value of the reference curve for the same duration $T$. For an event on the reference curve, the voltage-dip severity is exactly equal to 1; for an event above the reference curve the index is less than 1; for an event below the curve the index is greater than 1. For events with a residual voltage above the voltage-dip threshold the voltage-dip severity is set equal to zero.

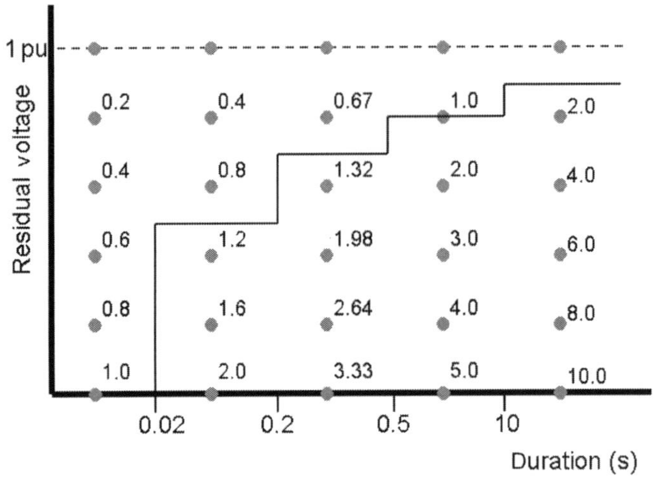


**Figure 8.46** Voltage-dip severity with reference to SEMI curve (solid curve). The values with the dots indicate the voltage-dip severity for voltage dips with that residual voltage and duration.

These events are normally not considered as dips when obtaining statistics. This implies that there is again a lowest nonzero value for this index.

The method is further explained in Figure 8.46. The longer the event duration and the lower the residual voltage, the higher the voltage-dip severity. But contrary to the voltage-dip energy, the resulting value is automatically limited when the reference curve starts to become horizontal. The SEMI curve does not exceed 90% of nominal so that the highest value of the voltage-dip severity is equal to 10.

Using the SEMI curve as a reference gives the following algorithm for calculating the voltage-sag severity:

$$
\begin{array}{rl}
T \leq 20\,\text{ms} & S_e = 1 - V \\
20\,\text{ms} < T \leq 200\,\text{ms} & S_e = 2(1 - V) \\
200\,\text{ms} < T \leq 500\,\text{ms} & S_e = 3.3(1 - V) \\
500\,\text{ms} < T \leq 10\,\text{s} & S_e = 5(1 - V) \\
T > 10\,\text{s} & S_e = 10(1 - V)
\end{array}
$$

## 8.7 SINGLE-EVENT INDICES: VOLTAGE SWELLS

Voltage swells are the opposite event to voltage dips. A voltage swell is a short-duration increase in rms voltage. The basic measurement is the same for dips as for swells: the one-cycle rms voltage updated every half-cycle (for class A compliance with IEC 61000-4-30). The starting of the event is detected when at least one of the three rms voltages exceeds a swell threshold. When all three voltages are back below the threshold, the end of the event is detected. The duration is defined as the difference between the start and end of the event. The threshold value is set as a percentage of the nominal or sliding-reference voltage. The discussion held for voltage dip on this issue also applies to voltage dips. A typical value for the swell-starting threshold is 110% of nominal, with the swell-ending threshold typically being 2% lower.

Comparing the definitions of swell, dip, and interruptions shows that the same event may be classified as a voltage dip, as a voltage swell, and as an interruption. An example of such an event is the one that was shown in Figure 7.2. Using the classification defined in IEC 61000-4-30 leads to the conclusion that this event is a voltage swell with duration $T_1$, a voltage dip with duration $T_1 + T_2 + T_3$, and an interruption with duration $T_3$. The standard document does not give any recommendation on how to solve this obvious case of double counting. This will be up to the user of the monitor and depend strongly on the application.

## 8.8 SINGLE-EVENT INDICES BASED ON THREE-PHASE CHARACTERISTICS

The characteristics versus time for three-phase unbalanced voltage dips can be used as the basis for single-event indices. The number of options is rather large, but we will give a characterization method along the lines of IEC 61000-4-30:

- The characteristic voltage, PN factor, and dip type are calculated over a one-cycle window. Either of the two methods discussed before may be used for this. The six-phase algorithm is part of the current draft of an IEEE standard on voltage-dip indices (IEEE task force P1564). The characteristic voltage is given in that document as one of the alternatives for calculating the magnitude of voltage dips in three-phase systems.
- The calculation of the characteristics versus time is updated every half-cycle between the transition segments.
- The residual voltage is the lowest absolute value of the characteristic voltage during the event (i.e., between the transition segments).
- The PN factor of the event is the lowest value of the PN factor during the event.
- The dip type of the event is the dip type at the instant of the lowest absolute value in characteristic voltage.
- The duration of the event is the time during which the absolute value of the characteristic voltage is below the dip threshold.

## 8.9 ADDITIONAL INFORMATION FROM DIPS AND INTERRUPTIONS

Several examples of voltage dips and interruptions were presented in this chapter and in Chapter 6. The single-event indices that were presented were mainly aimed at obtaining statistics to quantify the performance of the supply at a certain location. Most of the indices in use do not incorporate any information about the cause of the event. This is justifiable if one considers that the impact of an event on end-user equipment is only determined by its waveform and not by its cause. Thus to quantify the performance of the supply the origin of the event does not need to be considered.

There are however a number of reasons why information about the origin of a power quality event is of importance. Those reasons are listed below. In addition, the voltage and current recordings in some cases also provide diagnostic information on the power system.

- During "troubleshooting" the origin of an event leading to equipment maltrip or damage will give essential information about the mitigation method. For example, if equipment trips are due to the starting of a large motor from the same bus, the problem may be solved by moving the motor to another bus.
- This same argument can be extended to the presentation of statistics. When deciding about methods to improve the voltage quality at a given location, it is important to know what are the predominant causes of the events. An important issue in collecting statistics is whether the events originate at the customer premises or elsewhere in the system.
- Voltage dips are due to short circuits and earth faults. Faults are not only the cause of troublesome voltage dips for customers, they are also a serious threat to reliable and secure operation of the power system as a whole. Reducing the number of faults is thus an ongoing concern for all network operators. Voltage-dip recordings contain information about these faults that may be used to help reduce the number of faults.
- The duration of voltage dips is mainly determined by the speed of operation of the power system protection. Information on the voltage recovery instants thus provides information on the operation of the protection.

An expert system is presented in [282] in which voltage dips and interruptions are classified based on their origin. The expert system is based on the segmentation method presented in Section 7.4 and on a comparison between different event segments and between the event segments in different phases. The voltage recording is divided into transition segments and event segments. A transition segment corresponds with a change in the voltage waveform, in most cases a change in voltage magnitude (rms voltage). Event segments are bordered by transition segments and by the beginning and end of the recording. The number of event segments is always one more than the number of transition segments. A subdivision is made in the following classes of events, where the description has been somewhat generalized as compared to [282]:

*Energizing (End of an Interruption)* The recording consists of two event segments (one transition segment). The voltage is zero at the start of the recording (the first event segment) and within the normal voltage range or tending toward the normal voltage range at the end of the recording (the second event segment).

*Fault Interruption (Start of Interruption)* The recording consists of three or more event segments. The voltage is within the normal operating range in the first event segment and zero or tending toward zero in the last event segment. The voltages in the middle event segments are characteristic of a voltage dip; this may be a balanced dip or an unbalanced dip. In the latter case an rms overvoltage may be present in one or two phases during the middle event segment. The event can be further classified based on the voltage dip in the middle event segments.

*Nonfault Interruption (Start of Interruption)* The recording consists of two event segments. The voltage is within the normal operating range in the first event segment and zero or tending to zero in the last event segment. There are some cases when an automatic classification algorithm has difficulties in distinguishing between a fault interruption and a non–fault interruption. This is, for example, the case for a three-phase fault at the distribution level leading to an interruption. Due to the radial character of the network the rms voltage already drops to zero during the fault.

The transition segment for non-fault interruption may be rather long when the event is associated with an overvoltage.

*Transformer Energizing* The recording consists of two event segments. The second event segment shows an exponentially recovering voltage that is unbalanced between the phases. The voltage is also associated with a high even-harmonic distortion.

*Induction Motor Starting* Similar to transformer energizing this event consists of two event segments. The second segment again shows a gradually recovering rms voltage, but in this case the rms voltage is the same in the three phases. No significant even-harmonic distortion is present during any of the event segments.

*Step Change* The recording consists of two event segments. The transition segment is not a dip for any of the three phases. The voltage is in the normal operating range during both event segments and is higher or lower in the second event segment than in the first event segment.

*Transformer Saturation and Protection Operation* The recording consists of three event segments. The voltages during the first and last event segments are within the normal operating range. The voltages in the middle event segment show the exponential recovery, the unbalance, and the even-harmonic distortion associated with transformer energizing.

*Single-Stage Voltage Dip Due to a Fault* The recording consists of three event segments. The magnitude of the voltage in at least one of the three phases is below the dip threshold during the middle segment. The first transition segment is associated with fault initiation. The second transition segment is associated with fault clearing. The voltage during the middle transition segment (i.e., during the fault) may be balanced or unbalanced.

*Multistage Voltage Dip Due to a Fault* The recording consists of more than three event segments. The first and last event segments show an rms voltage within the normal operating range. The voltage during at least one of the middle segments is below the dip threshold in at least one of the three phases. The change in magnitude during the dip may be due to a change in fault type or due to clearing of the fault by one circuit breaker only ("change in system"). To distinguish between these two changes, the three-phase classification introduced in Section 6.2.3 may be used. If the dip changes type during a transition segment, that segment is associated with a change in fault. If the dip changes magnitude but not type, the transition segment is associated with an opening of a circuit breaker.

The expert system has been applied to 962 recordings that were obtained in a MV network. The results of the main classification are presented in Table 8.13. About half of the recordings are single-stage dips due to faults. The total number of events due to faults (i.e., including the multistage ones) is almost 60% of all recordings. Interruptions are responsible for about 20% of the events and transformer saturation for about 15%.

Some additional results from the analysis are presented in Tables 8.14 and 8.15. Of the 962 events, 196 were associated with an overvoltage, either a transient or a steady-state overvoltage. Most overvoltages were associated with non-fault interruptions and with fault-related swells. For the single-stage dips due to faults, the voltage-dip type was calculated, resulting in Table 8.14. Of the 455 single-stage dips due to faults, 50% were classified as type D, 38% as type C, and 12% as type A.

The reader should note that these statistics are only given as an example of the kind of information that can be extracted. It is not possible to draw any general conclusions from these statistics. The kind of system, the location and setting of the monitors, and random circumstances have a strong influence on the statistics. The

**TABLE 8.13 Classification Results Obtained from Expert System**

| Event | Number of Events |
|---|---|
| Energizing (end of interruption) | 104 |
| Fault interruption | 13 |
| Non-fault interruption | 88 |
| Transformer energizing during autoreclosure | 37 |
| Other transformer energizing | 82 |
| Transformer energizing followed by protection operation | 6 |
| Step change, increase | 15 |
| Step change, decrease | 21 |
| Single-stage dip due to a fault | 455 |
| Multistage dip due to fault, fault change | 56 |
| Multistage dip due to fault, system change | 56 |
| Nonclassified, short-duration dip | 16 |
| Nonclassified, overvoltage | 13 |

**TABLE 8.14 Further Analysis Results: Overvoltages**

| Overvoltage Associated With | Number of Events |
|---|---|
| Energizing (switching transient) | 17 |
| Non-fault interruption | 49 |
| Transformer energizing (harmonic overvoltage) | 7 |
| Step change, increase | 13 |
| Step change, decrease | 6 |
| Fault related, fault initiation | 10 |
| Fault related, swell | 86 |
| Fault related, fault clearing | 8 |

**TABLE 8.15 Further Analysis Results: Voltage Dip Type (Single Stage Dips Only)**

| Dip type | A | Ca | Da | Cb | Db | Cc | Dc |
|---|---|---|---|---|---|---|---|
| Number of events | 55 | 55 | 59 | 55 | 104 | 64 | 63 |

high number of non-fault interruptions is related to the location of some of the monitors. Most of these interruptions were due to switching actions in the distribution system in preparation of preventive maintenance. Not all of these interruptions resulted in interruptions for low-voltage customers. The high percentage of non–fault interruptions with overvoltages is also related to this. The number of swells is relatively low because a large fraction of the monitors are connected phase to phase. These measurements were obtained in a resistance-earthed distribution system. The result is that single-phase faults lead to shallow dips. The setting of the monitors was such that most of these dips were captured. Removing the shallow dips from the results would have reduced the number of type C and type D dips but not the number of type A dips. The fraction of balanced dips would thus have increased. Note that Table 8.13 does not include motor-starting dips. The measurements were performed at the MV level, where motor-starting dips are very uncommon. Motor-starting dips only occur at LV and in industrial MV networks.

As mentioned before, the duration of a voltage dip due to a fault is mainly due to the fault-clearing time by the protection. If voltage dips have durations that are longer than expected, this points to an incorrect setting of the protection. Also a duration shorter than expected would have to be treated with suspicion because it may point to a problem with protection coordination. Several network operators have used information from voltage-dip recordings as a basis to improve the setting of their protection. The result has been a reduction in the number of long dips and a subsequent reduction in the number of equipment trips due to dips.

The fault-clearing time (i.e., the dip duration) may also be used as an indication of the voltage level at which the fault takes place. Each voltage level has a certain range of durations, which can be used to estimate the voltage level where the dip originates. No general rules can be set for this as these fault-clearing times are different in different countries. In the United States instantaneous tripping is used for many distribution feeders, resulting in very short dips. This practice is less common in Europe, where dips due to distribution system faults are in general of longer duration. Durations over 1 s have been observed but are fortunately not common. When a dip shows a two-stage voltage recovery, this points to an origin at a higher voltage level. Two-stage voltage recovery is due to the difference in fault-clearing time for the breakers on both sides of a cable or line. Such multiple infeed is normally only used for higher voltage levels. (A small number of network operators also use this practice for their MV networks.) A large delay between the two recovery instants may again point to a problem with the protection setting. A large delay in fault-clearing instant in different phases may point to a mechanical problem with a circuit breaker.

Information on the location of the fault can also be obtained from the phase-angle jump and from the point on wave of the voltage recovery instant. However, the faulted feeder is in most cases known to the network operator from the protection operation. An analysis of the dip recording may provide some additional information on the location of the fault. Information on the speed of the protection is however generally not available from the protection relays. Voltage-dip recordings may prove a useful source of information for this.

Also information on the fault type and on its development is often not available from protection records, whereas such information can be easily extracted from voltage-dip records when the faulted voltage level and the transformer-winding connections are known. Especially information on developing faults may be of importance for planning of mitigation methods.

Often it is important to know if a voltage-dip event has an upstream or a downstream origin. With a downstream origin of the dip, the local overcurrent is the cause of the drop in voltage. To determine the origin of the event it is thus important to measure both voltage and current. If the drop in voltage is associated with a significant rise in current, one may reasonably assume that the origin of the event is located downstream of the monitor location. Note that the concepts of downstream and upstream are directly linked to the location of the current measurement. If, for example, the current is measured in one of the MV feeders in an MV substation, a fault in any of the other feeders will be classified as an upstream event. For balanced dips any of the three phases can be used to determine the direction of the dip, but for unbalanced dips the positive-sequence voltage should be used. As the negative-sequence voltage is close to zero in normal operation, it will always show a rise during an unbalanced dip. The presence of large motor loads downstream of the monitor location may lead to a rise in current magnitude even for an upstream fault. In such a case it may be somewhat more difficult to determine the direction of the dip. The question that should always be asked in case of doubt is: Is it reasonable to assume that the measured rise in current has resulted in the measured drop in voltage? If the overcurrent lasts for only one cycle but the dip lasts for several cycles, the current cannot be the cause of the dip. Even if an overcurrent lasts for the whole duration of the dip, it is worth verifying if the relation between voltage drop

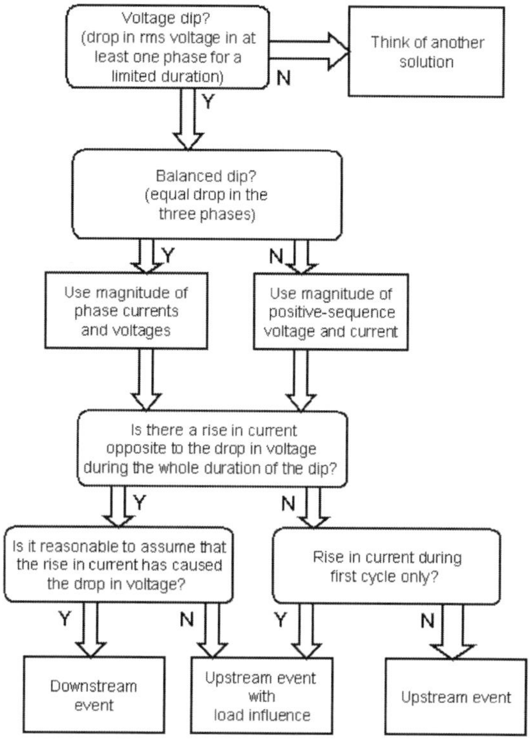


**Figure 8.47** Algorithm for determining whether voltage dip has upstream or downstream origin.

and current rise corresponds with a reasonable value of the source impedance. A general algorithm is shown in Figure 8.47 that can be used to determine whether a voltage dip is due to an upstream or a downstream event.

## 8.10 TRANSIENTS

The characterization of transients in voltages is somewhat different from that of interruptions, swells, and sags. However, the general method in Figure 8.1 still applies here. The first choice to be made is the choice of "characteristic versus time". Two principally different approaches can be distinguished here:

- The original measured (sampled) waveform is used as a characteristic versus time.
- Characterization is based on the deviation from the normal waveform, that is, on the actual transient.

The choice of characterization method depends also on the way in which the transient is detected. In theory the detection method and the characterization method are independent. For individual events any characterization method can be used, but especially when extracting statistics the characteristics should be related to the method used for detecting the event (the triggering method). When the triggering method and characteristics do not fit, the statistics may be irrelevant.

### 8.10.1 Extracting Transient Component

***8.10.1.1 Test Signal*** To test the performance of methods for extracting the transient component a synthetic test signal has been created. The test signal consists of a double-frequency transient superimposed on a quasi-steady-state signal (a fundamental component with some minor harmonic distortion). The signal represents a typical voltage waveform. The random variations in the voltage during normal operation are included by using a random-walk model. The fundamental component at any sample instant is obtained from the value at the previous sample by adding a random number:

$$U_1(n+1) = U_1(n) + k(\rho - 0.5) \tag{8.55}$$

where $\rho$ is a random sample from the uniform distribution on the interval (0,1). The result is a slow drift in the fundamental component. The same model was used for the third, fifth, and seventh harmonics. The amount of variation in the signal depends on the value of the constant $k$. For the fundamental component we have chosen $k = 0.096$ V, and for the harmonic components $k = 0.024$ V. The initial values were 230 V for the fundamental component and 5.75, 6.9, and 3.45 V (2.5, 3, and 1.5%) for harmonics 3, 5, and 7, respectively. From the fundamental and harmonic components as a function of time, the quasi-steady-state voltage waveform as a function of time has been calculated as follows:

$$\begin{aligned} U_s(t) = {} & U_1\sqrt{2}\sin(2\pi 50t) + U_3\sqrt{2}\sin(2\pi 150t) \\ & + U_5\sqrt{2}\sin(2\pi 250t) + U_7\sqrt{2}\sin(2\pi 350t) \end{aligned} \tag{8.56}$$

This signal is calculated over a window with a length of 14 cycles with a sampling frequency of 128 samples per cycle, resulting in a total of 1792 voltage values.

Added to this signal is a transient signal with a duration of two cycles, starting five cycles after the start of the quasi-steady-state signal. The transient signal consists of 370 and 410-Hz components with amplitudes of 23 and 12 V, respectively:

$$U_{\mathrm{tr}}(t) = 23 \times \sqrt{2} \times \cos(2\pi 370t) - 12\sqrt{2}\cos(2\pi 410t) \tag{8.57}$$

The fundamental-voltage component further shows an increase of 8 V (2.5%) at the start of the transient. The total waveform is obtained by adding (8.56) and (8.57):

$$U(t) = U_s(t) + U_{\mathrm{tr}}(t) \tag{8.58}$$

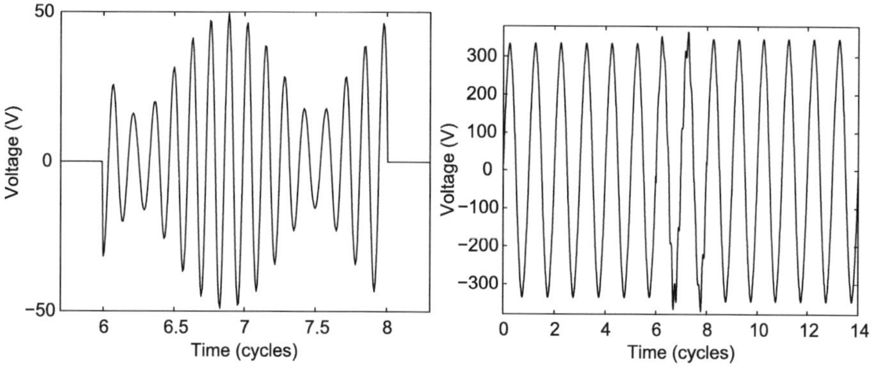


**Figure 8.48** Synthetic transient (left) and total (right) signal used to test methods for extracting transient component.

The transient waveform, as in (8.57), and the total waveform, as in (8.58), are shown in Figure 8.48. This waveform will be used to test various methods for extracting the transient component from a measured waveform.

***8.10.1.2 Cycle-by-Cycle Difference*** Comparing one cycle of the voltage or current waveform with the waveform one cycle earlier in time is a commonly used method to detect transients. This cycle-by-cycle difference could also be used for characterization purposes. One has to be careful, however, to prevent the same transient from coming back twice: One cycle after the transient there is again a difference with the previous cycle. One either has to limit the duration of the transient to less than one cycle or compare each cycle with the last cycle before the transient (alternatively the first cycle of the recording). The latter method has been used here.

The results of applying this method to the synthetic test signal are shown in Figure 8.49. The nonzero value before the event is due to the small variations in voltage between the first cycle (which has been used as a reference) and the sixth cycle (during which the transient occurred). The main difference between the original and extracted signals is however due to the change in the fundamental component. As the amplitude of the transient is at most 40 V, a change of 8 V in fundamental has a significant influence on the extracted transient.

To quantify the difference between the original and the extracted signal, and thus to quantify the performance of the extraction method, we define the rms error $\varepsilon$ as follows:

$$\varepsilon = \sqrt{\frac{1}{N_2 - N_1} \sum_{N_1}^{N_2} (U_{\text{tr}} - U_{\text{xt}})^2} \tag{8.59}$$

with $N_1$ and $N_2$ the starting and ending points of the synthetic transient and $U_{\text{xt}}$ the extracted signal, that is, the estimation of the original transient. The above calculations have been repeated 1000 times in order to get statistical information on the performance

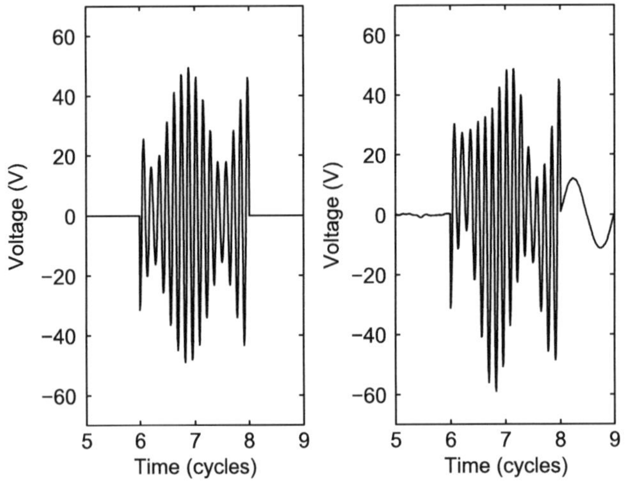


**Figure 8.49** Original transient (left) and extracted transient (right).

of the algorithm. The resulting probability distribution function is shown in Figure 8.50. The majority of values are between 7 and 9 V; the mean value of the error, over all 1000 runs, is 8.0 V, and the standard deviation is 0.76 V.

It is no surprise that the expected rms error is close to 8 V as this is the contribution of the change in the fundamental component. As a comparison the distribution of the error has also been calculated in case there is no step in fundamental voltage. The results are shown on the right in Figure 8.50. The error is significantly less, with the majority of values between 0.5 and 1.5 V. The expected value and standard deviation are 0.85 and 0.42 V, respectively.

When applying the method to a measured waveform, another problem occurs: The length of one cycle is typically not equal to an integer number of samples. Comparing the signal value with the value one cycle back in time requires

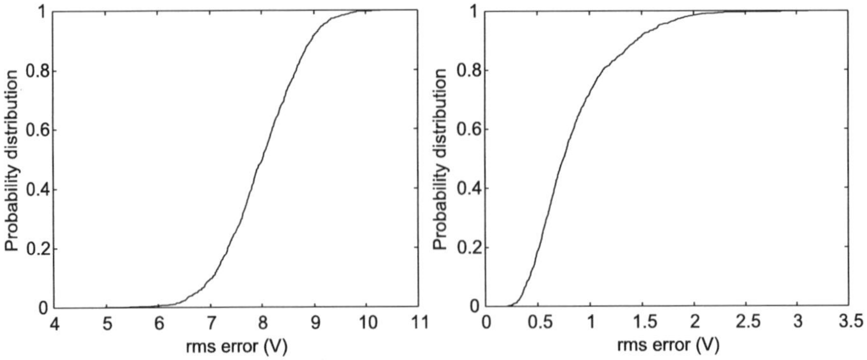


**Figure 8.50** Distribution of rms error when extracting a transient by subtracting preevent waveform: with 8-V step in fundamental voltage (left); no step in fundamental (right).

interpolation between samples. This will make the implementation of the method more complicated. However, as a transient rarely lasts longer than a few cycles, the total error in the extracted transient will in many cases be dominated by the change in fundamental voltage associated with the transient.

**8.10.1.3 Comparison with Average Fundamental Waveform** Instead of comparing the measured waveform with the value one cycle back in time, the waveform can be compared with the fundamental waveform. This is easier to implement as there is no need for interpolation. When the extraction of the transient takes place in the monitor, a phase-locked loop may be used to generate the reference signal with which the waveform is compared.

This method has been applied to the same synthetic test signal as before. The original transient and the extracted transient are compared in Figure 8.51. When comparing with Figure 8.49 we see that even before the transient the difference signal is not fully zero. This is understandable as the harmonic distortion is not removed from the signal. The harmonic distortion thus becomes part of the extracted transient. After the transient, the extracted signal is a combination of the harmonic distortion and the change in fundamental. As the harmonic distortion is rarely less than a few percent of the fundamental, this will introduce a significant error in the extracted transient, especially for less severe transients.

This method has been tested by applying it to 1000 different synthetic signals, as for the previous method. The results are shown in Figure 8.52. The calculations have been performed for the case with an 8-V step in the fundamental and without a step in the fundamental. The error is about twice as large when there is a step in the fundamental. The expected rms errors are 9.03 and 4.27 V with and without a step in the

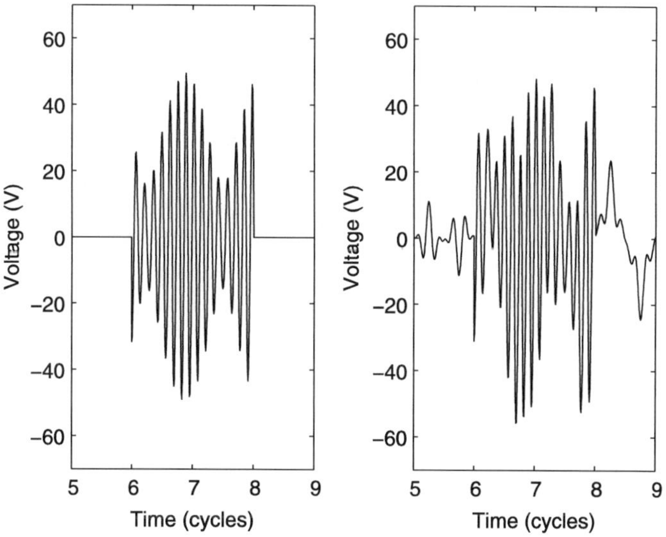


**Figure 8.51** Original transient (left) and extracted transient (right).

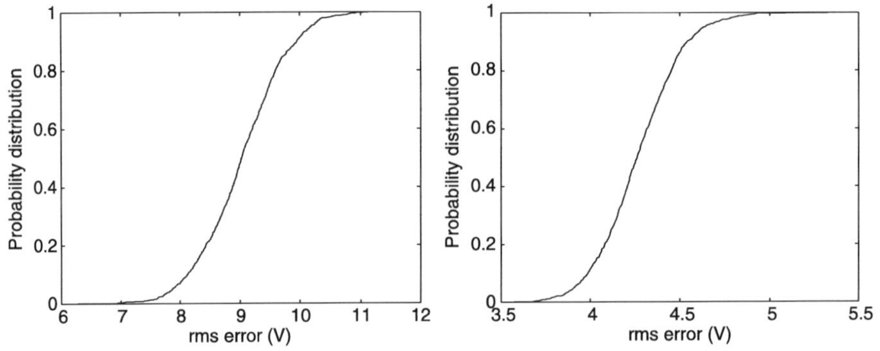


**Figure 8.52** Distribution of rms error for synthetic signal with (left) and without (right) step in fundamental component.

fundamental, respectively. The standard deviations are 0.70 and 0.23 V, respectively. For both cases the error is bigger than for extracting the preevent waveform.

***8.10.1.4 High-Pass Filter Output*** A straightforward way of removing the fundamental frequency is the use of a high-pass filter. The performance of the method depends to a large extent on the properties of the filter. The main aim of the filter is to remove the fundamental component, so that the stop band should include the power system frequency. Where possible also the dominant harmonics should be removed. The passband should include all the possible frequency components of the transient. As these are rarely known before, the choice of the stop-band edge and passband edge could be a problem. Capacitor-energizing transients as low as 250 Hz have been observed which would require a passband edge below 250 Hz. Unfortunately this will include the often dominant fifth- and seventh-harmonic components as well.

Where no lower frequency transients are present or expected, the passband edge should be around 300 to 400 Hz and the stop-band edge around 500 to 600 Hz. In case a passband edge of 250 Hz or lower has to be chosen, this method could be combined with the subtraction of the preevent waveform. In that case, the high-pass filter removes the fundamental component and the subtraction removes the harmonics. What remains are the changes in harmonic distortion associated with the transient.

This method has again been applied to the synthetic test signal. The results are shown in Figure 8.53. A type II Chebychev filter was chosen with a stop-band edge at 50 Hz. The left-hand figure has been obtained for a filter with passband edge at 250 Hz, passband ripple 0.1 dB, and stop-band attenuation 40 dB. For the right-hand filter the design criteria were passband edge 150 Hz, passband ripple 1 dB, and stop-band attenuation 80 dB.

The output of the signal before and after the transient is very similar; the steady-state output is slightly less for the one on the left due to the higher damping of the third-harmonic component. The initial oscillation is more severe in the first case. However, the initial oscillation can be easily handled by repeating the first cycle

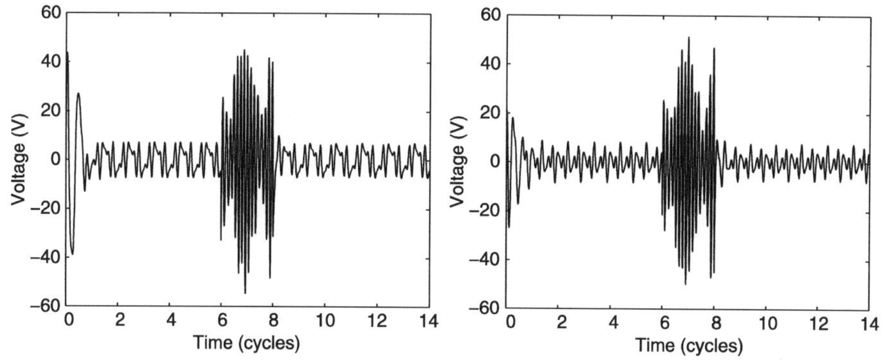


**Figure 8.53** Transient extracted by high-pass filter from synthetic waveform: 150 Hz passband edge (left); 250 Hz passband edge (right).

of the recording a number of times so that the oscillation is damped out well before the actual transient starts. But when the transient is associated with a significant change in fundamental component (this is, e.g., the case for current measurements during downstream switching actions), a similar oscillation will appear due to this change in fundamental component.

A problem that may occur with any digital filter is the introduction of phase errors or *waveform distortion*. The sharper the transition between the stop band and passband, the higher the risk of phase errors. To show the impact of the phase error, the extracted waveform and the original transient are compared in detail in Figure 8.54 for the second example (250 Hz, 1.0 dB, 80 dB).

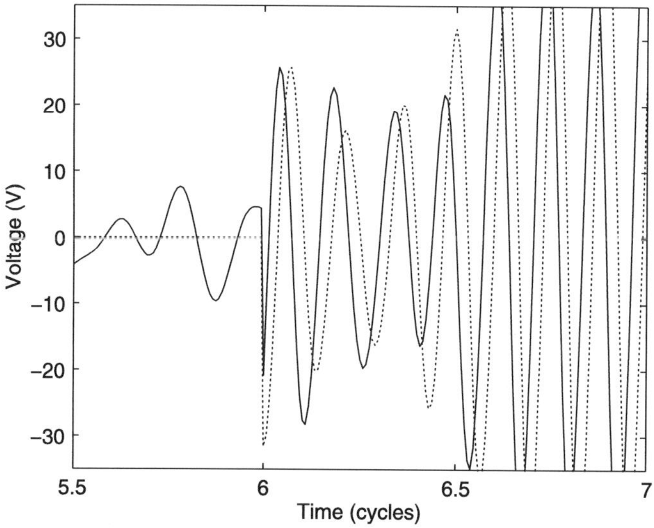


**Figure 8.54** Comparison of high-pass-filter output (solid line) and original signal (dotted line).

**TABLE 8.16 Performance of Different High-Pass Filters: Expected Value and Standard Deviation of rms Error When Extracting transient**

| Stop-band Edge (Hz) | Passband Edge (Hz) | Passband Ripple (dB) | Stop-Band Attenuation | Expected Value (V) | Standard Deviation (V) |
|---|---|---|---|---|---|
| 50 | 150 | 0.1 | 80 | 6.08 | 0.14 |
| 50 | 150 | 0.1 | 40 | 6.77 | 0.13 |
| 50 | 250 | 0.1 | 80 | 7.60 | 0.11 |
| 50 | 250 | 0.1 | 40 | 6.40 | 0.13 |
| 50 | 250 | 1 | 40 | 5.55 | 0.13 |
| 50 | 250 | 1 | 80 | 6.88 | 0.12 |
| 50 | 150 | 1 | 40 | 6.41 | 0.12 |
| 50 | 150 | 1 | 80 | 7.60 | 0.11 |

Calculating the rms error for this filter in the same way as for the earlier methods would result in a rather large error. The error is however mainly due to the time delay introduced by the filter. As long as the extracted transient has the same shape and magnitude as the original transient, any delay should not be seen as an error. Therefore the extracted transient has been delayed by a number of samples before the rms error was calculated as in (8.59). Different delay times were chosen until the lowest expected value was obtained for the rms error. The results for a number of different high-pass filters are shown in Table 8.16. We can conclude from the table that the properties of the high-pass filter do not have any significant influence on the rms error. Compare this with the expected error of 8 V for subtracting the preevent waveform.

#### 8.10.1.5 *Output of Notch Filter*

An alternative filter to remove the fundamental component is a so-called notch filter. Such a filter is able to remove the fundamental frequency completely: The transfer function is zero for the center frequency of the filter.

A second-order digital notch filter is designed in [221, Section 7.2.2]. The design starts from a second-order analog notch filter with the following transfer function:

$$H_a(s) = \frac{s^2 + \Omega_0^2}{s^2 + B_w s + \Omega_0^2} \tag{8.60}$$

where $\Omega_0 = 2\pi f_0$, with $f_0$ the angular notch frequency, and $B_w = 2\pi f_w$, with $f_w$ the 3-dB bandwidth. Using a bilinear transformation, the following transfer function for the digital function, in the $z$ domain, is obtained:

$$G(z) = \frac{\frac{1}{2}(1+\alpha) - \beta(1+\alpha)z^{-1} + \frac{1}{2}(1+\alpha)z^{-2}}{1 - \beta(1+\alpha)z^{-1} + \alpha z^{-2}} \tag{8.61}$$

with

$$\alpha = \frac{1 - \tan\left(\frac{1}{2}b_w\right)}{1 + \tan\left(\frac{1}{2}b_w\right)} \tag{8.62}$$

and

$$\beta = \cos(\omega_0) \tag{8.63}$$

where $b_w = 2\pi f_w / f_s$ is the *normalized angular notch frequency* and $\omega_0 = 2\pi f_0 / f_s$ the *normalized angular 3-dB bandwidth*.

A second-order notch filter centered around 50 Hz with a 3-dB bandwidth of 20 Hz has been used to extract the transient. The results are shown in Figure 8.55 for the same synthetic signal used before. In right-hand figure, the original transient is shown as a dotted line. The difference between the extracted transient and the actual transient is almost exclusively due to the harmonic distortion which is not removed by the filter. But just like the high-pass filter the output of the notch filter shows oscillations for a change in fundamental and at the start of the transients. The oscillations at the start of the signal are significant: with an amplitude of 75% of the step in fundamental and they take about four cycles to decay. This could be a problem for transients associated with a large change in the fundamental component.

The initial oscillation can be reduced by increasing the bandwidth of the filter. This however leads to a larger error in amplitude and phase angle. The performance of the notch filter, expressed as an expected rms error, has been calculated as a function of the 3-dB bandwidth. The result is shown in Figure 8.56. Each point is the average of 1000 stochastically independent runs. The standard deviation is small however (around 0.13 V for most values of the bandwidth); the error is mainly due to the distortion in the waveform. The rms error has been calculated for a delay of two samples and one sample and for no delay; the lowest rms error has been chosen for the curve. The minimum around 150 Hz corresponds to a time

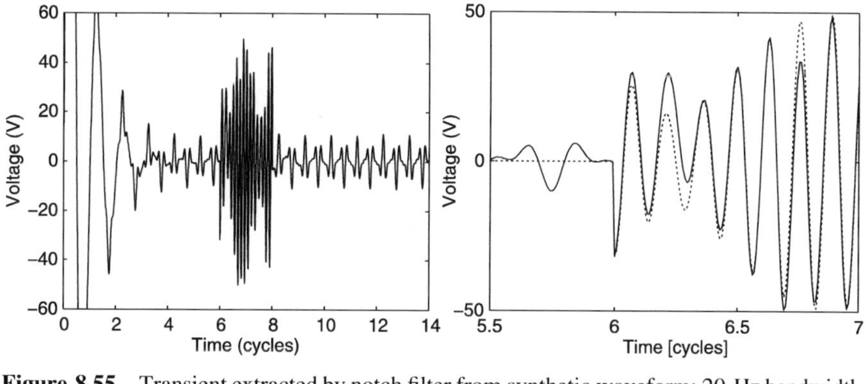


**Figure 8.55** Transient extracted by notch filter from synthetic waveform; 20-Hz bandwidth.

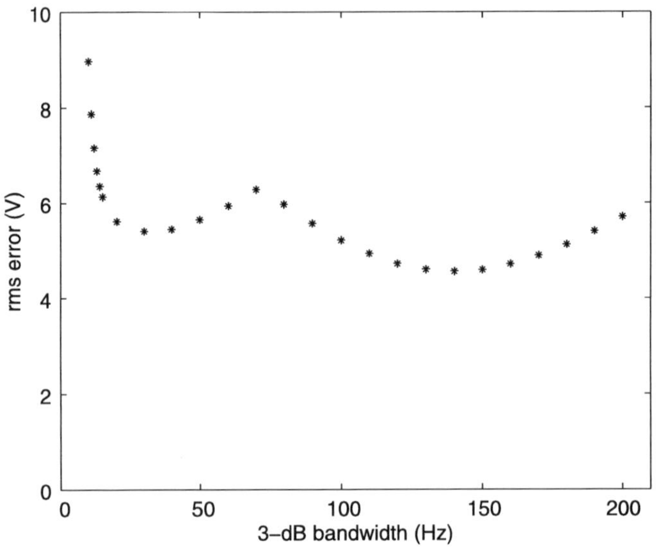


**Figure 8.56** Expected rms error for notch filters with different bandwidth.

shift between input and output of the filter of close to one sample. At 70 Hz bandwidth the error is the same for no delay and one-cycle delay. For a bandwidth less than 30 Hz, the oscillations due to the step in fundamental start to affect the result. This explains the steep rise in rms error for smaller bandwidths.

### 8.10.2 Transients: Single-Event Indices

The characterization of transients is in most cases based on magnitude and duration: the same two terms used for voltage dips. However, the terms have a different definition than for voltage dips. The use of the same terminology means that dips and transients are sometimes plotted in the same magnitude–duration curve. The standard example is the CBEMA curve [33, Section 5.2.4], where the time scale ranges from less than 1 ms through several seconds. The shortest duration of a voltage dip is 10 ms; the short-duration part of the curve refers to transients. Although the use of the curve is mathematically incorrect, it remains a very useful tool for presenting a wide range of power quality events.

The main problem, however, is not the presentation of dips and transients in one curve but the difficulties in defining the magnitude and duration of transients. A number of different methods exist that lead to significantly different results. We will discuss a number of methods below, including some of their advantages and disadvantages. The choice of method to use will depend among others on the application. When characterizing the performance of the supply, it is in general recommended to use a number of different methods so as to maximize the application range.

The method for determining the single-event indices is determined from the characteristics versus time, which are in the case of transients either the original waveform or the extracted transients by using any of the methods discussed in Section 8.10.1. We will introduce a number of methods below, present the problems in definition, and apply the method to a number of measurement examples.

**8.10.2.1 Peak Value** A simple method of characterizing a transient is by its highest absolute value. A distinction may be made between the positive and the negative peaks when more detail is required. The main disadvantage of this method is that its result may strongly depend on the bandwidth of the measurement equipment. For very fast transients, the transient amplitude gets lower for a smaller bandwidth. As a rule of thumb, if the peak value is far outside the range of the other values, the actual peak value is likely to have been much higher. Such a recording may still have its value for trouble shooting purposes and for understanding the power system behavior, but it should not be used for characterization purposes. For pulses with a short rise time the peak value is also very much affected by the anti-aliasing filter. We will study the effect of the anti-aliasing filter in the next section. Some monitors contain a special (often mainly analoge) circuit to detect the peak value of voltage and current transients.

An advantage of using the peak value as a single-event index is that it is not affected by the setting of a threshold. Most of the indices discussed below are affected in different ways.

The resulting peak value depends strongly on whether the actual waveform is considered or only the extracted transient. The peak value has been calculated for three examples. The same examples will be used for the other single-event indices. For each example only one phase has been considered. The first example is an impulsive transient in the current. The actual waveform and the extracted transient are shown in Figure 8.57. A notch filter with a bandwidth of 30 Hz has been used to extract the transient.

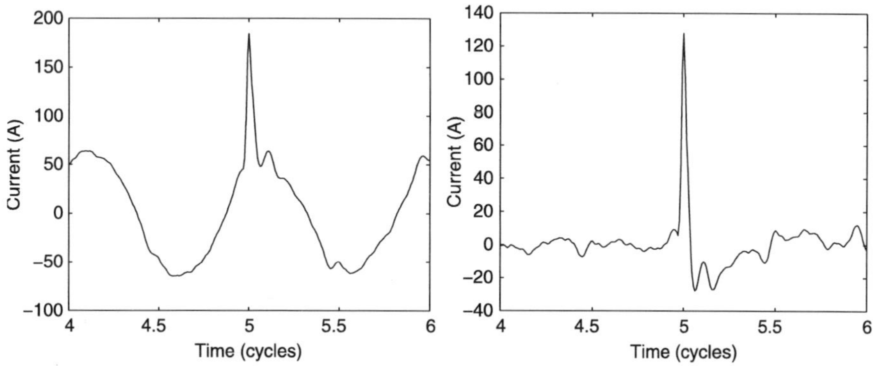


**Figure 8.57** Impulsive current transient: original waveform (left) and extracted transient (right).

The highest absolute value of the original waveform is 184 A; the highest absolute value of the extracted transient is 128 A. In this case the difference between these two values (56 A) is close to the amplitude of the preevent waveform (65 A). Note, however, that this only holds for transients near the maximum or minimum of the steady-state waveform. Also the initial direction of the transient should correspond with the sign of the steady-state waveform. We will see from the forthcoming examples that there is no general relation between the peak of the original waveform and the peak of the extracted transient.

The peak value can be expressed in a number of different ways, each leading to a different result. Some examples are given below, together with the resulting values. When quoting a value for any single-event index it is always very important to indicate how this value has been calculated. Future standards may simplify this by prescribing the way to calculate the indices. In the list below, $I_p$ is the highest absolute value of the original waveform, $I_{\mathrm{tr},p}$ the highest absolute value of the extracted transient, $I_{\mathrm{ss},p}$ the highest absolute value of the steady-state (preevent) waveform, and $I_{\mathrm{nom}}$ the (rms value of the) nominal current. For the calculation we assumed a nominal current of 55 A.

- The peak of the original waveform, $I_p = 184$ A.
- The absolute overshoot of the original waveform compared to the preevent waveform, $I_p - I_{\mathrm{ss},p} = 119$ A.
- The absolute overshoot of the original waveform compared to the nominal waveform, $I_p - \sqrt{2}I_{\mathrm{nom}} = 106$ A.
- The relative peak of the original waveform compared to the preevent waveform, $I_p/I_{\mathrm{ss},p} = 283\%$.
- The relative overshoot of the original waveform compared to the preevent waveform, $(I_p - I_{ss,\mathrm{p}})/I_{ss,\mathrm{p}} = 183\%$.
- The relative peak of the original waveform compared to the nominal waveform, $I_p/\sqrt{2}\,I_{\mathrm{nom}} = 236\%$.
- The relative overshoot of the original waveform compared to the preevent waveform, $(I_p - I_{\mathrm{nom}})/I_{\mathrm{nom}} = 136\%$.
- The peak of the extracted transient, $I_{\mathrm{tr},p} = 128$ A.
- The relative peak of the extracted transient compared to the preevent waveform, $I_{\mathrm{tr},p}/I_{\mathrm{ss},p} = 197\%$.
- The relative peak of the extracted transient compared to the nominal waveform, $I_{\mathrm{tr},p}/\sqrt{2}I_{\mathrm{nom}} = 165\%$.

The second example is an oscillatory transient due to capacitor energizing: measured in a 60-Hz, 120-V system. The original waveform and the extracted transient are shown in Figure 8.58. The figure shows a typical capacitor-energizing transient, with a first step in voltage toward zero (i.e., in opposite direction to the sign of the voltage), followed by an exponentially damped oscillation. The oscillation is especially clear in the extracted transient. Also note that the oscillation is of rather low frequency, about 300 Hz in this case.

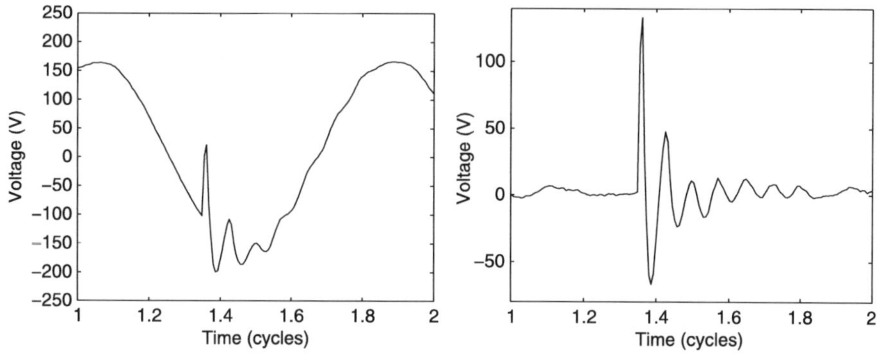


**Figure 8.58** Oscillatory voltage transient: original waveform (left); extracted transient (right).

The following peak values are obtained for this transient:

- The peak of the original waveform, $U_p = 199$ V.
- The relative overshoot of the original waveform compared to the preevent waveform, $(U_p - U_{\mathrm{ss},p})/U_{\mathrm{ss},p} = 21\%$.
- The peak of the extracted transient, $U_{\mathrm{tr},p} = 133$ V.
- The relative peak of the extracted transient compared to the preevent waveform, $U_{\mathrm{tr},p}/U_{\mathrm{ss},p} = 81\%$.

For voltages the preevent peak is close to the nominal peak, so there is no need to distinguish between them. Note that the overvoltage in the original waveform is only 21%, despite the peak of the extracted transient being 81% of the steady-state peak.

The third example is shown in Figure 8.59: an oscillatory voltage transient measured in a 10-kV, 50-Hz system. The transient is associated with a drop in rms voltage of about 1.5%. Using the relation between voltage step and resonance

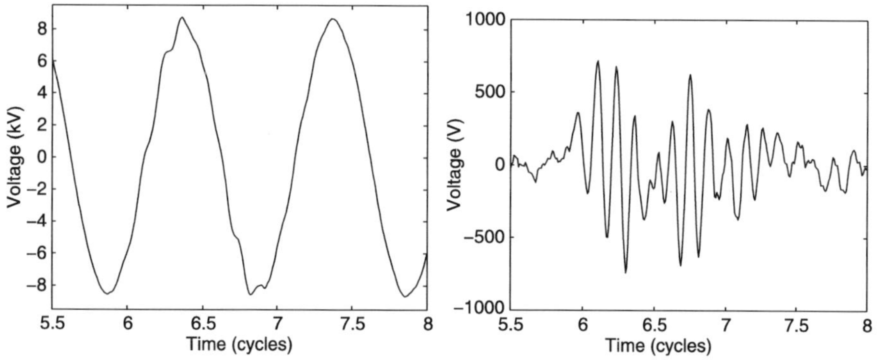


**Figure 8.59** Double-frequency oscillatory voltage transient: original waveform (left); extracted transient (right).

frequency for capacitor energizing, this would correspond to an oscillation frequency of about 400 Hz, which is in agreement with the measurement. The extracted transient is of a different character than the one shown in Figure 8.58. This is due to the presence of a second oscillation frequency. The modulation frequency of the amplitude of the oscillation equals the difference frequency, about 80 Hz in this case. A likely explanation for this transient is restrike during capacitor deenergizing at the next voltage level (132 kV), which excites a double-frequency oscillation together with the capacitor at the local 10-kV bus.

The peak of the extracted transient is 740 V, 8.5% of the steady-state peak. The original transient shows almost no overvoltage, only 0.5%. Without extracting the transient, this event would have been very difficult to detect.

***8.10.2.2 Impact of Filtering on Peak Value*** When a short-duration pulse appears at the terminals of a monitor, its high-frequency components will be removed by the anti-aliasing filter. This will typically lead to a reduction in the amplitude (peak value) of the pulse. Sampling takes place after the anti-aliasing filter, so that the peak value as obtained by the processing of the digital waveform is not a good estimation for the peak value of the actual waveform.

As an example we consider a monitor using a sampling rate of 12.8 kHz (256 samples per cycle in a 50-Hz system). We choose a passband edge at 3 kHz and the corresponding stop-band edge at 9.8 kHz. (See Section 3.2.3 for a discussion of the design criteria for anti-aliasing filters.) These requirements are met by an analog Butterworth filter of order 7 and natural frequency 4300 Hz.

Test pulses were created with a zero rise time and decay time constants of 70 and 350 μs. The test pulses were applied to the above-mentioned anti-aliasing filter. The results are shown in Figure 8.60, where the solid line gives the test waveform and the dotted line the output of the anti-aliasing filter. We see that the filter causes a significant drop in peak voltage as well as a delay of about 0.3 ms. Note that both input and output of the anti-aliasing filter are analog signals; sampling takes place after the

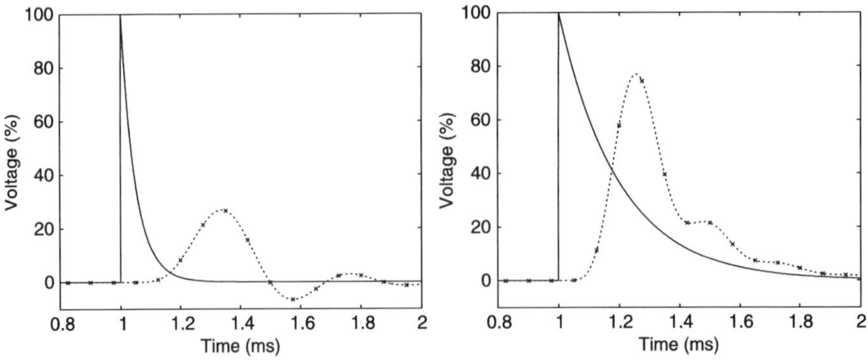


**Figure 8.60** Input pulse (solid) and output (dotted) of anti-aliasing filter with sampled waveform (crosses). Left: 70 μs time constant; right: 350 μs time constant.

anti-aliasing filter. The crosses indicate the sampled waveform. Note that the position of the first sample can be taken at random. This implies that there is some uncertainly in the highest value of the sampled waveform. If a sample instant corresponds with the peak of the analog waveform, the highest value of the sampled waveform is about 27% in this example. But if the peak of the analog waveform comes just between two sampling instants, the observed peak value is about 25%.

The peak value as observed by the monitor depends on the time constant of the original impulse. This effect is quantified in Figure 8.61, where the peak value after the anti-aliasing filter is shown as a function of the time constant of the impulse. The time constant has been varied between 10 and 500 μs to represent overvoltages due to lightning strokes. (See [146] for a discussion of the range of time constants of lightning strokes.) As we can see in Figure 8.60, the shape of the waveform around the peak is not affected by the time constant. There are some differences in the tail of the signal, but these will be hard to detect as the transient is superimposed on a distorted steady-state signal. When using the peak voltage to quantify impulsive transients, care should be taken that the bandwidth of the complete measurement chain is sufficiently wide.

When using the extracted transient to determine the peak value, it is further important to choose the filter parameters in an appropriate way. We saw before that the choice of filter and its parameter values affect the waveform of the extracted transient. To illustrate this, the impact of the bandwidth for a notch filter on the estimated peak value has been determined. The results are shown in Figure 8.62 for the impulsive current transient shown in Figure 8.57. We see a steady decrease of the peak value with increasing bandwidth. The larger the bandwidth, the more the higher frequencies are affected, so that a low bandwidth would be preferable.

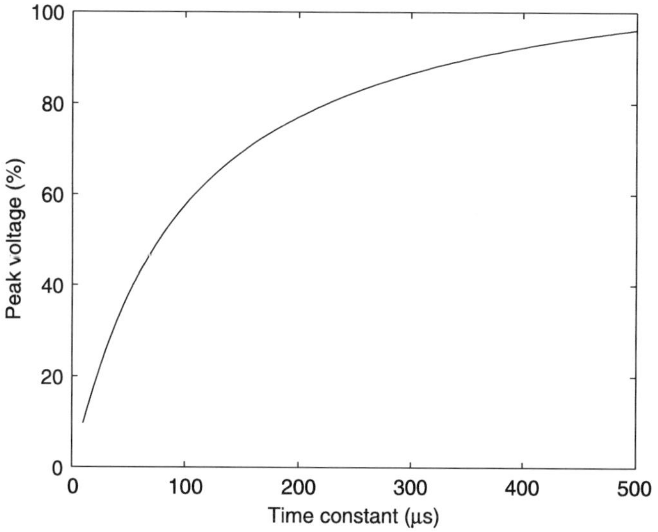


**Figure 8.61** Attenuation of peak of impulsive transient by anti-aliasing filter.

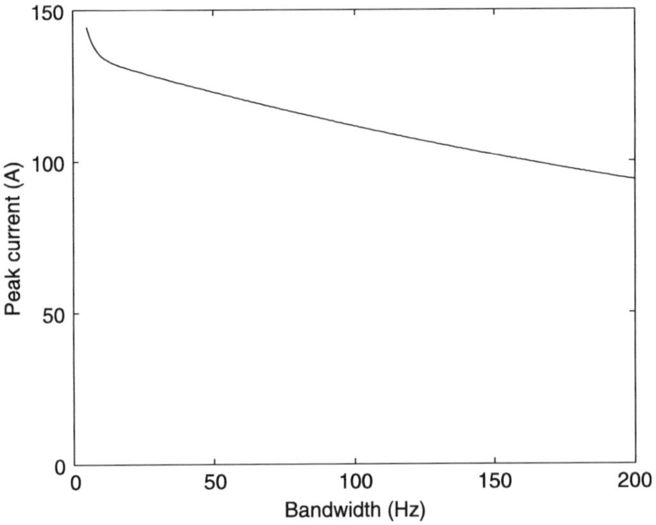


**Figure 8.62** Impact of notch filter bandwidth on estimated peak value of extracted transient.

However, for a bandwidth less than about 20 Hz, the change in steady state associated with the transient causes additional oscillations at the output of the notch filter. This results in an apparent increase in peak value.

***8.10.2.3 Duration*** The duration of a transient can be defined in a number of ways. However, the most appropriate way is to define the duration of a transient as the amount of time during which the voltage or current is above a threshold. This method is very similar to the method used for defining the duration of dips, swells, and interruptions. The more complicated character of transients requires however a number of additional issues to be defined:

- The absolute value of voltage or current should be compared with the threshold level.
- Either the original waveform or any of the methods to obtain the extracted transient can be used to determine the duration of the event. Obviously the choice of waveform compared with the threshold affects the resulting value.
- For oscillating transients (the majority of transients), multiple threshold crossings may occur, especially when the extracted transient is used. The duration should be defined as the time between the first upward threshold crossing and the last downward threshold crossing.

The main problem in quantifying the duration of a transient is the choice of the threshold. As the duration is defined as the time above threshold, the threshold level will affect the value of the duration. The higher the threshold level, the lower the resulting value for the duration. A suitable choice for the threshold used for calculating the duration would be the same threshold as the one used for detecting the transient. This is especially recommended when the single-event indices are used to obtain statistics.

To show how the choice of threshold affects the apparent duration of the event, the impulsive current transient shown in Figure 8.57 is plotted again in Figure 8.63. The absolute value of both the original waveform and the extracted transient are shown, together with a number of suitable threshold levels. The threshold levels for the original waveform correspond to 110, 120, and 130% of the nominal peak current ($\sqrt{2} \times 55$ A in this case). The crosses indicate the actual voltage samples; the duration equals the time step (0.1563 ms in this case) times the number of sampled voltages above the threshold. A threshold setting equal to 110% of the nominal peak corresponds to the common setting used for detection of swells: 110% of nominal rms voltage. The resulting duration is 1.09 ms. The same value of the duration would be obtained for a 120% threshold setting; for a 130% threshold the duration becomes 0.78 ms.

The threshold setting has more impact when the extracted transient is used, as shown in the right-hand plot in Figure 8.63. The threshold setting of 10% would somewhat correspond to a 110% threshold for the original waveform. However, the harmonic distortion after the event results in a steady-state high-pass component with an amplitude exceeding 10%. Such a severe distortion is uncommon for voltage waveforms but occurs occasionally for current waveforms, especially measurements close to distorting equipment. In this case a 10% threshold setting would lead to an infinite duration of the threshold. But even for a 20 or 30% threshold setting determining the duration is not straightforward. What looks like an impulsive transient in the original waveform turns out to be an oscillatory transient from the extracted transient. This is rather common; pure impulsive transients are very rare. The result of the oscillatory character of the event is that the threshold is crossed multiple times. In this case, where the transient is dominated by the first peak, the duration could be defined as the time between the first upward and the first downward threshold crossing. We will soon see examples where such a definition makes no sense.

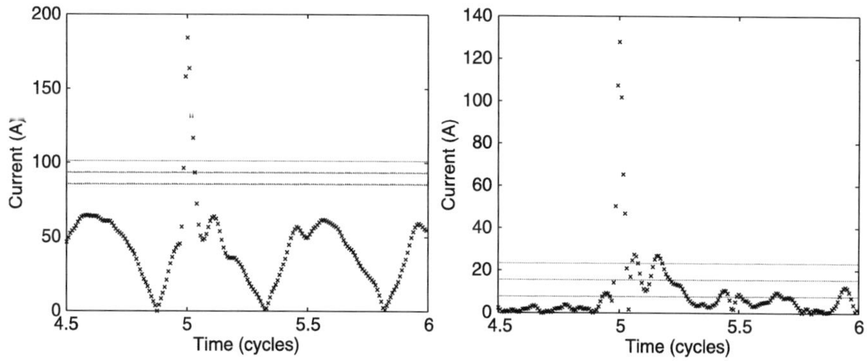


**Figure 8.63** Absolute value of original waveform (left) and of extracted transient (right) compared with a number of thresholds (dotted lines); waveform in Figure 8.57.

**TABLE 8.17 Duration for Different Threshold Settings and Different Duration Definitions: Impulsive Current Transient**

| Threshold Setting | 20/120% (ms) | 30/130% (ms) |
|---|---|---|
| Original waveform | 1.09 | 0.78 |
| Extracted transient | | |
| First downward to first upward zero crossing | 1.09 | 0.94 |
| First downward to last upward zero crossing | 4.69 | 4.06 |
| Total time above threshold | 3.75 | 2.34 |

Instead the duration should be defined either as the time between the first upward and the last downward threshold crossing or as the total time above threshold. The values resulting from the three different definitions are summarized in Table 8.17 for two threshold settings. We again see that it is very important to clearly define the way in which the index is defined. The resulting duration varies from less than 1 ms up to almost 5 ms, even for a simple transient.

The second example is the oscillatory voltage transient due to capacitor energizing, as shown in Figure 8.58. The absolute value of the voltage waveform and the extracted transient are shown in Figure 8.64. The thresholds are given for the same percentage values as before: 10/110, 20/120, and 30/130% of the nominal voltage of 120 V. The first observation is that the original waveform only just exceeds the 110% threshold. Setting the threshold to 120% of nominal or higher would lead to missing the threshold when the original transient would be used for detection. The extracted transient on the right exceeds even the 30% threshold without any problem.

The resulting duration values are shown in Table 8.18 together with the values for other single-event indices, to be discussed below.

The transient in the third example (Figure 8.59) has a very small amplitude compared to the steady-state signal. This means that the output of the notch filter will be

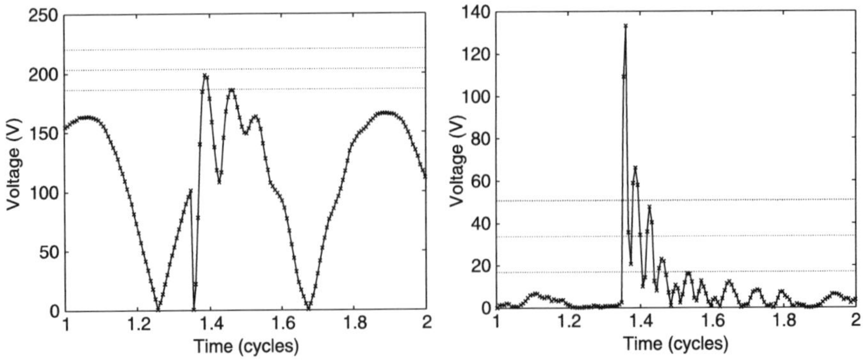


**Figure 8.64** Absolute value of original waveform (left) and of extracted transient (right) compared with a number of thresholds (dotted lines); waveform in Figure 8.58.

**TABLE 8.18 Overview of Single-Event Indices for Transient in Figure 8.64 at Threshold Settings of 10, 20, and 30%**

| | Duration (ms) | | | $V_t$ Integral (ms V) | | | Transient energy ($V^2s$) | | |
|---|---|---|---|---|---|---|---|---|---|
| Duration Definition | 10% | 20% | 30% | 10% | 20% | 30% | 10% | 20% | 30% |
| Time between first upward and last downward zero crossing | 2.34 | 1.69 | 0.91 | 98.2 | 87.1 | 63.2 | 6.65 | 6.44 | 5.56 |
| Total time above threshold | 1.82 | 1.30 | 0.65 | 91.9 | 80.9 | 55.5 | 6.58 | 6.34 | 5.34 |

dominated by harmonic components. The absolute value of the extracted transient is plotted in Figure 8.65 together with thresholds of 3% and 5% of the steady-state peak. It is clear that this event is close to the detection limit for this method (i.e., using the output of a notch filter). Subtracting the preevent waveform may do slightly better in this case, but even for that method a threshold less than 2% will cause difficulties due to the 1.5% drop in rms voltage associated with the transient. The double-frequency character further means that the beginning and end of the transient are difficult to detect. The result is that the apparent duration depends very much on the threshold setting. The impact on the $V_t$ integral and transient energy is less.

The single-event indices are given in Table 8.19 for threshold settings of 3 and 5% of the steady-state peak, corresponding to the two lines in Figure 8.65. We see that the transient energy is again the least affected by the threshold setting

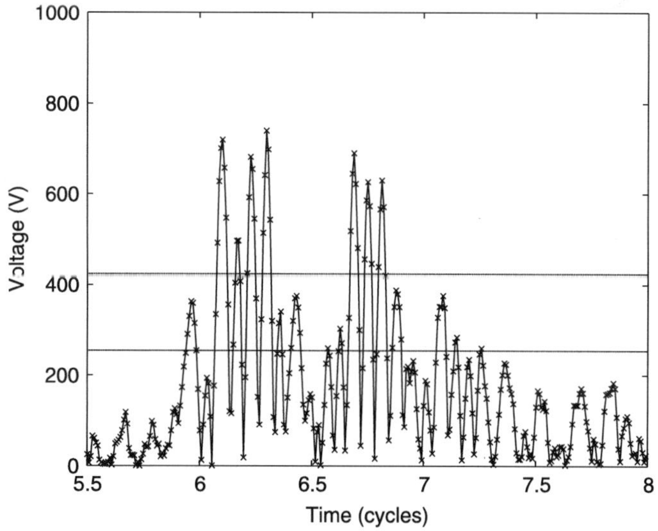


**Figure 8.65** Absolute value of extracted transient compared with a number of thresholds (dotted lines); waveform in Figure 8.59.

**TABLE 8.19 Overview of Single-Event Indices for Transient in Figure 8.59 at Threshold Settings of 3 and 5%**

| | Duration (ms) | | $V_t$ Integral (V s) | | Transient energy ($V^2$ s) | |
|---|---|---|---|---|---|---|
| Duration Definition | 3% | 5% | 3% | 5% | 3% | 5% |
| Time between first upward and last downward zero crossing | 26.4 | 15.0 | 7.1 | 4.9 | 2810 | 2249 |
| Total time above threshold | 11.5 | 5.0 | 5.1 | 2.9 | 2453 | 1734 |

and the integration window. The ratios between the highest and lowest values are 5.3 for the duration, 2.4 for the $V_t$ integral, but only 1.6 for the transient energy.

**8.10.2.4 $V_t$ Integral** Due to the problems in determining the peak value, an integration of the voltage or current over the duration of the event is used to characterize the severity of the transient. The $V_t$ integral is defined as

$$V_t = \int_0^T v(t)\,dt \tag{8.64}$$

In case of current transients the $I_t$ integral is defined in the same way. When using the $V_t$ or $I_t$ integral it is important to define the window over which the integral is taken. The most obvious choice is to take the integral over the time during which the voltage or current is above the threshold. In that case, the threshold level affects the value of the integral. This should be considered when comparing surveys: The setting of the threshold affects both the duration and the integral of the event. The lower the threshold, the longer the duration and the higher the integral; in other words, the severity of the event appears to increase with a reduction of the threshold. The resulting values for the impulsive current transient are summarized in Table 8.20.

From the duration and the $V_t$ integral it is possible to calculate an equivalent voltage:

$$V_{\mathrm{eq}} = \frac{V_t}{T} \tag{8.65}$$

**TABLE 8.20 $I_t$ Integral for Different Threshold Settings and Different Duration Definitions: Impulsive Current Transient, Figure 8.57**

| Duration Definition | 20/120% (mA s) | 30/130% (mA s) |
|---|---|---|
| Original waveform | 156.5 | 133.0 |
| Extracted transient | | |
| First upward to first downward zero crossing | 83.7 | 80.4 |
| First upward to last downward zero crossing | 152.5 | 141.2 |
| Total time above threshold | 140.5 | 114.3 |

**TABLE 8.21 Average Current for Different Threshold Settings and Different Duration Definitions: Impulsive Current Transient, Figure 8.57**

| Duration Definition | 20/120% (A) | 30/130% (A) |
|---|---|---|
| Original waveform | 144 | 171 |
| Extracted transient | | |
| First upward to first downward zero crossing | 77 | 86 |
| First upward to last downward zero crossing | 33 | 35 |
| Total time above threshold | 37 | 52 |

This value is equal to the average voltage during the integration period. An increase in threshold will thus result in an increase in equivalent voltage. The equivalent current can be calculated in the same way for a current transient. The results for the impulsive current transient are shown in Table 8.21, again for two values of the threshold setting: 20% and 30% of nominal peak.

#### 8.10.2.5 Transient Energy

An alternative definition for the severity of the transient is the so-called 'transient energy'. The transient energy is defined as

$$E = \int_0^T v^2(t)\,dt \tag{8.66}$$

The transient energy, when calculated from the original waveform, is a measure of the amount of energy absorbed by equipment due to an overvoltage. The overvoltage part of the well-known CBEMA curve [33, Section 5.2.4] is based on the fact that equipment can absorb a certain amount of transient energy. Also the performance of fuses and current-limiting circuit breakers is based on the transient energy of the fault current through the interrupting device. This method has again been applied to the impulsive current transient, with the results summarized in Table 8.22.

**TABLE 8.22 Transient Energy for Different Threshold Settings and Different Duration Definitions: Impulsive Current Transient, Figure 8.57**

| Duration Definition | 20/120% ($A^2$ s) | 30/130% ($A^2$ s) |
|---|---|---|
| Original waveform | 21.6 | 19.7 |
| Extracted transient | | |
| First upward to first downward zero crossing | 7.5 | 7.4 |
| First upward to last downward zero crossing | 9.0 | 8.8 |
| Total time above threshold | 8.8 | 8.3 |

### 8.10.3 Transients in Three Phases

Most power quality measurements take place in three-phase systems. A voltage or current transient rarely affects only one phase, so a three-phase approach is needed to fully understand the event. The event can be studied by comparing the transients in the respective phases. An alternative approach will be developed here. The approach is similar to the classification of three-phase unbalanced voltage dips, as discussed in Section 6.2.3.

***8.10.3.1 Theory*** The start of the method is the so-called Clarke transform as introduced by E. Clarke [68] to simplify calculations of traveling waves along overhead transmission lines. The method is discussed among others in [279, Section 5.4; 320, Section 6.2.1]. The Clarke transform relates phase voltages and component voltages through the following matrix expression:

$$\begin{pmatrix} V_a \\ V_b \\ V_c \end{pmatrix} = \begin{bmatrix} -1 & 0 & 1 \\ -\frac{1}{2} & \frac{1}{2}\sqrt{3} & 1 \\ -\frac{1}{2} & -\frac{1}{2}\sqrt{3} & 1 \end{bmatrix} \begin{pmatrix} V_\alpha \\ V_\beta \\ V_0 \end{pmatrix} \tag{8.67}$$

The components are referred to as the *alpha component*, *beta component* and *zero-sequence component*. The advantage of this method over the symmetrical-component transformation is that the transformation matrix only contains real elements. This makes it possible to do the transformation in the time domain.

The alpha component gives a voltage in phase a and the opposite voltage of half the amplitude in phases $b$ and $c$. [Fill in $V_\alpha = 1$, $V_\beta = 0$, $V_0 = 0$ in (8.67).] This corresponds to the type Da dip that was introduced in Tables 6.2 and 6.4. The beta-component gives a voltage in phase $b$ and the opposite voltage in phase $c$. This corresponds to a type Ca dip. The zero-sequence component results in the same voltage in the three phases.

Note that expressions for the voltage dips in three phases were given for the residual complex voltages. With transients, however, we are interested in the changes compared to the preevent voltages. Rewriting the expressions for dip types C and D in terms of changes in complex voltage results for type D in

$$\begin{aligned} \Delta U_a &= E - V \\ \Delta U_b &= -\tfrac{1}{2}(E - V) \\ \Delta U_c &= -\tfrac{1}{2}(E - V) \end{aligned} \tag{8.68}$$

and for type C in

$$\Delta U_a = 0$$
$$\Delta U_b = +\tfrac{1}{2}j(E - V)\sqrt{3} \tag{8.69}$$
$$\Delta U_c = -\tfrac{1}{2}j(E - V)\sqrt{3}$$

Substituting $V_\alpha = E - V$ in (8.68) and $V_\beta = j(E - V)$ in (8.69) results in (8.67).

To calculate the component voltages from the phase voltages, the inverse transformation of (8.69) is needed:

$$\begin{pmatrix} V_\alpha \\ V_\beta \\ V_0 \end{pmatrix} = \frac{1}{3}\begin{bmatrix} 2 & -1 & -1 \\ 0 & \sqrt{3} & -\sqrt{3} \\ 1 & 1 & 1 \end{bmatrix}\begin{pmatrix} V_a \\ V_b \\ V_c \end{pmatrix} \tag{8.70}$$

We may thus refer to the alpha and beta components as the Da and Ca components, respectively. The other dip types are obtained by cyclic rotation of (8.67);

$$\begin{pmatrix} V_a \\ V_b \\ V_c \end{pmatrix} = \begin{bmatrix} -\frac{1}{2} & -\frac{1}{2}\sqrt{3} & 1 \\ 1 & 0 & 1 \\ -\frac{1}{2} & -\frac{1}{2}\sqrt{3} & 1 \end{bmatrix}\begin{pmatrix} V_{\mathrm{Db}} \\ V_{\mathrm{Cb}} \\ V_0 \end{pmatrix} \tag{8.71}$$

with as inverse transformation

$$\begin{pmatrix} V_{\mathrm{Db}} \\ V_{\mathrm{Cb}} \\ V_0 \end{pmatrix} = \frac{1}{3}\begin{bmatrix} -1 & 2 & -1 \\ -\sqrt{3} & 0 & \sqrt{3} \\ 1 & 1 & 1 \end{bmatrix}\begin{pmatrix} V_a \\ V_b \\ V_c \end{pmatrix} \tag{8.72}$$

and for the Dc and Cc components

$$\begin{pmatrix} V_a \\ V_b \\ V_c \end{pmatrix} = \begin{bmatrix} -\frac{1}{2} & \frac{1}{2}\sqrt{3} & 1 \\ -\frac{1}{2} & -\frac{1}{2}\sqrt{3} & 1 \\ 1 & 0 & 1 \end{bmatrix}\begin{pmatrix} V_{\mathrm{Dc}} \\ V_{\mathrm{Cc}} \\ V_0 \end{pmatrix} \tag{8.73}$$

with as inverse transformation

$$\begin{pmatrix} V_{\mathrm{Dc}} \\ V_{\mathrm{Cc}} \\ V_0 \end{pmatrix} = \frac{1}{3}\begin{bmatrix} -1 & -1 & 2 \\ \sqrt{3} & -\sqrt{3} & 0 \\ 1 & 1 & 1 \end{bmatrix}\begin{pmatrix} V_a \\ V_b \\ V_c \end{pmatrix} \tag{8.74}$$

Note that the zero-sequence component is the same in all three cases, so only seven different components result. All seven components can be obtained from the phase voltages by combining (8.70), (8.72), and (8.74), resulting in the following matrix expression:

$$\begin{pmatrix} V_{\mathrm{Da}} \\ V_{\mathrm{Db}} \\ V_{\mathrm{Dc}} \\ V_{\mathrm{Ca}} \\ V_{\mathrm{Cb}} \\ V_{\mathrm{Cc}} \\ V_0 \end{pmatrix} = \frac{1}{3}\begin{bmatrix} 2 & -1 & -1 \\ -1 & 2 & -1 \\ -1 & -1 & 2 \\ 0 & \sqrt{3} & -\sqrt{3} \\ -\sqrt{3} & 0 & \sqrt{3} \\ \sqrt{3} & -\sqrt{3} & 0 \\ 1 & 1 & 1 \end{bmatrix}\begin{pmatrix} V_a \\ V_b \\ V_c \end{pmatrix} \tag{8.75}$$

At this stage it is very important to realize that this is not the standard decomposition into orthogonal modal components: These seven components are not mutually independent; there will be cross-coupling between the components. Consider, for example, a Da-type transient: $V_\alpha = V_{\mathrm{Da}} = 1$, $V_\beta = V_{\mathrm{Ca}} = 0$, $V_0 = 0$ in (8.67). This results, after calculating the phase voltages from (8.67) and substituting these into (8.75), in the following values for the seven components:

$$\begin{aligned} V_{\mathrm{Da}} &= 1 \\ V_{\mathrm{Db}} &= -\tfrac{1}{2} \\ V_{\mathrm{Dc}} &= -\tfrac{1}{2} \\ V_{\mathrm{Ca}} &= 0 \\ V_{\mathrm{Cb}} &= -\tfrac{1}{2}\sqrt{3} \\ V_{\mathrm{Cc}} &= \tfrac{1}{2}\sqrt{3} \\ V_0 &= 0 \end{aligned} \tag{8.76}$$

The Da component is the dominating component but four other components also have a nonzero value. Only the Ca and zero-sequence components are zero: Da, Ca, and $V_0$ form an orthogonal base. The consequence of the choice to use seven components is that there will be multiple solutions: After all there are only three phase voltages, which can be reproduced in an infinite number of ways from seven components. The aim is to reproduce a measured transient with the minimum number of components.

The cross-coupling between the seven components is given in the cross-coupling matrix:

$$\begin{pmatrix} V_{\mathrm{Da}} \\ V_{\mathrm{Db}} \\ V_{\mathrm{Dc}} \\ V_{\mathrm{Ca}} \\ V_{\mathrm{Cb}} \\ V_{\mathrm{Cc}} \\ V_0 \end{pmatrix} = \begin{bmatrix} 1 & -\frac{1}{2} & -\frac{1}{2} & 0 & -\frac{1}{2}\sqrt{3} & \frac{1}{2}\sqrt{3} & 0 \\ -\frac{1}{2} & 1 & -\frac{1}{2} & \frac{1}{2}\sqrt{3} & 0 & -\frac{1}{2}\sqrt{3} & 0 \\ -\frac{1}{2} & -\frac{1}{2} & 1 & -\frac{1}{2}\sqrt{3} & \frac{1}{2}\sqrt{3} & 0 & 0 \\ 0 & \frac{1}{2}\sqrt{3} & -\frac{1}{2}\sqrt{3} & 1 & -\frac{1}{2} & -\frac{1}{2} & 0 \\ -\frac{1}{2}\sqrt{3} & 0 & \frac{1}{2}\sqrt{3} & -\frac{1}{2} & 1 & -\frac{1}{2} & 0 \\ \frac{1}{2}\sqrt{3} & -\frac{1}{2}\sqrt{3} & 0 & -\frac{1}{2} & -\frac{1}{2} & 1 & 0 \\ 0 & 0 & 0 & 0 & 0 & 0 & 1 \end{bmatrix} \begin{pmatrix} V_{\mathrm{Da}} \\ V_{\mathrm{Db}} \\ V_{\mathrm{Dc}} \\ V_{\mathrm{Ca}} \\ V_{\mathrm{Cb}} \\ V_{\mathrm{Cc}} \\ V_0 \end{pmatrix} \tag{8.77}$$

We see from the cross-coupling matrix that $V_0$ has no cross-coupling with any of the other components; also is there no cross-coupling between Da and Ca, between Db and Cb, and between Dc and Cc. Another conclusion that can be drawn from the cross-coupling matrix is that there are only three combinations that form an orthogonal base: ($V_0$, Da, Ca), ($V_0$, Db, Cb), and ($V_0$, Dc, Cc). In the next section we will use the decomposition method introduced here to analyze voltage and current transients.

***8.10.3.2 Practical Implementation*** The method introduced in the previous section has been implemented as follows:

- The transient is extracted in each of the three phases. Any of the methods discussed in Section 8.10.1 can be used for this.
- The seven modes are calculated from the extracted transients by using (8.75). This results in seven waveforms as a function of time. The zero-sequence component can be treated separately because it has no correlation with any of the other components. The other six components are correlated according to the cross-correlation matrix in (8.77). The assumption to be made is that one of the six components is dominating. In that case the dominating component will have the largest amplitude and its orthogonal component (Ca for Da, etc.) will have an amplitude close to zero.
- The transient is classified by its dominating component. The single-event characteristics introduced before can be calculated from the dominating component.

***8.10.3.3 Measurement Examples*** The method introduced in the two previous sections has been applied to a number of measured transients. In all examples the transient is extracted by using a second-order notch filter centered at the nominal power system frequency (50 or 60 Hz) with a bandwidth of 30 Hz. The first example is the transient due to the energizing of one phase of a three-phase capacitor bank. The results are shown in Figure 8.66. In addition to to the seven components, the original waveform and the extracted transients for the three phases are shown. The method used for characterizing voltage dips, identifying the highest of six rms voltages, cannot be directly applied here. The amplitudes of Dc, Ca, and Cb are very similar: 108.6, 97.7, and 90.4 V, respectively. However, Cc clearly has the lowest amplitude. Also the six components show the same pattern as in the cross-coupling matrix:

$$U_{\text{Da}} = U_{Db} = -\tfrac{1}{2}U_{\text{DC}}$$
$$U_{\text{Ca}} = -U_{\text{Cb}}$$
$$U_{\text{Cc}} = 0$$

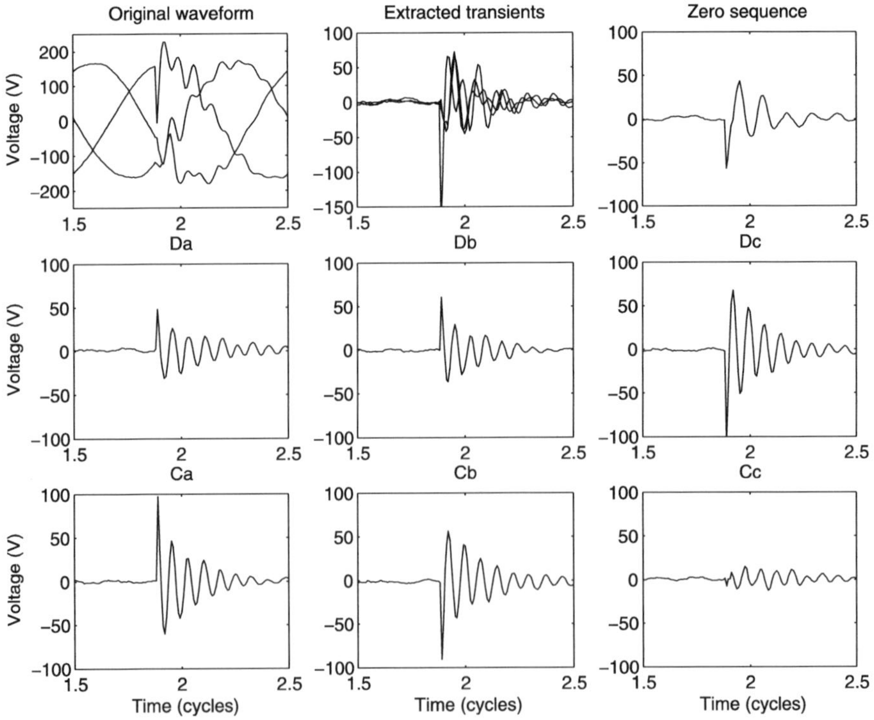


**Figure 8.66** Original waveform, extracted transients in three phases, and seven components for measured transient: example 1.

Identifying this pattern is a good method to find the dominating component, which is Dc in this case. This event would thus be classified as a type Dc transient, with a peak-voltage of 108.6 V. The other characteristics, such as transient energy, oscillation frequency, and damping time constant, can be obtained in the same way as for a single-phase recording.

A second example is given in Figure 8.67; it is again the transient due to closing one phase of a three-phase capacitor bank. In this case the transient is classified as type Da: The pattern fits best with Da as dominating component. Note that the zero-sequence component is much higher than in the first example. Whereas the zero-sequence component was about half the dominating component in the first example, it is even somewhat higher than the dominating component in this case. Strictly speaking the zero-sequence component should be referred to as the dominating component in this case. But in the same way as for the classification of three-phase unbalanced voltage dips, the zero-sequence component will be treated separately. Another interesting observation from the plots is that $V_{Da}$ and $V_0$ have the most smoothest waveform. Methods for extracting the oscillation frequency and time constant will be most successful when applied to these components.

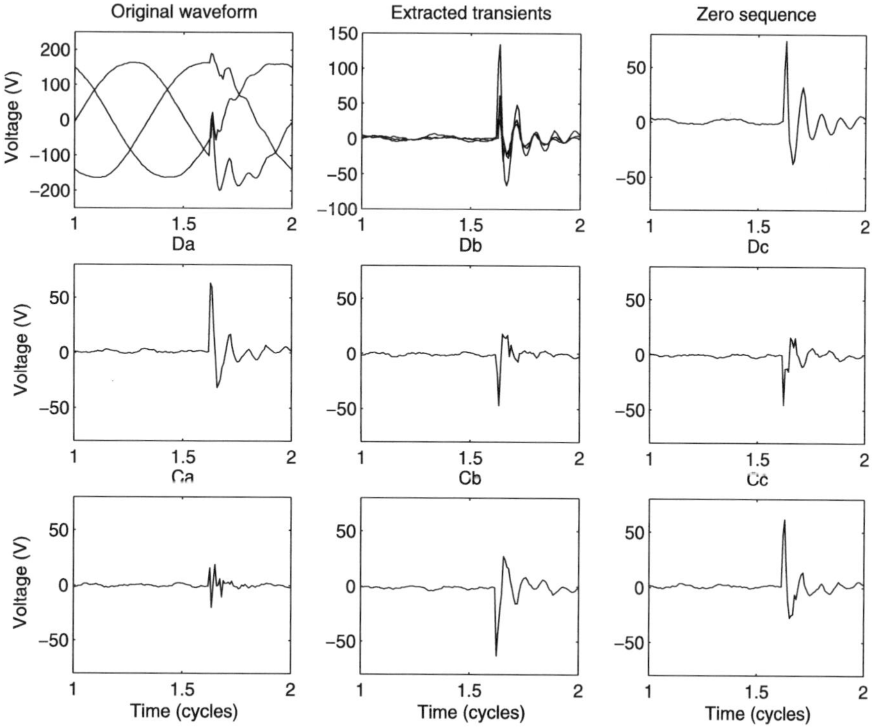


**Figure 8.67** Original waveform, extracted transients in three phases, and seven components for measured transient: example 2.

Figure 8.68 shows the results of applying the method to a third example. This recording contains two transients, about one and a half cycle apart in time. The first transient is classified as a type Cc transient. A Cc transient corresponds to a switching action between phases *a* and *b*. With such a switching action one would not expect a zero-sequence component, in the same way as there is no zero-sequence component associated with type C voltage dips. However, closer inspection of the zero-sequence component shows that it is damped very quickly. It is probably due to a small difference in closing instant between phases *a* and *b*. The second transient is of type Dc and is probably caused by the closing of the third phase.

An example of a current transient is shown in Figure 8.69. The recording is obtained in a low-voltage network close to a medium-size industrial company. The transient is due to load switching on the customer premises. Based on the relation between the waveforms in the six components, the event is classified as a type Ca transient, being a switching action between phases *b* and *c*. The event is probably due to the switching of a drive. The current transient caused a voltage transient that was also classified as a type Ca transient.

Example 5 is the voltage transient due to synchronized capacitor energizing. As the aim of synchronized capacitor energizing is to limit the impact of the switching

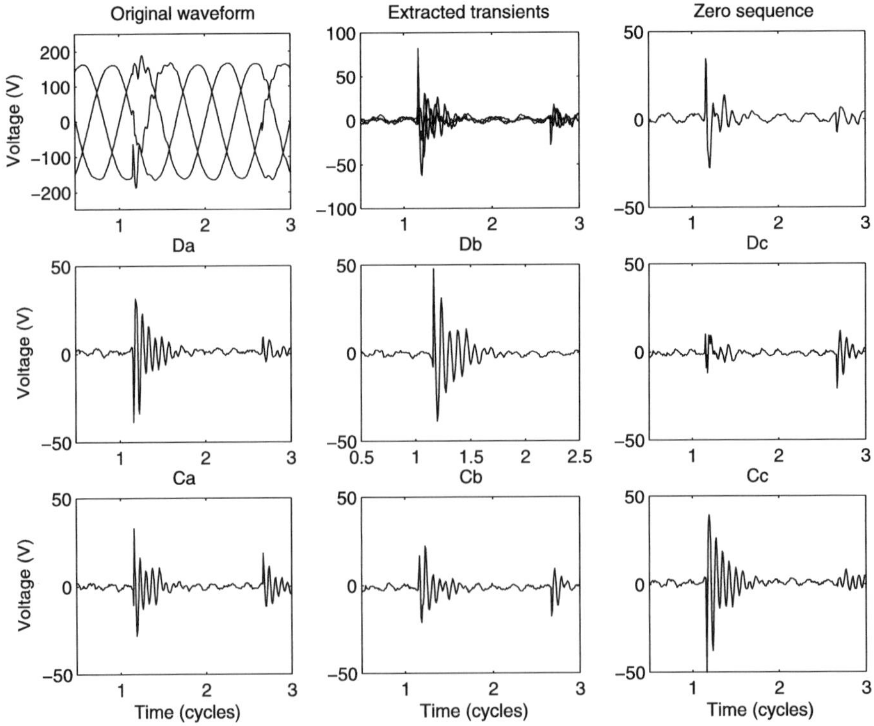


**Figure 8.68** Original waveform, extracted transients in three phases, and seven components for measured transient: example 3.

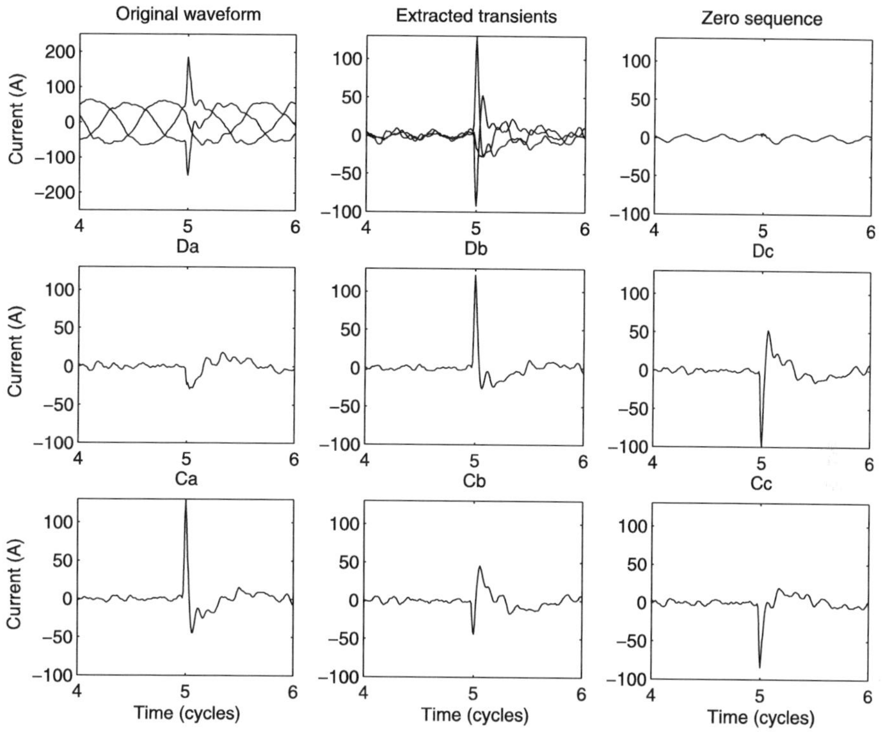


**Figure 8.69** Original waveform, extracted transients in three phases, and seven components for measured transient: example 4.

action, the transient is of limited magnitude. It is however still large enough to be detected, and it should be noted that most switching actions went unnoticed even by the monitor. The transient shown is thus one of the more severe ones due to synchronized switching. Comparing this with the previous examples clearly shows the mitigating effect of synchronized switching. The results of applying the decomposition method are shown in Figure 8.70. At first impression the classification method cannot be applied to this transient. None of the six components are small, so that it is not possible to find the dominant component. However, a closer look at components Dc and Cc shows that Cc increases about one quarter-cycle earlier than Dc. This transient can thus be interpreted as the superposition of a type Cc transient and a type Dc transient, where the latter one starts one quarter-cycle after the first one. The capacitor bank in this case is star-connected but not grounded. Two phases are energizing at the zero crossing of their voltage difference, the third phase one quarter-cycle later. This corresponds with the Cc–Dc classification found. Note that Cc and Dc are the only components that show a smooth damped waveform; the other ones are all a superposition of two components. The oscillation frequency and other single-event indices should be extracted from Cc and Dc.

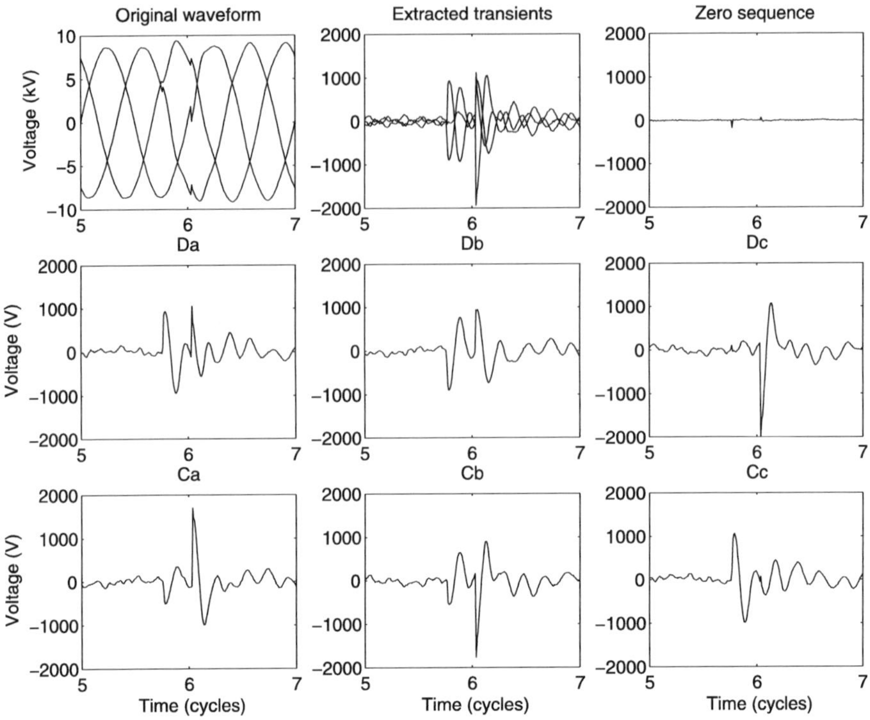


**Figure 8.70** Original waveform, extracted transients in three phases, and seven components for measured transient: example 5.

***8.10.3.4 Limitations of Method*** The methods for characterization and classification of transients in three-phase systems give a satisfactory result for most of the measured transients studied by the authors. The method is however based on the underlying assumption that no two switching actions occur at the same time. In the last example the two switching actions could still be distinguished because their time difference was somewhat more than one cycle of the transient oscillation. Any smaller difference in time will mean that the classification does not apply anymore. Such a transient should be labeled "nonclassified." The same holds when three switching actions occur shortly after each other.

The final example is such a nonclassifiable transient. Its component waveforms are shown in Figure 8.71. The origin of this transient is unknown, but it is probably due to restrike during capacitor deenergizing. Both Cc and Da are small compared to the other components, and both Dc and Ca are large. Thus both Dc and Ca could be classified as the dominant component. A close visual inspection of the waveforms for the six components shows Cb–Dc–Ca as a possible sequence, but the pattern is not very clear. An automatic pattern recognition algorithm may help in identifying such a sequence.

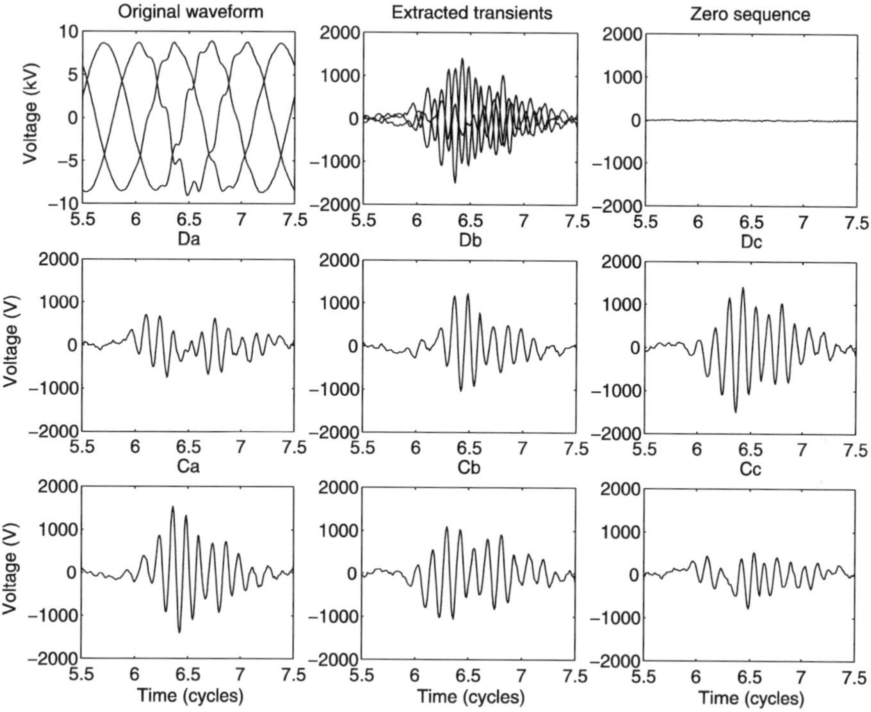


**Figure 8.71** Original waveform, extracted transients in three phases, and seven components for measured transient: example 6.

There are a number of reasons why the classification algorithm works less well for transients than for voltage dips. At first there are the measurement problems: The harmonic distortion in the original waveform causes an error in the extracted transient. For small transients this may make the classification more difficult. The second cause of error is in the method used for extracting the transient. A change in fundamental component leads to an oscillation in the output of a notch filter that adds to the actual transient. When extracting the fundamental or the preevent waveform, the difference in fundamental adds to the transient.

Another source of error is in the model used for the system. The Clarke transform is based on a balanced model of the power system: All three phases are equal and the coupling between each pair of phases is equal. Whereas this is a reasonable model for fundamental frequency, it may no longer be acceptable for higher frequencies. The nonbalanced character of the system manifests itself as a coupling between the components; thus Ca, Da, and zero sequence are no longer fully decoupled. A method similar to the one presented here is used for selection of the faulted phase with some algorithms for traveling-wave-based protection [29, 295]. The coupling between the components is also mentioned in [29, 295] as making the selection more difficult. Note that the "superimposed components" mentioned in [295] correspond to our extracted transients.

The characterization method introduced here for transients on three-phase systems is comparable to the characterization of voltage dips under the assumption that the PN factor remain equal to 1 pu. This is an acceptable assumption for single switching events, either between phase and ground or between two phases. We saw however, when analyzing voltage dips in three phases, that a second characteristic is needed, the PN factor, to cover dips due to two-phase-to-ground and three-phase faults. Such an extension may also be needed for the characterization of transients.

### 8.10.4 Additional Information from Transients

Most of the discussion on the previous pages was directed toward characteristics of transients for statistical purposes and was thus aimed at quantifying the supply performance. When interested in the origin of a transient, information about its magnitude and duration is of limited value. Some other characteristics may provide additional information that may help in identifying the origin of the event.

***8.10.4.1 Change in Steady-State Voltage or Current*** A transient is in many cases associated with a transition between two system states, for example, the energizing of a capacitor bank. This change in system state often also impacts the steady-state voltage or current magnitude and waveform. For a downstream switching event it is obvious that a change in steady-state current can be expected. But even an upstream switching event may cause a change in voltage magnitude. Capacitor energizing is the best example of a transient that affects the voltage magnitude and waveform. As shown in Section 2.2.3 the switching of a capacitor bank causes a change in voltage magnitude due to the injection of reactive power into the system. The switching of the capacitor bank may at the same time cause a significant change in the harmonic impedance and thus in the harmonic voltage distortion.

Consider as an example the voltage and current transient shown in Figure 6.70. This event is due to the synchronized energizing of a 6.6-Mvar capacitor bank just downstream of the measurement location. The three-phase characteristics of the voltages were shown in Figure 8.70 and the phase angle of the current versus time in Figure 8.22. The latter is especially interesting as it gives information about the underlying event without looking into the actual transient. The fact that the current advances 40° in phase angle indicates the presence of a large downstream source of reactive power. Some more characteristics are shown in Figure 8.72. The transient is associated with a rise in voltage of 3.4%, a rise in fifth-harmonic voltage from 0.5 to 1.2%, a rise in active power consumption by 4.5%, and a change in reactive power flow from 2.85 Mvar reactive to 3.95 Mvar capacitive. The rise in fundamental voltage is common for capacitor bank energizing. It is due to the injection of reactive power into the system (see Section 2.2.3). The rise in fifth-harmonic voltage is due to a harmonic resonance between the fifth and the sixth harmonics (see Section 2.5.5). If no other capacitance would be present at the substation, a 3.4% rise in voltage would correspond to a resonance frequency of 270 Hz. The additional capacitance, mainly due to underground cables, will even further reduce this frequency somewhat. Also note the significant rise in active power consumption.

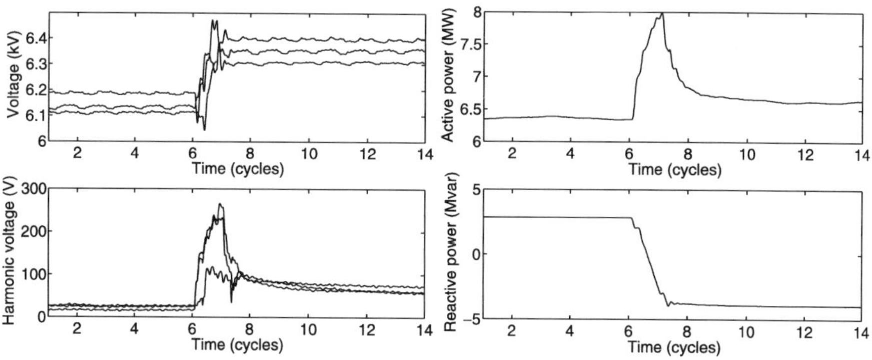


**Figure 8.72** Changes in steady-state parameters for capacitor-energizing transient: rms voltage (upper left); fifth-harmonic voltage (lower left); active power (upper right); reactive power (lower right).

This is partly due to the losses in the capacitor bank, but mainly due to the voltage dependence of the load.

#### *8.10.4.2 Frequency Spectrum*

Additional information may also be obtained from the spectral components of the transient. The dominant frequencies may be obtained by applying the discrete Fourier transform, as shown in Section 6.3, or alternatively from applying an ESPRIT or MUSIC method, as introduced in Section 3.5.2 and 3.5.3. The latter methods give more accurate estimates of the resonance frequency and even give estimates for the damping factors. In this case the spectrum only reveals harmonic components. From the step in voltage, the estimated resonance frequency is 270 Hz, which is too close to the fifth harmonic (250 Hz) to be detected by a simple DFT.

Figure 8.73 shows the voltage and current for a transient recorded at 132-kV subtransmission. The current measured is the total current through three 132/10-kV transformers. The current is lagging the voltage by 90°, which indicates a reactive load, in this case mainly formed by the three transformers. Plotting the product of the extracted voltage and current transient waveforms (the *transient power*, not shown here) shows that there is no net power flow outside the power system frequency. In the plot of the extracted voltage transient we see that the voltage contains a higher frequency component which is not present in the voltage. Only two currents were available, the transients parts of which are in phase, just like the transient part of the three voltages. The cause of this transient is not known, but from the measurements we can conclude that the source is upstream of the measurement location. The event is further associated with a 1% rise in positive-sequence voltage. We also see an increase in seventh-harmonic voltage (from 0.25 to 0.45%) and a minor increase in most of the other harmonic voltages (fifth harmonic rises from 0.66 to 0.78%). The latter may be associated with the change in harmonic currents, possibly related to the rise in voltage magnitude.

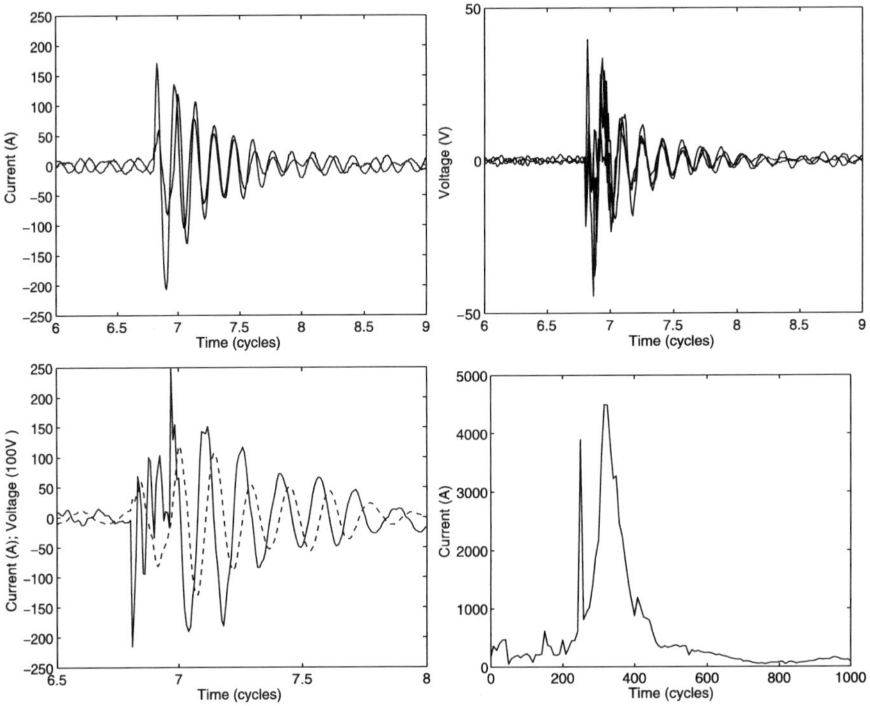


**Figure 8.73** Additional information from transient recorded at 132 kV: current transient (top left); voltage transient (upper right); voltage and current transients (bottom left); spectrum of current transient (bottom right).

Both the rise in positive-sequence voltage and the change in harmonic spectrum point to capacitor energizing as the cause of the transient.

The spectrum of transient current (only one of the phases is shown) possesses a sharp peak around 250 Hz (due to the fifth harmonic) and a wider peak between 300 and 350 Hz. The latter is the resonance frequency of the capacitor bank being energized with the source impedance at the bank's location. As this is between the fifth- and seventh-harmonic voltages, it is understandable that both show a rise after the switching. We will come back to the voltage transients below. As the three voltage waveforms are available, it is recommended to first perform a component transformation.

***8.10.4.3 Three-Phase Components*** The three-phase classification of transients, as introduced in the previous section, has been applied to the extracted voltage transients of this event. As only two currents were available, it was not possible to apply the method to the currents. The results are shown in Figure 8.74. The zero-sequence character, which was already visible in the previous figure, is better quantified now. The amplitudes of the seven components are similar, however, the non-zero-sequence components damp out in about one half-cycle whereas the

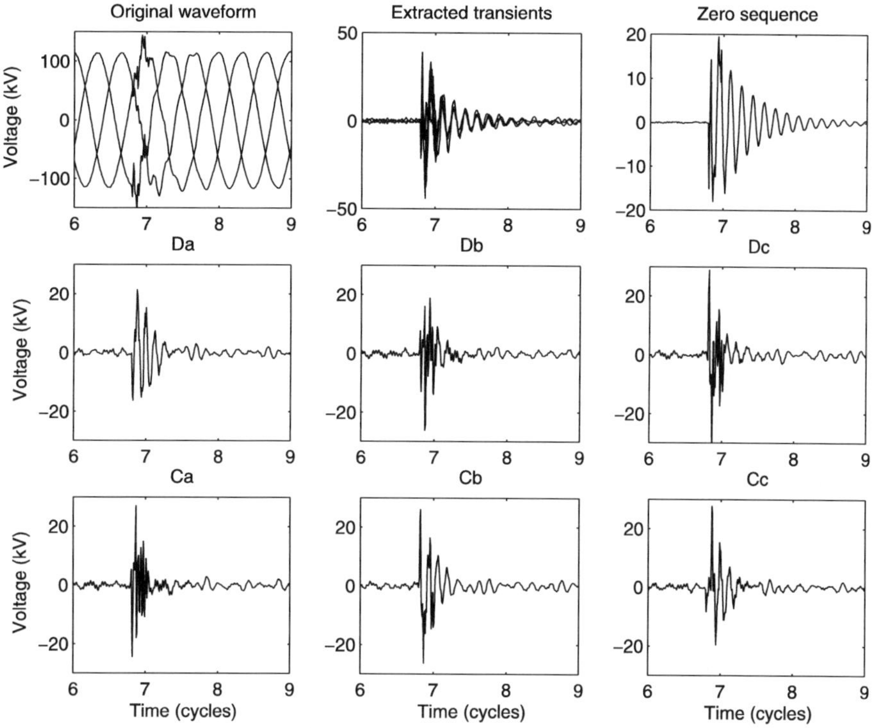


**Figure 8.74** Three-phase characteristics of voltage transient shown in Figure 8.73.

zero-sequence component remains for more than one cycle. A close inspection of the different components shows that component Cc rises about 0.1 cycle after Dc. The initial transient is thus of Dc character (i.e., mainly in phase *c*); after the second phase is switched, the character of the transient becomes difficult to determine, in part because the dominant frequencies are relatively high compared to the sampling frequency being used.

The frequency spectra have been obtained for four of the components shown in Figure 8.74 by applying a DFT over a 10-cycle window. The resulting magnitude spectra for components $V_0$, Da, Cb, and Cc are shown in Figure 8.75. The zero-sequence component has a clear maximum near 320 Hz. The third-harmonic component at 150 Hz is also clearly visible. As shown in Section 2.5.3, the third-harmonic component has a mainly zero-sequence character in a balanced system. The spectra of the non-zero-sequence components have a completely different character: The fifth harmonic is dominating as well as a wide peak around 400 Hz which corresponds to the oscillations visible in Figure 8.74. A resonance frequency of 400 Hz would correspond with a voltage step of about 1.5%, which is close enough to the 1% step in positive-sequence voltage that is measured. Note that it is important to extract the three-phase components before the spectrum is obtained.

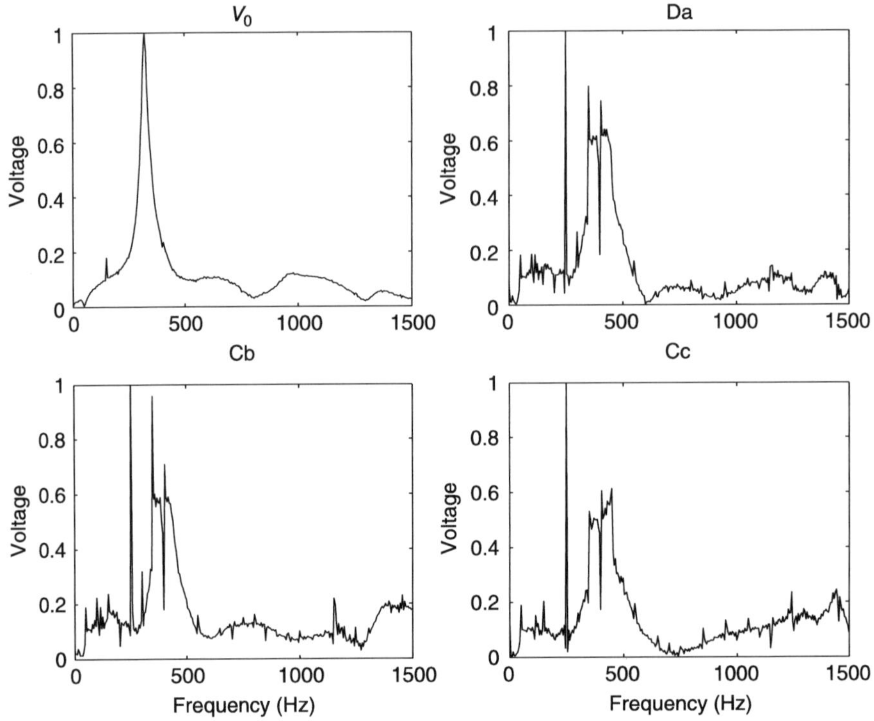


**Figure 8.75** Frequency spectra for four of the three-phase components of voltage shown in Figure 8.73.

The oscillations in the zero-sequence component typically involve other capacitances and inductances than the ones associated with the step in voltage.

**8.10.4.4 *Voltage and Currents*** By comparing voltages and currents, information can be obtained about the direction of the transient, that is, whether the transient's origin is downstream or upstream of the measurement location. For a downstream origin, the current transient is the cause of the voltage transients, whereas the voltage transient causes the current transient for an upstream origin. With transients this is however not always obvious, especially when a capacitor bank is present downstream of the monitor location. Some examples of downstream and upstream events were shown earlier in this chapter as well as in Section 6.3.

**8.10.4.5 *Extracting Spectral Components*** Instead of using the DFT to obtain information on the spectral components in the transient, a model-based method can be used. A model-based method will result in a much more accurate estimate of the oscillation frequencies and will additionally give information on the initial phase and the damping time constants of the various spectral components. The ESPRIT and MUSIC methods, introduced in Section 3.5.3 can both be used.

For both methods it is essential that the power system component is removed. In [44] the ESPRIT method is used to decompose transient recordings in a number of damped sinusoids. Here we will apply this method to the components obtained from the three-phase classification.

The ESPRIT method introduced in [44] and Section 3.5.3 has been applied to the Clarke components as shown in Figures 8.74 and 8.75. In Figure 8.74 we see that the transients do not completely damp out due to the presence of waveform distortion in the original waveform. This waveform distortion has been removed by subtracting a periodic continuation from the last cycle of the recording. The last cycle has been chosen instead of the first cycle because the harmonic spectrum is typically different before and after a switching action leading to a transient. For the example discussed here, this resulted in a reduction of the remaining waveform distortion by about a factor of 5. This removal of the waveform distortion took place after the fundamental component was removed by the notch filter and before the Clarke components were calculated. Next the spectral components were estimated by using ESPRIT over a window starting at the initiation of the transient. The latter instant was obtained by visual inspection of the waveforms. The window length was taken equal to four cycles for the zero-sequence component. However, as the non-zero-sequence components are damped mostly within one cycle, the cycle length has been reduced to 1.5 cycles. The results for components $V_0$ and Da are shown in Figure 8.76. Both original and reconstructed transients are shown. The high-frequency part of the transient is not easy to reproduce in this case. The three phases involved in the transient close at different instants in time. With every closing instant a new high-frequency oscillation is generated. The occurrence of multiple transients starting at different time instants is not part of the model used in ESPRIT. By increasing the model order, it will be possible to obtain a reconstructed signal that is close to the original transient, but the resulting frequency components will not have physical meaning. In this example the model order was limited to 3 for $V_0$ and to 4 for Da.

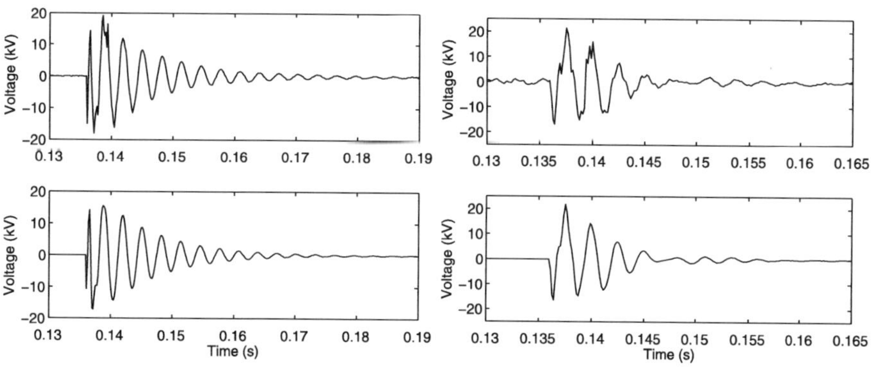


**Figure 8.76** Original transients (top) and reconstructed transients (bottom) from frequency components estimated using ESPRIT: zero-sequence component $V_0$ (left); non-zero-sequence component Da (right).

The characteristics for the Clarke components are summarized in Table 8.23. Model order 3 has been used for $V_0$, model order 5 for Dc and Ca, and model order 4 for the other components. For $V_0$ the 320-Hz component is clearly the dominating one; it has a high initial magnitude and a long time constant. However, during the first few milliseconds the single-frequency model does not hold and additional frequencies are needed to describe the signal. Note that the 303-Hz component is almost in opposite phase with the 320-Hz component. The result is that the 320-Hz component is de facto not present during the first millisecond. Instead the 1-kHz component dominates there.

For Da the two dominating oscillations are 370 and 449 Hz. The other components either are of smaller magnitude or possess higher damping. For Cb we find that the 381-Hz component is dominating, whereas Cc has again two dominating components (366 and 451 Hz). A dominating frequency around 400 Hz is

**TABLE 8.23 Characteristics for Clarke Components Estimated using ESPRIT**

| Component | Frequency (Hz) | Time Constant (ms) | Magnitude (kV) | Initial Phase (deg) |
|---|---|---|---|---|
| $V_0$ | 303 | 1.1 | 25.3 | −138 |
| | 320 | 8.7 | 25.3 | 23 |
| | 1064 | 1.0 | 21.9 | 108 |
| Da | 127 | 4.6 | 4.7 | −89 |
| | 370 | 5.4 | 16.7 | −164 |
| | 449 | 4.5 | 15.5 | 54 |
| | 1315 | 1.7 | 8.5 | −28 |
| Db | 424 | 4.0 | 19.1 | −72 |
| | 538 | 0.5 | 68.4 | 92 |
| | 1257 | 1.1 | 57.3 | −148 |
| | 1433 | 1.8 | 44.4 | −8 |
| Dc | 89 | 0.8 | 60.1 | 66 |
| | 378 | 6.0 | 10.5 | −14 |
| | 1226 | 1.0 | 55.5 | 87 |
| | 1394 | 2.0 | 43.1 | −152 |
| | 1838 | 5.7 | 4.1 | −13 |
| Ca | 417 | 0.9 | 26.2 | 161 |
| | 1190 | 1.3 | 38.4 | −102 |
| | 1416 | 2.4 | 33.0 | 11 |
| | 1817 | 3.3 | 6.7 | −165 |
| | 2326 | 1.6 | 8.3 | −54 |
| Cb | 243 | 2.0 | 13.8 | −26 |
| | 381 | 4.4 | 25.1 | −10 |
| | 738 | 0.2 | 62.5 | 115 |
| | 1415 | 2.8 | 11.6 | 171 |
| Cc | 366 | 2.8 | 33.1 | −135 |
| | 451 | 2.6 | 37.6 | 35 |
| | 1157 | 1.0 | 27.0 | 69 |
| | 1432 | 2.6 | 16.5 | −164 |

**TABLE 8.24 Characteristics for Four Clarke Components in First-Order Model**

| Component | Frequency (Hz) | Time Constant (ms) | Magnitude (kV) | Initial Phase (deg) |
|---|---|---|---|---|
| Da | 402 | 7.3 | 16.7 | 134 |
| Cb | 397 | 5.9 | 17.7 | −42 |
| Cc | 410 | 8.2 | 12.4 | 123 |
| $V_0$ | 320 | 13.9 | 15.2 | 22 |

present for most non-zero-sequence components, except for Ca (the 417-Hz component is damped too fast to be of significance). Note also that a frequency component around 1.3 to 1.4 kHz is present with all non-zero-sequence components. By combining the information from all the components, we may conclude that two main frequencies are present in the transient: one around 400 Hz and another one around 1.3 to 1.4 kHz.

For comparison, Table 8.24 gives the same characteristics when a first-order model is assumed. The results are given for four of the seven Clarke components. For the other components, no suitable first-order model could be found. We may conclude from the results that the dominating frequency is 320 Hz for the zero-sequence component and around 400 Hz for the non-zero-sequence components.

It is important that the ESPRIT method is applied to a window in which the relevant parameters of the signal are stationary, that is, where the basic model is valid. In the basic model used the signal consists of the sum of a finite number of damped sinusoids. This requires that the starting and ending points of the transient are found. Here these have been chosen through visual inspection of the waveform. Alternative methods for automatic detection of such an interval may include using wavelets to find the corresponding singular points or a segmentation method that exploits Kalman filter residues. Choosing the starting point too early may result in the capture of traveling waves or the difference in switching instants in different phases. This may lead to unreliable results. Delaying the starting point will only reduce the signal-to-noise ratio but not the frequency and damping of the components. Further development of the method is needed before it can be used for automatic analysis of voltage and current transients.

## 8.11 SUMMARY AND CONCLUSIONS

The characterization of individual power quality events is generally a two-step process. The first step consists of the calculation of suitable characteristics versus time from the voltage or current waveforms. In the second step the characteristics versus time form a base for the calculation of single-event indices. The most commonly used characteristic versus time for voltage dips, swells, and interruptions is the one-cycle rms voltage (as defined in IEC 61000-4-30). It is shown that a sudden step in voltage magnitude at the start and end of the dip may lead to artifacts that could result in a wrong value for the single-event indices. A number of alternative and additional characteristics are introduced in this chapter. Instead of the one-cycle rms voltage, the half-cycle rms voltage or the voltage amplitude (peak voltage) may be used to quantify the voltage magnitude. The half-cycle rms voltage gives a faster transition after a step in voltage but is also sensitive to even-harmonic distortion in the waveform. The peak voltage (half-cycle or one-cycle) could be an appropriate characteristic to quantify the impact of dips on electronic equipment. However, the method is rarely used. Some further discussion and additional standard development may be needed to propagate the use of the peak voltage instead of or in addition to the rms voltage.

Methods are discussed to extract the phase angle versus time of a voltage or current waveform. It is shown that accurate estimation of the frequency is needed and that the values obtained during transition segments are not reliable. Any automatic method for extracting phase-angle information from voltage dips thus requires a frequency estimation method as well as a method for detecting transition segments. Further development and implementation work are needed here.

Voltage-dip characteristics that consider the three-phase character of the power system are discussed in detail in this chapter. Two methods are discussed, both related to the three-phase classification of voltage dips as presented in Section 6.2.3. The symmetrical-component method is more accurate but somewhat more difficult to implement than the six-phase algorithm. Both methods result in a dip type plus three characteristics versus time. One of these characteristics, the characteristic voltage, is a generalized magnitude of the voltage that can be used to replace the rms voltage versus time as introduced for single-phase events. Currently, the IEC standards do not consider the three-phase character of the power system in the prescribed voltage-dip indices.

For voltage or current transients, the extracted transient is the most suitable characteristic versus time. Only for insulation-related problems is it appropriate to use the actual waveform. A number of methods have been introduced to obtain the extracted transient: subtracting the preevent steady-state waveform, subtracting the preevent fundamental voltage or current, the output of a high-pass filter, and the output of a notch filter centered at the power system frequency. Although each method has advantages and disadvantages, subtracting the preevent steady-state waveform and the notch filter appear the two most appropriate methods. Further development work is needed here, where a combination of these two methods (used in one of the examples in this chapter) should be investigated. No standard document discusses methods to extract the transient, which means that measurements obtained with different instruments may be difficult to compare.

Single-event indices are presented for voltage dips, swells, and interruptions and for voltage and current transients. The standard method prescribes the use of a duration for dips, swells, and interruptions; a residual voltage for voltage dips; and a maximum swell voltage for voltage swells. All these indices are obtained from the one-cycle rms voltage as a function of time. Single-event indices can also be obtained from the other characteristics versus time: the phase angle, the (three-phase) characteristic voltage, and the PN factor. A set of single-event indices is proposed based on these characteristics. The lowest characteristic voltage is a suitable alternative for the residual voltage in a three-phase system. A perceived disadvantage of introducing additional indices is that this further increases the dimensionality of a voltage-dip event. The opposite tendency is to quantify the severity of an event by a single value. Two such indices are discussed here: the voltage-dip energy and the voltage-dip severity, where the voltage-dip severity has preferable properties. It should be clearly noted, however, that the authors are not propagating the use of single-index methods for anything other than as an intermediate step in the calculation of site and system indices, and even in that case it should be used with care. The use of a set of indices is preferred as this will provide additional information on the event and on the relation between event severity and equipment performance. Further work is needed to develop standard methods to calculate the additional indices. Further research is needed to establish relations between the different indices and equipment performance.

Also voltage and current transients are generally characterized by a magnitude and a duration. However, contrary to voltage dips, there is no clarity on the method used and different methods give significantly different results. The different methods are presented here, without giving an opinion about which method is preferred. It is seen as a task for standard-setting organizations to define a set of single-event indices for transients. There is simply not enough information available on the relation between transients and equipment performance to recommend the use of just one index. Such a set of well-defined indices could form a starting point for further study of equipment performance during transients.

A method has been introduced for the characterization of transients in a three-phase system. The method is based on the Clarke transform and results in seven components. It is shown that the method is a generalization of the method for characterization of three-phase unbalanced voltage dips. In a three-phase system the single-event indices should be obtained from the Clarke components. To understand this method better, it should be applied to a sufficient number of simulated and measured transients.

The material on transients presented in this chapter is limited in at least two aspects. The frequency range has been limited to a few kilohertz due to the availability of measurement data. However, most of the methods presented can be applied to higher frequency ranges as well. The other limitation is that we only consider phase-to-ground and phase-to-phase voltages. With transient measurements the neutral-to-ground voltage is as important to equipment performance as the phase voltages. Systematic research on this subject has not even started but will be important in order to understand voltage transients.

Another subject on which a wide research effort is needed concerns the extraction of additional information from event recordings. Such information can be important when deciding on mitigation methods against power quality disturbances. However, event recordings also provide useful information on system performance that often cannot be obtained in any other way. A set of expert-system rules are presented that have been successfully used in an automatic system for classification of voltage-dip recordings based on the origin of the event. Whereas the rules are rather well understood for voltage dips, such a system for transients is still far from complete. Some general considerations are presented here, including a method to identify the main frequency components from the extracted waveform.

## CHAPTER 9

# EVENT CLASSIFICATION

## 9.1 OVERVIEW OF MACHINE LEARNING METHODS FOR EVENT CLASSIFICATION

In Chapters 3 and 4 we described signal decomposition and various parametric models of signals, where the extraction of signal characteristics (or features, attributes) becomes easier in some analysis domain as compared with directly using signal waveforms in the time domain. For example, one may extract features such as the locations of transition and harmonic magnitudes from the subband components of signal or estimate model parameters such as frequencies, magnitudes, and phasors of principal harmonics. These features can be used as the input of a classification system whose output would be, for example, the underlying causes of the disturbances. For many real-world problems we may have very little knowledge about the characteristics of events and some incomplete knowledge of the systems. Hence, learning from data is often a practical way of analyzing the power system disturbances. Among numerous methodologies of machine learning, we shall concentrate on a few methods that are commonly used or are potentially very useful for power system disturbance analysis and diagnostics. These methods include (a) learning machines using linear discriminants, (b) probability distribution–based Bayesian classifiers and Neyman–Pearson hypothesis tests, (c) multilayer neural networks, (d) statistical learning theory–based support vector machines, and (e) rule-based expert systems. There are many other classification methods; readers are referred to [98, 87, 114, 219, 110, 136, 176] for details.

*Signal Processing of Power Quality Disturbances.* By Math H. J. Bollen and Irene Yu-Hua Gu

For an input vector $\mathbf{x} \in \mathbb{R}^{m_0}$ (which can be features or the original data) and desired output $\mathbf{y} \in Y$, a machine-learning process can roughly be described as follows:

- Preprocessing a signal (or its attributes) for decorrelating or reducing the dimension of features. This can be described as finding a mapping function from an $m_0$-dimensional *input space* onto an $M$-dimensional *feature space*, that is, $\Phi: \mathbb{R}^{m_0} \rightarrow \mathcal{F}$.
- Selecting the topologies of the classification system and determining a set of mapping functions $\{f_i(\cdot)\}$, $f_i: \mathcal{F} \rightarrow Y$, that project an $M$-dimensional feature space onto a $L$-dimensional *decision space*. The mapping functions are chosen according to some criterion that results in the best feature separation among classes. The classification can be described as finding a decision tree that contains a set of hypotheses each associated with an individual class.
- Evaluating the performance of the trained system by using *cross-validation*.
- Choosing the best *hypothesis* for each input vector from the test set according to the machine that is learned from the training.

*Choosing a Good Criterion for Designing a Classifier* It is worth noting that many classical learning machines behave like *a rote learner*, that is, for each training set (regardless of whether the class boundaries are linear or nonlinear) a classifier will be able to yield a hypothesis that correctly classifies the data in the training set after sufficient training. Such a classifier may result in an overfit to the training set and may have a poor *generalization performance* as it can make uncorrelated predictions on unseen test data. Cross-validation is a method for estimating the generalization error of a classifier to the unseen data. It is also a method for choosing the parameter settings of a classifier such that the estimated generalization error is minimized. This can be achieved by choosing a best subset of the training data.

A modern learning machine tries to minimize the generalization error, that is, the error on the test data set rather than on the training data set. For example, a support vector machine (SVM) is an approximate implementation of structural risk minimization such that the generalization error is bounded by the sum of training error and a term that is dependent on the Vapnik–Chervonenkis (VC) dimension [308, 79] (where a VC dimension is associated with the complexity of the learning machine), without requiring to incorporate the knowledge in the problem domain. It is worth mentioning that artificial neural networks, frequently used in the classification of power system disturbances, are inferior to SVMs. Neural networks can be associated with a poor generalization performance for the test data. Since a SVM is designed to minimize the error on the test set with an upper error bound, it is clearly a more attractive option.

When there exists some prior information on different classes of data, employing such information would generally lead to increased performance of the classifier. A Bayesian classifier uses both the prior information and the information from measurement data to minimize the overall risk. It requires the prior and the conditional probabilities, which can often be estimated if a large training data set is available.

## 9.2 TYPICAL STEPS USED IN CLASSIFICATION SYSTEM

A typical classification system, as depicted in Figure 9.1, consists of the following steps: feature extraction and optimization, selection of classifier topologies (or architecture), supervised/unsupervised learning, and finally the testing. These steps are briefly described below. It is worth noting that for each particular system some of these steps may be combined or replaced.

### 9.2.1 Feature Extraction

More frequently, features extracted from the signals are used as the input of a classification system instead of the signal waveform itself, as this usually leads to a much smaller system input. Selecting a proper set of features is thus an important step toward successful classification. It is desirable that the selected set of features may characterize and distinguish different classes of power system disturbances (e.g., associated with different underlying causes). This can roughly be described as selecting features with a large interclass (or between-class) mean distance and a small intraclass (or within-class) distance. Furthermore, it is desirable that the selected features are uncorrelated and that the total number of features is small. Other issues that could be taken into account include mathematical definability, numerical stability, insensitivity to noise, invariability to affine transformations, and physical interpretablility.

Features can be divided into *static features* whose statistics are stationary (or time invariant) and *dynamic features* whose statistics are nonstationary (or time varying). For example, features for harmonic distortion are static in most cases, and features for voltage variations and dips are likely to be dynamic. This distinction is strongly related to the distinction between variations and events that has been used as a basis for several discussions throughout this book. However, it is worth noting that some of the features may still be static. For example, in oscillatory transient events (e.g., caused by capacitor switching, line/cable energizing, or restriking during capacitor deenergizing) the feature related to the damping factor in the damped sinusoid

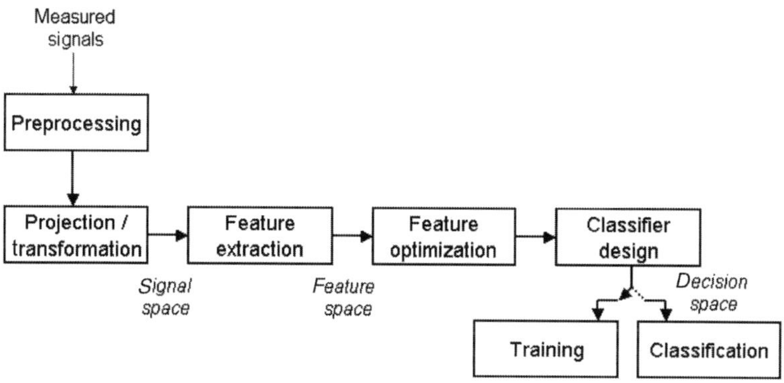


**Figure 9.1** Typical steps in pattern classification system.

transient model [44] is static. There is no direct link between the stationarity of a signal and the static/dynamic nature of a feature from the signal; it all depends on how one selects the feature from the signal. If we apply the concept of event segmentation (see Chapter 7), then the features in each event segment are static, while the features in a transition segment are often dynamic. The classification methods in this chapter mainly handle the static features. However, if the features are dynamic, then one may consider features in a smaller time interval (or take a "snapshot") within which the features can be approximated as static. Classification systems that handle dynamic features (with time-evolving statistics) are beyond the scope of this chapter.

Some typical features used for power quality disturbances are listed below:

- Harmonic and interharmonic components and their frequencies, magnitudes, and phasors
- Triggering points of events (start, end, transition)
- Principal components of a distorted waveform
- Residual voltage, phase angle, duration, and unbalance of voltage dips
- Principal components (amplitude, frequency, initial phase, damping factor of oscillatory transients)
- Spectral distributions of events
- Total harmonic distortion and its frequency range
- Three-phase balance/unbalance information
- Information on synchronization, propagation of voltage/current disturbances at different locations, and levels of a power system
- Statistics, probability distributions, or histograms of disturbances

Note that this list is only a small subset of possible feature candidates and is far from complete. From the earlier chapters it is obvious that the list of possible features can be much longer.

If there is no prior knowledge on which features could "best" characterize the disturbance data class (or type), a common way is to select many features even though these features may be highly correlated. The redundancy among the features can be removed through feature optimization (see the description below).

### 9.2.2 Feature Optimization

**_9.2.2.1 Features Associated with Signal and Averaging Features Associated with Class of Signals_** For each signal (e.g., a disturbance recording), we may extract a set of features that characterize the signal. A set of signals (e.g., disturbance recordings) may contain different classes (e.g., for disturbances they can be categorized into classes such as voltage dips, capacitor switching, transformer energizing, interruptions, and many more). For a training set, we often assume that we can prelabel each signal to one of these classes (see supervised training in Section 9.2.4). Assuming a total of $N$ signals in the training set, of which $N_j$ signals belong to the $j$th class, $N = \sum_{j=1}^{L} N_j$ for a total of $L$ classes. We can compute the mean features and the (co)variance of the features for each class. Obviously, two classes of features can easily be distinguished if the "distance" between the mean features of these classes is large. Further, if the variances of features for each class are small, then less "confusion" would be caused during the classification.

***9.2.2.2 Feature Normalization*** Feature normalization is usually applied to ensure that each feature in a feature vector is unbiased and properly scaled so that different features are equally weighted in a classifier if there is no a priori information on which feature is more important than the others. Let $C_1, C_2, \ldots, C_L$ denote $L$ classes of features, $\mathbf{x} \in \mathcal{C} = \bigcup_{i=1}^{L} C_i$, and $\mathbf{x}$ be a $m_0$-dimensional feature vector $\mathbf{x} = [x_1, x_2, \ldots, x_{m_0}]^{\mathrm{T}}$ extracted from a signal (e.g., a block of data samples or a disturbance recording). Feature normalization is done by applying

$$\tilde{x}_i = \frac{x_i - \mu_i}{\sigma_i}, \qquad i = 1, 2, \ldots, m_0 \tag{9.1}$$

where $\mu_i$ and $\sigma_i$ are the mean and standard deviation of the $i$th feature of a signal and $\mathbf{x} \in \mathcal{C}$. It is worth noting that normalization is applied to features over all classes. After normalization each feature has a zero mean and unit variance, that is, $\tilde{\mu}_i = (1/N)\sum_{k=1}^{N} \tilde{x}_i^{(k)} = 0$, $\tilde{\sigma}_i^2 = (1/N)\sum_{k=1}^{N} \left(\tilde{x}_i^{(k)} - \tilde{\mu}_i\right)^2 = 1$, where a total of $N$ feature vectors is assumed.

***9.2.2.3 Feature Optimization and Dimension Reduction*** Feature optimization is achieved through feature transformations. The purpose of feature optimization can be different, for example, to minimize numerical ill-conditioning, to reduce redundancy and feature dimensions (or achieve orthogonality), to rank features according to their importance, and to increase class separability. Furthermore, optimization may also be dependent on the topologies of a classifier; for example, an optimization process may increase class separability. This process can be considered as finding a mapping function that projects the features from the $m_0$-dimensional input space to a $M$-dimensional feature space $f$: $\mathbb{R}^{m_0} \rightarrow \mathcal{F}$.

*(a) Transformation for Decorrelating Features and Reducing Feature Dimensions* One application of feature transformation is to decorrelate features and reduce the size of the feature vector. This can be very beneficial in terms of computation, generalization performance, and preventing overfit of a classifier when using a high-dimensional feature vector. One usually tries to keep the dimensions of the input space (or the size of the feature vector) low, since a high input dimension drastically increases the difficulties and complexity of designing a classifier—a phenomenon sometimes referred to as the *curse of dimensionality*; that is, when the number of dimensions increases, the number of patterns that are required to properly sample the space increases exponentially.

Since different features often exhibit some degree of correlation and highly correlated features contain little new information, a common practice is to decorrelate features through a transformation followed by retaining only the principal features. Insignificant features result in little performance improvement but significantly increase the complexity of the classifier.

To measure the degree of correlation, the normalized cross-correlation between any two features is computed,

$$\rho_{ij} = \frac{E(x_i x_j)}{\sqrt{E(x_i^2)E(x_j^2)}} \qquad i,j = 1, 2, \ldots, m_0 \tag{9.2}$$

where $E(\cdot)$ is the expectation computed from the features over all classes. It is worth mentioning that two features are highly correlated if their cross-correlation value is large. To decorrelate feature components, one can apply the KL (Karhunen–Loéve) transformation [223],

$$\mathbf{y} = \mathbf{U}^{\mathrm{H}}\mathbf{x} \tag{9.3}$$

where H denotes the Hermitian operator (i.e., complex conjugate and transpose), $\mathbf{y}$ is the transformed feature vector whose components are uncorrelated,

$$E(\mathbf{y}\mathbf{y}^{\mathrm{H}}) = \mathbf{\Lambda} \tag{9.4}$$

and the matrices $\mathbf{U}$ and $\Lambda$ are obtained from

$$\mathbf{R}_{\mathbf{x}} = \mathbf{U}\mathbf{\Lambda}\mathbf{U}^{\mathrm{H}} \tag{9.5}$$

The columns of $\mathbf{U}$ contain the normalized eigenvectors of $\mathbf{R}_{\mathbf{X}}$, $\mathbf{\Lambda}$ is a diagonal matrix containing the corresponding eigenvalues (or variances), and $\mathbf{R}_{\mathbf{X}} = E(\mathbf{x}\mathbf{x}^{\mathrm{H}})$ is the autocorrelation matrix of $\mathbf{x}$.

It is worth noting that a KL transformation is class independent and hence does not take class separability into account. The KL transformation can be explained as performing a coordinate rotation such that the transformed coordinate axes are aligned with the direction of maximum variance.

Assuming that the transformed feature components are sorted according to their eigenvalues $\lambda_1 \geq \lambda_2 \geq \cdots \geq \lambda_{m_0}$ and $\mathbf{\Lambda} = \mathrm{diag}(\lambda_1, \lambda_2, \ldots, \lambda_{m_0})$, reducing feature dimensionality is equivalent to retaining the first $M$ columns of $\mathbf{U}$ associated with the $M$ largest eigenvalues in the transformation, or the transformation

$$\mathbf{y} = \widetilde{\mathbf{U}}^{\mathrm{H}}\mathbf{x} \tag{9.6}$$

where $\widetilde{U} = [\mathbf{u}_1 \ \mathbf{u}_2 \ \cdots \ \mathbf{u}_M]$, $\mathbf{u}_i$ is the $i$th column vector of $\mathbf{U}$, and $m_0 \geq M$. It is worth noting that features with large eigenvalues are the significant features (or principal features) and features after the transformation are uncorrelated.

It is also worth mentioning that dimension reduction is not always used in designing a classifier; for example, in a SVM the input features are (nonlinearly) mapped onto a higher dimension space through the use of a kernel function (see Section 9.6).

*(b) Transformation for Increasing Feature Separability* Another purpose for feature transformation is to maximize the feature discrimination, or the separability between different classes of features. For each feature vector $\mathbf{x} \in \mathcal{F}$, where $\mathbf{x}$ may belong to a class $C_i$, $i \in \{1, 2, \ldots, L\}$, apply the transformation

$$\mathbf{y} = \mathbf{W}^{\mathrm{T}}\mathbf{x} \tag{9.7}$$

where $\mathbf{W}$ is a matrix of size $M \times (L-1)$ that projects features from an $M$-dimensional feature (or input) space to a $(L-1)$dimensional decision space, $M > L$. The optimum matrix $\mathbf{W}$ is chosen such that class separability is maximized.

In order to determine the optimum $\mathbf{W}$, let us first define the within-class (or intraclass) scatter matrix $\mathbf{S}_w$ and the between-class (or interclass) scatter matrix $\mathbf{S}_b$ of $\mathbf{x}$,

$$\mathbf{S}_w = \sum_{i=1}^{L} \mathbf{S}_i \qquad \mathbf{S}_b = \sum_{i=1}^{L} N_i(\boldsymbol{\mu}_i - \boldsymbol{\mu})(\boldsymbol{\mu}_i - \boldsymbol{\mu})^{\mathrm{T}} \tag{9.8}$$

where $\mathbf{S}_i$ is the covariance matrix for the $i$th class,

$$\mathbf{S}_i = \sum_{\mathbf{x} \in \mathbf{C}_i} (\mathbf{x} - \boldsymbol{\mu}_i)(\mathbf{x} - \boldsymbol{\mu}_i)^{\mathrm{T}} \tag{9.9}$$

and $\boldsymbol{\mu}$ is the total mean vector,

$$\boldsymbol{\mu} = \frac{1}{N} \sum_{\mathbf{x} \in C} \mathbf{x} = \frac{1}{N} \sum_{i=1}^{L} N_i \boldsymbol{\mu}_i \tag{9.10}$$

Here, $\boldsymbol{\mu}_i$ is the mean vector for the $i$th class, $N_i$ is the number of feature vectors in the $i$th class, and $N = \sum_{i=1}^{L} N_i$ is the total number of feature vectors. Using (9.7) we obtain the corresponding scatter matrices for the transformed features $\mathbf{y}$ as

$$\tilde{\mathbf{S}}_w = \mathbf{W}^{\mathrm{T}}\, \mathbf{S}_w\, \mathbf{W} \qquad \tilde{\mathbf{S}}_b = \mathbf{W}^{\mathrm{T}}\, \mathbf{S}_b\, \mathbf{W} \tag{9.11}$$

*The criterion of class separability* is then defined as the ratio of the two scatter matrices of $\mathbf{y}$,

$$J(\mathbf{W}) = \frac{\mathbf{W}^{\mathrm{T}}\, \mathbf{S}_b\, \mathbf{W}}{\mathbf{W}^{\mathrm{T}}\, \mathbf{S}_w\, \mathbf{W}} \tag{9.12}$$

The transformation matrix $\mathbf{W}$ is chosen such that $J(\mathbf{W})$ is maximized. It can be shown that the optimal solution of $\mathbf{W}$ is obtained by solving the generalized eigenproblem where the columns of $\mathbf{W}$, $\mathbf{w}_i$, are the *generalized eigenvectors* corresponding to the $L-1$ largest eigenvalues,

$$\mathbf{S}_b \, \mathbf{w}_i = \lambda_i \, \mathbf{S}_w \, \mathbf{w}_i \tag{9.13}$$

where $\lambda_1 \geq \lambda_2 \geq \cdots \geq \lambda_{L-1}$. For the two-class case ($L = 2$), this reduces to the Fisher linear discriminant where the optimum $\mathbf{W}$ is a vector of size $M \times 1$,

$$\mathbf{W} = \mathbf{S}_w^{-1}(\boldsymbol{\mu}_1 - \boldsymbol{\mu}_2) \tag{9.14}$$

It is also possible to use other definitions to maximize class separability. Readers are referred to [98, 114] for more details.

### 9.2.3 Selection of Topologies or Architectures for Classifiers

Mathematically, designing a classifier is equivalent to finding a mapping function between the input space and the feature space followed by applying some decision rules that map the feature space onto the decision space. For example, a discriminant function $f(\cdot)$ can be employed to map a transformed feature $\mathbf{y} \in \{C_1, C_2\}$ in the feature space onto the decision space,

$$f(\mathbf{y}) = \begin{cases} \geq \eta & \textbf{if} \;\; \mathbf{y} \in C_1 \\ < \eta & \textbf{if} \;\; \mathbf{y} \in C_2 \end{cases} \tag{9.15}$$

Another example is to employ a neural network as a classifier. In such a case one needs to determine the structure of the classifier, for example, the number of layers and the number of nodes in the hidden layer. A third example is to select a SVM as a classifier. A SVM (nonlinearly) maps the input space onto a higher dimensional feature space where the classes are likely to be linearly separable, and hence linear discriminant functions can be employed. In such a case, choosing different kernel functions will lead to different types of SVMs.

There are many different ways of categorizing the classifiers. A possible distinction is between statistical-based classifiers (e.g., Bayes, Neyman–Pearson, SVMs) and non-statistical-based classifiers (e.g., syntactic pattern recognition methods, expert systems). A classifier can be linear or nonlinear. The mapping function can be parametric (e.g., kernel functions) or nonparametric (e.g., $K$ nearest neighbors) as well. Some frequently used classifiers are listed below:

- Linear machines
- Artificial neural networks
- Bayesian classifiers
- Kernel machines
- Support vector machines
- Rule-based expert systems
- Fuzzy classification systems

### 9.2.4 Supervised/Unsupervised Learning

Before choosing the topologies of classifiers, one should consider whether or not the given training data are prelabeled according to their classes. Supervised learning refers to training a classifier using prelabeled input vectors. Contrary to this, for unsupervised learning the class labels of input vectors are unknown during the training process.

For unsupervised training, clustering methods such as $k$ means (or $C$ means), mean shift, or mixture of Gaussians are frequently employed. Validation of classifiers is necessary after unsupervised training in order to determine whether these clusters correspond to the actual classes. Although unsupervised training is frequently required in real-world problems, the theories are less well developed as compared with the supervised training. Since we can reasonably assume that the training data can be prelabeled by power system experts, we shall limit ourselves to classifiers with supervised training.

### 9.2.5 Cross-Validation

Cross-validation is a method that can be used for estimating the generalization error (i.e., the error on the test set) by "resampling" the training set data [252]. A $k$-fold cross-validation divides the training data into $k$ subsets, of which $k$-1 subsets are used for the training and the remaining one subset is used for predicting the classification error. This process is repeated $k$ times until each single subset has been used once for the prediction. The averaging error can be used to predict the generalization error of the classification system. Viewing $k$-fold cross-validation as a "leave-$k$-out" method, a special case of the cross-validation is the so-called leave-one-out method.

Cross-validation is also used to select a subset of the training set so that the generalization performance of the classifier can be improved. This is equivalent to selecting a classifier (among many possible ones) that is trained by the "best" subset of training data. For example, outliers from the training set can be removed by using leave-one-out cross-validation: A small number of samples that lead to very large prediction error could be removed from the training set.

For most classifiers described in this chapter, cross-validation is required in order to estimate the generalization error as well as to improve the performance of a classifier.

### 9.2.6 Classification

Once a classifier is trained, classification can be performed by inputting vectors from the test data set into the system, where the "setting" of the classifier (e.g., parameters) is "frozen" from the training.

An important issue of concern is that for real-world problems a classifier should be designed to allow updating/adapting the settings/parameters dynamically; for example, if a new class of events is obtained from the measurements, the information should be added to enrich the existing classifier.

## 9.3 LEARNING MACHINES USING LINEAR DISCRIMINANTS

Linear methods are one of the most basic and commonly used methods for classification, where linear discriminant functions are the fundamental tool [114].

Let the total number of classes be $L$. A linear machine can mathematically be described as finding a set of linear discriminant functions $f_i(\cdot)$ that can best project each $M$-dimensional feature onto one of the $L$ decision regions. Let an $M$-dimensional feature vector be $\mathbf{x} \in \mathcal{F}$, $L$ decision regions be $\mathcal{R}_i$, $i = 1,2,\ldots,L$, $\mathcal{R}_i \in Y$, and $L$ feature classes be $C_1,C_2,\ldots,C_L$. The linear mapping (or discriminant) functions perform $f_i(\cdot)$: $\mathcal{F} \rightarrow Y$ in the following way:

$$f_i(\mathbf{x}) = \mathbf{w}_i^{\mathrm{T}}\mathbf{x} + b_i \qquad i = 1, 2, \ldots, L \tag{9.16}$$

where $\mathbf{w}_i$ is the weight vector of size $M$ and $b_i$ is the bias or threshold weight for the $i$ th class. The weight vectors $\mathbf{w}_i$, $i = 1,2,\ldots, L$, are determined according to some given criterion, for example, to maximize the class separability criterion in (9.12). Once the weight vector $\mathbf{w}_i$ for each class is determined, the decision that a given feature vector $\mathbf{x}$ from a test set belongs to the class $C_i$ will be made if the following conditions are satisfied:

$$\text{Choose } \mathbf{x} \in C_i \qquad \text{if } f_i(\mathbf{x}) > f_j(\mathbf{x}) \qquad \text{for all } j \neq i \tag{9.17}$$

The classification can be explained as partitioning the decision space to $L$ regions where the linear boundaries are defined by

$$\mathbf{w}_i^{\mathrm{T}}\mathbf{x} + b_i = \mathbf{w}_j^{\mathrm{T}}\mathbf{x} + b_j, \qquad i \neq j \tag{9.18}$$

The decision regions for a linear learning machine are convex. It is worth noting that (9.16) can easily be generalized, for example, to quadratic discriminants,

$$f_i(\mathbf{x}) = b_i + \sum_{m=1}^{M} w_{i,m}x_m + \sum_{m=1}^{M}\sum_{n=1}^{M} w_{i,mn}x_m x_n \qquad i = 1,2,\ldots,L \tag{9.19}$$

## 9.4 LEARNING AND CLASSIFICATION USING PROBABILITY DISTRIBUTIONS

In Chapters 5 and 10 we characterize or quantify power quality data in terms of their statistics. Mathematically speaking this is equivalent to estimating the statistics of some random variables (e.g., the triggering location, phasor, magnitude, harmonic frequency) and determining the estimation accuracy (e.g., mean, bias, standard deviation, or variance, among others). In this section we will describe two statistical-based methods for classification of disturbances according to their classes, for example, the underlying causes of disturbances.

*Estimate Probability Distributions for Different Event Classes* The feature statistics as discussed in Chapters 5 and 10 are not complete in uniquely characterizing/quantifying different classes of power quality disturbances. For a complete statistical characterization, one requires to know (or estimate) the probability distribution for each class of feature vectors. In reality we often do not know the distributions, as follows:

- The probability distribution itself is unknown. The shape of the probability distribution is unknown. For example, we do not know whether a feature vector is Gaussian or Laplacian distributed or whether the shape of a distribution can be approximated by some other existing function.
- The type of distribution function is known but its parameters are unknown. In this case, we assume that the type of distribution function is known (e.g., Gaussian distribution). However, one or more parameters in the distribution are unknown (e.g., the mean value $\mu$ and the variance $\sigma^2$ of a Gaussian distribution).

If the probability distribution is unknown, we can estimate the distribution by using the existing data. This can be done by using the normalized histogram as the estimated joint probability distribution for each class of feature vectors. We thus assume that the estimated probability distribution for each feature class is a true estimate of the actual distribution that would be obtained from an infinite amount of data.

Next, it is worth checking whether the shape of the estimated probability distribution fits (or is close to) the shape of a particular probability distribution function in the existing set. If so, the problem is converted to estimating the unknown parameters of a given distribution. Table 9.1 lists some frequently used probability distributions. Among these, the Gaussian (or normal) distribution is widely used due to the laws of large numbers and the central limit theorem; the exponential and

**TABLE 9.1 Some Frequently Used Probability Distributions in Power Quality Analysis**

| | | |
|---|---|---|
| Gaussian | $p_{\mu,\sigma}(x) = \dfrac{1}{\sqrt{2\pi}\sigma} e^{(x-\mu)^2/(2\sigma^2)},$ | $x \in (-\infty, \infty)$ |
| Laplacian | $p_\alpha(x) = \dfrac{\alpha}{2} e^{-\alpha\|x\|},$ | $x \in (-\infty, \infty),\ \alpha > 0$ |
| Poisson | $p_\lambda(x) = \dfrac{\lambda^x}{x!} e^{-\lambda},$ | $x = 0, 1, 2, \ldots$ |
| Exponential | $p_\lambda(x) = \lambda e^{-\lambda x},$ | $x \geq 0$ |
| Weibull | $p_{\lambda,\beta}(x) = \lambda^\beta \beta x^{\beta-1} e^{-(\lambda x)^\beta},$ | $x \geq 0$ |

Weibull distributions are commonly used for reliability calculations. It is also recommended to use a physical-based reasoning leading to a type of probability distribution in all cases.

Alternatively, one can first select a particular probability distribution function (e.g., Gaussian) based on some assumptions. It is important to mention that validation is usually needed afterward to be sure that the selected probability distribution function is close to the actual one. This can be done by comparing the shape between the histogram of data/features and the selected distribution. If they are similar, then the assumption holds and the problem is converted to estimating the unknown parameters of the distribution. However, if the shapes do not resemble one another, one should use the histogram as the estimated distribution rather than enforce an existing distribution function that is unfit.

Likewise, we can learn the unknown parameters if the distribution is known. Assuming that we have sufficient data, these parameters can be estimated in principle. Details of estimating unknown parameters is beyond the scope of this book. Readers are referred to [98, 114] for further reading.

*Formulation of Statistical-Based Classification Problems Using Selected Features* When a large amount of measurement data is available, it is a clear advantage to design a classifier based on the statistics of features that characterize the data. We assume that the classification system uses some features that characterize the signal, and the features are optimized through preprocessing. For the probability distribution–based methods that will be described in this section, we need to consider the following probabilities and parameters:

*A Priori Probabilities* From the knowledge and prior information for each class of features, one can determine the prior-probabilities associated with each class $P(C_i)$, $i = 1, 2, \ldots, L$. If there is no a priori information on the classes, we shall assume an equal prior $P(C_i) = P(C_j) = 1/L$; that is, a feature vector $\mathbf{x}$ has an equal prior probability of being in any of the $L$ classes.

*Conditional Probabilities (or Likelihood)* Another probability of concern is the conditional probability $p(\mathbf{x}|C_i)$, also referred to as the likelihood. Here, $p(\mathbf{x}|C_i)$ is the probability of feature vector $\mathbf{x}$ to be in the given class $C_i$, or the likelihood of $C_i$ being the true class for $\mathbf{x}$.

*Posterior Probabilities* The posterior probability $P(C_i|\mathbf{x})$ is the probability that a given feature vector $\mathbf{x}$ belongs to class $C_i$. The posterior probability can be computed using the Bayes formula:

$$P(C_i|\mathbf{x}) = \frac{p(C_i)P(\mathbf{x}|C_i)}{\sum_{j=1}^{L} P(C_j)p(\mathbf{x}|C_j)} \tag{9.20}$$

The posterior probability can be viewed as the combination of the likelihood of $C_i$ being the true class for $\mathbf{x}$ and the prior class probability indicating how frequent the class $C_i$ will appear among all $L$ classes. The denominator in (9.20) can be viewed as a scale factor that guarantees that the posterior probabilities sum to 1.

*False-Alarm Rate* For a given feature vector **x**, if it is classified as belonging to class $C_i$ (e.g., a voltage dip due to fault) while the true class of **x** is $C_j$ (e.g., a dip due to transformer energizing), then **x** is called a false alarm to the class $C_i$ (a false alarm of fault-induced dip). It is often required that the false-alarm rate be kept low in order to avoid unnecessary cost (e.g., a protection action in the powersystem).

In the two approaches described below, the Neyman–Pearson approach uses the conditional probabilities (or the likelihood) with the false-alarm rate as a parameter of constraint, while the Bayesian-based approach employs both the prior probability and the likelihood.

In all these approaches, we use $P(\cdot)$ to denote the probability and $p(\cdot)$ the probability density function (pdf).

### 9.4.1 Hypothesis Tests and Decision Trees

Assume that there are $L$ classes of disturbances $C_0, C_1, \ldots, C_{L-1}$ and $L$ is known in advance. For each data recording a vector of features is extracted, followed by some preprocessing such as feature normalization, transformation, and dimension reduction (see Section 9.2), yielding an $M$-dimensional feature vector $\mathbf{x} = [x_1 \ x_2 \ \cdots \ x_M]^{\mathrm{T}}$. Further, assume that the probability distribution of **x** for each class is known (or is estimated).

Let $L$ hypotheses be $\{\mathcal{H}_i, \ i = 0,1,\ldots, L-1\}$. Let the $i$th hypothesis $\mathcal{H}_i$ denote that a given feature vector **x** belongs to the class $C_i$, that is,

$$\mathcal{H}_i\colon \mathbf{x} \in C_i \qquad i = 0,1,\ldots,L-1 \tag{9.21}$$

The classification problem under the framework of hypothesis testing can be regarded as, given a feature vector **x**, deciding the best $\mathcal{H}_j$ among the $L$ possible hypotheses $\{\mathcal{H}_i, \ i = 0,1,\ldots, L-1\}$ under a preselected criterion.

Assuming the probability of **x** being in each class is known, (9.21) can also be written as

$$\mathcal{H}_i\colon \ \mathbf{x} \sim p(\mathbf{x};\, C_i) \qquad i = 0, 1, \ldots, L-1 \tag{9.22}$$

where $p(\mathbf{x};\, C_i)$ is the probability of **x** being in the class $C_i$. The classification process can be depicted in Figure 9.2 by a decision tree under the multiple hypothesis test.

### 9.4.2 Neyman–Pearson Approach

Assume the conditional *probability density functions* $\{p(\mathbf{x}|C_i), \ i = 0, 1, \ldots, L-1\}$ are known. Further assume the prior probabilities of all classes are equal, that is, $P(C_i) = P(C_j)$. For a given feature vector **x**, (9.22) can thus be expressed by

$$\mathcal{H}_i\colon \mathbf{x} \sim p(\mathbf{x}|C_i) \tag{9.23}$$

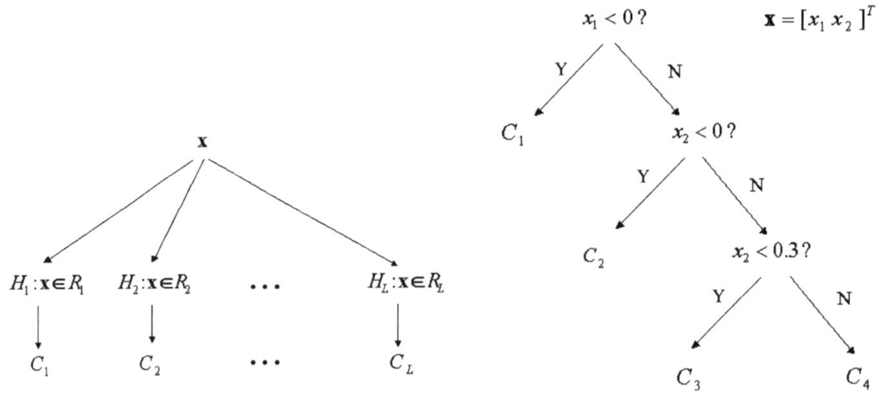


**Figure 9.2** Multiple hypothesis test and decision tree. (*a*) Decision tree by using $L$ hypotheses for $L$ classes. $R_i$ is the region for class $i$. (*b*) Example of multilevel binary decision tree, where $\mathbf{x} = [x_1, x_2]^{\mathrm{T}}$ and there are four classes in total.

Under the Neyman–Pearson (NP) approach, the best classification is determined according to the maximum-likelihood (ML) criterion under the constraint that the probability of false alarm $P_{\mathrm{FA}}$ (for choosing some particular class $C_i$) is below a predetermined threshold $\eta$. This is equivalent to deciding $\mathcal{H}_i$ such that

$$\begin{aligned}&\text{Maximize} \quad p(\mathbf{x}|C_i) = \max \{p(\mathbf{x}|C_j) \qquad j = 0,1,\ldots,L-1\}\\&\text{Subject to} \quad P_{\mathrm{FA}} = \sum_{j:\, j\neq i} P(\mathcal{H}_i;\, C_j) = \int_{\mathcal{R}_i}\left[\sum_{j:\, j\neq i} p(\mathbf{x}|C_j)\ dx\right] \leq \eta\end{aligned} \tag{9.24}$$

where $p(\mathbf{x}|C_i)$ in the first equation implies deciding $\mathcal{H}_i$ based on the ML criterion, $P(\mathcal{H}_i;\, C_j)$ is the probability of deciding $\mathcal{H}_i$ when the true class of $\mathbf{x}$ is $C_j$, and $\mathcal{R}_i$ is the decision region for $\mathbf{x} \in C_i$.

***9.4.2.1 Two-Class Case*** Consider the classification of two classes of disturbances. We will show later how the approach can be extended to a multiclass case.

Let the hypothesis $\mathcal{H}_0$ denote that a given feature vector $x$ belongs to class $C_0$ and $\mathcal{H}_1$ to class $C_1$, that is,

$$\mathcal{H}_0\colon \mathbf{x} \in C_0 \qquad \mathcal{H}_1\colon \mathbf{x} \in C_1 \tag{9.25}$$

Similar to (9.23), the above equation can be expressed as

$$\mathcal{H}_0\colon \mathbf{x} \sim p(\mathbf{x}|C_0) \qquad \mathcal{H}_1\colon \mathbf{x} \sim p(\mathbf{x}|C_1) \tag{9.26}$$

where $p(\mathbf{x}|C_i)$, $i = 0, 1$, are the conditional pdf's of $\mathbf{x}$. The task is to "best" classify each incoming data based on its feature vector $\mathbf{x}$ under the constraint that the false alarm of classifying **x**, for example, as class $C_1$ (i.e., deciding $\mathcal{H}_1$ when the actual class is $C_0$) is smaller than a predetermined threshold. One may, for example, choose the hypothesis $\mathcal{H}_1$ as being the presence of a voltage dip and $\mathcal{H}_0$ as without a voltage dip (note, this can also be regarded as a detection problem or a binary classification problem on whether a voltage dip exists). In some other case one may classify the cause of a disturbance, for example, choosing $\mathcal{H}_0$ as a dip caused by transformer energizing, $\mathcal{H}_1$ as a dip caused by a fault, and $\mathcal{H}_2$ as a disturbance caused by capacitor switching. The latter class can easily be distinguished when a postprocessing is applied (i.e., when the waveform of an entire event is available). But during real-time processing, when only part of the waveform in the event is available, distinguishing between these two events is no longer trivial. Capacitor-energizing transients may lead to erroneous detection of a voltage dip (a so-called false alarm).

For the two-class case, we can apply the *likelihood ratio test*,

$$\text{Decide } \mathcal{H}_1 \qquad \text{if} \quad L(\mathbf{x}) = \frac{p(\mathbf{x}|C_1)}{p(\mathbf{x}|C_0)} > \gamma \tag{9.27}$$

or the log-likelihood ratio test,

$$\text{Decide } \mathcal{H}_1 \qquad \text{if} \quad \tilde{L}(\mathbf{x}) = \log \frac{p(\mathbf{x}|C_1)}{p(\mathbf{x}|C_0)} > \log(\gamma) \tag{9.28}$$

where the threshold $\gamma$ is determined by the constraint of the false-alarm probability,

$$P_{\mathrm{FA}} = \int_{\mathcal{R}_1} p(\mathbf{x}|C_0)\, d\mathbf{x} \le \eta \tag{9.29}$$

and the decision region $\mathcal{R}_1$ is defined as $\mathcal{R}_1 = \{\mathbf{x}\colon L(\mathbf{x}) > \gamma\}$. The threshold $\gamma$ can be determined analytically when the exact mathematical expression of the pdf's is known. Otherwise $\gamma$ has to be determined numerically (see the sequential classification example in Section 9.4.5). Finally, to evaluate the performance, one may examine the classification (or detection) probability,

$$P_D = \int_{\mathcal{R}_1} p(\mathbf{x}|C_1)\, d\mathbf{x} \tag{9.30}$$

The above binary classifier can be regarded as a detector (e.g., detecting whether there exists a voltage dip), where one may encounter some frequently used terminologies in detection theory:

| | |
|---|---|
| Detection (or a hit) | Decide $\mathcal{H}_1$ when the actual class is $C_1$ |
| Type I error (or a false alarm) | Decide $\mathcal{H}_1$ when the actual class is $C_0$ |
| Type II error (or a miss) | Decide $\mathcal{H}_0$ when the actual class is $C_1$ |
| A correct rejection | Decide $\mathcal{H}_0$ when the actual class is $C_0$ |

The performance of the binary classifier or a detector is often evaluated by a *receiver operating characteristic* (ROC) curve where $P_{\mathrm{FA}} \sim P_D$ curves are shown under different parameter settings.

**Example 9.1** Hypothesis Tests for Detecting Whether Fault-Induced Dip Originated at Transmission Level or at Distribution Level. In this example we shall use the NP approach to classify whether a fault-induced voltage dip originated at a transmission level or at a distribution level of a power system. Fault-induced dips from these two different levels have different range of dip duration. For the fault-induced dips at the transmission level, the mean duration value is roughly around $d_1 = 100$ ms, while for the dips at the distribution level, the mean duration value is roughly around $d_0 = 200$ ms.

Let the random variable $x$ denote the measured duration of a fault-induced dip. It can be viewed as the true value (deterministic) plus the white noise (include the measurement noise and the feature variation). We formulate the two hypotheses as follows:

$$\begin{array}{lll} \mathcal{H}_1 \text{ (fault-induced dip at transmission level):} & x = w + d_1 \\ \mathcal{H}_0 \text{ (fault-induced dip at distribution level):} & x = w + d_0 \end{array} \qquad (9.31)$$

where the white noise $w$ is assumed to be Gaussian distributed, $p(w) = N(0, \sigma^2)$, and $d_i$, $i = 0{,}1$, are constant values representing the mean dip durations at the corresponding levels. The standard deviation is set to be $\sigma = 30$ ms and is empirically determined. Figure 9.3 shows the two pdf's $p(x|C_1)$ and $p(x|C_0)$ and the decision boundary $\gamma$ that is to be decided below.

Under both hypotheses, $x$ is also Gaussian distributed, but with different mean values. For the hypothesis $\mathcal{H}_0$ the conditional probability is $p(x|C_0) = N(d_0, \sigma^2)$, and for the hypothesis $\mathcal{H}_1$ the conditional probability is $p(x|C_1) = N(d_1, \sigma^2)$. The log-likelihood ratio test in (9.28) (here, $\log_e$ is used instead of $\log_{10}$ for mathematical convenience in the Gaussian pdf without influencing the classification result) can be applied in this occasion,

$$\tilde{L}(x) = \ln \frac{1/(\sqrt{2\pi}\sigma)e^{-(x-d_1)^2/(2\sigma^2)}}{1/(\sqrt{2\pi}\sigma)e^{-(x-d_0)^2/(2\sigma^2)}} = \frac{(x-d_0)^2 - (x-d_1)^2}{2\sigma^2} > \ln(\gamma) \qquad (9.32)$$

To classify in which level of the power system the fault-induced dip happened, under the constraint that the probability of wrongly classifying a dip in the distribution level as in the transmission level does not exceed $\eta$,

$$P_{\mathrm{FA}} = P\{x \in \mathcal{R}_1; C_0\} = \frac{1}{\sqrt{2\pi}\sigma}\int_{-\infty}^{\gamma} e^{-(x-d_0)^2/(2\sigma^2)}\, dx \le \eta \qquad (9.33)$$

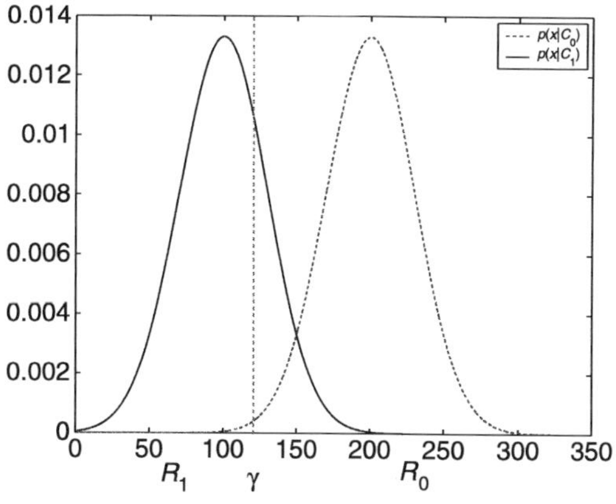


**Figure 9.3** Gaussian distributions for $x$ under two hypotheses $\mathcal{H}_1$ and $\mathcal{H}_0$. The vertical dashed line indicates the threshold $\gamma$ (being selected by the NP method) that separates the two decision regions $\mathcal{R}_1$ (the left region) and $\mathcal{R}_0$ (the right region). Note that the vertical line in the figure does not correspond to the actual position that is selected.

where $\mathcal{R}_1 = \{x\colon x < \gamma\}$. After manipulation we can obtain the equivalence

$$P_{\text{FA}} = \frac{1}{\sqrt{2\pi}\sigma}\int_{-\infty}^{\gamma} e^{-(x-d_0)^2/(2\sigma^2)}\,dx = \frac{1}{\sqrt{2\pi}}\int_{-\infty}^{-(d_0-\gamma)/\sigma} e^{-\frac{1}{2}\xi^2}\,d\xi \le \eta \qquad (9.34)$$

where $(d_0 - \gamma)/\sigma$ is a variable larger than zero. For a given $\eta$ (e.g., set $\eta = 0.05$), we can look up the mathematical table for the Gaussian distribution $P(x) = (1/\sqrt{2\pi})\int_{-\infty}^{-x} e^{-(1/2)\xi^2}\,d\xi = 0.05$ where $x = (d_0 - \gamma)/\sigma$. We obtain $(d_0 - \gamma)/\sigma = 1.6449$. Substituting $\sigma = 30$ ms and $d_0 = 200$ ms, we obtain the threshold $\gamma \le 150.653$ ms as the boundary that separates the regions of $\mathcal{R}_1$ and $\mathcal{R}_0$ under the specified false-alarm constraint.

To determine the classification probability for fault-induced dips in the transmission level, the following integration is applied:

$$P_{D_{\text{trans}}} = P\{x \in \mathcal{R}_1;\, C_1\} = \frac{1}{\sqrt{2\pi}\sigma}\int_{-\infty}^{\gamma} e^{-(x-d_1)^2/(2\sigma^2)}\,dx \qquad (9.35)$$

Note that $\mathcal{R}_1 = \{x\colon x < \gamma = 150.653\}$, $d_1 = 100$ ms, and $\sigma = 30$ ms. From the table of Gaussian distribution, we obtain the detection probability $P_{D_{\text{trans}}} = 0.9545$ for this case.

The classification probability for fault-induced dips in the distribution level can be obtained straightforwardly as

$$P_{D_{\text{dist}}} = P\{x \in \mathcal{R}_0;\, C_0\} = \frac{1}{\sqrt{2\pi}\sigma}\int_{\gamma}^{\infty} e^{-(x-d_0)^2/(2\sigma^2)}\,dx \ge 1 - 0.05 = 0.95 \quad (9.36)$$

It is worth mentioning that this method is not limited to Gaussian distributions. Any distribution can be used (e.g., distributions obtained from measurement data). However, for most practical cases it is no longer possible to use analytical expressions, and numerical solutions should be used to find the threshold setting.

#### 9.4.2.2 *Multiclass Case*

For $L$ classes, $L > 2$, $C_0, C_1, \ldots, C_{L-1}$, classification using the likelihood ratio test (or log-likelihood ratio test) in (9.27) is replaced by the criterion

$$\text{Decide } \mathcal{H}_i \qquad \text{if } L(\mathbf{x}) = \frac{p(\mathbf{x}|C_i)}{p(\mathbf{x}|C_j)} > \gamma, \qquad \text{for all } j \neq i \tag{9.37}$$

subject to the constraint of the false-alarm probability in (9.24).

### 9.4.3 Bayesian Approach

Instead of using (9.24) as the criterion in the NP approach (i.e., the ML criterion under the false-alarm constraint where no a priori information is assumed), the Bayesian approach minimizes the overall risk by combining the prior knowledge of each class. The corresponding multiple hypotheses $\mathcal{H} = \{\mathcal{H}_0, \ldots, \mathcal{H}_{L-1}\}$ are defined as

$$\mathcal{H}_i\colon \mathbf{x} \sim P(C_i|\mathbf{x}) \qquad i = 0,1,\ldots, L-1 \tag{9.38}$$

where $P(C_i|\mathbf{x})$ is the *posterior probability* of $\mathbf{x}$ being in the class $C_i$. The conditional risk of deciding $\mathcal{H}_i$ for a given feature vector $\mathbf{x}$ is

$$R(\mathcal{H}_i|\mathbf{x}) = \sum_j \lambda_{ij} P(C_j|\mathbf{x}) \tag{9.39}$$

where $\lambda_{ij}$ is the *cost incurred for deciding* $\mathcal{H}_i$ *when* $\mathbf{x}$ *actually belongs to* $C_j$. When we make some error, for example, classify $\mathbf{x}$ as being in the class $i$ when it actually belongs to class $j$, there is a cost related to this. Naturally, if we make a correct classification, there is no cost (or loss). Therefore, $\lambda_{ij} - \lambda_{jj}$ is always positive for all $i \neq j$. The overall risk is

$$R = \int R(\mathcal{H}(\mathbf{x})|\mathbf{x}) p(\mathbf{x})\, d\mathbf{x} \tag{9.40}$$

The Bayes decision rule is to minimize the overall risk $R$ (or Bayes risk) in (9.40). This is equivalent to choosing the hypothesis $\mathcal{H}_j$ that has the minimum $R(\mathcal{H}_j|\mathbf{x})$,

$$\text{Decide } \mathcal{H}_j\colon R(\mathcal{H}_j|\mathbf{x}) = \min\{R(\mathcal{H}_i|\mathbf{x}), \quad i = 0, 1, \ldots, L-1\} \tag{9.41}$$

***9.4.3.1 Two-Class Case*** For simplicity, we again consider the two-class case. The hypotheses in (9.38) become

$$\mathcal{H}_0\colon \mathbf{x} \sim P(C_0|\mathbf{x}) \qquad \mathcal{H}_1\colon \mathbf{x} \sim P(C_1|\mathbf{x}) \tag{9.42}$$

The conditional risks corresponding to (9.39) are

$$\begin{aligned} R(\mathcal{H}_0|\mathbf{x}) &= \sum_j \lambda_{0j} P(C_j|\mathbf{x}) = \lambda_{00} P(C_0|\mathbf{x}) + \lambda_{01} P(C_1|\mathbf{x}) \\ R(\mathcal{H}_1|\mathbf{x}) &= \sum_j \lambda_{1j} P(C_j|\mathbf{x}) = \lambda_{10} P(C_0|\mathbf{x}) + \lambda_{11} P(C_1|\mathbf{x}) \end{aligned} \tag{9.43}$$

To choose a hypothesis with the minimum risk, one performs thefollowing:

$$\text{Choose } \mathcal{H}_1 \qquad \text{if } R(\mathcal{H}_0|\mathbf{x}) > R(\mathcal{H}_1|\mathbf{x}) \tag{9.44}$$

Substituting (9.43) to (9.44) and noting the Bayes formula

$$P(C_i|\mathbf{x}) = \frac{p(\mathbf{x}|C_i)P(C_i)}{\sum_j p(\mathbf{x}|C_j)P(C_j)}$$

it follows,

$$\text{Choose } \mathcal{H}_1 \qquad \text{if } \frac{p(\mathbf{x}|C_1)}{p(\mathbf{x}|C_0)} > \frac{(\lambda_{10} - \lambda_{00})P(C_0)}{(\lambda_{01} - \lambda_{11})P(C_1)} \tag{9.45}$$

where $\lambda_{ij} - \lambda_{jj} > 0$ is assumed. The overall risk in (9.40) becomes

$$\begin{aligned} R = &\int_{\mathcal{R}_0} [\lambda_{00} P(C_0) p(\mathbf{x}|C_0) + \lambda_{01} P(C_1) p(\mathbf{x}|C_1)]\, d\mathbf{x} \\ &+ \int_{\mathcal{R}_1} [\lambda_{10} P(C_0) p(\mathbf{x}|C_0) + \lambda_{11} P(C_1) p(\mathbf{x}|C_1)]\, d\mathbf{x} \end{aligned} \tag{9.46}$$

where $\mathcal{R}_i$ is the decision region that the classifier decides $\mathbf{x} \in C_i$, $i = 0,1$.

***9.4.3.2 Multiclass Case*** For multiple classes, $L > 2$, one simply extends (9.44) by choosing the hypothesis $\mathcal{H}_i$ that is associated with the minimum conditional risk (hence minimizes the overall risk),

$$\text{Choose } \mathcal{H}_i \qquad \text{if } R(\mathcal{H}_j|\mathbf{x}) > R(\mathcal{H}_i|\mathbf{x}), \qquad \text{for all } j \neq i \tag{9.47}$$

#### 9.4.3.3 Discussion

*(a) Minimum Error Rate Classifier* It is interesting to notice that if one chooses

$$\lambda_{ij} = \begin{cases} 1 & \text{for } i \neq j \\ 0 & \text{otherwise} \end{cases} \tag{9.48}$$

then a Bayesian classifier becomes a *minimum-error-rate-classifier*. It is worth noting that in a minimum-error-rate classifier one would make an equal cost (or risk) whether or not a feature $\mathbf{x}$ is wrongly classified to the class $i$ or any other class. In such a case, the decision can be made from comparing the posterior probabilities and choosing the hypothesis that is associated with the maximum posterior probability,

$$\text{Choose } \mathcal{H}_i \qquad \text{if } P(C_i|\mathbf{x}) > P(C_j|\mathbf{x}) \qquad \text{for all } j \neq i \tag{9.49}$$

*(b) Relations to Neyman–Pearson Approach* It is also interesting to notice that if all priors $P(C_i)$ are equal and if we choose to minimize the overall error in (9.40) subject to the constraint that the conditional risk $\int R(\mathcal{H}_i|\mathbf{x})\,d\mathbf{x} < \eta$ for some particular class $i$, a Bayesian classifier becomes the classifier under the NP criterion.

### 9.4.4 Bayesian Belief Networks

A Bayesian belief network is a graphical representation of the dependencies of system variables and the relative probabilities of these variables by means of directed acyclic graph. A Bayesian belief network can be used to form a Bayesian classifier, or a maximum a posteriori classifier if the Bayes rule is carefully applied.

***9.4.4.1 Definition of Bayesian Belief Network*** A Bayesian belief network, or simply a belief network, consists of *nodes that represent system variables and directional links that represent the conditional probabilities;* for example, the link from node $x_1$ to node $x_2$ represents the conditional probability $P(x_2|x_1)$. It is worth noting that variables have links only when they have some dependencies. Independent variables are not linked to each other.

***9.4.4.2 Feature Dependencies*** Let us assume that we assign features as the system variables of a belief network. When not all features are correlated, the expression of their joint probabilities can be simplified. For example, if the feature vector consists of three features, $\mathbf{x} = [x_1, x_2, x_3]^T$, of which $x_3$ is dependent on $x_1$ and $x_2$ while $x_1$ and $x_2$ are independent, then the joint probability of $\mathbf{x}$ can be written as $P(\mathbf{x}) = P(x_3|x_1, x_2)P(x_1)P(x_2)$. The dependencies of these features and their probabilities and conditional probabilities can be graphically described by the belief network in Figure 9.4*a*. More general, for a given feature vector $\mathbf{x} = [x_1, \ldots, x_M]^T$ some features are correlated and others are uncorrelated. If choosing features as the system variables in the belief network, we can graphically describe their dependencies.

To determine whether features $x_i$ and $x_j$ are dependent, one may compute their cross-correlation [see (9.2)]. If the cross-correlation value equals zero (or close to zero), then the two features are uncorrelated (or weakly correlated). Weakly correlated features can be approximated as uncorrelated in order to simply the belief network. If the cross-correlation value is close to 1, then they are highly correlated, which is undesirable. If this happens, one may consider decorrelating features or selecting some principal features through feature optimization (see Section 9.2) in order to simplify the belief network.

#### 9.4.4.3 Assign System Variables

Clearly, assigning system variables (or, nodes) in the belief network is essential and is dependent on a classification problem. For example, to classify two classes of voltage dips, one may set three system variables: The first variable $x_1$ is used for the dip class, where $x_1 = x_{1,1}$ denotes the variable taking a discrete value as the fault-induced dip, $x_1 = x_{1,2}$ denotes the variable taking a discrete value as the non-fault induced dip. The second variable $x_2$ describes the shape of the rms voltage drop with the discrete value $x_{2,1}$ being a sharp drop and $x_{2,2}$ being a smooth drop. The third variable $x_3$ describes the shape of the rms voltage recovery with the discrete value $x_{3,1}$ being a sharp recovery and $x_{3,2}$ being a smooth recovery.

#### 9.4.4.4 Classification Using Bayesian Belief Network

Assuming we have the complete conditional probabilities and some prior probabilities of the variables at some states in the belief network [e.g., $P(x_{1,1}) = P(x_{1,1}|e^{x_1}) = 0.3$ as the prior probability of fault-induced dips and $P(x_{1,2}) = P(x_{1,2}|e^{x_1}) = 0.7$ as the prior probability of non-fault-induced dips, together with all other required conditional probabilities]. If we carefully apply the Bayes rule, we will be able to determine the posterior probabilities [e.g., $P(x_{1,1}|\mathbf{e})$ as the probability of classifying the dip as being a fault given the evidence **e** from the belief network and $P(x_{1,2}|\mathbf{e})$ as the probability of classifying the dip as being a non-fault given the evidence **e** from the belief network]. We can then select a class that has the maximum posterior probability. For example, selecting the dip as the fault-induced class if $P(x_{1,1}|\mathbf{e}) > P(x_{1,2}|\mathbf{e})$.

#### 9.4.4.5 Compute Probability of Variable from Evidence in Network

Since a Bayesian belief network describes a set of relative probabilities of system variables for the given evidence **e** from the rest of the network, it is important to establish the method to compute such a relative probability of $P(x_i|\mathbf{e})$ for $x_i$. For a given node in the network assigned with the variable $x_i$, the relative probability can be computed from

$$P(x_i|\mathbf{e}) \propto P(\mathbf{e}^c|x_i)P(x_i|\mathbf{e}^{\mathscr{P}}) \tag{9.50}$$

where $\mathbf{e}^c$ denotes the evidence from the children and $\mathbf{e}^{\mathcal{P}}$ the evidence from the parents of $x_i$, a child node of $x_i$ is defined as the node that receives the directional link from $x_i$, and a parent node of $x_i$ is defined as the node that originates the directional link toward $x_i$. The two probabilities on the right-hand side of (9.50) can be computed as

$$P(\mathbf{e}^c|x_i) = \prod_{c_j} P(e^{c_j} \mid x_i) \tag{9.51}$$

where $c_j$ denotes the $j$th child of $x_i$, and

$$\begin{aligned} P(x_i \mid \mathbf{e}^{\mathcal{P}}) &= P(x_i \mid e^{\mathcal{P}_1}, \ldots, e^{\mathcal{P}_{|\mathcal{P}|}}) \\ &= \sum_{i,\ldots,k} P(x_i \mid \mathcal{P}_{1,i}, \ldots, \mathcal{P}_{|\mathcal{P}|,k}) P(\mathcal{P}_{1,i} \mid e^{\mathcal{P}_1}) \cdots \\ &\quad \times P(\mathcal{P}_{|\mathcal{P}|,k} \mid e^{\mathcal{P}_{|\mathcal{P}|}}) \end{aligned} \tag{9.52}$$

where $\mathcal{P}_l$ denotes the $l$th parent of $x_i$ and $\mathcal{P}_{l,k}$ denotes the variable $\mathcal{P}_l$ taking a value $\mathcal{P}_{l,k}$.

To illustrate this, a hypothetical example is shown in Figure 9.4*b* where the belief network consists of six variables $x_i$, $i = 1, \ldots, 6$. There are dependencies between the variables $x_1$ and $x_3$, $x_2$ and $x_3$, $x_3$ and $x_4$, $x_4$ and $x_5$, $x_4$ and $x_6$, and $x_5$ and $x_6$, and the corresponding conditional probabilities are $P(x_3|x_1)$, $P(x_3|x_2)$, $P(x_4|x_3)$, $P(x_5|x_4)$, $P(x_6|x_4)$, and $P(x_6|x_5)$. Further, the prior probabilities of two variables $x_1$ and $x_2$ are given as $P(x_1) = P(x_1|e^{x_1})$ and $P(x_2) = P(x_2|e^{x_2})$. For computing the relative probability of variable $x_4$ from the network evidence $\mathbf{e}$, we apply (9.50). The first term on the right-hand side of (9.50), that is, the evidence from the children of $x_4$ in the belief network, can be computed using (9.51) as

$$\begin{aligned} P(\mathbf{e}^c|x_4) &= P(e^{x_5} \mid x_4) P(e^{x_6} \mid x_4) \\ &= \left( \sum_j P(e^{x_5} \mid x_{5,j}) P(x_{5,j} \mid x_4) \right) \\ &\quad \times \left( \sum_k P(e^{x_6} \mid x_{6,k}) P(x_{6,k} \mid x_4) \right) \end{aligned} \tag{9.53}$$

where $x_{l,j}$ denotes the variable $x_l$ taking a discrete value $x_{l,j}$. The second term on the right-hand side of (9.50), that is, the evidence from the parents of $x_4$ in the belief network, can be computed by using (9.52), which results in

$$P(x_4|\mathbf{e}^{\mathcal{P}}) = \sum_j P(x_4 \mid x_{3,j}) P(x_{3,j} \mid e^{x_3}) \tag{9.54}$$

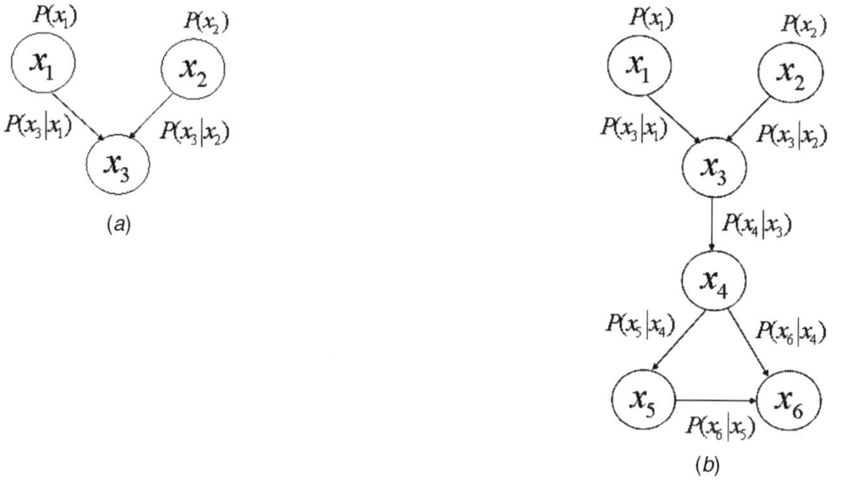


**Figure 9.4** Example of Bayesian belief network. (*a*) The belief network describes the probabilities of a feature vector consisting of three components, of which $x_1$ and $x_2$ are independent and $x_3$ is dependent on both $x_1$ and $x_2$. The joint probability of three features is $P(\mathbf{x}) = P(x_3|x_1, x_2)P(x_1)P(x_2)$. (*b*) In this application example the feature vector contains six components, $\mathbf{x} = [x_1, x_2, \ldots, x_6]^T$; $P(x_1)$ and $P(x_2)$ are the prior probabilities.

Hence, the relative probability of $x_4$ from the evidence $\mathbf{e}$ throughout the entire belief network can be obtained by multiplying the results from (9.53) and (9.54),

$$P(x_4 \mid \mathbf{e}) \propto P(\mathbf{e}^c|x_4)P(x_4|\mathbf{e}^{\mathcal{P}})$$

### 9.4.5 Example of Sequential Classification of Fault-Induced Voltage Dips

In this example, we shall describe an application that uses the NP approach for the fast detection of fault-induced voltage dips from three classes of disturbance recordings [127]. The three classes considered here are dips caused by faults, dips caused by transformer energizing, and disturbances caused by capacitor switching. This example concerns an online processing where a decision should be made as soon as possible. The classification process is performed sequentially, enabling a classification within the time interval from a minimum $\frac{1}{8}$ cycle to $\frac{1}{4}$, $\frac{1}{2}$, $\frac{3}{4}$, up to a maximum of one cycle.

The first step is to compute the rms voltage sequence from a given recording using a half-cycle sliding window and to detect the triggering point of voltage drop from the rms voltages using a threshold of 0.95 pu. A sequential processing is then applied, starting using features from a short analysis window of $\frac{1}{8}$ cycle. If the probabilities of detection and false alarm cannot meet some predetermined value, then the features from larger analysis windows ($\frac{1}{4}$, $\frac{1}{2}$, $\frac{3}{4}$ cycle) are subsequently applied for the detection until the required probabilities are met or otherwise a fault-induced event is rejected.

Two features are extracted from each rms sequence to form a feature vector $\mathbf{x} = [x_1, x_2]^T$. The first feature $x_1$ is dip-related. It is obtained by selecting one rms sequence among the three phases which has the lowest rms value within the analysis window and assigning the rms value at the end of the analysis window to $x_1$. The second feature $x_2$ is swell related. It is obtained by selecting the rms sequence among the three phases which has the highest rms value within the analysis window and assigning the rms value at the beginning of the analysis window to $x_2$. The motivation of defining $x_1$ is based on the observation that fault-induced dips are more severe than the other classes of dips. The motivation of defining $x_2$ is based on the observation that fault-induced dips might be associated with an increase of rms voltage in the nonfaulted phases (voltage swell), while such an increase is rare in a transformer saturation.

The NP approach is used for detecting whether or not the measured disturbance is a fault-induced dip. Two hypotheses are used:

$$\begin{array}{ll} \mathcal{H}_1 \text{ (fault-induced voltage dip):} & x \sim p(x|C_1) \\ \mathcal{H}_0 \text{ (non-fault-induced dip or event):} & x \sim p(x|C_0) \end{array}$$

It is assumed that the conditional pdf's $p(x|C_i)$, $i = 0,1$, can be obtained from data learning (empirically estimated from the histograms) and that the non-fault-induced event further consists of two subclasses: One is the dip caused by transformer energizing and another is the disturbance caused by capacitor switching. To decide $\mathcal{H}_1$, (9.27) is applied where the likelihood ratio $\gamma$ is determined by the constraint of false alarm in (9.29).

To train the detector, histograms of features are prelabeled according to the event classes. For a one-dimensional $x_1$ (or $x_2$), detector design can be described as: Let $\gamma$ be the threshold separating the two decision regions $\mathcal{R}_i$, $i = 0,1$. For a given $\gamma$, the detection rate $P_D$ is estimated as the ratio between the number of fault-induced dips in $\mathcal{R}_1$ and the total number of fault-induced dips in $\mathcal{R}_1 \cup \mathcal{R}_0$. The false-alarm $P_{FA}$ is estimated as the ratio between the number of non-fault-induced events in $\mathcal{R}_1$ and the total number of events in $\mathcal{R}_1 \cup \mathcal{R}_0$. For each given false-alarm threshold $\eta$, the detector shifts its decision boundary $\gamma$ such that the false alarm satisfies $P_{FA} \leq \eta$ meanwhile $P_D$ is selected to have the maximum possible value. The discrete values were set to $\eta = \eta_i$ by taking a step size 1% from 1 to 100%. A table for the trained detector is then made containing $\eta_i$ and the resulting $\gamma_i$, $P_D(i)$, and $P_{FA}(i)$. Further, the ROC (i.e., $P_D \sim P_{FA}$) curves are obtained by plotting all pairs of ($P_D(i)$, $P_{FA}(i)$), each corresponding to a particular sized analysis window. For a two-dimensional feature vector $[x_1\ x_2]^T$, the threshold is set to be $\eta = \eta^{(1)} + \eta^{(2)}$ as an approximation, where $\eta^{(1)}$ and $\eta^{(2)}$ are thresholds associated with $x_1$ and $x_2$, respectively.

For each new disturbance measurement from the test set a feature vector is extracted. Normalized histograms of features are computed. The corresponding values in the normalized histograms are used as $p(x|\text{faults})$, $p(x|\text{TS})$, $p(x|\text{CS})$, that is, the probability of faults, nonfaults including transformer saturation (TS), and capacitor switching (CS) transients. Then (9.27) is applied; that is,

$p(x|\text{faults})/p(x|\text{TS})$ and $p(x|\text{faults})/p(x|\text{CS})$ are calculated, respectively. If both values are larger than $\gamma_0$ (for a predetermined $\eta_0$, $\gamma_0$ is found from the saved table associated with the trained detector), then it is classified as a fault-induced dip. Meanwhile, for each selected false-alarm threshold $\eta_0$, the accuracy of the decision and the estimated $P_D$ and $P_{\text{FA}}$ can be found from looking up the saved table obtained from the training process.

The sequential detector is a simple extension of the single detector. Once a voltage drop and the triggering point are detected, a detector is applied using the smallest analysis window ($w_1 = \frac{1}{8}$ cycle) and the normalized histograms are estimated. If (9.27) is satisfied, then the data recording is classified as a fault-induced dip and the procedure is terminated. Otherwise, the detector is applied to the features obtained from a larger analysis window ($w_2 = \frac{1}{4}$ cycle and so forth) until the event is either accepted as a fault-induced dip or classified as a non-fault-induced disturbance if it is not accepted from using the largest analysis window ($w_4 = \frac{3}{4}$ cycle). If in any analysis window the rms voltage is above 0.95 pu, then the event is no longer considered. This is typical for a dip due to transformer saturation that produces a rms signature of short-duration repeating voltage drops [127].

The sequential detection scheme is trained using the recordings containing dips due to faults and transformer saturation measured in an 11-kV network over a one-month period. The false-alarm threshold was set to $\eta = 6.5\%$. The disturbances caused by capacitor switching were found to be completely separable from the fault-induced dips using the first feature element $x_1$. Therefore, only the detection results from the two classes of dips are included.

The training results showed that about 70 and 95% of the fault-induced dips were correctly classified within the $\frac{1}{4}$- and $\frac{3}{4}$-cycle analysis window, respectively, under the specified false alarm $\eta$. Figure 9.5*a* shows the resulting ROC curves for four different analysis windows sized $w_i = \frac{1}{8}, \frac{1}{4}, \frac{1}{2}, \frac{3}{4}$ cycle. It can be seen that the detection probability $P_D$ increases significantly with the increase of the analysis window size.

For the test set, however, data recordings measured from the same network over a different period of time were used. Figure 9.5*b* shows the resulting ROC curves from the test data set using four different analysis window sizes $w_i$. Compared to the training results in Figure 9.5*a*, Figure 9.5*b* shows a decrease in the detection probability or an increase in the false-alarm rate. To reduce the degradation of performance on the test set, the training data set can be collected from recordings spread over all different months rather than concentrated in a few months of a year. Further, cross-validation (see Section 9.6.6) can be applied for selecting a subset of training data that may yield better generalization performance.

The sequential detection method was able to distinguish fault-induced dips from transformer saturation and capacitor switching at $\frac{1}{8}, \frac{1}{4}, \frac{1}{2}$, or $\frac{3}{4}$-cycle analysis window, with a detection probability $P_D > 95\%$ with a (user-specified) maximum false-alarm level $\eta = 6.5\%$. The time required for a decision is found to be a trade-off between detection accuracy and speed and is dependent on the class of the dips/disturbances. It was found that long analysis windows were mainly required for shallow dips due to faults. Such dips are less severe than deep ones;

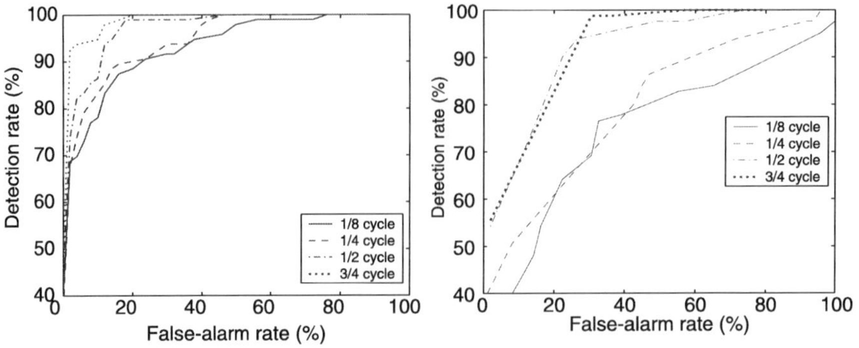


**Figure 9.5** Performance of detector: ROC curves for detecting fault-induced dips in presence of dips due to transformer saturation: (*a*) from training data; (*b*) from test data.

hence a longer detection time is justified. Deep dips are found in almost all cases being detected by the $\frac{1}{8}$-cycle analysis window.

## 9.5 LEARNING AND CLASSIFICATION USING ARTIFICIAL NEURAL NETWORKS

In this section we shall briefly describe two types of layered feedforward neural networks (NNs) that are frequently used in power system applications: the multilayer perceptrons (MLPs) and the radial-basis function (RBF) networks. The former tries to construct global approximations while the latter tries to construct local approximations to nonlinear input–output mappings. The former usually uses nonlinear mappings for all layers, while the latter only uses nonlinear mapping in the hidden layer. The design of the former corresponds to solving the stochastic approximation problem and the latter the surface-fitting problem. Despite all these differences, they both try to minimize the empirical error on the prelabeled training set of data. The generalization performance of these NNs is not guaranteed and could be poor depending on the selection of the training data set. There is also a risk of overfitting to the training data. Despite these limitations, artificial neural networks (ANNs) provide some attractive aspects, especially when the underlying signal models or the physical systems associated with the signals are largely unknown. Artificial neural networks offer us a way for signal classification under "black-box" models but without guaranteeing the performance.

### 9.5.1 Multilayer Perceptron Classifiers

A MLP [110, 133, 136] is a feedforward NN typically consisting of an input layer, one or several hidden layers, and one output layer, as shown in Figure 9.6. Each layer of a MLP contains several neurons that are connected by synaptic weights.

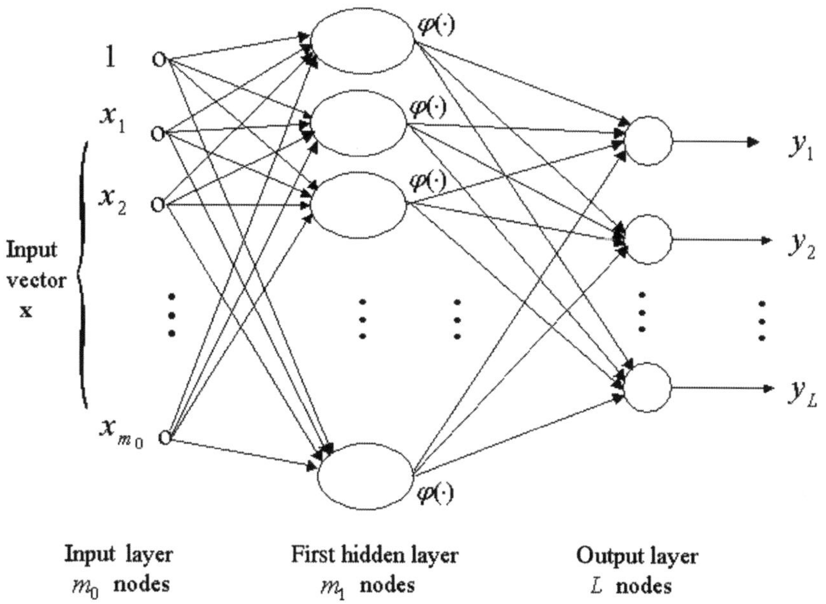


**Figure 9.6** Example of MLP neural network that consists of one hidden layer.

An activation function $\varphi(\cdot)$ is employed on the induced local field associated with each neuron. This is equivalent to applying a nonlinear function, for example, a sigmoidal function, whose input is the weighted sum of neuron outputs from the previous layer, which yields the output of the neuron. For an $L$-class classification problem, we need a total of $L$ neurons in the output layer to generate the outputs $y_j^{(O)}$, $j = 1, \ldots, L$, to represent all possible decision classes. The number of neurons in the input layer is usually the same as the size of the feature vector (or the size of the input signal vector). The number of neurons in a hidden layer as well as the total number of hidden layers are determined by the structure of the NNs and may vary significantly for different applications.

Let $y_j^{(l)}(n)$ denote the output from the $j$th neuron of layer $l$ at time $n, j = 1, \ldots, m_l$. Let the synaptic weight from neurons $i$ at layer $l-1$ to neuron $j$ at layer $l$ at time $n$ be $w_{ji}^{(l)}(n)$, the induced local field associated with the $j$th neuron be $v_j^{(l)}(n)$, the activation function on the local induced field of neuron $j$ be $\varphi_j^{(l)}(\cdot)$, and the bias of weights be $w_{j0}^{(l)} = b_j^{(l)}$ (see Fig. 9.6). Then, the output from neuron $j$ at layer $l$ can be obtained as

$$y_j^{(l)}(n) = \varphi_j^{(l)}(v_j^{(l)}(n)) \tag{9.55}$$

where the induced local field of this neuron is

$$v_j^{(l)}(n) = \sum_{i=1}^{m_{l-1}} w_{ji}^{(l)}(n) y_i^{(l-1)}(n) + w_{j0}^{(l)}(n) \tag{9.56}$$

and $\varphi(\cdot)$ is the activation function. An example of an activation function is the sigmoidal function that can achieve the nonlinearity,

$$\varphi_j^{(l)}(v_j^{(l)}(n)) = \frac{1}{1+\exp[-av_j^{(l)}(n)]} \qquad a > 0 \tag{9.57}$$

Some other logistic functions can also be used for achieving the nonlinearity. For the input layer ($l = 0$) $[y_1^{(0)}(n), \ldots, y_{m_0}^{(0)}(n)] = [x_1, x_2, \ldots, x_{m_0}]$, which is equal to the feature vector $\mathbf{x}$.

***9.5.1.1 Training Phase*** A MLP is trained using some recursive techniques known in statistics as stochastic approximation. A most commonly used approach for supervised training of a MLP is to use the back-propagation algorithm. Each iteration in a training process consists of a two-pass process. First, a feedforward pass is conducted when an input training sample $\mathbf{x}$ from the prelabelled training set $\{(\mathbf{x}_i, d_i), i = 1, \ldots, N\}$ is fed into the network where all synaptic weights are fixed, and the errors associated with the outputs of neurons are computed. A backward pass is then performed usually by a back-propagation algorithm where synaptic weights are updated according to the error gradient.

More specifically, during the feedforward pass, a vector of input data of a prelabeled class is fed into the NN starting from the input layer, the hidden layer(s), to the output layer with fixed synaptic weights. The corresponding error $e_j(n)$ in the $j$th neuron of the output layer is the difference between the desired output $d_j$ and the actual output $y_j^{(O)}(n)$,

$$e_j(n) = d_j - y_j^{(O)}(n) \tag{9.58}$$

Hence, the instantaneous error energy at time $n$ is

$$\varepsilon(n) = \frac{1}{2}\sum_{j=1}^{m_O} e_j^2(n) \tag{9.59}$$

where $m_O = L$ is the number of neurons in the output layer.

During the feedback pass, the errors at the outputs of neurons are propagated backward through the layers. The synaptic weights are updated in a manner similar to the LMS (least mean square) algorithm [221],

$$w_{ji}^{(l)}(n+1) = w_{ji}^{(l)}(n) - \eta\frac{\partial\varepsilon(n)}{\partial w_{ji}^{(l)}(n)} \tag{9.60}$$

where $\eta$ is a parameter controlling the learning rate. Applying the chain rule of calculus to $\partial\varepsilon(n)/\partial w_{ji}^{(l)}(n)$ yields

$$\frac{\partial\varepsilon(n)}{\partial w_{ji}^{(l)}(n)} = -\delta_j^{(l)}(n)y_i^{(l-1)}(n) \tag{9.61}$$

where

$$\delta_j^{(l)}(n) = -\frac{\partial \varepsilon(n)}{\partial v_j^{(l)}(n)} = -\frac{\partial \varepsilon(n)}{\partial y_j^{(l)}(n)} \frac{\partial y_j^{(l)}(n)}{\partial v_j^{(l)}(n)}$$

is the local gradient, which is equivalent to

$$\delta_j^{(l)}(n) = \begin{cases} \varphi_j'(v_j^{(l)}(n)) \sum_k \delta_k^{(l+1)}(n) w_{kj}^{(l+1)}(n) & \text{neuron } j \text{ is in hidden layer} \\ \varphi_j'(v_j^{(O)}(n)) e_j(n) & \text{neuron } j \text{ is in output layer} \end{cases} \tag{9.62}$$

where $\varphi_j'(\cdot)$ is the 1st derivative of $\varphi_j(\cdot)$. The iterations of the above forward pass and backward pass continue until the stop criterion is met. An example of a stop criterion is to minimize the following cost function (the average squared error energy over all training samples) at the output layer, or, when the value of the cost function is smaller than a prespecified small number $\varepsilon_0$,

$$\varepsilon_{\text{average}} = \frac{1}{N} \sum_{n=1}^{N} \varepsilon(n) \tag{9.63}$$

where $\varepsilon(n)$ is obtained from (9.59).

It is worth noting that the back-propagation training algorithm is a suboptimal technique, and it may get stuck in a local minimum instead of converging to the desired global minimum.

***9.5.1.2 Classification Phase*** In the classification phase, the feature vector belonging to a signal of unknown class is feedforward to the neurons in the input layer and propagates through the layers to generate output signals, where the synaptic weights from the training process are employed. The outputs $y_j^{(O)}, j = 1, \ldots, L$, are then compared with the target value of each class. For example, if the network is trained with binary target values, then for the class $C_k$, the target value is 1 for the $k$th neuron and zeros for the remaining neurons in the output layer. Hence, a target vector for $C_k$ can be set as the $L$-dimensional vector

$$\begin{bmatrix} 0 \\ \vdots \\ 1 \\ \vdots \\ 0 \end{bmatrix} \longleftarrow k\text{th element} \tag{9.64}$$

An input signal from the test set is classified to the class $C_k$ if the $k$th neuron in the output layer has the highest value or if the output vector has the shortest distance to the target vector in (9.64).

It is worth noting that a MLP classifier can lead to an asymptotic approximation of the underlying a posteriori class probabilities, provided that the training set is large enough and the back-propagation learning does not get stuck to a local minimum. It is also worth mentioning that such a NN is only designed to minimize the empirical error on the training set.

### 9.5.2 Radial-Basis Function Networks

A RBF network [110, 133, 136] is another type of feedforward NN where the network is designed by means of the so-called curve-fitting problem: finding a surface in a multidimensional space that best fits the training data.

A RBF network usually consists of three layers (see Fig. 9.7). The input layer of the network receives the feature vector extracted from the signal. The second layer (the only hidden layer) applies a nonlinear transformation from the input space to the feature space, usually having a higher dimension than that of the input space. The outputs of the network are the weighted sum from the neurons in the hidden layer.

***9.5.2.1 Cost Function, Regularization Solution of Mapping Function, and Weight Vector*** Supervised learning using RBFs is an ill-posed hyperspace reconstruction problem for a given set of data that may be sparse. We shall consider here the *generalized RBF networks*. Without loss of generality, let us consider a one-dimensional output space; that is, $y_j$ is a scalar, $y_j \in Y = \mathbb{R}$. A generalized RBF

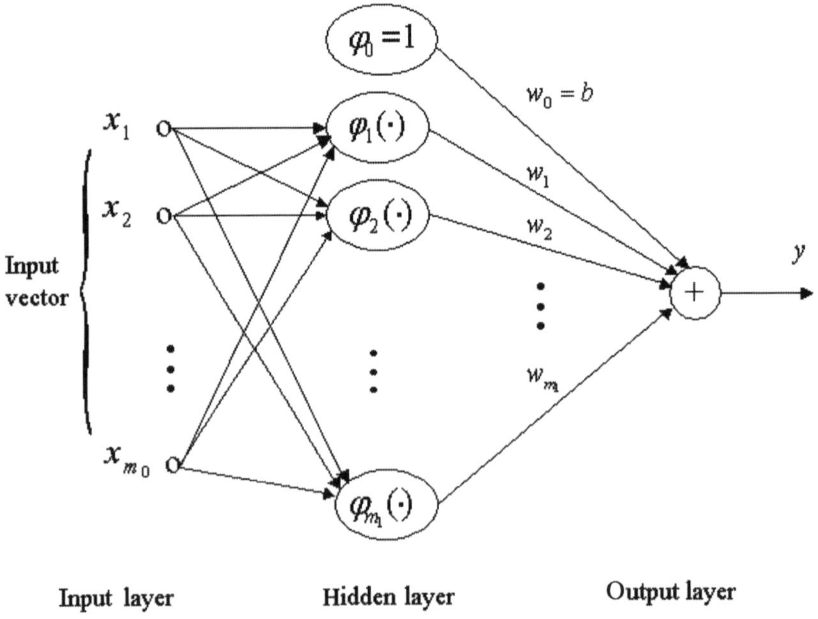


**Figure 9.7** Generalized RBF network.

network is designed to minimize the cost function

$$\varepsilon(F) = \frac{1}{2}\sum_{i=1}^{N} [d_i - F(\mathbf{x}_i)]^2 + \frac{1}{2}\lambda\|\mathbf{D}F\|^2 \tag{9.65}$$

where the first term on the right-hand side is the standard empirical error using $N$ training samples (or the cost function for a conventional RBF network) and the second term is the regularization term (as a constraint on the smoothness of $F$), $\mathbf{D}$ is a linear differential operator, $F(\mathbf{x}_i) = y_i$, and $d_i$ is the desired output. If the number of neurons is set to be equal to the number of training samples $N$, $F$ satisfies the interpolation condition $F(\mathbf{x}_i) = d_i$. It follows that the optimal solution to the minimization problem in (9.65) is

$$F(\mathbf{x}) = \sum_{i=1}^{N} w_i G(\mathbf{x},\mathbf{x}_i) \tag{9.66}$$

where $G(\cdot)$ is the Green's function. The solution in (9.66) requires one to use the same number of neurons as the size of the training set (i.e., $m_1 = N$). This is undesirable as it will lead to a high computational cost. Therefore, one often seeks *an approximated solution* that requires significantly fewer neurons $m_1$ in the hidden layer, $m_1 < N$. The interpolation $F^*(\mathbf{x}_i) = y_i$ may not exactly be equal to $d_i$ but is close to $d_i$. It is worth mentioning that in such cases the number of neurons $m_1$ and their center positions will affect the outcome of the approximated solution. For $m_1$ neurons in the hidden layer, $m_1 \leq N$, the corresponding cost function to (9.65) is

$$\varepsilon(F^*) = \frac{1}{2}\sum_{i=1}^{N}\left(d_i - \sum_{j=1}^{m_1} w_j G(\|\mathbf{x}_i - \mathbf{t}_j\|)\right)^2 + \lambda\|\mathbf{D}F^*\|^2 \tag{9.67}$$

where $\mathbf{t}_j$ is the center of the Green's function $G$. Minimizing the above function leads to the approximation of the regularized solution,

$$F^*(\mathbf{x}) = \sum_{i=1}^{m_1} w_i G(\|\mathbf{x} - \mathbf{t}_i\|_{\mathbf{C}}) \tag{9.68}$$

where $\mathbf{C}$ is a norm weighting matrix of size $m_0 \times m_0$. A specific case is the Gaussian RBF, with the following Green's function,

$$G(\|\mathbf{x} - \mathbf{t}_i\|_{\mathbf{C}}) = \exp\left[-\tfrac{1}{2}(\mathbf{x} - \mathbf{t}_i)^{\mathrm{T}}\boldsymbol{\Sigma}^{-1}(\mathbf{x} - \mathbf{t}_i)\right] \tag{9.69}$$

where $\frac{1}{2}\boldsymbol{\Sigma}^{-1} = \mathbf{C}^{\mathrm{T}}\mathbf{C}$.

It can be shown that minimizing (9.67) with respect to the weight vector $\mathbf{w}$ yields

$$(\mathbf{G}^{\mathrm{T}}\mathbf{G} + \lambda\mathbf{G}_0)\mathbf{w} = \mathbf{G}^{\mathrm{T}}\mathbf{d} \tag{9.70}$$

where

$$\begin{aligned}
\mathbf{d} &= [d_1, d_2, \ldots, d_N]^{\mathrm{T}} \\
\mathbf{w} &= [w_1, w_2, \ldots, w_{m_1}]^{\mathrm{T}} \\
\mathbf{G} &= \begin{bmatrix} G(\mathbf{x}_1, \mathbf{t}_1) & \cdots & G(\mathbf{x}_1, \mathbf{t}_{m_1}) \\ \vdots & \ddots & \vdots \\ G(\mathbf{x}_N, \mathbf{t}_1) & \cdots & G(\mathbf{x}_N, \mathbf{t}_{m_1}) \end{bmatrix} \\
\mathbf{G}_0 &= \begin{bmatrix} G(\mathbf{t}_1, \mathbf{t}_1) & \cdots & G(\mathbf{t}_1, \mathbf{t}_{m_1}) \\ \vdots & \ddots & \vdots \\ G(\mathbf{t}_{m_1}, \mathbf{t}_1) & \cdots & G(\mathbf{t}_{m_1}, \mathbf{t}_{m_1}) \end{bmatrix}
\end{aligned} \tag{9.71}$$

The weight vector $\mathbf{w}$ can be obtained by solving (9.70). If the regularization parameter is set to zero, $\lambda = 0$, then the solution corresponds to that of a conventional RBF network, which is associated with an overdetermined least-squares data-fitting problem whose weight vector converges to

$$\mathbf{w} = (\mathbf{G}^{\mathrm{T}}\mathbf{G})^{-1}\mathbf{G}^{\mathrm{T}}\mathbf{d} \tag{9.72}$$

where $(\mathbf{G}^{\mathrm{T}}\mathbf{G})^{-1}\mathbf{G}^{\mathrm{T}} = \mathbf{G}^{\dagger}$ is the pseudoinverse.

***9.5.2.2 Structure of Generalized RBF Network*** As shown in Figure 9.7, a generalized RBF network contains an input layer consisting of $m_0$ nodes that receive the input feature vector $\mathbf{x}$. The number of nodes $m_0$ is equal to the size of $\mathbf{x}$. For the training process, $\mathbf{x} \in \{\mathbf{x}_1, \mathbf{x}_2, \ldots, \mathbf{x}_N\}$. The hidden layer contains $m_1$ neurons (usually $m_1 \leq N$, $N$ is the number of training samples), and each neuron contains a nonlinear activation function, that is, a RBF,

$$\varphi_i(\mathbf{x}) = G(\|\mathbf{x} - \mathbf{t}_i\|_C) \qquad i = 1, \ldots, m_1 \tag{9.73}$$

with its center location $\mathbf{t}_i$. In (9.73) $G(\cdot)$ is often chosen to be the Gaussian RBF defined in (9.69). The output of the network and the neurons in the hidden layer are linked by *linear weights* $w_i$ as

$$y_j = \sum_{i=1}^{m_1} w_i \varphi_i(\mathbf{x}_j) + w_0 \tag{9.74}$$

where $w_0$ is the bias.

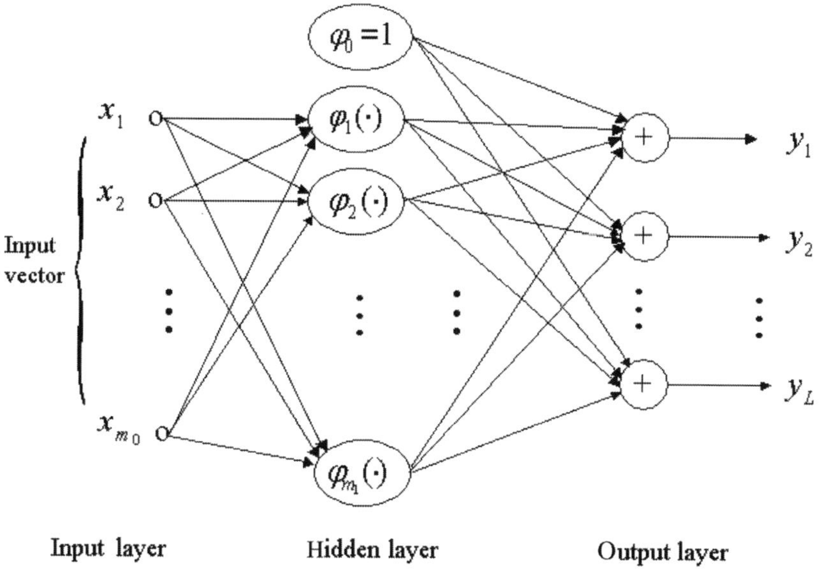


**Figure 9.8** Generalized RBF network that contains $L$ output nodes.

The output layer may contain multiple nodes; for example, one may use $L$ nodes for an $L$-class classification problem, as shown in Figure 9.8.

***9.5.2.3 RBF Network Learning*** The leaning process for a RBF network includes learning the linear weights associated with the output node(s) and the parameters in the RBFs associated with the hidden layer. Since these two sets of parameters are evolving in different rates, they are usually learned separately by using different techniques. We assume that the training set contains $N$ prelabeled samples, that is, $\{(\mathbf{x}_i, d_i), i = 1, \ldots, N\}$.

- *Learning Parameters of RBFs* For Gaussian RBFs, there are two sets of parameters to be estimated. They are the center locations $\mathbf{t}_i$ and the standard deviation $\sigma$.

  (a.1) A simple approach is to set some fixed RBF center values. The center locations $\mathbf{t}_i$ can be chosen randomly from the training set, that is, $\mathbf{t}_i \in \{\mathbf{x}_1, \mathbf{x}_2, \ldots, \mathbf{x}_N\}$, $i = 1, \ldots, m_1$. The standard deviation of Gaussian RBFs can be chosen as

$$\sigma = \frac{d}{\sqrt{2m_1}} \tag{9.75}$$

where $d$ is the maximum distance between the centers of RBFs,

$$d = \max\{\|\mathbf{t}_i - \mathbf{t}_j\|\} \qquad \forall \mathbf{t}_i, \mathbf{t}_j \in \{\mathbf{t}_1, \ldots, \mathbf{t}_{m_1}\} \tag{9.76}$$

This way of selection ensures that RBFs are not too peaked or too flat; hence the resulting RBFs are satisfactory provided that the training set is large. An improved way of setting the centers of RBFs is to use a self-organized center selection method, for example, using $k$-means clustering on training samples $\mathbf{x}_i$, $i = 1, \ldots, N$.

(b.1) Another approach is to adaptively update the RBF parameters by employing supervised learning. This can be performed by using gradient descent similar to the LMS algorithm. First, define a cost function at time $n$ as

$$\varepsilon(n) = \frac{1}{2}\sum_{i=1}^{N} e_i^2(n) \tag{9.77}$$

where

$$e_i(n) = d_i - \sum_{j=1}^{m_1} w_j G(\|\mathbf{x}_i - \mathbf{t}_j(n)\|_{\mathbf{C}_j}) \tag{9.78}$$

The positions of RBF centers are then updated as follows,

$$\mathbf{t}_i(n+1) = \mathbf{t}_i(n) - \alpha \frac{\partial \varepsilon(n)}{\partial \mathbf{t}_i(n)} \qquad i = 1, \ldots, m_1 \tag{9.79}$$

where $\partial\varepsilon(n)/\partial\mathbf{t}_i(n)$ is obtained by applying the chain rule of calculus,

$$\frac{\partial \varepsilon(n)}{\partial \mathbf{t}_i(n)} = 2w_i(n)\sum_{l=1}^{N} e_l(n) G'(\|\mathbf{x}_l - \mathbf{t}_i\|_{\mathbf{C}_i}) \boldsymbol{\Sigma}_i^{-1}[\mathbf{x}_l - \mathbf{t}_i(n)] \tag{9.80}$$

Similarly, the covariance matrices of Gaussian RBFs can be updated by

$$\boldsymbol{\Sigma}_i^{-1}(n+1) = \boldsymbol{\Sigma}_i^{-1}(n) - \beta \frac{\partial \varepsilon(n)}{\partial \boldsymbol{\Sigma}_i^{-1}(n)} \qquad i = 1, \ldots, m_1 \tag{9.81}$$

where

$$\frac{\partial \varepsilon(n)}{\partial \boldsymbol{\Sigma}_i^{-1}(n)} = -w_i(n)\sum_{l=1}^{N} e_l(n) G'(\|\mathbf{x}_l - \mathbf{t}_i\|_{\mathbf{C}_i})[\mathbf{x}_l - \mathbf{t}_i(n)][\mathbf{x}_l - \mathbf{t}_i(n)]^{\mathrm{T}} \tag{9.82}$$

and $\alpha$, $\beta$ in (9.79) and (9.81) are parameters controlling the learning rates.

- *Learning Linear Weights* (a.2) The simplest method to determine the weight vector $\mathbf{w}$ is to use the pseudoinverse described in (9.72), that is, $\mathbf{w} = (\mathbf{G}^{\mathrm{T}}\mathbf{G})^{-1}\mathbf{G}^{\mathrm{T}}\mathbf{d}$.

Using the method described in (a.1) for estimating RBF parameters, components in **G** can be written as

$$g_{ji} = \exp\left(-\frac{m_1 \|\mathbf{x}_j - \mathbf{t}_i\|^2}{d^2}\right) \qquad i = 1, \ldots, m_1 \;\; j = 1, \ldots, N \tag{9.83}$$

Then, **w** can be obtained by using singular-value decomposition (SVD).

(b.2) Another method is to employ supervised learning by adaptively updating the linear weights $w_i$, $i = 1, 2, \ldots, m_1$, using the gradient descent,

$$w_i(n+1) = w_i(n) - \gamma \frac{\partial \varepsilon(n)}{\partial w_i(n)} \qquad i = 1, \ldots, m_1 \tag{9.84}$$

where $\gamma$ is a parameter controlling the learning rate and

$$\frac{\partial \varepsilon(n)}{\partial w_i(n)} = \sum_{j=1}^{N} e_j(n) G(\|\mathbf{x}_j - \mathbf{t}_i(n)\|_{\mathbf{C}_i}) \tag{9.85}$$

***9.5.2.4 Discussion: Multidimensional Output Space*** In the above description, the interpolation function $F^*(\cdot)$ in (9.68) maps an $m_0$-dimensional input vector to a one-dimensional output, that is, $F^*\colon \mathbb{R}^{m_0} \to Y = \mathbb{R}^1$, or $\mathbf{x}_i \mapsto y_i = F^*(\mathbf{x}_i)$. It is straightforward to extend the above function $F^*(\cdot)$ that can map an $m_0$-dimensional input space onto a $L$-dimensional output space. This is done by replacing the function $F^*(\cdot)$ with $F_k^*(\cdot)$, $k = 1, \ldots, L$.

### 9.5.3 Applications to Classification of Power System Disturbances

Artificial neural networks for classification of power system disturbances and faults have been extensively studied; some of the previous work can be found in [269, 288, 82, 144, 116, 147]. We shall describe an example of ANNs for the classification of power system disturbances proposed by [269], where multiple NNs are applied in the wavelet domain.

The type of NN that was used in the system is a self-organizing map (also known as a *Kohonen map*) [136]. A self-organizing map is characterized by the formation of topologically ordered maps of the input patterns. It is based on competitive learning where there is only one winning neuron in the output at each time.

The system includes the following steps. First, the disturbance data are decomposed into subband components through a dyadic wavelet transform with five scale levels. This decomposes the signal into six subbands including a low-pass band. Then, the outputs from these six subband filters are used as the inputs of multiple NNs, each of which is a self-organizing map network. Finally, the outputs from the NNs are integrated using decision-making schemes, for example, a simple voting scheme or a scheme based on the Dempster–Shafer theory of evidence [275].

The NN has been designed to recognize six classes of power quality disturbances, including high-frequency capacitor-switching disturbance, low-frequency capacitor-switching disturbance, a perfect 60-Hz sinusoidal waveform (for North America), impulsive transient, voltage dip, and momentary interruption. The test set contained 675 recordings. For the simple voting scheme, the classifier was reported to have obtained a 91.8% recognition rate at the expense of rejecting 7.8% of the waveforms as ambiguous. Further, the classifier is able to reject about 60% of the outliers.

## 9.6 LEARNING AND CLASSIFICATION USING SUPPORT VECTOR MACHINES

### 9.6.1 Why Use a Support Vector Machine for Classification?

Real-world problems often require hypothesis spaces that are more complex than those using linear discriminants. Support vector machines are able to find nonlinear boundaries if the classes are linearly nonseparable.

A natural question will arise before we study SVMs: Why should one be interested in a SVM when there are many other classification methods? The main issues of interest in using SVMs for classification are the generalization performance and the complexity of the classifier, which is a practical implementation concern.

More specifically, when we design a classification system, it is natural to want the classifier to have good generalization performance, that is, good performance on the test set rather than on the training set. We want the classifier to converge to a global optimum rather than some local optimal positions. We also consider the problem of overfitting: Do we offer too many or too few training samples? If we use too many training samples, we might have a classifier that is overfitted to the training samples. However, if we have too few training samples, we might not be able to obtain sufficient statistical coverage to most possible situations. Both cases will lead to poor generalization performance, that is, poor performance to the test data set. Is there any guarantee (e.g., an upper bound) of the error on the test set?

Another issue of concern is computational complexity. As discussed in Section 9.4, a Bayesian classifier is an elegant method using the posterior probability distributions. However, the computation cost is very high when the dimension of features becomes large even for a Gaussian pdf. This often hinders the practical application of Bayesian classifiers.

An answer to these issues by using a SVM as a classifier is positive! From the description below, we will see that a SVM classifier minimizes the generalization error on the test set under the *structural risk minimization* (SRM) principle. The design of a SVM can also be viewed as a constrained optimization problem.

### 9.6.2 SVMs and Generalization Error

A special "feature" in designing a SVM is that, instead of the dimension reduction commonly employed in conventional pattern recognition systems (see Section 9.2), the input space in a SVM is nonlinearly mapped by $\mathbf{\Phi}(\cdot)$ onto a high-dimensional feature space. The result is that the classes are more likely to be linearly separable than in a low-dimensional feature space. Although one may cast doubts on whether it is a good idea to go to a high-dimensional feature space for machine learning due to the *curse of dimensionality* (which states that the number of patterns required to properly sample the space is exponentially increased when the number of dimensions increases), there is fortunately a good statistical learning theory that guarantees the error bound and complexity of a SVM when using a high-dimensional feature space. Another important attraction of SVMs is the use of kernels. Kernel representations offer an alternative solution by nonlinearly projecting the input space onto a high-dimensional feature space. A kernel function $k(\mathbf{x}_i, \mathbf{x})$ is an inner product $\langle \mathbf{\Phi}(\mathbf{x}_i), \mathbf{\Phi}(\mathbf{x})\rangle$. Instead of directly using vectors $\mathbf{\Phi}(\mathbf{x}_i)$ in the feature space to solve the primal problem, only inner products of feature vectors are required in a kernel function to solve the dual problem in SVM learning. A SVM classifier employs a kernel function $k(\mathbf{x}_i, \mathbf{x})$ relating the *support vectors* $\mathbf{x}_i$ (which is a small subset of training samples) and the input vectors $\mathbf{x}$ drawn from the test data. Hence one can often avoid handling a large training set. Further, using different kernel functions leads to different SVMs. Figure 9.9 depicts the spaces and the mappings in a SVM.

Let the input–output training data pairs be described as $(\mathbf{x}_i, d_i)$, $i = 1, 2, \ldots, N$, $\mathbf{x}_i \in \mathbb{R}^{m_0}$ (i.e., the input space) and $d_i \in Y$ (i.e., the decision space). First, we apply a nonlinear function $\mathbf{\Phi}(\cdot)$ that maps the input space $\mathbb{R}^{m_0}$ onto a high-dimensional feature space $\mathcal{F}$,

$$\mathbf{\Phi}: \mathbb{R}^{m_0} \longrightarrow \mathcal{F} \qquad \mathbf{x}_i \longmapsto \Phi(\mathbf{x}_i) \tag{9.86}$$

We will see later that choosing $\mathbf{\Phi}$ is associated with selecting a kernel function leading to different types of SVMs. Once a high-dimensional feature space $\mathcal{F}$ is chosen, another function $f(\cdot)$ is applied to map the feature space onto a decision space,

$$f: \mathcal{F} \longrightarrow Y \qquad \mathbf{\Phi}(\mathbf{x}_i) \longmapsto f(\mathbf{\Phi}(\mathbf{x}_i)) \tag{9.87}$$

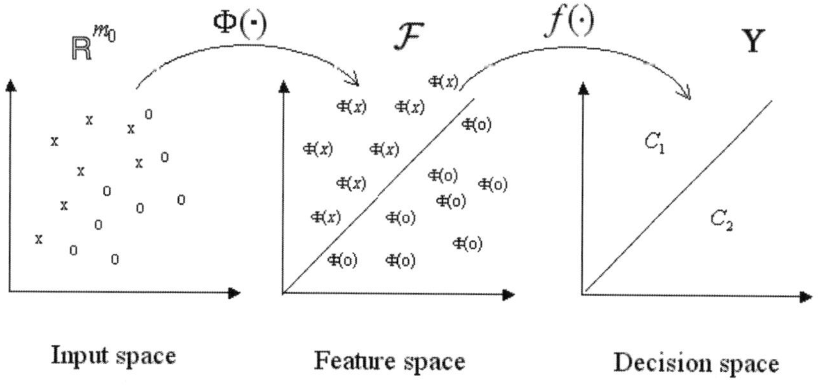


**Figure 9.9** Different spaces and mappings in SVM.

The best function $f(\cdot)$ that may correctly classify a unseen example $(\mathbf{x}, d)$ *on a test set* is the one minimizing the expected error or the generalization error,

$$R(f) = \int l[f(\mathbf{\Phi}(\mathbf{x})), d] \, dP(\mathbf{\Phi}(\mathbf{x}), d) \tag{9.88}$$

where $l(\cdot)$ is the loss function and $P(\mathbf{\Phi}(\mathbf{x}), d)$ is the probability of $(\mathbf{\Phi}(\mathbf{x}), d)$ which can be obtained if the probability of generating the input–output pair $(\mathbf{x}, d)$ is known. A loss (or error) occurs if $f(\mathbf{\Phi}(\mathbf{x})) \neq d$. Since $P(\mathbf{x}, d)$ is unknown, we cannot directly minimize (9.88). Instead we try to estimate the function $f(\cdot)$ that is close to the optimal one from the function class $\mathbb{F}$ using the training set. It is worth mentioning that *there exist many $f(\cdot)$ that give perfect classification on the training set, but they give different results on the test set.*

According to VC theory [308, 79, 276, 52], what we need is to choose a function $f(\cdot)$ that fits the necessary and sufficient conditions for consistency of empirical risk minimization,

$$\lim_{n \to \infty} P\left\{ \sup_f [R(f) - R_{\text{emp}}(f)] > \varepsilon \right\} = 0 \qquad \text{for all } \varepsilon > 0 \tag{9.89}$$

where the empirical risk (i.e., the training error) is defined on the training set,

$$R_{\text{emp}}(f) = \frac{1}{N} \sum_{i=1}^{N} l(f(\mathbf{\Phi}(\mathbf{x}_i)), d_i) \tag{9.90}$$

and $N$ is the total number of samples (or feature vectors) in the training set. A specific way to control the complexity of function class $\mathbb{F}$ is given by VC theory and the SRM principle [79, 224].

Under the SRM principle one chooses the function class $\mathbb{F}$ (and the function $f$) such that the upper bound of the generalization error in (9.88) is minimized. For all $\delta > 0$ and $f \in \mathbb{F}$, it follows that the bound of the generalization error

$$R(f) \leq R_{\text{emp}}(f) + \sqrt{\frac{h[\ln(2N/h) + 1] - \ln(\delta/4)}{N}} \tag{9.91}$$

holds with a probability of at least $1 - \delta$ for $N > h$, where $h$ is the VC dimension for the function class $\mathbb{F}$. *The VC dimension, roughly speaking, measures the maximum number of training samples that can be correctly separated (or shattered) by the class of functions* $\mathbb{F}$. For example, $N$ data samples can be labeled to $2^N$ possible ways in a binary class $Y = \{1, -1\}$ case, of which there exists at least one set of $h$ samples that can be correctly classified to their class labels by the chosen function class $\mathbb{F}$. As one can see from (9.91), minimization of $R(f)$ is obtained by yielding a small training error $R_{\text{emp}}(f)$ (the first term) while keeping the function class as small as possible (the second term). *Hence, SVM learning can be viewed as seeking the best function $f(\cdot)$ from the possible function set $\mathbb{F}$ according to the SRM principle, which results in the lowest upper bound in (9.91).*

### 9.6.3 Case 1: SVMs for Linearly Separable Patterns

Let prelabeled training sample pairs be $(\mathbf{x}_1, d_1), \ldots, (\mathbf{x}_N, d_N)$, where $\mathbf{x}_i$ are the input vectors and $d_i$ are the corresponding labels in a binary class $Y = \{-1, +1\}$ that are linearly separable. The conditions for perfect classification in the feature space $\mathcal{F}$ without training error [i.e., the first term in the right-hand side of (9.91) is zero] are, for $i = 1, 2, \ldots, N$,

$$\begin{aligned} \langle \mathbf{w}, \mathbf{\Phi}(\mathbf{x}_i)\rangle + b \geq 0 \qquad & \text{for } d_i = +1 \\ \langle \mathbf{w}, \mathbf{\Phi}(\mathbf{x}_i)\rangle + b < 0 \qquad & \text{for } d_i = -1 \end{aligned} \tag{9.92}$$

Rescaling (9.92) yields the following equivalence,

$$d_i(\langle \mathbf{w}, \mathbf{\Phi}(\mathbf{x}_i)\rangle + b) \geq 1 \qquad i = 1, 2, \ldots, N \tag{9.93}$$

It is worth noting that $\mathbf{x}_i$ can be either the data samples or the attributes (or features) of data samples, depending on each individual application and the choice of the designers. The goal of learning is to find the best $f(\cdot)$ for a given $\mathbf{\Phi}(\cdot)$ such that $R(f)$ in (9.91) is minimized. Since $R(f)$ cannot be directly computed, one should minimize the bound on the right-hand side of (9.91). It is known that for a linear classifier in the feature space $\mathcal{F}$ the VC dimension $h$ is bounded according to

$$h \leq \min\left\{\left\lceil \tfrac{1}{4}\|\mathbf{w}\|^2 r^2 \right\rceil, m_0\right\} + 1 \tag{9.94}$$

where $r$ is the radius of the smallest sphere enclosing the training data that is fixed for a giving training set, $m_0$ is the dimension of input space, and $\lceil\cdot\rceil$ is the ceiling sign (e.g., the smallest integer greater than or equal to the number enclosed within $\lceil 4.2 \rceil = 5$). Hence, we can minimize the complexity term [i.e., the second term of (9.91)] by minimizing $\|\mathbf{w}\|^2$. This can be formulated as a quadratic optimization problem

$$\min_{\mathbf{w}, b} \tfrac{1}{2}\|\mathbf{w}\|^2 \tag{9.95}$$

It is worth mentioning that minimizing $\|\mathbf{w}\|^2$ also corresponds to selecting the maximum margin of a classifier, where a margin is defined as the shortest distance between the separating boundary and an input vector that can be correctly classified.

Combining (9.93) and (9.95), we can formulate the primal form of the optimization problem as a Lagrangian optimization,

$$L(\mathbf{w}, b, \boldsymbol{\alpha}) = \tfrac{1}{2}\|\mathbf{w}\|^2 - \textstyle\sum_{i=1}^{N} \alpha_i \left[d_i \left(\langle \mathbf{w}, \boldsymbol{\Phi}(\mathbf{x}_i)\rangle + b\right) - 1\right] \tag{9.96}$$

where $\alpha_i \geq 0$, $i = 1, 2, \ldots, N$, are the Lagrangian multipliers. To find optimal $\alpha_i$ values, the so-called Karush–Kuhn–Tucker (KKT) conditions for (constrained) optimization problems must be satisfied [79]. The KKT conditions are associated with the necessary and in some cases sufficient conditions for a set of variables to be optimal. Only those $\alpha_i$ in (9.96) having nonzero values satisfy the KKT conditions,

$$\alpha_i[d_i(\langle \mathbf{w}, \boldsymbol{\Phi}(\mathbf{x}_i)\rangle + b) - 1] = 0 \qquad i = 1, 2, \ldots, N \tag{9.97}$$

The corresponding vectors $\mathbf{x}_i$ satisfying the above KKT conditions are called *support vectors*, that is, all support vectors $\mathbf{x}_i$ lie on the margins, and they satisfy,

$$d_i(\langle \mathbf{w}, \boldsymbol{\Phi}(\mathbf{x}_i)\rangle + b) = 1 \tag{9.98}$$

Once the $\alpha_i$ are found, the optimal solution and the bias $b$ can be obtained as

$$\mathbf{w} = \sum_{i=1}^{N_s} \alpha_i d_i \boldsymbol{\Phi}(\mathbf{x}_i) \qquad b = 1 - \langle \mathbf{w}, \boldsymbol{\Phi}(\mathbf{x}_i^{(s)})\rangle \tag{9.99}$$

where $N_s$ is the number of support vectors, $\mathbf{x}_i$ are support vectors, $\mathbf{x}_i \in \mathrm{SV}$, and $\mathbf{x}_i^{(s)}$ is a support vector for which $d_i^{(s)} = +1$. *This implies that training samples that are not support vectors will not contribute to the decision boundaries of the classifier and hence can be removed from the training set!*

The optimization problem formulated in the above primal form in (9.96) can equivalently be expressed from its dual form by introducing the kernel function,

$$\begin{aligned}
\max_{\alpha}: \quad & \sum_{i=1}^{N} \alpha_i - \frac{1}{2}\sum_{i,j=1}^{N} \alpha_i \alpha_j d_i d_j k(\mathbf{x}_i, \mathbf{x}_j) \\
\text{Subject to:} \quad & \sum_{i=1}^{N} \alpha_i d_i = 0 \\
& \alpha_i \geq 0 \qquad i = 1, 2, \ldots, N
\end{aligned} \tag{9.100}$$

where $k(\mathbf{x}_i, \mathbf{x}_j) = \langle \boldsymbol{\Phi}(\mathbf{x}_i), \boldsymbol{\Phi}(\mathbf{x}_j)\rangle$ is the kernel function. It is worth noting that $\mathbf{w}$ has disappeared in the above dual form. The decision function for the classifier is shown to be

$$f(\mathbf{x}) = \mathrm{sgn}\left(\sum_{i=1}^{N_s} \alpha_i d_i k(\mathbf{x}, \mathbf{x}_i) + b\right) \tag{9.101}$$

where $\mathbf{x}$ is from the test set, $N_s$ is the number of support vectors $\mathbf{x}_i$, and $\mathbf{x}_i \in \mathrm{SV}$ is drawn from the training set.

### 9.6.4 Case 2: Soft-Margin SVMs for Linearly Nonseparable Patterns

For a linearly nonseparable case, a classifier with zero training error may lead to overfitting. Further, achieving a high training accuracy may not be the best way to minimize the error on the test set [i.e., minimize the right-hand side of (9.91)]. A better strategy is therefore to allow some training error in order to achieve less error on the test set. A soft-margin SVM is designed for this purpose.

As is known, a *margin* is defined as the shortest distance between the separating boundary and an input vector that can be correctly classified (e.g., classified as being $d_1 = +1$ and $d_2 = -1$). For linearly separable cases, the training error is equal to zero, and all support vectors from the training set are located exactly on the margins. In the soft-margin SVMs one is allowed to have some training error, where support vectors lie both on the margins and inside the margins: Those support vectors that lie between the separating boundaries and the margins are the misclassified samples from the training set. In a soft-margin SVM, a soft-margin constraint is added. This can be described as replacing (9.93) and (9.95) in the linearly separable case by the equations

$$d_i(\langle \mathbf{w}, \mathbf{\Phi}(\mathbf{x}_i)\rangle + b) \geq 1 - \xi_i \qquad \xi_i \geq 0 \qquad i = 1, 2, \ldots, N \tag{9.102}$$

$$\min_{\mathbf{w},b,\xi} \tfrac{1}{2}\|\mathbf{w}\|^2 + C\sum_{i=1}^{N} \xi_i \tag{9.103}$$

where $C \geq 0$ is a user-specified *regularization parameter* that determines the trade-off between the upper bound on the complexity term and the empirical error (or the training error) and $\xi_i$ are the slack variables. For example, $C$ can be determined experimentally through the use of a cross-validation process. The primal form of the Lagrangian optimization problem can be formed as

$$L(\mathbf{w}, b, \boldsymbol{\alpha}, \boldsymbol{\xi}, \boldsymbol{\mu}) = \tfrac{1}{2}\|\mathbf{w}\|^2 + C\sum_{i=1}^{N} \xi_i - \sum_{i=1}^{N} \alpha_i[d_i(\langle \mathbf{w}, \mathbf{\Phi}(\mathbf{x}_i)\rangle + b) - 1 + \xi_i] - \sum_{i=1}^{N} \mu_i \xi_i \tag{9.104}$$

where $\mu_i$ is the Lagrangian multiplier to enforce the positivity of $\xi_i$. The KKT conditions for the primal optimization problem are, for $i = 1, 2, \ldots, N$,

$$\begin{aligned} \alpha_i[d_i(\langle \mathbf{w}, \mathbf{\Phi}(\mathbf{x}_i)\rangle + b) - 1 + \xi_i] &= 0 \\ \mu_i \xi_i &= 0 \end{aligned} \tag{9.105}$$

The Lagrangian multiplies $\alpha_i$ are nonzero only when the KKT conditions are met. The corresponding vectors $\mathbf{x}_i$ (for nonzero $\alpha_i$) satisfying the following equation are support vectors:

$$d_i(\langle \mathbf{w}, \boldsymbol{\Phi}(\mathbf{x}_i)\rangle + b) = 1 - \xi_i$$

This is equivalent to

$$d_i(\langle \mathbf{w}, \boldsymbol{\Phi}(\mathbf{x}_i)\rangle + b) \leq 1 \tag{9.106}$$

It is worth mentioning that in a soft margin SVM support vectors may lie on the margins as well as within the margins, as compared to the SVMs for linearly separable cases where support vectors may only lie on the margins. Notice that those support vectors that lie between the margin and the decision boundary are those training samples that are erroneously classified [since a SVM minimizes the generalization error (9.88), it allows some nonzero empirical error].

Similar to (9.99), once the $\alpha_i$ are determined, the optimal solution $\mathbf{w}$ can be obtained as

$$\mathbf{w} = \sum_{i=1}^{N_s} \alpha_i d_i \boldsymbol{\Phi}(\mathbf{x}_i) \tag{9.107}$$

where $\mathbf{x}_i \in \mathrm{SV}$ and the bias $b$ can be determined by the KKT conditions in (9.105).

The equivalent dual problem corresponding to the primal optimization problem in (9.104) can be described as follows:

$$\begin{aligned} \max_{\boldsymbol{\alpha}}: \quad & \sum_{i=1}^{N} \alpha_i - \frac{1}{2}\sum_{i,j=1}^{N} \alpha_i\alpha_j d_i d_j k(\mathbf{x}_i, \mathbf{x}_j) \\ \text{Subject to:} \quad & \sum_{i=1}^{N} \alpha_i d_i = 0 \\ & C \geq \alpha_i \geq 0 \qquad i = 1, 2, \ldots, N \end{aligned} \tag{9.108}$$

Note that the slack variables $\xi_i$ and the weight vector $\mathbf{w}$ do not appear in dual form.

Finally, the decision function for linearly nonseparable cases is shown to be the same as that for linearly separable SVM classifiers in (9.101),

$$f(\mathbf{x}) = \mathrm{sgn}\left(\sum_{i=1}^{N_s} \alpha_i d_i k(\mathbf{x}, \mathbf{x}_i) + b\right) \tag{9.109}$$

where $N_s$ is the number of support vectors $\mathbf{x}_i$, $\mathbf{x}_i \in \mathrm{SV}$, and $k(\mathbf{x}, \mathbf{x}_i) = \langle \boldsymbol{\Phi}(\mathbf{x}), \boldsymbol{\Phi}(\mathbf{x}_i)\rangle$ is the selected kernel function.

***9.6.4.1 Discussion: When Should One Choose a Linear or a Nonlinear Classifier?*** From the above two case studies, a natural question arrives: For a given classification problem, can we decide beforehand if the patterns are linearly separable? There are no common rules of knowing whether one should choose a linear or a nonlinear classifier.

One way is to make an initial guess from the physical problem. From the guess one may decide to proceed with the design of a linear classifier. However, if it appears that the linear classifier has large training errors even after removing the outliers, then the patterns are linearly nonseparable and a nonlinear classifier should be used instead.

Another way is to use a nonlinear classifier design regardless of whether or not the patterns are linearly separable. If the patterns appear to be linearly (or close to linearly) separable, then the learned classifier would have linear (or close to linear) class boundaries.

***9.6.4.2 ν-SVMs: Modified Version of Soft-Margin SVMs*** Several modifications have been proposed to the basic SVMs described above. One particular useful modification is the so-called $\nu$-SVM where the regularization constant $C$ is replaced by another constant $\nu \in (0,1]$ and is found to be associated with a clearer interpretation. Instead of solving (9.108), one solves

$$\begin{aligned} \max_{\boldsymbol{\alpha}}: \quad & -\frac{1}{2}\sum_{i,j=1}^{N} \alpha_i\alpha_j d_i d_j k(\mathbf{x}_i, \mathbf{x}_j) \\ \text{Subject to:} \quad & \sum_{i=1}^{N} \alpha_i d_i = 0 \\ & \frac{1}{N} \geq \alpha_i \geq 0 \qquad \sum_i \alpha_i \geq \nu \qquad i = 1, 2, \ldots, N \end{aligned} \tag{9.110}$$

It is shown that a $\nu$-SVM has the same decision function as (9.109) for a C-SVM.

***9.6.4.3 Solving Quadratic Programming (QP) Problem*** To solve the SVM problem, which is a constrained optimization problem [see (9.96) and (9.104)], one has to solve the (convex) quadratic programming (QP) problem. There exists a huge body of literature on this subject and some free or commercial software packages that can be used to solve this problem. Readers are referred to [94].

### 9.6.5 Selecting Kernels for SVMs and Mercer's Condition

One of the important issue in SVMs is the use of kernel functions. A kernel function $k(\mathbf{x}_i, \mathbf{x}_j)$ corresponds to a dot product of mappings in some feature space $\mathcal{F}$. A mapping function $\boldsymbol{\Phi}$ is typically chosen as a (nonlinear) mapping between a lower dimensional input space $\mathbb{R}^{m_0}$ and a higher dimensional feature space $\mathcal{F}$. The reason for such a mapping is that it is more likely to obtain a linearly separable region in a higher dimensional space than in a lower dimensional space. It is worth noting that *the choice of the kernel has a crucial effect on the SVM performance.*

Since SVMs only depend on the dot products between various patterns, it suffices to use $k(\mathbf{x}_i, \mathbf{x}_j) = \langle \mathbf{\Phi}(\mathbf{x}_i), \mathbf{\Phi}(\mathbf{x}_j) \rangle$ instead of $\mathbf{\Phi}(\mathbf{x})$ explicitly. Depending on how a kernel is generated, one may construct different learning machines.

*Kernels used in SVMs must satisfy Mercer's condition.* Roughly speaking, Mercer's condition states for which kernel $k(\mathbf{x}_i, \mathbf{x}_j)$ there exists (or not exists) a pair $\{\mathcal{F}, \mathbf{\Phi}\}$ or whether a kernel is a dot product in some feature space $\mathcal{F}$. However, it does not tell how one can construct $\mathbf{\Phi}$ or in which exact space $\mathcal{F}$. Details on how to select/construct a kernel for a SVM is beyond the scope of this book, and readers are referred to [276, 139] for further details. In most engineering applications a common way is to choose a kernel from the existing list of SVM kernels. Table 9.2 lists some of these kernels. Once a kernel function is selected, one may still need to determine the best parameter values associated with the selected kernel, for example, for a Gaussian RBF kernel there is a parameter associated with the spread $\gamma$. A common way to select the parameter values is to use *cross-validation* on the training set. Also note that for some types of kernel the choice of parameter values may be limited for a SVM; for example, for a sigmoidal kernel, only certain $k$ and $\theta$ values can satisfy Mercer's condition.

**TABLE 9.2 Some Frequently Used Kernels in SVMs**

| | |
|---|---|
| Polynomial | $k(\mathbf{x}, \mathbf{y}) = (\langle \mathbf{x}, \mathbf{y} \rangle + \theta)^d$ |
| Gaussian RBF | $k(\mathbf{x}, \mathbf{y}) = \exp\left(\frac{-\lVert \mathbf{x} - \mathbf{y} \rVert^2}{\gamma}\right)$ |
| Exponential RBF | $k(\mathbf{x}, \mathbf{y}) = \exp\left(\frac{-\lVert \mathbf{x} - \mathbf{y} \rVert}{\gamma}\right)$ |
| Sigmoidal (or MLP) | $k(\mathbf{x}, \mathbf{y}) = \tanh(\kappa \langle \mathbf{x}, \mathbf{y} \rangle + \theta)$ |
| Inverse multiquadratic | $k(\mathbf{x},\mathbf{y}) = \frac{1}{\sqrt{\lVert \mathbf{x} - \mathbf{y} \rVert^2 + \gamma^2}}$ |
| B-spline | $k(\mathbf{x}, \mathbf{y}) = B_{2N+1}(\lVert \mathbf{x} - \mathbf{y} \rVert)$ |
| Additive kernel | $k(\mathbf{x}, \mathbf{y}) = \sum_i k_i(\mathbf{x}, \mathbf{y})$ |
| Tensor product of kernels | $k(\mathbf{x}, \mathbf{y}) = \prod_i k_i(x_i, y_i)$ |

### 9.6.6 Implementation Issues and Practical Examples of SVMs

Despite many model selection techniques that provide principal ways of selecting a proper kernel [276, 139], we shall only discuss one model selection method that is based on cross-validation. Further, we shall limit ourselves to using RBF kernels as a practical example in this section.

Figure 9.10 shows the block diagram of a SVM classifier where $\mathbf{x}$ is the input vector drawn from the test set and $\mathbf{x}_i$, $i = 1, \ldots, N_s$, are the support vectors drawn from the training set.

It is worth mentioning that although the block diagram structure of this SVM looks similar to a two-class RBF neural network, they are fundamentally different. First, the weights $\alpha_i d_i$ to the output of a SVM contain Lagrangian multipliers $\alpha_i$ which are associated with the solutions of the constrained optimization problem in the SVM learning such that the generalization error (or the classification error on the test set) is minimized. In RBF neural networks, the weights are selected for minimizing the empirical error (i.e., the error between the desired outputs $d$ and the network outputs from the training set). Second, for a general multilayer NN activation functions are used in the hidden layer to achieve the nonlinearity, while a SVM employs a kernel function $k(\cdot)$ which is a function dependent only on the difference of $\mathbf{x}$ from the test set and the support vectors from the training set.

***9.6.6.1 Normalization of Input Features*** To avoid having some feature components with large dynamic range dominating other small components,

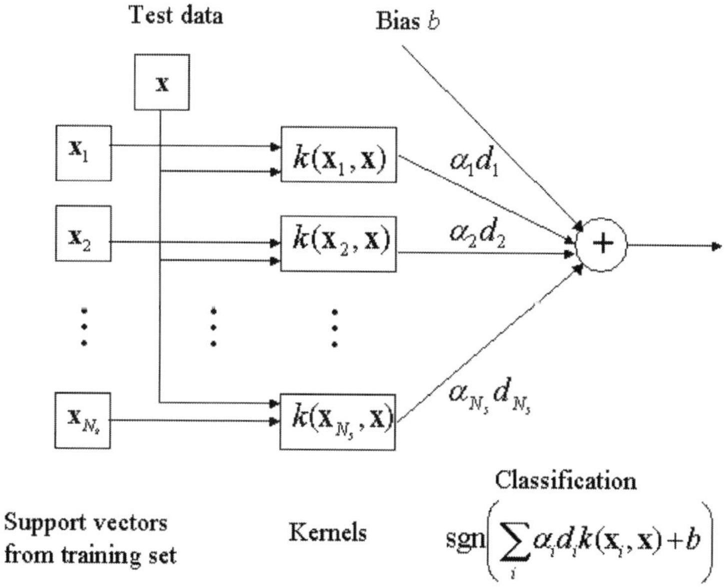


**Figure 9.10** Block diagram of SVM classifier for two-class case.

normalization should be applied (see Section 9.2). Note that the same normalization approach should be applied on the test set as well.

**9.6.6.2 *Selection of SVM Kernels and Parameters*** We assume that a kernel function is selected, for example, from the list in Table 9.2. Among these kernel functions, a RBF kernel $k(\mathbf{x}, \mathbf{y}) = \exp(-\|\mathbf{x} - \mathbf{y}\|^2/\gamma)$ is one of the most frequently used ones to perform the nonlinear mapping from a low-dimensional input space to a high-dimensional feature space.

Using a RBF kernel for a C-SVM (i.e., a soft-margin SVM with $C$ as the regularization parameter), there are two parameters to be selected: the regularization parameter $C$ in (9.104) and the kernel parameter $\gamma$. It is known that a RBF kernel behaves like a linear kernel under certain parameter settings $(C, \gamma)$ and like a sigmoid kernel under some other parameter settings. A RBF is also known to have less numerical difficulties since the kernel matrix $\mathbf{K} = [k_{ij}]$, $i, j = 1, \ldots, N$, is symmetric positive definite.

Cross-validation can be used here to find the best parameters $(C,\gamma)$, for example, by using a simple grid search. First, a coarse grid search can be applied (e.g., $C = \{2^{-5}, 2^{-3}, \ldots, 2^{15}\}$, $\gamma = \{2^{15}, 2^{13}, \ldots, 2^{-3}\}$). Then the search is tuned to a finer grid in the region where the predicted error rate from the cross-validation is the lowest in the coarse search.

**9.6.6.3 *Cross-Validation*** As one usually does not know in advance which parameters are the best for a given application, these parameters must be determined during the machine learning.

Since the goal is to best predict unknown test data, achieving high training accuracy may not be most useful. A commonly used procedure is to apply *cross-validation*. A $k$-fold cross-validation can be described as follows: For a given training set, one first divides the data set into $k$ subsets of equal size. The classifier is then trained $k$ times: In the $l$th iteration, $l = 1, 2, \ldots, k$, the classifier is trained using all subsets except the $l$th subset. The trained classifier is then tested using the $l$th subset, and the classification error for this subset is then computed. In such a way, each training subset is tested (or predicted) once. The cross-validation accuracy is the percentage of data which are correctly classified.

**9.6.6.4 *SVM Learning and Classification*** After finding the suitable parameters $(C, \gamma)$ for a RBF kernel, the whole training set is used for the training. The trained SVM is then ready to be used as a classifier for samples $\mathbf{x}$ drawn from the test set. The whole process is depicted in the block diagram of Figure 9.11.

**9.6.6.5 *Discussion: SVM Learning Versus Bayesian Learning*** Support vector machine learning uses the statistical learning (VC) theory that takes into account the complexity of the functions a learning machine can implement. A Bayesian learning machine uses probability distributions, however, without taking into account the implementation complexity. When the distributions are

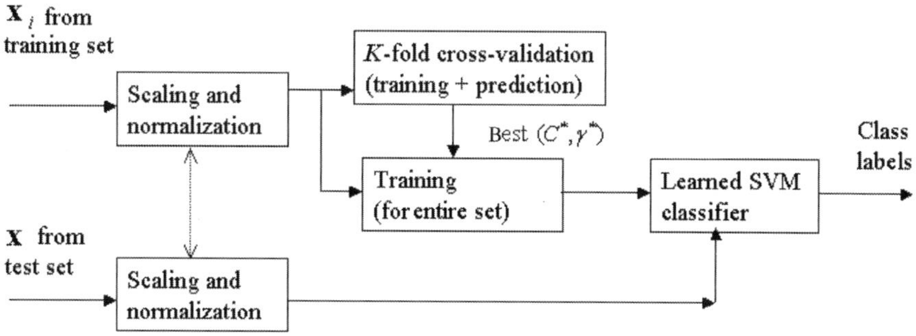


**Figure 9.11** Block diagram for learning a RBF kernel SVM. The SVM is subsequently used as a classifier for input **x** from a test set.

non-Gaussian, the computation of a Bayesian learning machine becomes rather intensive and difficult. A support vector learning machine directly uses feature vectors **x** as the inputs while a Bayesian learning machines uses probabilities of the feature vectors as the inputs. A SVM is based on statistical learning (VC) theory that minimizes the classification errors on the test set with an upper bound and is equivalent to solving a constrained optimization problem, while a Bayesian learning machine is elegant and entirely based on using the probability distributions. A Bayesian classifier selects a class that maximizes the posterior probability of input feature vector **x**, which is the likelihood weighted by the prior class probability $P(C_i)$. A SVM guarantees the generalization performance by using the SRM principle, while a Bayesian approach requires cross-validation to choose a best subset of training data for achieving good generalization. As a result, a SVM is often more attractive than a Bayesian classifier from the practical implementation point of view.

### 9.6.7 Example of Detecting Voltage Dips Due to Faults

To show how a SVM can be used for classification of power system disturbances, we shall describe a very simple example of classifying fault-induced voltage dips.

A voltage dip is defined when the power system voltage drops to a certain percentage (e.g., 0.9 pu) of its nominal value (230 V in most of Europe). Furthermore, for dips caused by faults, rms voltage sequences (rms voltage versus time) are close to a rectangular shape. However, if a rms voltage sequence contains a dip with sharp drop followed by slow recovery, the dip is caused by transformer saturation or induction motor starting depending on whether they have a similar voltage drop in all three phases [282]. In the ideal case, classification of fault-induced voltage dips can be described as an AND problem. Let each feature vector $\mathbf{x} = [x_1, x_2]^T$ consist of two components, where:

$x_1 \in \{0, 1\}$ indicates whether there is a fast drop in rms voltage to below 0.9 pu (i.e., rectangular-shaped drop). If there is a sharp voltage drop to below 0.9 pu, then $x_1$ is set to 1.

**TABLE 9.3 AND Problem for Classifying Voltage Dips Due to Faults**

| Input Vector **x** | Desired Response $d$ |
|---|---|
| (1,1) | +1 |
| (1,0) | −1 |
| (0,0) | −1 |
| (0,1) | −1 |

$x_2 \in \{0, 1\}$ indicates whether there is a quick recovery in rms voltage (i.e., rectangular-shaped recovery in the rms sequence). When the recovery is fast, then $x_2$ is set to 1.

Also, define the desired output for dips due to faults as $d = 1$ and otherwise as $d = -1$. Table 9.3 describes such an AND operation.

However, a voltage drop or recovery will never be absolutely instantaneous but instead always take a finite time. The closeness to rectangularity is a fuzzy quantity that can be defined as a continuous function ranging between 0 and 1.0 (e.g., by defining a fuzzy function, see [189]), where 0 implies a complete flat shape while 1.0 an ideal rectangular shape ($\theta = 90°$). Consequently, each component of the feature vector **x** takes a real value within [0, 1].

In this simple example, the training set contains the following 12 vectors,

$$\mathbf{x}_1 = (1.0, 1.0) \quad \mathbf{x}_2 = (1.0, 0.0) \quad \mathbf{x}_3 = (0.0, 0.0) \quad \mathbf{x}_4 = (0.0, 1.0)$$
$$\mathbf{x}_5 = (0.8, 0.7) \quad \mathbf{x}_6 = (0.9, 0.4) \quad \mathbf{x}_7 = (0.2, 0.8) \quad \mathbf{x}_8 = (0.4, 0.8)$$
$$\mathbf{x}_9 = (0.8, 0.9) \quad \mathbf{x}_{10} = (0.9, 0.5) \quad \mathbf{x}_{11} = (0.3, 0.2) \quad \mathbf{x}_{12} = (0.3, 0)$$

and their corresponding desired outputs are

$$d_1 = d_5 = d_9 = +1 \qquad d_2 = d_3 = d_4 = d_6 = d_7 = d_8 = d_{10} = d_{11} = d_{12} = -1$$

A Gaussian RBF kernel (see Table 9.2) is employed in the SVM with the above training set. Figure 9.12 shows the decision boundaries obtained from training a SVM using Gaussian RBF kernels with two different $\gamma$ values (related to the spread of kernels). One can see that the bend in the decision boundary (i.e., the solid line) changes depending on the choice of the parameter related to Gaussian spread. In Figure 9.12*a* the RBF kernels have a relatively large spread and the decision boundary becomes very smooth, which is close to a linear boundary, while in Figure 9.12*b* the RBFs have a smaller spread and the decision boundary has a relatively large bending (nonlinear boundary). If the features of event classes are linearly nonseparable, a smoother boundary would introduce more training errors. Interestingly enough, the decision boundary obtained from the training set (containing feature vectors with soft values of rectangularity in voltage drop and recovery) divides the feature space into two regions where the upright region

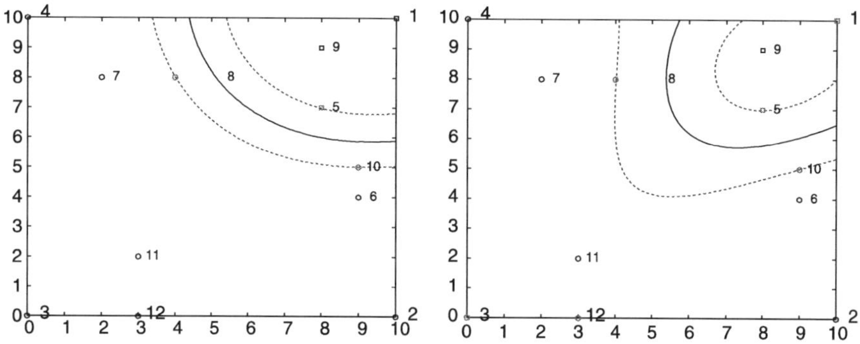


**Figure 9.12** Decision boundaries and support vectors obtained by training SVM with Gaussian RBF kernel functions. Solid line: decision boundary; dashed lines: margins. (*a*) The parameter for RBF kernels is $\gamma = 36m_0$. (*b*) The parameter for RBF kernels is $\gamma = 8m_0$. For both cases, the regularization parameter $C = 10$ and $m_0 = 2$ is the size of the input vector. Solid line is the decision boundary. Support vectors are the training vectors that are located on the margins (the dotted lines) and between the boundary and the margin (misclassified samples). For the given training set, there is no misclassified samples. The two sets of feature vectors are marked with squares and circles, respectively. Notice that only support vectors are contributed to the location of the decision boundary and margins. In the figure, all feature values are scaled by a factor of 10.

(notice that the shape of the region makes good sense!) is associated with the fault-induced dip events. Notice that all training vectors are enlarged by 10 times in the figure.

***9.6.7.1 Discussion*** If the training set contains more samples, one may expect an improved classifier. Using more training data will lead to changes in the decision boundary, the margins, and the distance between the margins and the boundary. We might also find some support vectors that are lying between the boundary and the margin, indicating training errors. However, for achieving a good performance on the test set, some training errors are allowed.

***9.6.7.2 Some Useful Software for SVMs*** There exist many software packages of SVMs that can be download free from the Internet. Some examples are as follows:

- LIBSVM from Chih-Wei Hsu and Chih-Jen Lin, http://www.csie.ntu.edu.tw/cjlin/libsvm/
- MATLAB SVM Toolbox from the Technical University of Graz, http://www.igi.tugraz.at/aschwaig/software.html
- MATLAB SVM Toolbox by Steve Gunn, http://www.isis.ecs.soton.ac.uk/resources/svminfo/

- SVM and Kernel Methods MATLAB Toolbox from Insa de Rouen, http://asi.insa-rouen.fr/%7Earako tom/toolbox/index

## 9.7 RULE-BASED EXPERT SYSTEMS FOR CLASSIFICATION OF POWER SYSTEM EVENTS

Expert systems are meant to solve real problems which normally would require specialized human experts in the area [176, 78]. Building an expert system for the classification or diagnostics of power system disturbances requires understanding the "rules of thumb" from human experts and involves automatically extracting the relevant information by computers. Such knowledge is often heuristic in nature rather than absolute certainty. Furthermore, "converting" such knowledge in a way that can be used by a computer is generally a difficult task, requiring its own expertise.

### 9.7.1 Structure and Rules of Expert Systems

***9.7.1.1 Structure of Typical Expert System*** As shown in the block diagram of Figure 9.13, a rule-based expert system usually consists of the following blocks:

*User Interface* It is the interface where the data are fed as the input into a system (e.g., from the output of a power system monitor); the classification or diagnostic results are the output through the interface (e.g., a computer terminal).

*Inference Engine* An inference engine performs the reasoning between the expert system knowledge (or rules) and the data from a particular problem.

*Explanation System* An explanation system allows the system to explain the reasoning to a user.

*Knowledge Base Editor* The system may sometimes include this block so as to allow a human expert to update or check the rules.

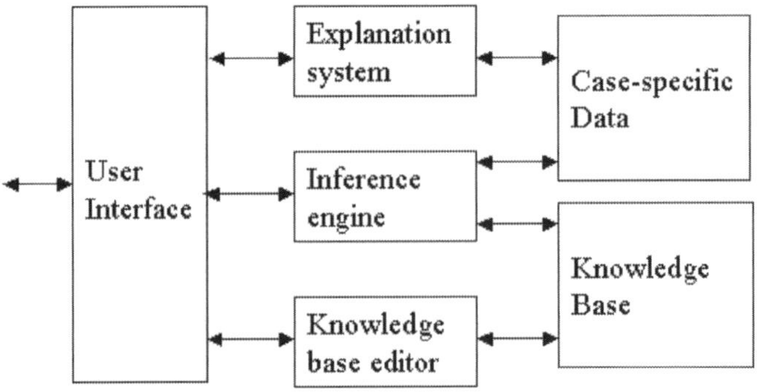


**Figure 9.13** Block diagram of an expert system.

*Knowledge Base* It contains all the rules, usually a set of IF–THEN rules.

*Case-Specific Data* This block includes data provided by the user and can also include partial conclusions or additional information from the measurements.

Note that the methodology used in the classification of power system disturbances in an expert system is very different from those described in previous sections of this chapter. Here, we consider and handle the measurement data (or their features) as deterministic signals.

**9.7.1.2 Examples of Rules** A disturbance in a power system can be associated with a certain class of events, for example, overvoltage, interruption, transformer saturation, capacitor switching, step change, single-stage dip due to faults, multistage dip due to faults, and many more. To recognize which event a disturbance recording is associated with, good power system knowledge from human experts is required. The knowledge associated with these events can usually be analyzed using the fundamental-voltage magnitudes where a set of rules can be deduced [284, 282].

When there is a sudden voltage drop, depending on its level and duration, it can be classified as an interruption, step change, or voltage dip. Possible "rules" to distinguish these three classes of events are as follows:

*Rule 1. Interruption* IF at least two consecutive rms voltages are less than 0.01 pu, THEN the event is an interruption.

*Rule 2. Voltage Swell* IF at least two consecutive rms voltages are more than 1.10 pu AND the rms voltage drops below 1.10 pu within 1 min, THEN the event is a voltage swell.

*Rule 3. Sustained Overvoltage* IF the rms voltage remains above 1.06 pu for 1 min or longer AND the event is not a voltage swell, THEN the event is a sustained overvoltage.

*Rule 4. Voltage Dip* IF at least two consecutive rms voltages are less than 0.9 pu AND the rms voltage rises above 0.90 pu within 1 min AND the event is not an interruption, THEN the event is a voltage dip.

*Rule 5. Sustained Undervoltage* IF the rms voltage remains below 0.94 pu for 1 min or longer AND the event is not a voltage dip AND the event is not an interruption, THEN the event is a sustained undervoltage.

*Rule 6. Voltage Step* IF the rms voltage remains between 0.90 and 1.10 pu AND the difference between two consecutive rms voltages remains within 0.0025 pu of its average value for at least 20 s before and after the step AND the event is not a sustained overvoltage AND the event is not a sustained undervoltage, THEN the event is a voltage step.

Note that these rules do allow for an event being both a voltage swell and a voltage dip. There are also events that possibly do not fall in any of the event classes. The inference engine should be such that both combined events and nonclassifiable events are recognized. Alternatively, two additional event classes may be defined as "combined dip–swell" and "other events."

To further classify the underlying causes of a voltage dip, we can apply the following rules:

*Rule 4.1. Voltage Dip Due to Fault* IF the rms sequence has a fast recovery after a dip (rectangular-shaped recovery), THEN the dip is due to a fault (rectangular shape is caused by protection operation).

*Rule 4.2. Induction Motor Starting* IF the rms sequences for all three phases of voltage recover gradually but with approximately the same voltage drop (i.e., symmetric recovery), THEN it is caused by induction motor starting.

*Rule 4.3. Transformer Saturation* IF the rms sequences for all three phases of voltage recover gradually but with different voltage drop (i.e., nonsymmetric recovery), THEN it is caused by transformer saturation.

Figure 9.14 shows the rms sequences corresponding to these classes of events.

### 9.7.2 Application of Expert Systems to Event Classification

There exist many expert systems for power system applications [284, 198, 62, 81]. As a practical example, we shall describe an expert system from [284, 282] that is able to classify the corresponding event classes based on the underlying causes and whose input is the power system disturbance recordings. For each measurement recording containing a power system disturbance, the expert system uses rules from the knowledge base and outputs the event class to which the disturbance belongs.

As shown in the block diagram of Figure 9.15, the expert system consists of the following processing blocks:

1. *Segmentation* For each recording, voltage waveform samples are partitioned into event segments and transition segments (see Chapter 7).
2. *Feature Extraction* For each segment, features (e.g., rectangularity, swell, dip depth and duration, etc.) are extracted.
3. *Event Classification* Rules drawn from the knowledge base are used as the inference engine for the classification. Results are the event classes for the recordings.
4. *Further Analysis* Further detailed rules from the knowledge base are used for analysis and further classification.

The sample expert system is designed to classify nine classes of events:

- Energizing
- Non–fault interruption
- Fault interruption
- Transformer saturation due to fault
- Induction motor starting
- Step change
- Transformer saturation followed by protection operation

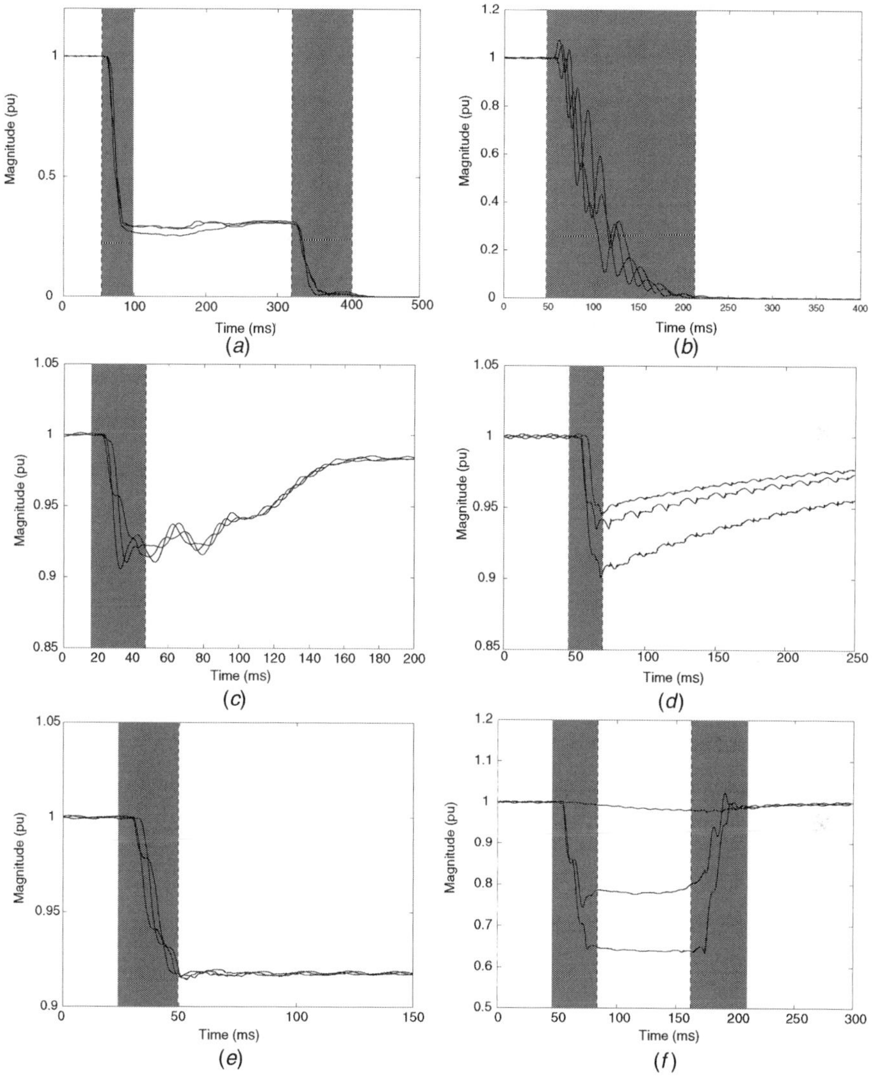


**Figure 9.14** The rms voltage values versus time (shadowed parts: transition segments): (*a*) interruption due to fault; (*b*) nonfault interruption; (*c*) induction motor starting; (*d*) transformer saturation; (*e*) step change; (*f*) single-stage voltage dip due to fault.

- Single-stage dip due to fault
- Multistage dip due to fault

Some further analysis and classification are then applied: for example:

- Seven classes of dip (i.e., class A, Ca, Cb, Cc, Da, Db, Dc) [33]
- Step change associated with voltage increase/decrease

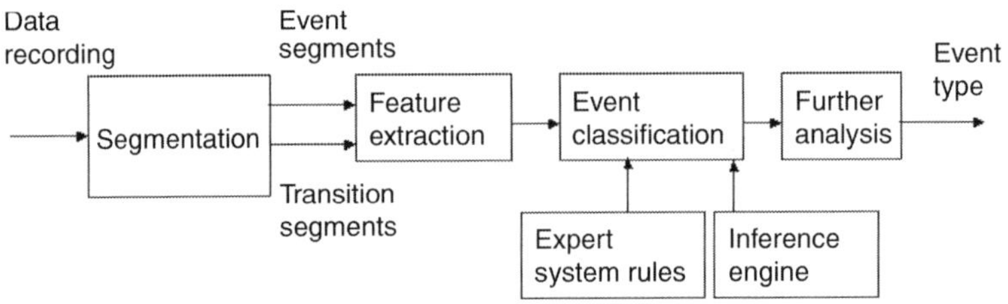


**Figure 9.15** Block diagram of sample rule-based expert system for event classification of power system disturbances.

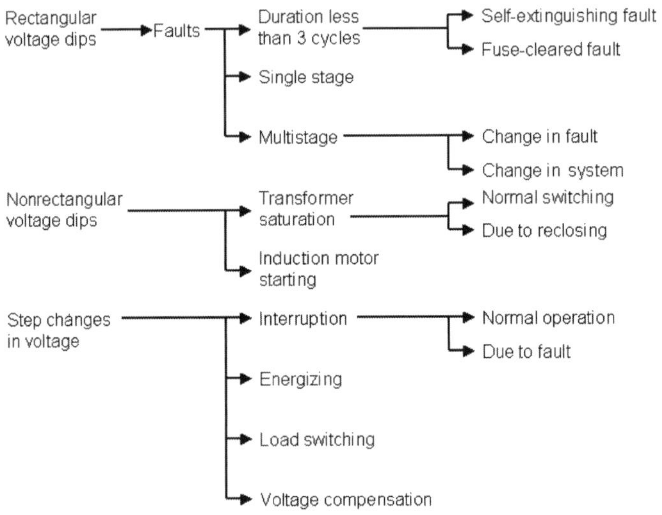


**Figure 9.16** Tree-structured inference process for event classification from classified disturbance recordings.

- Overvoltage associated with energizing/transformer saturation/stepchange/faults
- Fault related to starting/voltage swell/clearing

The tree-structured inference process for classifying events from disturbance recordings is shown in Figure 9.16 [284, 282].

Table 9.4 shows the results from the expert system. Compared to the ground truth (manually classified results from power system experts), the expert system has achieved approximately 97% of classification rate for a total of 962 disturbance recordings.

## 9.8 SUMMARY AND CONCLUSIONS

In this chapter, we have discussed methods for further characterizing power system disturbances in terms of classification, for example, classifying the types (or classes) or the underlying causes of power system disturbances. Although in Chapters 3 and

**TABLE 9.4 Classification Results for Total of 962 Recordings**

| Name of Events | Numbers | | |
|---|---|---|---|
| Energizing | 104 | | |
| | Fault | | Nonfault |
| Interruption | 13 | | 88 |
| | Reclosure | Other class | Protection operation |
| Transformer saturation | 37 | 82 | 6 |
| | Increase | | Decrease |
| Step change | 15 | | 21 |
| Single-stage dip due to fault | 455 | | |
| | Fault change | | System change |
| Multistage-dip due to fault | 56 | | 56 |
| | Short duration dip | | Overvoltage |
| Not classified | 16 | | 13 |

4 we discussed some useful signal-processing methods, they mainly serve to better examine, exploit, and quantify some individual characteristics of the disturbances in some domains (or under certain models), but without systematic classification of their types.

Several classification methods have been addressed in this chapter. These classifiers can roughly be categorized according to whether the designing method is statistical or deterministic in nature. Examples of statistical classifiers discussed in this chapter include classifiers using the simplest linear discriminants, Bayesian belief networks and NP hypotheses, ANNs, and SVMs, while rule-based expert systems are an example of a deterministic classifier.

It is obvious that a statistics-based classification method requires that we have a large amount of data available for training the classifier. This is often the case since an increasing number of power quality monitors are employed to measure the power quality. However, if one only has a small amount of measurement data, a deterministic-based method such as an expert system is often the best choice.

A classification system based on linear discriminant functions is the simplest and most fundamental tool. The disadvantage is that the performance of the classifier is limited by whether or not the different classes of power system disturbances are linearly separable. Poor performance may appear if they are linearly nonseparable.

An ANN is another choice which could offer relatively better performance if the classes of disturbances are linearly nonseparable. Especially if one has little knowledge about the power system disturbances (i.e., they appear as a black box), "blind" learning through training is a reasonably good choice. However, one should be aware that such a system may behave as *a rote learner*, having little guarantee on the generalization performance (despite a very small training error) and on the convergence to the desired global extrema. The performance is heavily dependent on how the training data set is drawn and whether it has a large coverage over all classes of disturbances, how large is the available training set, and the structure (or topologies) of the neural networks (e.g., the number of hidden layers and neurons and the interconnection of subneural networks if employed).

To overcome the above potential shortcomings, a better strategy is to design a classifier by exploiting the probabilities (or probability distributions) associated with each class of disturbances. This is achievable, as either we may know in advance the pdf's conditioned for each class of disturbances or we can use the estimated pdf's or the histograms from a large training data set.

If we do not know the prior probability for the different classes of disturbances, it is reasonable to assume an equal prior to all classes of disturbances. In such a case, we may choose to design a ML classifier (corresponding to a Bayesian classifier with equal priors). We are often interested in the type of classifiers that can maximize the classification rate under the constraint that the false-alarm rate is below a prespecified small value. This leads to the choice of the NP approach.

If we have the knowledge about the prior probability of each class in addition to the conditional probabilities it is worth exploiting this information by using a Bayesian classifier. The method is elegant in theory; however, when the dimension of the features is large or the probabilities are non-Gaussian distributed, the computation becomes difficult, if not impossible. A Bayesian belief network describes graphically the probabilities of each feature in relation to other features while taking into account that not all features in the feature vector are dependent on each other in real world problems. Carefully applying the Bayes rule to a belief network one may design a Bayesian classifier with reduced computational complexity. Despite all the elegance, the use of Bayesian classifiers is often "prohibited" by either its computational cost/complexity or the unknown or non-Gaussian pdf's.

As an alternative option, SVMs, armed with the statistical learning (VC) theory, may provide a good solution and compromise. Support vector machines are designed to minimize the classification error on the test set rather than on the training set under the SRM principle. The classification error on the test set is guaranteed by an upper bound which is dependent on the choice of the mapping function. Meanwhile, SVMs take into account the complexity of the learning machines, which is associated with a VC dimension. Choosing appropriate kernels in a SVM that are suitable to each particular application problem is essential. Different kernels lead to different classifiers and classification performance. Compared with ANNs both from the theoretical and implementation view point, a SVM is a superior choice for designing a classification system. Although there is little reported work on applications of SVMs to classify power system disturbances, this is clearly a good direction for future research and development in power system applications.

A rule-based expert system can be regarded as a deterministic approach that is parallel to these statistics-based classifiers. If one does not have many measurement recordings at hand but has much expert knowledge, a rule-based expert system is a wise choice that can translate a human expert classification/diagnostic process to a machine classification/diagnostic system. In such a system, it is essential that good knowledge is gained from human experts and subsequently translated into, for example, a set of IF–THEN rules. It also shows that by first segmenting each disturbance recording into event and transition segments, followed by extracting features from each segment, and subsequently applying expert system rules, one may obtain an efficient expert system for classifying events from power system disturbance recordings.

CHAPTER 10

# EVENT STATISTICS

In this chapter we reach the end of the analysis of measurement data from power quality events. In the previous chapters we discussed a number of methods to obtain single-event indices from sampled voltage or current waveforms. The single-event indices quantify the severity of the event and they form the basis for the quantification of the voltage quality at a certain monitor location. The resulting values are referred to as *site indices*. As a final step the site indices are used to calculate *system indices*, a way to quantify the voltage quality for a whole system, for example, all measurement sites at one voltage level or over a whole country.

In this chapter we will present site and system indices for two common voltage events: interruptions and voltage dips. Indices for interruptions (commonly known as *reliability indices*) will be presented in Section 10.1, site and system indices for voltage dips in Sections 10.2 and 10.4, respectively. An important issue in calculating site and system indices, time aggregation, is discussed in detail in Section 10.3. The calculation of system indices is matured for long interruptions in distribution systems. For voltage dips the discussion is still ongoing and competing indices are in use.

## 10.1 INTERRUPTIONS

### 10.1.1 Interruption Statistics

Interruption statistics have been in use for many years already. The number and duration of interruptions are generally accepted as a way of quantifying the performance of the power supply. With the deregulation of the electricity market, performance assessment became suddenly much more important [25, Chapter 6]. This has led to fast developments in the way in which interruption statistics are being obtained, presented, and used.

The reliability of the supply is quantified by means of so-called reliability indices, or interruption indices, normally referring to a whole system. We will use the term *interruption indices* as other events are sometimes also considered as part of reliability. There are two different ways of obtaining interruption indices:

- One method is to obtain indices for each site in the system for which information can be obtained. This will typically be a medium- or low-voltage feeder. For each location the number and duration are collected for the interruptions that occurred during a certain period, typically one year. These are referred to as site indices. The system indices are obtained as a weighted average over all sites for which information is available. Weighting is typically done based on the number of customers represented by the site but may also be based on maximum, rated, or average power consumption or on importance of the load.
- An alternative method is to obtain information for each interruption that occurred during a period anywhere in the system. For each interruption the duration and the number of customers affected are recorded. The annual statistics are obtained as a weighted average of all interruptions.

The second method is most commonly used; the former method would be more suitable from a power quality viewpoint. This distinction lies at the heart of some of the more serious discussions on power quality indices: whether an interruption should be defined as a situation in which the voltage for a customer becomes zero (a *voltage interruption*) or as a situation in which an interrupting device removes a customer from the main system (a *current interruption*). The result for the customer is in almost all cases the same, as we saw in Section 6.1.2. However, when defining indices to quantify the system performance, the difference becomes essential. An important criterion in defining a practical index is that it should be possible to obtain a value for that index without excessively high costs. Defining an interruption as a disconnection of a customer from the main supply makes it possible for the network operator to record the interruption. It is after all a device operated by the network that causes the interruption.

Defining an interruption as a voltage zero would require measurement of the voltage at many locations in the power system. This would require heavy investment. Therefore all interruption indices in practical use are based on current interruptions. For distribution systems the difference between the two methods is small; however, for transmission systems some serious difficulties occur, as we will discuss in Section 10.1.3. For voltage dips all discussed indices are based on the second method: measuring the event experienced by the customer. This has been shown to be a serious limitation in introducing voltage-dip indices.

Most interruption indices in practical use also distinguish between short and long interruptions. Other terminologies are in use as well, as we will see later. An important reason for making this distinction is again in the costs for implementing the data collection. As mentioned before, an interruption may be terminated by the automatic closure of a device. Such an interruption would typically be of shorter duration than one terminated by manual switching. In the latter case somebody has to actually make the closing action. This will make record keeping relatively easy. The person restoring the supply can make a record of the duration and size of the interruption. With automatic reclosure there is no human intervention and often the network operator is not even aware that there was an interruption. This is one of the reasons why short interruptions are only slowly becoming accepted as being part of interruption indices. Collecting data on short interruptions with the same accuracy as for long interruptions would require a major investment. Another reason is that, according to many network operators, short interruptions are minor events which the customer should accept as a normal part of the power supply. This point of view is however being replaced more and more by the opinion that all disturbances are worth recording.

### 10.1.2 IEEE Standard 1366

The IEEE standard 1366 defines a number of system indices for quantifying the reliability of distribution systems [169]. For the purpose of this standard an interruption is defined as "the loss of service to one or more customers connected to the distribution portion of the system." [169, page 2] The following system indices are defined by the standard:

*SAIFI, System Average Interruption Frequency Index* This is the most commonly used index. It is also often referred to as the *interruption frequency*. It is defined as the ratio between the total number of customer interruptions and the total number of customers served from the system. The number of customer interruptions is defined as the sum, over all the interruptions during one year, of the number of customers affected by each interruption. Let there be $N$ interruptions during a year, with interruption $i$ affecting $C_i$ customers. The value of SAIFI is calculated from the expression

$$\mathrm{SAIFI} = \frac{\sum_{i=1}^{N} C_i}{N_c} \tag{10.1}$$

with $N_c$ the total number of customers served.

*SAIDI, System Average Interruption Duration Index* This index is defined as the ratio of the total interrupted customer minutes and the number of customers:

$$\mathrm{SAIDI} = \frac{\sum_{i=1}^{N} D_i C_i}{N_c} \tag{10.2}$$

with $D_i$ the duration of interruption $i$.

*CAIDI, Customer Average Interruption Duration Index* This index gives the average duration of an interruption as seen from a customer:

$$\text{CAIDI} = \frac{\sum_{i=1}^{N} D_i C_i}{\sum_{i=1}^{N} C_i} \tag{10.3}$$

The three indices SAIFI, SAIDI, and CAIDI are related by the expression

$$\text{CAIDI} \times \text{SAIFI} = \text{SAIDI} \tag{10.4}$$

which can be easily derived from (10.1) through (10.3). Most network operators quantify their supply performance with two of these three indices, although they do not all use the same terminology.

*CTAIDI, System Total Average Interruption Duration Index* This index only considers those customers that actually experienced an interruption during a certain year. The definition is the same as for CAIDI with the exception that the value in the denominator is no longer the total number of customers. Instead $N_{ci}$ in the denominator is equal to the number of customers that experienced at least one interruption during a given year:

$$\text{CTAIDI} = \frac{\sum_{i=1}^{N} D_i C_i}{N_{ci}} \tag{10.5}$$

*CAIFI, Customer Average Interruption Frequency Index* This index quantifies the number of interruptions experienced by those customers that actually experienced interruptions:

$$\text{CAIFI} = \frac{\sum_{i=1}^{N} C_i}{N_{ci}} \tag{10.6}$$

This index has a value of at least 1, as each of the customers considered for this index experiences at least one index. The value of the index indicates how much interruptions are experienced by a small group of customers. This could be an important index as customers will in most cases accept one interruption during a year but often not more than that.

*ASAI, Average Service Availability Index* This index gives the average percentage of time during which a customer has electricity available:

$$\text{ASAI} = \frac{8760 \times N_c - \sum_{i=1}^{N} r_i N_i}{8760 \times N_c} \tag{10.7}$$

where 8760 represents the number of hours per year (for a leap year 8784 should be used instead). This additional multiplication is needed because $r_i$ is expressed in hours whereas the number of events are obtained over one year. The ASAI is not very commonly used. It always results in a value close to 1, which could easily give the wrong impression of a very reliable supply.

*ASIFI, Average System Interruption Frequency Index* This index is sometimes used instead of SAIFI. Whereas for SAIFI the interruptions are weighted by the number of customers affected, they are now weighted by the amount of load affected. The amount of load is expressed in kilovolt-amperes. Let $L_i$ be the load interrupted by interruption $i$ and $L_T$ the total load supplied from the system:

$$\mathrm{ASIFI} = \frac{\sum_{i=1}^{N} L_i}{L_T} \tag{10.8}$$

*ASIDI, Average System Interruption Duration Index* This index is related to SAIDI in the same way as ASIFI is related to SAIFI—The weighting is based on the amount of load affected, not on the number of customers affected:

$$\mathrm{ASIDI} = \frac{\sum_{i=1}^{N} D_i L_i}{L_T} \tag{10.9}$$

*CEMIn Customers Experiencing Multiple Interruptions* This index gives the fraction or percentage of customers that experience more than $n$ interruptions.

The latest version of IEEE 1366 also defines indices for short interruptions, referred to as *momentary interruptions* in the standard document. Three indices are defined: MAIFI, MAIFIe, and CEMSMIn. To understand these indices it is important to distinguish between a momentary interruption and a *momentary interruption event.* A momentary interruption is a single opening–closing sequence of an interrupting device. The time between opening and closing should not exceed a maximum value, which is currently chosen as 5 min. A momentary interruption event consists of one or more momentary interruptions. The total time between the first opening and the last closing should not be more than 5 min. Any interruption or interruption event lasting longer than 5 min duration is referred to in IEEE 1366 as a *sustained interruption.*

The following indices are recommended based on momentary interruptions and momentary interruption events:

*MAIFI, Momentary Average Interruption Frequency Index* Just as SAIFI gives the count of sustained interruptions, MAIFI gives the count of momentary interruptions.

*MAIFIe, Momentary Average Interruption Event Frequency Index* This index counts the number of momentary interruption events. MAIFIe is generally considered as a more suitable index because momentary interruptions come in clusters. Many reclosing schemes allow for multiple attempts before a permanent trip is given.

*CEMSMIn, Customers Experiencing Multiple Sustained Interruptions and Momentary Interruption Events* This index gives the fraction of customers that experience more than $n$ sustained interruptions and momentary interruption events. Note that this is the only index that considers both momentary and sustained interruptions. The value of this index may be calculated for different values of $n$.

Four examples of interruption sequences are given in Figure 10.1. Based on the above-given rules (as in IEEE 1366) these sequences would classify as follows for the purpose of obtaining interruption indices:

- In sequence (*a*) the time between the first opening and the last closing is less than 5 min. This sequence classifies as two momentary interruptions and one momentary interruption event.
- In sequence (*b*) the time between the first opening and the last closing is more than 5 min. However, each individual interruption is of less than five min. duration. This sequence classifies as two momentary interruptions and two momentary interruption events.
- The same holds for sequence (*c*): two momentary interruptions and two momentary interruption events.
- The second interruption in sequence (*d*) has a duration longer than 5 min. This sequence classifies as one momentary interruption, one momentary interruption event, and one long interruption.

The concept of counting a number of momentary interruptions as one event is often referred to as *time aggregation*. In IEEE 1366 the maximum duration of a momentary interruption event is the same as the maximum duration of a momentary interruption. It would be possible to merge the individual interruptions over a

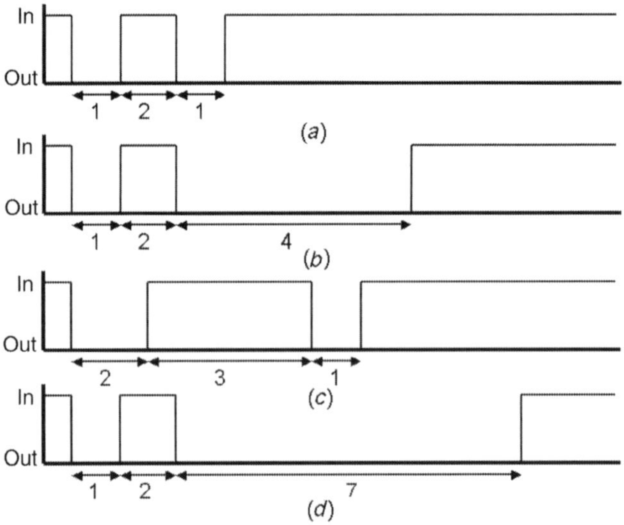


**Figure 10.1** Different opening and closing sequences. The time scale is given in minutes.

longer period of time. But in that case an *interruption event* could also contain sustained interruptions. The concept of time aggregation will be treated in more detail when discussing voltage-dip indices below.

The maximum duration of 5 min has been the issue of much discussion. Different documents use different values. The main importance is that the concept of maximum duration is used to distinguish between momentary (or short) and sustained (or long) interruptions. In line with the discussion above, an alternative would be to distinguish based on the closing mechanism: An interruption terminated by the automatic closing of the interrupting device would be a short interruption; an interruption terminated by manual closing would be referred to as a long interruption. But this definition is, as far as the authors are aware, not used anywhere for power quality or performance quantification purposes.

For collecting component failure data the restoration-type approach is used. In IEEE 859 [163] a distinction is made between three types of *forced outages*. A forced outage refers to the unintentional and unwanted removal of component from service. The distinction is made based on the way in which the component is restored to service:

*Transient Forced Outage* The component is restored automatically.

*Temporary Forced Outage* The component is restored through manual switching actions.

*Permanent Forced Outage* The component is restored through repair or replacement.

It would be equally possible to use such a criterion to distinguish between short and long interruptions. This would make it somewhat easier to collect statistics, as only manual restorations need to be counted. On the other hand, the duration-based classification is more reasonable from a customer viewpoint as the customer has no way of knowing if the supply is restored manually or automatically.

In several of the interruption indices defined above, the duration of the interruption is used. The duration of the interruption is defined as the time elapsed between the opening of the interrupting device and restoration of the supply. The moment of restoring the supply can be easily recorded because this is done manually. That leaves us with knowing when the interruption started. For interruptions that originate at the transmission level or at higher distribution levels, any (automatic or manual) switching action is recorded. This is however a minor part of all interruptions. For interruptions at the distribution and certainly at the low-voltage level, there is no record of the operation of an interruption device. In fact, the first information a network operator receives of an interruption is in many cases a phone call from a customer that is interrupted. There is no well-defined method yet for determining the start of an interruption. Some network operators take the time of the first phone call as the start of the interruption. The argument for doing this is, apart from the fact that no other information is available, that the restoration process can only start when it is known that there is an interruption.

For calculation of indices based on momentary interruptions, the interruption duration does not play any role. It is only the number of momentary interruptions and momentary interruption events that matters. Thus the moment at which the interruption starts or ends does not need to be recorded. However, any self-respecting network operator would record the information anyway because the extra costs are small. An automatic recording mechanism is needed anyway for collecting the data on short interruptions.

### 10.1.3 Transmission System Indices

Attempts have been made to extend the definition for distribution systems (as, e.g., in IEEE 1366) to transmission systems. There are however a number of pitfalls which make it difficult to immediately apply the distribution indices.

A first pitfall is that transmission systems are typically operated meshed. This implies that the opening of an interrupting device (a current interruption) does not necessarily lead to a voltage interruption. In most cases there is a parallel path, which means the supply is not interrupted. Interruption definitions based solely on the opening of interrupting devices will not be possible. It is still possible to give availability statistics for individual components, typically transmission lines, which can next be averaged over all components in the system. There is no direct relation between this *component availability* and the reliability of supply as experienced by individual customers. However, this should not be interpreted as a reason for not publishing component availability data. A distinction may further be needed between planned unavailability (for maintenance) and forced unavailability. Planned unavailability typically takes place during off-peak periods and is aimed at increasing the supply reliability. A high forced unavailability of components points to a low system reliability.

The term *transmission incident* is often used for an event (such as a fault) at the transmission level that leads to an interruption for one or more customers. In some cases the term is reserved for events that lead to an interruption for end users, but here we will use the term more generally, including also those cases where the connection with another network operator is lost. In some countries, the availability of transmission system components is used as a performance index [72]. In addition to the average annual system availability (in percent) and the winter-peak system availability of the English system, the availability of the interconnections with France and Scotland is reported separately (96.2 and 99.7%, respectively, for the 2000 to 2001 reporting period). A distinction is further made between planned and unplanned unavailability. Whereas the planned availability (reported on a monthly basis) is up to 8.4%, the unplanned unavailability varied between 0.20 and 0.36% (long-term averages; 2000 to 2001 data).

A second pitfall is that the number of customers is no longer as easily determined as for distribution system indices. The transmission system operator does not have any record of the number of customers that may be affected by an interruption. The loss of voltage on a transmission substation may not have any effect for a large industrial customer connected to this substation when there is an automatic transfer to another substation. The decision on whether this should be counted as an interruption or not is made even more complicated when the backup supply may not in all cases be able to feed the total load. Would such a case be counted as an interruption in the normal supply or in the back-up supply?

A third pitfall is that some large customers may settle for a less reliable supply to save connection costs. Should an interruption in such a case be attributed to the transmission operator or not?

The issue of reliability indices for transmission networks is discussed in detail in the final report by the CIGRE working group C4.07 [64]: An overview is given of the differences between the indices as used in different countries. Based on existing practice and on perceived ambiguities in existing practice, three philosophies are proposed by the working group:

- End-customer interruption performance
- Connection point interruption performance
- System-interrupted energy performance

The end-customer load interruption performance gives the best quantification of the transmission system performance as experienced by the end customer. The reliability indices are the same as those for distribution reliability and are calculated in the same way, but only events originating in the transmission system are included in the statistics. It is not specifically stated in the document, but "end customers" should be read as consumers of electrical energy. Thus large power stations are not considered as end customers here. How small power stations (*distributed generation*) should be treated remains unclear.

The resulting indices quantify how much the transmission system unreliability affects the end customers. The indices are a measure of the performance of the transmission system, but the data collection cannot be done solely by the transmission company. Data have to be provided by the distribution companies on the number of customers affected by each *transmission system incident.* Alternatively the data can be collected by the distribution companies as part of their own reporting on reliability indices. This will also automatically result in consistency between transmission and distribution data, which is an important advantage of reporting end-customer interruption performance.

An alternative approach is the so-called connection point interruption performance. Instead of counting the number of interruptions experienced by end customers, the number of interruptions of connections points is counted. A connection point, or *transmission connection point*, is defined as "the interface (border) between the transmission network and another transmission network, a distribution company, large end-customer, or producer. It is the point up to which the transmission company has responsibility and control over" [64, Section 4.5.2.1]. A connection point may be a complete busbar, an individual section of a busbar, or a single circuit breaker field. Note that the exact choice of connection point can significantly affect the statistics. This is one of the reasons why it is extremely important to very clearly define all details of the way in which the statistics are collected before any comparisons can be made.

An interruption is defined as a situation in which the connection point is not available for transport of energy. Frequency and duration of supply point interruptions are reported in pretty much the same way as for customer interruptions at the distribution level. One may collect site indices first (where each connection point becomes one site) and average those over the system. Alternatively one may record the number of connection points that experience an interruption for each transmission system incident. The results are presented per connection point in the same way as the results for distribution networks are presented per customer.

The way in which an end customer experiences a transmission system incident is affected very much by the design of the distribution system. The determining factor is often the location of the interface between transmission and distribution systems. When the distribution network operator only covers up to medium voltage, the network will typically be operated completely radial. Two extreme cases are shown in Figure 10.2. In the left-hand drawing, the *subtransmission network* is operated by the transmission company. The connection points are located at MV; it may be either the 30-kV busbar or the 30-kV side of the 30/10-kV transformer. But in either case a connection point interruption will lead to an interruption for all distribution customers. The connection point interruption philosophy is equivalent to the end-customer interruption philosophy. Note, however, that the resulting values will still be different as the number of end customers is not the same for each connection point. There may also be some difference concerning the duration of interruptions. It is often possible to feed distribution customers from another transformer, and thus from another transmission connection point, through switching actions in the distribution network.

The situation becomes completely different for the situation on the right-hand side in Figure 10.2. The border between the transmission and distribution network is now located at 200 kV. The connection point is at the 200-kV bus. A connection point interruption will not always lead to an interruption for end customers. In this very simple network there are four connection points. Loss of one connection point should not affect any end customers; loss of two connection points in the same substation could be a problem, however. The situation can become even

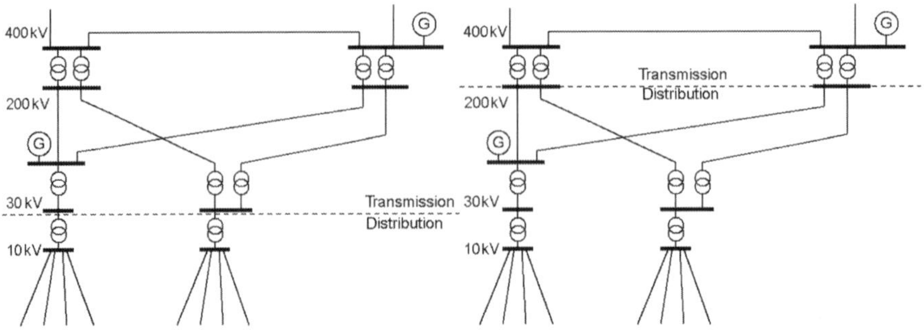


**Figure 10.2** Two different division lines between transmission and distribution networks.

more complicated when the subtransmission network is owned by a third company. Such a subtransmission company has four types of connection points:

- Connection points with the transmission network operator
- Connection points with neighboring subtransmission operators
- Connection points with distribution companies
- Connection points with generators

### 10.1.4 Major Events

An old discussion of the performance assessment of any system is the treatment of extreme external circumstances. The discussion can be traced back to the basic design philosophy of the system. Some technical systems are designed such that they should never experience a major breakdown. The financial costs involved in obtaining such an extreme reliability are only of secondary importance.

For other technical systems, in fact for most systems, the design only considers so-called normal operation under "normal circumstances." Considering "exceptional circumstances" would lead to much higher costs than what are considered worthwhile.

The same discussion can be applied to electric power systems. Even though such a discussion has never been held publicly, the outcome can be observed when looking at the resulting design. For generation and transmission systems the design rules are such that interruptions should be very unlikely. Large-scale blackouts (e.g., recently in the United States and Canada, August 14, 2003; in the South of Sweden and Copenhagen, September 23, 2003; and in Italy, September 28, 2003) are seen as unacceptable. Small-scale blackouts are acceptable when occurring with a low frequency and preferably not in major urban areas (the one affecting parts of South London and Kent on August 28, 2003, was treated in the press as a major blackout even though it only affected 500,000 customers for 34 min). In the design of transmission networks the amount of redundancy is such that these large-scale blackouts are very unlikely. A single short circuit or loss of a generator station should not affect any customer. Even multiple contingencies will in most cases not lead to a large-scale interruption.

At the distribution level the situation is completely different. The system is designed such that it will operate as long as no fault occurs. After a fault the supply will be restored as soon as possible. Important aspects of design and operation include the limitation of the number of faults and speeding up the restoration time. However, every fault will still lead to an interruption of the supply to some customers. There are some exceptions where even at the distribution level the design is such that interruptions are very unlikely. Examples are the downtown areas of New York and most of Singapore. A complete discussion of the relation between distribution system design and interruptions is outside the scope of this text. For more information the reader is referred to [33, Chapter 7]. What is important for this discussion is that the design and operation of distribution networks are based on normal circumstances. When suddenly a lot of damage occurs to the distribution network (e.g., due to an earthquake or snowstorm), it is no longer possible to restore the supply within a few hours as would have been possible for a single fault during normal circumstances. Including those interruptions in the reliability statistics would not provide a good indicator of the normal performance of the system. Therefore interruptions related to exceptional circumstances are normally excluded from reliability statistics.

However, such exceptions should be treated with care. The reasoning is that the system is not designed to withstand such extreme circumstances. This is obviously true, but it is equally true that the system is not designed to not get an interruption after a fault. Using the same reasoning one could conclude that interruptions due to distribution system faults should not be included in the statistics. Any interruption can be prevented by proper design. It is just a matter of including sufficient redundancy in the design and operation. It has happened that a network operator explained an interruption as "due to lightning" and thus outside of his or her control.

There are good reasons for treating exceptional circumstances with care before including them in performance statistics. A good reason is that from a regulatory viewpoint one is not interested in encouraging investment of such a level that the system can operate even under extreme circumstances. If the regulatory feedback is only aimed at reducing the number of faults and the restoration time during normal circumstances, only normal circumstances should be included in the statistics. The regulator has the task of defining what should be considered as normal circumstances.

An acceptable compromise may be to give separate statistics for exceptional circumstances. The interpretation of these statistics may be a separate issue. Currently there is at simply not enough experience with these statistics to be able to interpret them. That alone would be a good reason to collect them.

Having decided that interruptions during exceptional circumstances have to be treated differently, the next question immediately comes up: What are exceptional circumstances? There are three different approaches to this, all three of which are in use:

- The decision could be made on a case-by-case basis. When collecting interruption statistics, typically at the end of a year, it is decided whether any periods of the year count as exceptional circumstances. Exceptional circumstances are rare and as such easy to recognize. This method is used by the regulatory bodies in some countries.
- A predefined rule is agreed upon where the exceptional circumstances are listed. Examples of well-defined rules are [169] "national weather service declares severe weather watch or warning for the area," "named storms and tornados verified by the national weather service," and "winds in excess of 140 km/h or 1.2 cm of ice combined with winds in excess of 60 km/h." In most cases the description is more prone to interpretation, for example, "a catastrophic event that exceeds the design limits of the electric power system." Apart from the problem of exactly defining what are extreme circumstances (when snow fall

becomes a snow storm), there is a risk of forgetting something in the listing (snow is not mentioned by any of the examples in [169] although ice storms are).

- A predefined rule based on the impact on the system. The impact is typically measured in terms of the percentage of customers affected and the time it takes to restore the supply. Examples of such rules are [169] "more than 10% of customers affected and more than 1% not restored in 24 hours," "more than 15000 customers out," "more than 15 separate interruptions and more than 200 000 interrupted customer minutes within a 24 hour period."

An interesting method of determining exceptional circumstances is given in IEEE 1366 [169]. A so-called major-event day is defined by assuming a lognormal distribution for the daily SAIDI values. Those days for which the daily SAIDI value is outside the normal range is classified as a major-event day.

It is well known from stochastic theory that the sum of a number of independent stochastic variables forms a normal distribution. Equally but less well-known, the product of a number of independent stochastic variables forms a so-called lognormal distribution [200]. The density function for the lognormal distribution reads as follows:

$$f(y) = \frac{1}{\sqrt{2\pi} s y} \exp\left\{ -\frac{1}{2s^2} \left[ \ln\left(\frac{y}{y_0}\right)^2 \right] \right\} \tag{10.10}$$

where $y_0$ is the median value of the distribution and $s$ the standard deviation of $\log y$. The mean $\mu$ of the lognormal distribution is not equal to $y_0$ but

$$\mu = y_0 \exp\left(\tfrac{1}{2} s^2\right) \tag{10.11}$$

Expressions (10.10) and (10.11) are rarely used. In most cases, when studying lognormally distributed variables, the logarithms of values are used instead of the values of the stochastic variable. If a stochastic variable $y$ is lognormally distributed, then $\log y$ is normally distributed and all the standard methods for the analysis of normally distributed variables can be applied.

Coming back to the approach for detecting major-event days as defined in IEEE 1366, the logarithm of the daily SAIDI value is used for further analysis. The daily SAIDI value is defined in the same way as in (10.2) with the difference that only those interruptions are counted that are starting within the given day. Note that it is the starting time of the event that counts. The whole duration of the interruption is considered even if it ends outside of the given day. The mean $\mu$ and standard deviation $\sigma$ of log SAIDI are determined next. A day is classified as a major-event day when its SAIDI value is more than 2.5 standard deviations above the mean:

$$\text{SAIDI} > \mu + 2.5\sigma \tag{10.12}$$

Two issues immediately become apparent. The first is that a day without any interruption cannot be considered in the distribution (the logarithm of zero is minus infinity). The standard recommends that any zero SAIDI value be made equal to the lowest nonzero value measured. The effect of this on the result is not clear as there is no information available yet on the application of this method.

The second issue is the choice of the threshold, in this case 2.5 standard deviations above the mean. If we assume that the SAIDI values show a lognormal distribution under normal circumstances, there is a nonzero probability that the value will be higher than the threshold. This could be seen as incorrect detection of a major-event day and would call for a higher threshold. However, a too high threshold would result in the risk that a major-event day is not detected. The probability that a normally distributed variable exceeds its mean value by more than 2.5 standard deviations is equal to 0.62%. This would correspond to about two days per year. But this assumes that the distribution of SAIDI is lognormal even for such high values. More experience with the method is needed before this can be verified.

## 10.2 VOLTAGE DIPS: SITE INDICES

In the previous chapter a number of indices were introduced to characterize individual voltage dips. These so-called single-event indices or single-event characteristics include duration and residual voltage (or magnitude), the two most commonly used. In fact, only these two are defined in standard documents. In this section we will therefore mainly discuss the further processing of duration and residual voltage. In the discussion we will in most cases assume that residual voltage and duration are known for all dips at one site or at a number of sites. Most of the discussion in the remainder of this chapter is independent of the way in which the residual voltage and duration are calculated.

The calculation of voltage-dip indices takes place in two steps. From the residual voltage and duration values collected at one location over a period of time, a site index is calculated. From the site indices of a number of sites in one system, a system index is calculated.

Voltage-dip site indices are aimed at quantifying the voltage quality at a certain location. Most indices simply give the number of events per year within a certain range of magnitude and duration. The difference between the various proposed methods is mainly in the grouping of the individual events, although some methods are more fundamentally different.

### 10.2.1 Residual Voltage and Duration Data

Consider as an example the voltage dips recorded during a six-year period at an MV site (in the voltage range 10 to 35 kV). During these six years, a total of 124 events were recorded at this site. The duration and residual voltage for all 124 events are presented in Table 10.1. The residual voltage is calculated as the lowest of the three phase-to-ground voltages. Table 10.1 is a way of describing the performance of the

**TABLE 10.1 Voltage Dips Recorded at One Site During Six-Year Period**

| Duration (ms) | Residual Voltage (%) | Duration (ms) | Residual Voltage (%) | Duration (ms) | Residual Voltage (%) |
|---|---|---|---|---|---|
| 60 | 72.86 | 50 | 73.56 | 80 | 4.93 |
| 120 | 7.82 | 300 | 74.46 | 150 | 6.11 |
| 60 | 75.34 | 70 | 84.56 | 130 | 15.50 |
| 20 | 124.18 | 70 | 83.10 | 70 | 83.04 |
| 40 | 87.86 | 180 | 2.64 | 70 | 87.50 |
| 60 | 87.80 | 80 | 71.20 | 60 | 76.76 |
| 50 | 88.78 | 140 | 82.15 | 60 | 86.76 |
| 60 | 85.97 | 130 | 82.97 | 70 | 86.91 |
| 140 | 80.21 | 50 | 74.73 | 40 | 83.87 |
| 120 | 22.44 | 80 | 87.26 | 100 | 61.01 |
| 70 | 85.62 | 50 | 85.62 | 130 | 63.94 |
| 940 | 79.87 | 810 | 80.73 | 850 | 80.90 |
| 110 | 80.70 | 380 | 80.52 | 310 | 80.81 |
| 80 | 66.56 | 60 | 87.03 | 130 | 1.70 |
| 14,940 | 0.0 | 90 | 0.0 | 130 | 12.44 |
| 110 | 76.42 | 70 | 53.79 | 100 | 79.68 |
| 70 | 85.69 | 50 | 87.30 | 130 | 1.67 |
| 80 | 65.29 | 190 | 10.22 | 70 | 80.21 |
| 40 | 89.14 | 130 | 1.86 | 20 | 89.53 |
| 300 | 85.44 | 360 | 47.53 | 120 | 2.31 |
| 170 | 2.16 | 80 | 74.07 | 570 | 85.34 |
| 30 | 89.78 | 620 | 9.20 | 460 | 89.50 |
| 550 | 83.82 | 80 | 86.53 | 100 | 70.48 |
| 450 | 70.41 | 130 | .91 | 180 | 61.05 |
| 330 | 86.17 | 960 | 86.11 | 150 | 64.76 |
| 240 | 9.16 | 40 | 88.90 | 60 | 88.78 |
| 170 | 6.49 | 90 | 10.59 | 120 | 82.92 |
| 120 | 84.02 | 120 | 83.17 | 130 | 80.70 |
| 70 | 86.23 | 310 | 88.13 | 530 | 4.45 |
| 660 | 86.48 | 120 | 76.48 | 540 | 76.61 |
| 60 | 89.68 | 120 | 85.49 | 120 | 5.50 |
| 120 | 6.43 | 300 | 86.60 | 310 | 86.93 |
| 310 | 86.79 | 150 | 80.55 | 1,020 | 85.47 |
| 110 | 82.82 | 1,020 | 84.80 | 1,020 | 83.73 |
| 70 | 80.39 | 570 | 83.96 | 230 | 87.73 |
| 50 | 84.22 | 50 | 86.54 | 130 | 46.40 |
| 70 | 70.79 | 70 | 75.34 | 110 | 63.09 |
| 290 | 86.06 | 60 | 79.66 | 90 | 79.93 |
| 80 | 82.69 | 120 | 11.36 | 40 | 88.51 |
| 50 | 88.09 | 60 | 88.46 | 160 | 79.14 |
| 140 | 15.30 | 130 | 14.41 | 120 | 14.26 |
| 120 | 11.21 | | | | |

system at this specific site. This is in fact the way in which many power quality monitors store the data: simply as a list of single-event indices, in almost all cases together with date and time stamps. The latter are of interest to correlate voltage dips with events in the system and for time aggregation, as will be discussed in Section 10.3.

The site indices are calculated from the residual voltage and duration values in Table 10.1. It is in theory possible to calculate average values of these two indices, but that is rarely ever done and difficult to interpret. Especially the average residual voltage depends very strongly on the threshold setting. When only voltage dips below 70% residual voltage are considered, the average residual voltage will be much lower than when all voltage dips below 90% residual voltage are considered. The average duration is not so much affected by the threshold setting, but one or two long events can seriously affect the average value. A discussion on using the average residual voltage and how to interpret the results is started in [237], where the average residual voltage is compared with the value for a radial system.

Next to an upper threshold, some index methods use a lower threshold, (e.g., 10% of nominal). An event with a residual voltage below the lower threshold is in that case classified not as a voltage dip but as a short interruption. Statistics on short interruptions are given by one of the methods discussed in Section 10.1. This will prevent double counting but may leave some events uncounted. A short circuit near a substation may lead to a voltage dip with a residual voltage close to zero. Such an event would be classified as a voltage interruption but not as a current interruption and thus not be counted in the interruption indices. To prevent events becoming uncounted, a comparison of time stamps would be needed, but this is not always feasible.

### 10.2.2 Scatter Plot

A straightforward graphical method to present the performance of a site is by means of *scatter plot*. Each event is represented as one point in a plot of residual voltage versus duration. The scatter plot is not generally considered as a site index because it does not quantify anything, but it is included here for completeness and because it is very commonly used.

The scatter plot for the dips shown in Table 10.1 is shown in the top left plot in Figure 10.3. Only events with a duration up to 2 s are included in the plot. For this site one event was recorded with a duration of almost 15 s. Including this event would push all other events to the left-hand side of the plot. All other events had a duration of less than 2 s.

The advantage of the scatter plot is that it gives a graphical view of the voltage quality of a site. The user can rather quickly draw conclusions about residual voltage and duration of dips at a given site. In Figure 10.3 one can see some "clusters" of dips: for the MV site we observe a number of very deep dips (residual voltage less than 30%) and with durations up to 200 ms. There is a second cluster of dips with moderate depths (residual voltage down to 50%) and durations up to 200 ms and a third cluster of shallow dips (residual voltage down to 75% and durations

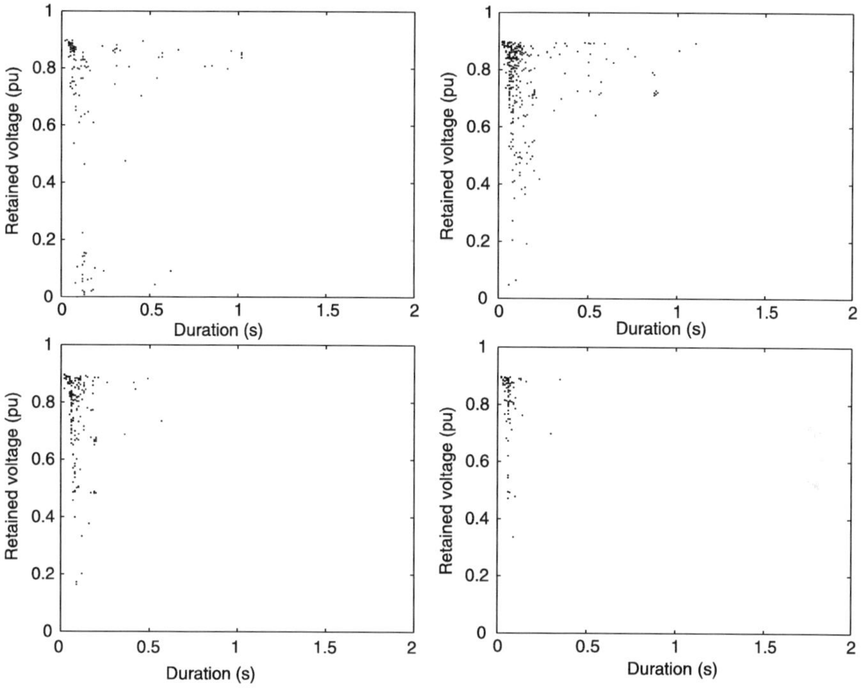


**Figure 10.3** Scatter plot showing voltage dips appearing at four different sites during six-year period: MV (top left); HV1 (top right); HV2 (bottom left); EHV (bottom right).

up to 1 s. These three clusters point to three different origins for the dips. The measurement was performed in an impedance-earthed system. The short, deep dips are due to single-phase faults. A single-phase fault in an impedance-earthed system leads to a low voltage in one phase and overvoltages in the nonfaulted phases. After a Dy transformer, however, the fault only results in a minor voltage dip. We will come back to this in Section 10.2.8.

Disadvantages of the scatter plot are that patterns may be observed that are not really present, different events with the same characteristics show up as only one point, and quantification of the voltage quality is not possible. In the forthcoming sections we will present a number of methods that are better at quantifying the supply performance but in general less able to present the dips occurring at a site in one single glance.

For comparison a number of scatter plots are given for sites at other voltage levels in Figure 10.3. The site labelled HV1 is in the voltage range 35 to 100 kV, the site labelled HV2 is in the voltage range 100 to 230 kV, and the site labelled EHV is in the voltage range above 230 kV. Again all events with a duration exceeding 2 s are not presented in the plots. The scatter plot for the HV1 site (top right) is similar to the one for the MV site (top left). The HV1 site shows a higher dip frequency than the MV site, resulting in more dots. But the distribution of the dips over the plot is similar, with the exception of the short, deep dips. These are only present for the MV site, not for the HV1site.

The biggest difference between the HV2 site (bottom left) and the MV and HV1 sites is that in the former the dips are shorter in duration. This reflects the faster protection used in the HV2 network compared with the networks at lower voltages. At EHV the dip duration is even shorter, as shown in the bottom right plot.

### 10.2.3 Density and Distribution Functions

Another way of presenting the results is through one-dimensional probability density and distribution functions. The probability distribution function for the event duration gives the relative number of events with a duration less than a given value. This function is shown in Figure 10.4 for the four sites mentioned earlier. From this figure one can conclude, for example, that 75% of the dips have a duration less than 200 ms at the HV1 site (dash-dot curve), whereas over 95% of the dips have a duration less than 200 ms at the EHV site (dotted curve). By comparing the four curves, one can conclude that the dip duration at EHV and HV2 is shorter than at HV1 and MV. We already drew that conclusion from the scatter plots in Figure 10.3, but from the distribution function it is possible to quantify the reduction in dip duration.

The same kind of probability distribution can be plotted for the residual voltage, resulting in Figure 10.5 for the four sites. For the MV site we see again the large number of events with small values for the residual voltage. We also see that a

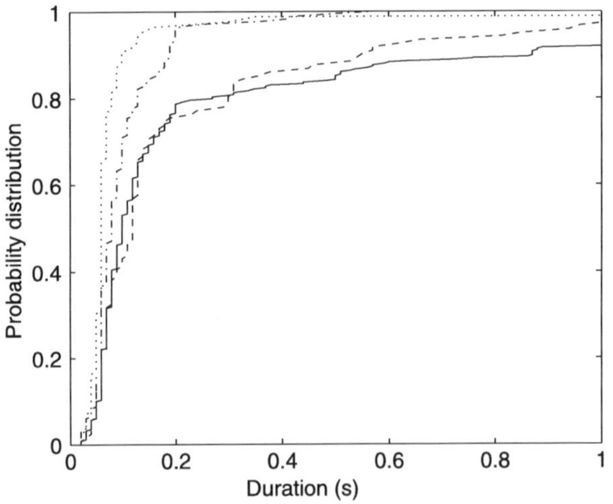


**Figure 10.4** Probability distribution function of dip duration for four different sites: MV (solid curve), HV1 (dashed), HV2 (dash-dot), and EHV (dotted).

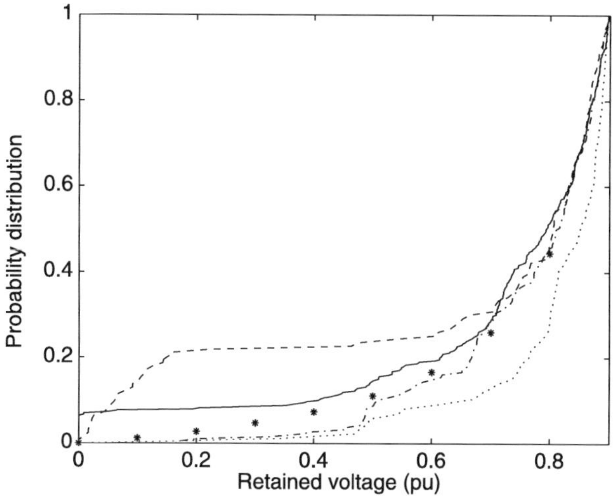


**Figure 10.5** Probability distribution function of residual voltage for four different sites: MV (solid curve), HV1 (dashed), HV2 (dash-dot), and EHV (dotted).

substantial fraction of the events at the HV1 site have zero residual voltage. These events can be classified as short interruptions as well.

We can also conclude from Figure 10.5 that the number of events increases very quickly with increasing residual voltage. The threshold setting will thus have a significant influence on the number of events recorded. In [33, Section 6.5] it is shown that under rather strict conditions (infinitely long radial feeders with uniform fault frequency) the number of dips with a residual voltage less than $V$ (in per-unit) is proportional to

$$N_{\text{dips}} = N_0 \frac{V}{1 - V} \tag{10.13}$$

This relation is shown as asterisks in Figure 10.5. The approximation holds rather well, despite its simplicity and strict conditions. The same conclusion has been drawn in other studies [31, 245].

In some cases it is more appropriate to give the absolute number of events more severe than a certain residual voltage, instead of the probability distribution function. This allows a direct comparison with equipment performance. Such plots are shown in Figure 10.6 for the four sites.

Instead of the distribution function it is rather common to use a density like function. The (absolute or relative) number of dips within certain ranges of residual voltage or duration are given. Figure 10.7 gives the absolute number of dips for nine different ranges in residual voltage: 0 to 0.1 pu through 0.8 to 0.9 pu. The label with each of the bars gives the upper limit of the range. The figure

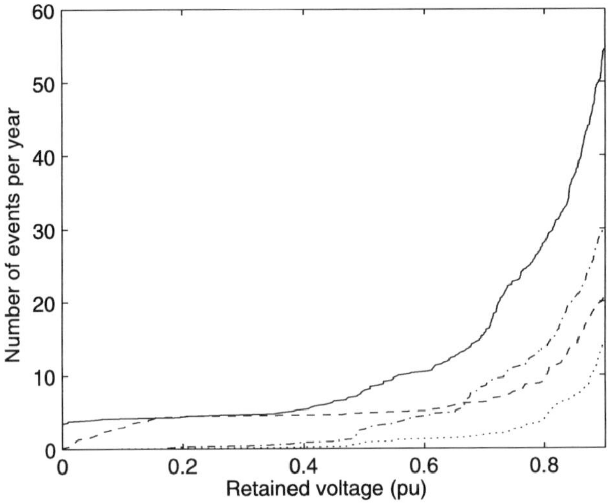


**Figure 10.6** Number of events more severe than a certain residual voltage for four different sites: MV (solid curve), HV1 (dashed), HV2 (dash-dot), and EHV (dotted).

again confirms the relatively large number of events with very low residual voltage for the MV site. For the HV2 site we only see an increase in the event frequency for higher values of the residual voltage.

Note that we give a curve for the distribution function but a bar chart for the density function. The distribution function is by nature a rather smooth, monotonously increasing function, whereas the density function gets a very noisy appearance when we attempt to make the resolution very small. The result of using a too small step size is shown in Figure 10.8. This figure gives basically the same information as Figure 10.7, but instead of a resolution of 0.1 pu a resolution of 0.01 pu has been used.

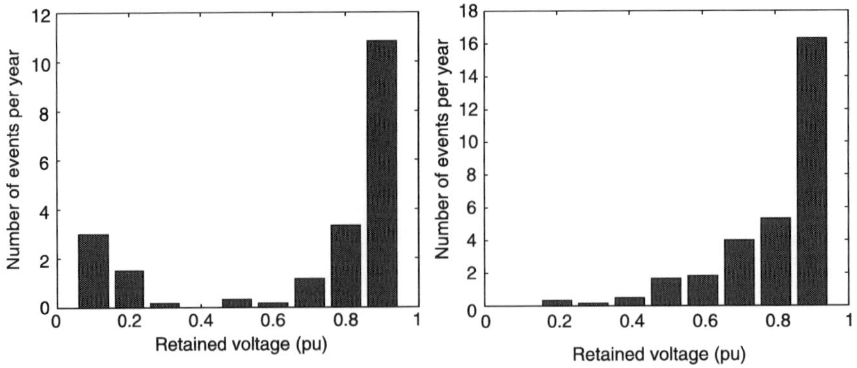


**Figure 10.7** Number of events for different ranges of residual voltage: MV site (left) and HV2 site (right).

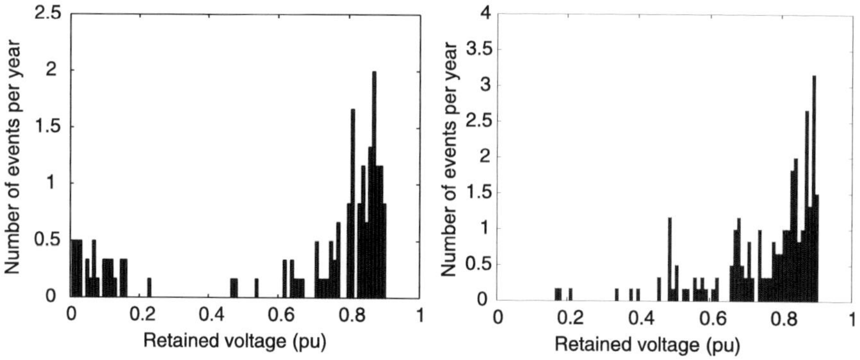


**Figure 10.8** Resulting bar charts for the MV (left) and HV2 (right) sites when a voltage resolution of 0.01 pu is used.

### 10.2.4 Two-Dimensional Distributions

As voltage dips are characterized by two independent parameters—residual voltage (magnitude) and duration—a two-dimensional distribution will give more information that two one-dimensional distributions. We saw already that a scatter plot gives more information than a distribution function. However, those were one-dimensional distribution functions. A two-dimensional distribution could quantify the voltage-dip frequency in both dimensions.

Such a two-dimensional distribution is shown in Figure 10.9 for the same four sites presented earlier. From the residual voltage and duration of the voltage dips recorded during a six-year period, a two-dimensional distribution function is calculated. The function value is equal to the number of events per year more severe than the given residual voltage and duration. This function is represented by means of a contour plot, where points with equal function values are connected. The differences in probability distribution are clearly visible from comparing the plots for the different sites. Contour plots are given for 1, 2, 5, 10, 15, 20, and 30 events over the six-year period. The MV site has a significantly higher number of short, deep events than the other sites. The HV1 site has a higher number of long, shallow events than the other sites.

The use of contour plots to quantify the performance of a site is recommended in two IEEE standards that were written in the first half of the 1990s: IEEE 493 and IEEE 1346 [161, 168]. However, recently the shift has been very much toward indices (i.e., numbers) instead of charts, preferably simple indices. We will come across a number of those indices in the remainder of this section. Methods such as using the contour chart and the scatter plot should however not be forgotten as they give a graphical overview of the performance of the supply. The contour chart, as in Figure 10.9, also has the advantage that it allows a quantitative comparison of supply performance and equipment immunity against voltage dips [74].

Note that a contour plot is only appropriate for a cumulative function, not for a density function. Attempting to create a contour plot for a two-dimensional

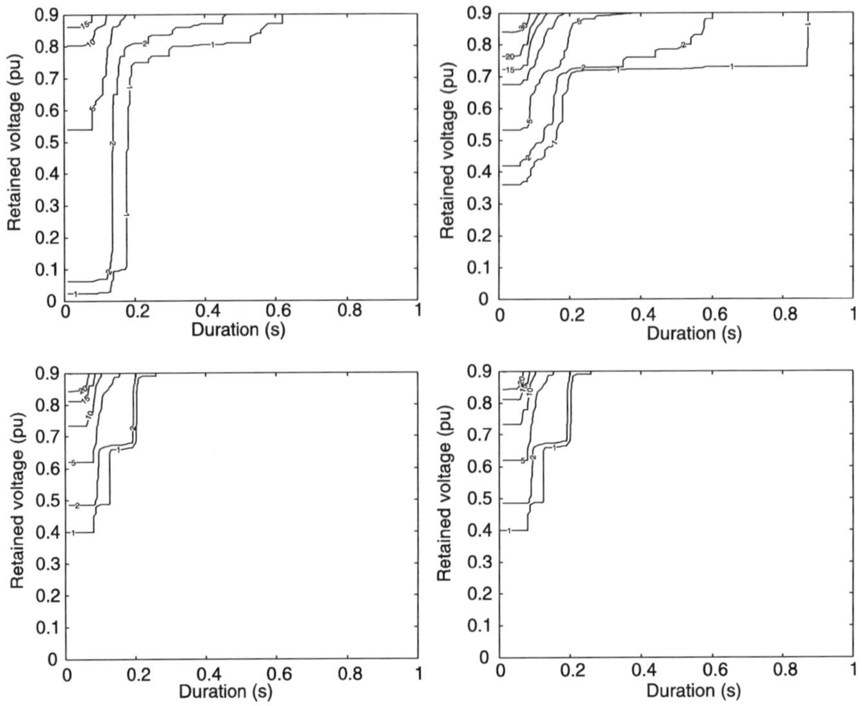


**Figure 10.9** Contour plot of voltage-dip probability distribution for four different sites: MV (top left); HV1 (top right); HV2 (bottom left); EHV (bottom right).

density function would result in a graph that is heavily dependent on the resolution used. A similar problem was presented before with one-dimensional density functions. Two contour plots for the density function are shown in Figure 10.10 for the MV site. The left-hand plot was created using a resolution of 20 ms in duration and 0.02 pu in residual voltage. The result shown is similar to the scatter plot in

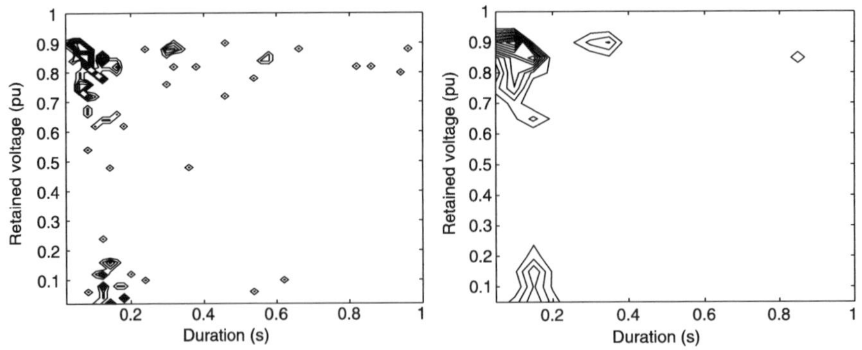


**Figure 10.10** Contour plot of density function for MV site with high resolution (left) and low resolution (right).

Figure 10.3. The resolution is so high that there is seldom more than one dip within a cell (of size 20 ms by 0.02 pu). The resolution used for the right-hand figure was 50 ms in duration and 0.05 pu in residual voltage. This plot lacks the detail to draw quantitative conclusions, although it indicates where the main concentrations of dips can be expected. The contour chart of the density function does not add much information compared to the scatter plot.

**10.2.4.1 Voltage-Dip Tables** The calculation of voltage-dip site indices and system indices can be seen as a way of compressing the information in Table 10.1. The choice of index, in fact of any compression method, is a compromise between limiting the resulting amount of data and maximizing the amount of information remaining. The method with the least compression, and thus with the largest amount of information remaining, is the *voltage-dip table.* Both the residual voltage and the duration range are split in a number of intervals. Each event belongs in one residual voltage interval and in one duration interval. For each interval in two-dimensional space the number of events is counted, resulting in a table of residual voltage against duration. The discussion on the use of these tables concerns mainly the choice of intervals. No consensus has been reached to date, although a recommendation is made in an IEC technical report (IEC 61000-2-8). A number of examples will be given below.

Voltage-dip tables for the sites discussed before are presented in Tables 10.2 through 10.5. The so-called UNIPEDE table has been used as proposed in [301].

The voltage-dip table for the MV site is given in Table 10.2. We see the majority of events in the upper row (referring to shallow events) and in the duration ranges between 20 ms and 1 s. Note that the values presented are averages over a six-year period, hence the noninteger numbers. No events have been recorded with a duration exceeding 20 s.

The voltage-dip table for the HV1 site is given in Table 10.3. Again the majority of events are in the 20- to 100-ms and 100- to 500-ms columns. This is obviously related to the typical range of fault-clearing time. The longest duration values occur for events in the 0 to 10% residual voltage range. These are often referred to as *short interruptions* and separate statistics are sometimes presented. Note, however, that short interruptions are also measured from the opening and closing of interrupting devices (as quantified in the reliability indices).

**TABLE 10.2 Voltage-Dip Table for MV Site**

| | 0–20 ms | 20–100 ms | 100–500 ms | 0.5–1 s | 1–3 s | 3–20 s | 20–60 s | 60–180 s |
|---|---|---|---|---|---|---|---|---|
| 85–90% | 0.2 | 4.8 | 2 | 0.6 | 0.2 | 0 | 0 | 0 |
| 70–85% | 0 | 3.5 | 2.83 | 1 | 0.33 | 0 | 0 | 0 |
| 40–70% | 0 | 0.67 | 1 | 0 | 0 | 0 | 0 | 0 |
| 10–40% | 0 | 0.17 | 1.5 | 0 | 0 | 0 | 0 | 0 |
| ≤10% | 0 | 0.33 | 2.17 | 0.33 | 0 | 0.17 | 0 | 0 |

**TABLE 10.3 Voltage-Dip Table for HV1 Site**

| | 0–20 ms | 20–100 ms | 100–500 ms | 0.5–1 s | 1–3 s | 3–20 s | 20–60 s | 60–180 s |
|---|---|---|---|---|---|---|---|---|
| 85–90% | 0.8 | 10.8 | 7.6 | 1 | 0.4 | 0 | 0.2 | 0 |
| 70–85% | 0 | 13 | 5.83 | 2.5 | 0 | 0 | 0 | 0 |
| 40–70% | 0 | 5.5 | 4.67 | 0.17 | 0 | 0 | 0 | 0 |
| 10–40% | 0 | 0.5 | 0.67 | 0 | 0 | 0 | 0 | 0 |
| ≤ 10% | 0 | 0.33 | 0 | 0 | 0 | 0.17 | 2 | 1.83 |

The voltage-dip table for the HV2 site is given in Table 10.4. The statistics are very similar to those for the HV1 site, with the exception of the short interruptions, which are absent in the HV2 site. This will be related to the operation of the system. Higher voltage levels are operated more meshed than lower voltage levels, thus experiencing less interruptions.

The voltage-dip table for the EHV site is given in Table 10.5. The voltage-dip frequency is clearly less than for the lower voltage levels. The fault frequency at this voltage level is very low, so that despite the high exposed area, the dip frequency is lower than at lower voltage levels. Another contributing factor is that dips propagate almost undamped from high to low voltage levels, whereas a fault at a lower voltage level causes at most a shallow dip at a higher voltage level. The result is that each voltage level experiences dips due to faults at that voltage level and at all higher levels. Naturally the dip frequency will increase when moving from high to low voltage levels. The relation between system parameters and voltage-dip frequency is discussed among others in [33, Chapter 6; 228].

**TABLE 10.4 Voltage-Dip Table for HV2 Site**

| | 0–20 ms | 20–100 ms | 100–500 ms | 0.5–1 s | 1–3 s | 3–20 s | 20–60 s | 60–180 s |
|---|---|---|---|---|---|---|---|---|
| 85–90% | 1.2 | 6.8 | 3.6 | 0 | 0 | 0 | 0 | 0 |
| 70–85% | 0.17 | 9.33 | 2.33 | 0.17 | 0 | 0 | 0 | 0 |
| 40–70% | 0 | 4.83 | 2.67 | 0 | 0 | 0 | 0 | 0 |
| 10–40% | 0 | 0.5 | 0.5 | 0 | 0 | 0 | 0 | 0 |
| ≤10% | 0 | 0 | 0 | 0 | 0 | 0 | 0 | 0 |

**TABLE 10.5 Voltage-Dip Table for EHV Site**

| | 0–20 ms | 20–100 ms | 100–500 ms | 0.5–1 s | 1–3 s | 3–20 s | 20–60 s | 60–180 s |
|---|---|---|---|---|---|---|---|---|
| 85–90% | 0.6 | 7.2 | 1 | 0 | 0 | 0 | 0 | 0 |
| 70–85% | 0 | 5 | 0.17 | 0 | 0 | 0 | 0 | 0 |
| 40–70% | 0 | 1.5 | 0.17 | 0 | 0 | 0 | 0 | 0 |
| 10–40% | 0 | 0.17 | 0 | 0 | 0 | 0 | 0 | 0 |
| ≤10% | 0 | 0 | 0 | 0 | 0 | 0 | 0.17 | 0 |

From Table 10.5 we might conclude an event frequency of 0.2 per year for short interruptions with a duration in the 20- to 60-s interval. One should be aware, however, that the value in this cell is due to one individual event. It is not possible to draw any conclusions from this. It may have been a very exceptional event, which is in fact confirmed by the fact that none of the neighboring cells show any event. One of the dangers of presenting voltage-dip statistics in this way is that the values are sometimes used to predict the future performance of a certain site. One should realize, however, that the average value is only an estimation of the expected value. Consider, for example, the five events per year with duration between 20 and 100 ms and residual voltage between 70 and 85%. This value is based on the measurement of $N = 30$ dips during a six-year period. But the number of events during any arbitrary six-year period is a stochastic variable as well, with an expected value and a standard deviation. The average value observed (30) is a so-called pure estimate of the expected value. If we further assume that the time between events is exponentially distributed (thus that the occurrence of dips is a Poisson process), the standard deviation of this stochastic variable can be estimated as well. A pure estimation of the standard deviation is $\sqrt{N}$ and the 95 percentile of this stochastic variable is (estimated again) between $N - 2\sqrt{N}$ and $N + 2\sqrt{N}$. With $N = 30$ this confidence interval is from 19 to 41. Thus, under the assumption that the occurrence of dips is a Poisson process, we can state with 95% confidence that the number of dips during the next six-year interval will be between 20 and 40. Thus, although we measured over a six-year period, the uncertainty is still a factor of 2. The uncertainty for the number of dips in one year is even higher. Using a Poisson distribution with expected value 5, we find (e.g., by using a table for the Poisson distribution) that the number of dips in any given year will be between 2 and 9, with 95% confidence. Those familiar with statistics will realize that a number of simplifications have been made in the calculation of the confidence interval. The main simplification is, however, the assumption that the occurrence of voltage dips is a Poisson process. Most measurements indicate that dips are clustered more than according to a Poisson process. The result is an even higher standard deviation and thus even less predictive value from the monitoring results, as we will see in Section 10.2.7.

In the voltage-dip tables presented above the residual voltage was expressed in percent, but it may also be expressed in volts or per-unit. The duration should be expressed in milliseconds, seconds, or minutes, with the exception of the first column, where the maximum duration may be expressed as one cycle. A distinction between short interruptions and other voltage dips is made already when defining the range of retained voltages. If short interruptions are treated differently from other voltage dips, the lower row of the table should start at 1 or 10%.

Measured voltage dips may have a duration or residual voltage that corresponds exactly with the border between two cells. These events could in theory be placed in any of these two cells. A recommendation for this is made in IEEE 493[161]: A voltage dip with characteristics on the border between two cells will be added to the cell with the most severe dips. Thus a dip with 500 ms duration will be added to the 500-ms to 1-s duration range; a dip with a residual voltage of 85% will be added to the 70 to 85% range.

We already mentioned the UNIPEDE table as a method for dividing residual voltage and duration ranges in suitable intervals. The intervals in residual voltage become smaller for more shallow dips. This compensates for the fact that shallow dips are more common than deep dips. The same occurs in the horizontal direction, where the interval length increases from 20 ms through 2 min. The examples above showed that many of the cells do not have any event count at all. There remains, however, a need for the longer duration intervals as this is where short interruptions due to fast auto-reclosing are found. The merging of the long-duration cells should still be handled with care. From the fact that there are rarely any (noninterruption) events longer than 1 s one could conclude that these cells can be merged. However, the rarety of the events does not reduce their importance. In the rare case that such an event occurs, it is important to know if it was in the range 1 to 3 s or in the range 1 to 3 min. The possible causes, the impact on equipment, and even the impact on the system will be completely different. Therefore it is recommended to keep the long-duration cells even if no events occur that will fill them.

The voltage-sag table proposed in IEC technical report 61000-2-8 is shown as Table 10.6. The main difference with the UNIPEDE table is in the higher resolution in the residual voltage ranges. Further, an additional duration range is added with 250 ms as a limit. Note that the lowest duration interval starts at one cycle. The result for this is that events with a duration of only one half-cycle are not included in the voltage-dip statistics. (When measuring according to IEC 61000-4-30 the duration of a voltage dip is an integer multiple of one half-cycle.)

Some utilities have developed approaches to defining specific areas on a magnitude/duration plane that attempt to provide generalized guidelines on areas where sags are likely to occur and areas that customers are likely to be affected by. The aim of these generalized areas is to reduce the number of indices that need to be reported and managed, based on the most "appropriate" grouping of sag events. The categorization method adopted by NRS-048-2:2003 (South Africa) is shown in Figure 10.11. The Y-type area reflects sags that are expected to occur frequently on typical HV and MV systems and against which customers should protect their plant. The X-type areas (X1 and X2) reflect "normal" HV protection clearance

**TABLE 10.6 Voltage Dip Table Recommended by IEC 61000-2-8**

| | 1 cycle–0.1 s | 0.1–0.25 s | 0.25–0.5 s | 0.5–1 s | 1–3 s | 3–20 s | 20–60 s | 1–3 min |
|---|---|---|---|---|---|---|---|---|
| 80–90% | | | | | | | | |
| 70–80% | | | | | | | | |
| 60–70% | | | | | | | | |
| 50–60% | | | | | | | | |
| 40–50% | | | | | | | | |
| 30–40% | | | | | | | | |
| 20–30% | | | | | | | | |
| 10–20% | | | | | | | | |
| <10% | | | | | | | | |

<table>
<tr><th></th><th>20-150 ms</th><th>150-600 ms</th><th>0.6-3 s</th></tr>
<tr><td>85-90%</td><td colspan="3">Y</td></tr>
<tr><td>80-85%</td><td colspan="2">Y</td><td rowspan="2">Z1</td></tr>
<tr><td>70-80%</td><td>Y</td><td rowspan="3">S</td></tr>
<tr><td>60-70%</td><td>X1</td><td rowspan="3">Z2</td></tr>
<tr><td>40-60%</td><td>X2</td></tr>
<tr><td>0-40%</td><td colspan="2">T</td></tr>
</table>

**Figure 10.11** Voltage-dip table as defined by NRS-048-2:2003.

**TABLE 10.7 Voltage-Dip Table Based on IEC 61000-4-11**

| | <1 cycle | 1 cycle − 200 ms | 0.2–0.5 s | 0.5–5 s | >5 s |
|---|---|---|---|---|---|
| 70–80% | | | | | |
| 40–70% | | | | | |
| 10–40% | | | | | |
| ≤10% | | | | | |

times and hence a significant number of events are expected to occur in this area. Customers with sensitive equipment should attempt to protect against at least X1-type sags, which are more frequent. The T-type area reflects close-up faults, which are not expected to happen too regularly and which a utility should specifically address if excessive. The S-type sags are not as common as X- and Y-type events but may occur where impedance protection schemes are used or where voltage recovery is delayed. The Z-type dips are very uncommon in HV systems (particularly Z2-type events), as this generally reflects problematic protection operation. These may be more common in MV systems [64, 187].

Another example of a voltage-dip table is shown as Table 10.7. The intervals in duration and residual voltage are based on IEC 61000-4-11 Ed. 2 [156]. This standard recommends the duration values 0.5 cycle, 1 cycle, 10/12 cycles, 25/30 cycles, and 250/300 cycles and the magnitude values 0%, 40%, 70%, and 80% for the testing of equipment against voltage dips and short interruptions. In the table 0% has been replaced by the more practical value of 10%. The use of these values would allow for an easy comparison between system performance and performance of equipment tested under IEC 61000-4-11.

### 10.2.5 SARFI Indices

An alternative approach to counting voltage dips is proposed in [50]. The method is recommended by the CIGRE/CIRED working group on voltage quality and one of the methods in the document by the CIGRE working group on power quality indices

**TABLE 10.8 Ranges of Voltage-Dip Residual Voltage and Duration Based on IEEE 1159**

| | Up to 60 s | Up to 0.5 s | 0.5–3 s | 3–60 s |
|---|---|---|---|---|
| 140% and above | $SARFI_{140}$ | $SIARFI_{140}$ | $SMARFI_{140}$ | $STARFI_{140}$ |
| 120% and above | $SARFI_{120}$ | $SIARFI_{120}$ | $SMARFI_{120}$ | $STARFI_{120}$ |
| 110% and above | $SARFI_{110}$ | $SIARFI_{110}$ | $SMARFI_{110}$ | $STARFI_{110}$ |
| Up to 90% | $SARFI_{90}$ | $SIARFI_{90}$ | $SMARFI_{90}$ | $STARFI_{90}$ |
| Up to 80% | $SARFI_{80}$ | $SIARFI_{80}$ | $SMARFI_{80}$ | $STARFI_{80}$ |
| Up to 70% | $SARFI_{70}$ | $SIARFI_{70}$ | $SMARFI_{70}$ | $STARFI_{70}$ |
| Up to 50% | $SARFI_{50}$ | $SIARFI_{50}$ | $SMARFI_{50}$ | $STARFI_{50}$ |
| Up to 10% | $SARFI_{10}$ | $SIARFI_{10}$ | $SMARFI_{10}$ | |

and objectives. The SARFI indices also play a prominent role in the current draft of IEEE 1564 [172].

The method was originally based on the duration values as used in IEEE 1159 [165] to distinguish between instantaneous, momentary, and temporary interruption: 0.5, 3, and 60 s. The abbreviation SARFI stands for System Average rms-variation Frequency Index and it should be seen as the voltage-dip equivalent of SAIFI as defined for interruptions. The term *rms variations* refers to voltage dips, swells, and short interruptions. Just like SAIFI considers all interruptions equal, independent of their duration, so are all voltage dips considered equal when determining SARFI. However, the residual voltage remains as a parameter in SARFI. For example, $SARFI_{80}$ is the number of voltage-dip events per year with a residual voltage of 80% or less. Although the abbreviation stands for "system average," the term SARFI is also used for site indices.

The original proposal contained three more indices, SIARFI, SMARFI, and STARFI, giving the frequency of instantaneous, momentary, and temporary rms variations, respectively. However, these indices are rarely used. An overview of the indices is shown in Table 10.8. Some selected SARFI indices have been calculated for the four sites presented before. The results are summarized in Table 10.9. The possibility to immediately compare sites is one of the reasons for the popularity of the method [277].

The SARFI indices in the second column of Table 10.8 are referred to as $SARFI_X$; they give the number of dips below a voltage threshold. In addition, $SARFI_{CURVE}$

**TABLE 10.9 SARFI Indices for Four Sites**

| | MV | HV1 | HV2 | EHV |
|---|---|---|---|---|
| $SARFI_{90}$ | 21.8 | 58.0 | 32.1 | 16.0 |
| $SARFI_{85}$ | 14.0 | 37.2 | 20.5 | 7.2 |
| $SARFI_{70}$ | 6.3 | 15.8 | 8.5 | 2.0 |
| $SARFI_{40}$ | 4.7 | 5.5 | 1.0 | 0.3 |
| $SARFI_{10}$ | 3.0 | 4.3 | 0.0 | 0.2 |

indices have been introduced that give the number of dips below a voltage tolerance curve. For example, $SARFI_{SEMI}$ gives the number of dips per year below the curve defined in the SEMI F47 standard. Also $SARFI_{ITIC}$ and $SARFI_{CBEMA}$ are commonly used [172, 244].

### 10.2.6 Single-Index Methods

The previous sections have presented the calculation of site and system indices based on residual voltage and duration. Some of the problems discussed are due to the fact that one event is characterized by two numbers. This makes it impossible to give a ranking of the severity of the events. A number of methods have been proposed in the literature to characterize a single event with only one characteristic.

Two single-index methods are currently dominating the discussion: the voltage-dip energy and the voltage-dip severity. The single-event indices for both methods were defined in Section 8.6. Having only one dimension it is straightforward to give a probability distribution of the index. We will give some examples below. The site indices calculated are as follows:

- The sum of the index values for all events recorded during a given period (typically one year)
- The total number of events recorded during that period
- The average index value for the events recorded during that period

Especially for the second and third indices it is important that the threshold value be defined. A reduction in threshold (e.g., from 90 to 80% of nominal) will lead to a large reduction in the number of events and a large change in the average index value.

***10.2.6.1 Voltage-Dip Energy Indices*** The dip energy method of characterization uses three site indices: number of events per site, total lost energy per site, and average lost energy per event [172]. The *dip energy index* (SEI, for *sag energy index*) equals the sum of the voltage dip energy values for all qualified events recorded at a given site during a given period, typically one year:

$$\mathrm{SEI} = \sum_{i=1}^{N} E_{VS-i} \tag{10.14}$$

where $i$ is the sag event number and $N$ is the number of qualified events during the given period at the given site. The dip energy index, when expressed in units of time, can be interpreted as the length of the equivalent interruption with the same lost energy as all dips together that occurred during the observation period.

The *average dip energy index* (ASEI, for *average sag energy index*) is the average of the voltage sag energies for all events measured at a given site during a given period:

$$\text{ASEI} = \frac{1}{N}\sum_{i=1}^{N} E_{VS-i} \tag{10.15}$$

The average dip energy is dependent on the triggering of the monitor. A sensitive setting will result in a large number of shallow events (with a low dip energy) and this in a lower value for the average. The dip energy index, on the other hand, will increase for sensitive setting of the monitor.

The number of events $N$ recorded at the site during the observation period is the third site index. The SARFI-90 index can be used for this when a 90% threshold value is used. Note that only two of the three indices are needed as they are related according to

$$\text{SEI} = \text{ASEI} \times \text{SARFI}_{90} \tag{10.16}$$

The voltage-dip energy for all events in Table 10.1 have been calculated. The voltage-dip energy varies over a wide range of values, from 4 ms to almost 15 s. This is a disadvantage when calculating site indices as one event may dominate the index for a whole year. To overcome this problem, some authors propose to use a maximum value for the voltage-sag energy per event. A more appropriate alternative is to define classes of events and to calculate separate indices for each class. A simple option is to exclude all interruptions from the statistics. Interruptions may be quantified as a separate class of voltage dips or through the reliability indices. The voltage-dip energy has been calculated for the same four sites as in the earlier examples. The probability distribution function for the voltage-dip energy is shown for each site in Figure 10.12.

The site indices for the four sites are shown in Table 10.10. The number of dips varies by a factor of 4 between the sites, but the voltage-dip-energy index (the sum

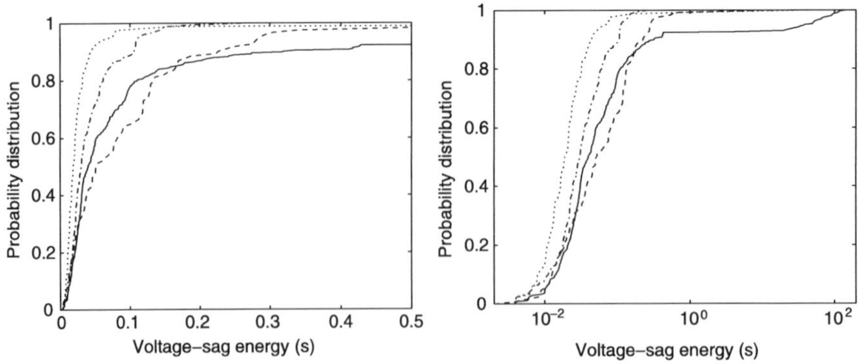


**Figure 10.12** Probability distribution of voltage-sag energy values at four different sites: MV (dashed curve), HV1 (solid), HV2 (dash-dot), and HV2 (dotted); linear scale (left) and logarithmic scale (right).

**TABLE 10.10 Voltage-Dip Energy Indices for Four Sites**

| Site | Number of Events | Event Index | | | | No Interruptions | | |
|---|---|---|---|---|---|---|---|---|
| | | Sum | Average | Median | Maximum | Number | Sum | Average |
| MV | 124 | 26.1 | 0.21 | 0.052 | 14.94 | 106 | 7.89 | 0.07 |
| HV1 | 327 | 1595 | 4.9 | 0.040 | 143 | 305 | 25.7 | 0.08 |
| HV2 | 181 | 7.74 | 0.043 | 0.030 | 0.26 | 181 | 7.74 | 0.043 |
| EHV | 87 | 43.0 | 0.49 | 0.019 | 40.9 | 86 | 2.06 | 0.024 |

of the single-event values) varies by a factor of 200, and the average value by a factor of 100. Note, however, that the median value is similar for the four sites, with only a factor of 3 between the lowest and the highest values. The high voltage-dip energy index for the MV site is due to a number of short interruptions; see, for example, Table 10.3. These short interruptions are also visible as the tail of the probability distribution on the right of the logarithmic plot in Figure 10.12. Note that the voltage-dip energy index for the EHV site is higher than for the MV and HV2 site, despite the lower number of events. This is due to only one event, with a voltage-dip energy equal to 40.9. This one event stands for 47% of the site index. Also the value for the MV site is highly inflated, with one value standing for 57% of the site index. The last three columns of Table 10.10 have been obtained by removing all events from the statistics with a residual voltage less than 10%. The HV2 site is not affected, as no short interruptions were recorded at this site. The other three sites show a significant reduction in voltage-dip energy index. Note that the resulting average values are of the same order of magnitude again.

***10.2.6.2 Voltage-Dip Severity Indices*** The calculation of site indices for the voltage-dip severity method is very similar to the calculation of site indices based on the voltage-dip energy. Three site indices are introduced to characterize the site performance [172]:

- Total voltage-dip severity: $S_{\text{site}} = \sum_{i=1}^{N} S_{e-i}$
- Average voltage-dip severity: $S_{\text{average}} = S_{\text{site}}/N$
- Number of events for site: $N$

When a 90% threshold setting is used, $N$ is again equal to $\text{SARFI}_{90}$ and the same relation between the indices holds as for the voltage-sag energy method:

$$S_{\text{site}} = S_{\text{average}} \times \text{SARFI}_{90} \tag{10.17}$$

The probability distribution functions for the four sites at four different voltage levels are shown in Figure 10.13. Most of the properties visible for the voltage-dip energy index can also be found here. The site indices are shown in Table 10.11.

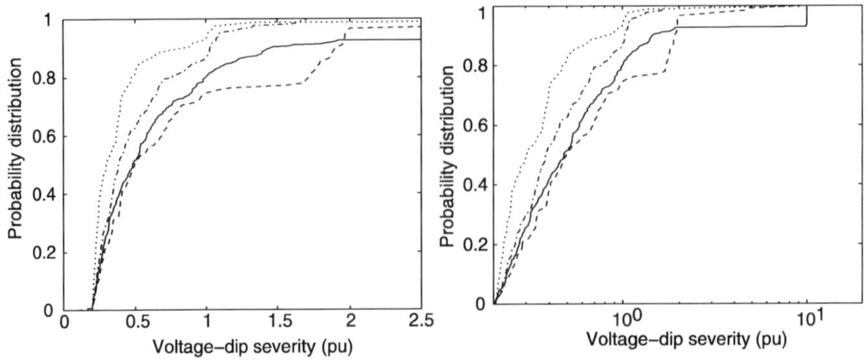


**Figure 10.13** Probability distribution of voltage-dip severity values at four different sites: MV (dashed curve), HV1 (solid), HV2 (dash-dot), and HV2 (dotted); linear scale (left) and logarithmic scale (right).

**TABLE 10.11 Voltage-Dip Severity Indices for Four Sites**

| | | Event Index | | | |
|---|---|---|---|---|---|
| Site | Number of Events | Sum | Average | Median | Maximum |
| MV | 124 | 113.2 | 0.91 | 0.493 | 10 |
| HV1 | 327 | 412.0 | 1.26 | 0.490 | 10 |
| HV2 | 181 | 94.0 | 0.52 | 0.384 | 1.70 |
| EHV | 87 | 43.6 | 0.50 | 0.303 | 10 |

A comparison with the voltage-dip energy is natural here as both methods aim to achieve the same thing: characterize the performance of the supply with one value, being the sum of the severities of each individual event. Comparing Figure 10.12 with Figure 10.13 we see that the voltage-dip energy values are mainly concentrated in the range between 0 and 0.1 s, with a long tail up to 100 s. The voltage-dip severity values show much less of a spread: The mean concentration is between 0.2 and 1 pu with a tail up to 10 pu. The high number of values at 10 pu. for the HV1 site are interruptions lasting more than 10 s. Comparing Table 10.10 with Table 10.11 we see that the heavy dominance of one event in the site indices, as observed for the voltage-dip energy index, is no longer present with the voltage-dip severity index. This is a direct consequence of the smaller range in value for the voltage-dip severity. For the EHV site we still see that one event (an interruption lasting more than 10 s) accounts for 23% of the site index. When removing interruptions from the statistics, a much smaller spread of values was obtained for the voltage-dip energy. The same may be done for the voltage-dip severity.

### 10.2.7 Year-to-Year Variations

***10.2.7.1 Poisson Process*** The simplest statistical model of the occurrence of voltage dips is the *Poisson process*. The underlying assumption of the

Poisson process is that the occurrence of an event is independent of the occurrence of the previous event. If the expected number of events per year is 365, then the expected number of events is 1 for every day of the year, $\frac{1}{24}$ for every hour of the year, and so on. The mathematics for the Poisson process are straightforward and commonly used, so it is worth repeating them here and using them as a reference, even though we will see later that the underlying assumptions do not hold for voltage dips.

Assume that the expected number of dips per year is equal to $\mu$. For a very long measuring period we would record on average $\mu$ events per year. The number of events recorded in any individual year will show a Poisson distribution [200]. The probability of observing $n$ events in a given year equals

$$p_n = \frac{\mu^n}{n!} e^{-\mu} \tag{10.18}$$

The resulting probability density and distribution functions are plotted in Figure 10.14 for two values of the expected value: $\mu = 4.5$ and $\mu = 15$. We see that the probability that the number of events in a given year is equal to the expected value is very low. For an expected value of 4.5, the probability that 4 events occur in a given year is only 19%; the probability of 5 events is 17%. For an expected value of 15, the probability of 15 events is only slightly more than 10%, but the probability that there are 20 or more events is 12%.

To indicate the uncertainty in the results, the *confidence interval* is used. The 95% confidence interval gives the range of values in which a stochastic variable is found in 95% of the cases. The 95% confidence interval is between 1 and 10 for an expected value of 4.5 and between 9 and 23 for an expected value of 15. In relative values, the uncertainty is about 100% of the expected value in the first case and about 50% of the expected value in the second case. We will come back to the consequences of this later; first we will verify if the measurement statistics agree with a Poisson distribution.

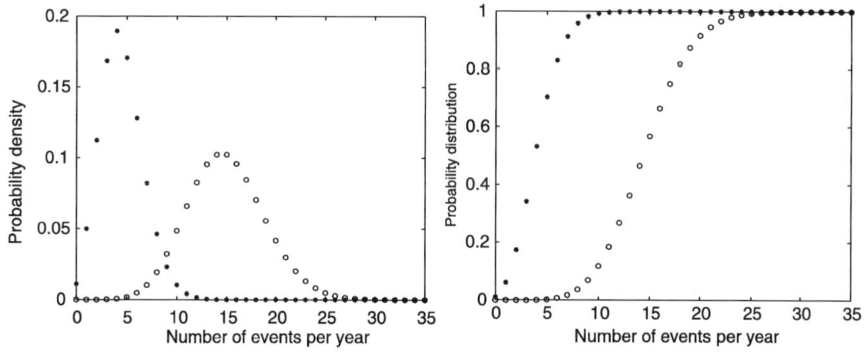


**Figure 10.14** Probability density function (left) and probability distribution function (right) for Poisson distributions with expected value 4.5 (asterisks) and 15 (circles).

***10.2.7.2 Comparing Statistics with Poisson Process*** There are a number of reasons why the occurrence of voltage dips is not a Poisson process. The first reason is that voltage dips show clustering at a time scale of several seconds. Some voltage dips come *in* groups, as we will see later when we discuss time aggregation. We will also see later that the probability that a voltage dip occurs shows a seasonal variation. It is thus to be expected that the number of dips is not distributed in a Poisson distribution.

To verify if the occurrence of voltage dips behaves as a Poisson process, the number of recorded voltage dips for a number of sites over a six-year period has been used to calculate the relation between the mean and the standard deviation of the yearly values. For example, for the MV site used in the earlier examples, the dip frequency was 9, 12, 14, 11, 16, and 22 events per year for the six years during which measurements were available. The dip frequency is defined here as the number of events with a residual voltage of 85% or less ($SARFI_{85}$).

For this site the average dip frequency over the six-year period was 14.0 events per year, whereas the standard deviation was 4.6 events per year. For a Poisson distribution the standard deviation is the square root of the mean value: $\sigma = \sqrt{\mu}$. In other words, for a Poisson distribution we expect that $\sigma/\sqrt{\mu} = 1$. This ratio equals $4.6/\sqrt{14.0} = 1.23$ for this site. The standard deviation if thus higher than would be expected for a Poisson process. These calculations have been repeated for a number of sites at four voltage levels. The results are shown in Table 10.12. The first column gives the voltage level; columns 2 through 7 give the recorded dip frequency for six consecutive years. The last three columns give the average dip frequency $\mu$ over the six-year period; the standard deviation of the six yearly values, $\sigma$; and the ratio $\sigma/\sqrt{\mu}$.

Only 5 of the 42 sites have a standard deviation less than that according to the Poisson distribution. We may thus state that the spread of the dip frequency between consecutive years is more than would be expected for a Poisson process. The average value of the last column is 1.6; the standard deviation is on average 1.6 times the value for a Poisson process.

The average and standard deviation of the dip frequency are presented in a graphical way in Figure 10.15. Each circle or square represents one of the 42 sites in Table 10.12. The dotted line is the theoretical relation for a Poisson process: $\sigma = \sqrt{\mu}$. The least-squares line for all the sites is represented by the solid straight line.

There appears to be some clustering of points around the Poisson curve and along two lines. There is, however, not enough statistical data available to further study this. The interpretation of the least-squares approximation is also not possible at this stage. Further study of the underlying statistical process is needed before semi-empirical relations can be obtained.

***10.2.7.3 Required Monitor Duration*** In [33, Section 6.3.2] an estimation is made of the length of time needed before a given accuracy is reached with voltage-dip monitoring. The estimation was based on the assumption that the occurrence of voltage dips is a Poisson process and thus that the standard deviation is the square

**TABLE 10.12 Year-to-Year Variation of Dip Frequency at Selected Sites**

| Voltage Level | 1996 | 1997 | 1998 | 1999 | 2000 | 2001 | Mean | Standard Deviation | $\sigma/\sqrt{\mu}$ |
|---|---|---|---|---|---|---|---|---|---|
| MV | 10 | 15 | 13 | 2 | 0 | 0 | 6.7 | 6.8 | 2.6 |
| | 11 | 11 | 6 | 5 | 4 | 9 | 7.7 | 3.1 | 1.1 |
| | 9 | 12 | 14 | 11 | 16 | 22 | 14.0 | 4.6 | 1.2 |
| | 16 | 15 | 11 | 15 | 14 | 22 | 15.5 | 3.6 | 0.9 |
| | 15 | 16 | 9 | 26 | 13 | 22 | 16.8 | 6.2 | 1.5 |
| | 8 | 3 | 6 | 3 | 12 | 4 | 6.0 | 3.5 | 1.4 |
| HV1 | 31 | 47 | 30 | 35 | 46 | 34 | 37.2 | 7.5 | 1.2 |
| | 5 | 19 | 10 | 14 | 8 | 13 | 11.5 | 4.9 | 1.5 |
| | 8 | 20 | 11 | 15 | 8 | 13 | 12.5 | 4.6 | 1.3 |
| | 10 | 19 | 14 | 22 | 28 | 15 | 18.0 | 6.4 | 1.5 |
| | 12 | 20 | 9 | 21 | 28 | 18 | 18.0 | 6.8 | 1.6 |
| | 12 | 16 | 13 | 20 | 16 | 12 | 14.8 | 3.1 | 0.8 |
| | 24 | 54 | 11 | 22 | 21 | 27 | 26.5 | 14.5 | 2.8 |
| | 9 | 20 | 10 | 19 | 11 | 6 | 12.5 | 5.7 | 1.6 |
| | 11 | 11 | 14 | 14 | 21 | 12 | 13.8 | 3.8 | 1.0 |
| | 31 | 60 | 37 | 65 | 91 | 61 | 57.5 | 21.5 | 2.8 |
| | 27 | 59 | 41 | 45 | 88 | 58 | 53.0 | 20.8 | 2.9 |
| | 11 | 17 | 23 | 28 | 28 | 35 | 23.7 | 8.7 | 1.8 |
| | 9 | 16 | 7 | 14 | 15 | 7 | 11.3 | 4.1 | 1.2 |
| | 15 | 15 | 11 | 14 | 14 | 11 | 13.3 | 1.9 | 0.5 |
| | 9 | 8 | 7 | 2 | 17 | 11 | 9.0 | 4.9 | 1.7 |
| HV2 | 21 | 27 | 8 | 21 | 30 | 16 | 20.5 | 7.8 | 1.7 |
| | 8 | 17 | 6 | 11 | 5 | 8 | 9.2 | 4.4 | 1.4 |
| | 13 | 10 | 3 | 7 | 6 | 4 | 7.2 | 3.8 | 1.4 |
| | 0 | 1 | 0 | 1 | 0 | 1 | 0.5 | 0.6 | 0.8 |
| | 49 | 10 | 5 | 10 | 8 | 12 | 15.7 | 16.5 | 4.2 |
| | 2 | 0 | 16 | 10 | 9 | 15 | 8.7 | 6.6 | 2.2 |
| | 0 | 6 | 10 | 15 | 10 | 6 | 7.8 | 5.1 | 1.8 |
| | 6 | 14 | 6 | 16 | 5 | 5 | 8.7 | 5.0 | 1.7 |
| | 6 | 17 | 0 | 10 | 10 | 3 | 7.7 | 6.0 | 2.2 |
| | 7 | 10 | 7 | 10 | 10 | 2 | 7.7 | 3.1 | 1.1 |
| | 18 | 14 | 4 | 7 | 10 | 5 | 9.7 | 5.5 | 1.8 |
| | 8 | 13 | 7 | 11 | 16 | 8 | 10.5 | 3.5 | 1.1 |
| | 8 | 17 | 14 | 5 | 11 | 11 | 11.0 | 4.2 | 1.3 |
| | 6 | 15 | 11 | 7 | 16 | 5 | 10.0 | 4.7 | 1.5 |
| | 5 | 12 | 8 | 8 | 30 | 2 | 10.8 | 10.0 | 3.0 |
| EHV | 5 | 8 | 5 | 3 | 4 | 6 | 5.2 | 1.7 | 0.8 |
| | 9 | 10 | 5 | 11 | 6 | 2 | 7.2 | 3.4 | 1.3 |
| | 8 | 7 | 0 | 7 | 6 | 2 | 5.0 | 3.2 | 1.4 |
| | 4 | 8 | 2 | 3 | 7 | 5 | 4.8 | 2.3 | 1.1 |

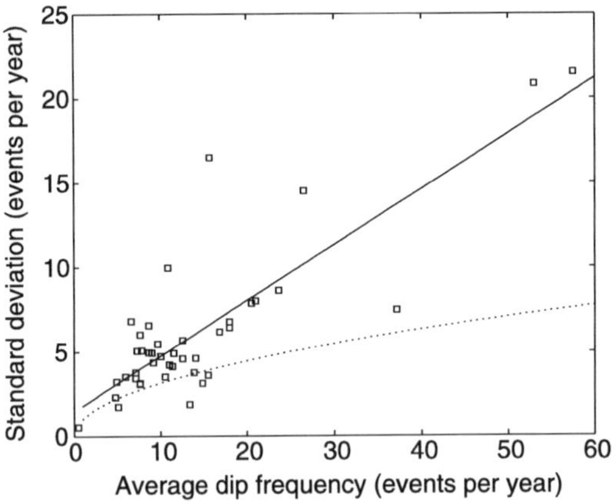


**Figure 10.15** Average and standard deviation of dip frequency over six-year period for 42 sites. The solid line gives the least-squares approximation, the dotted line the relation for a Poisson process.

root of the number of events. From Table 10.12 we find the following empirical relation between the standard deviation $\sigma$ and the number of events $N$:

$$\sigma = 1.6 \times \sqrt{N} \tag{10.19}$$

This results in the following expression for the 95% confidence interval:

$$\delta_{95} = 1.96 \times 1.6 \times \sqrt{N} \tag{10.20}$$

where it has been assumed that the confidence interval is the same as for a normal distribution with the same standard deviation, which is a reasonable assumption. Finally we obtain for the relative error in the estimated expected value

$$\varepsilon = \frac{3.1}{\sqrt{N}} \tag{10.21}$$

To obtain an error less than $\varepsilon_{max}$, at least $N_{min}$ events have to be recorded, with

$$N_{min} = \frac{9.8}{N^2} \tag{10.22}$$

To obtain an accuracy of 10%, about 1000 events need to be recorded. This corresponds with a monitoring period of about 80 years when the dip frequency is one event per year. (The corresponding value for a Poisson distribution is 30 years.) The table in [33, Section 6.3.2] has been reproduced with the standard deviation estimated from Table 10.12, that is, according to (10.19). The results are shown in Table 10.13.

When interpreting this table it is very important to realize that the relation between the average and standard deviation has been based on measurements in four systems only. Different systems may result in a different relation. Also all sites within one system may not obey the same rules. The standard deviation may

**TABLE 10.13 Minimum Monitoring Period Needed to Obtain Given Accuracy**

| Event Frequency | 50% Accuracy | 10% Accuracy | 2% Accuracy |
|---|---|---|---|
| 1 per day | 1 month | 2.5 years | 65 years |
| 1 per week | 9 months | 20 years | 450 years |
| 1 per month | 3 years | 80 years | 2000 years |
| 1 per year | 40 years | 1000 years | 24,000 years |

be larger for some sites than for others, for example, sites with mainly overhead lines versus sites with mainly underground cables. Fortunately it is always possible to calculate the standard deviation from a set of measurements, as we have done here, and thus estimate the uncertainty in the estimation for the expected number of events. Consider the HV1 site with an average dip frequency of 26.5 events per year and a standard deviation of 14.5 events per year. The standard deviation in the average value can be estimated as the standard deviation in the yearly values divided by the number of years of monitoring:

$$\sigma_{\mu} = \frac{\sigma}{\sqrt{N}} \tag{10.23}$$

In this case we find: $\sigma_{\mu} = 14.5/\sqrt{6} = 5.9$. The 95% confidence interval is approximately $(\mu - 2\sigma_{\mu},\ \mu + 2\sigma_{\mu})$, in this case (15,38). Thus the dip frequency per year is with 95% confidence between 15 and 38, that is, an uncertainty of a factor of 2. Even this value may be optimistic as we have made a number of implicit assumptions, one of them being that these six years are representative for the stochastic process. A once-in-ten-year storm with a large number of dips may not have been in the statistics yet.

In many cases the conclusion will be that the uncertainty is so large that no accurate prediction can be drawn about the expected event frequency in the future unless monitoring results over many years are available or the dip frequency is very high. The alternative, stochastic prediction methods, is outside the scope of this book but should be seriously considered when accurate predictions for dip frequency are needed.

### 10.2.8 Comparison Between Phase–Ground and Phase–Phase Measurements

From the discussion of unbalanced dips in Section 6.2.3 it has become clear that phase-to-phase and phase-to-ground connected monitors will measure different phase voltages. This was after all one of the reasons for introducing the classification of unbalanced dips. In this section we will present the results for two MV sites at which monitors were connected both phase to phase and phase to ground. This allows us to make a comparison of the effect of the monitor connection on the recorded dip frequency. In all of the figures in this section we have combined the measurements from the two monitor locations and treated them as if they were one site. Even though this is not correct for describing site performance, all the conclusions concerning monitor connection remain valid.

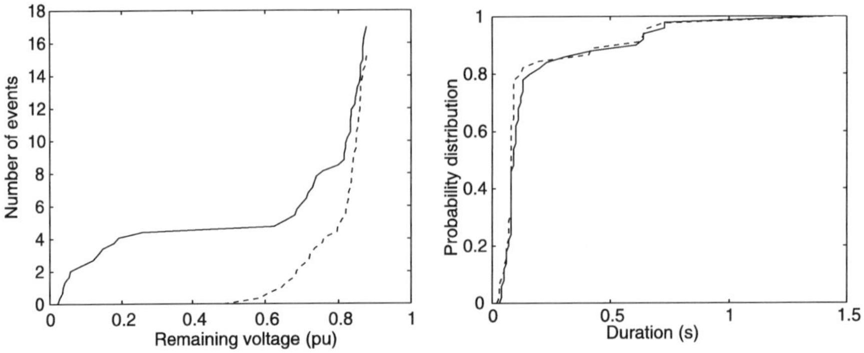


**Figure 10.16** Dip frequency (events per year) as function of residual voltage (left) and as function of duration (right) for phase-to-phase connected monitor (dashed line) and phase-to-ground connected monitor (solid line) at same two sites.

Figure 10.16 compares the number of events as a function of the residual voltage and as a function of the duration. The number of deep events is rather large for the phase-to-ground connected monitors, but for the phase-to-phase connected monitors there are no events with a residual voltage less than 0.45 pu. On the other hand, the duration is not significantly affected by the monitor connection.

The scatter plots for the two monitor connections are compared in Figure 10.17. This figure again shows that the phase-to-phase connected monitors did not record any deep events.

Comparing individual events showed that a total of 58 events was recorded during a three-year period at the two sites. Of these, 38 events were recorded by both monitors, 7 by the phase-to-phase connected monitor only, and 13 by the phase-to-ground connected monitor only. The latter all show the characteristic signature of single-phase-to-ground faults in an impedance-grounded system, with a deep dip in one phase and a significant overvoltage in the other two phases. The

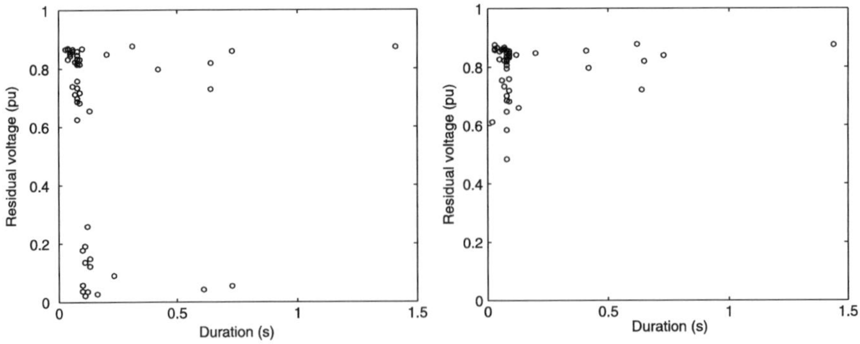


**Figure 10.17** Scatter plots for phase-to-ground connected monitors (left) and phase-to-phase connected monitors (right).

residual voltage of the three-phase event is, in accordance with IEC 61000-4-30, defined as the lowest of the three voltages in the individual phases. This gives a low value for the phase-to-ground connected monitor, but the phase-to-phase connected monitor does not even experience a drop below 90% in any of the phases.

The seven events recorded by the phase-to-phase connected monitor only all showed a drop slightly below the dip threshold of 90% in only one phase. These were most likely type D dips (drop in one phase) for the phase-to-phase voltages. The corresponding type C dip (drop in two phases) for the phase-to-ground voltages did not result in any of the three voltages being below the dip threshold.

The residual voltage of individual events (i.e., the lowest of the three phase voltages) is compared for the two monitor connections in Figure 10.18. The figure also contains three theoretical relations, indicated by dashed lines. For type A dips (due to three-phase faults), both monitors will observe about the same residual voltage:

$$V_{PP}^{(A)} = V_{PN}^{(A)} \tag{10.24}$$

A type C dip (drop in two phases) observed by the phase-to-ground connected monitor will be a type D dip (drop in one phase) observed by the phase-to-phase connected monitor, and the other way around. The residual voltages for a type D and a type C dip are related as follows [33, Section 4.6] when the characteristic phase-angle jump is assumed to be zero:

$$V_C^2 = \tfrac{1}{4} + \tfrac{3}{4} V_D^2 \tag{10.25}$$

For a type D event in phase-to-ground voltages, we get the following relation between the residual voltages for the two monitors:

$$V_{PP} = \sqrt{\tfrac{1}{4} + \tfrac{3}{4} V_{PN}^2} \tag{10.26}$$

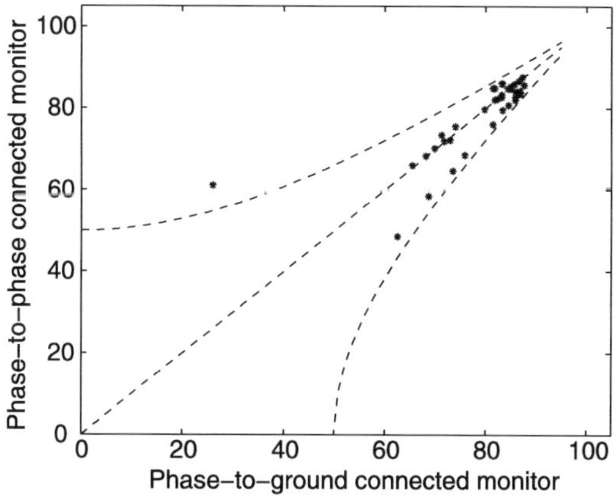


**Figure 10.18** Comparison of residual voltage for individual events, phase-to-ground or phase-to-phase connected monitors, including theoretical relations.

For a type C event in the phase-to-ground voltages, the relation reads as follows:

$$V_{PP} = \sqrt{\tfrac{4}{3}V_{PN}^2 - \tfrac{1}{3}} \tag{10.27}$$

These expressions are plotted as dashed lines in Figure 10.18. The upper dashed curve represents type D events in phase-to-ground voltages (drop in one phase-to-ground voltage): The phase-to-phase connected monitor observes a higher residual voltage. The diagonal represents balanced dips, which are observed in the same way for the two monitors. The lower curve represents type C events in phase-to-ground voltages (drop in one phase-to-phase voltage): The phase-to-phase connected monitor observes a lower residual voltage. The measurements fit reasonably well with the theoretical results. The differences are most likely due to the approximations made in the theoretical model: The characteristic phase-angle jump, assumed to be zero, and the effect of load on the voltages during a fault. Measurements and simulations have shown that the load effects cause the dip to become more like a balanced dip. This explains the fact that the measurement results are all toward the left of the theoretical curve, that is, closer to the theoretical curve for a balanced dip (the diagonal of the plot). A more detailed study of the influence of the monitor connection on the residual voltage is presented in [197].

The main conclusion is that the monitor connection seriously affects the observed number of dips and the residual voltage of individual dips. The CIGRE working group on power quality indices and objectives [64] recommends that measurements at MV and higher voltage levels be performed by means of phase-to-phase connected monitors. Alternatively the phase-to-phase voltages may be calculated from the phase-to-ground waveforms or from the phase-to-ground rms voltage. In the latter case an algorithm similar to the one proposed in [38] should be used. The use of phase-to-ground connected monitors does provide additional information to the network operator, for example, on local single-phase faults, that would be lost when the monitors were connected phase to phase. However, for quantifying the supply performance, the phase-to-phase voltages are more relevant.

Note, however, that the phase-to-phase voltages do not in all cases give a correct representation of the voltages as experienced by the end-user equipment. This depends on the winding connections of transformers between the monitoring location and the end-user equipment as well as on the connection of the end-user equipment. The classification method for three-phase unbalanced dips as described in Section 6.2.3 does not suffer from this problem.

An interesting conclusion for these specific sites is that there do not appear to be many type D events in phase-to-ground voltages. This will be related to the structure of the power system at higher voltage levels, including the transformer winding connections and the failure rate at individual voltage levels. Note that this conclusion does not hold generally; it is a specific conclusion for this location.

## 10.3 VOLTAGE DIPS: TIME AGGREGATION

### 10.3.1 Need for Time Aggregation

Even though obtaining site indices may be described as "just a matter of counting," reality is not as simple is it may sound. A very important question is: What is one event? Consider, for example, the rms voltage versus time shown in Figure 10.19. The dip threshold is indicated by a dashed line. The rms voltage "dips" below the threshold three times. According to the definition of single-event indices as given in the previous chapter, this would result in three events with residual voltages $V_1$, $V_2$, $V_3$ and durations $T_1$, $T_2$, $T_3$.

However, there is some kind of general understanding that this gives an overly pessimistic picture of the actual voltage quality. The interest in voltage dips is obviously due to its effect on the operation of equipment connected to this site. Any site index should have at least some kind of relation to the performance of equipment. If a piece of equipment will trip on the first dip in Figure 10.19, it will still be out of operation the moment the second and first dips occur. Even if the equipment itself will be back in operation within a second (e.g., a computer), it will take a while before the user will have come back to using it. If the equipment is part of a production process, it will take several minutes up to hours before the process can start again. The second and third dips will not have any effect in that case.

On the other hand, if the first dip does not affect the equipment, the second one will probably also not affect the equipment. Thus these three events will have the same effect for the customer as a single event. Therefore the three events can be counted as one. The process of merging such a "multiple event" into one event for statistical purposes is called *time aggregation.*

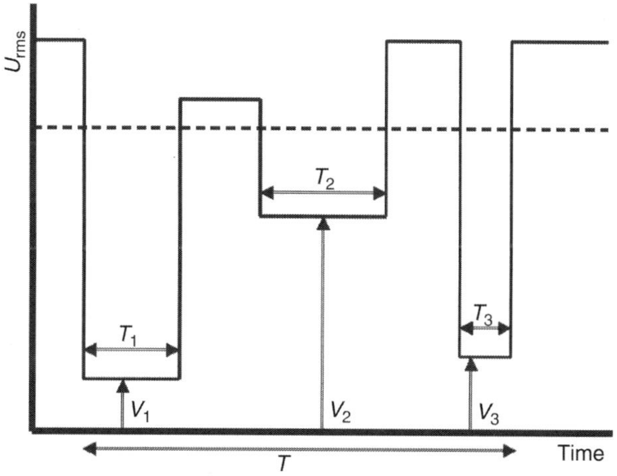


**Figure 10.19** Hypothetical example of voltage dip consisting of different stages. The dashed line indicates the dip threshold.

Another argument for time aggregation is that multiple voltage-dip events often have the same underlying cause in the power system. For example, an unsuccessful autoreclosure will lead to two dips separated in time by the autoreclosing interval. Even a successful autoreclosure may give a second dip when it involves a distribution feeder. The autoreclosure may lead to saturation of the distribution transformers leading to a long, shallow dip.

There are also arguments against time aggregation. The first is that the combined effect of two events may be more than the effect of each event individually. In other words, the equipment may survive the first event but still be affected somewhat. When the second one occurs, it may not have recovered fully from the first event. The combination of two events would have to be considered as an event somewhat more severe than the first event. Another phenomenon that speaks against time aggregation is that the consequences of the second event may aggravate those of the first event. The first event may lead to tripping of a piece of equipment (e.g., a computer or a drive). While this piece of equipment is restarting, a second dip occurs. There are indications that the second event may actually lead to damage to the equipment, whereas the first one only led to malfunction.

When we decide to use time aggregation for multiple events, as in Figure 10.19, two questions come up immediately. The first one concerns the choice of the time aggregation window. Currently this is still a point of discussion. The choice of window is based, among others, on the reasoning leading to the choice for time aggregation. If time aggregation is used to remove multiple events due to autoreclosure from the statistics, a time aggregation window up to a few seconds would be sufficient. However, if time aggregation is used to present a more accurate picture of the effect on equipment, a window of several minutes up to hours would be acceptable.

The second question concerns the indices for the aggregated event. With reference to Figure 10.19 several options are possible:

- Duration $T$, residual voltage $V_1$
- Duration $T_1 + T_2 + T_3$, residual voltage $\min(V_1, V_2, V_3)$
- Duration and residual voltage of the first event
- Duration $\max(T_1, T_2, T_3)$, residual voltage $\min(V_1, V_2, V_3)$

The last option is the one most commonly used: The residual voltage is the lowest value and the duration the highest value for the individual events. This is based on the expected effect of the event on equipment. Ideally one would like to choose the indices for the most severe event. But as there are two indices involved, it is not always possible to decide which event is more severe. Therefore the worst case is used for both residual voltage and duration. This could lead to cases where the aggregated event is much more severe than any of the individual events. This occurs when a short, deep event (e.g., due to a nearby fault) is followed by a long, shallow event (e.g., due to transformer saturation associated with the autoreclosure). Neither of the individual events would be of concern for most equipment, but the aggregated event would be classified as a long, deep dip. If these events occur often, it could seriously disrupt the statistics. It is not known to the authors how common such sites are in practice. It is in any case a good idea to calculate site indices with and without time aggregation to allow for the detection of such anomalies.

### 10.3.2 Time Between Events

The effect of time aggregation on the site and system indices is strongly related to the distribution of the time between events. The more events that occur close to each other in time, the more the effect of time aggregation on the indices will be.

The time between events is defined here as the time elapsed between the end of an event and the start of the next event. If this difference is less than the time aggregation window, the two events will be treated as one event. Residual voltage and duration are obtained for the aggregated event. We will discuss the aggregated-event indices later, for now we will only discuss the time between events. The distribution of the time between events for four sites is given in Figure 10.20. The percentage of events close together in time is clearly different for different voltage levels. The vertical lines indicate values of (from left to right) 1 s, 1 min, and 1 h. For MV and EHV there is a step in the distribution function for around 1 s. This is probably due to automatic reclosure, a phenomenon often mentioned as being responsible for multiple events. However, for none of the voltage levels do we observe more than a few percent of the events with a time between events around 1 s. For HV1 and HV2 the distribution of the time between events is even a very smooth curve.

The same curves are repeated in Figure 10.21 in a Weibull plot. The Weibull plot is commonly used to determine the parameters of probability distributions.

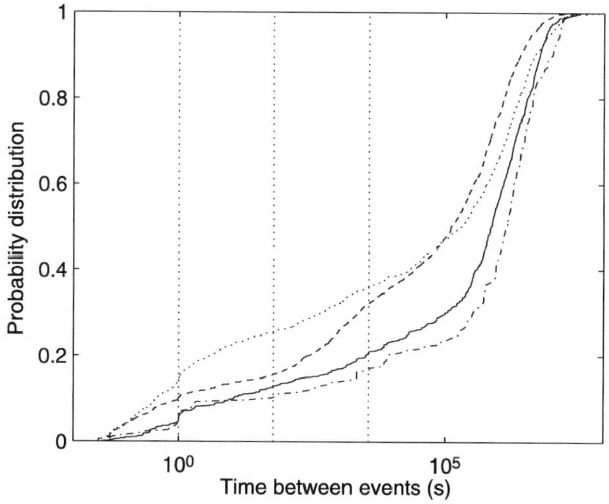


**Figure 10.20** Probability distribution function for time between events recorded at four sites: EHV (dash-dot curve); HV2 (dotted curve); HV1 (dashed curve); MV (solid curve).

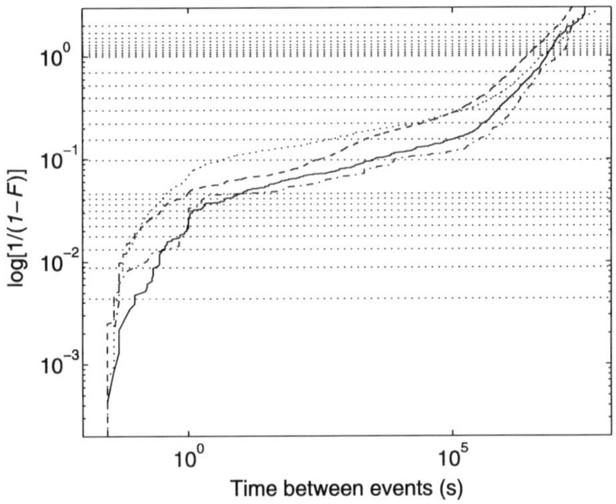


**Figure 10.21** Probability distribution function (presented as Weibull plot) for time between events recorded at four sites: EHV (dash-dot curve); HV2 (dotted curve); HV1 (dashed curve); MV (solid curve).

A Weibull distribution produces a straight line in a Weibull plot [200, pp. 140–142]. The horizontal lines in Figure 10.21 correspond to probabilities of 1%, 2%, ..., 9%, 10%, 20%, ..., 90%, 91%, ..., 99%.

A Weibull distribution is defined as

$$F(t) = 1 - \exp\left[-\left(\frac{t}{\theta}\right)^m\right] \tag{10.28}$$

From (10.28) we can write

$$\log\log\left(\frac{1}{1 - F(t)}\right) = m \log t - m \log\theta \tag{10.29}$$

Plotting $\log\log\{1/[1 - F(t)]\}$ versus $\log t$ results in a straight line with slope $m$. Applying this to Figure 10.21 we see a shape factor around 0.5 for time between events up to a few seconds. For time between events between a few seconds and several days (10 days corresponds to $10^{5.94}$ seconds) the shape factor is somewhere around 0.2. For time between events of several days and longer the shape factor moves toward 1. A possible interpretation of this behavior is as follows:

- Time between events up to a few seconds correspond to repetitive events, for example, due to automatic reclosing actions. From Figure 10.21 we conclude that these are up to about 10% of events. Note that this category also includes events in which the rms voltage stays close to the voltage-dip threshold for a longer duration; see Section 10.3.3.

- Time between events between a few seconds and a few days corresponds to events that occur close in time but that are not related to each other. This is most likely due to periods of adverse weather such as lightning storms. The shape factor of the Weibull distribution is lowest in this range, corresponding to the highest degree of clustering.
- Time between events of several days and longer corresponds to events that occur randomly throughout the year. But even those are not fully randomly distributed as the shape factor remains less than unity.

To study the latter phenomenon, part of the previous plot has been repeated in Figure 10.22. Only time-between event values of one day and longer are included. The asterisks indicate an exponential distribution (shape factor unity) with an expected value of 60 days. The curves obtained from the observation are in parallel with the curve for the exponential curve. The difference between the curves is only in the expected (average) value, not in the shape factor. The horizontal lines in Figure 10.22 correspond to probabilities of 30%, 40%..., 90%, 91%,..., 99%. We can thus conclude from this figure that about half of the voltage dips occur completely random, whereas the other half shows strong clustering.

The middle part of the distribution of the time between events is shown in Figure 10.23. The line of asterisks represents a Weibull distribution with a shape factor of 0.125 and a characteristic time of 100 days.

Note that the characteristic time is not the same as the expected or average time between events. The low shape factor points to a high degree of clustering.

Another conclusion from the above figures is that correlated events with a time between events of more than a few seconds are rare. The impact of time aggregation windows exceeding a few seconds on the total number of events is thus limited. A number of different time aggregation windows will be studied later when discussing system indices.

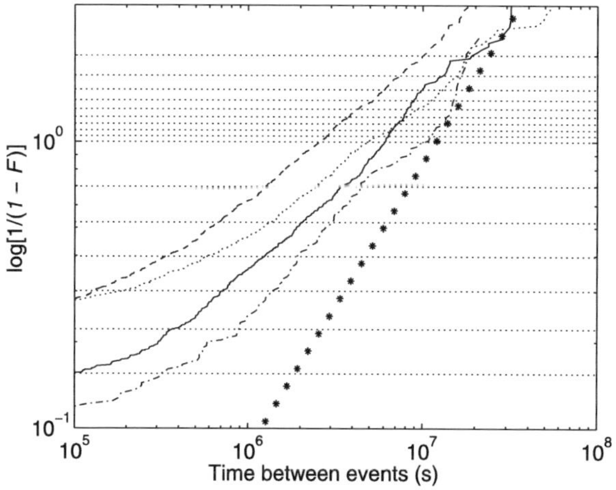


**Figure 10.22** Enlargement of Figure 10.21 for time between events of one day and longer.

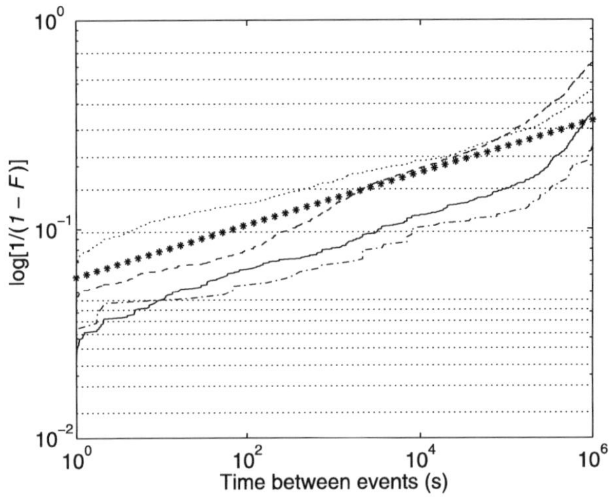


**Figure 10.23** Enlargement of Figure 10.21 for time between events of 1 s up to one day.

### 10.3.3 Chains of Events for Four Different Sites

To show the extent to which time aggregation affects the site indices, Figure 10.24 plots the distribution of the time between events for the MV site. We see from the figure that about 10% of the 124 events recorded at this site occur less than 1 min in time after another event.

The characteristics of the events at this site that are close together in time are presented in Table 10.14. We see that the events are similar in residual voltage, which indicates that the same fault has led to both events. In all but one case, the time

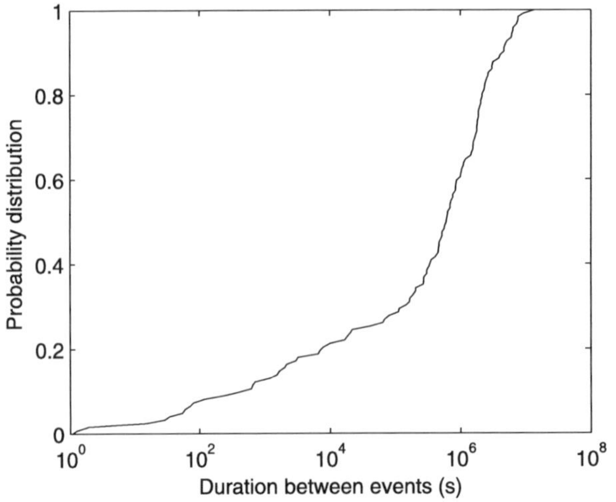


**Figure 10.24** Probability distribution for time between events at MV site.

**TABLE 10.14 Events Less Than 100 s Separated in Time for MV Site**

| Number | Time Since Start of Previous Event (s) | Duration (s) | Residual Voltage (pu) | Aggregated Event | |
|---|---|---|---|---|---|
| | | | | Duration (s) | Voltage |
| 1 | | 0.07 | 0.8304 | 0.07 | 0.8304 |
| | 1.00 | 0.04 | 0.8786 | | |
| 2 | | 0.07 | 0.8562 | 0.07 | 0.8562 |
| | 1.87 | 0.05 | 0.8562 | | |
| 3 | | 0.81 | 0.8073 | 0.85 | 0.8073 |
| | 58.16 | 0.85 | 0.809 | | |
| 4 | | 0.11 | 0.807 | 0.38 | 0.8052 |
| | 54.10 | 0.38 | 0.8052 | | |
| 5 | | 14.94 | 0 | 14.94 | 0 |
| | 0.04 | 0.09 | 0 | | |
| 6 | | 0.33 | 0.8617 | 0.96 | 0.8611 |
| | 34.01 | 0.96 | 0.8611 | | |
| 7 | | 0.12 | 0.8292 | 0.12 | 0.7993 |
| | 28.87 | 0.12 | 0.8402 | | |
| | 80.04 | 0.12 | 0.8317 | | |
| 8 | | 0.09 | 0.7993 | 0.08 | 0.7993 |
| | 1.20 | 0.08 | 0.8269 | | |
| 9 | | 0.13 | 0.1441 | 0.13 | 0.1426 |
| | 71.17 | 0.12 | 0.1426 | | |

separation between the individual events is much larger than the duration of the individual events. The exception is aggregated event 5: an event with zero residual voltage of 14.94 s duration followed 40 ms later by another zero-residual-voltage event with a duration of 90 ms. For this event one could argue that the sum of the durations is a more suitable index for the aggregated event. We will come back to this later.

The aggregated events for the HV1 site are presented in Tables 10.15 and 10.16. The situation for this site is significantly more complicated than for the MV site. Not only are there more aggregated events, there are also more individual events within one aggregated event. The most individual events are found in event chain 5. The total chain lasts about 4 min, during which the monitor triggered 19 times. Looking closer at the values for residual voltage, time between events, and event duration shows that the voltage stays around 90% of nominal for about 3 min, after which the voltage becomes zero for 22 s. Such a chain of events can be prevented from affecting the statistics by choosing a triggering threshold for the monitor which is a few percent higher than the upper limit of the highest residual voltage range in the voltage-dip table. Another solution is to use a hysteresis in the threshold setting, where the dip-ending threshold is 1 or 2% higher than the dip-starting threshold.

Event chain 19 contains five individual events of two different residual voltage values. It appears that there are two distinctive system events behind this. The 100 s time aggregation interval may have been too long for this case.

**TABLE 10.15 Events Less Than 100 s Separated in Time, for HV1 Site**

| | | | | Aggregated Event | |
|---|---|---|---|---|---|
| Number | Time Since End of Previous Event (s) | Duration (s) | Residual Voltage (pu) | Duration (s) | Voltage |
| 1 | | 0.09 | 0.5015 | 0.16 | 0.5015 |
| | 0.79 | 0.16 | 0.5098 | | |
| | 0.14 | 0.07 | 0.8313 | | |
| 2 | | 0.06 | 0.6785 | 0.06 | 0.6681 |
| | 5.80 | 0.06 | 0.6681 | | |
| 3 | | 0.06 | 0.7357 | 0.13 | 0.7331 |
| | 85.48 | 0.06 | 0.7331 | | |
| 4 | | 0.13 | 0.8524 | 0.13 | 0.841 |
| | 2.14 | 0.10 | 0.8599 | | |
| | 38.59 | 0.07 | 0.841 | | |
| | 2.10 | 0.09 | 0.841 | | |
| 5 | | 0.51 | 0.8926 | 28.23 | 0 |
| | 0.17 | 0.48 | 0.8953 | | |
| | 0.10 | 0.50 | 0.8938 | | |
| | 0.14 | 0.50 | 0.8941 | | |
| | 0.04 | 0.59 | 0.8926 | | |
| | 0.11 | 1.10 | 0.8938 | | |
| | 0.04 | 0.53 | 0.8938 | | |
| | 46.94 | 28.23 | 0.8785 | | |
| | 0.22 | 0.12 | 0.8948 | | |
| | 0.13 | 0.06 | 0.8946 | | |
| | 0.42 | 0.03 | 0.8946 | | |
| | 5.75 | 0.05 | 0.8943 | | |
| | 0.07 | 0.02 | 0.8958 | | |
| | 0.40 | 0.32 | 0.8943 | | |
| | 0.05 | 0.02 | 0.8985 | | |
| | 0.27 | 0.37 | 0.8928 | | |
| | 0.43 | 0.03 | 0.8973 | | |
| | 1.81 | 0.07 | 0.8958 | | |
| | 93.25 | 22.38 | 0 | | |
| 6 | | 0.31 | 0.6587 | 0.31 | 0.6587 |
| | 5.08 | 0.08 | 0.6631 | | |
| 7 | | 0.20 | 0.8056 | 0.16 | 0.6408 |
| | 4.92 | 0.09 | 0.8071 | | |
| | 88.77 | 0.16 | 0.6432 | | |
| | 38.77 | 0.14 | 0.6408 | | |
| 8 | | 0.12 | 0.8779 | 0.12 | 0.8779 |
| | 67.27 | 0.06 | 0.8796 | | |
| 9 | | 0.15 | 0.6378 | 0.54 | 0.6378 |
| | 4.94 | 0.54 | 0.6423 | | |
| 10 | | 0.20 | 0.7257 | 0.2 | 0.7257 |
| | 4.16 | 0.20 | 0.7302 | | |

**TABLE 10.16 Events Less Than 100 s Separated in Time for HV1 Site**

| Number | Time Since End of Previous Event (s) | Duration (s) | Residual Voltage (pu) | Aggregated Event | |
|---|---|---|---|---|---|
| | | | | Duration (s) | Voltage |
| 11 | | 0.20 | 0.7202 | 0.20 | 0.7202 |
| | 3.67 | 0.20 | 0.7232 | | |
| 12 | | 143.02 | 0 | 143.02 | 0 |
| | 44.29 | 110.02 | 0.0089 | | |
| 13 | | 52.37 | 0 | 52.37 | 0 |
| | 69.40 | 82.02 | 0 | | |
| 14 | | 0.10 | 0.7858 | 0.87 | 0.7213 |
| | 0.41 | 0.87 | 0.7213 | | |
| 15 | | 0.08 | 0.8456 | 0.44 | 0.8456 |
| | 0.14 | 0.27 | 0.8635 | | |
| | 0.57 | 0.44 | 0.8649 | | |
| 16 | | 0.51 | 0.856 | 0.21 | 0.7028 |
| | 58.80 | 0.21 | 0.7028 | | |
| 17 | | 0.05 | 0.8928 | 0.12 | 0.8824 |
| | 0.34 | 0.12 | 0.8824 | | |
| 18 | | 0.06 | 0.6968 | 0.11 | 0.6968 |
| | 0.95 | 0.11 | 0.739 | | |
| 19 | | 0.05 | 0.8423 | 0.18 | 0.7966 |
| | 2.04 | 0.05 | 0.8528 | | |
| | 90.73 | 0.18 | 0.7996 | | |
| | 98.04 | 0.17 | 0.7966 | | |
| | 2.69 | 0.09 | 0.7966 | | |
| 20 | | 0.05 | 0.88 | 0.05 | 0.8785 |
| | 7.10 | 0.05 | 0.8785 | | |
| 21 | | 0.08 | 0.2723 | 0.08 | 0.2053 |
| | 1.23 | 0.08 | 0.2053 | | |
| 22 | | 0.13 | 0.8702 | 0.15 | 0.806 |
| | 88.62 | 0.15 | 0.806 | | |
| 23 | | 0.38 | 0.8841 | 0.38 | 0.8794 |
| | 0.83 | 0.31 | 0.8794 | | |
| 24 | | 0.04 | 0.8645 | 0.04 | 0.8645 |
| | 0.96 | 0.04 | 0.8659 | | |
| | 0.07 | 0.03 | 0.8777 | | |
| 25 | | 0.08 | 0.7803 | 0.08 | 0.759 |
| | 5.12 | 0.08 | 0.759 | | |
| 25 | | 0.09 | 0.7934 | 0.86 | 0.7922 |
| | 0.42 | 0.86 | 0.7922 | | |
| 26 | | 0.10 | 0.7458 | 0.51 | 0.7268 |
| | 0.77 | 0.51 | 0.7268 | | |

Another interesting issue coming from the study of these events is whether to aggregate all events within a certain window or to aggregate all events closer in time than a certain window. For example, event chain 19 in Table 10.16 could be aggregated in two ways (when using a 100-s aggregation interval): as one event or as two events, as shown in Figure 10.25. When the time between individual events is used as an aggregation criterion, the chain will be aggregated into one event as none of the individual events occur more than 100 s after the end of the previous event. This chain of events will however result in two aggregated events when aggregation is based on the time elapsed since the start of the chain. Interestingly, the latter method would still result in the third individual event becoming part of the "wrong" chain. The occurrence of cases such as this is unavoidable with any standardized or automatic method.

The HV2 site in Table 10.17 shows a less complex behavior than the HV1 site. The only event chains worth mentioning are 3, 9, and 13. In these three cases the individual events show significantly different residual voltages. Chain 3 may be due to three system events, chain 9 to two system events, and chain 13 to three system events. The latter chain appears to consist of three individual events that have no correlation with each other. They just appear close to each other in time, possibly during a period of adverse weather. One of the conclusions from Table 10.17 is that there is no time aggregation value such that only those dips due to the same underlying system event will be aggregated. More complicated aggregation methods are needed for that, including both magnitude and duration information as well as system information. Such methods could provide very important information on the operation of the system but would be of limited value for quantifying the performance of the supply.

Finally, the aggregated events are given for the EHV site in Table 10.18. The number of aggregated events is limited in this case, but they are of a number of different characters. Event chains 2, 5, 6, 7, and 9 show two or three events only

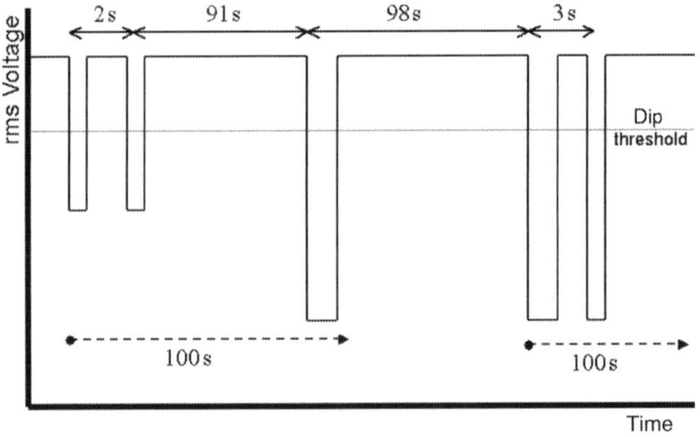


**Figure 10.25** Chain of event that can be aggregated in two different ways: corresponding to event chain 19 in Table 10.16.

**TABLE 10.17 Events Less Than 100 s Separated in Time for HV2 Site**

| Number | Time Since End of Previous Event (s) | Duration (s) | Residual Voltage (pu) | Aggregated Event Duration (s) | Aggregated Event Voltage |
|---|---|---|---|---|---|
| 1 | | 0.07 | 0.824 | 0.07 | 0.8315 |
| | 0.83 | 0.06 | 0.8315 | | |
| 2 | | 0.06 | 0.7074 | 0.06 | 0.6945 |
| | 5.80 | 0.06 | 0.6945 | | |
| 3 | | 0.13 | 0.8388 | 0.13 | 0.8277 |
| | 2.14 | 0.10 | 0.8465 | | |
| | 38.59 | 0.07 | 0.8277 | | |
| | 2.10 | 0.11 | 0.8277 | | |
| | 22.51 | 0.02 | 0.895 | | |
| | 19.49 | 0.02 | 0.8964 | | |
| 4 | | 0.13 | 0.868 | 0.13 | 0.868 |
| | 67.27 | 0.06 | 0.8697 | | |
| 5 | | 0.20 | 0.6761 | 0.20 | 0.6761 |
| | 4.16 | 0.20 | 0.6761 | | |
| 6 | | 0.20 | 0.6712 | 0.20 | 0.6712 |
| | 3.67 | 0.20 | 0.6769 | | |
| 7 | | 0.08 | 0.8527 | 0.41 | 0.8527 |
| | 0.15 | 0.26 | 0.8697 | | |
| | 0.60 | 0.41 | 0.8697 | | |
| 8 | | 0.11 | 0.881 | 0.11 | 0.8841 |
| | 2.05 | 0.08 | 0.8841 | | |
| 9 | | 0.05 | 0.8213 | 0.18 | 0.7728 |
| | 2.04 | 0.05 | 0.8322 | | |
| | 90.73 | 0.18 | 0.7744 | | |
| | 98.04 | 0.17 | 0.7728 | | |
| | 2.69 | 0.09 | 0.7728 | | |
| 10 | | 0.05 | 0.8644 | 0.05 | 0.8644 |
| | 7.10 | 0.05 | 0.8658 | | |
| 11 | | 0.05 | 0.8738 | 0.05 | 0.8736 |
| | 2.05 | 0.05 | 0.8736 | | |
| 12 | | 0.11 | 0.8775 | 0.11 | 0.8775 |
| | 44.78 | 0.10 | 0.8828 | | |
| 13 | | 0.09 | 0.6015 | 0.13 | 0.556 |
| | 80.22 | 0.08 | 0.556 | | |
| | 52.63 | 0.13 | 0.7173 | | |
| 14 | | 0.11 | 0.7503 | 0.19 | 0.4832 |
| | 56.90 | 0.19 | 0.4832 | | |
| 15 | | 0.08 | 0.8755 | 0.08 | 0.8755 |
| | 40.11 | 0.09 | 0.8847 | | |
| 16 | | 0.04 | 0.865 | 0.04 | 0.865 |
| | 1.96 | 0.04 | 0.8698 | | |
| | 0.07 | 0.03 | 0.8809 | | |
| 17 | | 0.06 | 0.8742 | 0.06 | 0.7392 |
| | 67.83 | 0.06 | 0.7392 | | |

**TABLE 10.18 Events Less Than 100 s Separated in Time for EHV Site**

| | | | | Aggregated Event | |
|---|---|---|---|---|---|
| Number | Time Since End of Previous Event (s) | Duration (s) | Residual Voltage (pu) | Duration (s) | Voltage |
| 1 | | 0.07 | 0.836 | 0.09 | 0.836 |
| | 0.15 | 0.09 | 0.8761 | | |
| | 19.64 | 0.09 | 0.8599 | | |
| 2 | | 0.06 | 0.8095 | 0.06 | 0.805 |
| | 1.74 | 0.06 | 0.805 | | |
| 3 | | 0.06 | 0.4945 | 0.06 | 0.4945 |
| | 70.68 | 0.06 | 0.5441 | | |
| 4 | | 0.06 | 0.8744 | 0.06 | 0.8744 |
| | 55.09 | 0.06 | 0.8807 | | |
| 5 | | 0.08 | 0.8098 | 0.08 | 0.8098 |
| | 1.79 | 0.07 | 0.8142 | | |
| 6 | | 0.12 | 0.8908 | 0.12 | 0.8908 |
| | 2.01 | 0.05 | 0.8917 | | |
| 7 | | 0.06 | 0.5522 | 0.06 | 0.4716 |
| | 1.01 | 0.06 | 0.4716 | | |
| 8 | | 0.04 | 0.8955 | 0.07 | 0.8729 |
| | 0.67 | 0.04 | 0.8941 | | |
| | 2.17 | 0.07 | 0.8729 | | |
| 9 | | 0.06 | 0.8838 | 0.06 | 0.8838 |
| | 0.68 | 0.03 | 0.8872 | | |

a few seconds apart in time. Unsuccessful autoreclosure is a likely cause of these events. The first two individual dips in event chain 9 are likely due to the same fault with the residual voltage close to the dip threshold. For event chains 3 and 4 the time between individual dips is about 1 min. But as the residual voltages of the two individual events are very similar, they are probably due to the same fault.

The indices for the aggregated events are given in the final two columns of Tables 10.14 through 10.18. The time between individual events has been used as an aggregation criterion. The residual voltage of the aggregated event is equal to the lowest residual voltage of the individual events. The duration of the aggregated event is equal to the longest duration of the individual events.

### 10.3.4 Impact on Site Indices

The time aggregation algorithm described above has been applied to voltage dips recorded at the four sites used in the earlier examples. All individual events that start less than 100 s after the previous event have been aggregated into one event. The effect of time aggregation on the total number of events is shown in Table 10.19. The resulting voltage-dip tables are shown as Tables 10.20 through 10.23. Note that Table 10.19 gives the total number of events over the six-year monitoring period, whereas the other tables give number of dips per year. The resulting

**TABLE 10.19 Total Number of Events Using 100-s Time Aggregation Window**

| Voltage Level | Number of Events | Number of Aggregated Events | Number of Events After Aggregation |
|---|---|---|---|
| MV | 124 | 9 | 114 |
| HV1 | 327 | 27 | 273 |
| HV2 | 181 | 17 | 154 |
| EHV | 87 | 9 | 76 |

**TABLE 10.20 Voltage-Dip Table for MV Site Using 100-s Time Aggregation Window**

| | 0–20 ms | 20–100 ms | 100–500 ms | 0.5–1 s | 1–3 s | 3–20 s | 20–60 s | 60–180 s |
|---|---|---|---|---|---|---|---|---|
| 85–90% | 0.2 | 4.4 | 1.8 | 0.6 | 0.2 | 0 | 0 | 0 |
| 70–85% | 0 | 3.3 | 2.33 | 0.83 | 0.33 | 0 | 0 | 0 |
| 40–70% | 0 | 0.67 | 1 | 0 | 0 | 0 | 0 | 0 |
| 10–40% | 0 | 0.17 | 1.33 | 0 | 0 | 0 | 0 | 0 |
| ≤10% | 0 | 0.17 | 2.17 | 0.33 | 0 | 0.17 | 0 | 0 |

voltage-dip table for the MV site is given in Table 10.20. Comparing this with Table 10.2 shows that the effect of time aggregation on the voltage-dip table is small for this site. The main impact is for events in the range 20 to 500 ms/70 to 90%, the dips due to ordinary faults, partly because this is where the majority of dips occur.

The results for the HV1 site are shown in Table 10.21. The values in this table should be compared with the nonaggregated values in Table 10.3. The difference is significant, especially for shallow events. Almost all events in the range 85 to 90%/0.5 to 60 s have disappeared after time aggregation. It is interesting to note that the main difference between the two tables is due to one chain of events, as was shown in Table 10.15.

The voltage-dip table after time aggregation for the HV2 site is given in Table 10.22. Comparison with Table 10.4 shows that also for this site the main difference is for shallow events, especially for those due to ordinary faults (i.e., in the time range 20 to 100 ms).

**TABLE 10.21 Voltage-Dip Table for HV1 Site Using 100-s Time Aggregation Window**

| | 0–20 ms | 20–100 ms | 100–500 ms | 0.5–1 s | 1–3 s | 3–20 s | 20–60 s | 60–180 s |
|---|---|---|---|---|---|---|---|---|
| 85–90% | 0.4 | 8.4 | 5.4 | 0.2 | 0.2 | 0 | 0 | 0 |
| 70–85% | 0 | 11 | 5.17 | 2.67 | 0 | 0 | 0 | 0 |
| 40–70% | 0 | 4.8 | 4.5 | 0.17 | 0 | 0 | 0 | 0 |
| 10–40% | 0 | 0.33 | 0.67 | 0 | 0 | 0 | 0 | 0 |
| ≤10% | 0 | 0.33 | 0 | 0 | 0 | 0.17 | 1.83 | 1.67 |

**TABLE 10.22 Voltage-Dip Table for HV2 Site Using 100-s Time Aggregation Window**

| | 0–20 ms | 20–100 ms | 100–500 ms | 0.5–1 s | 1–3 s | 3–20 s | 20–60 s | 60–180 s |
|---|---|---|---|---|---|---|---|---|
| 85–90% | 0.8 | 4.8 | 3.4 | 0 | 0 | 0 | 0 | 0 |
| 70–85% | 0.17 | 8.17 | 1.67 | 0.17 | 0 | 0 | 0 | 0 |
| 40–70% | 0 | 4.5 | 2.5 | 0 | 0 | 0 | 0 | 0 |
| 10–40% | 0 | 0.5 | 0.5 | 0 | 0 | 0 | 0 | 0 |
| ≤10% | 0 | 0 | 0 | 0 | 0 | 0 | 0 | 0 |

**TABLE 10.23 Voltage-Dip Table for EHV Site Using 100-s Time Aggregation Window**

| | 0–20 ms | 20–100 ms | 100–500 ms | 0.5–1 s | 1–3 s | 3–20 s | 20–60 s | 60–180 s |
|---|---|---|---|---|---|---|---|---|
| 85–90% | 0.6 | 5.8 | 1 | 0 | 0 | 0 | 0 | 0 |
| 70–85% | 0 | 4.7 | 0.17 | 0 | 0 | 0 | 0 | 0 |
| 40–70% | 0 | 1.17 | 0.17 | 0 | 0 | 0 | 0 | 0 |
| 10–40% | 0 | 0.17 | 0 | 0 | 0 | 0 | 0 | 0 |
| ≤10% | 0 | 0 | 0 | 0 | 0 | 0 | 0.17 | 0 |

The resulting voltage-dip table for the EHV site is given in Table 10.23. The difference with the nonaggregated values in Table 10.5 is small and only in the time range 20 to 100 ms.

## 10.4 VOLTAGE DIPS: SYSTEM INDICES

Having obtained site indices for all monitored sites in a system, the next step is to calculate system indices. System indices quantify the voltage quality of a whole system. A system may be one voltage level, the network operated by one company or a whole country. In any case a system consists of a (large) number of monitoring sites.

These are different ways of obtaining system indices. They fall into two principally different approaches:

- The individual events collected over all sites are combined into one index or set of indices. The whole system is treated as if it were one site. Examples are the scatter plot and the average voltage-dip table.
- The system index or indices are obtained from the indices for each individual site. Examples are the median and 95 percentile values of a site index.

There are advantages with both methods, but the authors are of the opinion that the second method gives a more accurate picture of the voltage quality for a whole system. Under the first approach one site can significantly affect the results, especially when a smaller number of sites are monitored. Also the resulting value does not always give a good indication of what an individual customer can

**TABLE 10.24 Systems Used for Calculating System Indices**

| Voltage Level | Number of Sites | Number of Monitor-Years | Number of Events | Average Dip Frequency (events/monitor/yr) |
|---|---|---|---|---|
| EHV | 4 | 24 | 236 | 9.8 |
| HV2 | 21 | 114 | 1832 | 16.1 |
| HV1 | 33 | 110 | 3678 | 33.4 |
| MV | 17 | 74 | 982 | 13.3 |

expect. The discussion on which method to use is related to the discussion on the aim of system indices.

In the remainder of this section a number of methods are discussed to present and/or quantify the system performance. The examples given are based on monitoring data obtained in four different systems, at four voltage levels, that were monitored over a period of up to six years. Some basic data on the systems are given in Table 10.24.

### 10.4.1 Scatter Plots

In addition to being obtained for one site, a scatter plot can be obtained for a whole system. Every dip recorded at any of the sites results in a dot in the scatter plot. The results for the four voltage levels are presented in Figure 10.26. In all figures, only

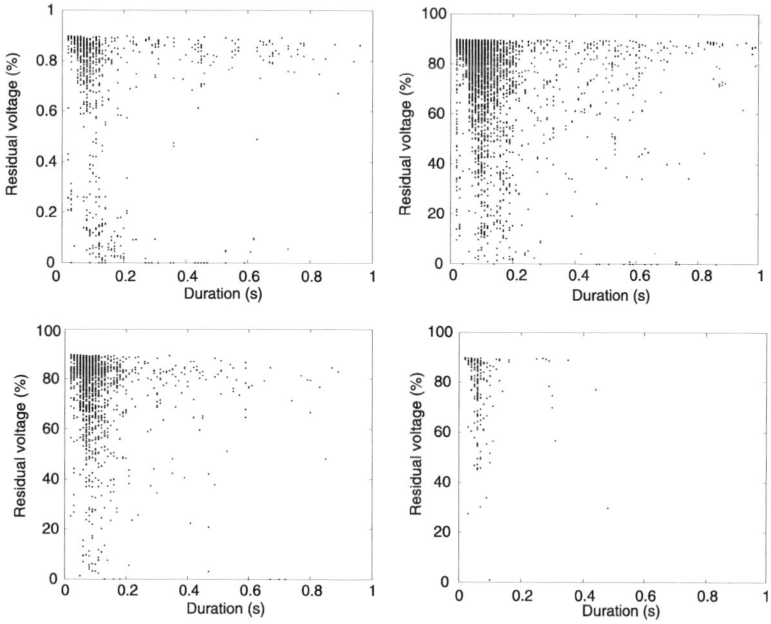


**Figure 10.26** Scatter plot for all sites obtained during six years of measurements: MV sites (top left); HV1 sites (top right); HV2 sites (bottom left); EHV sites (bottom right).

dips with durations up to 1 s have been plotted, even though longer duration events had been recorded.

The scatter plot obtained from recordings at four EHV sites (bottom right) shows that the majority of dips have a duration of up to 150 ms. There are some isolated dips of longer duration, including some rather deep ones, but the majority of dips are short and shallow.

The scatter plot for HV2 (bottom left) looks rather different. A first comparison reveals that there are many more long and deep events present than at EHV. However, one should not forget that the number of events recorded at HV2 is more than seven times the number recorded at EHV, in part due to the larger number of sites, but also due to a higher dip frequency.

The HV1 scatter plot (top right) contains again twice as many dots as the one for HV2. Interestingly there are a surprisingly large number of very short events, with one-cycle duration. Some of these events are very deep, which rules out a voltage transient as an origin. Further analysis of the individual events is needed to reveal the origin of these events.

The scatter plot for the MV sites (top left) looks similar to the one for HV1 with the exception of the large number of events with low residual voltage. As we saw earlier (in the discussion on single-phase versus phase-to-phase measurements in Section 10.2.8), these are due to single-phase faults in an impedance-earthed network.

### 10.4.2 Distribution Functions

Another way of presenting the performance of a system is through probability distribution functions. These are not used to quantify the performance, but rather are used to graphically show the statistics of individual events. In the remainder of this section we will show some examples, similar to those used when discussing site indices in Section 10.2.3.

The probability distribution function of the dip duration is shown in Figure 10.27 for the four voltage levels. As with the scatter diagram the system has been treated as one big site. The probability distribution is obtained from the duration values for all events recorded at the same voltage level. No correction has been made for the fact that the monitoring period is different for different sites or for the fact that the dip threshold setting has been changed from 85 to 90% after the first year of monitoring.

From Figure 10.27 one can conclude that voltage dips for lower voltage levels are of somewhat longer duration that for higher voltage levels. This is obviously related to the faster protection used at higher voltage levels. The figure also shows that a few percentage of events have a duration exceeding 1 s.

Also for the residual voltage a probability distribution function can be obtained. The way to calculate this is the same as for the duration of the event. The results are shown in Figure 10.28 for the four voltage levels.

The distribution functions are very similar, with the exception of MV (solid line). For the MV sites the number of dips with a low residual voltage is significantly higher than with other voltage levels. As mentioned before this is due to the high-impedance earthing used at this voltage level.

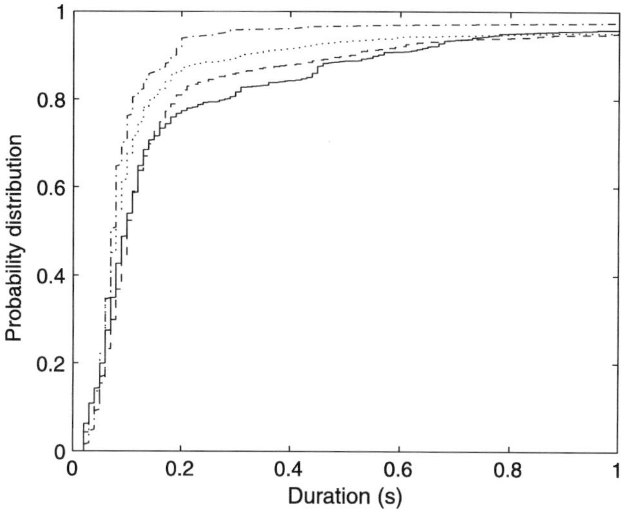


**Figure 10.27** Comparison of probability distribution function for dip duration at four different voltage levels: MV (solid line); HV1 (dashed line); HV2 (dotted line); EHV (dash-dot line).

About 5% of the dips have a residual voltage close to zero. No dips have been recorded with a residual voltage above 90%. This refers to the dip detection threshold of the monitors, which was set to 90%. Note that the function is very steep toward the right-hand side of the plot. A small change in threshold setting will give a significant change in the frequency of recorded voltage dips.

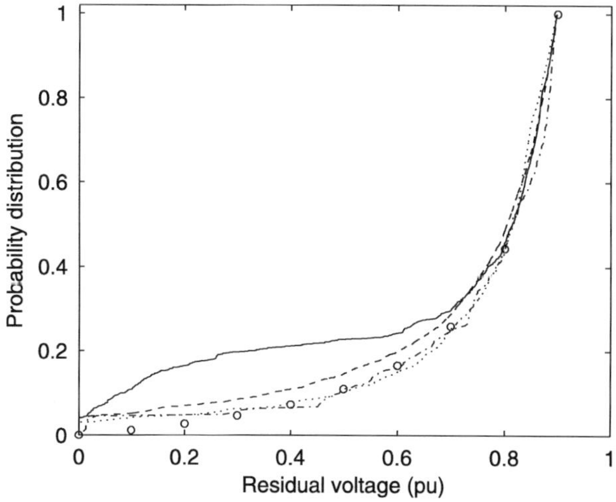


**Figure 10.28** Comparison of probability distribution function for residual voltage at four different voltage levels: MV (solid line); HV1 (dashed line); HV2 (dotted line); EHV (dash-dot line). The circles indicate the theoretical relation for a radial system.

The different voltage levels are also compared with the theoretical curve. As explained in [33, Section 6.5] the number of voltage dips with a residual voltage less than $V$ (in per-unit) is proportional to $V/(1 - V)$. As can be seen in the figure the measured distribution is close to the theoretical one for all voltage levels with the exception of MV, where the number of deep dips is higher than expected. This can also be concluded from Figure 10.26 when comparing the scatter plot for this voltage level with the scatter plots for other voltage levels.

### 10.4.3 Contour Charts

The two-dimensional distribution of residual voltage and duration is represented in the *contour chart*. The absolute number of events is given, not the relative number, so that this is (strictly speaking) not a probability distribution function. The contour charts for the four voltage levels are presented in Figure 10.29. It is hard to draw general conclusions about the charts. They are mainly used to compare equipment performance with the performance of the network at individual sites. One of the few general conclusions from the charts is that the dip frequency reduces when

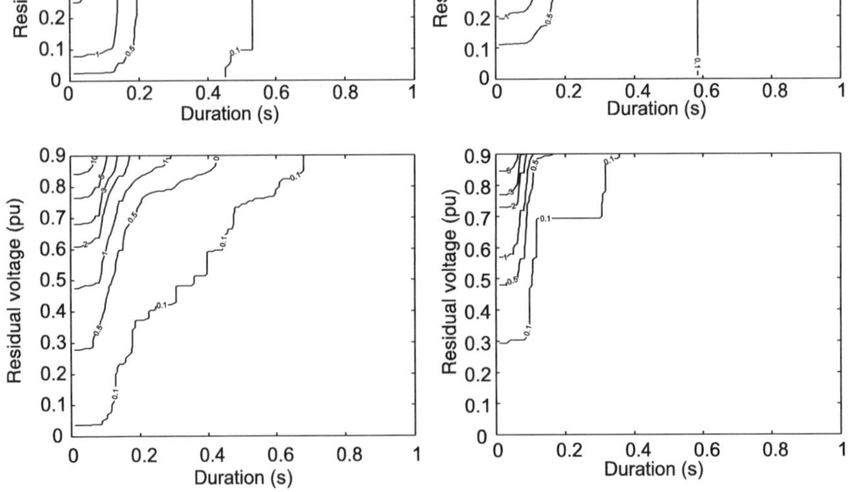


**Figure 10.29** Contour chart (events per site per year) obtained for all sites: MV (top left); HV1 (top right); HV2 (bottom left); EHV (bottom right).

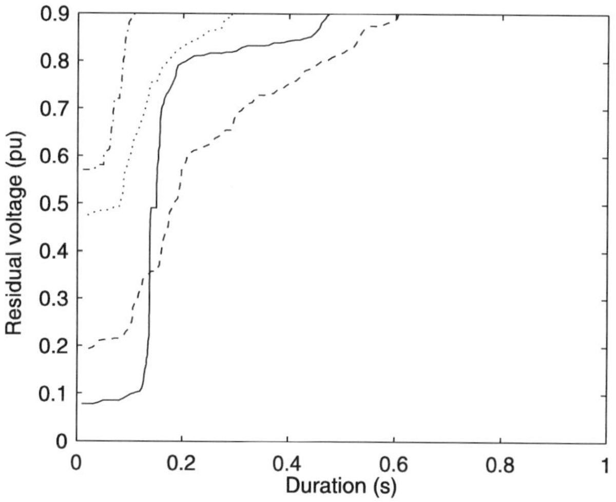


**Figure 10.30** One-event-per-site-per-year contours for four voltage levels: MV (solid line); HV1 (dashed line); HV2 (dotted line); EHV (dash-dot line).

moving to higher voltage levels. This is understandable as voltage dips propagate downward (i.e., to lower voltage levels) with only a small damping whereas the damping is very big when moving upward. In other words, a fault at HV2 gives a deep dip (a low residual voltage) for HV1 and MV but only a shallow dip (a high residual voltage) for EHV.

The four voltage levels are compared in Figure 10.30, where the one-event-per-site-per-year contours are plotted. Equipment with a voltage tolerance curve toward the right of the contour will experience less than one malfunction per year (due to voltage dips). This figure clearly confirms that the dip frequency increases with decreasing voltage level.

### 10.4.4 Seasonal Variations

The number of voltage dips recorded at a certain location is related to the number of faults in the network. Therefore any seasonal variation in fault frequency will likely show up as a seasonal variation in dip frequency. Figure 10.31 shows the number of dips recorded for each month of the year. Data from all voltage levels and all sites have been used. The dip frequency is significantly higher during the summer months (May through August) than during the winter months. This is probably related to the higher lightning activity during summer. The seasonal pattern is most pronounced for the EHV and HV1 sites. The high contribution of December and January to the number of dips is due to a very high dip frequency during a few days. The seasonal pattern is less visible with the MV sites, probably because a higher part of the MV network is underground.

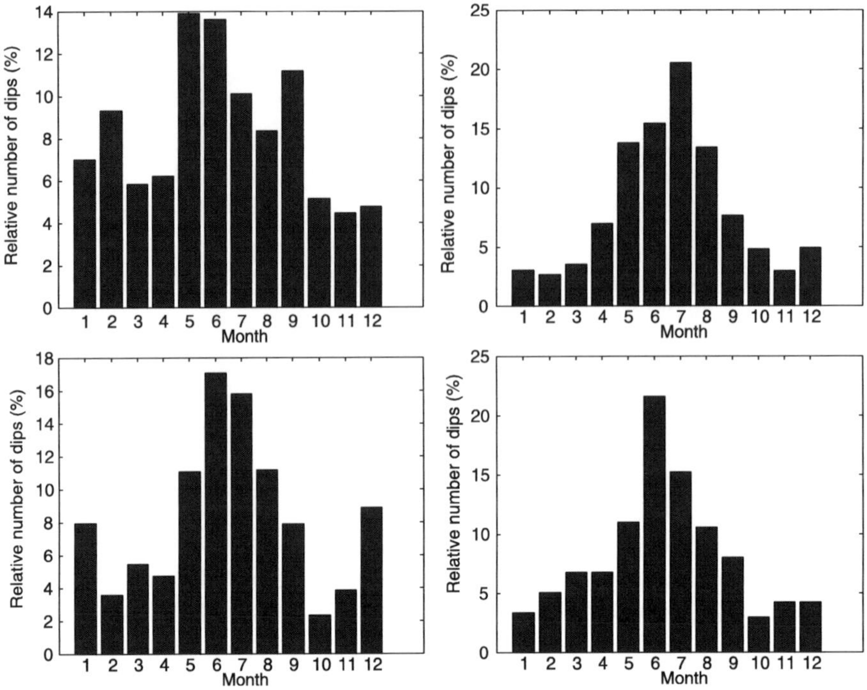


**Figure 10.31** Seasonal distribution of voltage-dip frequency: MV sites (top left); HV1 sites (top right); HV2 sites (left bottom); EHV sites (right bottom).

### 10.4.5 Voltage-Dip Tables

All the ways of presenting the system performance discussed before are strictly speaking not indices as they do not result in individual numerical values. This poses limits on the calculation and interpretation of the results. Voltage-dip tables do not suffer from these disadvantages. The voltage-dip tables for a whole system are calculated from the tables for individual sites. A number of methods are being used for this.

From the voltage-dip tables for the individual sites, an average voltage-dip table can be obtained by averaging the values for each cell over the different sites. This is a straightforward way of obtaining system indices, but it does not give the same result as when the whole system is treated as one site. The method does require the calculation of indices for the individual sites.

Consider a system with $K$ sites. Site $k$ has been monitored over a period $T_k$ during which $N_k$ events were recorded within a certain range of residual voltage and duration. The value in the corresponding cell of the voltage-dip table is thus

$$I_k = \frac{N_k}{T_k} \tag{10.30}$$

The system index obtained as the average of the site indices equals

$$I_{\text{system}} = \frac{1}{K}\sum_{k=1}^{K}\frac{N_k}{T_k} \tag{10.31}$$

By treating the whole system as one site, the following value is obtained:

$$I_{\text{site}} = \frac{\sum_{k=1}^{K} N_k}{\sum_{k=1}^{K} T_k} \tag{10.32}$$

The two methods only lead to the same result when the monitoring period is the same for all sites, that is, when

$$T_1 = T_2 = \cdots = T_K \tag{10.33}$$

The voltage-dip table obtained as the average over the HV2 sites is reproduced in Table 10.25.

An advantage of calculating the system indices from the site indices is that different weighting factors can be given to different sites. The weighting factors may be a representation of the importance of a load connected to a certain site or simply be proportional to the amount of load supplied from the site. Another way of weighting is to give lower weighting factors to a site when a number of nearby sites are also monitored. The weighting factor will then be proportional to the number of sites (or the amount of load) represented by this monitored site.

Instead of the average value per cell, the 95 percentile for each cell can be calculated. This is only relevant when there is a sufficient number of sites. For surveys with less than 10 sites the 95 percentile has no meaning. The 95 percentile for 21 values (the number of HV1 sites in this survey) is the highest value but one. In this case the highest value for this cell is 26.0 dips per year and the highest-but-one value is 9.3 dips per year. The value in the corresponding cell of the 95% table becomes 9.3 dips per year. The whole 95% table is shown in Table 10.26. The 95 percentile table should not be confused with the table for the 95 percentile site. The 95 percentile site is obtained by ranking the total dip frequency (i.e., the sum of all elements of the table) and obtaining the 95 percentile for this.

Alternatively, the 95 percentile may be estimated from the average and standard deviation of the site indices. For the site indices the average $\mu$ and standard

**TABLE 10.25 Average Voltage-Dip Table (Dips per Site per Year) for All HV2 Sites**

| | 0–20 ms | 20–100 ms | 100–500 ms | 0.5–1 s | 1–3 s | 3–20 s | 20–60 s | 60–180 s |
|---|---|---|---|---|---|---|---|---|
| 85–90% | 0.29 | 3.67 | 0.71 | 0.03 | | | | |
| 70–85% | 0.33 | 5.10 | 2.52 | 0.27 | 0.24 | 0.10 | 0.02 | |
| 40–70% | 0.02 | 1.65 | 1.04 | 0.04 | 0.04 | 0.02 | | |
| 10–40% | 0.01 | 0.25 | 0.24 | | 0.01 | 0.01 | 0.01 | |
| 0–10% | | 0.11 | 0.08 | 0.03 | 0.02 | 0.05 | 0.10 | 0.28 |

**TABLE 10.26 The 95% Voltage-Dip Table (Events per Site per Year) for All HV2 Sites**

| | 0–20 ms | 20–100 ms | 100–500 ms | 0.5–1 s | 1–3 s | 3–20 s | 20–60 s | 60–180 s |
|---|---|---|---|---|---|---|---|---|
| 85–90% | 1.0 | 6.6 | 1.5 | 0.2 | | | | |
| 70–85% | 1.5 | 9.3 | 4.3 | 0.2 | | | | |
| 40–70% | 0.2 | 3.3 | 2.7 | 0.2 | | | | |
| 10–40% | | 0.7 | 0.7 | | | 0.2 | | |
| 0–10% | | 0.6 | 0.3 | 0.2 | | 0.3 | 0.3 | 1.0 |

deviation $\sigma$ are estimated using the standard expressions. Assuming that the site indices are normally distributed, the 95 percentile interval is between $\mu - 2\sigma$ and $\mu + 2\sigma$. The method is appropriate when only a small number of sites is available.

The most commonly used system indices for voltage-dip tables are the average table and the 95 percentile table. It is important to realize the difference between these two ways of characterizing the system performance. When using the average voltage-dip table, a bad performance in part of the system (i.e., a large number of dips) can be compensated for by a good performance in other parts of the system. With the 95 percentile table this is not possible. The values in the 95 percentile table give the performance (or voltage quality) as experienced by 95% of the customers, when a sufficient number of sites are monitored. In other words, the average table may give a performance value that is much better than is experienced by a substantial part of the customers, but the 95 percentile table gives a performance value that is much worse than is experienced by the majority of the customers. A system index will never be able to give the power quality for individual customers; it only gives a measure of the performance of the system toward all customers. When the aim of the system is to satisfy the majority of the customers, the 95 percentile value is an appropriate index.

### 10.4.6 Effect of Time Aggregation on Voltage-Dip Tables

The voltage-dip tables for all HV21 sites have been recalculated using time aggregation. The time aggregation algorithm used is as follows:

- A time aggregation window of 10 s or 10 min is used.
- The time between events is defined as the time between the end of an event and the start of the next event.
- If the time between events is less than the time aggregation window, the two events belong to the same "chain of events".
- A chain of events is represented by one "aggregated event" with residual voltage equal to the lowest residual voltage in the chain and duration equal to the longest duration in the chain.

The impact of the 10-s aggregation will be given below for the average and for the 95 percentile table. The impact of the 10-min aggregation will only be given for the 95 percentile table.

**TABLE 10.27 Average Voltage-Dip Table (Dips per Year) for All HV1 Sites**

| | 0–20 ms | 20–100 ms | 100–500 ms | 0.5–1 s | 1–3 s | 3–20 s | 20–60 s | 60–180 s |
|---|---|---|---|---|---|---|---|---|
| 85–90% | 0.86 | 6.75 | 3.40 | 0.70 | 0.20 | | | |
| 70–85% | 0.16 | 6.12 | 4.98 | 0.69 | 0.18 | | | |
| 40–70% | 0.17 | 1.99 | 2.89 | 0.31 | 0.02 | | | |
| 10–40% | 0.09 | 0.56 | 1.14 | 0.03 | 0.01 | | | |
| 0–10% | 0.01 | 0.07 | 0.20 | 0.13 | 0.20 | 0.23 | 0.14 | 0.44 |

**TABLE 10.28 The 95 Percentile Voltage-Dip Table (Dips per Year) for All HV1 Sites**

| | 0–20 ms | 20–100 ms | 100–500 ms | 0.5–1 s | 1–3 s | 3–20 s | 20–60 s | 60–180 s |
|---|---|---|---|---|---|---|---|---|
| 85–90% | 3.2 | 12.2 | 7.6 | 2.8 | 1.0 | | | |
| 70–85% | 0.7 | 13.0 | 11.0 | 2.5 | 0.7 | | | |
| 40–70% | 1.5 | 6.7 | 8.7 | 2.0 | | | | |
| 10–40% | 0.2 | 1.8 | 4.0 | 0.2 | | | | |
| 0–10% | — | 0.3 | 0.7 | 1.0 | 1.2 | 1.0 | 0.3 | 2.0 |

The average and the 95 percentile voltage-dip tables for the HV1 sites are shown as Tables 10.27 and 10.28, respectively. The impact of a 10-s time aggregation on the HV1 indices is shown in Tables 10.29 and 10.30. The impact of a 10-min time aggregation on the 95 percentile table is shown in Table 10.31. The impact of time aggregation is rather large here. It should be noted, however, that the impact is

**TABLE 10.29 Average Voltage-Dip Table (Events per Site per Year) for HV1 Sites Using 10-s Time Aggregation Window**

| | 0–20 ms | 20–100 ms | 100–500 ms | 0.5–1 s | 1–3 s | 3–20 s | 20–60 s | 60–180 s |
|---|---|---|---|---|---|---|---|---|
| 85–90% | 0.50 | 5.68 | 3.02 | 0.57 | 0.16 | 0.01 | 0.01 | |
| 70–85% | 0.13 | 5.76 | 4.92 | 0.66 | 0.17 | | | |
| 40–70% | 0.16 | 1.78 | 2.89 | 0.33 | 0.02 | | | |
| 10–40% | 0.09 | 0.54 | 1.14 | 0.03 | 0.01 | | | |
| 0–10% | 0.01 | 0.07 | 0.17 | 0.13 | 0.26 | 0.24 | 0.14 | 0.44 |

**TABLE 10.30 The 95 Percentile Voltage-Dip Table (Events per Site per Year) for HV1 Sites Using 10-s Time Aggregation Window**

| | 0–20 ms | 20–100 ms | 100–500 ms | 0.5–1 s | 1–3 s | 3–20 s | 20–60 s | 60–180 s |
|---|---|---|---|---|---|---|---|---|
| 85–90% | 1.0 | 8.6 | 5.8 | 2.6 | 1.0 | | | |
| 70–85% | 0.3 | 11.5 | 10.5 | 2.5 | 0.7 | | | |
| 40–70% | 1.3 | 6.0 | 8.8 | 2.0 | 0.2 | | | |
| 10–40% | 0.2 | 1.8 | 4.0 | 0.2 | | | | |
| 0–10% | — | 0.3 | 0.7 | 1.0 | 1.3 | 1.0 | 0.3 | 2.0 |

**TABLE 10.31 The 95 Percentile Voltage-Dip Table (Events per Site per Year) for HV1 Sites Using 10-min Time Aggregation Window**

| | 0–20 ms | 20–100 ms | 100–500 ms | 0.5–1 s | 1–3 s | 3–20 s | 20–60 s | 60–180 s |
|---|---|---|---|---|---|---|---|---|
| 85–90% | 1.0 | 8.2 | 5.3 | 2.4 | 0.7 | | | |
| 70–85% | 0.3 | 10.3 | 9.7 | 2.7 | 0.5 | | | |
| 40–70% | 1.3 | 5.5 | 7.8 | 1.8 | 0.3 | | | |
| 10–40% | 0.2 | 1.8 | 3.3 | 0.2 | 0.2 | | | |
| 0–10% | — | 0.3 | 0.7 | 1.0 | 1.2 | 1.0 | 0.3 | 1.7 |

mainly due to a small number of sites with a large number of events closely together in time. Time aggregation remains difficult to interpret as part of a system index unless the number of sites is sufficiently large. It is difficult to give a lower limit, but a small number of sites should not have a significant impact on the results. As the 95 percentile excludes 1 in 20 sites, a minimum value of 50 to 100 sites would seem reasonable. At this moment, there is simply not sufficient experience on extracting indices for such large systems, with the exception of the EPRI Distribution Power Quality (DPQ) survey [92, 277].

The effect on the 95 percentile values is larger, as can be concluded by comparing the different tables. The impact of time aggregation tends to be bigger for sites with a large number of recorded dips. This is understandable as the occurrence of a large number of dips in a short period will lead to a high voltage-dip count and will also mean that time aggregation has a severe impact. In this case the 10-min aggregation window has a large influence whereas the 10-s window has only minor influence.

The 95 percentile tables have been calculated for three systems for different time aggregation windows. The results are shown schematically in Figures 10.32 through 10.34. The different bars in each group correspond to the following time aggregation windows (left to right): no time aggregation, 1 s, 5 s, 10 s, 20 s, 1 min, 5 min, 10 min, and 15 min.

The effect of time aggregation on the elements in the 95 percentile table for MV sites, is shown in Figure 10.32. The impact is small with the exception of the (10–40%, 20–100 ms) cell in the table where heavy clustering appears in the range 5 s to 1 min.

For the HV1 system the main impact is found for a 1-s time aggregation window; some cells show a gradual decrease with increasing time aggregation window.

The effect of time aggregation on the 95 percentile table is small for the HV2 system, as can be concluded from Figure 10.32. Again the dip frequency is shown as a function of the time aggregation window for those cells with a dip frequency exceeding one event per year when no time aggregation is used. The effect of time aggregation is much bigger for the MV sites. The effect on the 95 percentile table is shown in Figure 10.33.

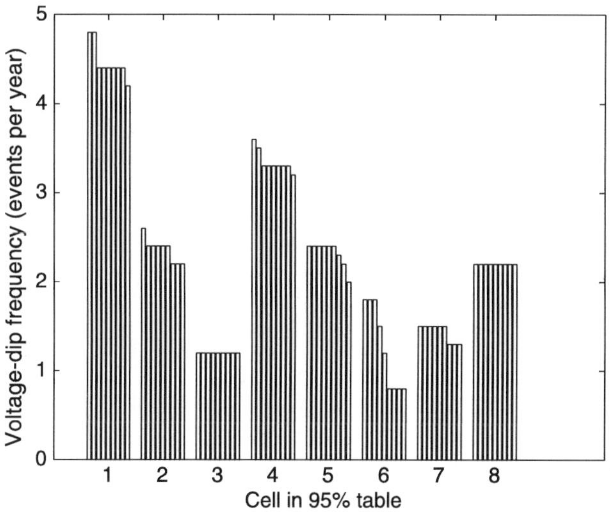


**Figure 10.32** Effect of time aggregation on selected elements of 95 percentile table for MV sites: (1) 85–90%, 20–100 ms; (2) 85–90%, 100–500 ms; (3) 85–90%, 500–1000 ms; (4) 70–85%, 20–100 ms; (5) 70–85%, 100–500 ms; (6) 10–40%, 20–100 ms; (7) 10–40%, 100–500 ms; (8) 0–10%, 100–500 ms.

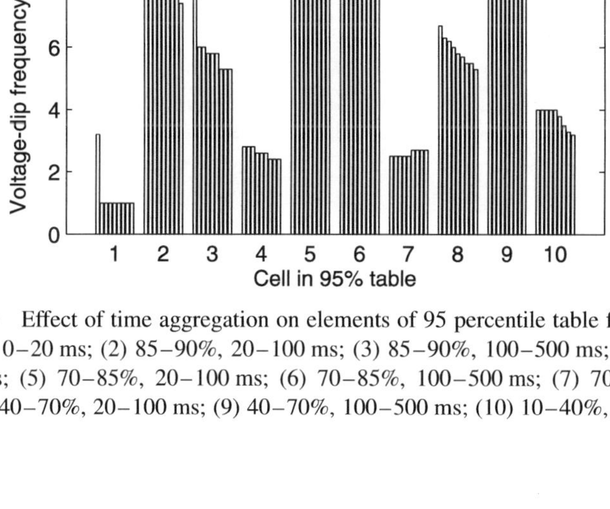


**Figure 10.33** Effect of time aggregation on elements of 95 percentile table for HV1 sites: (1) 85–90%, 0–20 ms; (2) 85–90%, 20–100 ms; (3) 85–90%, 100–500 ms; (4) 85–90%, 500–1000 ms; (5) 70–85%, 20–100 ms; (6) 70–85%, 100–500 ms; (7) 70–85%, 500–1000 ms; (8) 40–70%, 20–100 ms; (9) 40–70%, 100–500 ms; (10) 10–40%, 100–500 ms.

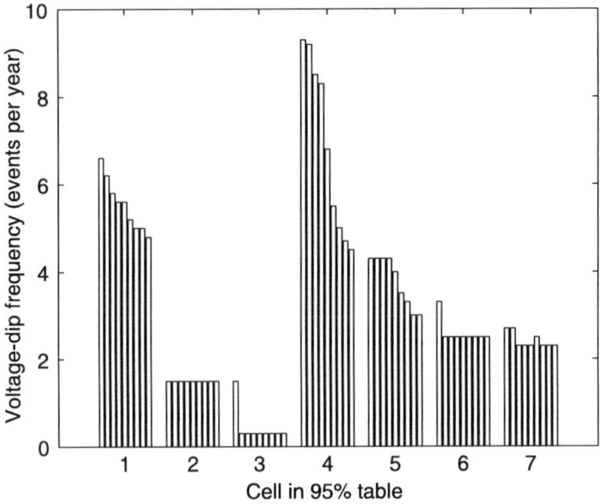


**Figure 10.34** Effect of time aggregation on elements of 95 percentile table for HV2 sites: (1) 85–90%, 20–100 ms; (2) 85–90%, 100–200 ms; (3) 70–85%, 0–20 ms; (4) 70–85%, 20–100 ms; (5) 70–85%, 100–500 ms; (6) 40–70%, 20–100 ms; (7) 40–70%, 100–500 ms.

### 10.4.7 SARFI Indices

The SARFI indices have been introduced as a system index from the beginning; SARFI stands for System Average rms-variation Frequency Index and was intended as the voltage-dip equivalent of SAIFI (System Average Interruption Frequency Index). Let $N_{90}(s)$ be the number of events with residual voltage below 90% at site $s$; $\mathrm{SARFI}_{90}$ is then defined as

$$\mathrm{SARFI}_{90} = \sum_{s=1}^{S} \omega_s N_{90}(s) \tag{10.34}$$

with $\omega_s$ a weighting factor quantifying the importance of site $s$ such that $\sum_{s=1}^{S} \omega_s = 1$. Using the equivalence between SARFI and SAIFI the number of customers affected by an event is an obvious choice for a weighting factor. With voltage dips it is however not straightforward to determine the number of customers affected by each event. Instead the number of customers "represented" by the monitor site is used. This results in the following expression for the weighting factor:

$$\omega_s = \frac{C_s}{\sum_{s=1}^{S} C_s} \tag{10.35}$$

with $C_s$ the number of customers represented by site $s$. The use of such a weighting method would be suitable when, for example, all distribution substations were equipped with a monitor. Each monitor would then represent all the customers supplied from that substation. In practice only part of the substations are equipped with a power quality monitor. For each nonmonitored substation a monitored substation has to be found that is likely to experience the same number of voltage dips. This requires the use of stochastic prediction methods to estimate the expected number of dips at each substation. The prediction methods are easy to implement and apply but unfortunately still rarely used. Alternatively it is simply assumed that the monitored substations are picked at random so that the observed number of dips is a good estimator of the number of dips at the nonmonitored stations. This method is more commonly used, although it is not clear how accurate the method is.

The total number of customers supplied from a substation is mainly determined by the number of domestic customers. Industrial and commercial customers are a small fraction of the total number of customers, even though they make up a substantial part of the total power. Voltage dips are however mainly a concern for industrial customers and to a lesser extent for commercial customers. Domestic customers are in most cases only affected by the more severe voltage dips and by short and long interruptions. This reasoning would lead to the conclusion that the weighting factors in (10.34) should be based on the number of industrial customers or on the combined rated power of the industrial customers connected to a substation.

For monitoring at higher voltage levels (subtransmission, transmission) it is often very difficult to determine the number of end customers, so that the total rated power can be used as a weighting factor. Alternatively the same weight is given to each substation, so that the weighting factor for each monitor location is proportional to the number of nonmonitored sites for which a similar number of dips is expected. In practice it is assumed that nearby substations experience a similar number of dips.

Determining weighting factors is not straightforward, and no matter which method is used, there will always be arguments for using another method. One of the main obstacles is that only a fraction of the substations are monitored so it is very difficult to include all customers in the statistics. Installing monitor equipment at all substations will be expensive. Methods for estimating the dip frequency at nonmonitored substations are under development [229], but it may take some time before these are generally accepted. Therefore the system indices are in most cases calculated as a simple average of the site indices for all monitored sites. The weighting factor is equal for all sites and (10.34) becomes

$$\mathrm{SARFI}_{90} = \frac{1}{S}\sum_{s=1}^{S} N_{90}(s) \tag{10.36}$$

Five different SARFI indices have been calculated for the MV, HV1, and HV2 sites. The results are shown in Figures 10.35 through 10.37. The left-hand plots give the value of $\mathrm{SARFI}_{10}$, $\mathrm{SARFI}_{40}$, $\mathrm{SARFI}_{70}$, $\mathrm{SARFI}_{85}$, and $\mathrm{SARFI}_{90}$ for each site. In the right-hand plots each of the SARFI values have been sorted. Note that a column in the right-hand plot typically refers to different actual sites. For example, site 4 in Figure 10.35 has the highest value for $\mathrm{SARFI}_{90}$ but a rather low value for $\mathrm{SARFI}_{10}$.

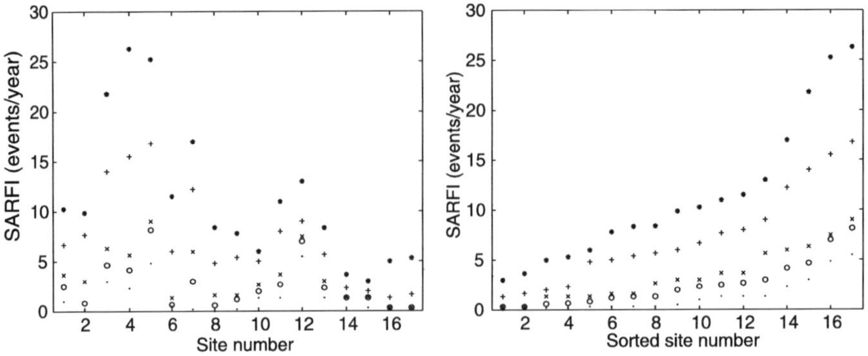


**Figure 10.35** SARFI indices for all MV sites: $SARFI_{10}$ (dot); $SARFI_{40}$ (circle); $SARFI_{70}$ (cross); $SARFI_{85}$ (plus); $SARFI_{90}$ (star). The left-hand plot gives the values per site; the right-hand plot gives the sites sorted after SARFI value.

The results for the MV sites (Fig. 10.35) show a rather uniform distribution of the index values. There are no extreme values, neither on the low side nor on the high side. The results for the HV1 and HV2 sites show a different behavior. Especially for the HV2 sites we see a number of extreme sites: some with a very high value and some with a very low value. It is not possible to give an explanation for this without going into more detail of the system structure and the kind of events that caused voltage dips during the monitoring period. It should further be noted that no time aggregation has been applied to the data. Time aggregation tends to especially reduce the high values so that the spread in values becomes less.

From the site indices, three system indices have been calculated: the average value, the median (50%) value, and the 95% value. The results are shown in Table 10.32.

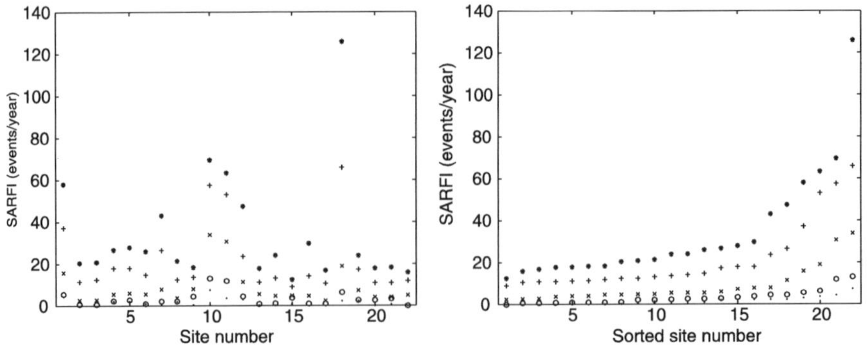


**Figure 10.36** SARFI indices for all HV1 sites: $SARFI_{10}$ (dot); $SARFI_{40}$ (circle); $SARFI_{70}$ (cross); $SARFI_{85}$ (plus); $SARFI_{90}$ (star). The left-hand plot gives the values per site; the right-hand plot gives the sites sorted after SARFI value.

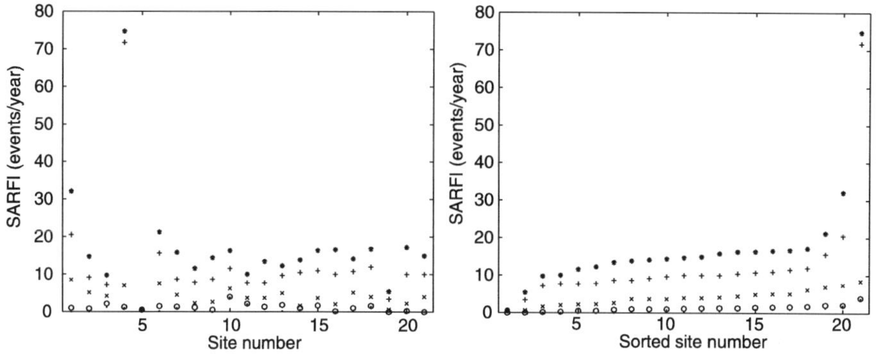


**Figure 10.37** SARFI indices for all HV2 sites: $SARFI_{10}$ (dot); $SARFI_{40}$ (circle); $SARFI_{70}$ (cross); $SARFI_{85}$ (plus); $SARFI_{90}$ (star). The left-hand plot gives the values per site; the right-hand plot gives the sites sorted after SARFI value.

**TABLE 10.32 SARFI Indices for MV, HV1, and HV2 Systems**

| | MV System | | | HV1 System | | | HV2 System | | |
|---|---|---|---|---|---|---|---|---|---|
| | Mean | 50% | 95% | Mean | 50% | 95% | Mean | 50% | 95% |
| $SARFI_{90}$ | 11.4 | 8.4 | 25.2 | 34.0 | 24.0 | 69.5 | 17.3 | 14.5 | 32.1 |
| $SARFI_{85}$ | 7.3 | 5.7 | 15.0 | 21.6 | 13.8 | 57.5 | 12.6 | 9.7 | 20.5 |
| $SARFI_{70}$ | 3.4 | 2.7 | 7.5 | 8.9 | 5.5 | 30.8 | 4.0 | 3.7 | 7.5 |
| $SARFI_{40}$ | 2.5 | 1.3 | 7.0 | 3.4 | 2.3 | 12.0 | 1.2 | 1.0 | 2.2 |
| $SARFI_{10}$ | 1.4 | 0.3 | 4.8 | 1.5 | 0.7 | 4.3 | 0.7 | 0.3 | 2.0 |

### 10.4.8 Single-Index Methods

Characterization methods such as voltage-dip energy and voltage-dip severity result in only one value for the site index. This makes it straightforward to calculate system indices. Here we will only discuss the voltage-dip energy. The calculation of system indices associated with the voltage-dip severity proceeds in a similar way.

The system index associated with the voltage-dip energy is the System Average Sag Energy Index, or SASEI, the average over all sites of the dip energy index:

$$\text{SASEI} = \frac{1}{S}\sum_{s=1}^{N} \text{SEI}_s \tag{10.37}$$

with $\text{SEI}_s$ the dip energy index (i.e., the sum of the dip energy for all events) at site $s$. The dip energy index has been calculated for the HV1 and HV2 sites. The results are shown in Figure 10.38 and Table 10.33. The variation among sites is very large, even when excluding the short interruptions (note the logarithmic vertical scale). The result of this large variation is that a system average does not represent the majority of the sites.

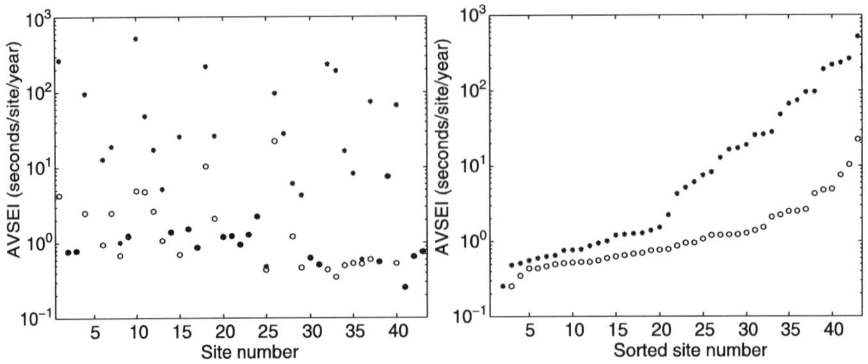


**Figure 10.38** Dip energy Site Index for all HV1 and HV2 Sites: including all events (stars); excluding short interruptions (circles). The left-hand plot gives the values per site; the right-hand plot gives the sites sorted after SEI value.

**TABLE 10.33 System Indices Associated with Voltage-Dip Energy (s/year/site)**

| | HV1 Sites | | | HV2 Sites | | |
|---|---|---|---|---|---|---|
| | Mean | 50% | 95% | Mean | 50% | 95% |
| All events | 57.7 | 1.53 | 265.9 | 35.5 | 2.24 | 192.3 |
| No interruptions | 2.16 | 1.22 | 4.93 | 2.03 | 0.53 | 7.54 |

## 10.5 SUMMARY AND CONCLUSIONS

This chapter presents methods of quantifying the performance of the power network as experienced by its customers. The discussion of power quality indices addresses an essential question in the design and operation of a power system: Which performance do we expect from a power system? The answer to this question concerns much more of customer relations than just power quality. However, the discussion in this chapter has concentrated on the power quality part of system performance.

### 10.5.1 Interruptions

Reliability indices are needed as feedback to the network operator. As the operator of a transmission or distribution network possesses a so-called natural monopoly, the normal market mechanism cannot be relied upon to guarantee an optimal reliability (the *product quality* in economic terms). Instead an artificial feedback is introduced where a regulatory body defines reliability indices and objectives. Also, before deregulation many network operators collected interruption statistics to obtain a measure of the performance of their system. As the use of those statistics was mainly internal, there was less emphasis on exact index definitions.

Many interruption indices for distribution systems are defined in IEEE 1366, where the ones most commonly used are: SAIFI (system average interruption frequency index, number of interruptions per year), SAIDI (system average interruption duration index, interrupted minutes per year), and CAIDI (customer average interruption duration index, minutes per interruption). These indices are calculated from the number of customers affected by each interruption and the duration of each interruption. The interruption indices are defined from the opening of interrupting devices, not from the voltage experienced by the customers. In the vast majority of cases, these two viewpoints would have resulted in the same statistics.

Data collection of interruptions relies with most network operators on information obtained from the service personnel responsible for restoring the supply. Automatic recording based on the status of interrupting devices is rarely used at the distribution level. A direct consequence of this is that the interruption statistics are limited in two aspects.

The first limitation is that the start of the interruption is in many cases not known. Automatic recording equipment is needed to obtain accurate information on this. Most network operators rely on information from customers that report an interruption, often with dedicated telephone numbers being available for this.

The second limitation is that interruptions terminated by automatic restoration of the supply (by means of fast autoreclosing) are not part of the statistics. The duration of such interruptions is typically short. To prevent a bias toward manual restoration, all interruptions shorter than a few minutes (short interruptions) are not considered in the reliability statistics. The border between short and long interruptions is placed somewhat higher than the longest automatic reclosing time in use (1 to 5 min). Data collection on short interruptions requires in all cases automatic measurements. The high costs associated with this mean that in the majority of cases only long interruptions are included in the reliability statistics. The definition of reliability indices for distribution networks is rather straightforward due to the radial operation of those networks. For transmission networks several complications occur, mainly related to the meshed operation of transmission networks. A number of different approaches have been discussed in this chapter, with their advantages and disadvantages.

An interpretational issue with interruption statistics for transmission networks is that interruptions are rare at the transmission level. The statistics will likely show large year-to-year variations. A serious consequence of this is that it will take many years before a deterioration in the transmission reliability becomes clear from interruption statistics. Transmission system statistics are thus less applicable as a feedback mechanism than those for distribution systems. Some network operators solve this by providing statistics on component availability, but those do not consider the component redundancy, and a decrease in component unavailability may be due to more maintenance or more failures.

### 10.5.2 Voltage Dips

The calculation of voltage-dip indices is a two-stage process. In the first stage, the performance of each monitored site is quantified through a site index. Next the system index is calculated from the site indices for all monitored sites. The choice of site index in many cases dictates the choice of system indices. In some cases, the choice of single-event indices already determines the choice of site and system indices. This holds especially for single-index methods.

A number of site indices have been discussed in this chapter. Scatter plots can be easily obtained and give a quick first impression of the voltage quality, but they do not provide any quantitative information. The probability distribution of residual voltage and duration can be presented in a number of ways: as a one-dimensional function or as a two-dimensional contour chart; as a density function or as a (cumulative) distribution function. Several examples have been presented in this chapter, with the contour chart of the two-dimensional distribution function having clear advantages when coordinating supply performance and equipment performance (the voltage dip coordination chart).

A voltage-dip table is a way of presenting the density function in a limited number of values. The residual voltage and duration range are divided into a small number of intervals. For each of the resulting cells the number of dips per year is given. The advantage of using a voltage-dip table with a high resolution is that the data can easily be further compressed into another index at a later stage. It is, for example, appropriate to use less resolution for the system indices than for the site indices.

The SARFI indices are another commonly used method for quantifying the performance of the supply. The SARFI indices, in the way they are commonly used, only consider the residual voltage. The duration of a dip does not impact the statistics. Such indices can be applied to systems where the fault-clearing time does not vary much or cannot be affected. Such would typically hold for transmission systems and even for some distribution systems. SARFI indices are also valid for equipment where the voltage tolerance curve is flat for the normal range of dip durations (somewhat less than 100 ms to somewhat longer than 1 s). The use of voltage-dip tables does however provide additional information on the duration of dips and on the fault-clearing time. It should be noted that a reduction in fault-clearing time may be easier to achieve for a distribution system operator than a reduction in the number of faults.

Two single-index methods have been presented: voltage-dip energy and voltage-dip severity. Especially with voltage-dip energy the values for individual events vary over a wide range. The result is that site indices (being the sum of all single-event indices over one year) are often dominated by one or two events.

Currently, a trend exists to make site and system indices simpler, with some authors proposing a "unified power quality index" combining all power quality variations and events into one single index [123, 124, 138, 206]. Such indices may have their application in a regulatory process, but the relation with equipment performance or customer concerns has completely disappeared. Where such indices are used for regulatory purposes, more detailed indices are needed to be able to interpret the results and to provide technical feedback to the network operators, allowing them to make the necessary improvements in their systems.

An important parameter in the calculation of system indices from site indices is the weighting of the different sites. With interruption indices the weighting is based on the number of customers affected by an interruption. However, with voltage dips such an approach would require a very large number of monitors, preferably one with every customer. In practice the number of monitoring sites is very limited so that it is difficult to link a number of customers with a monitor location. The result is that each monitor location is considered equal. The use of stochastic prediction methods (Section 10.5.4) in combination with a limited number of monitor locations will enable a better estimation of the system performance over all customers.

It was shown that the number of voltage dips per year is not a Poisson distribution. From a limited set of data it was concluded that the standard deviation is about 50% more than for a Poisson distribution. This immediately translates in a longer monitoring period being required before accurate statistics are obtained. However, not enough statistical data were available to get more information on this. Data from large power quality surveys need to be studied to obtain more statistical patterns, such as differences between overhead and underground networks and between countries with different weather patterns. A parallel approach is to use fault data, which is typically available over a longer period and for the whole system, as a starting point for the study.

Knowledge on the probability distribution of the number of events per year is of importance for interpreting the results of power quality monitoring. An often-made but rarely mentioned assumption in collecting voltage-dip statistics is that the average dip frequency in the past is equal to the expected value for the future. Not enough is known about the stochastic nature of dip statistics to assess their predictive value.

Collecting dip statistics in three-phase systems involves the choice of phase-to-phase or phase-to-ground measurements. The IEC power quality monitoring standard leaves this choice open. It is shown here that phase-to-phase measurements at medium and higher voltage levels give a better indication of the voltage experienced by the equipment than phase-to-ground measurements. For performance assessment it is therefore recommended to measure phase-to-phase. However, for system diagnostics, as discussed in Section 8.9, a phase-to-ground connected monitor is preferred. This dilemma can be solved by using the three-phase classification method for voltage dips as proposed in Section 6.2.3.

Some voltage-dip indices are very sensitive to the threshold setting of the monitor. When presenting the results of such a method it is important to indicate the threshold setting used. The current trend is toward using a standard value of 90% for the threshold setting.

The literature on voltage-dip indices mainly consists of applications of methods proposed elsewhere. An overview of several methods is found in IEC 61000-2-8, in the draft versions of IEEE 1564 [172], and in the report by the CIGRE working group C4.07 [64]. Other publications on voltage-dip indices include [50, 74, 76, 85, 88, 179, 192, 196, 237, 253].

### 10.5.3 Time Aggregation

Time aggregation is the process of merging events that occur close together in time into one aggregated event. Such a procedure is applied in many surveys, but a systematic study of time aggregation is missing. When implementing time aggregation two choices have to be made: which events will be aggregated (what is the maximum distance in time between them) and how the indices of the aggregated event are to be calculated.

Two types of time aggregation must be distinguished: events close together in time that are actually one event and events close together in time that are really two events but may have a common origin (such as an unsuccessful reclosure) or have the same impact on equipment as one event. In the first case, an appropriate choice for the duration of the aggregated event is the sum of the durations of the individual events. In the latter case it is more appropriate to choose the worst of the individual events.

Time aggregation requires an open discussion supported by quantitative results on system performance and on equipment susceptibility to multiple events before it can find a place in international standard documents. It has been shown in this chapter that the threshold setting can have a significant impact on the number of events. Also a difference between the dip-starting threshold and the dip-ending threshold can impact the number of events. A discussion on time aggregation further should include a discussion on threshold setting.

The examples in this chapter show that the impact of time aggregation is bigger on the 95 percentile indices than on average indices. The reason is that sites with a high number of multiple events are also more likely to have a high total number of events. It was shown how the impact of time aggregation was mainly due to a few chains of events at a limited number of sites. No similar information from other surveys is available to verify these conclusions.

### 10.5.4 Stochastic Prediction Methods

The process of calculating site and system indices was explained in this chapter for single-event indices obtained from measurements. There is however no reason to not apply the same methods to simulation results. In fact, simulations can lead to statistically relevant results much faster than measurements. For a discussion on stochastic prediction methods for voltage dips the reader is referred to [33, Chapter 6]. More recent publications include [15, 228, 240, 249].

Methods for stochastic prediction of interruptions can be found in the extensive literature on powersystem reliability [e.g., 26, 27, 28, 51].

Stochastic prediction methods may also be further investigated to solve some of the problems associated with collecting statistics for interruption and voltage-dip indices. Statistical patterns can be obtained from stochastic prediction methods. Limited sets of measurements (i.e., a limited number of monitor locations) can next be used to obtain absolute values. Such a method may be suitable to obtain statistics on short interruptions from a limited number of monitor locations.

Stochastic methods may also be used to determine weighting factors for monitor locations when collecting voltage-dip statistics. From stochastic prediction studies the correlation can be obtained between the voltage measured at a monitor location and the value experienced by customers. This information can next be used to give a value for the number of customers "represented" by every monitor. An alternative method is to calculate the voltage-dip statistics at all nonmonitored locations in the system [229].

Reliability techniques can be used to calculate the interruption risk for transmission systems from observed component outages and faults. Such a stochastic technique would reveal a deterioration in the system reliability much faster than pure statistical techniques.

### 10.5.5 Other Events

In this chapter, site and system indices have only been calculated for voltage dips and interruptions. The same approach can also be applied to other events. Voltage transients are an obvious choice, where duration and magnitude would be the two single-event indices to form the basis for the statistics. Much more than with voltage dips, the definitions of magnitude and duration will have a strong influence on the resulting statistics.

Site and system indices may also be obtained for "new" events such as frequency swings and bursts of harmonics. In those cases it is first important to define appropriate triggering methods and appropriate single-event indices. The range of possible power quality events that can be defined is almost unlimited. However, any new event should be justified based on its potential impact on either equipment or on the system.

# CHAPTER 11

# CONCLUSIONS

The chapters of this book contain conclusions that are specifically directed to the subjects discussed in those chapters. This final chapter contains a number of general conclusions. The general-conclusions part of this chapter contains a recapitulation of the general framework for power quality monitoring (the *events-and-variations-framework*) and a discussion of the signal-processing needs to implement this framework. As in the conclusion sections in the different chapters heavy emphasis is placed on future work.

## 11.1 EVENTS AND VARIATIONS

The structure of this book is heavily based on the distinction between power quality events and power quality variations. The latter are also referred to as *steady-state disturbances* or *normal operation*. Variations are small disturbances: small and slow deviations from a nominal or ideal value mainly due to the joined impact of all load connected to the system. On the other hand, events are large disturbances: sudden large deviations from the nominal or ideal value. Events are mainly due to step changes in the power system or with a load. This definition does not provide a clear distinction between variations and events: There is no unique border between "small" and "large." This dilemma can be solved in two ways. Power quality events may be defined as the consequence of distinguishable step changes in the power system (*underlying events* or *power system events*). For example, a voltage dip may be defined as the voltage waveform due to a fault, the starting of

*Signal Processing of Power Quality Disturbances.* By Math H. J. Bollen and Irene Yu-Hua Gu

a motor, or the energizing of a power transformer. Such a definition is applicable in power system analysis studies. Alternatively, a threshold may be agreed upon to define when a deviation should be treated as a large deviation. Using the same example, a voltage dip is defined as a situation in which the rms voltage at a certain location drops below 90% of the nominal voltage. This definition is most appropriate for power quality monitoring and is used in both IEEE and IEC power quality standards. In this book we use this method to distinguish between variations and events. The definition matter is thus simply taken care of by the way in which the sampled data are processed. For variations, the data are processed at predefined instants or (more commonly) over predefined measurement windows. For events, the data processing depends on a triggering mechanism. It is, however, worth emphasizing that the other approach is equally valid, although probably impossible to implement in a power quality monitoring program. Any short-duration drop in (rms) voltage may be referred to as a voltage dip, even if the voltage does not drop below 90% of nominal. Of course, minor drops in voltage are not of interest for most customers as they do not impact the end-user equipment. But they may contain information of importance to the network operator.

## 11.2 POWER QUALITY VARIATIONS

The previous chapters introduce a general framework for the processing of events and variations. Below we give a summary of the difference between the processing of events and the processing of variations. This difference is an essential part in understanding the processing of power quality monitoring data:

- A defining characteristic of power quality variations is that the measurements are performed at predefined instants or over predefined intervals. The first step in setting up measurements is to define the basic measurement instants or windows and the features that will be calculated. The choice of feature is determined by the kind of disturbance that is being studied. For example, when studying voltage unbalance the ratio between negative- and positive-sequence voltages is an appropriate feature.
- As a form of data compression, features obtained with a high time resolution are combined into features with a lower time resolution. A standard example is the time aggregation of 200-ms values via 3-s values into 10-min values. Another, more interesting example is the calculation of the short-time (10-min) flicker severity from the voltage waveforms through a complicated process. The window lengths and time scales for time aggregation of variations (200 ms, 3 s, 10 min, and 2 h) are based on a number of IEC standard documents, notably IEC 61000-4-30. There are, however, a number of reasons to add an additional value between 3 s and 10 min; a 1-min time scale is proposed in this book.
- The features obtained from the basic measurements (eventually after time aggregation) are next used to obtain site indices. When defining a site index

it is important to include in the definition the measurement period (e.g., one week), the basic measurement values (e.g., the 10-min total harmonic distortion), and the way in which a statistically relevant value is chosen (e.g., the 95 percentile). These three choices also provide three degrees of freedom so that the number of possible indices becomes quickly very large. Consider voltage unbalance as an example. The first choice is between the use of the 3-s or the 10-min values. Next the site indices can be calculated per day or per week. From the probability distribution the average, 50 percentile, 90 percentile, 95 percentile, 99 percentile, or highest value can be chosen. For a longer measurement survey (e.g., one year) the highest or average daily or weekly value can be taken. Even this simple example allows for about 30 different indices to quantify the performance of the supply over a one-year period. If we include other ways to quantify the unbalance of a waveform (e.g., the difference between the highest and the lowest rms voltage), the number of options easily becomes more than 100. This is a serious issue, not just because of the large number of options, but even more because there is no clear argumentation to favor one index above the other and because different indices may result in significantly different values. This issue cannot be solved by researchers only but requires a set of predefined standard indices. The work done in a number of IEC and CIGRE standards and working groups is moving in the right direction but is far from complete.

- Having calculated the site indices for a number of sites, the system indices are calculated as a statistically relevant value (e.g., the average or the 95 percentile) of the site indices. Some of the choices here are similar to the ones when defining the site indices. But a more fundamental problem is often hidden behind the discussion on how to calculate system indices: the choice of the monitoring sites. The closer the measurement location is to the terminals of end-user equipment, the more relevant the result and in many cases the higher the disturbance level. However, no statistical information is available on how significant this effect is. For example: Is the system index for voltage variations calculated from measurements on the low-voltage side of the distribution transformers and how relevant is this index for the voltage quality at the terminals of end-user equipment? These questions will have to be addressed before system indices can be used to quantify the system performance.

## 11.3 POWER QUALITY EVENTS

The processing of power quality events proceeds in a completely different way, although the end result is again a set of system indices:

- A triggering method is needed to distinguish an event from variations. Rather simple methods are used in practice to detect voltage dips and interruptions, but a number of more advanced methods are discussed in the literature. More advanced methods are especially needed for transients and for minor events. A basic issue with the development of triggering methods is the difficulty in testing them. The triggering method is often the definition of the event (e.g., a voltage dip is a drop in rms voltage below a threshold). In that case we cannot state that any method is better than another one. Therefore, before testing any new triggering method, it is important to define the event independent of a triggering method. One approach is to assess the ability of the triggering method to detect the underlying events in the system (in case of voltage dips, faults, motor starting, or transformer energizing). Another possible approach is to assess the ability of the method for predicting equipment behavior. The same fundamental problem exists when testing the other aspects of the analysis of power quality events: from characteristic through system events. This is a subject that has not received the attention it deserves.

- The actual processing of a power quality event starts after the triggering and is often performed off-line. The processing typically consists of two parts: From the sampled waveform one or more characteristics versus time are calculated. These characteristics versus time are used to calculate single-event indices. Complex events may be split in segments and the indices calculated per segment. In this book a distinction has been made between event segments and transition segments. The latter correspond to changes in the power system whereas the former are intervals during which the system is in quasi–steady state. The methods for calculating characteristics versus time and single-event indices in most cases assume quasi-stationarity of the signal, so that the distinction between event segments and transition segments becomes a basic aspect of the processing of events. The procedure for the calculation of single-event indices is standardized for dips and interruptions. However, the method prescribed by the standards only considers simple events and does not take into account the three-phase character of the system. No standardized method is available yet for transients or for current quality events.
- Time aggregation with events serves a completely different purpose than with power quality variations. Time aggregation of power quality events is mainly aimed at preventing double counting. Double counting may take place in different ways, which requires different algorithms for time aggregation. The triggering algorithm may lead to multiple triggering instants for the same underlying event (e.g., the rms voltage fluctuates around the dip threshold); related underlying events cause identical power quality events (e.g., reclosing of a distribution feeder); and a number of events close together in time only cause one production stoppage. The discussion on time aggregation is ongoing; most existing studies aggregate all events within a predefined time aggregation window. Time aggregation is currently applied after the single-event indices have been calculated, but it may be more logical to apply it as part of the calculation of the single-event indices.
- Site indices are a means of quantifying the performance of the supply at a given location in the system for one specific customer or even for one specific piece of

equipment. In most cases a site index gives the number of events per year for a specific range of single-event indices. The difference between the different site indices is mainly in the choice of this range. Despite this similarity, it is important to distinguish between two different philosophies in defining site indices. One school of thought is aimed at defining simple indices, preferable just one value, which allows an easy comparison between different sites, recording of trends, and optimization of the supply. The alternative philosophy is that site indices should allow for an assessment of the compatibility between equipment and supply. This leads to more complex indices, often a set of indices. Both types of indices have their applications. In case of doubt on the application it is recommended to calculate a set of indices (simple as well as complex ones) to allow the final user to make a choice based on the application. Different site indices for voltage dips are proposed by a number of standard-setting organizations, but no single method has come to dominate yet. The discussion is currently ongoing at a number of levels, with the difference between the two above-mentioned philosophies (simple versus complex indices) playing an important role in the discussion. Site indices for interruptions are in use or under development in many countries for regulatory purposes, although the emphasis is mainly on system indices.

- Whereas site indices quantify the performance of the supply at a given location, system indices aim at quantifying the performance of the supply of the system as a whole. A system index is typically calculated as a statistically relevant value from the site indices at all the monitor locations. System indices for interruptions (typically referred to as reliability indices) are in use or under development in many countries for regulatory purposes. Similar attempts have been made for voltage dips, but as yet without much consistency or success. A number of reasons can be pointed out for this lack of progress. Voltage dips are typically only measured at a limited number of sites, which makes it difficult to obtain results that are relevant for all locations. A related problem is the lack of knowledge of the statistical distribution of the site indices. A second reason for the lack of progress concerns the choice of the measurement location. Power quality monitors are rarely located at the terminals of the sensitive equipment but typically are located at medium-voltage substations. The difference between the voltage dip at the monitoring location and at the terminals of the sensitive equipment is relatively small and at least qualitatively understood, but the fact that there is an unknown difference limits the use of the measurements for regulatory purposes. A fundamental discussion is needed on these two limiting factors before voltage-dip indices can be used for regulatory purposes.

The above-discussed framework has been applied to the most commonly discussed power quality disturbances: variations in voltage frequency and magnitude; harmonic distortion; unbalance and flicker (voltage fluctuations) for variations; and interruptions, voltage dips, and transients for events. The framework can however be extended to other disturbances. Recently a discussion has started on electric and magnetic fields. Much of the discussion on this subject has to do with a lack of clear definitions. For example, limit values are proposed without defining an appropriate index. Applying the general approach for variations and events may simplify this discussion significantly.

## 11.4 ITEMIZATION OF POWER QUALITY

Most power quality studies start with a so-called itemization of the subject. The deviation from the ideal waveform is subdivided into a number of phenomena, for example (as in Chapter 2), frequency variations, voltage variations, waveform distortion, unbalance, and voltage fluctuations. It is next assumed that these phenomena are mutually exclusive and thus are independent phenomena. Next each phenomenon is characterized by one or more performance indicators (referred to as power quality indices in this book). This approach has been successful in describing the performance of the power supply but should not become a goal by itself. What is needed is continuous verification if the resulting set of performance indicators gives a complete and appropriate description of the supply performance. Some indicators may not (or no longer) have a relation with the way in which the supply impacts equipment (for voltage quality indices) or the other way around (for current quality indices). New indices may be needed to address either new phenomena at the equipment side or new disturbances. An example of a new equipment-related phenomenon is the generation of frequency components associated with the switching frequency in active converters, for example, in energy-saving lamps or solar power converters. Another example is the range in time scale between a few seconds and 10 min, which is covered by the very short variations introduced in this book. The range was not covered by any index beforehand and the introduction of increasing numbers of solar and wind power installations may cause an increase in voltage variations in this range of time scales.

## 11.5 SIGNAL-PROCESSING NEEDS

The processing of power quality measurements is an interesting signal-processing application, as the authors have hopefully managed to make clear in the preceding chapters. An important conclusion from our cooperation is that neither power engineering nor signal processing can solve the various problems on its own. New developments require knowledge from both areas. In a number of cases it turned out that what was considered general knowledge in one area was unknown in the other. We also found out that terminology was either defined differently or interpreted differently in the two areas. As a basis for further research it is important to bridge the gap between power engineering and signal processing. Joint research projects and joint courses are important approaches in bridging this gap.

### 11.5.1 Variations

The processing of power quality variations is well defined in standard documents and generally well understood. Well-understood signal-processing tools (e.g., rms and DFT) are sufficient for most applications. For studying variations only small and slow deviations from the ideal waveform are considered, where advanced methods do not have any significant advantage. In signal-processing terms, the quasi-stationary approximation holds. When this approximation does not hold due to an event taking place, other tools are needed.

There are however two areas where further development of signal-processing tools is still needed: fast voltage fluctuations and high-frequency variations. The DFT may also be applied to frequencies above a few kilohertz, but alternative methods are worth considering. Light flicker has been an issue for many years and has moved a huge step forward with the introduction of the flickermeter method for quantifying the severity of voltage fluctuations for causing light flicker. The method is a combination of an amplitude demodulator, a bandpass filter, and a weighted probability distribution function. The next step is the incorporation of other lamp types (fluorescent lights and energy-saving lamps being most urgent) in the algorithm. This poses some interesting signal-processing challenges.

Fast fluctuations in voltage magnitudes have been studied because of the resulting light flicker. Fast changes in voltage waveforms have not been studied at all. There is no direct need of this for their impact to equipment (no major impact is currently known), but there is an important indirect need for such studies. A problem in testing almost any signal-processing method for power quality is the lack of a stochastic model for voltage and current signals. As a consequence, measurements are needed to verify the performance of the algorithm. However, this has the disadvantage that it is not possible to compare the result of an algorithm with any "exact value." Another disadvantage is that it is difficult to study the algorithm under a range of disturbance conditions and to find limits to its correct performance. Some researchers use white noise superimposed onto a synthetic signal to study the performance of an algorithm. The use of white noise is common in signal processing but may not be suitable for power system signals. A realistic stochastic model of voltage and current signals would allow the testing of new algorithms under realistic conditions. The development of such a model would require a large collection of measurement data as well as an understanding of stochastic signal modeling.

Most of the studies presented in this book, elsewhere in the literature, and even in standard documents involve voltage signals. Current signals show a wider range of variations in magnitude and waveform so that the limitations of existing characterization methods may be reached earlier. The above-mentioned stochastic model would be a great help in assessing the suitability of standard methods for current variations.

An important application of signal-processing tools is the location of sources of power quality variations. Several studies have been performed after finding sources of harmonic distortion, and recently an interesting method has been developed for finding the source of voltage fluctuations in the flicker range [18]. Further signal-processing work in combination with the understanding of the spread of disturbances is needed here. A basic problem, of course, remains that power quality variations are in many cases due to the combination of a (large) number of sources. Those methods should therefore be able to distinguish between *background distortion* and *locally generated distortion.*

### 11.5.2 Variations and Events

Some requirements of signal processing appear at the border line between variations and events, all related to the problem of distinguishing between variations and events. In fact, we have two different questions at hand here.

The first question concerns when a small slow deviation becomes a large sudden deviation. This is currently being solved by introducing triggering levels to distinguish between "small" and "large." More advanced methods may be introduced to include "slow" and "sudden" as well. A discussion of power system criteria to distinguish between variations and events may be worthwhile.

The second question is determining when the measurement of a variation is no longer reliable or should not be included in the statistics for some other reasons. The "flagging concept" has been introduced for this in standard documents. This concept has received remarkably little attention from signal-processing experts and has only been discussed in standard-setting organizations. A more fundamental discussion is certainly important here.

An interesting approach to distinguishing between variations and events is to define a small number of characteristics to be tracked continuously. The same characteristic is to be used to quantify variations and detect events. An example is a generalized magnitude definition that on the one hand is the basis for flicker and voltage variation indices and on the other hand is used for triggering and quantification of dips, swells, and interruptions. The challenge is to include harmonics and transients as well. The *dq*-voltage is a characteristic worth looking into. The development of such characteristics would gain significantly from the above-mentioned discussion on triggering and flagging.

### 11.5.3 Events

It is in the processing of power quality events that advanced signal-processing tools have found their main applications. Tools such as wavelet transforms and Kalman filters provide little advantage over standard tools for the processing of variations but may give significant advantages for the study of the fast changes in signal amplitude and waveform that are associated with events. In this book we emphasize the distinction between event segments and transition segments. The quasi-stationarity assumption holds during event segments, but during the transition segments this assumption no longer holds. Note that the distinction between event and transition segments shows similarities with the distinction between variations and events, including the need for triggering methods and flagging.

A distinction has to be made between methods to characterize event segments and methods to detect transition segments. Characterization methods are often based on the quasi-stationarity assumption and they do not provide reliable results during transition segments. In fact, transition segments may be defined as time intervals during which the quasi-stationarity assumption does not hold. The actual transition (e.g., fault clearing due to the opening of a circuit breaker) is often very short. However, the limitation of the characterization methods makes the results unreliable for a longer period. What is needed are methods to accurately time allocate a transition and methods to characterize event segments close to the actual transition. Both will lead to short transition segments (and thus to longer event segments). All of the existing methods, including the ones presented in this book, move "forward in time," but with postprocessing of event recordings (as is often the case) it may be worth considering to apply methods moving "backward in time."

Methods are also needed to further investigate transition segments. This includes methods that characterize the signal properties within a transition segment. In fact, such methods lead to a reduction of the transition segment. Another important issue is to detect multiple transitions close together in time, such as different instants of fault clearing or different switching instants in different phases.

The processing of events such as voltage dips and interruptions is still rather straightforward, mainly because the event segments dominate for all but the shortest events. However, for voltage and current transients no quasi-stationary state is reached. In other words, the event only consists of a transition segment between two normal-operation states. Other methods are needed for analyzing transient events. This is an area that requires a large signal-processing input. The underlying problem is very similar to the further investigation of transition segments. Again the different instants in the three phases are an important issue. Further, appropriate methods for use in three-phase systems are needed.

### 11.5.4 Event Classification

An important application of signal-processing methods is the classification of power quality events. Both statistical methods and rule-based methods are presented in the literature. In both cases the choice of features is a very important first step in the development. It is also very important to understand the reason for developing a classification method. Some methods are only interested in distinguishing between the different events as experienced at the customer interface, whereas other methods are aimed at distinguishing between different underlying causes (or underlying events) in the power system. The latter is a more challenging application. Support vector machines provide a promising systematic approach for the development of a classification system based on statistical learning. Next to classifying a recorded event to a class (to an underlying cause), a confidence value should be given, with alternative classes or causes in case of a low-confidence outcome.

# APPENDIX A

# IEC STANDARDS ON POWER QUALITY

IEC 61000: Electromagnetic Compatibility (EMC) consists of six parts each with several sections. The list below gives the power quality and power-quality-related standards within the 61000 series as well as some documents currently (May 2006) in preparation.

- *Part 1: General*

  Section 1: Application and interpretation of fundamental definitions and terms.

  Section 2: Methodology for the achievement of functional safety of electrical and electronic equipment with regard to electromagnetic phenomena.

  Section 4: Rationale for limiting power–frequency conducted harmonic and interharmonic current emissions from equipment in the frequency range up to 9 kHz.

- *Part 2: Environment*

  Section 1: Description of the environment—Electromagnetic environment for low-frequency conducted disturbances and signalling in power supply systems.

  Section 2: Compatibility levels for low-frequency conducted disturbances and signaling in public supply systems.

  Section 3: Description of the environment—Radiated and non-network-frequency-related conducted disturbances.

*Signal Processing of Power Quality Disturbances.* By Math H. J. Bollen and Irene Yu-Hua Gu

Section 4: Compatibility levels in industrial plants for low-frequency conducted disturbances.

Section 5: Classification of electromagnetic environments.

Section 6: Assessment of the emission levels in the power supply of industrial plants as regards low-frequency conducted disturbances.

Section 7: Low-frequency magnetic fields in various environments.

Section 8: Voltage dips and short interruptions on public electric power supply systems with statistical measurement results.

Section 12: Compatibility levels for low-frequency conducted disturbances and signalling in public medium voltage power supply systems.

Section 14: Overvoltages on public electricity distribution networks (in preparation).

- *Part 3: Limits*

Section 1: Overview of emission standards and guides (in preparation).

Section 2: Limits for harmonic current emissions (equipment input current $\leq$16 A per phase).

Section 3: Limitation of voltage changes, voltage fluctuations, and flicker in public low-voltage supply systems for equipment with rated current $\leq$16 A per phase and not subject to conditional connection.

Section 4: Limits for harmonic current emissions (equipment with rated current greater than 16 A per phase).

Section 5: Limitation of voltage fluctuations and flicker in low voltage power supply systems for equipment with rated current greater than 16 A.

Section 6: Assessment of emission limits for distorting loads in MV and HV power systems.

Section 7: Assessment of emission limits for fluctuating loads in MV and HV power systems.

Section 8: Signalling on low voltage electrical installations—Emission levels, frequency bands, and electromagnetic disturbance levels.

Section 9: Limits for interharmonic current emissions (equipment with input power $\leq$16 A per phase and prone to produce interharmonics by design) (in preparation).

Section 10: Emission limits in the frequency range 2 to 9 kHz (in preparation).

Section 11: Limitation of voltage changes, voltage fluctuations, and flicker in public low-voltage supply systems—Equipment with rated current $\leq$75 A and subject to conditional connection.

Section 12: Limits for harmonic currents produced by equipment connected to public low voltage systems with input current $>$16 A and $\leq$75 A per phase and subject to restricted connection.

- *Part 4: Testing and Measurement Techniques*
  - Section 1: Overview of IEC 61000-4 series.
  - Section 2: Electrostatic discharge immunity test.
  - Section 3: Radiated, radio-frequency, electromagnetic field immunity test.
  - Section 4: Electrical fast transient/burst immunity test.
  - Section 5: Surge immunity test.
  - Section 6: Immunity to conducted disturbances, induced by radio-frequency fields.
  - Section 7: General guide on harmonics and interharmonics measurement and instrumentation for power supply systems and equipment connected thereto.
  - Section 8: Power frequency magnetic field immunity test.
  - Section 9: Pulse magnetic field immunity test.
  - Section 10: Damped oscillatory magnetic field immunity test.
  - Section 11: Voltage dips, short interruptions, and voltage variations immunity tests.
  - Section 12: Oscillatory waves immunity test.
  - Section 13: Harmonics and interharmonics, including mains signaling at ac power port, low-frequency immunity tests.
  - Section 14: Voltage fluctuation immunity test.
  - Section 15: Flickermeter—Functional and design specifications.
  - Section 16: Test for immunity to conducted, common-mode disturbances in the frequency range 0 Hz to 150 kHz.
  - Section 17: Ripple on dc input power port immunity test.
  - Section 18: Oscillatory wave immunity test (in preparation).
  - Section 27: Unbalance, immunity test.
  - Section 28: Variation of power frequency, immunity test.
  - Section 29: Voltage dips, short interruptions, and voltage variations on dc input power ports, immunity test.
  - Section 30: Power quality measurement methods.
  - Section 33: Measurement methods for high-power transient parameters.
  - Section 34: Voltage dips, short interruptions, and voltage variations immunity tests for equipment with input current more than 16 A per phase.
- *Part 5: Installation and mitigation guidelines*
  - Section 1: General considerations.
  - Section 2: Earthing and cabling.
  - Section 6: Mitigation of external EM influences.
  - Section 7: Degrees of protection provided by enclosures against electromagnetic disturbances (EM code).

- *Part 6: Generic standards*
  Section 1: Immunity for residential, commercial, and light-industrial environments.
  Section 2: Immunity for industrial environments.
  Section 4: Emission standard for industrial environments.
  Section 5: Immunity for power station and substation environments.

# APPENDIX B

# IEEE STANDARDS ON POWER QUALITY

- *141-1993* Recommended practice for electric power distribution for industrial plants, also known as the Red Book.
- *142-1991* Recommended practice for grounding of industrial and commercial power systems, also known as the Green Book.
- *213-1993* Standard procedure for measuring conducted emissions in the range of 300 kHz to 25 MHz from television and FM broadcast receivers to power lines.
- *241-1990* Recommended practice for electric power systems in commercial buildings, also known as the Gray Book.
- *281-1994* Standard service conditions for power system communication equipment.
- *299-1997* Standard method of measuring the effectiveness of electromagnetic shielding enclosures.
- *367-1996* Recommended practice for determining the electric power station ground potential rise and induced voltage from a power fault.
- *430-1991* Standard procedures for the measurement of radio noise from overhead power lines and substations.
- *446-1995* Recommended practice for emergency and standby power systems for industrial and commercial applications, also knows as the Orange Book.
- *449-1998* Standard for ferroresonance voltage regulators.

*Signal Processing of Power Quality Disturbances.* By Math H. J. Bollen and Irene Yu-Hua Gu

- *473-1991* Recommended practice for an electromagnetic site survey (10 kHz to 10 GHz).
- *493-1997* Recommended practice for the design of reliable industrial and commercial power systems, also known as the Gold Book.
- *519-1992* Recommended practice and requirements for harmonic control in electric power systems.
- *762-1987* Standard definitions for use in reporting electric-generating unit reliability, availability, and productivity.
- *859-1987* Standard terms for reporting and analyzing outage occurrences and outage states of electric transmission facilities.
- *998-1996* Guide for direct lightning stroke shielding of substations.
- *1057-1994* Standard for digitizing waveform recorders.
- *1100-1999* Recommended practice for powering and grounding sensitive electronic equipment, also known as the Emerald Book.
- *1159-1995* Recommended practice for monitoring electric power quality.
- *1159.1* Guide for recorder and data acquisition requirements for characterization of power quality events (in preparation).
- *1159.3* Recommended practice for the transfer of power quality data.
- *1184-1994* Guide for the selection and sizing of batteries for uninterruptible power systems.
- *1250-1995* Guide for service to equipment sensitive to momentary voltage disturbances.
- *1325-1996* Recommended practice for reporting field failure data for power circuit breakers.
- *1346-1998* Recommended practice for evaluating electric power system compatibility with electronics process equipment.
- *1366-2001* Full-use guide on electric power distribution reliability indices.
- *1409* Guide for application of power electronics for power quality improvement on distribution systems rated 1 through 38 kV (in preparation).
- *1453-2004* Recommended practice for measurement and limits of voltage fluctuations and associated light flicker on ac power systems.
- *1459-2000* Trial-use standard definitions for the measurement of electric power quantities under sinusoidal, nonsinusoidal, balanced, or unbalanced conditions.
- *1504* Guide for the specification and application of active harmonic filters (in preparation).
- *1547-2003* Standard for interconnecting distributed resources with electric power systems.
- *1547.1-2005* Standard for conformance test procedures for equipment interconnecting distributed resources with electric power systems.

- *1547.2* Application guide for IEEE 1547 standard for interconnecting distributed resources with electric power systems (in preparation).
- *1547.3* Guide for monitoring, information exchange, and control of distributed resources interconnected with electric power systems (in preparation).
- *1547.4* Guide for design, operation, and integration of distributed resource island systems with electric power systems (in preparation).
- *1547.5* Technical guidelines for interconnection of electric power sources greater than 10 MVA to the power transmission grid (in preparation).
- *1560-2005* Standard for methods of measurement of radio frequency power line interference filter in the range of 100 Hz to 10 GHz.
- *1564* Recommended practice for the establishment of voltage-sag indices (in preparation).
- *1585-2002* Guide for the functional specification of medium voltage (1 to 35 kV) electronic series devices for compensation of voltage fluctuations.
- *1651* Guidelines for reducing bird-related outages (in preparation).
- *C.57.110-1998* Recommended practice for establishing transformer capability when supplying nonsinusoidal load currents.
- *C62.41.1-2002* Guide on the surge environment in LV (1000-V and less) ac power circuits.
- *C62.41.2-2002* Recommended practice on characterization of surges in LV (1000-V and less) ac power circuits.
- *C62.45-2002* Guide on surge testing for equipment connected on LV ac power circuits.
- *C62.48-1995* Guide on interactions between power system disturbances and surge-protective devices.

# BIBLIOGRAPHY


1. ABB Switchgear manual, ABB, 1993.
2. E. Acha, and M. Madrigal. *Power System Harmonics—Computer Modelling and Analysis*, Wiley, Chichester, 2001.
3. R. A. Adams, S. W. Middlekauf, E. H. Camm, and J. A. McGee. Solving customer power quality problems due to voltage magnification. *IEEE Transactions on Power Delivery*, 13(4): 1515–1520, October 1998.
4. E. Agneholm, cold load pick-up. PhD thesis, Department of Electric Power Engineering, Chalmers University of Technology, Gothenburg, Sweden, 1998.
5. J. Amentegui, PQ Measurements Procedure in Spain, Paper presented at the Seventeenth International Conference on Electricity Distribution, Round Table on Power Quality at the Interface Transmission and Distribution, Barcelona, Spain, May 2003.
6. P. M. Anderson, and A. A. Fouad. *Power System Control and Stability*. IEEE Press, New York, 2003.
7. P. M. Andersson, and R. G. Farmer. *Series Compensation of Power Systems*. PBLSH!, Encinitas, CA, 1996
8. L. Angrisani, P. Daponte, M. D.' Apuzzo, and A. Testa. A measurement method based on the wavelet transform for power quality analysis. *IEEE Transactions on Power Delivery*, 13(4): 990–998, October 1998.
9. J. Arrillaga, M. H. J. Bollen, and N. R. Watson. Power quality following deregulation. *Proceedings IEEE*, 88(2): 46–261, February, 2000.
10. J. Arrillaga, D. A. Bradley, and P. S. Bodger. *Power System Harmonics*. Wiley, Chichester, 1985.



*Signal Processing of Power Quality Disturbances.* By Math H. J. Bollen and Irene Yu-Hua Gu

11. J. Arrillaga, B. C. Smith, N. R. Watson, and A. R. Wood. *Power System Harmonic Analysis*. Wiley, Chichester, 1997.

12. J. Arrillaga, and N. R. Watson. *Power System Harmonics*, 2nd ed. Wiley, Chichester, 2003.

13. J. Arrillaga, N. R. Watson, and S. Chen. *Power System Quality Assessment*. Wiley, Chichester, 2000.

14. R. Arseneau, and M. Ouellette. The effects of supply harmonics on the performance of compact fluorescent lamps. *IEEE Transactions on Power Delivery*, 8(2): 473–479, April 1993.

15. M. T. Aung, J. V. Milanovic, and C. P. Gupta, Propagation of asymmetrical sags and the influence of boundary crossing lines on voltage sag prediction. *IEEE Transacions on Power Delivery*, 19(4): 1819–1827, October 2004.

16. H. Awad. Vector control of static series compensators for mitigation of voltage dips, Licentiate thesis, Department of Electric Power Engineering, Chalmers University of Technology, Gothenburg, Sweden, 2002.

17. H. Awad, J. Svensson, and M. H. J. Bollen. Tuning software phase-locked loop for static series compensator. *IEEE Transactions on Power Delivery*, 20(1):300–308, January 2005.

18. P. Axelberg. Measurement Methods for Calculating the Direction to a Flicker Source, Licentiate thesis. Chalmers University of Technology, Gothenburg, Sweden, March 2003.

19. Y. Baghzouz (chair), Time-varying harmonics: Part II—harmonic summation and propagation. *IEEE Transactions on Power Systems*, 17(1): 279–285, January 2002.

20. G. Ban, L. Prickler, and G. Banfai, 750 kV reactive power control, automatic reclosing and overvoltage protection. Paper presented at the IEEE Bologna Power Tech Conference, Bologna, Italy, June 2003.

21. P. P. Barker, Determining the impact of distributed generation on power systems. Paper presented at the IEEE Power Engineering Society Summer Meeting, Seattle, WA, July 2000.

22. J. Barros, and J. M. Drake. Realtime fault detection and classification in power systems using microprocessors. *IEE Proceedings—Generation, Transmission, Distribution*, 141(4): 315–322, July 1994.

23. L. F. Beites, J. G. Mayordomo, A. Hernadez, and R. Asensi. Harmonics, interharmonics and unbalances of arc furnaces: A new frequency domain approach. *IEEE Transactions on Power Delivery*, 16(4): 661–668, October 2001.

24. L. Berthet, D. Boudou, X. Mamo, P. Eyrolles, and J. Martinon. State of play of harmonic levels in the French low-voltage networks. Paper presented at the Seventeenth International Conference on Electricity Distribution, Barcelona, Spain, May 2003.

25. K. Bhattacharya, M. H. J. Bollen, and J. E. Daalder. *Operation of Restructured Power Systems*, Kluwer Academic, Boston, 2001.

26. R. Billinton, and R. N. Allan. *Reliability Assessment of Large Electric Power Systems*. Kluwer Academic, Boston, 1988.

27. R. Billinton, and R. N. Allan. *Reliability Evaluation of Power Systems*, 2nd ed. Plenum, New York, 1996.

28. R. Billinton, and W. Li. *Reliability Assessment of Electric Power Systems Using Monte Carlo Simulation*. Plenum, New York, 1994.

29. M. H. J. Bollen. Extensive Testing of Algorithms for Travelling-Wave Based Protection. Paper presented at the Fourth International Conference on Developments in Power System Protection, London, 1989.

30. M. H. J. Bollen. The influence on motor reacceleration on voltage sags. *IEEE Transactions on Industry Applications*, 31(4): 667–674, July 1995.

31. M. H. J. Bollen. Fast assessment methods for voltage sags in distribution systems. *IEEE Transactions on Industry Applications*, 31(6): 1414–1423, November 1996.

32. M. H. J. Bollen. Characterisation of voltage sags experienced by three-phase adjustable-speed drives. *IEEE Transactions on Power Delivery*, 12(4): 1666–1671, October 1997.

33. M. H. J. Bollen. *Understanding Power Quality Problems—Voltage Sags and Interruptions*. IEEE Press, New York, 2000.

34. M. H. J. Bollen. Definitions of Voltage Unbalance. *IEEE Power Engineering Review*, 22(11): 49–50, November 2002.

35. M. H. J. Bollen. Algorithms for characterizing measured three-phase unbalanced voltage dips. *IEEE Transactions on Power Delivery*, 18(3): 937–944, July 2003.

36. M. H. J. Bollen. Additions to the method of critical distances for stochastic assessment of voltage sags. *Paper presented at the IEEE Power Engineering Society Winter Meeting*, New York, February 1999.

37. M. H. J. Bollen, and R. A. A. de Graaff. Behavior of ac and dc drives during voltage sags with phase-angle jump and three-phase unbalance. Paper presented at the IEEE Power Engineering Society Winter Meeting, New York, February 1999.

38. M. H. J. Bollen, Ph. Goossens, and A. Robert. Assessment of Voltage Dips in hv-networks: Deduction of complex voltages from the measured rms-voltages. *IEEE Transactions on Power Delivery*, 19(2): 783–790, April 2004.

39. M. H. J. Bollen, M. Häger, and C. Roxenius. Effect of induction motors and other load on voltage dips: Theory and measurements. Paper presented at the IEEE Power Tech Conference, Bologna, June 2003.

40. M. H. J. Bollen, and M. Häger. Power quality: Interactions between distributed energy resources, the grid, and other customers. Paper presented at the First International Conference on Renewable Energy Sources and Distributed Energy Resources, Brussels, December 2004.

41. M. H. J. Bollen, M. Häger, and C. Roxenius. Voltage dips in distribution systems: Load effects, measurements and theory. Paper presented at the Seventh International Conference on Electricity Distribution (CIRED), Barcelona, May 2003.

42. M. H. J. Bollen, and A. Sannino. Voltage control with inverter-based distributed generation. *IEEE Transactions on Power Delivery*, 20(1): 519–520, January 2005.

43. M. H. J. Bollen, and E. Styvaktakis. Characterization of three-phase unbalanced dips (as easy as one two three?). Paper presented at the IEEE Power Summer Meeting, Seattle, WA, July 16–20, 2000.

44. M. H. J. Bollen, E, Styvaktakis, and I. Y. H. Gu. Categorization and analysis of power-system transients. *IEEE Transactions on Power Delivery*, 20(3):2298–2306, July 2005.

45. M. H. J. Bollen, and P. T. M. Vaessen, Frequency spectra for admittances and voltage transfers measured on a three-phase power transformer. EUT-Report 87-E-191, Eindhoven University of Technology, Faculty of Electrical Engineering, The Netherlands, October 1987.

46. M. H. J. Bollen, and L. D. Zhang. Analysis of voltage tolerance of ac adjustable-speed drives for three-phase balanced and unbalanced sags. *IEEE Transactions on Industry Applications*, 36(3): 905–910, May/June 2000.

47. E. Bompard, E. Carpaneto, G. Chicco, P. Ribaldone, and C. Vercellino. The impact of public lighting on voltage distortion in low voltage distribution systems. *IEEE Transactions on Power Delivery*, 16(4): October 2001, 752–757.

48. The New Encyclopaedia Britannica. 1989.

49. Encyclopaedia Britannica DVD. 2001

50. D. L. Brooks, R. C. Dugan, M. Waclawiak, and A. Sundaram. Indices for assessing utility distribution system RMS variation performance. *IEEE Transactions on Power Delivery*, 13(1): January 1998, 254–259.

51. R. E. Brown. *Electric Power Distribution Reliability*. Marcel Dekker, New York, 2002.

52. C. J. C. Burges. A tutorial on support vector machines for pattern recognition. *Data Mining and Knowledge Discovery*, 2, 1998, 121–167,

53. J. J. Burke. *Power Distribution Engineering—Fundamentals and Applications*, Marcel Dekker, New York, 1994.

54. R. Carbone, D. Mennitini, R. E. Morrison, and A. Testa. Harmonic and interharmonic distortion modeling in multiconverter system. *IEEE Transactions on Power Delivery*, 10(3): July 1995, 1685–1692.

55. A. B. Carlson. *Communication Systems*. 2nd ed., McGraw-Hill Kogakuska, Tokyo, 1975.

56. R. Carmiona, W.-L. Hwang, and B. Torresani. *Practical Time-Frequency Analysis*, Academic, San Diego, CA, 1998.

57. D. Carnovale, and D. Ellis. Mind your P's and Q's. *IEEE Industry Applications Magazine*. March–April 2003, pp. 55–63.

58. A. Cavallini, G. C. Montanari, L. Pitti, and D. Zaninelli. ATP simulation for arc-furnace flicker investigation. *European Transactions on Electric Power*, 5(3): May/June 1995, 165–172.

59. X. X. Chen. A wavelet-based method to detect, locate, quatify and identify slight deviations of amplitude and phase in power systems. Paper presented at the IEEE Power Engineering Society General Meeting, Toronto, July 2003.

60. R. F. Chu, and J. J. Burns. Impact of Cycloconverter Harmonics. *IEEE Transactions on Industry Applications*, 25(3): May/June 1989, 427–435.

61. J. Chung, E. J. Powers, W. M. Grady, and S. C. Bhatt. New robust voltage sag disturbance detector using and adaptive prediction error filter. Paper presented at the IEEE Power Engineering Society Summer Meeting, Edmonton, Alta 1999.

62. J. Chung, E. J. Powers, W. M. Grady, and S. C. Bhatt. Power disturbance classifier using a rule-based method and wavelet packet-based hidden Markov model. *IEEE Transactions on Power Delivery*, 17(1): January 2002, 233–241.

63. CIGRE Working Group 36-05. Harmonics, characteristic parameters, methods of study, estimates of existing values in the network. Paper presented at the International Conference on Large High Voltage Networks (CIGRE), Paris 1981.

64. Power quality indices and objectives. Final report by CIGRE working group C4.07, CIGRE, Paris February 2004.

65. Specification of radio disturbance and immunity measuring apparatus and methods. Part 1: Radio disturbance and immunity measuring apparatus. CISPR 16-1, International Electrotechnical Commission, Geneva, October 1999.

66. C. K. Chui. *An Introduction to Wavelets*. Academic, Boston, 1992.
67. C. K. Chui. *Wavelets: A Tutorial in Theory and Applications*, Academic, Boston, 1992.
68. E. Clarke. Circuit analysis of AC systems. *Symmetrical and Related Components*, General Electric Co., Schenectady, NY, 1950.
69. R. Collantes-Bellido, and T. Gomez. Identification and modeling of a three-phase arc furnace for voltage disturbance simulation. *IEEE Transactions on Power Delivery*, 12(4): 1812–1817, October 1997.
70. E. R. Collins, and M. A. Bridgwood. The impact of power system disturbances on ac-coil contactors. Paper presented at the IEEE Annual Technical Conference on Textile, Fiber and Film Industry, Greenville, SC, 1997.
71. E. R. Collins, and F. Zapardiel. An experimental assessment of ac contactors during voltage sags. Paper presented at the IEEE International Symposium on Industrial Electronics, Guimaraes, 1997.
72. Report to the director of the office of gas and electricity markets. National Grid Company, Warwick, May 2001.
73. L. E. Conrad. Proposed chapter 9 for prediction voltage sags (dips) in revision to IEEE Std 493, the gold book. *IEEE Transactions on Industry Applications*, 30(3): May 1994, 805–821.
74. L. E. Conrad, and M. H. J. Bollen. Voltage sag coordination for reliable plant operation. *IEEE Transactions on Industry Applications*, 33(6): November/December 1997, 1465–1471.
75. L. E. Conrad, K. Little, and C. Grigg. Predicting and preventing problems associated with remote fault-clearing voltage dips. *IEEE Transactions on Industry Applications*, 27(1): January 1991, 167–172.
76. D. De Macado Correia, and D. De Oliveira Campones. Probability distribution of the sag performance of some Brazilian network busbars. Paper presented at the CIGRE Sessions, Paris, 2004.
77. Working group on quality of electricity supply Council of European Energy Regulators. Quality of electricity supply: Initial benchmarking on actual levels, standards and regulatory strategies, April 2001.
78. R. G. Cowell, S. L. Lauritzen, and D. J. Spiegelhater. *Probabilistic Networks and Expert Systems*, 2nd printing. Springer, 2003.
79. N. Cristianini, and J. Shawe-Taylor. *An Introduction to Support Vector Machines*. Cambridge University Press, Cambridge, 2000.
80. C. Dai, and Y. Baghzouz. On the voltage profile of distribution feeders with distributed generation. Paper presented at the IEEE Power Engineering Society General Meeting, Toronto, Canada, July 2003.
81. P. K. Dash, S. Mishra, M. M. A. Salama, and A. C. Liew. Classification of power system disturbances using a fuzzy expert system and Fourier linear combiner. *IEEE Transactions on Power Delivery*, 15(2): April 2000, 472–477.
82. A. De, and E. E. Mombello. Recognition of impulse fault patterns in transformers using Kohonen's self-organizing feature map. *IEEE Transactions on Power Delivery*, 17(2): April 2002, 489–494.
83. M. M. de Oliveira. Power electronics for mitigation of voltage sags and improved control of ac power systems. PhD thesis, Department of Electric Power Engineering, Royal Institute of Technology, Stockholm, Sweden, 2000.

84. P. Delarue, P. Bartholomeus, F. Minne, and E. de Jaeger. Study of harmonic currents introduced by three-phase PWM-converters connected to the grid. Paper presented at the Seventeenth International Conference on Electricty Distribution (CIRED), Barcelona, Spain, May 2003.

85. G. Desquilbet, C. Foucher, and P. Fauquembergue. Statistical analysis of voltage dips. International Conference on Power Quality. End-Use Applications and Perspectives, PQA '94, Amsterdam, 1994.

86. J. Deuse. Impact of DER on power systems. Paper presented at the EU-DEEP Workshop, March 2004, http://www.eudeep.com.

87. L. Devroye, L. Gyorfi, and G. Lugosi. *A Probabilistic Theory for Pattern Recognition*. Springer, New York, 1996.

88. L. Ruiz Dias. Regulators indices on power quality. Paper presented at the CIRED, Session 2—Power Quality and EMC, Barcelona, May 2003.

89. E. P. Dick, D. Fisher, R. Marttila, and C. Mulkins. Point-on-wave capacitor switching and adjustable-speed drives. *IEEE Transactions on Power Delivery*, 11(3): 1367–1372, July 1996.

90. M. Didden. Techno-economic analysis of methods to reduce damage due to voltage dips. PhD thesis, Department Electrical Engineering, Katholieke Universiteit Leuven, Leuven, Belgium, December 2003.

91. R. Dohorty, E. Denny, and M. O'Malley. System operation with a significant wind power penetration. Paper presented at the IEEE Power Engineering Society, General Meeting, Denver, CO, 2004.

92. D. S. Dorr, M. B. Hughes, T. M. Gruzs, R. E. Juewicz, and J. L. McClaine. Interpreting recent power quality surveys to define the electrical environment. *IEEE Transactions on Industry Applications*, 33(5): 1480–1487, November 1997.

93. J. A. K. Douglas. Voltage quality levels. Paper presented at the CIRED 2003, Session 2—Power Quality and EMC, Barcelona, 2003.

94. D. P. Bertsekas. *Nonlinear Programming*. Athena Scientific, Belmont, MA, 1995.

95. *Handbook of Power Signatures*, 2nd ed. Dranetz BMI, Edison, NJ, 1997.

96. C. Q. Du. The control of VSC-HVDC and its use for large industrial power systems. Licentiate report, Department of Electric Power Engineering, Chalmers University of Technology, Gothenburg, Sweden, December 2003.

97. L. H. S. Duarte and M. F. Alves. The degradation of power capacitors under the influence of harmonics. Paper presented at the International Conference on Harmonics and Quality of Power, Rio de Janeiro, 2002.

98. R. O. Duda, P. E. Hart, and D. G. Stork. *Pattern Classification*, 2nd ed. Wiley, New York, 2001.

99. R. C. Dugan, M. F. McGranaghan, S. Santosa, and H. W. Beaty. *Electric Power Systems Quality*, 2nd ed. McGraw-Hill, New York, 2003.

100. D. M. Dunsmore, E. R. Taylor, B. F. Wirtz, and T. L. Yanchula. Magnification of transient voltages in multi-voltage-level, shunt-capacitor-compensated circuits. *IEEE Transactions on Power Delivery*, 7(2): April 1992, 644–673.

101. L. L. Grisby (Ed.). *The Electric Power Engineering Handbook*. CRC Press, Boca Raton, FL, 2001.

102. S. Haykin (Ed.) *Kalman Filtering and Neural Networks*. Wiley, New York, 2001.

103. W. A. Elmore, C. A. Kramer, and S. E. Zocholl. Effect of waveform distortion on protective relays. *IEEE Transactions on Industry Applications*, 29(2): March/April 1993, 404–411.

104. A. E. Emanuel. IEEE standard 1459: Questions and future direction—Quo Vadis? *Paper presented at the IEEE Power Engineering Society General Meeting*, Toronto, Canada, 2003.

105. A. E. Emanuel and J. A. Orr. Six-pulse converter atypical harmonics caused by second harmonic voltage. Paper presented at the International Conference on Harmonics and Quality of Power, Rio de Janeiro, 2002.

106. Voltage characteristics of electricity supplied by public distribution systems, EN 50160. CENELEC: European Committee for Electrotechnical Standardization, Brussels, Belgium, November 1999. Also available as a national standard from most European national standard organizations.

107. J. Endreyni. *Reliability Modeling in Electric Power Systems*. Wiley, New York, 1978.

108. J. H. R. Enslin, W. T. J. Hulshorst, A. M. S. Atmadji, P. J. M. Heskes, A. Kotsopoulos, J. F. G. Cobben, and P. van der Sluijs. Harmonic interaction between large numbers of photovoltaic inverters and the distribution network. Paper presented at the IEEE Bologna Power Tech Conference, Bologna, Italy, June 2003.

109. M. Etezadi-Amoli and T. Florence. Power factor and harmonic distortion characteristics of energy efficient lamps. *IEEE Transactions on Power Delivery*, 4(3): July 1989, 1965–1969.

110. L. Fausett. *Fundamentals of Neural Networks*. Prentice-Hall, Englewood Cliffs, NJ, 1994.

111. W. E. Feero and W. B. Gish. Overvoltages caused by DSG operation: Synchronous and induction generators. *IEEE Transactions on Power Delivery*, 1(1): January 1986, 258–264.

112. R. Flores. Signal processing tools for power quality event classification. Licentiate report, Department of Signals and Systems Chalmers University of Technology, Gothenburg, Sweden, November 2003.

113. R. A. Flores, I. Y. H. Gu, and M. H. J. Bollen. Positive and negative sequence estimation for unbalanced voltage dips. Paper presented at the IEEE Power Engineerings Society, General Meeting, Toronto, Canada, July 2003.

114. K. Fukunaga. *Introduction to Statistical Pattern Recognition*, 2nd ed. Academic Press Professional, San Diego, CA, 1990.

115. J. F. Fuller, E. F. Fuchs, and D. J. Roesler. Influence of harmonics on power distribution system protection. *IEEE Transactions on Power Delivery*, 3(2): April 1988, 549–557.

116. Z.-L. Gaing. Implementation of power disturbance classifier using wavelet-based neural networks. Paper presented at the IEEE Bologna Power Tech Conference, Bologna, Italy, June 2003.

117. M. Galassi. Determinazione degli instanti iniziali e finali dei buchi di tensione: Analisi teoretica e sperimentale. Master's thesis, Università Degli Studi di Bologna, 2001.

118. A. Ghosh and G. Ledwich. *Power Quality Enhancement Using Custom Power Devices*. Kluwer Academic, Boston, 2002.

119. A. A. Girgis, C. M. Fallon, J. C. P. Rubino, and R. C. Catoe. Harmonics and transient overvoltages due to capacitor switching. *IEEE Transactions on Industry Applications*, 29(6): November/December 1993, 1184–1188.

120. A. A. Girgis, D. G. Hart, and W. B. Chang. An adaptive scheme for digital protection of power transformers. *IEEE Transactions on Power Delivery*, 7(2): April 1992, 546–553.

121. A. A. Girgis and J. Qui. Measurement of the parameters of slowly time varying high frequency transients. *IEEE Transactions on Instrumentation and Measurement*, 38(6): December 1989, 1057–1063.

122. I. F. Gonos, M. B. Kostic, and F. V. Topalis. Harmonic distortion in electric power systems introduced by compact fluorescent lamps. Paper presented at the IEEE Power Tech Conference, Budapest, 1999.

123. V. J. Gosbell, B. S. P. Perera, and H. M. S. C. Herath. Unified power quality index (UPQI) for continuous disturbances. Paper presented at the International Conference on Harmonics and Quality of Power (ICHQP 2002), Rio de Janeiro, October 2002.

124. V. J. Gosbell, D. Robinson, and S. Perera. The analysis of utility voltage sag data. Paper presented at the International Power Quality Conference (IPQC 2002), Singapore, October 2002.

125. A. Greenwood. *Electrical Transients in Power Systems*, 2nd ed. Wiley, New York, 1991.

126. R. Grünbaum, M. Noroozian, and A. Palesjö. Connecting the London underground to the public grid: Solutions for safeguarding of power quality. Paper presented at the Seventeenth International Conference on Electricity Distribution (CIRED), Barcelona, Spain, May 2003.

127. I. Y. H. Gu, N. Ernberg, E. Styvaktakis, and M. H. J. Bollen. Statistical-based sequential method for fast online detection of fault-induced voltage dips. *IEEE Transactions on Power Delivery*, 19(2): April 2004, 497–504.

128. I. Y. H. Gu, E. Styvaktakis, and M. H. J. Bollen. Analyzing power disturbances using the residuals of AR models. *IEEE Power Engineering Review*, 20(4): April 2000, 60–62.

129. Y. H. Gu and M. H. J. Bollen. Time-frequency and time-scale domain analysis of voltage disturbances. *IEEE Transactions on Power Delivery*, 15(4): October 2000, 1279–1284.

130. L. Guash, L. Corcoles, and J. Pedra. Effects of voltage sags on three-phase three-legged transformers. Paper presented at the Tenth International Conference on Harmonics and Quality of Power, Rio de Janeiro, 2002.

131. L. Guash, L. Corcoles, and J. Pedra. Effects of symmetrical and unsymmetrical voltage sags on induction motors. *IEEE Transactions on Power Delivery*, 19(2): April 2004, 774–782.

132. E. W. Gunther. Preliminary analysis of measurement data from New York blackout of 2003. http://www.enernex.com/special/blackout.htm, August 2003.

133. K. Gurney. *An Introduction to Neural Networks*. UCL Press, London, 1997.

134. R. Gustavsson. *Praktisk Elkvalitet*. NORBO KraftTeknik AB, Sweden, 2003.

135. S. M. Halpin. Harmonic modeling and simulation requirements for the revised IEEE standard 519–1992. Paper presented at the Seventeenth IEEE Power Engineering Society General Meeting, Denver, CO, June 2004.

136. S. Haykin. *Neural Networks—A Comprehensive Foundation*. Prentice-Hall, Upper Saddle River, NJ, 1999.

137. S. Haykin. *Kalman Filtering and Neural Networks*. Wiley, New York, 2001.

138. C. Herath, V. Gosbell, S. Perera, and D. Robinson. A transient index for reporting power quality surveys. Paper presented at the Seventeenth International Conference on Electricity Distribution, Barcelona, May 2003.

139. R. Herbrich. *Learning Kernel Classifiers, Theory and Algorithms.* MIT Press, Cambridge, MA, 2002.

140. S. Herraiz Jaramillo, G. T. Heydt, and E. O'Neill-Carrillo. Power quality indices for aperiodic voltages and currents. *IEEE Transactions on Power Delivery*, 15(2): April 2000, 784–790.

141. G. T. Heydt. *Electric Power Quality*. Stars in a circle publications, West LaFayette, IN, 1991.

142. W. A. Hill and S. C. Kapoor. Effect of two-level PWM sources on plant power system harmonics. Paper presented at the IEEE Industry Applications Conference, St. Louis, MO, 1998.

143. M. Hird, N. Jenkins, and P. Taylor. An active 11 kV voltage controller: Practical considerations. Seventeenth International Conference on Electricity Distribution (CIRED), Barcelona, May 2003.

144. J. Huang, M. Negnevitsky, and D. T. Nguyen. A neural-fuzzy classifier for recognition of power quality disturbances. *IEEE Transactions on Power Delivery*, 17, 2002, 609–616.

145. S.-J. Huang, C.-T. Hsieh, and C.-L. Huang. Application of Morlet wavelets to supervise power system disturbances. *IEEE Transactions on Power Delivery*, 14(1): January 1999, 235–243.

146. M. Ianoz. Lightning and EMC. V. Cooray (Ed.), *The Lightning Flash*. Institution of Electrical Engineers, London, 2003.

147. W. R. A. Ibrahim and M. M. Morcos. Artificial intelligence and advanced mathematical tools for power quality applications: A survey. *IEEE Transactions on Power Delivery*, 17(2): April 2002, 668–673.

148. Electromagnetic compatibility (EMC), Part 1, Section 1: Application and interpretation of fundamental defintions and terms, IEC 61000-1-1.

149. Electromagnetic compatibility (EMC), Part 2, Section 2: Compatibility levels for low-frequency conducted disturbances and signalling in public low-voltage power supply systems, IEC 61000-2-2.

150. Electromagnetic compatibility (EMC), Part 3, Section 2: Limits for harmonic current emissions (equipment input current $\leq$ 16 A per phase), IEC 61000-3-2.

151. Electromagnetic compatibility (EMC), Part 3, Section 4: Limitation of emission of harmonic currents in low-voltage supply systems for equipment with rated current $>$16 A, IEC 61000-3-4.

152. Electromagnetic compatibility (EMC), Part 3, Section 5: Limitation of voltage fluctuations and flicker in low-voltage power supply systems for equipment with rated current greater than 16 A, IEC 61000-3-5.

153. Electromagnetic compatibility (EMC), Part 3, Section 6: Assessment of emission limits for distorting loads in MV and HV power systems, IEC 61000-3-6.

154. Electromagnetic compatibility (EMC), Part 3, Section 11: Limitation of voltage changes, voltage fluctuations and flicker in public low-voltage supply systems—equipment with rated current $\leq$75 A and subject to conditional connection, IEC 61000-3-11.

155. Electromagnetic compatibility (EMC), Part 4, Section 7: General guide on harmonics and interharmonics measurements and instrumentation, for power supply systems and equipment connected thereto, IEC 61000-4-7.

156. Electromagnetic Compatibility (EMC), Part 4, Section 11: Voltage dips, short interruptions and voltage variations immunity tests, IEC 61000-4-11, Edition 2.

157. Electromagnetic compatibility (EMC), Part 4, Section 15: Flickermeter—functional and design specifications, IEC 61000-4-15, Edition 1.1.
158. Electromagnetic compatibility (EMC), Part 4, Section 30: Power quality measurement methods, IEC 61000-4-30.
159. The IEEE standard dictionary of electrical and electronics terms, 6th ed., IEEE Std. 100-1996.
160. Recommended practice for electric power distribution for industrial plants, IEEE Std.141-1993.
161. Recommended practice for the design of reliable industrial and commercial power systems. IEEE Std. 493-1997.
162. Recommended practices and requirements for harmonic control in electrical power systems, ANSI/IEEE Std. 519-1992.
163. Standard terms for reporting and analyzing outage occurrences and outage states of electrical transmission facilities, ANSI/IEEE Std. 859-1987.
164. Recommended practice for powering and grounding electronic equipment, IEEE Std. 1100-1999.
165. Recommended practice for monitoring electric power quality, IEEE Std. 1159-1995.
166. IEEE project group P1159.2, voltage-sag characteristics. The work from this group only resulted in a draft document, unfortunately never in a final document.
167. Guide for service to equipment sensitive to momentary voltage disturbances, IEEE Std. 1250-1995.
168. Recommended practice for evaluating electric power system compatibility with electronic process equipment, IEEE Std. 1346-1998.
169. Guide for electric power distribution reliability indices, IEEE Std. 1366-2003.
170. Draft recommended practice for measurement and limits of voltage flicker on AC power systems, IEEE Std. 1453, Draft 9, June 2004.
171. Standard definitions for the measurement of electric power quantities under sinusoidal, nonsinusoidal, balanced and unbalanced conditions, IEEE Std. 1459-2000.
172. Draft recommended practice for the establishment of voltage sag indices, IEEE Std. 1564, Draft 6, June 2004.
173. Recommended practice for establishing transformer capability when supplying nonsinusoidal load currents, IEEE Std. C57.110-1986.
174. The INSPEC database is administered by the Institution of Electrical Engineerings. It was accessed through the AXIOM online research service (http://axiom.iop.org/) as subscribed to by Chalmers University of Technology.
175. I. Y. H. Gu and E. Styvaktakis. Bridge the gap: Signal processing for power quality applications. *Electric Power System Research Journal*, 66(1): March 2003, 83–96.
176. P. Jackson. *Introduction to Expert Systems*. Addison-Wesley, Harlow, United Kingdom, 1998.
177. D. A. N. Jacobson. Examples of ferroresonance in a high voltage power system. Paper presented at the IEEE Power Engineering Society General Meeting, Toronto, Canada, July 2003.
178. N. Jenkins, R. Allan, P. Crossley, D. Kirschen, and G. Strbac. *Embedded Generation*. Institution of Electrical Engineers, London, 2000.

179. N. Kagan, E. L. Ferrera, N. M. Matsuo, S. X. Duarte, J. L. Cavaretti, A. Tenorio, and L. R. Souza. A methodology for the assessment of short duration voltage variations in electric power distribution systems. In Paper presented at the International Conference on Harmonics and Quality of Power, Rio de Janeiro, 2002.

180. K. Kahle, J. Pedersen, T. Larson, and M. M. de Oliveira. The new 18 Mvar, 18 kV static var compensator at CERN: Background, design and commissioning. Paper presented at the Seventeenth International Conference on Electricity Distribution (CIRED), Barcelona, Spain, May 2003.

181. P. G. Kaminski, A. E. Bryson, and S. F. Schmidt. Discrete square root filtering: A survey of current techniques. *IEEE Transactions on Automatic Control*, 16(6): December 1971, 727–736.

182. I. Kamwa, R. Grondin, and D. McNabb. On-line tracking of changing harmonics in stressed power systems: Application to Hydro-Quebec network. *IEEE Transactions on Power Delivery*, 11(4): October 1996, 2020–2027.

183. S. M. Kay. *Fundamentals of Statistical Signal Processing: Estimation Theory*. Prentice-Hall, Englewood Cliffs, NJ, 1993.

184. S. M. Kay. *Fundamentals of Statistical Signal Processing: Vol.II Detection Theory*. Prentice-Hall, Upper Saddle River, NJ, 1998.

185. W. E. Kazibwe and M. H. Sendaula. *Electric Power Quality Control Techniques*. Van Nostrand Reinhold, New York, 1993.

186. A. E. Kiprakis and A. R. Wallace. Hybrid control of distributed generators connected to weak rural networks to mitigate voltage variation. Paper presented at the Seventeenth International Conference on Electricity Distribution (CIRED), Barcelona, May 2003.

187. R. Koch, R. Balgobind, P. Jonsson, I. Sigwebela, R. McCrach, D. Bhana, and J. Wilson. Power quality management in a regulated environment: The South African experience. Paper presented at the CIGRE Sessions, Paris, 2004.

188. L. Kojovic. Impact of DG and voltage regulator interaction on distribution system voltage regulations. Paper presented at the Seventeenth International Conference on Electricity Distribution (CIRED), Barcelona, May 2003.

189. B. Kosko. *Fuzzy Engineering*. Prentice-Hall, Upper Saddle River, NJ, 1997.

190. D. O. Koval. How long should power system disturbance site monitoring be to be significant? *IEEE Transactions on Industry Applications*, 26(4): July/August 1990, 705–710.

191. D. O. Koval and C. Carter. Power quality characteristics of computer load. *IEEE Transactions Industry Applications*, 33(3): 613–621, May/June 1997.

192. D. O. Koval and M. B. Hughes. Canadian national power quality survey: Frequency of industrial and commercial voltage sags. *IEEE Transactions on Industry Applications*, 33(3): 622–627, May/June 1997.

193. E. Lakervi and E. J. Holmes. *Electricity Distribution Network Design*. Institution of Electrical Engineers, London, 1995.

194. G. Lalor, J. Ritchie, S. Rourke, D. Flynn, and M. O'Malley. Dynamic frequency control with increasing wind generation. Paper presented at the IEEE Power Engineering Society, General Meeting, Denver, CO, 2004.

195. A. Larsson. Flicker emission of wind turbines during continuous operation. *IEEE Transactionson Energy Conversion*, 17(1): March 2002, 114–118.

196. R. C. Leborgne, J. M. Carvalho Filho, J. Policarpo, T. C. Oliveira, A. A. Postal, and L. H. Zaparoli. Alternative methodology for characterization of industrial process sensitity to voltage sags. Paper presented at the IEEE Bologna Power Tech Conference, Bologna, Italy, June 2003.

197. R. C. Leborgne, G. Olguin, and M. H. J. Bollen. The influence of PQ-monitor connection on voltage dip measurements. Paper presented at the Mediterranean Power Engineering Conference (MEDPOWER), Lemesos, Cyprus, 2004.

198. H.-J. Lee, D.-Y. Park, B.-S. Ahn, Y.-M. Park, J.-K. Park, and S. S. Venkata. A fuzzy expert system for the integrated fault diagnosis. *IEEE Transactions on Power Delivery*, 15:833–838, 2002.

199. H. Leite, H. Y. Li, N. Jenkins, and P. F. Gale. Real-time voltage regulation of distribution networks with distributed generation. Paper presented at the seventeenth International Conference on Electricity Distribution (CIRED), Barcelona, May 2003.

200. E. E. Lewis. *Introduction to Reliability Engineering*. Wiley, New York, 1987.

201. D. Lin, T. Batan, E. F. Fuchs, and W. M. Grady. Harmonic losses of single-phase induction motors under nonsinusoidal voltages. *IEEE Transactions on Energy Conversion*, 11(2):273–286, June 1996.

202. J. Lundquist. On harmonic distortion in power systems. Licentiate thesis, Department of Electric Power Engineering, Chalmers University of Technology, Gothenburg, Swedem, 2001.

203. J. Machowski, J. W. Bialek, and J. R. Bumby. *Power System Dynamics and Stability*. Wiley, Chichester, 1997.

204. K. Macken. Analysis and control of renewable generation units used to improve power quality. PhD thesis, Department Electrical Engineering, November 2003, Leuven, Belgium, Katholieke Universiteit Leuven.

205. S. Mallat and W. L. Hwang. Singularity detection and processing with wavelets. *IEEE Transactions on Information Theory*, 38(2):617–643, March 1992.

206. X. Mamo and J.-L. Javerzac. Power quality indicators. Paper presented at the IEEE Porto Power Tech, Porto, Portugal, September 2001.

207. D. G. Manolakis, V. K. Ingle, and S. M. Kogon. *Statistical and Adaptive Signal Processing*. McGraw-Hill, New York, 2000.

208. A. Mansoor. Low-order harmonic cancellation: Impact of low-voltage network topology. Paper presented at the IEEE Power Engineering Society Winter Meeting, New York, 1999.

209. A. Mansoor, W. M. Grady, A. H. Chowdhury, and M. J. Samotyj. An investigation of harmonics attenuation and diversity among distributed single-phase power electronic loads. *IEEE Transactions on Power Delivery*, 10(1): 467–473, January 1995.

210. A. Mansoor, W. M. Grady, P. T. Staats, R. S. Thallam, M. T. Doyle, and M. J. Samotyj. Predicting the net harmonic currents produced by large numbers of distributed single-phase computer loads. *IEEE Transactions on Power Delivery*, 10(4): 2001–2006, October 1995.

211. R. Marconato. *Electric Power Systems,* Vol. 1: *Background and Basic Components*. Italian Electrotechnical Committee, Milan, 2002.

212. R. M. Mathur and R. J. Varma. *Thyristor-Based FACTS Controllers for Electrical Transmission Systems*. IEEE Press, Piscataway, NJ, 2002.

213. C. E. McCoy and B. L. Floryancic. Characteristics and measurement of capacitor switching at medium voltage distribution level. *IEEE Transactions on Industry Applications*, 30(6): 1480–1489, November/December 1994.

214. M. McGranaghan, A. Mansoor, and C. Miller. LFEIC survey of residential harmonic voltage distortion levels in Europe—discussion of survey implications. Paper presented at the CIRED 2003—Proceedings of Session 2—Power Quality and EMC, Barcelona, 2003.

215. M. F. McGranaghan, T. E. Grebe, G. Hensley, M. Samotyj, and T. Singh. Impact of utility-switched capacitors on customer systems—II: Adjustable-speed drive concerns. *IEEE Transactions on Power Delivery*, 6(4): 1623–1628, October 1991.

216. M. F. McGranaghan, D. R. Mueller, and M. J. Samotyj. Voltage sags in industrial power systems. *IEEE Transactions on Industry Applications*, 29(2): 397–403, March/April 1993.

217. C. Melhorn, A. Maitra, A. Sundaram, and A. Mansoor. Power quality predictive maintenance. Paper presented at the International Power Quality Conference, Singapore, October 2002.

218. C. J. Melhorn, R. J. Schwabe, and M. Samotyj. Effect of high efficiency lighting on power quality in public buildings. Paper presented at the IEEE Industry Applications Society Annual Meeting, Orlando, FL, 1995.

219. E. Micheli-Tzanakou. *Supervised and Unsupervised Pattern Recognition*. CRC Press, Boca Raton, FL, 1999.

220. M. Misiti, Y. Misiti, G. Oppenheim, and J.-M. Poggi. Matlab wavelet toolbox user's guide. MathWorks, Natick, MA, 1996.

221. S. K. Mitra. *Digital Signal Processing—A Computer-Based Approach*. McGraw-Hill, Boston, 2001.

222. G. C. Montanari, M. Loggini, A. Cavallini, L. Pitti, and Z. Zaninelli. Arc furnace model for the study of flicker compensation in electrical networks. *IEEE Transactions on Power Delivery*, 9(4):2026–2036, October 1994.

223. T. K. Moon and W. C. Stirling. *Mathematical Methods and Algorithms for Signal Processing*. Prentice-Hall, Upper Saddle River, NJ, 2000.

224. K. R. Müller, S. Mika, G. Rätsch, K. Tsuda, and B. Schölkopf. An introduction to kernel-based learning algorithms. *IEEE Transactions on Neural networks*, 12(2): 181–202, March 2001.

225. Electricity supply—quality of supply. Part 2: Voltage characteristics, compatibility levels, limits and assessment methods. South African National Standard NRS 048-2:2002, South African Bureau of Standards, Pretoria, RSA.

226. Network of experts for standardisation. Power quality in european electricity supply networks. Eurelectric, Brussels, February 2002.

227. The Mathematical Division of the Institute of Chinese Academy of Science. *Mathematical Methods for Filtering of Discrete Time Systems*. Defence Industries Publisher, China, 1975 (in Chinese).

228. G. Olguin. Stochastic assessment of voltage dips caused by faults in large transmission systems. Licentiate thesis, Department of Electric Power Engineering, Chalmers University of Technology, Gothenburg, Sweden, September 2003.

229. G. Olguin. Stochastic prediction and state estimation of voltage dips (preliminary title). PhD thesis, Department of Electric Power Engineering, Chalmers University of Technology, Gothenburg, Sweden, May 2005.

230. A. V. Oppenheim and R. W. Schafer. *Discrete-Time Signal Processing*, 2nd ed. Prentice-Hall, Upper Saddle River, NJ, 1999.

231. J. A. Orr and A. E. Emanuel. On the need for strict second harmonic limits. *IEEE Transactions on Power Delivery*, 15(3): 967–971, July 2000.

232. D. A. Paice. *Power Electronic Converter Harmonics—Multipulse Methods for Clean Power*. IEEE Press, Piscataway, NJ, 1996.

233. M. P. Papadopoulos, S. A. Papathanassiou, S. T. Tentzerakis, and N. G. Boulaxis. Investigation of the flicker emission by grid connected wind turbines. Paper presented at the International Conference on Harmonics and Quality of Power, Athens, 1998.

234. S. Pathanassiou, D. Georgakis, N. Hatziargyriou, A. Engler, and Ch. Hardt. Operation of a prototype microgrid system based on microsources equipped with fast-acting power electronic interfaces. Paper presented at the IEEE Power Electronic Specialist Conference, 2004.

235. F. Z. Peng. Application issues of active power filters. *IEEE Industry Applications Magazine*, pp. 21–30, September/October, 1998.

236. F. Z. Peng. Harmonic sources and filtering approaches. *IEEE Industry Applications Magazine*, pp. 18–25, July/August 2001.

237. C. Di Perna, G. Olguin, P. Verde, and M. H. J. Bollen. On probabilistic system indices for voltage dips. Paper presented at the International Conference on Probabilistic Methods applied to Power Systems (PMAPS), Ames, IA, 2004.

238. P. Pillay and M. Manyage. Definitions of voltage unbalance. *IEEE Power Engineering Review*, 21(5): 50–51, May 2001.

239. P. Pohjanheimo and M. Lehtonen. Equipment sensitivity to voltags sags—test results for contactors, PCs and gas discharge lamps. Paper presented at the Tenth International Conference on Harmonics and Quality of Power, Rio de Janeiro, 2002.

240. P. Pohjanheimo and M. Lehtonen. Introducing prob-a-sag—a probabilistic method for voltage sag management. Paper presented at the Eleventh International Conference on Harmonics and Quality of Power, Lake Placid, NY, 2004.

241. O. Poisson, P. Rioual, and M. Meunier. New signal processing tools applied to power quality analysis. *IEEE Transactions on Power Delivery*, 14(2):561–566, April 1999.

242. O. Poisson, P. Rioual, and M. Meunier. Detection and measurement of power quality disturbances using wavelet transform. *IEEE Transactions on Power Delivery*, 15(3): 1039–1044, July 2000.

243. J. E. Potter. A matrix equation arising in statistical filter theory. NASA Contractor Report, NASA CR-270, August 1965.

244. PQ view website, http://www.pqview.com, accessed July 2004.

245. M. R. Qader, M. H. J. Bollen, and R. N. Allan. Stochastic prediction of voltage sags in a large transmission system. *IEEE Transactions on Industry Applications*, 35(1): 152–162, January 1999.

246. L. R. Rabiner and R. W. Schafer. *Digital Processing of Speech Signals*. Prentice-Hall, Upper Saddle River, NJ. 1978.

247. S. J. Ranade (Chair). Modeling and simulation of the propagation of harmonics in electric power networks part II: Sample systems and examples. *IEEE Transactions on Power Delivery*, 11(1): 452–465, January 1996.

248. S. J. Ranade (Chair). Modeling and simulation of the propagation of harmonics in electric power networks part I: Concepts, models and simulation techniques. *IEEE Transactions on Power Delivery*, 11(1):466–474, January 1996.

249. H. Renner and M. Sakulin. Calculation of voltage dips in meshed grids. Paper presented at the Eleventh International Conference on Harmonics and Quality of Power, Lake Placid, NY, 2004.

250. P.-F. Ribeiro, R. Celio, and M. J. Samotyj. Future analysis tools for power quality. Paper presented at the EPRI Conference on Power Quality and Applications (PQA93), Amsterdam, 1993.

251. D. E. Rice. Adjustable-speed drive and power rectifier harmonics—their effect on power system components. *IEEE Transactions on Industry Applications*, 22(1):161–167, January 1986.

252. B. D. Ripley. *Pattern Recognition and Neural Networks*. Cambridge University Press, Cambridge, 1996.

253. A. Robert and E. De Jaeger. Round table on power quality at the interface T and D. Paper presented at the CIRED 2003, Barcelona, May 2003.

254. A. Robert and J. Marquet. Assessing voltage quality in relation to harmonics, flicker and unbalance. Paper presented at the International Conference on Large Electric Power Systems (CIGRE), paper 36-203, Paris, 1992.

255. R. Robinson (Chair). Power system harmonics—what's being done against them. *Power Quality Assurance*, pp. 26–32, November/December 1995.

256. R. Rüdenberg. *Transient Performance of Electric Power Systems—Phenomena in Lumped Networks*. MIT Press, Cambridge, MA, 1950.

257. Z. Saad-Saoud and N. Jenkins. Models for predicting flicker induced by large wind turbines. *IEEE Transactions on Energy Conversion*, 14(3): 743–748, September 1999.

258. D. D. Sabin, D. L. Brooks, and A. Sundaram. Indices for assessing harmonic distortion from power quality measurements: Definitions and benchmark data. *IEEE Transactions on Power Delivery*, 14(2):489–496, April 1999.

259. D. D. Sabin, T. E. Grebe, D. L. Brooks, and A. Sundaram. Rules-based algorithm for detecting transient overvoltages due to capacitor switching and statistical analysis of capacitor switching in distribution systems. Paper presented at the IEEE Power Engineering Society, Transmission and Distribution Conference, New Orleans, LA, 1999.

260. D. S. Sabin. Analysis of harmonic measurement data. Paper presented at the IEEE Power Engineering Society, Summer Meeting, Chicago, IL, 2002.

261. F. Saccomanno. *Electric Power Systems—Analysis and Control*. IEEE Press, New York, 2003.

262. V. M. M. Saiz and J. B. Guadalupe. Application of Kalman filtering for continuous real-time tracking of power system harmonics. *IEE Proceedings—Generation, Transmission, Distribution*, 144(1): 13–20, January 1997.

263. C. Sankaran. *Power Quality*. CRC Press, Boca Raton, FL, 2002.

264. A. Sannino. Static transfer switch: Analysis of switching conditions and actual transfer time. Paper presented at the IEEE Power Engineering Society Winter Meeting, Columbus, OH, 2001.

265. A. Sannino. STS and induction motors. *IEEE Industry Applications Magazine*, 9(4): 50–57, July/August 2003.

266. A. Sannino, M. H. J. Bollen, and J. Svensson. Voltage tolerance testing of three-phase voltage-source converters. *IEEE Transactions on Power Delivery*, in press.

267. S. Santoso, W. M. Grady, E. J. Powers, J. Lamoree, and S. C. Bhatt. Characteristics of distribution power quality events with Fourier and wavelet transforms. *IEEE Transactions on Power Delivery*, 15(1): 247–254, January 2000.

268. S. Santoso, E. J. Powers, and W. M. Grady. Power quality disturbance data compression using wavelet transform methods. *IEEE Transactions on Power Delivery*, 12(3): 1250–1257, July 1997.

269. S. Santoso, E. J. Powers, W. M. Grady, and A. C. Parsons. Power quality disturbance waveform recognition using wavelet-based neural classifier—part 1: Theoretical foundation. *IEEE Transactions on Power Delivery*, 15(1): 222–228, January 2000.

270. S. Santoso, E. J. Powers, W. M. Grady, and A. C. Parsons. Power quality disturbance waveform recognition using wavelet-based neural classifier—Part 2: Application. *IEEE Transactions on Power Delivery*, 15(1): 229–235, January 2000.

271. J. Schlabbach, D. Blume, and T. Stephanblome. *Voltage Quality in Electrical Power Systems*. Number 36 in IEE Power and Energy Series. Institution of Electrical Engineers, London, 2001.

272. C. Schwaegerl, M. H. J. Bollen, K. Karoui, and A. Yagmur. Voltage control in distribution systems as a limitation of the hosting capacity for distributed energy resources. Paper presented at the International Conference on Electricity Distribution (CIRED), Turin, Italy, June 2005.

273. H. Seljeseth and K. Sand. Krav till spenningskvalitet [Requirements on voltage quality]. Report No. TR A5856, SINTEF Energiforskning AS, Trondheim, Norway, October 2003 (in Norwegian).

274. O. Sen, Z.-X. Song, D. Chen, and J. H. Wang. Application of wavelet soft-detection threshold denoising technique in power quality detection. Paper presented at the International Power Quality Conference, Singapore, October 2002.

275. G. Shafer. *A Mathematical Theory of Evidence*. Princeton University Press, Princeton, NJ, 1976.

276. J. Shawe-Taylor and N. Cristianini. *Kernel Methods for Pattern Analysis*. Cambridge University Press, Cambridge, 2004.

277. T. A. Short, A. Mansoor, W. Sunderman, and A. Sundaram. Site variation and prediction of power quality. *IEEE Transactions on Power Delivery*, 18(4): 1369–1375, October 2003.

278. J. C. Smith, J. Lamoree, P. Vinett, T. Duffy, and M. Klein. The impact of voltage sags on industrial plants. Paper presented at the International Conference on Power Quality. End-Use Aplications and Perspectives, Paris, 1991.

279. G. W. Stagg and A. H. El-Abaid. *Computer Methods in Power System Analysis*. McGraw-Hill, Auckland, 1968.

280. H. Stoltz. Kartläggning av elkvaliteten påSvenska kraftnätet [Mapping the power quality in the Swedish power system]. MSc dissertation, Department of Power Electronics, Royal University of Technology, Stockholm, Sweden, March 1995 (in Swedish).

281. G. Strang and T. Nguyen. *Wavelet and Filter Banks*. Wellesley-Cambridge Press, 1996.

282. E. Styvaktakis. Automating power quality analysis. PhD thesis, Department of Signals and Systems, Chalmers University of Technology, Gothenburg, Sweden, 2002.

283. E. Styvaktakis, M. H. J. Bollen, and I. Y. H. Gu. Expert system for classification and analysis of power system events. *IEEE Transactions on Power Delivery*, 17(2): 423–428, April 2002.

284. E. Styvaktakis, M. H. J. Bollen, and I. Y. H. Gu. Expert system for voltage dip classification and analysis. Paper presented at the IEEE Power Engineering Society Summer Meeting, Vancouver, Canada, July 2001.

285. T. Sun. Power quality of grid-connected wind turbines with DFIG and their interaction with the grid. PhD thesis, Aalborg University of Technology, Institute of Energy Technology, Aalborg, Denmark, May 2004.

286. D. J. Swinden. The arc furnace, Report EC4600/2.88. The Electricity Council, London.

287. L. Tang, D. Hall, M. S. Amotyj, and J. Randolph. Analysis of DC arc furnace application and flicker caused by 187 Hz voltage distortion. *IEEE Transactions on Power Delivery*, 9(2): 1098–1107, April 1994.

288. M. M. Tawfik and M. M. Morcos. ANN-based techniques for estimating fault location on transmission lines using Prony method. *IEEE Transactions on Power Delivery*, 16: 219–224, 2001.

289. C. W. Taylor. *Power System Voltage Stability*. McGraw-Hill, New York, 1994.

290. A. Testa, D. Gallo, and R. Langella. On the processing of harmonics and interharmonics: Using Hanning window in standard framework. *IEEE Transactions on Power Delivery*, 19(1): 28–34, January 2004.

291. T. Thiringer. Power quality measurements performed on a low-voltage grid equipped with two wind turbines. *IEEE Transactions on Energy Conversion*, 11(3): 601–606, September 1996.

292. T. Thiringer and J. A. Dahlberg. Periodic pulsations from a three-bladed wind turbine. *IEEE Transactions on Energy Conversion*, 16(2): 128–133, June 2001.

293. T. Thiringer, T. Petru, and C. Liljegren. Power quality impact of a sea located hybrid wind park. *IEEE Transactions on Energy Conversion*, 16(2): 123–127, June 2001.

294. T. Thiringer, T. Petru, and S. Lundberg. Flicker contribution from wind turbine installations. *IEEE Transactions on Energy Conversion*, 19(1): 157–163, March 2004.

295. D. W. P. Thomas, M. S. Jones, and C. Christopoulos. Phase selection based on superimposed components. *IEE Proceedings—Generation, Transmission and Distribution*, 143(3): 295–299, May 1996.

296. L. Tihanyi. *Electromagnetic Compatibility in Power Electronics*. IEEE Press, Piscataway, NJ, 1995.

297. A. Tjäder. High frequency conducted disturbances. PhD thesis, Department of Electric Power Engineering, Chalmers University of Technology, Gothenburg, Sweden, 2004.

298. T. Tran-Quoc, C. Andrieu, and N. Hadjsaid. Technical impacts of small distributed generation units on LV networks. Paper presented at the IEEE Power Engineering Society General Meeting, Toronto, Canada, July 2003.

299. N. S. Tunaboylu and E. R. Collins. The wavelet transform approach to detect and quantify voltage sags. Paper presented at the International Conference on Harmonics and Quality of Power, Las Vegas, NV, October 1996.

300. A. E. Turner and E. R. Collins. The perfomance of ac contactors during voltage sags. Paper presented at the Seventh International Conference on Harmonics and Quality of Power (ICHQP), Las Vegas, NV, October 1996.

301. Guide to Quality of Electrical Supply for industrial installation. Part 2: Voltage Dips and Short Interruptions. International Union for Electricity Applications, Paris, 1996.

302. Guide to Quality of Electrical Supply for Industrial Installation. Part 4: Voltage Unbalance. International Union for Electricity Applications, Paris, 1998.

303. Guide to Quality of Electrical Supply for Industrial Installation. Part 5: Flicker and Voltage Fluctuations. International Union for Electricity Applications, Paris, 1999.

304. P. C. van der Sluijs and F. Provoost. Results of long term power quality monitoring in the nuon grid. Paper presented at the Seventeenth International Conference in Electricity Distribution (CIRED 2003), Barcelona, May 2003.

305. L. van der Sluis. *Transients in Power Systems*. Wiley, Chichester, 2001.

306. A. van Jouanne and B. B. Banerjee. Assessment of voltage unbalance. *IEEE Transactions in Power Delivery*, 16(4): 782–790, October 2001.

307. S. J. van Zyl and C. T. Gaunt. Control strategies for distributed generators operating on weak distribution networks. Paper presented at the IEEE Bologna Power Tech Conference, Bologna, Italy, June 2003.

308. V. Vapnik. *The Nature of Statistical Learning Theory*. Springer, New York, 1995.

309. R. R. Verderber, O. C. Morse, and W. R. Alling. Harmonics from compact fluorescent lamps. *IEEE Transactions on Industry Applications*, 29(3): 670–674, May/June 1993.

310. C. Vilar, J. Usaola, and H. Amaris. A frequency domain approach to wind turbines for flicker analysis. *IEEE Transactions on Energy Conversion*, 18(2): 335–341, June 2003.

311. V. E. Wagner, J. P. Staniak, and T. L. Orloff. Utility capacitor switching and adjustable-speed drives. *IEEE Transactions on Industry Applications*, 27(4): 645–651, July/August 1991.

312. V. E. Wagner (Chair). Effects of harmonics on equipment—report of the IEEE task force on the effects of harmonics on equipment. *IEEE Transactions on Power Delivery*, 8(2): 672–680, April 1993.

313. G. J. Wakileh. *Power System Harmonics—Fundamentals, Analysis and Filter Design*. Springer, Berlin, 2001.

314. G. J. Wakileh. Harmonics in rotating machines. *Electric Power Systems Research*, 66(1): 31–37, July 2003.

315. M. Waller. *Harmonics*. Howard Sams, Indianapolis, IN 1994.

316. F. Wang. Power quality disturbances and protective relays—component switching and frequency deviation. PhD thesis, Department of Electric Power Engineering, Chalmers University of Technology, Gothenburg, Sweden, 2003.

317. F. Wang and M. H. J. Bollen. Frequency response characteristics and error estimation in RMS measurement. *IEEE Transactions on Power Delivery*, 19(4):1 569–1578, October 2004.

318. Y.-N. Wang, J.-Y. Gu, and C.-M. Chen. Real-time tracking of the torsional vibration of an induction motor supplied by distorted voltage sources. *Electric Power Systems Research*, 57, 2001, 205–215.

319. Y. L. Wang, L. Pierrat, and R. Feuillet. An analytical method for predicting current harmonics produced by an ac/dc converter under unbalanced supply voltage. *European Transactions on Electric Power*, 2(4): 237–244, July/August 1992.

320. N. Watson and J. Arrillaga. *Power Systems Electromagnetic Transients Simulation*. Institution of Electrical Engineers, London, 2003.

321. N. R. Watson and J. Arrillaga. Harmonics in large systems. *Electric Power Systems Research*, 66(1): 15–29, July 2003.

322. B. M. Weedy and B. J. Cory. *Electric Power Systems*, 4th ed. Wiley, Chichester, 1998.

323. W. A. Wilkinson and M. D. Cox. Discrete wavelet analysis of power system transients. *IEEE Transactions on Power Systems*, 11(4): 2038–2044, November 1996.

324. H. L. Willis. *Power Distribution Planning Reference Book*. Marcel Dekker, New York, 1997.

325. A. Woyte. Design issues of photovoltaic systems and their grid interaction. PhD thesis, Department Electrical Engineering, Katholieke Universiteit Leuven, Leuven, Belgium, December 2003.

326. A. Woyte, T. Vu Van, K. Purchala, R. Belmans, and J. Nijs. Quantifying the occurrence and duration of power fluctuations introduced by photovoltaic systems. Paper presented at the IEEE Bologna Power Tech, Bologna, Italy, June 2003.

327. W. Xu (Chair). Characteristics and modeling of harmonic sources—power electronic devices. *IEEE Transactions on Power Delivery*, 16(4): 791–800, October 2001.

328. R. Yacamini. Power system harmonics—Part 3. Problems caused by distorted supplies. *Power Engineering Journal*, pp. 233–238, October 1995.

329. R. Yacamini. Power system harmonics—Part 4. Interharmonics. *Power Engineering Journal*, pp. 185–193, August 1996.

330. G. Yalcinkaya, M. H. J. Bollen, and P. A. Crossley. Characterisation of voltage sags in industrial distribution systems. *IEEE Transactions on Industry Applications*, 34(4): 682–688, July 1998.

331. H.-T. Yang and C.-C. Liao. A de-noising scheme for enhancing wavelet-based power quality monitoring scheme. *IEEE Transactions on Power Delivery*, 16(3): 353–360, July 2001.

332. D. Yildirim and E. F. Fuchs. Measured transformer derating and comparison with harmonic loss factor approach. *IEEE Transactions on Power Delivery*, 15(1): 186–191, January 2000.

333. L. D. Zhang. Three-phase unbalance of voltage dips. Licentiate thesis, Department of Electric Power Engineering, Chalmers University of Technology, Gothenburg, Sweden, November 1999.

334. L. D. Zhang and M. H. J. Bollen. Characteristics of voltage dips (sags) in power systems. *IEEE Transactions on Power Delivery*, 15(2): 827–832, April 2000.

335. J. Zhong. On some aspects of design of electric power ancillary service markets. PhD thesis, Department of Electric Power Engineering, Chalmers University of Technology, Gothenburg, Sweden, February 2003.

336. *Field Handbook for Power Quality Analysis*. Dranetz BMI, Edison, NJ, 1998.

# INDEX



*Signal Processing of Power Quality Disturbances.* By Math H. J. Bollen and Irene Yu-Hua Gu

Printed and bound by CPI Group (UK) Ltd, Croydon, CR0 4YY